INTRODUCTION TO STRUCTURAL DYNAMICS

This textbook provides the student of aerospace, civil, or mechanical engineering with all the fundamentals of linear structural dynamics and scattered discussions of nonlinear structural dynamics. It is designed to be used primarily for a first-year graduate course. This textbook is a departure from the usual presentation of this material in two important respects. First, descriptions of system dynamics throughout are based on the simpler-to-use Lagrange equations of motion. Second, no organizational distinction is made between single and multiple degree of freedom systems. In support of those two choices, the first three chapters review the needed skills in dynamics and finite element structural analysis. The remainder of the textbook is organized mostly on the basis of first writing structural system equations of motion, and then solving those equations. The modal method of solution is emphasized, but other approaches are also considered. This textbook covers more material than can reasonably be taught in one semester. Topics that can be put off for later study are generally placed in sections designated by double asterisks or in endnotes. The final two chapters can also be deferred for later study. The textbook contains numerous example problems and end-of-chapter exercises.

Bruce K. Donaldson was first exposed to aircraft inertia loads when he was a carrier-based U.S. Navy antisubmarine pilot. He subsequently worked in the structural dynamics area at the Boeing Co. and at the Beech Aircraft Co., both in Wichita, Kansas, before returning to school and then embarking on an academic career in the area of structural analysis. At the University of Maryland he became a professor of aerospace engineering and then a professor of civil engineering. Professor Donaldson is the recipient of numerous teaching awards and has maintained industrial contacts, working various summers at government agencies and for commercial enterprises, the last being Lockheed Martin in Fort Worth, Texas.

INTRODUCTION TO

Structural Dynamics

BRUCE K. DONALDSON, Ph.D.

Department of Civil and Environmental Engineering
University of Maryland, College Park

CAMBRIDGE
UNIVERSITY PRESS

CAMBRIDGE
UNIVERSITY PRESS

32 Avenue of the Americas, New York NY 10013-2473, USA

Cambridge University Press is part of the University of Cambridge.

It furthers the University's mission by disseminating knowledge in the pursuit of education, learning and research at the highest international levels of excellence.

www.cambridge.org
Information on this title: www.cambridge.org/9780521865746

First published 2006
First paperback edition 2011

A catalogue record for this publication is available from the British Library

Library of Congress Cataloguing in Publication data

Donaldson, Bruce K.
Introduction to structural dynamics / Bruce K. Donaldson.
p. cm.
Includes bibliographical references and index.
ISBN 0-521-86574-3 (hardback)
1. Structural dynamics – Textbooks. 2. Linear systems – Textbooks. I. Title.
TA654.D65 2006
624.1′71 – dc22 2005037990

ISBN 978-0-521-86574-6 Hardback

To Matteo, Olivia, and Bridget

Spiego, così imparo

Contents

Preface for the Student

No actual structure is rigid. All structures deform under the action of applied loads. When the applied loads vary over time, so, too, do the deflections. The time-varying deflections impart accelerations to the structure. These accelerations result in body forces[1] called inertial loads. Since these inertia loads affect the deflections, there is a feedback loop tying together the deflections and at least the inertial load part of the total loads. When the applied loads result from the action of a surrounding liquid, then the deflections determine all the applied dynamic loads. Therefore, unlike static loads (i.e., slowly applied loads), differential equations based on Newton's laws are required to mathematically describe time-varying load–deflection interactions. Inertial loads can also have the importance of being the largest load set acting on parts of a structure, particularly if the structure is quite flexible.

In order to appreciate how significant time-varying forces can be, consider, for example, the time-varying loads that act on a typical large aircraft. After the aircraft starts its engines, it generally must taxi along taxiways to a runway and then travel along the runway during its takeoff run. Taxiways and runways are not perfectly flat. They have small alternating hills and valleys. As will be examined in a simplified form later in this book, these undulations cause the aircraft to move up and down and rock back and forth on its landing gear, that is, its suspension system. Since the aircraft structure is not rigid, this vibratory motion of the aircraft as a whole leads to the flexing of the major parts of the aircraft, particularly the wings. The relative deformations between various parts of the wing structure are, of course, conveniently described as strains. The strains go hand in hand with stresses, and these stresses can be the maximum stresses for the aircraft structure. For example, the maximum in-flight gross weight of many large aircraft is greater than the maximum takeoff gross weight. (In-flight refueling makes possible these different gross weights.) The up-and-down inertial loads induced by the design values for the anticipated waviness of the taxiway are often responsible for the lesser value of the takeoff gross weight.

[1] All structural engineering forces are either contact forces or body forces. Contact forces are simply the result of one mass system abutting another. Body forces are the result of a force field, such as a gravity or magnetic field.

Once the aircraft has taken off, it generally climbs to the desired altitude by using full power or at least higher values of engine thrust. This type of power plant operation often produces the worst-case acoustical (high-frequency) loading on the aircraft structure adjacent to the power plant. That noise, a high-frequency vibration, is of concern because it can induce acoustical fatigue, as well as be bothersome to passengers and crew. Each time the aircraft maneuvers during its flight, the control system alters the so-called g-loads (another name for inertia loads) distributed over the aircraft structure. For those types of aircraft, such as fighter aircraft, for which rapid maneuvers are important, it is easy to imagine that the maneuver loads could be, for the most part, the dominant load set. It is also possible that the critical loads occur when a large aircraft is flying straight and level if the aircraft is subjected to substantial vertically directed wind gusts. Such gusts can add considerable bounce to the flight, with considerable flexing of the aircraft's major components. If the flight goes well, eventually the aircraft will land, and that landing will create another set of important dynamic loads as a result of the impact of the aircraft's landing gear with the runway. A landing on an aircraft carrier in particular requires careful estimation of the distributed inertial loads along the wing as the wing tips bend toward the carrier flight deck immediately after the landing gear impacts on the flight deck. All the above-described situations generally result in an initial structural motion and a snap-back motion; that is, a *vibration* that is now defined as any back-and-forth motion of the structure. Again, those motions result in inertial loads that, when combined with other loads, can cause the critical stresses within the aircraft structure. Thus the importance of vibrations for aircraft structural engineers is clear. Similar scenarios are possible for other types of vehicles: land, sea, air, or space. The structure does not have to be that of a vehicle to be endangered by time-varying loads. Time-varying wind gust and earthquake loads must be considered in the design and analysis in many civil engineering structures.

If the possibility of dynamic loads providing the maximum stresses is not enough of a reason for structural engineers to study vibrations, then there is the matter of the dynamic instabilities that are possible. In bridges and aircraft, these critical instabilities are grouped mostly under the heading "flutter." The general concept of flutter is familiar to anyone who has ever watched a ribbon tied to a fan or observed Venetian blinds lowered over an open window in a mild breeze. There are two possible sources of difficulty. One type of problem is where, perhaps because of a nonlinearity, the vibration amplitude is limited but nevertheless maintained at large deflection amplitudes. In such circumstances, there is the threat of a rapid fatigue failure. The second type of problem is where the combination of aerodynamic, inertial, and elastic loads produces vibrations whose amplitudes continue to increase. When the amplitudes of the vibration steadily increase, the strains and stresses also steadily increase until structural failure occurs. These dynamic instabilities generally result from the same combination of elastic, inertial, and applied loads that are present in any structural dynamics problem. Moreover, there is also the aircraft phenomenon called *propeller whirl* that, in addition to depending on aerodynamic, inertial, and elastic forces, depends on the gyroscopic forces of the rotating propeller.

The above brief discussion is intended to support two facts of engineering practice. The first fact is that particularly for land, sea, air, and space vehicles, the dynamics of the structure are important, often critically important. The second fact is that for an engineering analyst to prepare an adequate mathematical description of a structural dynamics problem, that analyst needs a certain understanding of dynamics as well as of structural analysis. This textbook first focuses on providing the student with all the information on the dynamics of solids that the student needs for such analyses. The textbook then explains how to use the commonplace finite element stiffness method to create those matrix differential equations that adequately describe the structural dynamics problem. The remainder of the textbook discusses solution techniques, principally the technique called the modal method.

For a student to succeed in using this book, he or she should have already studied some applications of Newton's laws and have studied structural mechanics to the point of being reasonably comfortable with elementary beam theory. Chapter 3 provides a sufficient and self-contained explanation of structural modeling using the finite element method to the extent of structures composed of such structural elements as beams, bars, and springs. An attempt has been made to illustrate all aspects of the presented theory by providing numerous example problems and exercises at the end of each chapter. The answers to the exercises are found in Appendix I.

Preface for the Instructor

This textbook is designed to be the basis for a one-semester course in structural dynamics at the graduate level, with some extra material for later self-study. Using this text for senior undergraduates is possible also if those students have had more than one semester of exposure to rigid body dynamics and are well versed in the basics of the linear, stiffness finite element method. This textbook is suitable for structural dynamics courses in aerospace engineering and mechanical engineering. It also can be used in civil engineering at the graduate level when the course focus is on analysis rather than earthquake design. The first two chapters on dynamics should be particularly helpful to civil engineers.

This textbook is a departure from the usual presentation of this material in two important ways. First, from the very beginning, descriptions of system dynamics are based on the simpler-to-use Lagrange equations. To this end, the Lagrange equations are derived from Newton's laws in the first chapter. Second, no organizational distinctions are made between multidegree of freedom systems and single degree of freedom systems. Instead, the textbook is organized on the basis of first writing structural system equations of motion and then solving those equations mostly by means of a modal transformation. Beam and spring stiffness finite elements are used extensively to describe the structural system's linearly elastic forces. If the students are not already confident assemblers of element stiffness matrices, Chapter 3 provides a brief explanation of that material. One of the advantages of this textbook is that it provides practice in the hand assembly of system stiffness matrices. Otherwise the student is expected only to bring to this study topic the usual calculus and differential equation skills developed in an accredited undergraduate curriculum. The one exception with respect to math skills occurs in Chapter 8, a wholly optional chapter, which deals with continuous mass models. There a couple of Bessel equations are used to describe nonuniform, vibratory systems. These tapered-beam examples just push that topic to its limits and thus easily can be skipped.

The traditional textbook and course material organization starts with an exhaustive study of single degree of freedom systems and only then proceeds to multidegree of freedom systems. The author's departure from this customary organization is prompted by his experience that this usual material organization leaves little time at the end of the semester for students to obtain a comfort level with the use of the modal transformation. Furthermore, the present organization provides more and better opportunities to apply the superior Lagrange equations and beam stiffness matrices to multidegree of freedom systems. This advantage in turn allows consideration of example structural systems that actually look like models for small structures as opposed to collections of rigid masses on wheels connected by elastic springs. Thus the vital link between structural analysis and structural dynamics is both maintained and evident.

The following are details of the textbook's organization. Chapters 1 and 2 provide a brief overview, or review, of only that portion of rigid body dynamics that is necessary to understand structural dynamics. Chapters 3 and 4 deal with writing the matrix equations of motion for undamped, discrete mass structural systems. Again, the elastic forces are described using mostly beam stiffness finite elements. Chapters 6 and 7 focus primarily on the modal method for solving those equations. Chapter 8 considers continuous mass structural systems as a means for providing further insight into discrete systems and as a means for demonstrating the serious difficulties often associated with continuous mass models. Numerical integration techniques, with or without modal transformations, are presented in Chapter 9.

Acknowledgments

The organization of this textbook mainly reflects the author's experience working at the Boeing Co., Beech Aircraft, and Lockheed Martin at various times. I acknowledge all the many people at those organizations from whom I have learned, as well as the helpful people at the places I have consulted and the universities where I was a student. I also thank my students, who, having endured earlier versions of this textbook, challenged me to prepare this material in a clear and concise manner.

I thank Professor Jewel B. Barlow for reviewing the first two chapters on dynamics. I thank Jack A. Ellis, now retired from Lockheed Martin, for reviewing almost the entire draft. I thank Dr. Suresh Chander of Network Computing Services for obtaining commercial finite element program solutions for some of the more extensive example problems. I thank the anonymous reviewers and those involved in the publication process. Of course, any errors are solely my responsibility, and I would appreciate being notified of them.

B.D.

List of Symbols

$\cdot$	each dot placed above a symbol indicates one total differentiation with respect to time.
,	a comma as part of a subscript indicates partial differentiation with respect to all the variables that follow the comma.
$'$	each prime indicates one differentiation with respect to the single variable of the equation, usually a spatial variable.
[ ]	a square or rectangular matrix.
{ }	a column matrix.
⌊ ⌋	a row matrix; i.e., the transpose of a column matrix.
[\ \]	a diagonal (square) matrix.
a, b	with a single subscript, a coefficient of a power series expansion of (usually) a deflection function for a structural element.
a, b, c	general lengthwise dimensions or proportionality factors.
$\mathbf{a}$	general acceleration vector. A subscript indicates the acceleration of a particular mass particle.
c	a general damping coefficient such that the damping coefficient multiplied by the corresponding velocity produces a damping force. In brackets, the damping matrix.
c	a flexibility coefficient, which in general terms, is the inverse of a stiffness coefficient; with square brackets, a flexibility matrix; and with two subscripts, the row and column entry of that matrix identified by the subscripts.
c	an airfoil chord length; i.e., the streamwise distance between the airfoil leading edge and the airfoil trailing edge.
d	various distances.
e	subscript or superscript refers to an individual structural finite element.
e	eccentricity of an ellipse or an offset distance of a lumped mass from a finite element model node.

$\boldsymbol{e}_j$ — position vector of the jth mass particle relative to the center of mass of the mass system.

f — the frequency of a vibration measured in hertz (Hz; cycles per second). This is not to be confused with the circular frequency, ω, which has units of radians per second.

f — a general mathematical function of engineering interest.

f — a force per unit length acting along the length of a beam or a friction force.

g — the acceleration of gravity.

g — with an argument, the step response function. See Section 7.5.

g — a fictitious material damping factor distinguished from the actual material damping factor symbolized by γ.

h — the vertical translation of a wing segment.

h — with an argument, the impulse response function. A subscript indicates the associated natural mode.

h, k — with subscripts, increments in deflection and velocity for various numerical methods of integrating differential equations.

$\boldsymbol{i}, \boldsymbol{j}, \boldsymbol{k}$ — fixed unit vectors aligned with the Cartesian coordinate system.

i, j, k — positive integer indices.

k — a stiffness coefficient for a single coiled spring or, more generally, an entry in the stiffness matrix of a spring or a more complicated structural element such as a beam or plate. In square brackets, a stiffness matrix.

l, ℓ — generally a beam segment length.

m — mass. A subscript indicates a particular mass particle or mass at the ith finite element node. In square brackets, a mass matrix.

m, n — positive integer indices.

p — in braces, the vector of modal deflections; with a subscript, an entry of such a vector.

$\boldsymbol{p}, \boldsymbol{q}$ — orthogonal unit vectors in the z plane that, depending on the subscripts, rotate (positive counterclockwise) with either the center of mass of the mass system or a particular mass particle of the mass system. See Figure 1.4.

q — general symbol for a generalized coordinate (degree of freedom). In braces, a vector of generalized coordinates.

$\boldsymbol{r}$ — a position vector; i.e., a vector that locates the position $[x(t), y(t), z(t)]$ of a mass or mass particle. In the latter case there is a subscript that indicates which mass particle.

r — a radial polar coordinate.

r — with a subscript, the ramp response function for the mode indicated by the subscript.

r	in brackets, a rotation matrix that is part of the Jacobi method or a variation on the Jacobi method.
s	entries in the "sweeping" matrix that relates (1) the generalized coordinate vector constrained to be orthogonal to the lower numbered mode shapes to (2) the unconstrained generalized coordinate vector.
s	with a subscript, the sine response function for the mode indicated by the subscript.
sgn( )	a function that has the value positive 1.0 when the argument is positive and the value negative 1.0 when its argument is negative.
stp( )	the Heaviside step function. See Section 7.5.
t	time.
t	the thickness of a thin beam cross section.
u, v, w	translational deflections in the x, y, z directions, respectively. With subscripts, such translations at the nodes of a finite element model.
v	general velocity vector.
x, y, z	Cartesian coordinates.
A	a beam cross-sectional area.
A	in braces or within row matrix symbols, an eigenvector of the amplitudes of the natural vibration.
A_0	an aerodynamic coefficient equal to $\frac{1}{2}\, C_{l\alpha}\, \rho S$.
A, B, C	constants of integration or unknown amplitudes of a vibratory motion.
B	a general matrix or a coefficient matrix for the column matrix of generalized coordinates that yields the strains appropriate to the finite element.
BCs	abbreviation for boundary conditions.
C_l	an airfoil lift coefficient. See Eq. (7.16).
$\mathcal{C}(\kappa)$	The Theodorsen function. See the explanation following Eq. (7.20).
CG	abbreviation for center of gravity, which here is the same as center of mass.
D	a plate stiffness factor. See Example 8.8.
D	with a single subscript, a term associated with an initial deflection vector.
D	in brackets, a material stiffness matrix for an elastic material; i.e., a coefficient matrix for strains that yields the corresponding stresses.
D	in brackets, the system dynamic matrix; i.e., the product of the inverse of the stiffness matrix premultiplying the mass matrix when it is nonsymmetrical or the result of a transformation using a Cholesky decomposition when it is symmetrical.
DOF	abbreviation for degrees of freedom.

E Young's modulus; i.e., the slope of the straight-line portion of the stress–strain curve for a structural material loaded in tension or compression.

E error term.

$\mathscr{E}$ total mechanical energy of a structural system.

$\boldsymbol{F}$ general force vector. An *ex* superscript indicates forces external to the mass system under consideration. An *in* superscript indicates forces internal to the mass system. A subscript indicates a force acting on a particular mass particle.

$\mathcal{F}$ the magnitude of an impulse; i.e., the integral of a short duration force over time.

F, G general mathematical functions of engineering interest.

G the shear modulus; i.e., the slope of the straight-line portion of the stress–strain curve for a structural material subjected to a shear loading.

G the universal gravitational constant.

H a general symbol for mass moment of inertia of a mass system about a point or an axis indicated, respectively, by the single or double subscript. As a "second moment," it is the sum of each mass particle or differential sized mass multiplied by the square of the distance from the point or axis indicated to the mass particle or differential mass.

H with a subscript, the complex frequency response function associated with the mode indicated by the subscript.

I a general symbol for the area moment of inertia of a beam cross section. Double subscripts indicate the centroidal axis about which the second moment of area is calculated. A p subscript indicates a polar moment of inertia.

I the value of an integral that is to be optimized or evaluated.

I in brackets, the identity matrix.

J for a beam cross section, the St. Venant constant for uniform torsion. It is equal to the cross-sectional area polar moment of inertia only in the case of circular and annular cross sections.

J with a subscript and an argument, a Bessel function of the first kind. A subscript indicates the order.

K in brackets, the stiffness matrix for an entire structural system that is composed of the compatible sum of the stiffness matrices of the individual structural elements. With subscripts, a submatrix of the total stiffness matrix.

K a torsional spring constant; i.e., the proportionality factor multiplying the twist in the spring that yields the moment necessary to achieve that twist.

L a general symbol for length, usually the length of a beam or beam segment.

L in brackets, the lower (or left) triangular matrix of a Cholesky decomposition.

$\boldsymbol{L}$	general angular momentum (moment of momentum) vector. Subscripts can indicate the point about which the moment arm is measured, or the mass particle under consideration.
$\mathcal{L}$	an aerodynamic lift force.
M	an externally applied moment, or the internal moment stress resultant for a beam cross section.
M	In brackets, a mass matrix.
$\mathcal{M}$	an aerodynamic moment; i.e., the moment acting on a wing segment because of the surrounding airflow.
$\boldsymbol{M}$	general moment vector.
N	the axial force in a beam or a bar of a truss.
N	with a single subscript, a "shape function" that, together with a generalized coordinate of a finite element, describes the deflections of the finite element associated with that generalized coordinate.
$O[\,]$	order of magnitude of the quantity within the brackets.
P	in braces, the vector of applied modal forces; with a subscript, an entry of such a vector.
$\boldsymbol{P}$	general momentum vector equal to the scalar mass value multiplied by the velocity vector.
Q	general symbol for a generalized force. Subscripts often indicate the corresponding generalized coordinate.
R	general symbol for a support reaction, either a force reaction or a moment reaction.
R	in brackets, a right triangular matrix of a Cholesky decomposition.
R	in brackets, a coefficient matrix that relates one set of generalized coordinates to a second set of generalized coordinates that is rotated through one or more angles relative to the first set of generalized coordinates.
R	a principal radius of curvature of a curved beam.
S	the planform area of a wing segment.
S	in brackets, a "sweeping" matrix that removes the presence of lower numbered modes from the system dynamic matrix.
T	kinetic energy.
T	the time period of one vibratory cycle.
T	a matrix that transforms one set of generalized coordinates into another set of generalized coordinates.
U	strain energy; i.e., the recoverable energy stored in an elastic system because of the deformation of that system.
U, V, W	with an argument, a vibration amplitude function of a system with a continuous mass distribution.

V	a general potential energy other than strain energy.
V	the shearing force acting on a beam cross section. Subscripts indicate the direction and lengthwise position of the shearing force.
V	with a single subscript, a term associated with a system initial velocity vector.
V	the airspeed of the airfoil.
W	work, either the product of force acting through a translational displacement or moment acting through a rotational displacement. Superscripts and subscripts indicate the type of forces or moments doing the work, such as internal or external to the mass system, or energy conservative or nonconservative.
X	the x Cartesian coordinate nondimensionalized by division by the length of a beam segment.
X, Y	the horizontal and vertical components of an internal bar force.
Y	with a subscript and argument, a Bessel function of the second kind.
α	an alternate polar coordinate.
α, β	various angles, parameters, or proportionality factors.
γ	general symbol for angular or shearing strain, positive when the reference right angle decreases. Two differing Cartesian coordinate subscripts indicate the two coordinate axes forming the original right angle.
γ	nondimensional material damping factor.
γ	when used as a multiplier of an area term, an area correction factor that attempts to account for the variation of shearing stresses and strains over a beam cross section.
δ	the variational operator which (here) always precedes a function of deflections. When applied to quantities of engineering interest, such as work W, the resulting engineering interpretation is that of a "virtual" quantity, which in this case is virtual work.
δ	the Dirac delta function, which always has an associated spatial or temporal argument. The argument is always the difference between the variable and a related parameter. The latter is sometimes zero, in which case the argument contains only the variable. See Section 7.5.
ϵ	general symbol for normal strains (changes in length because of deformation divided by the original length). Two repeated subscripts indicate the direction in which the strain is measured. Elongations are positive.
ζ	a nondimensionalized value of the damping coefficient.
θ	various angles or an angular generalized coordinate at the node of a finite element model.
κ	the "reduced frequency"; a nondimensional frequency or nondimensional airspeed equal to $c\,\omega/(2\,V)$, where c is the airfoil chord length.

λ	a general symbol for an eigenvalue or, occasionally, a parameter having the same units as the system eigenvalues.
μ	coefficient of Coulomb friction, a mass ratio, or a coefficient of Duffing's equation.
ν	Poisson's ratio.
ρ	mass density; i.e., mass divided by volume.
ρ	an alternate polar coordinate.
σ	general symbol for both normal and shearing stresses. The same double subscripts indicate a normal stress in that Cartesian coordinate direction, whereas the first of two unlike subscripts indicates the plane on which the shearing stress acts, and the second of the two unlike subscripts indicates the direction in which the shearing stress acts.
τ	a value of the time variable different from another measure of time, t. Usually used as a parametric value.
ϕ, ψ	various angles or finite element nodal rotations.
ω	a given angular velocity, or more commonly here, a circular frequency of vibration having the units of radians per second, and therefore equal to $2\pi f$, where f is the frequency of the vibration measured in units of cycles per second.
Γ	with a subscript, the participation factor for the mode shape indicated by the subscript.
Δ	an operator that indicates a small increment in the quantity to which it is applied.
Θ	the fixed amplitude of the vibratory generalized coordinate θ.
Λ	a general symbol for the matrix of eigenvalues of another matrix or the vibratory system.
Π	the magnitude of an impulsive force expressed in modal coordinate terms.
Υ	an amplitude of a forced motion.
Φ	a matrix of all, or selected, eigenvectors.
Ω	the ratio of a forcing frequency to a natural frequency where the subscript indicates which natural frequency. Without a subscript, the natural frequency is the first natural frequency.

1 The Lagrange Equations of Motion

1.1 Introduction

A knowledge of the rudiments of dynamics is essential to understanding structural dynamics. Thus this chapter reviews the basic theorems of dynamics without any consideration of structural behavior. This chapter is preliminary to the study of structural dynamics because these basic theorems cover the dynamics of both rigid bodies and deformable bodies. The scope of this chapter is quite limited in that it develops only those equations of dynamics, summarized in Section 1.10, that are needed in subsequent chapters for the study of the dynamic behavior of (mostly) elastic structures. Therefore it is suggested that this chapter need only be read, skimmed, or consulted as is necessary for the reader to learn, review, or check on (i) the fundamental equations of rigid/flexible body dynamics and, more importantly, (ii) to obtain a familiarity with the Lagrange equations of motion.

The first part of this chapter uses a vector approach to describe the motions of masses. The vector approach arises from the statement of Newton's second and third laws of motion, which are the starting point for all the material in this textbook. These vector equations of motion are used only to prepare the way for the development of the scalar Lagrange equations of motion in the second part of this chapter. The Lagrange equations of motion are essentially a reformulation of Newton's second law in terms of work and energy (stored work). As such, the Lagrange equations have the following three important advantages relative to the vector statement of Newton's second law: (i) the Lagrange equations are written mostly in terms of point functions that sometimes allow significant simplification of the geometry of the system motion, (ii) the Lagrange equations do not normally involve either external or internal reaction forces and moments, and (iii) the Lagrange equations have the same mathematical form regardless of the choice of the coordinates used to describe the motion. These three advantages alone are sufficient reasons to use the Lagrange equations throughout the remaining chapters of this textbook.

1.2 Newton's Laws of Motion

Newton's three laws of motion can be paraphrased as (Ref. [1.1]):

1. Every particle continues in its state of rest or in its state of uniform motion in a straight line unless it is compelled to change that state by forces impressed upon it.

2. The time rate of change of momentum is proportional to the impressed force, and it is in the direction in which the force acts.

3. Every action is always opposed by an equal reaction.

These three laws are not the only possible logical starting point for the study of the dynamics of masses. However, (i) these three laws are at least as logically convenient as any other complete basis for the motion of masses, (ii) historically, they were the starting point for the development of the topic of the dynamics, and (iii) they are the one basis that almost all readers will have in common. Therefore they are the starting point for the study of dynamics in this textbook.

There are features of this statement of Newton's laws that are not immediately evident. The first of these is that these laws of motion are stated for a single particle, which is a body of very, very small spatial dimensions, but with a fixed, finite mass. The mass of the jth particle is symbolized as m_j. The second thing to note is that momentum, which means rectilinear momentum, is the product of the mass of the particle and its instantaneous velocity. Of course, mass is a scalar quantity, whereas velocity and force are vector quantities. Hence the second law is a vector equation. The third thing to note is that the second law, which includes the first law, is not true for all coordinate systems. The best that can be said is that there is a *Cartesian* coordinate system "in space" for which the second law is valid. Then it is easy to prove (see the first exercise) that the second law is also true for any other Cartesian coordinate system that translates at a constant velocity relative to the valid coordinate system. The second law is generally not true for a Cartesian coordinate system that rotates relative to the valid coordinate system. However, as a practical matter, it is satisfactory to use a Cartesian coordinate system fixed to the Earth's surface *if* the duration of the motion being studied is only a matter of a few minutes. The explanation for this exception is that the rotation of the Cartesian coordinate system fixed at a point on the Earth's surface at the constant rate of one-quarter of a degree per minute, or 0.0007 rpm, mostly just translates that coordinate system at the earth's surface in that short period of time. See Figure 1.1(a).

As is derived below, when Newton's second law is extended to a mass m of finite spatial dimensions, which is subjected to a net external force of magnitude[1] $\boldsymbol{F}$, then Newton's second law can be written in vector form as follows:

$$\boldsymbol{F} = \frac{d\boldsymbol{P}}{dt} = m\frac{d\boldsymbol{v}}{dt} = m\boldsymbol{a}, \tag{1.1}$$

where $\boldsymbol{P} = m\boldsymbol{v}$ is the momentum vector, $\boldsymbol{v}$ is the velocity vector of the total mass m relative to the valid coordinate origin, t is time, and $\boldsymbol{a}$ is the acceleration vector, which of course is the time derivative of the velocity vector. The velocity vector is not the

[1] Vector quantities are indicated by the use of italic boldface type.

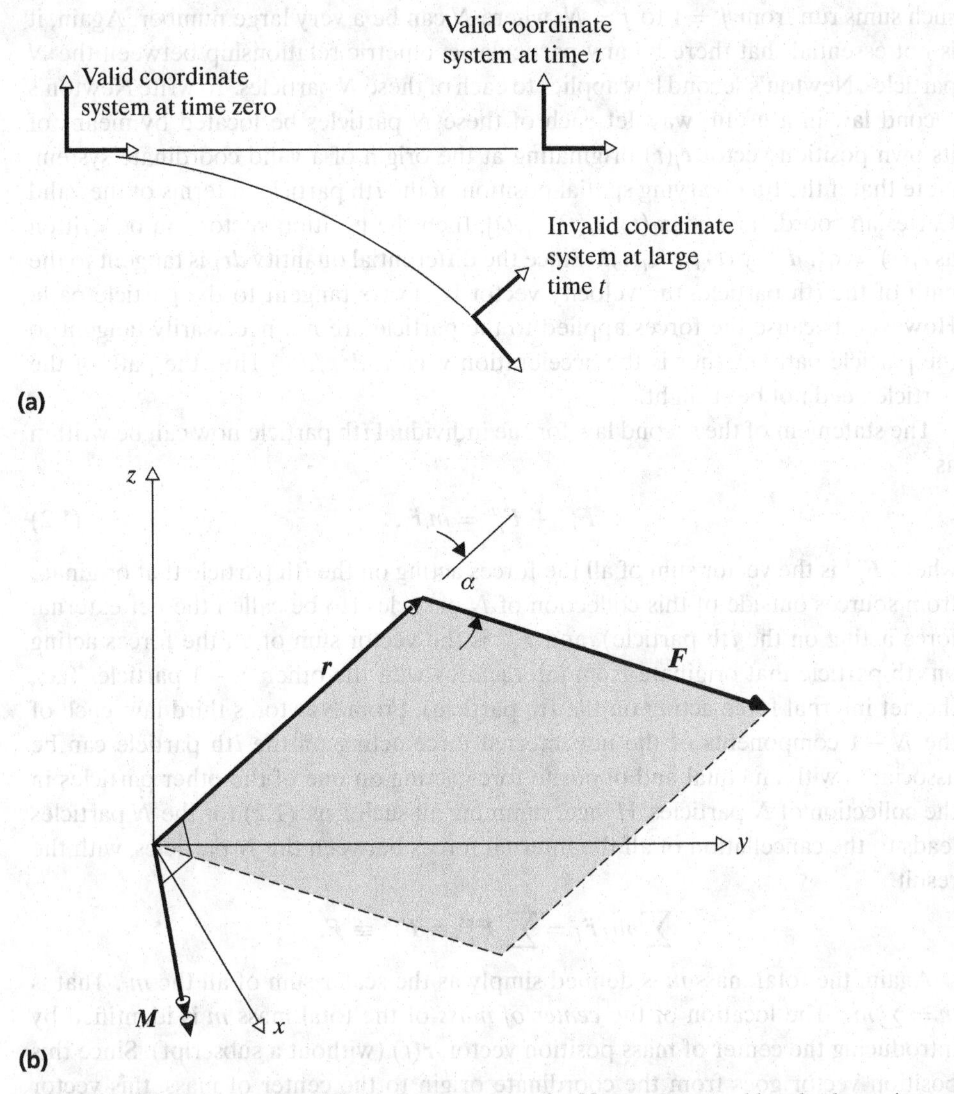

Figure 1.1. (a) Valid and invalid coordinate systems for Newton's second law, both moving at constant speed. (b) Illustration of the right-hand rule for $\boldsymbol{r} \times \boldsymbol{F} = \boldsymbol{M} = rF \sin \alpha \boldsymbol{n}$.

velocity of all points within the mass m relative to the valid coordinate system. Rather, it is the velocity of the one point called the *center of mass*, which is defined below. Further, note that the mass of the system of particles whose motion is described by this equation is the mass of a fixed collection of specific mass particles. Hence, even though the boundary surface that encloses these specified mass particles may change considerably over time, the mathematical magnitude of the mass term is a constant. Those mass particles that are included within the mass, or alternately, enclosed by the boundary surface of the mass system, are defined by the analyst as the "mass system under study."

The above basic result, Eq. (1.1), can be derived as follows. Consider a collection of, that is, a specific grouping of, N particles of total mass $m = \sum m_j$, where all such sums run from $j = 1$ to $j = N$, where N can be a very large number. Again, it is not essential that there be any particular geometric relationship between the N particles. Newton's second law applies to each of these N particles. To write Newton's second law in a useful way, let each of these N particles be located by means of its own position vector $\boldsymbol{r}_j(t)$ originating at the origin of a valid coordinate system. Note that if the time-varying spatial position of the ith particle in terms of the valid Cartesian coordinates is $[x_i(t), y_i(t), z_i(t)]$, then the position vector can be written as $\boldsymbol{r}_i(t) = x_i(t)\boldsymbol{i} + y_i(t)\boldsymbol{j} + z_i(t)\boldsymbol{k}$. Since the differential quantity $d\boldsymbol{r}_i$ is tangent to the path of the ith particle, the velocity vector is always tangent to the particle path. However, because the forces applied to the particle are not necessarily tangent to the particle path, neither is the acceleration vector, $d^2\boldsymbol{r}/dt^2$. Thus the path of the particle need not be straight.

The statement of the second law for the individual ith particle now can be written as

$$\boldsymbol{F}_i^{ex} + \boldsymbol{F}_i^{in} = m_i\ddot{\boldsymbol{r}}_i, \tag{1.2}$$

where $\boldsymbol{F}_i^{ex}$ is the vector sum of all the forces acting on the ith particle that originate from sources outside of this collection of N particles (to be called the net external force acting on the ith particle), and $\boldsymbol{F}_i^{in}$ is the vector sum of all the forces acting on ith particle that originate from interactions with the other $N-1$ particles (i.e., the net internal force acting on the ith particle). From Newton's third law, each of the $N-1$ components of the net internal force acting on the ith particle can be associated with an equal and opposite force acting on one of the other particles in the collection of N particles. Hence, summing all such Eqs. (1.2) for the N particles leads to the cancellation of all the internal forces between the N particles, with the result

$$\sum m_j\ddot{\boldsymbol{r}}_j = \sum \boldsymbol{F}_j^{ex} \equiv \boldsymbol{F}^{ex} \equiv \boldsymbol{F}.$$

Again, the total mass m is defined simply as the scalar sum of all the m_i. That is $m = \sum m_j$. The location of the *center of mass* of the total mass m is identified by introducing the center of mass position vector, $\boldsymbol{r}(t)$ (without a subscript). Since this position vector goes from the coordinate origin to the center of mass, this vector alone fully describes the path traveled by the center of mass as a function of time. The center of mass position vector $\boldsymbol{r}$ at any time t is defined so that

$$m\boldsymbol{r} \equiv \sum m_i\boldsymbol{r}_i.$$

This definition means that the center of mass position vector is a mass-weighted average of all the mass particle position vectors. This definition can also be viewed as an application of the mean value theorem. Differentiating both sides of the definition of the center of mass position vector with respect to time twice and then substituting into the previous equation immediately yields Eq. (1.1): $\boldsymbol{F} = m\ddot{\boldsymbol{r}} \equiv m\boldsymbol{a}$. Again, the force vector $\boldsymbol{F}$, without superscripts and subscripts, is the sum of all the external forces. Note that external forces can arise from only one of two sources: (i) the direct contact of the boundary surface of the N particles under study with the boundary of other masses or (ii) the distant action of other masses, in which case they are called field forces. Gravitational forces are an example of the latter type of action.

1.3 Newton's Equations for Rotations

A knowledge of the motion of the center of mass can tell the analyst a lot about the overall motion of the mass system under study. However, that information is incomplete because it tells the analyst nothing at all about the rotations of the mass particles about the center of mass. Since rotational motions can be quite important, this aspect of the overall motion needs investigation.

Just as the translational motion of the center of mass can be viewed as determined by forces, rotational motions are determined by moments of forces. Recall that the mathematical definition of a *moment about a point*, when the moment center is the origin of the valid coordinate system, is

$$\boldsymbol{M} \equiv \boldsymbol{r} \times \boldsymbol{F}.$$

Recall that reversing the order of a vector cross product requires a change in sign to maintain an equality. Further note that it is immaterial where this position vector intercepts the line of action of the above force vector because the product of the magnitude of the $\boldsymbol{r}$ vector and the sine of the angle between the $\boldsymbol{r}$ and $\boldsymbol{F}$ vectors is always equal to the perpendicular distance between (i) the line of action of the force and (ii) the moment center.

Structural engineers are more familiar with moments about Cartesian coordinate axes than moments about points. The relation between a moment about a point and a moment about such an axis can be understood by reference to Figure 1.1(b). This figure illustrates that the moment resulting from the cross product of the $\boldsymbol{r}$ vector and the $\boldsymbol{F}$ vector, by the rules of vector algebra, is in the direction of the unit vector $\boldsymbol{n}$, which is perpendicular to the plane formed by the $\boldsymbol{r}$ and $\boldsymbol{F}$ vectors. The positive direction of $\boldsymbol{n}$ is determined by the thumb of the right hand after sweeping the other four fingers of the right hand from the direction of $\boldsymbol{r}$, the first vector of the cross product, through to the direction of $\boldsymbol{F}$. In terms of α, the angle between these two vectors in the plane formed by the two vectors

$$\boldsymbol{M} \equiv \boldsymbol{r} \times \boldsymbol{F} \equiv Fr \sin\alpha \, \boldsymbol{n}.$$

Like any other vector, the vector $\boldsymbol{M}$ has components along the Cartesian coordinate axes. In terms of the components of the force $\boldsymbol{F}$ and the position vector $\boldsymbol{r}$, the moment about a point can be written, using vector algebra, as follows:

$$\begin{aligned} \boldsymbol{M} = \boldsymbol{r} \times \boldsymbol{F} &= (x\boldsymbol{i} + y\boldsymbol{j} + z\boldsymbol{k}) \times (F_x\boldsymbol{i} + F_y\boldsymbol{j} + F_z\boldsymbol{k}) \\ &= (yF_z - zF_y)\boldsymbol{i} + (zF_x - xF_z)\boldsymbol{j} + (xF_y - yF_x)\boldsymbol{k} \\ &= M_x\boldsymbol{i} + M_y\boldsymbol{j} + M_z\boldsymbol{k}. \end{aligned}$$

Considering the last equation, it is clear that moments about axes are simply components of moments about points.

When describing the rotation of the mass m, it is often convenient to consider a reference point P that is other than the valid coordinate origin, which is here called the point O. See Figure 1.2. Let the this new reference point P move in an arbitrary fashion relative to the coordinate origin, point O, in a fashion defined by the position

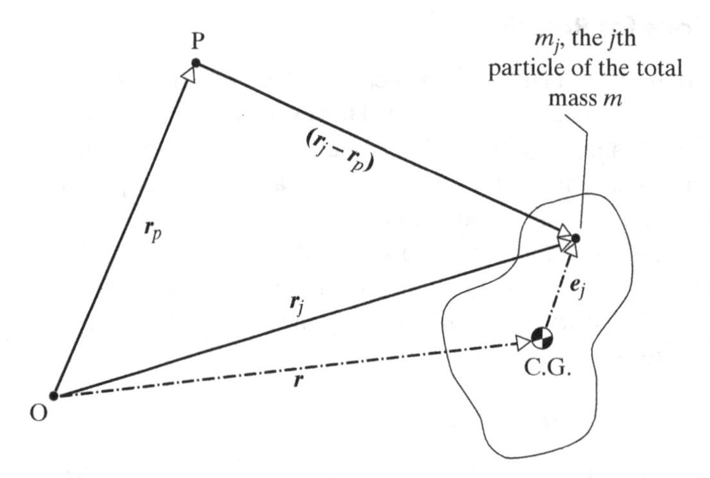

Figure 1.2. Vectors relevant to the rotational motion of a mass. Point P has an arbitrary motion relative to point O.

vector $\boldsymbol{r}_P(t)$. Introduce the vector quantity $\boldsymbol{L}_{Pj}(t)$ which is to be called the *angular momentum* about point P, or, more descriptively, the *moment of momentum* of the mass particle m_j about the arbitrary point P. That is, the angular momentum about point P of the jth mass particle is defined as the vector cross product of (i) the position vector from point P to the particle m_j and (ii) the momentum vector of m_j where the associated velocity vector is that relative to point P rather than the origin of the coordinate system, point O. Thus, in mathematical symbols, relative to point P, the angular momentum of the jth particle, and the angular momentum of the total mass m are, respectively,

$$\boldsymbol{L}_{P_j} \equiv (\boldsymbol{r}_j - \boldsymbol{r}_P) \times m_j(\dot{\boldsymbol{r}}_j - \dot{\boldsymbol{r}}_P) \quad \text{and} \quad \boldsymbol{L}_P \equiv \sum \boldsymbol{L}_{P_j}.$$

Differentiating both sides of the total angular momentum with respect to time, and noting that the cross product of the relative velocity vector $(\dot{\boldsymbol{r}}_j - \dot{\boldsymbol{r}}_P)$ with itself is zero, yields the following result:

$$\frac{d\boldsymbol{L}_P}{dt} = 0 + \sum [(\boldsymbol{r}_j - \boldsymbol{r}_P) \times m_j(\ddot{\boldsymbol{r}}_j - \ddot{\boldsymbol{r}}_P)].$$

From the original statement of Newton's second law, it is possible to substitute in the above equation the net external and internal forces on the jth particle for $m_j(d^2/dt^2)\boldsymbol{r}_j$. The result is

$$\frac{d\boldsymbol{L}_P}{dt} = \sum \left[(\boldsymbol{r}_j - \boldsymbol{r}_P) \times \left(\boldsymbol{F}_j^{ex} + \boldsymbol{F}_j^{in}\right) - m_j(\boldsymbol{r}_j - \boldsymbol{r}_P) \times \ddot{\boldsymbol{r}}_P\right].$$

The term involving the net internal forces sums to zero because all the component internal forces are not only equal and oppositely directed, but, by the strong form of Newton's third law, they are also collinear. See Exercise 1.1. The remaining portion of the first term, that involving the net external forces on the N particles, sums to

$\boldsymbol{M}_P$, called the moment about point P of all the external forces acting on the mass m. The last term in the above sum can be simplified by noting that

$$\sum m_i(\boldsymbol{r}_i - \boldsymbol{r}_P) \times \ddot{\boldsymbol{r}}_P = -\left[\boldsymbol{r}_P \sum m_i - \sum m_i \boldsymbol{r}_i\right] \times \ddot{\boldsymbol{r}}_P$$
$$= -[m\boldsymbol{r}_P - m\boldsymbol{r}] \times \ddot{\boldsymbol{r}}_P$$
$$= +m(\boldsymbol{r} - \boldsymbol{r}_P) \times \ddot{\boldsymbol{r}}_P.$$

Thus the final result for the time derivative of the angular momentum of the mass m is

$$\frac{d\boldsymbol{L}_P}{dt} = \boldsymbol{M}_P - m(\boldsymbol{r} - \boldsymbol{r}_P) \times \ddot{\boldsymbol{r}}_P. \tag{1.3a}$$

In other words, with reference to Figure 1.2,

$$d\boldsymbol{L}_P/dt = \boldsymbol{M}_P - m * \text{(position vector from P to the center of mass)}$$
$$* \text{(acceleration vector of point P relative to point O)}.$$

Clearly, if point P is coincident with the center of mass (called the center of mass or CG case, where $\boldsymbol{r}_P = \boldsymbol{r}$), or if the relative position vector $\boldsymbol{r}_P - \boldsymbol{r}$ and the acceleration vector $(d^2/dt2)\boldsymbol{r}_P$ are collinear (unimportant because it is unusual), or if point P is moving at a constant or zero velocity with respect to point O (called, for simplicity, the fixed point or FP case), then the rotation equation reduces to simply

$$\frac{d\boldsymbol{L}_P}{dt} = \boldsymbol{M}_P \quad \text{if P is a "fixed" point or located at the center of mass.} \tag{1.3b}$$

Note that the above vector equation is the origin of the static equilibrium equations, which state that "the sum of the moments about any *axis* is zero." That is, when the angular momentum relative to the selected *point* P is zero or a constant, then the three orthogonal components of the total moment vector of the external forces acting on the system about point P are zero. These three orthogonal components are the moments about any three orthogonal axes.

The above rotational motion equation, Eq. (1.3b) is not as useful as Eq. (1.1), the corresponding translational motion equation. In Eq. (1.1), the three quantities force, mass, and acceleration are individually quantifiable. In Eq. (1.3b), while the moment term is easily understood, the time rate of change of the angular momentum needs further refinement so that perhaps it too can be written as some sort of fixed mass type of quantity multiplied by some sort of acceleration. Recall that for the mass system m, the total angular momentum relative to point P, is defined as the sum of the moments of the momentum of all the particles that comprise the mass m. That is, again

$$\boldsymbol{L}_P = \sum(\boldsymbol{r}_i - \boldsymbol{r}_P) \times m_i(\dot{\boldsymbol{r}}_i - \dot{\boldsymbol{r}}_P).$$

From the previous development, that is, Eqs. (1.3a,b), there are two simplifying choices for the reference point P: the FP (so-called fixed point) case and the CG (center of mass) case, where the time derivative of the angular momentum is equal to just the moment about point P of all the external forces. First consider the FP case, where point P has only a constant velocity relative to the coordinate origin, point O. Then, from Exercise 1.1, either point P or point O is the origin of a valid Cartesian coordinate system. Since these two points are alike, for the sake of simplicity, let the reference point P coincide with the origin of the coordinate system, point O. Again, this placement of point P at point O does not compromise generality within the FP case because when point P is only moving at a constant velocity relative to point O, point P can also be an origin for a valid coordinate system. Then with $\boldsymbol{r}_P = 0$, and because the $\boldsymbol{e}_i$ vectors of Figure 1.2 originate at the center of mass, the total angular momentum becomes

$$\begin{aligned} \boldsymbol{L}_{FP} &= \sum \boldsymbol{r}_i \times m_i \dot{\boldsymbol{r}}_i = \sum (\boldsymbol{r} + \boldsymbol{e}_i) \times m_i (\dot{\boldsymbol{r}} + \dot{\boldsymbol{e}}_i) \\ &= \boldsymbol{r} \times \dot{\boldsymbol{r}} \left(\sum m_i\right) + \boldsymbol{r} \times \left(\sum m_i \dot{\boldsymbol{e}}_i\right) \\ &\quad + \left(\sum m_i \boldsymbol{e}_i\right) \times \dot{\boldsymbol{r}} + \sum (\boldsymbol{e}_i \times m_i \dot{\boldsymbol{e}}_i) \\ &= \boldsymbol{r} \times m \dot{\boldsymbol{r}} + \sum (\boldsymbol{e}_i \times m_i \dot{\boldsymbol{e}}_i). \end{aligned} \tag{1.4a}$$

To explain why the second and third terms of the above second line are zero, recall the definition of the center of mass position vector, $\boldsymbol{r}$. That mean value definition is $m\boldsymbol{r} \equiv \sum m_i \boldsymbol{r}_i$. Since $\boldsymbol{r}_i = \boldsymbol{r} + \boldsymbol{e}_i$, $m\boldsymbol{r} \equiv \sum m_i \boldsymbol{r} + \sum m_i \boldsymbol{e}_i$. Since $\boldsymbol{r}$ is not affected by the summation over the N particles, it can be factored out of the first sum on the above right-hand side. The result is $m\boldsymbol{r} \equiv m\boldsymbol{r} + \sum m_i \boldsymbol{e}_i$ or $0 = \sum m_i \boldsymbol{e}_i$. Furthermore, because the mass value of each particle is a constant, the time derivative of this last equation shows that $0 = \sum m_i \dot{\boldsymbol{e}}_i$. This is just an illustration of the general fact that first moments, that is, multiplications by distances raised to the first power, of mass or area, or whatever, about the respective mean point are always zero. Multiplications of mass by distances with exponents other than one lead to terms which are generally not zero.

In the above FP equation, Eq. (1.4a), for the angular momentum, the first term depends only on the motion of the center of mass relative to the Cartesian coordinate origin. Even if the mass is not rotating relative to the Cartesian coordinate origin, this term is generally not zero. The second part of the angular momentum exists even if the center of mass is not moving. This second part accounts for the spin of the mass about its own center of mass.

The CG case is where the reference point P is located at the center of mass, point C, rather than at the coordinate origin, point O, as in the FP case. In this CG case, $\boldsymbol{r} = \boldsymbol{r}_P$ and $\boldsymbol{r}_i - \boldsymbol{r}_P = \boldsymbol{e}_i$. Substituting these vector relationships into the expression for $\boldsymbol{L}_P$ immediately leads to the same result for the angular momentum, as was obtained for the FP case, except that the first of those two terms is absent. Hence the mathematics of the CG case are included within that of the FP case, and therefore the CG case does not need a parallel development.

1.4 Simplifications for Rotations

Since Newton's second law is a vector equation, it has been convenient to derive its rotational corollaries by use of vector algebra in three-dimensional space. However, it is no longer convenient to pursue the subject of rotations using three-space vector forms because, in general, the rotations themselves about axes in three dimensions (as opposed to moments about axes in three dimensions) are not vector quantities.

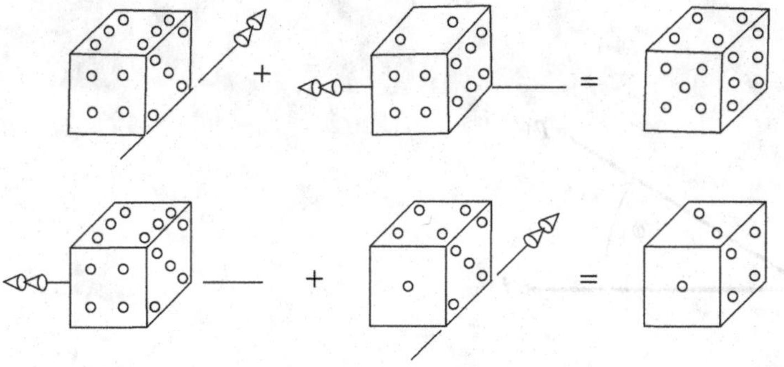

Figure 1.3. Proof that, generally, rotations are not vectors because the order of the rotations is not irrelevant.

For a quantity be classified as a vector, the order of an addition has to be immaterial; that is, it is necessary that $\boldsymbol{A}+\boldsymbol{B}=\boldsymbol{B}+\boldsymbol{A}$, which is called the commutative law for vector addition. In contrast, as Figure 1.3 illustrates, the order of addition of rotations in three-space can greatly change the final orientation of the mass whenever the rotational angles involved are large, like the 90° angles selected for Figure 1.3. There are two simple ways of circumventing this difficulty. The first simplifying approach is to restrict the rotational motion equations to a single plane. In a single plane, all rotations simply add or subtract as scalar quantities. This is a wholly satisfactory approach for most of the illustrative pendulum problems considered in the next chapter. The second option for simplification is to retain rotations about more than one orthogonal axis but limit all those rotations to being small. Here "small" means that the tangent of the angle is closely approximated by the angle itself.[2] As is explained in Ref. [1.2], p. 271, in contrast to larger angles, angles about orthogonal axes of these small magnitudes can be added to each other as vector quantities. This approach of restricting the rotations to either being small or lying in a single plane would not be adequate for formulating a general analysis of the motion of bodies of finite size, which is not a present concern. However, this is a satisfactory approach for almost all structural dynamics problems because structural rotations due solely to the vibrations of a flexible structure are almost always less than 10° or 12°. Therefore, to repeat and thus underline this important point, for the present purposes of structural dynamics, it is often satisfactory only to look at rotations in a single plane or restrict the analysis to small rotations, which can be added vectorially.

To further the discussion, consider all rotations confined to a single plane that, for the sake of explicitness, is identified as the z plane. To reflect the change from three to two dimensions, the notation FP for a fixed point in three-dimensional space, transitions to FA for a fixed axis perpendicular to the z plane. This simplification from a general state of rotations to those only about an axis paralleling the z axis allows the introduction of a pair of convenient unit vectors in the z plane called $\boldsymbol{p}_1$ and $\boldsymbol{q}_1$ such that $\boldsymbol{p}_1$ is directed from the origin toward the center of mass and $\boldsymbol{q}_1$ is rotated 90° counterclockwise from $\boldsymbol{p}_1$. These two unit vectors rotate in the z plane as the center

[2] For example, 10° (expressed in radians) and the tangent of 10° differ by only 1%.

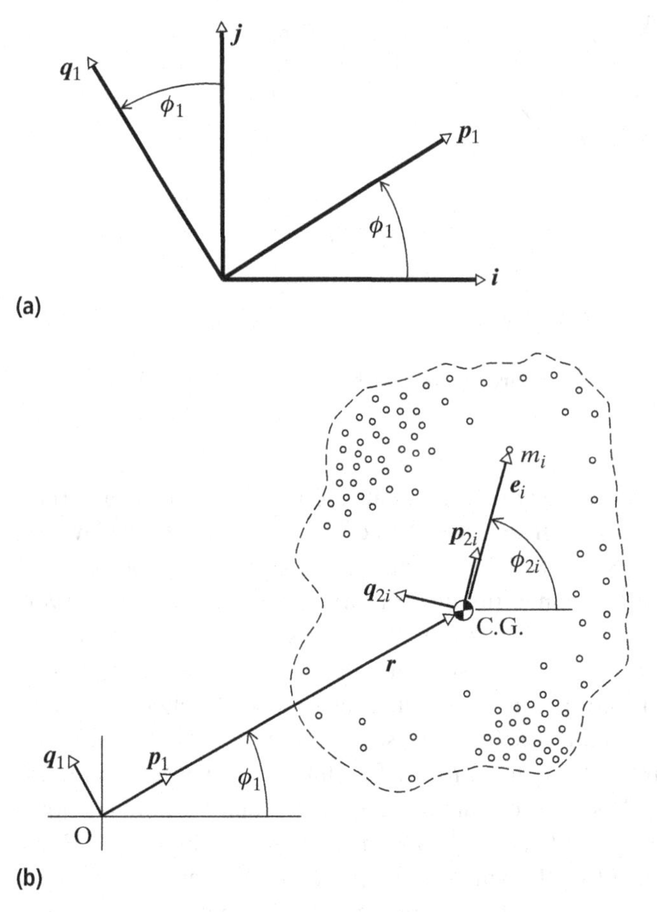

Figure 1.4. (a) The relationship between the rotating unit vectors and the fixed unit vectors, $\boldsymbol{i}$ and $\boldsymbol{j}$. (b) Use of unit vectors to locate the ith mass particle.

of mass moves in that plane. In terms of the fixed-in-space Cartesian coordinate unit vectors, $\boldsymbol{i}$, $\boldsymbol{j}$, as shown in Figure 1.4(a),

$$\begin{aligned} \boldsymbol{p}_1 &= +\boldsymbol{i}\cos\phi_1 + \boldsymbol{j}\sin\phi_1 \\ \boldsymbol{q}_1 &= -\boldsymbol{i}\sin\phi_1 + \boldsymbol{j}\cos\phi_1. \end{aligned}$$

Again, even though $\boldsymbol{p}_1$ and $\boldsymbol{q}_1$ have a fixed unit length, they have time derivatives because their orientation in the z plane varies with time as the angle ϕ changes with time. The above equations show that the time derivatives of these rotating unit vectors are

$$\dot{\boldsymbol{p}}_1 = \dot{\phi}_1\boldsymbol{q}_1 \qquad \dot{\boldsymbol{q}}_1 = -\dot{\phi}_1\boldsymbol{p}_1.$$

This unit vector pair $\boldsymbol{p}, \boldsymbol{q}$ can be used with both the position vector for the center of mass and the vector from the center of mass to the ith mass particle. That is, as illustrated in Figure 1.4(b),

$$\boldsymbol{r} = r\boldsymbol{p}_1 \quad \text{and} \quad \boldsymbol{e}_i = e_i\boldsymbol{p}_{2i}.$$

As the final limitation on the dynamics equations to be developed, let the geometry of the total mass be restricted to small changes in overall shape so that the rotation angle for the jth mass about the center of mass differs so little from that average rotation that the average rotation ϕ_2 can be used as the rotation angle about the center of mass for all the mass particles that are included within the boundary of the total mass. This is a rather minor limitation, if any at all, for almost all structures.

With this preparation for the general FA case (i.e., when point P is at point O), it is now possible to write the expression for the angular momentum, Eq. (1.4), for this special case of rotations only about the z axis, as follows:

$$\begin{aligned}
\boldsymbol{L}_{FA} &= \boldsymbol{r} \times m\dot{\boldsymbol{r}} + \sum(\boldsymbol{e}_i \times m_i \dot{\boldsymbol{e}}_i) \\
&= r\boldsymbol{p}_1 \times m\frac{d}{dt}(r\boldsymbol{p}_1) + \sum e_i \boldsymbol{p}_{2i} \times m_i \frac{d}{dt}(e_i \boldsymbol{p}_{2i}) \\
&= r\boldsymbol{p}_1 \times m(\dot{r}\boldsymbol{p}_1 + r\dot{\phi}_1 \boldsymbol{q}_1) + \sum e_i \boldsymbol{p}_{2i} \times m_i(\dot{e}_i \boldsymbol{p}_{2i} + e_i \dot{\phi}_2 \boldsymbol{q}_{2i}) \\
&= (mr^2)\dot{\phi}_1 \boldsymbol{k} + \left(\sum m_i e_i^2\right)\dot{\phi}_2 \boldsymbol{k} \\
&= [(mr^2)\dot{\phi}_1 + H_{CG}\dot{\phi}_2]\boldsymbol{k},
\end{aligned}$$

where the definition of H_{CG}, the second moment of the mass about the center of mass, is readily apparent. If the values of r and H_{CG} are near to being constants, where the latter restriction is consistent with the previous approximation that the shape of the mass changes only slightly as the mass moves, then differentiating with respect to time leads to the good approximation that

$$M_{FA} = (mr^2)\ddot{\phi}_1 + H_{CG}\ddot{\phi}_2. \tag{1.4b}$$

The above result is valid for any type of body undergoing no changes or only negligibly small changes in shape and distance from the fixed axis. Again, although H_{CG} has a fixed value for a rigid body, this quantity varies with the displacements of a deformable body. However, the displacements of a flexible structural body are usually quite small relative to the overall dimensions of the body. In such cases, it is a good approximation to regard H_{CG} as a constant, using the undeformed value. Indeed, in the case of many vibrations, the undeformed value is an average value. Moreover, when the distance r from the coordinate system origin to the center of the mass under consideration is very close to being a fixed distance, *and* when ϕ_1 equals ϕ_2, then the two right-hand side terms in Eq. (1.4b) may be combined. The sum $mr^2 + H_{CG}$ can be defined as H_{FA}. This definition is the *parallel axis theorem* for mass moments of inertia. Thus in the case where the distances r and e_i are very nearly constants, and when the two rotational angles are equal, the rotation in a plane is governed by the following simple equation:

$$H_{FA}\ddot{\theta} = M_{FA}, \tag{1.5a}$$

where point P equals point O. This equation provides the desired form that parallels the original form of Newton's second law; that of a mass term multiplied by an acceleration term equaling an applied load term.

Turning now from the FP/FA case where point P is the same as point O ($\boldsymbol{r}_P = 0$), and turning to the CG case where point P is at the center of mass, C, so that $\boldsymbol{r} = \boldsymbol{r}_P$ and $\boldsymbol{r}_i - \boldsymbol{r}_P = \boldsymbol{e}_i$, the angular momentum about point P or C is simply

$$\boldsymbol{L} = \sum \boldsymbol{e}_i \times m_i \dot{\boldsymbol{e}}_i = \sum \left(m_i e_i^2\right) \boldsymbol{p}_2 \times \dot{\phi}_2 \boldsymbol{q}_2 = H_{CG} \dot{\phi}_2 \boldsymbol{k}.$$

Substituting into Eq. (1.3) yields the simple result for the mass of a structural body

$$H_{CG} \ddot{\phi} = M_{CG}, \tag{1.5b}$$

where point P equals point C. Equations (1.5a) and (1.5b) are particularly useful in the analysis of pendulums, which are studied in the next chapter for the sole purpose of practicing using the equations derived in this chapter.

1.5 Conservation Laws

Note that if the sum of all the forces external to the mass system, $\boldsymbol{F}$, is zero throughout the time period of study, then $\boldsymbol{M}_P$ is also zero. Then, in this case, there are two immediate corollaries of Eqs. (1.1) and (1.3b), which are that $\boldsymbol{P}$ and $\boldsymbol{L}_P$, the momentum and angular momentum respectively, are constants. These theorems of constant (or conservation of) momentum and constant (or conservation of) angular momentum can be quite useful when dealing with isolated system models such as those of some spacecraft. In general, when they apply, conservation of energy equations have an advantage over Newton's second law in its original or equivalent forms because these conservation equations essentially accomplish the first and more difficult integration of the otherwise required two integrations for a solution. However, the conservation of energy theorem is generally not recommended for use in structural dynamics problems because it is often more useful to employ all energy quantities within the context of the Lagrange equations (soon to be derived) and certainly more likely to avoid those deceptive situations where, despite the fact that there is no dissipation of energy incorporated in the system analytical model, the energy is nevertheless not a constant except in an average sense. An example of such a deceptive situation is considered in the next chapter.

1.6 Generalized Coordinates

Before deriving the Lagrange equations, it is necessary to introduce the concept of *generalized coordinates*. Generalized coordinates, also referred to as *degrees of freedom* (DOF), are instantaneous measures of the deflected position of the dynamical system relative to a reference or datum position. In the case of structures subjected to static loads, the datum position is usually the undeflected position. In the case of loads that vary significantly with time, the datum is usually the static equilibrium position (to be explained later). Regardless of the datum,

> Generalized coordinates are the independent time functions that specify the instantaneous position of all the mass particles of the system without ambiguity and without redundancy.

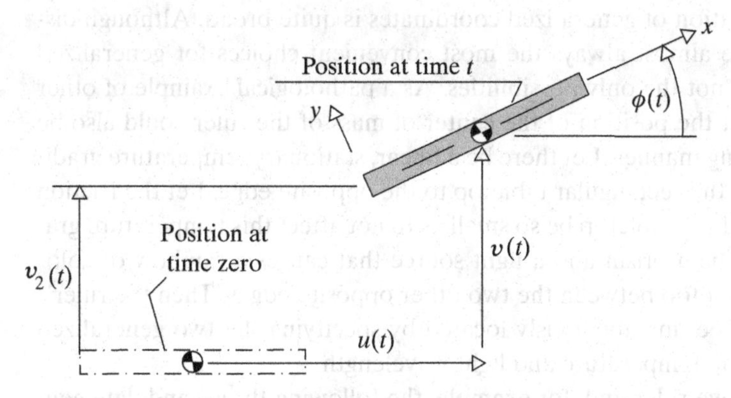

Figure 1.5. Generalized and other coordinates for a flat ruler moving in the plane of the paper.

It is important to note from this definition that it is the motions of the system mass, not those of, say, elastic elements, that are described by the generalized coordinates.

In contrast to generalized coordinates, spatial coordinates only perform the duty of merely distinguishing between different material points (particles) within the system under study. In other words, the spatial coordinates may merely name the particle by naming the point where the particle is located, say, before deformation or movement of the whole body. The following example clarifies the concept of generalized coordinates and the distinction between generalized coordinates and ordinary spatial coordinates.

Consider a flat ruler sliding on a tabletop. See Figure 1.5. Model the flat ruler as being a rigid body because its deformations due to its motion[3] are very small compared to its overall motion, and those deformations are of no present interest. Therefore a Cartesian system attached to, and thus moving with, the ruler is a coordinate system that identifies material points in the ruler by identifying their fixed location relative to the material point where this Cartesian coordinate system originates in the ruler. However, let the quantities u, v, and θ be as indicated in the figure. Since the ruler is modeled as rigid, the position of every mass particle in the ruler is exactly specified relative to its starting position when the values of u, v, and θ are specified and vice versa. Thus u, v, and θ are one possible set of generalized coordinates for all of the many mass particles contained in the ruler. There are many other possible sets of generalized coordinates for this ruler. For example, instead of using the Cartesian type coordinates u, v to locate the center of mass, two polar coordinate type coordinates would be equally effective. However, no matter what specific choices are made for this ruler, any valid set of generalized coordinates will consist of three, and only three, such coordinates. Therefore this dynamic system is called a three degree of freedom system. It is important to understand that the number of degrees of freedom is an inherent characteristic of the system mathematical model.

[3] Any motion of a mass that involves accelerations, by virtue of Newton's second law, creates (inertial) forces that in turn produce deformations. Thus there is feedback because the deformations affect the accelerations that may be thought of as starting the chain of consequences. The structural dynamics feedback loop must be described by differential equations.

Note that the definition of generalized coordinates is quite broad. Although distances and angles are almost always the most convenient choices for generalized coordinates, they are not the only possibilities. As a pathological example of other possibilities, note that the position of the center of mass of the ruler could also be located in the following manner. Let there be a linear, stationary temperature gradient from one edge of the rectangular tabletop to the opposite edge. Let the friction between the ruler and the tabletop be so small as to not affect this temperature gradient. Let there also be a prism and a light source that causes a rainbow of color to spread over the tabletop between the two other opposite edges. Then the ruler's center of mass could be unambiguously located by specifying the two generalized coordinates of tabletop temperature and light wavelength.

Reconsider the above ruler and, for example, the following three candidate generalized coordinates: u, v as previously discussed and a coordinate v_2 for the zero length end of the ruler, as is also shown in Figure 1.5. This set of three coordinates is not a valid set of generalized coordinates. The reason that these three are not valid is that specifying these coordinates does not distinguish between two possible positions for the ruler. One possible location for the ruler is the original position shown in Figure 1.5. The other possible location is that where the ruler is pivoted about a fixed center of mass (i.e., fixed u, v) so as to place the zero-inch mark to the right of the center of mass and, of course, a distance v_2 above the horizontal datum. By their definition, such an ambiguity is not permissible for generalized coordinates. Thus u, v, v_2 are not a valid set of generalized coordinates.

The following are some further examples of identifying generalized coordinates and establishing the number of DOF associated with a mass system. In the case of the arbitrary motion in the z plane of a single mass particle, only two generalized coordinates, such as the polar coordinates r and θ, are required to totally specify the position of the mass particle. This is so because geometrically the particle is only a point. Six generalized coordinates are necessary to specify the position of all the mass particles of a rigid body of finite size moving arbitrarily in three-dimensional space. A possible choice for those six DOF are three Cartesian type coordinates to locate the center of mass and pitch, roll, and yaw angles in a prescribed order. A butterfly with a rigid body to which are hinged two rigid wings (like a door is hinged to its frame) is an eight-DOF system. That is, first of all, the butterfly body would have the same six DOF that any such rigid body has. There would be an additional (angular) DOF for the location of each wing relative to fixed axes in the body for the total of eight DOF. A circular cylinder rolling in one direction without slipping on a flat tabletop has only one DOF. This is so because the zero-slip condition mandates that the distance moved by the line of contact between the cylinder and tabletop is the same as the length of the circumferential arc between the original line of contact (at zero distance) and the instantaneous line of contact. A more complicated rigid body situation is that of a thin hoop that remains vertical while rolling without slipping on a flat tabletop. A total of four DOF are required to specify the position of all the mass particles of the hoop. Two DOF can be used on the tabletop to locate the point of contact between the hoop and the tabletop. Another DOF is needed to locate the rotation of the plane of the hoop relative to a fixed plane perpendicular to the tabletop. A fourth DOF is needed to locate on the hoop the point of contact between the hoop and the tabletop. This latter DOF could be the arc length of the path taken by the nonslipping hoop or it could be more simply a circumferential arc length in a specified diection between a fixed point on the hoop and the point of contact. To understand why this latter DOF is necessary to avoid an ambiguity, picture the point of contact and any other point on the hoop. This other point can be made into the point of contact, with the first three DOF values unchanged in the final configuration, by rolling the hoop around a circle whose circumference is the arc length between these two circumferential points, the original point of contact and the newly desired point of contact. Examples and exercises provide other examples.

1.7 Virtual Quantities and the Variational Operator

Before deriving the Lagrange equations, it is also necessary to briefly review the concept of virtual quantities and introduce the variational operator. A *virtual displacement* may be defined as any displacement of arbitrarily small magnitude that does not violate the constraints of the system. A virtual displacement may be a real displacement (a rare choice) or entirely a figment of the imagination of the analyst. An arbitrary, and therefore imaginary, small displacement is the usual case for a virtual displacement because generally it is a more useful choice than an actual displacement.

Again consider the ruler of Figure 1.5, all of whose mass particles are completely located by the three generalized coordinates u, v, and θ. These three generalized coordinates are to be viewed as the real displacements of the ruler since the ruler began its travels in the z plane from the location where all three generalized coordinates have zero values. A virtual displacement in the vertical direction, symbolized as δv, would be positive in the upward vertical direction because v is positive in the upward direction.[4] Thus a positive δv would be any small, imagined, exclusively upward deflection beyond the real upward deflection v. Of course the ruler (in the analyst's imagination) would remain parallel to the position it had before this virtual displacement because at this point there are no virtual changes in the other two generalized coordinates. If a positive $\delta\theta$ were the only virtual displacement, the ruler would in the analyst's imagination, rotate through a small angle, counterclockwise about the point located by the u and v generalized coordinates. See Figure 1.6(a). Of course, all three generalized coordinates, symbolized generally as q_j, where $j = 1, 2, 3$, could be simultaneously augmented by virtual displacements, symbolized in general as δq_j. The lowercase δ is the prefix that identifies all types of virtual quantities. Finally, an example of a prohibited displacement, imagined or real, would be any movement of the ruler off the z plane because such a motion would violate the constraints of the mathematical model. Thus there generally would not be a virtual deflection away from the z plane.

If there are (real) forces and moments[5] acting on the mass system, then, just as real increments in the displacements, dq_i, give rise to real work done on the mass

[4] In contrast, the symbol dv is always to be interpreted as a small and real increment to the vertical displacement.

[5] Except for one endnote in Chapter 3, there is no need to consider virtual forces or virtual moments in this textbook.

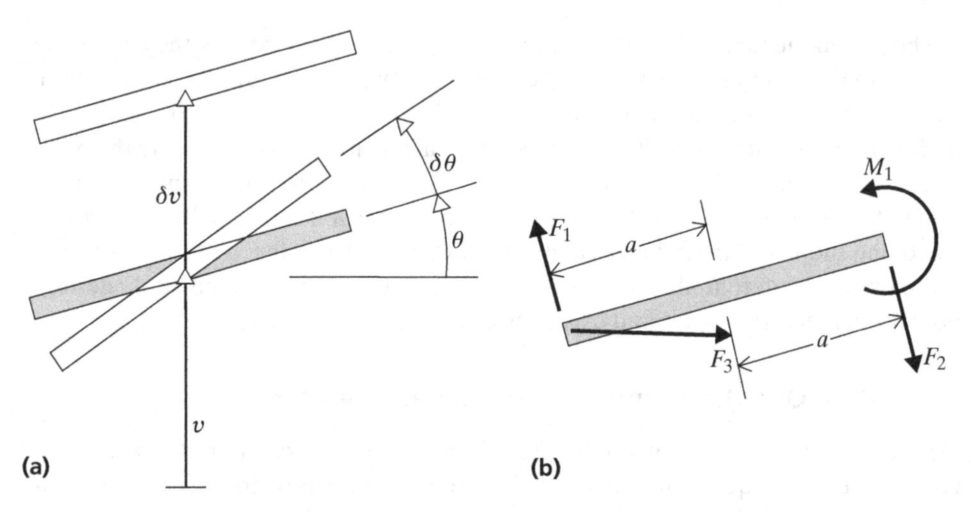

Figure 1.6. (a) Illustration of two (much exaggerated) virtual displacements. (b) Loads acting on the flat ruler moving in the plane of the paper.

system, virtual displacements give rise to virtual work done on the mass system. *Virtual work* is defined as real forces and moments moving through virtual displacements that are, respectively, distances and rotations. In mathematical terms, just as the standard definition for real work is $dW = \boldsymbol{F} \cdot d\boldsymbol{r}$, the definition for virtual work is $\delta W = \boldsymbol{F} \cdot \delta \boldsymbol{r}$. Again the δ as opposed to the d, is used to distinguish between virtual and real quantities. It is vital to understand that in the expression for virtual work, the magnitudes and directions of the real forces are, of course, entirely unaffected by the virtual displacements and therefore to be regarded as constants when calculating the virtual work.

Consider, as an example for calculating virtual work, Figure 1.6(b), where the real forces and moments acting on the ruler are displayed. Following the definition above, the virtual work done by these (external) forces and moments is obtained by simply considering each virtual displacement in turn and paying attention to whether the force components move in the positive or negative direction as a result of the virtual displacement. The result for this example is

$$\begin{aligned}\delta W = (F_3 - F_1 \sin\theta + F_2 \sin\theta)\delta u + (F_1 \cos\theta - F_2 \cos\theta)\delta v \\ + (M_1 - aF_1 - aF_2 + aF_3 \sin\theta)\delta\theta.\end{aligned}$$

As is true in any virtual work expression, the coefficients of the virtual displacements are called *generalized forces*. The generalized force corresponding to the generalized displacement δq_i (i.e., the coefficient of) is symbolized as Q_i. In the above expression the generalized forces are the quantities within the parentheses. That is, in general

$$\delta W \equiv \sum Q_i \delta q_i$$

so here

$$\begin{aligned}Q_u &= (F_3 - F_1 \sin\theta + F_2 \sin\theta) \quad Q_v = (F_1 \cos\theta - F_2 \cos\theta) \\ Q_\theta &= (M_1 - aF_1 - aF_2 + aF_3 \sin\theta).\end{aligned}$$

Note that when the virtual displacement has the units of radians, the generalized force has the units of moment.

Calculating the virtual work of external forces acting on flexible bodies is just as straightforward as it is for rigid bodies. If, for example, a simply supported beam whose elastic axis is in the x direction is loaded by a single concentrated force of magnitude F acting in the positive z direction at $x = a$, then the total external virtual work is $\delta W_{ex} = F\delta w(a)$, where $w(x)$ is the beam's elastic axis deflection in the positive z direction.[6] The vertical support reactions do no virtual (or real) work because the constraints on this beam mathematical model are that there are no lateral deflections at the beam support reactions. If in addition to the above-cited concentrated force there were an additional loading consisting of an externally applied force per unit length, $f(x)$, acting in the negative z direction, then the additional external virtual work would be

$$\delta W_{ex} = -\int_{\text{left end}}^{\text{right end}} f(x)\,\delta w(x)\,dx.$$

The virtual work or real work done by the *internal* forces of the above-discussed rigid ruler or any other rigid mass is always zero because the rigidity of the body requires all of those equal and opposite internal force pairs to move together in lockstep. That is, because these paired internal rigid body forces move through the same distances, together they do virtual work or real work of equal magnitude but opposite sign, and thus all such work cancels. However, if the body is elastically or plastically deformable, the virtual work or real work of the internal forces is generally not zero. This is so because, the virtual or real deformations of the body allow those equal and opposite forces to move closer together to, or further away from, each other. In either case, the net work is generally neither zero nor insubstantial.

Before introducing the variational operator, it is important to recall that a *point function* is a mathematical function whose value only depends on the instantaneous values of its argument. Perhaps the most important aspect of point functions is that the integral of the differential of a point function is not path dependent. Examples of point functions are (i) a generalized coordinate, $q_i(t)$; (ii) for a continuum, a temperature change distribution, $\Delta T(x, y, z)$; and (iii) for a continuum, the vector field of x-component displacements from an original location to another location, $u(x, y, z, t)$. Examples of a path-dependent function are (i) $W(s)$, the work done on a wood block by the friction force as it slides over a tabletop; (ii) $W(s)$, the work done by a force whose point of application and line of action is fixed relative to a body that translates and rotates; and (iii) $\sigma(\epsilon)$, the stress, as a function of strain, in a metal after the yield point has been passed and plastic strains and deformations begin. A work function is a point function if and only if the forces contributing to the work are "energy conservative"; that is, when the forces themselves are point functions.

[6] The symbols for the Cartesian components of the total displacement or deflection vector used here are u for the x direction, v for the y direction, and w for the z direction.

The distinction between point functions and path-dependent functions is necessary because they require different mathematical treatment. For example, if $f(x)$ is a point function, as before, then the value of its integral depends on only the initial and final points of the integration interval. If $f(x)$ is path dependent, then its integration between initial and final points must be line integration where a specific path is specified. The value of the line integral between the fixed points changes when the path changes.

Let f be a point function. Then, in this case, the d of the symbol df may be regarded separately from the remainder of the differential symbol, f. The d is called the *differential operator*. That is, $df = d(f)$. (The chain rule result of the application of the differential operator to a point function is reviewed below.) If f were not a point function, then the d and the f cannot be separated; that is, df is merely a single, infinitesimal quantity.

It is also useful to permit the δ of δf to be separated from the point function f. The separated δ is called the *variational operator*. In brief, the rules for the application of the variational operator are the same as those for the differential operator (meaning that the calculus already learned by the reader is fully applicable here), with just one exception. That exception is that the result of applying the variational operator to any independent variable is zero. This selectivity with regard to dependent and independent functions arises from the fact that, for example, the virtual displacement $\delta v(t)$ can happen without any sort of change in the time variable, and similarly for some function $G(x)$, there can be a $\delta G(x)$ without there being a corresponding δx. Therefore the variational or δ operator is more selective than the differential operator in somewhat the same way as partial differentiation is more selective than total differentiation.

In general, which variables are to be classified as independent, and which are to be classified as dependent, follows from the analyst's choice among a force, a displacement, or a hybrid-type analysis. All routine structural dynamics analyses are displacement-type analyses. Thus in this textbook all displacement-type quantities (such as deflections and strains) are classified as dependent quantities; that is, dependent on the applied loads, time, and so on. As such, the displacements and strains always have nonzero variations that can be written, respectively, as δq and $\delta \epsilon$. However, all force-type quantities, such as forces, moments, and stresses, are classified as independent variables and their variations are zero. Furthermore, spatial coordinates, time, and temperature (if not used as a generalized coordinate) are always to be considered as independent variables and thus also always have zero variations. For example, consider a point function of the form $f(F, u, \partial u/\partial x, x, t)$, where F is a force, u is a displacement, x is a spatial coordinate, and t is time. By the calculus, where the combination of a comma and a variable subscript indicates partial differentiation with respect to that subscripted variable (e.g., $u,_x \equiv \partial u/\partial x$),

$$df = \frac{\partial f}{\partial F}dF + \frac{\partial f}{\partial u}du + \frac{\partial f}{\partial u,_x}du,_x + \frac{\partial f}{\partial x}dx + \frac{\partial f}{\partial t}dt,$$

but, by the rules for the variational operator, where only the dependent variables have nonzero variations,

$$\delta f = \frac{\partial f}{\partial u}\delta u + \frac{\partial f}{\partial u,_x}\delta u,_x. \tag{1.6}$$

A justification for the rule that the variational operator ignores independent variables while acting only on dependent variables is offered in Endnote (1). The reader may also consult Ref. [1.3], Chapter 2, or Ref. [1.2], Chapter 17, for further discussion and explanation of the variational operator.

An important pair of rules for the use of the variational operator are that the order of the application of variational operator with both derivatives and definite integrations is interchangeable. (When the order of multiplication of two quantities is interchangeable, those quantities are called multiplicatively *commutative*.) That is, it is always possible to write the following types of interchanges, where the derivative interchange is accepted as part of the definition of the variational operator, whereas the integration interchange is proven in Endnote (1).

$$\delta u_{,x} \equiv \delta\left(\frac{\partial u}{\partial x}\right) = \frac{\partial}{\partial x}(\delta u) \qquad \delta\iiint u_{,x}\, d(\mathrm{vol}) = \iiint \delta u_{,x}\, d(\mathrm{vol}). \tag{1.7}$$

1.8 The Lagrange Equations

Now that generalized coordinates, virtual work, and the variational operator have been introduced, the Lagrange equations can be derived, via Hamilton's principle, for a collection of particles. The choice of a collection of particles as the mass system to be studied is prompted by the original formulation of Newton's laws and the above development of its corollaries. The final step of the derivation of Hamilton's principle is to suggest that the collection of particles can be made so large and so dense as to be any continuum[7] model of interest, such as a structure composed of beams and other structural elements. This questionable approach to a limit is valid in this case because the exact same final result also can be obtained starting with an arbitrary continuum as the structural body of interest, and using the equations of solid mechanics. See Ref. [1.2], p. 514.

Consider a specific grouping of N particles that comprise a deformable body of arbitrary shape and, in the above sense of a limit, represent any structural material. Newton's second law applies to each of the N particles. Let $\boldsymbol{F}_i^{ex}$ be the sum of all the forces acting on the ith particle that originate from sources outside of this collection of particles (to be called the net external force acting on the ith particle). Let $\boldsymbol{F}_i^{in}$ be the sum of all the forces acting on the ith particle that originate from interactions with the other $N-1$ particles (i.e., the net internal force acting on the ith particle). Then, with $\boldsymbol{r}_i$ being the position vector for the mass m_i, Newton's second law applied to only the ith mass is

$$m_i\ddot{\boldsymbol{r}}_i = \boldsymbol{F}_i^{ex} + \boldsymbol{F}_i^{in}. \tag{1.2}$$

Transpose the right-hand side of Eq. (1.2), and dot-multiply the result by an (arbitrary) virtual displacement for the ith mass to obtain one form of *d'Alembert's principle*

$$\left(m_i\ddot{\boldsymbol{r}}_i - \boldsymbol{F}_i^{ex} - \boldsymbol{F}_i^{in}\right)\cdot\delta\boldsymbol{r}_i = 0. \tag{1.8a}$$

[7] A *continuum* is a material that can be endlessly subdivided, and regardless of how small the material sample, it will still exhibit the material properties of the material as a whole. Continuums are convenient fictions for structural analysis.

Again sum over all the N particles.

$$\sum \left(m_i \ddot{\boldsymbol{r}}_i - \boldsymbol{F}_i^{ex} - \boldsymbol{F}_i^{in}\right) \cdot \delta \boldsymbol{r}_i = 0. \tag{1.8b}$$

Note that the sum involving the net internal force terms is not zero in this equation because each of these equal and opposite forces that is a component of a net internal force is multiplied by a generally different virtual displacement. To illustrate this point, for simplicity, consider a deformable "body" that includes only two mass particles. In this case of just two particles, there is only one internal force acting on each particle. Of course, these forces are of equal magnitude, but are oppositely directed. Since the body is deformable, the position of the first particle can be fixed in space while the second particle moves toward the first particle. When the moving particle approaches the stationary particle, the internal force acting on the moving particle moves as well. The result of the internal force moving through an actual distance is (actual) work that is called the internal work. What can be true for the actual displacements also can be true for the virtual displacements because there are no rigid body constraints that prohibit the particles from virtual movement toward, or away from, each other.

From the definition of virtual work, the second and third sums in the above equation, respectively, are identified as the virtual work done on the mass system by the external forces, δW_{ex}, and the virtual work done on the mass system by the internal forces, δW_{in}. Hence the above equation can be rewritten as

$$\sum m_i \ddot{\boldsymbol{r}}_i \cdot \delta \boldsymbol{r}_i = \delta W_{ex} + \delta W_{in}. \tag{1.9}$$

Using the product rule for differentiation in the form $u Dv = D(uv) - v Du$, which is as valid for the variational operator as it is for the differential or derivative operator, rewrite the left-hand side as follows:

$$\sum m_i \ddot{\boldsymbol{r}}_i \cdot \delta \boldsymbol{r}_i = \sum \left[m_i \frac{d}{dt}(\dot{\boldsymbol{r}}_i \cdot \delta \boldsymbol{r}_i) - m_i \dot{\boldsymbol{r}}_i \cdot \delta \dot{\boldsymbol{r}}_i \right].$$

The second term on the right-hand side can be rewritten as

$$\sum m_i \dot{\boldsymbol{r}}_i \cdot \delta \dot{\boldsymbol{r}}_i = \delta \sum (\tfrac{1}{2}\, m_i \dot{\boldsymbol{r}}_i \cdot \dot{\boldsymbol{r}}_i) \equiv \delta \sum (T_i) \equiv \delta T,$$

where each of the time-varying, velocity-dependent functions T_i, defined above, is called the *kinetic energy* of the ith particle, and, of course, T is the (system or total) kinetic energy. It can be seen from the second of the above equations that (i) the kinetic energy is always a positive quantity whenever the velocity is not zero; and (ii) because $\boldsymbol{r}_i(t)$ is a point function (dependent only on spatial position and time), so too is the kinetic energy. Since each of the kinetic energy expressions is a point function, the variational operator can be separated from kinetic energy expressions. The knowledge that the kinetic energy of a system of masses is a point function, and thus dependent only on the current position of the masses relative to the datum, will make calculating kinetic energy expressions a much easier task than it would be otherwise. This is so because, as illustrated in the next chapter, it will not be necessary to determine the actual trajectories of the masses, but only their velocities at the initial and end points of those trajectories. Substituting the above results into Eq. (1.8b), that equation can be rewritten as

$$\delta T + \delta W_{ex} + \delta W_{in} = \sum m_i \frac{d}{dt}(\dot{\boldsymbol{r}}_i \cdot \delta \boldsymbol{r}_i).$$

Since each of the particle mass values is a constant, the derivative operator acting on the right-hand side of the above equation may be placed at the beginning of the summation. The next step of the derivation of Hamilton's principle is to integrate both sides of the above equation over a time interval defined by the arbitrary time limits t_1 and t_2. Recall that, as is true for real displacements over time, the virtual displacements of each of the mass particles, δr_i, must be a smooth function of time. That is, the virtual displacements must abide by the constraint applied to the real displacements that they be smoothly continuous over time. It is now convenient to require that all these arbitrary virtual displacements have zero values at the two end points of the arbitrary time interval of integration, t_1 and t_2. Then, because the right-hand side is the definite integral of an exact differential, it can be integrated immediately with a zero result at the two arbitrary time limits. The result is *Hamilton's principle*:

$$\int_{t_1}^{t_2} (\delta T + \delta W_{ex} + \delta W_{in})\, dt = 0. \tag{1.10}$$

An important point about Hamilton's principle is that in its derivation there are no restrictions placed on either the external forces or the material represented by the internal forces of the N particles. Specifically, these internal or external forces can be either energy conservative or energy nonconservative because nowhere has the variational operator, δ, been separated from the W on the presumption that work was a point function.

It is a short series of steps from Hamilton's principle to the Lagrange equations. Like the flat ruler used in a previous example in this chapter (and like the structural models that are developed in later chapters), the motion of a collection of N mass particles (or M rigid or near rigid masses) can be described in terms of n generalized coordinates, q_j, $j = 1, 2, \ldots, n$. Since the position vector and the generalized coordinates both locate the position of m_i, there must be a functional relationship between the position vectors and the generalized coordinates. Of course, like the position vectors, the generalized coordinates are also implicit functions of time. It is also possible for the position vector to be an explicit function of time, as can be seen from the following example. Consider a pea (a particle) moving on the upper surface of a computer's hard disk as the disk rotates at a constant angular velocity ω. Let the generalized coordinates of the pea be the polar coordinates[8] $r(t)$ and $\theta(t)$, where θ is measured from a radial line *fixed on the disk*. Since that rotating fixed line on the upper disk surface cannot be expected to be part of a valid coordinate system, let the rotated position of the hard disk be measured relative to the front of the computer, which is stipulated to move uniformly relative to a valid coordinate system whose

[8] Here scalar r is not the magnitude of the soon-to-be-discussed position vector, vector $\boldsymbol{r}$.

fixed unit vectors are $\boldsymbol{i}$ and $\boldsymbol{j}$. Then the position vector for the pea relative to the valid coordinate system is

$$\boldsymbol{r}(r,\theta;t) = [r\cos(\theta+\omega t)]\boldsymbol{i} + [r\sin(\theta+\omega t)]\,\boldsymbol{j}.$$

Now that, by illustration, the validity of the functional form $\boldsymbol{r}_i = \boldsymbol{r}_i(q_j, t)$ has been established, the chain rule for differentiation leads to the following conclusion:

$$\frac{d\boldsymbol{r}_i}{dt} \equiv \dot{\boldsymbol{r}}_i = \sum_{j=1}^{n} \frac{\partial \boldsymbol{r}_i}{\partial q_j}\dot{q}_j + \frac{\partial \boldsymbol{r}_i}{\partial t}.$$

Since the velocity vector $\dot{\boldsymbol{r}}_i$ is clearly a function of the time derivatives of the generalized coordinates, which are called the generalized velocities, so too is the kinetic energy T. This example illustrates that the kinetic energy may also, through the agency of the above partial derivatives, be a function of the generalized coordinates themselves and an explicit function of time itself. Thus, in general, the kinetic energy can be written in functional form as $T = T(\dot{q}_i, q_i, t)$. Then application of the variational operator as per Eq. (1.6), where, as always, $\delta t = 0$, yields

$$\delta T = \sum\left[\left(\frac{\partial T}{\partial \dot{q}_i}\right)\delta\dot{q}_i + \left(\frac{\partial T}{\partial q_i}\right)\delta q_i\right]. \tag{1.11}$$

As discussed above, unlike the kinetic energy, the work functions can be path dependent. Again, as previously illustrated, regardless of the nature of those associated actual forces, it is always possible to write

$$\delta W_{ex} = \sum Q_i^{ex}\delta q_i \quad \text{and} \quad \delta W_{in} = \sum Q_i^{in}\delta q_i. \tag{1.12a}$$

The general validity of, say, the first of these previously introduced equations can be now deduced by using the chain rule with the variational operator as follows:

$$\begin{aligned}\delta W_{ex} &= \sum_i \boldsymbol{F}_i^{ex}\cdot\delta\boldsymbol{r}_i = \sum_i \boldsymbol{F}_i^{ex}\cdot\sum_j \frac{\partial \boldsymbol{r}_i}{\partial q_j}\delta q_j \\ &= \sum_j\left(\sum_i \boldsymbol{F}_i^{ex}\cdot\frac{\partial \boldsymbol{r}_i}{\partial q_j}\right)\delta q_j = \sum_j Q_j^{ex}\delta q_j.\end{aligned}$$

Again, the quantity Q_j^{ex} is called the jth generalized external force in keeping with the concept that force times displacement equals work. Be sure to note that the above equation for generalized forces is not nearly as efficient a way of determining the values of generalized forces as that previously illustrated. Recall that the more efficient approach was simply taking, in turn, positive variations in the generalized coordinates and calculating the work done by the applied forces and moments with each such virtual displacement.

When a generalized force of the summation is known to be energy conservative, such as an external generalized force resulting from a gravitational field, it is usually convenient to go one step further by introducing a potential (point) function that is associated with the energy conservative work done by such a force. In the case of the external work, let the potential function be written as $V = -W_{ex}$. Then, by use of the chain rule for partial differentiation

$$\delta V(q_i) = \sum_{i=1}^{n} \frac{\partial V}{\partial q_i}\delta q_i = -\delta W_{ex} = -\sum_{i=1}^{n} Q_i \delta q_i. \tag{1.12b}$$

Since each of the generalized coordinates is independent of the other generalized coordinates, so too are the variations on the generalized coordinates. Then by setting all but one of the varied generalized coordinates equal to zero, it is possible to conclude that the external generalized force is obtained from this potential point function by differentiation; that is, $Q_j = -\partial V/\partial q_j$. Note that here and from here on, the superscript *ex* for the external forces will be dropped because the other forces, the internal forces, will generally be only expressed in terms of a similar potential function as explained in the next paragraph.

From this point on, consider that the N particles constitute a continuum, and, unless otherwise stated, let that continuum be an elastic body. An elastic body or an elastic material is an ideal where all work done on the body is stored without energy loss, and thus can be fully recovered, say, by allowing the body to snap back to its original size and shape. In other words, elastic bodies are idealizations that ignore the internal friction, and hence energy loss, that occurs when real bodies are deformed. Furthermore, elastic bodies have infinite viscosity. This means that elastic bodies do not over time continue to deform under load, the phenomenon called creep. (Since rubber does creep, it is not an elastic material.)

The work or energy stored in an elastic body is called the elastic strain energy. Unfortunately, the above derivation of Hamilton's principle based on a collection of N particles is utterly useless for deducing a mathematical description of the strain energy. It is necessary to resort to the ideas and descriptions of solid or continuum mechanics where it is possible to talk about stresses and strains. From Ref. [1.3], Chapter 1, or Ref. [1.2], Chapter 17, the strain energy for an elastic body of volume *vol* is

$$U = \frac{1}{2}\iiint [\sigma_{xx}\epsilon_{xx} + \sigma_{yy}\epsilon_{yy} + \sigma_{zz}\epsilon_{zz} + \sigma_{xy}\gamma_{xy} + \sigma_{xz}\gamma_{xz} + \sigma_{yz}\gamma_{yz}]d(\text{vol}),$$

where each stress is possibly a function of all the strains. Since the strains are functions of the relative displacements, and the displacements of a solid body system, even a flexible one, can be described by the generalized coordinates of the system, the strain energy can be expressed in terms of the generalized coordinates. The details of those steps in the general context of the finite element method (FEM) are explained, for example, in Ref. [1.2], p. 711ff. The application of the FEM is described in Chapter 3.

In summary, if the external virtual work is separated into its energy conservative and energy nonconservative components, and the energy conservative components are written in terms of appropriate potential energies, whereas the nonconservative components are left in terms of generalized forces, then it is possible to write

$$\begin{aligned}\delta W_{ex} + \delta W_{in} &= -\delta V - \delta U + \sum Q_j \delta q_j \\ &= \sum \left[\left(\frac{\partial V}{\partial q_j}\right)\delta q_j + \left(\frac{\partial U}{\partial q_j}\right)\delta q_j - Q_j \delta q_j \right],\end{aligned} \tag{1.13}$$

where Q_j now represents just those external generalized forces that are chosen not to be, or, because they are not energy conservative, cannot be, included in the external potential energy expression. Hence, substituting Eqs. (1.11) and (1.13) into Eq. (1.10), and interchanging the order of the finite integration and the finite summation, a step that is always possible because the sum is a finite sum and the limits of integration are also finite, yields the following elaborated form of Hamilton's principle:

$$\sum \int_{t_1}^{t_2} \left[\left(\frac{\partial T}{\partial \dot{q}_j} \right) \delta \dot{q}_j + \left(\frac{\partial T}{\partial q_j} \right) \delta q_j - \left(\frac{\partial V}{\partial q_j} \right) \delta q_j - \left(\frac{\partial U}{\partial q_j} \right) \delta q_j + Q_j \delta q_j \right] dt = 0. \tag{1.14}$$

The first term has as a factor the quantity $\delta(dq_j/dt)$, whereas all the other terms have as a factor the term δq_j. Again, because the variational operator and the derivative operator are commutative, that is, because

$$\delta \dot{q}_j \equiv \delta \left(\frac{dq_j}{dt} \right) = \frac{d}{dt}(\delta q_j),$$

these two terms, the varied generalized coordinate and varied generalized velocity, are not independent of each other. The first term can be changed to have simply δq_i as its factor by integrating by parts. The result for the typical ith term is

$$\int_{t_1}^{t_2} \left(\frac{\partial T}{\partial \dot{q}_i} \right) \delta \dot{q}_i dt = \int_{t_1}^{t_2} -\frac{d}{dt} \left(\frac{\partial T}{\partial \dot{q}_i} \right) \delta q_i dt + \left(\frac{\partial T}{\partial \dot{q}_i} \right) \delta q_i \Big|_{t_1}^{t_2}.$$

Recall the requirement stated before Eq. (1.10) that all virtual displacements, including the δq_i, are to be zero at the end points of this arbitrary time interval of integration. Thus the second term on the right-hand side disappears. Substituting this result into Eq. (1.14) yields

$$\sum \int_{t_1}^{t_2} \left[-\frac{d}{dt} \left(\frac{\partial T}{\partial \dot{q}_i} \right) + \frac{\partial T}{\partial q_i} - \frac{\partial V}{\partial q_i} - \frac{\partial U}{\partial q_i} + Q_i \right] \delta q_i dt = 0. \tag{1.15}$$

The final argument begins by noting that all the virtual displacements δq_i within the open interval (t_1, t_2) are smoothly varying over time but otherwise arbitrary. In other words, they are what the analyst wishes them to be. It is now convenient to choose the following values for the virtual displacements. Let all but one, say δq_j, of the virtual displacements be zero. The result of this choice is that the summation symbol in Eq. (1.15) disappears, and all the i summation subscripts change to a specific j subscript. The integral that is left has an integrand that has two factors. The first factor is enclosed within brackets, and the second factor is δq_j. Recall that inside the limits of integration the value of this single nonzero virtual displacement is not restricted in any way other than that it must be small and, of course, that it must be a continuous function of time. In other words, $\delta q_j(t)$ can be positive, negative, or both inside the time interval of integration. Whatever values are chosen for continuous time function δq_j, the value of the one remaining integral remains fixed at zero. The only way the integral can be zero, regardless of the choice for the values of the virtual displacement, is for the quantity in brackets to always be zero. Hence the conclusion is the jth *Lagrange equation*:

$$\frac{d}{dt}\left(\frac{\partial T}{\partial \dot{q}_j}\right) - \frac{\partial T}{\partial q_j} + \frac{\partial (U+V)}{\partial q_j} = Q_j. \tag{1.16}$$

There are n of these equations, one for each virtual displacement. Since there are also n unknown displacement functions of time, the q_i, there are the proper number of equations for the number of unknowns. Recall that this result began with Newton's second law. Hence the Lagrange equations are Newton's second law in energy and generalized coordinate form utilizing Euler's concept of virtual work.

The examples in the early chapters of this textbook illustrate various applications of the Lagrange equations. In particular, these chapters review the calculation of the system kinetic energy, the calculation of the system virtual work to obtain the generalized forces, Q_i, and the construction of the potential functions V and U. A guide for writing kinetic energy expressions is presented below. Most applications are straightforward. However, an occasional application requires a familiarity with the derivation of these equations and a knowledge of Newton's laws from which Hamilton's principle and thus the Lagrange equations are derived as above.

1.9 Kinetic Energy

To conveniently elaborate the kinetic energy expressions, consider a rigid body or a flexible body that is only undergoing negligibly small geometric changes over time. Let the body be moving in three-dimensional space as pictured in Figure 1.2. From the definition of the kinetic energy term in the derivation of the Lagrange equation:

$$\begin{aligned} T &= \tfrac{1}{2}\sum m_i \dot{\boldsymbol{r}}_i \cdot \dot{\boldsymbol{r}}_i = \tfrac{1}{2}\sum m_i(\dot{\boldsymbol{r}} + \dot{\boldsymbol{e}}_i)\cdot(\dot{\boldsymbol{r}} + \dot{\boldsymbol{e}}_i) \\ &= \tfrac{1}{2}\left(\sum m_i\right)\dot{\boldsymbol{r}}\cdot\dot{\boldsymbol{r}} + \dot{\boldsymbol{r}}\cdot\left(\sum m_i \dot{\boldsymbol{e}}_i\right) + \tfrac{1}{2}\sum m_i \dot{\boldsymbol{e}}_i \cdot \dot{\boldsymbol{e}}_i \\ &= \tfrac{1}{2}\, m(CG\ velocity)^2 + 0 + \tfrac{1}{2}\sum m_i \dot{\boldsymbol{e}}_i \cdot \dot{\boldsymbol{e}}_i. \end{aligned}$$

Let u, v, w be the three Cartesian components of the displacement vector for the system center of mass from, say, its at rest position to its time-varying position. In other words, let $\boldsymbol{r} = u\boldsymbol{i} + v\boldsymbol{j} + w\boldsymbol{k}$. Then the time derivative of the center of mass position vector is $\dot{\boldsymbol{r}} = \dot{u}\boldsymbol{i} + \dot{v}\boldsymbol{j} + \dot{w}\boldsymbol{k}$. Hence, one possible and often convenient way of writing the first part of the kinetic energy, that part directly related to the motion of the center of mass, for vibrating mass systems is

$$T_{cg} = \tfrac{1}{2}\, m\dot{\boldsymbol{r}}\cdot\dot{\boldsymbol{r}} = \tfrac{1}{2}\, m(\dot{u}^2 + \dot{v}^2 + \dot{w}^2).$$

This result can be viewed as a manifestation of the Pythagorean theorem; that is, the square of the total velocity is the sum of the squares of the three orthogonal components of the total velocity.

The second part of the kinetic energy is the part associated with the spin about the center of mass because it involves the time derivatives of the $\boldsymbol{e}_i$ vector. The length of the $\boldsymbol{e}_i$ vector changes very little as a structural body moves and flexes. Thus the great majority of the time change of the $\boldsymbol{e}_i$ vector is associated with its change of orientation,

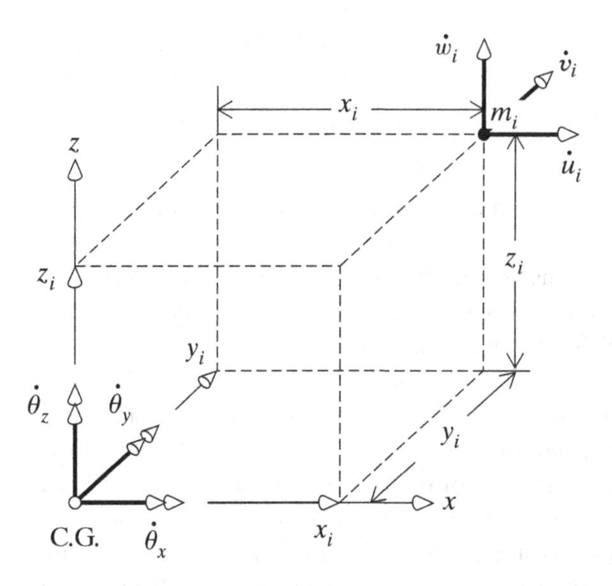

Figure 1.7. Setup for calculating rectilinear velocities $\dot{u}$, $\dot{v}$, $\dot{w}$ at m_i due solely to the three angular velocities at the CG, where $\boldsymbol{e}_i = x_i\boldsymbol{i} + y_i\boldsymbol{j} + z\boldsymbol{k}$.

that is, its spin. The kinetic energy associated with the spin of a structural body can be detailed in the following way. Consider Cartesian coordinate axes originating at the center of mass of the body and that have any fixed orientation with respect to the constant mass and (very closely) constant geometry of the body. Such a coordinate system is called *body fixed*. See Figure 1.7, where the ith mass particle is located by the position vector

$$\boldsymbol{e}_i = x_i\boldsymbol{i} + y_i\boldsymbol{j} + z_i\boldsymbol{k}.$$

Let the components of the angular velocity about the arbitrarily selected x axis be designated as $\dot{\theta}_x$, and so on,[9] which are positive according to the right-hand rule about the positive coordinate axes. The additional velocities at m_i because of the small rotations about the x, y, and z axes, labeled $\dot{u}_i$, $\dot{v}_i$, and $\dot{w}_i$ in the figure, can be deduced from Figure 1.7 by using the right-hand rule for the angular velocities $\dot{\theta}_x, \dot{\theta}_y, \dot{\theta}_z$, along with the perpendicular offset distances from the CG to m_i. Thus the total velocity vector at m_i is, again, for small rotations

$$\dot{\boldsymbol{e}}_i = \dot{u}_i\boldsymbol{i} + \dot{v}_i\boldsymbol{j} + \dot{w}_i\boldsymbol{k} = (z_i\dot{\theta}_y - y_i\dot{\theta}_z)\boldsymbol{i} + (x_i\dot{\theta}_z - z_i\dot{\theta}_x)\boldsymbol{j} + (y_i\dot{\theta}_x - x_i\dot{\theta}_y)\boldsymbol{k},$$

which is nothing but the cross product of the angular velocity vector and $\boldsymbol{e}_i$. It is valid to sum these three angular velocity terms in this way because, as previously discussed, small angles, and hence their time derivatives (the angular velocities), can be treated as vector quantities even though larger angles, in general, cannot be so described. When the "spin" kinetic energy of the ith particle is summed with that

[9] These body-fixed Cartesian axes rotate relative to the valid Cartesian coordinate system, and thus it is not valid to write Newton's laws or their corollaries in this coordinate system. However, it is valid to use relative velocities in this system.

of all other such particles to obtain the spin kinetic energy for the entire body, the result is

$$T_{spin} = \sum T_{i,spin} = \tfrac{1}{2}\sum m_i \dot{\boldsymbol{e}}_i \cdot \dot{\boldsymbol{e}}_i$$
$$= \tfrac{1}{2}\sum m_i \Big[\big(z_i^2\dot{\theta}_y^2 + y_i^2\dot{\theta}_z^2 - 2y_i z_i \dot{\theta}_y\dot{\theta}_z\big)$$
$$+ \big(x_i^2\dot{\theta}_z^2 + z_i^2\dot{\theta}_x^2 - 2x_i z_i \dot{\theta}_x\dot{\theta}_z\big) + \big(y_i^2\dot{\theta}_x^2 + x_i^2\dot{\theta}_y^2 - 2x_i y_i \dot{\theta}_x\dot{\theta}_y\big)\Big].$$

For this case of a rigid or a near rigid body, that is, where there are a great many particles packed into a fixed or very nearly fixed, closed, boundary surface, the above summations over the mass particles are conveniently restated as three-dimensional integrals that sum over differential masses $m_i \to dm$. This transition from a collection of particles to a continuum prompts the introduction of the mass density, ρ, where $dm = \rho d(\text{vol}) = \rho\, dx\, dy\, dz$. Then the summations found in the above spin kinetic energy expression can be written in terms of the angular velocities and the following integral forms suitable for any slightly deformed body

$$H_{xx} = \iiint \rho(y^2 + z^2)\, d(\text{vol}) \qquad H_{xy} = -\iiint \rho\, x\, y\, d(\text{vol})$$
$$H_{yy} = \iiint \rho(x^2 + z^2)\, d(\text{vol}) \qquad H_{xz} = -\iiint \rho\, x\, z\, d(\text{vol})$$
$$H_{zz} = \iiint \rho(x^2 + y^2)\, d(\text{vol}) \qquad H_{yz} = -\iiint \rho\, y\, z\, d(\text{vol}).$$

The next organizing step is to arrange the previously stated "spin" kinetic energy for this single finite body in matrix form as

$$T_{spin} = \tfrac{1}{2}\lfloor\dot{\theta}\rfloor[H]\{\dot{\theta}\},$$

where

$$\{\dot{\theta}\}^t = \lfloor \dot{\theta}_x \quad \dot{\theta}_y \quad \dot{\theta}_z \rfloor$$

is called the angular velocity vector (row matrix) and where

$$[H] = \begin{bmatrix} H_{xx} & H_{xy} & H_{xz} \\ H_{xy} & H_{yy} & H_{yz} \\ H_{xz} & H_{yz} & H_{zz} \end{bmatrix}$$

is called the mass moment of inertia matrix. Therefore the total kinetic energy of the single rigid or near rigid shaped body is as follows:

$$T = \tfrac{1}{2}\lfloor \dot{u} \quad \dot{v} \quad \dot{w} \rfloor \begin{bmatrix} m & 0 & 0 \\ 0 & m & 0 \\ 0 & 0 & m \end{bmatrix} \begin{Bmatrix} \dot{u} \\ \dot{v} \\ \dot{w} \end{Bmatrix} + \tfrac{1}{2}\lfloor\dot{\theta}\rfloor[H]\{\dot{\theta}\}. \tag{1.17}$$

Specifically, the second moments of mass for a continuum, such as H_{xx}, with two identical subscripts, are called mass moments of inertia. The quantity H_{xx} is spoken of as the mass moment of inertia "about the x axis" because its integrand factor $y^2 + z^2$ is the squared distance from the x axis to the ith particle or, in this case, the equivalent infinitesimal mass $dm = \rho d(\text{vol})$. Mass moments of inertia are measures of the dispersal of the body mass from the center of mass much like the statistical variance.

Second moments of mass such as H_{xy}, with subscripts that are not the same, are called mass products of inertia. Note that the order of the subscripts is immaterial. Products of inertia are measures on the nonsymmetry of the mass distribution. If, for example, for each x plane through the body, the mass is distributed symmetrically with respect to either the y or z axis, then $H_{yz} = 0$. The zero value is a result of the fact that for this symmetry, every dm that has a positive value of the product $yzdm$, there will be another dm with a equal but negative value of $yzdm$.

Note that when the mass moments of inertia are arranged as above in a three by three matrix where mass moment terms with an x subscript are placed in the first row and column, those with a y subscript are placed in the second row and column, and so on, the matrix $[H]$ is a symmetrical matrix just as is the three by three matrix $[m]$ that just has the value of the mass as its diagonal terms.

The symbol H was chosen to be very clear about the distinction among mass moments of inertia, area moments of inertia for beams, and the St. Venant constant for uniform torsion for beams (another area distribution measure related to the twisting of beams). Their respective basic symbols in this textbook are H, I, and J, with appropriate subscripts on the first two.

Recall that the triple matrix product involving the angular velocities and $[H]$ are calculated relative to arbitrarily chosen, body-fixed Cartesian coordinate axes. This calculation can be simplified by choosing that unique body-fixed Cartesian coordinate system where $[H]$ is a diagonal matrix; that is, where all the products of inertia are zero. The Cartesian coordinates for which $[H]$ is a diagonal matrix are called the principal mass axes, and the corresponding mass moments of inertial are called the principal mass moments of inertia. In many cases it is a simple matter to identify the principal mass axes. If the body has an axis of symmetry with respect to both body geometry and mass distribution, then that axis of symmetry is always one of the three principal axes. If the principal axes are not evident, then the location of the principal mass axes and the magnitudes of the principal mass moments of inertia can be calculated by solution of the same matrix eigenvalue problem used, for example, to extract the principal stresses at any point in a continuum from the three by three general stress matrix. See, for example, Ref. [1.3], Chapter 1. It is also worth noting that just as there are invariants associated with the stresses, there are invariants associated with moments of inertia. As an example of this parallel between stress and moment of inertia invariants (with respect to coordinate axis rotation), note that just as

$$\sigma_{xx} + \sigma_{yy} + \sigma_{zz} = \sigma_{XX} + \sigma_{YY} + \sigma_{ZZ}$$

where capital letter subscripts indicate principal axes, then

$$H_{xx} + H_{yy} + H_{zz} = H_{XX} + H_{YY} + H_{ZZ}.$$

Finally, from Eq. (1.17), when the motion is confined to a single plane, say the z plane, the kinetic energy in terms of the center of mass rectilinear and angular velocities is simply

$$T = \tfrac{1}{2}\, m(\dot{u}^2 + \dot{v}^2) + \tfrac{1}{2} H_{CG}\dot{\theta}_{CG}^2, \qquad (1.18)$$

where H_{CG} is the mass moment of inertia about an axis perpendicular to the z plane. Endnote (2) offers an immediate, although peripheral, example of the use of kinetic energy in a structural context.

1.10 Summary

For structural engineers to comprehend the rich and diverse history of structural analysis, a knowledge of the dynamics of rigid and flexible bodies is essential. For example, one of the four classical methods of analysis[10] in the theory of elastic stability is the dynamic method. See Chapter 1 in Ref. [1.4]. One example of the need for the dynamic method for solving an elastic stability problem is the seemingly static problem called Beck's problem, where a fixed magnitude compressive force acts at the end of a long beam and remains parallel to the beam axis (making it a non-conservative force) as the beam buckles. Newer developments such as the discrete element method, Refs. [1.5,1.6,1.7], which has been applied to the collapse and post-collapse histories of structures, also require a knowledge of structures and dynamics. However, because of the general limitation of the material discussed in this textbook to small angles of rotation in three-dimensional space, the knowledge of dynamics required for this textbook is limited to the basic equations summarized below, where the emphasis is on the use of the Lagrange equations.

1. Newton's second law of motion for a mass m of finite spatial dimensions, and velocity $\boldsymbol{v}$ at its center of mass, subjected to a net (external) force of magnitude $\boldsymbol{F}$ is

$$\boldsymbol{F} = m\frac{d\boldsymbol{v}}{dt}. \tag{1.1}$$

2. The rotational motion $\theta(t)$of a mass m about an axis parallel to a fixed z axis, or about a z axis passing through the center of mass, for both small and large angles in the single plane of motion, are respectively described in terms of external moments by

$$H_{FA}\ddot{\theta} = M_{FA} \tag{1.5a}$$

and

$$H_{CG}\ddot{\theta} = M_{CG}. \tag{1.5b}$$

3. The jth Lagrange equation in terms of the jth DOF q_j is

$$\frac{d}{dt}\left(\frac{\partial T}{\partial \dot{q}_j}\right) - \frac{\partial T}{\partial q_j} + \frac{\partial (U+V)}{\partial q_j} = Q_j. \tag{1.16}$$

4. The total kinetic energy of a rigid or near rigid body is

$$T = \tfrac{1}{2}\lfloor \dot{u} \quad \dot{v} \quad \dot{w} \rfloor [\backslash m \backslash] \begin{Bmatrix} \dot{u} \\ \dot{v} \\ \dot{w} \end{Bmatrix} + \tfrac{1}{2}\lfloor \dot{\theta} \rfloor [H] \{\dot{\theta}\}. \tag{1.17}$$

[10] The other three methods are the adjacent equilibrium method, the imperfection method, and the energy method.

The three angles in the above angular velocity vector are limited to being angles small enough that the angle (in radians) is a good approximation to the tangent of the angle because the second triple product of this kinetic energy expression involves the addition of angles from different orthogonal planes.

In this textbook, after preliminary use of the force and moment equations, Eqs. (1.1) and (1.5), the focus is entirely on the easier to use Lagrange equations, Eq. (1.16), using the description of the kinetic energy that is set forth in Eq. (1.17). The potential functions in the Lagrange equation, $V$and U, are discussed at greater length. The mathematical description of the strain energy U is further developed in Chapters 3 and 4. The following two example calculations hopefully will shed further light on the fourth component of the Lagrange equations, the generalized forces.

EXAMPLE 1.1 **(a)** Consider the system of two rigid bodies connected by a single, massless elastic spring, sketched in Figure 1.8(a). The rigid bodies roll on the rigid surface without friction. This is a two-DOF system. Select the generalized coordinates q_1 and q_2 to locate the position of each mass particle in the system. The task is to calculate the virtual work resulting from the presence of the externally applied force $F(t)$, and thereby deduce the values of the generalized forces associated with each of the generalized coordinates q_1 and q_2. Hint: Give each generalized coordinate a small variation in turn and calculate the virtual work that arises from those virtual displacements. Focus here only on the external force $F(t)$. The virtual work of the internal forces such as those associated with the elastic spring are taken up in part (c) of this example problem.

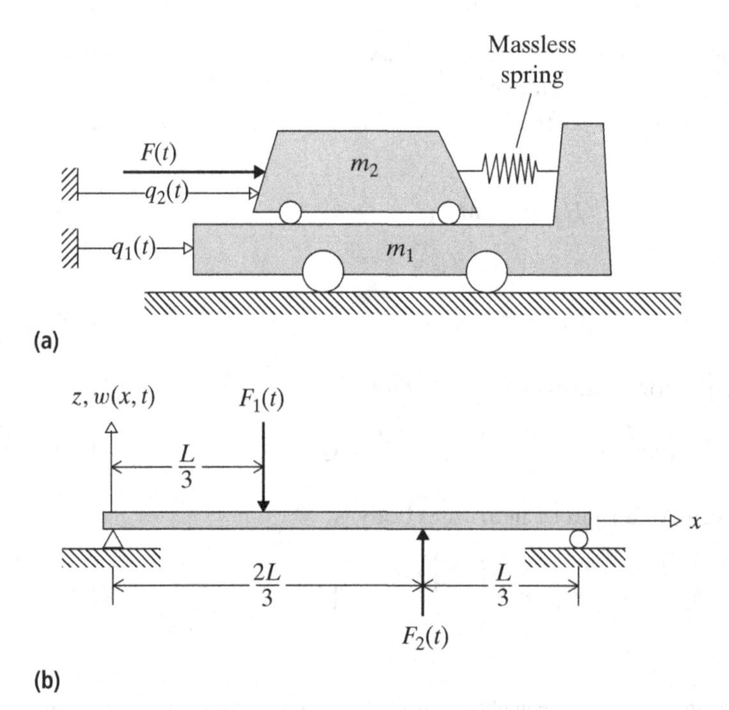

Figure 1.8. (a) Example 1.1. (b) Example 1.2.

(b) Repeat the above problem, but this time use as the two generalized coordinates q_1 and q_3, where q_3 specifies the relative motion between the two masses and is positive when the horizontal distance between the right-hand ends of the two masses increases.

(c) Repeat part (a), but this time include the internal forces of the elastic spring. (Including the internal elastic forces in the virtual work expression is an alternative to describing those same elastic forces using the strain energy potential function.) Recall that the spring force is the spring stiffness factor, k, multiplied by the relative deflection of the spring attachment points.

SOLUTION (a) Fix the value of the time, and give the nonzero value of q_1 a small, positive increment, δq_1. Observe what work is done by the external force $F(t)$. Since the nonzero value of q_2 has been unchanged, neither the upper mass moved nor the force $F(t)$ moved. Thus the virtual work associated with δq_1 is zero. Now increment q_2 and observe the work done. This time the force $F(t)$ moves in its positive direction a distance δq_2, and the virtual work is $F\delta q_2$. Thus the total virtual work is $\delta W = 0\delta q_1 + F\delta q_2$. Since the generalized forces are simply the coefficients of the virtual displacements, $Q_1 = 0$, and $Q_2 = +F(t)$.

(b) Again give q_1 a small increment δq_1, keeping q_3 fixed. With q_3 fixed, the upper mass moves through the same distance that the lower mass moves, δq_1. Hence the external force $F(t)$ moves through that same distance and does virtual work equal to $F\delta q_1$. Now increment q_3 while fixing q_1. Since any increase in q_3 when q_1 is fixed results in the externally applied force $F(t)$ moving opposite to is own direction, the virtual work done is $-F\delta q_3$. Therefore, the generalized forces are $Q_1 = +F(t)$, and $Q_3 = -F(t)$.

(c) All equal and opposite pairs of internal forces acting on the mass particles of the rigid bodies move together and thus do no virtual work. As previously discussed, such is not true for flexible bodies. From the original derivation of virtual work, note that the only forces to be considered are those that act on mass particles. Thus the only elastic forces to be considered are those acting on the two rigid masses, that is, those at the spring ends. Without loss of generality, let $q_1 > q_2$. In this case, the spring transmits a tensile force $k(q_1 - q_2)$ that acts to the right on mass two and to the left on mass one. Hence, following the above outlined procedure, the virtual change in the generalized coordinates produce virtual work equal to

$$\delta W = -k(q_1 - q_2)\delta q_1 + [F + k(q_1 - q_2)]\delta q_2.$$

The generalized forces are, as always, merely the coefficients of their respective virtual displacements. You should confirm that if the original choice was $q_1 < q_2$, the result still would be the same.

The generalized forces that are internal to an elastic mass system are much more easily obtained by resorting to their potential, the strain energy. In this textbook, that is what is done routinely in later chapters. ★

EXAMPLE 1.2 The time-varying deflection in the z direction of the axis of the simply supported beam shown in Figure 1.8(b), symbolized as $w(x, t)$, is (crudely) approximated by the following two-term series. The quantities q_1 and q_2 are generalized coordinates because their specification precisely locates (as far as this simple mathematical model is concerned) every mass particle along the length of the beam within the limits of engineering beam theory.

$$w(x,t) = q_1(t)\sin\frac{\pi x}{L} + q_2(t)\sin\frac{2\pi x}{L}.$$

Calculate the generalized forces associated with these two generalized coordinates, q_1 and q_2.

COMMENT These generalized coordinates are examples of what are called distributed coordinates, or sometimes Fourier coefficients, in that they provide an amplitude for a fixed spatial distribution. Later in this text, such coordinates are used extensively. Note that the two selected spatial distributions are the first two terms of a Fourier sine series. The use of only two such terms generally would not produce a reasonable approximation for the deflections along the length of the beam. However, commonly, six to ten such terms of a Fourier series would produce a very good approximation.

SOLUTION In this example problem, note that the deflections $w(x, t)$ are positive upwards, and simply write the basic virtual work expression $\delta W = -F_1\delta\ w(L/3, t) + F_2\delta w(2L/3, t)$. Note that the support reactions do no work or virtual work because they are constrained to have zero real deflections, and virtual deflections are not allowed to violate constraints. Now it is just a matter of substituting the given deflection approximation into the above virtual work expression. Thus $\delta W = -F_1[\delta q_1 \sin(\pi/3) + \delta q_2 \sin(2\pi/3)] + F_2[\delta q_1 \sin(2\pi/3) + \delta q_2 \sin(4\pi/3)]$. Hence, after regrouping terms around the virtual displacements, $Q_1 = \sqrt{3}/2(F_1 - F_2)$ and $Q_2 = -\sqrt{3}/2(F_1 + F_2)$.

Note that, as in the above virtual work expression, if there is a variational operator present in one term of an equation, then every term on both sides of the equation must also involve a variational operator. Note that in this problem there is a possibility of confusion regarding terms. Although δw is a varied deflection (a virtual deflection), the term virtual displacement is always reserved for variations on the generalized coordinates. In this case the virtual displacements are δq_1 and δq_2, and not δw. Generalized forces are associated only with virtual displacements.

Although generalized coordinates are always functions of time, generalized coordinates are also dependent on the time-varying loads acting on the mass system. Both the loads and time are independent variables and as such they have no variation. Therefore it is not possible to use, for example, the chain rule and the variational operator to write virtual displacements as functions of other varied (independent) quantities because those other variations must be zero. The virtual displacements are essentially indivisible, basic quantities. ★

It is emphatically *not* the purpose of this textbook to deal with rigid body dynamics. However, one nonnumerical, rigid body dynamics analysis of historic interest is presented as Endnote (3) just to present one problem that makes use of the above material. Although this example problem is another demonstration of the use of the Lagrange equations, its main purpose is to suggest that rigid body dynamics can be very challenging, which in turn underlines the simplicity of most of the dynamics associated with structural analyses, the focus of this textbook. This problem also provides a small insight into the genius of Newton.

REFERENCES

1.1 Symon, K. R., *Mechanics*, 2nd ed., Addison-Wesley, Reading, MA, 1961.

1.2 Donaldson, B. K., *Analysis of Aircraft Structures*, McGraw-Hill, New York, 1993.

1.3 Shames, I. H., and C. L. Dym, *Energy and Finite Element Methods in Structural Mechanics*, Hemisphere, Washington, D.C., 1985.

1.4 Ziegler, H., *Principles of Structural Stability*, Blaisdell, 1968.

1.5 Cundall, P. A., "A computer model for simulating progressive, large scale movements in blocky rock systems," *Proc. Sym. Int. Soc. Rock Mech.*, vol. 8, 1971.

1.6 Anandrajah, A., and N. Lu, "Structural analysis by distinct element method," *J. Eng. Mech., ASCE*, vol. 117, 9, 1991, pp. 2156–2163.

1.7 Hakuno, M., and K Meguro, "Simulation of concrete frame collapse due to dynamic loading," *J. Eng. Mech.*, ASCE, vol. 119, 9, 1993, pp. 1709–1723.

CHAPTER 1 EXERCISES (answers in Appendix I)

1.1 **(a)** Use a vector approach to prove that if $\boldsymbol{F} = m(d\boldsymbol{v}/dt)$ is true for a mass moving with an arbitrary (i.e., nonconstant) velocity relative to a "fixed" Cartesian coordinate system, the exact same form of force–acceleration equation applies in any other Cartesian coordinate system that translates at a constant velocity relative to the first Cartesian coordinate system.

(b) Show that the time rate of change of the angular momentum (i.e., moment of momentum) of a collection of particles is zero whenever all of the particles are moving at a constant velocity relative to the valid coordinate system.

(c) Show that the sum of the moments about an arbitrary point of two collinear forces of equal magnitudes, but oppositely directed, is zero.

(d) Show that if a rigid body is in static equilibrium, any virtual displacement leads to zero virtual work.

(e) Show that if, in connection with an arbitrary virtual displacement, the virtual work of a rigid body is zero, then that body is in static equilibrium.

1.2 Evaluate the mass moment of inertia at the center of mass about the z axis for a constant density, rectangular parallelepiped whose x, y, z uniform dimensions are, respectively, a, b, c. Hint: Multiple integration is expected.

1.3 How many degrees of freedom are necessary for the following rigid body systems?

(a) A cylinder rolling and slipping on a plane in one direction.

(b) A sphere rolling with or without slipping on a plane.

(c) A straight rod leaning against, and sliding down, a wall and across the floor, only up to the point where the long, thin rod becomes horizontal and in full contact with the floor. Let the motion of the rod be confined to a single plane perpendicular to the wall, and let the rod not rotate about its own longitudinal axis. Realize that at some point in its downward, rotating motion, the upper end of the rod will lose contact with the wall. Of course, the lower end of the rod will remain in contact with the floor.

(d) A circular pendulum; that is, a bar of finite dimensions that can pivot in all angular directions about its connection at one end to a universal joint.

(e) A bead moving along a wire of arbitrary shape in a three-dimensional space.

1.4 **(a)** The lateral deflection $w(x)$ (in the z direction) of a uniform, symmetric cross-section beam element of length ℓ subjected to a uniform lateral force per unit length, f_z, in terms of the applied load and the beam element end point generalized coordinates, w_1 through θ_2, is

$$w(x) = \frac{f_z \ell^4}{24EI}(X^4 - 2X^3 + X^2) + (2X^3 - 3X^2 + 1)w_1$$
$$+ \ell(X^3 - 2X^2 + X)\theta_1 + (-2X^3 + 3X^2)w_2 + \ell(X^3 - X^2)\theta_2,$$

where $X = x/\ell$ is the nondimensional beam spanwise, spatial coordinate. The use of the virtual strain energy for analyzing this beam segment requires the calculation of $\delta w(X)$. Determine that variation of the lateral deflection.

(b) A horizontal, rigid bar of length L is supported by a vertical spring at each of its ends. The bar can only move vertically, and while doing so, it can undergo only small rotations in the plane of the paper. Let the two DOF for this system be $q_1(t)$, which measures the upward deflection at the left end of the bar, and $q_2(t)$, which measures the upward deflection at the right end. The only applied load is an upwardly directed force $F(t)$ acting at the right end. Calculate the virtual work for this system.

(c) A simply supported (flexible) beam of length L is loaded over its span by a downwardly directed, uniform load per unit length, $f_0(t)$. Let the positive upward deflections of this beam be approximated by the two-term series $w(x, t) = q_1(t)\sin(\pi x/L) + q_2(t)\sin(2\pi x/L)$. Calculate the virtual work done on the beam by the applied distributed forces when this beam undergoes a virtual deflection in accordance with the given series approximation for the beam deflections. Determine the values of the generalized forces.

1.5 Use the Lagrange equations to derive the polar coordinate forms of Newton's second law that describe the planar (z-plane) motion of the center of mass of body of mass m. That is, let the two generalized coordinates of the CG of the mass m be r and θ, and let the forces acting at the CG be F_r and F_θ.

1.6 **(a)** Let X and Y be Cartesian coordinates in the Earth's equatorial plane and fixed to the Earth's surface. See Figure 1.9. With respect to the X, Y coordinate

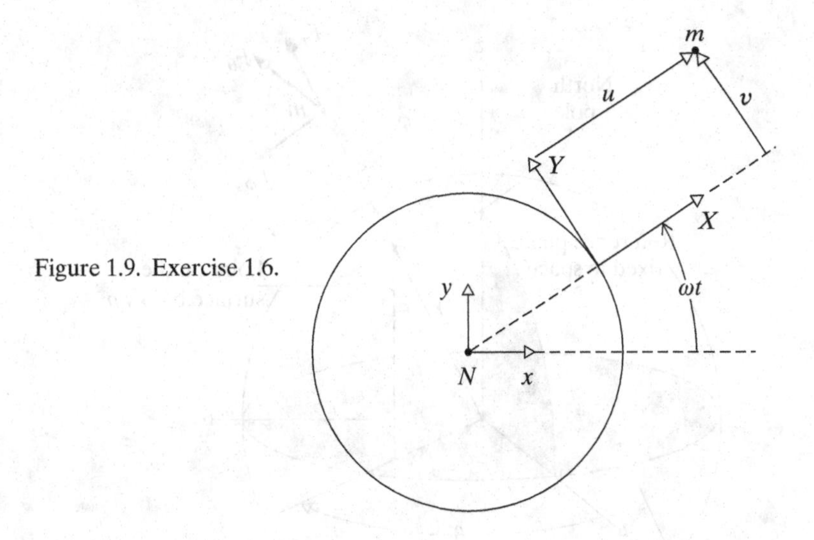

Figure 1.9. Exercise 1.6.

system, let u, v be the displacements of a particle whose motion is confined to that equatorial plane. Using a valid coordinate system fixed in space at the Earth's center, show that the equations of motion for the particle are

$$F_X = m(\ddot{u} + 2\dot{v}\omega - u\omega^2)$$
$$F_Y = m(\ddot{v} - 2\dot{u}\omega - \dot{v}\omega^2 - R\omega^2),$$

where ω is the Earth's angular velocity about its polar axis and R is the Earth's radius.

(b) In terms of the usual spherical coordinates ($q_1 = r$, the radial distance from the Earth's center; $q_2 = \phi$, the colatitude angle from the north pole; and $q_3 = \theta$, the east longitude angle from the prime meridian, where all three references, the Earth's center, the north pole, and the prime meridian are only translating, approximately, with respect to the valid coordinate axes), use the Lagrange equations to write the equations of motion of a particle of mass m moving above the Earth's surface. The particle is subjected to the total force components $\boldsymbol{F}_r$, $\boldsymbol{F}_\phi$, and $\boldsymbol{F}_\theta$. Hint: To obtain $\boldsymbol{r}(t)$, which can be differentiated to obtain the total velocity vector necessary for describing the kinetic energy of the particle, as per Figure 1.10, set up the following three orthogonal unit vectors corresponding to the three spherical coordinates: $\boldsymbol{p}(t)$ in the r direction, $\boldsymbol{q}(t)$ in the direction of increasing ϕ, and $\boldsymbol{s}(t)$ in the direction of increasing θ. (These are the same directions as the three force components mentioned above.) Then write $\boldsymbol{r} = r\boldsymbol{p}$. Also introduce Earth-centered, space-fixed Cartesian unit vectors such that $\boldsymbol{i}, \boldsymbol{j}$ are in the Earth's equatorial plane such that $\boldsymbol{i}$ is also in the plane of the space-fixed prime meridian and $\boldsymbol{k}$ points in the direction of the north pole. Since these Cartesian unit vectors are "fixed in space," they are not functions of time. In terms of the Cartesian unit vectors, the spherical unit vectors are

$$\boldsymbol{p} = (\sin\phi\cos\theta)\boldsymbol{i} + (\sin\phi\sin\theta)\boldsymbol{j} + (\cos\phi)\boldsymbol{k}$$
$$\boldsymbol{q} = (\cos\phi\cos\theta)\boldsymbol{i} + (\cos\phi\sin\theta)\boldsymbol{j} - (\sin\phi)\boldsymbol{k}$$
$$\boldsymbol{s} = -(\sin\theta)\boldsymbol{i} + (\cos\theta)\boldsymbol{j},$$

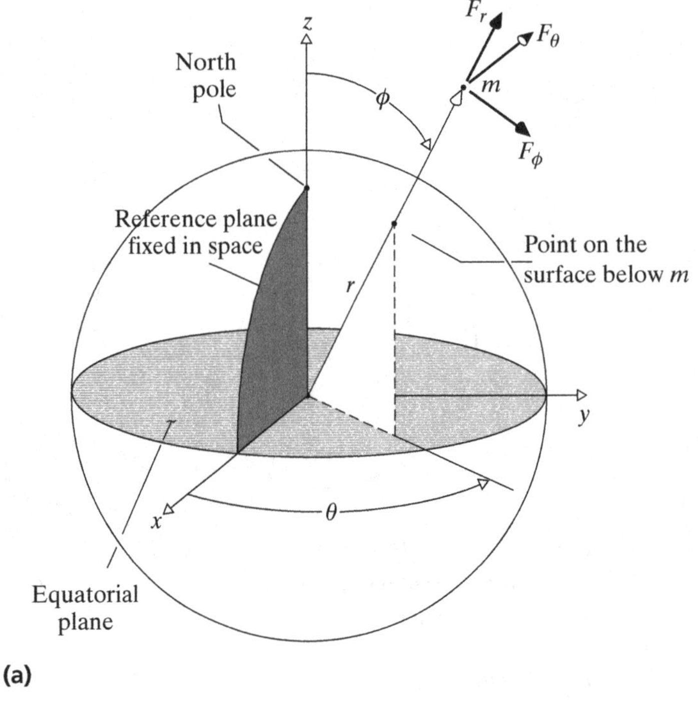

(a)

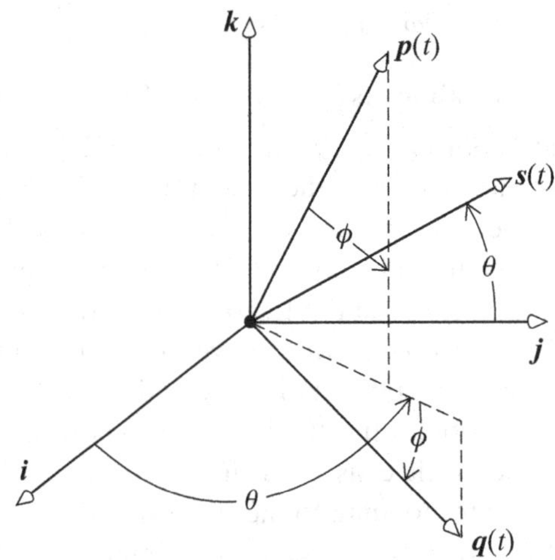

(b)

Figure 1.10. (a) The genaralized coordinates r, ϕ, and θ for the particle of mass m. (b) The orientation of the moving spherical coordinate system unit vectors relative to the fixed Cartesian unit vectors such that $\boldsymbol{r}(t) = r\,\boldsymbol{p}$.

hence

$$\dot{\boldsymbol{p}} = \dot{\phi}\,\boldsymbol{q} + \dot{\theta}\sin\phi\boldsymbol{s}.$$

(c) Repeat part (b), but this time, instead of using polar coordinate references that are fixed in space, use polar coordinate references that are fixed with respect to the Earth rotating about its polar axis.

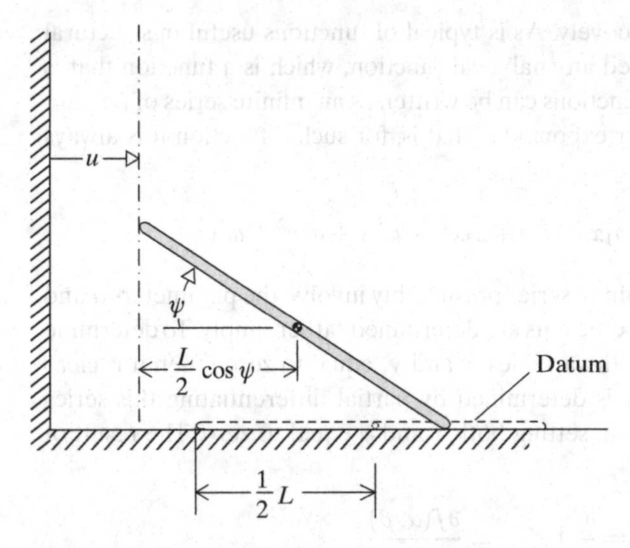

Figure 1.11. Exercise 1.3(c) and Exercise 1.7.

1.7 Using the generalized coordinates $q_1 = \psi$, and $q_2 = u$, where ψ is the angle between the rod and the floor and u is the horizontal motion of the top end of the rod of length L, as shown in Figure 1.11, write the equations of motion of the rod. Let both the wall and the floor be frictionless.

1.8 Show that when $T = T(\dot{q}, q, t)$, the order of differentiation in the first term of the Lagrange equation cannot be reversed. That is, prove

$$\frac{d}{dt}\frac{\partial T}{\partial \dot{q}} \neq \frac{\partial}{\partial \dot{q}}\frac{dT}{dt}.$$

COMMENT As any calculus book will explain, it is perfectly permissible to reverse the order of partial differentiation with respect to two different variables whenever either (1) the first-order derivatives are differentiable, or (2) the first- and second-order derivatives are continuous. However the implied qualification for these two theorems is that the two variables of differentiation are independent of each other, such as are the x and y Cartesian coordinates in two-space. When the two coordinates are not independent of each other, the reversal of the order of differentiation is not always valid. For example, with x, y being Cartesian coordinates and r, θ being polar coordinates, it is not generally correct to say

$$\frac{\partial^2 F}{\partial r \partial x} = \frac{\partial^2 F}{\partial x \partial r}. \quad \text{(not so)}$$

ENDNOTE (1): FURTHER EXPLANATION OF THE VARIATIONAL OPERATOR

Preparatory to further discussion of the variational operator, it is useful to review Taylor's series in terms of more than one variable. As representative of the multivariable case, let α and β be any two parameters and consider the function $f(\alpha + x, \beta + y)$. In this form it is convenient to consider the parameters α and β to be fixed values of the variables x and y, respectively. As is typical of functions useful in structural modeling, let f be what is called an analytical function, which is a function that is infinitely differentiable. Such functions can be written as an infinite series of polynomial terms with positive integer exponents. That is, for such a function it is always possible to write

$$f(\alpha + x, \beta + y) = a_0 + a_1 x + a_2 y + a_3 x^2 + a_4 xy + a_5 y^2 + a_6 x^3 + \cdots,$$

where the coefficients of this infinite series presumably involve the parameters α and β. The precise values of those coefficients are determined rather simply. To determine a_0, it is necessary only to set both variables, x and y, equal to zero. Then it is clear $a_0 = f(\alpha, \beta)$. The coefficient a_1 is determined by partial differentiating this series with respect to x and then, again, setting both x and y equal to zero. The result of that pair of steps is

$$a_1 = \left.\frac{\partial f}{\partial x}\right|_{x=y=0} \equiv \frac{\partial f(\alpha, \beta)}{\partial x}.$$

The remaining coefficients can be determined in exactly the same way. For example,

$$a_2 = \left.\frac{\partial f}{\partial y}\right|_{x=y=0} \equiv \frac{\partial f(\alpha, \beta)}{\partial y}$$

$$a_3 = \frac{1}{2}\frac{\partial f^2(\alpha, \beta)}{\partial x^2}$$

$$a_4 = \frac{\partial f^2(\alpha, \beta)}{\partial x \partial y}$$

and so on. Since α, β, x, and y can have any values whatever, choose the set x, y, dx, dy for α, β, x, and y, respectively. For this choice, the above series expansion for the function f becomes

$$\begin{aligned} f(x + dx, y + dy) &= f(x, y) + \frac{\partial f(x, y)}{\partial x} dx + \frac{\partial f(x, y)}{\partial y} dy \\ &\quad + \frac{1}{2}\frac{\partial f^2(x, y)}{\partial x^2}(dx)^2 + \frac{\partial f^2(x, y)}{\partial x \partial y}(dx)(dy) \\ &\quad + \frac{1}{2}\frac{\partial f^2(x, y)}{\partial y^2}(dy)^2 + \frac{1}{3!}\frac{\partial f^3(x, y)}{\partial x^3}(dx)^3 + \cdots \\ &= f(x, y) + \sum_{n=1}^{\infty}\left[\frac{1}{n!}\left(\frac{\partial}{\partial x} + \frac{\partial}{\partial y}\right)^n f(x, y)\right]. \end{aligned}$$

Having achieved the above, the various differentials of the function f can now be defined as follows, where $d^{(T)} f = f(x + dx, y + dy) - f(x, y)$ is called the total differential of f, and $d^{(1)} f$ is called the first differential of f, and so on.

$$d^{(T)} f = d^{(1)} f + d^{(2)} f + d^{(3)} f + \cdots$$

or, respectively,

$$
\begin{aligned}
&[f(x+dx, y+dy) - f(x, y)] \\
&\quad = \left[\frac{\partial f(x, y)}{\partial x}dx + \frac{\partial f(x, y)}{\partial y}dy\right] \\
&\qquad + \left[\frac{1}{2}\frac{\partial f^2(x, y)}{\partial x^2}(dx)^2 + \frac{\partial f^2(x, y)}{\partial x \partial y}(dx)(dy) + \frac{1}{2}\frac{\partial f^2(x, y)}{\partial y^2}(dy)^2\right] \\
&\qquad + \left[\frac{1}{3!}\frac{\partial f^3(x, y)}{\partial x^3}(dx)^3 + \cdots\right] \ldots.
\end{aligned}
$$

Since differentials are only useful when they are very small, the second differential, third differential, and so on, can be discarded as inconsequential. Thus the total differential is very well approximated by the first differential. Since $d^{(T)} f = d^{(1)} f$, there is no point in qualifying the sole remaining differential with a superscript. Thus all that remains is the chain rule for differentials

$$
df = \frac{\partial f}{\partial x}dx + \frac{\partial f}{\partial y}dy.
$$

The same thing that was done for differentials can now be done for variations. Going back to the above general form for Taylor's series, let α, β, x, and y have the respective values u, v, δu, and δv, where u and v can be thought of as deflections. Then, from Taylor's series

$$
\delta^{(T)} f = \delta^{(1)} f + \delta^{(2)} f + \delta^{(3)} f + \cdots \tag{1.19a}
$$

or

$$
\begin{aligned}
&[f(u+\delta u, v+\delta v) - f(u, v)] \\
&\quad = \left[\frac{\partial f(u, v)}{\partial u}\delta u + \frac{\partial f(u, v)}{\partial v}\delta v\right] \\
&\qquad + \left[\frac{1}{2}\frac{\partial f^2(u, v)}{\partial u^2}(\delta u)^2 + \frac{\partial f^2(u, v)}{\partial u \partial v}(\delta u)(\delta v) + \frac{1}{2}\frac{\partial f^2(u, v)}{\partial v^2}(\delta v)^2\right] \\
&\qquad + \left[\frac{1}{3!}\frac{\partial f^3(u, v)}{\partial u^3}(\delta u)^3 + \cdots\right].
\end{aligned}
$$

Unlike the differential, the higher order variations such as $\delta^{(2)} f$ can have importance because in certain circumstances the lower ordered variations are zero valued. For example, the higher ordered variations can be useful in energy-based stability analyses. Let the symbol δf is to be understood to mean simply $\delta^{(1)} f$, the first variation of f. From the above, after comparison of the quantities df and δf, the very important conclusion can be drawn that the differential operator and the variational operator follow the same rules of calculus. However, they are not quite the same. To understand the difference between the two, consider a function with a greater variety of terms in its argument. Consider, in the light of virtual work, the function $V(u, \epsilon, \sigma, F, t)$, where u is any deflection, ϵ is any strain (a derivative of a deflection), σ is any stress (an internal force dependent quantity), M is any internal stress resultant (like a moment), and t is time. Then, first, the differential of V, also known as the total change in V, depends on the changes in its argument. That is, from the above derived chain rule

$$dV(u, \epsilon, \sigma, M, t) = \frac{\partial V}{\partial u}du + \frac{\partial V}{\partial \epsilon}d\epsilon + \frac{\partial V}{\partial \sigma}d\sigma + \frac{\partial V}{\partial \epsilon}d\epsilon + \frac{\partial V}{\partial M}dM + \frac{\partial V}{\partial t}dt.$$

However, in this case where the argument of the function v involves more than just deflection quantities such as u and ϵ the variation of V is only

$$\delta V = \frac{\partial V}{\partial u}\delta u + \frac{\partial V}{\partial \epsilon}\delta\epsilon.$$

The reason for this difference goes back to the original definition of virtual work as real forces moving through virtual displacements (virtual deflections). In virtual work, only the deflections are given (possibly imaginary) variations. Although actual forces generally change when a body undergoes actual deflections, actual forces are always wholly unchanged by imaginary deflections. Since (i) there are no variations for force type quantities, (ii) time is an arbitrary fixed value for virtual displacements, and (iii) constant geometric and material properties will not be varied here, then the rule for variations to be used here is

> The variational operator follows the same rules of calculus as the differential operator except that it only acts on deflections and deflection-type quantities, such as strains.

In this textbook, except for one endnote, forces, time, and the usual constants of a structural analysis have zero variations. The above rule can be generalized for all textbooks by saying only those quantities that the analyst designates as dependent functions have nonzero variations. All quantities the analyst designates as independent quantities have zero variations. In other words, in this textbook the viewpoint is that the forces cause the deflections, so the deflections are dependent and the forces are arbitrary and hence independent. To see why, for example, strains have variations, whereas stresses do not, consider the simple example of a uniform bar of length L and cross-sectional area A subjected to a uniform tensile force F. Let the total extension of the bar be called u_0. Then, from their basic definitions, the stress in the bar is (F/A), and the strain in the bar is (u_0/L). Thus it is clear that stress is directly dependent on the external force, and if the force has a zero variation, so too does the stress. However, because the strain is directly dependent on the deflection, the strain has a nonzero variation because the deflection has a nonzero variation. Note that if the bar material is unidentified, as is the case here, it is not possible to write a mathematical relationship between stress and strain, a relationship that might cause confusion.

EXAMPLE 1.3 Prove that the variational operator is commutative with definite integration typified by an integral over a fixed volume. (Recall the commutativity of the variational operator and a derivative operator is part of the definition of the variational operator.)

SOLUTION Let $F(u, \partial u/\partial x)$ represent a typical point function of deflection type quantities. Consider the following integral over a fixed volume:

$$I = \iiint F\left(u, \frac{\partial u}{\partial x}\right) dx\, dy\, dz.$$

Then, from the first part of Eq. (1.19b) where $f(u+\delta u, v+\delta v) = f(u,v)+\delta f(u,v)$,

$$I+\delta I = \iiint F\left[u+\delta u, \frac{\partial u}{\partial x}+\delta\left(\frac{\partial u}{\partial x}\right)\right] dx\,dy\,dz.$$

Subtracting I from $I+\delta I$ simply yields δI on the left-hand side. On the right-hand side, the two volume integrals have the same limits. Therefore the difference of the two integrals is the integral of the difference of their two integrands. That is, again using the first part of Eq. (1.19b),

$$\delta I = \iiint\left[F\left(u+\delta u, \frac{\partial u}{\partial x}+\delta\frac{\partial u}{\partial x}\right) - F\left(u, \frac{\partial u}{\partial x}\right)\right] dx\ dy\ dz$$

$$= \iiint \delta F\left(u, \frac{\partial u}{\partial x}\right) dx\ dy\ dz \qquad \bigstar$$

ENDNOTE (2): KINETIC ENERGY AND ENERGY DISSIPATION

One direct use of the concept of kinetic energy in structural applications is in combination with energy dissipation in crash safety analyses. The basic idea is simply that the kinetic energy of the vehicle at crash impact needs to be dissipated through plastic deformations of a buffer structure before, say, the passenger compartment undergoes plastic deformations. Consider the greatly simplified situation where the entire structure, buffer structure, passenger compartment, and whatever else is a single, short rod. The rod is restricted to be short so that buckling of the rod need not be considered here. Let the rod move with a constant velocity v_0 in a direction paralleling its longitudinal axis. Let the rod impact, head on, a nearly rigid mass of such a relatively large mass that this target mass hardly moves in response to the impact of the rod. A single short rod is chosen here because the sole stress and sole strain are both longitudinal and simply related and can be approximated as uniform over the rod's cross section. A rigid mass representing the passenger compartment could be attached to the end of the rod without complicating the analysis. However, this additional mass would neither add anything insightful to the explanation nor make the model much more realistic.

If the rod's mass density, cross-sectional area, and length are, respectively, ρ, A, and L, then the kinetic energy to be dissipated is $\frac{1}{2}\rho A L v_0^2$. Since significant energy dissipation is only possible when the stresses due to the impact reach the yield stress of the rod's material, σ_y, let that be the case. Consider the simplified, bilinear stress strain curve shown in Figure 1.12. The area beneath the load path curve represents work done on the material. This is so because stress is equal to force over cross-sectional area, whereas strain is equal to change in length over length where this length is perpendicular to the cross-sectional area. Thus the product of a unit of stress and a unit of strain, which represents a unit of area under the stress–strain curve, is

$$\frac{F}{A}\frac{\Delta\ell}{\ell} = \frac{\text{work}}{\text{volume}}.$$

Again, when the strain is a fully plastic strain as represented by the horizontal portion of the stress–strain curve, the work done on the material is not recoverable because of the unloading curve being as shown. That is the same as saying that the work

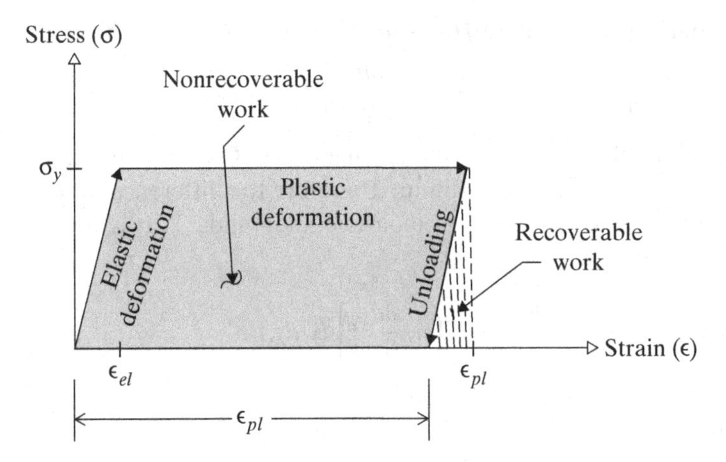

Figure 1.12. Endnote (1) simplified material behavior model.

done on the material is energy dissipated in deforming the material. In this case, the energy dissipated is the product of the yield stress, the magnitude of the plastic strain, and the volume of the rod. That is, the total dissipated energy under these simplifying assumptions is equal to $\sigma_y \epsilon_p AL$. Equating the kinetic energy to the energy dissipated allows a solution for the magnitude of the required plastic strain, which is $\epsilon_p = (\rho v_0^2)/(2\sigma_y)$. The permanent shortening of the rod axis is approximated by the product of this plastic strain and the rod length.

Of course, the above analysis is quite crude. In addition to the presumptions of a perfect in-line impact and no rod bending, the rod's entire end cross-sectional area is assumed to make an instantaneous flat contact with the immovable obstruction producing a uniform state of stress and strain over the end cross section. Furthermore, that initially uniform stress state is presumed to propagate down the length of rod. In fact, the friction of contact between the rod and the immovable object will restrict the lateral, Poisson-effect expansion of the end cross section, causing the rod to deform into a barrel shape that indicates a more complex stress state.

A more realistic crash safety analysis would involve the plastic bending of not necessarily straight beams and beam-columns. In this case, it would be necessary to identify those segment lengths of the beam that were undergoing fully elastic bending (no energy dissipation), partially plastic bending, and fully plastic bending. Although the bookkeeping is more complicated and requires a digital computer, the basic concepts are the same.

ENDNOTE (3): A RIGID BODY DYNAMICS EXAMPLE PROBLEM

EXAMPLE 1.4 **(a)** Use the Lagrange equations to write the equations of motion that determine the planar path traced by a single planet orbiting a much more massive star.

(b) Solve those equations and demonstrate that the path of the planet is elliptical.

SOLUTION (a) As shown in Figure 1.13, choose the polar coordinates $r(t)$ and $\theta(t)$ as the two generalized coordinates of the planet CG relative to the star CG. It is assumed that the star is moving at a constant velocity (at least approximately so for the time duration of one planetary orbit) and, thus, from Exercise 1.1(a), can be considered to be stationary for the application of the equations of motion. Recall from physics that the gravitational attraction between two bodies with masses M and m is equal to the force GMm/r^2, where G is the universal gravitational constant. Since the planet has orthogonal velocity components $\dot{r}$ and $r\dot{\theta}$

$$T = \frac{1}{2}m(\dot{r}^2 + r^2\dot{\theta}^2) \quad U = V = 0 \quad \delta W = -\frac{GMm}{r^2}\delta r.$$

Hence

$$Q_r = -\frac{GMm}{r^2} \qquad Q_\theta = 0.$$

Substituting the kinetic energy and the generalized forces into the Lagrange equations yields the following coupled, nonlinear, ordinary differential equations for the generalized coordinates:

$$\ddot{r} - r\dot{\theta}^2 + \frac{GM}{r^2} = 0 \qquad \frac{d}{dt}(r^2\dot{\theta}) = 0.$$

(b) The second of the above equations is integrated immediately to obtain $r^2\dot{\theta} = C_1$ where C_1 is simply a constant of integration. This result that $(r^2 d\theta)/dt$ is a constant is Kepler's second law, which is often stated as the radius vector of each planet passes over equal areas ($\frac{1}{2}r^2 d\theta$) in equal intervals of time (dt). Substitution of this result into the first equation of motion to eliminate the θ variable leads to the following:

$$\ddot{r} + \frac{GM}{r^2} - \frac{C_1^2}{r^3} = 0.$$

The first integration of this second-order ordinary differential equation is accomplished by first rewriting the second derivative term using the chain rule for derivatives

$$\ddot{r} = \frac{d\dot{r}}{dt} = \frac{d\dot{r}}{dr}\frac{dr}{dt} = \dot{r}\frac{d\dot{r}}{dr}.$$

Now it is possible to separate the radial velocity and radius variables so as to obtain, after carrying out the integration,

$$\dot{r}^2 = -\frac{C_1^2}{r^2} + \frac{2GM}{r} + C_2.$$

Since a direct relationship between the two polar coordinates in the form of $r(\theta)$ or $\theta(r)$ is desired rather than the parametric solution pair $r(t)$ and $\theta(t)$, divide the above equation by the previous result $r^2\dot{\theta} = C_1$, squared, to obtain

$$\frac{(dr)^2}{(d\theta)^2} = r^2\left[-1 + \frac{2GM}{C_1^2}r + \frac{C_2}{C_1^2}r^2\right].$$

```
In[1]: = <<Graphics 'Graphics'

In[2]: = r[t_]=1/(2 + Cos[t])

Out[2] =      1
         ----------
         2 + Cos[t]

In[3]:= PolarPlot[r[t],{t,0,2 Pi}]
```

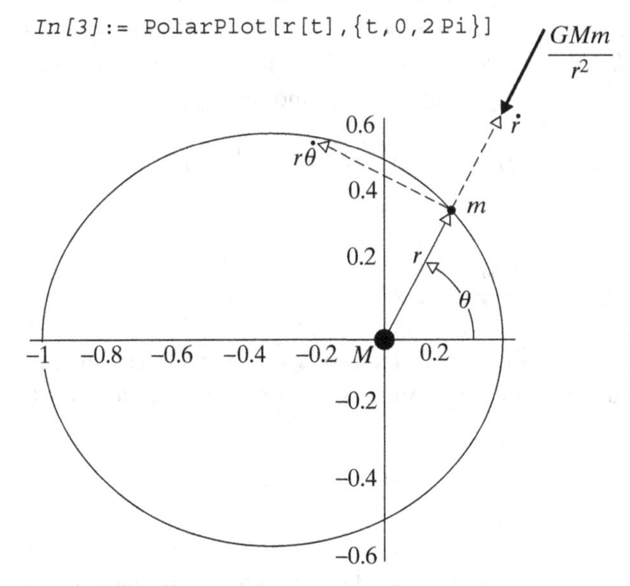

(a)
```
Out[3] =  -Graphics-
```

```
In[7]:= PolarPlot[r [z],{z,0,2 Pi}]
```

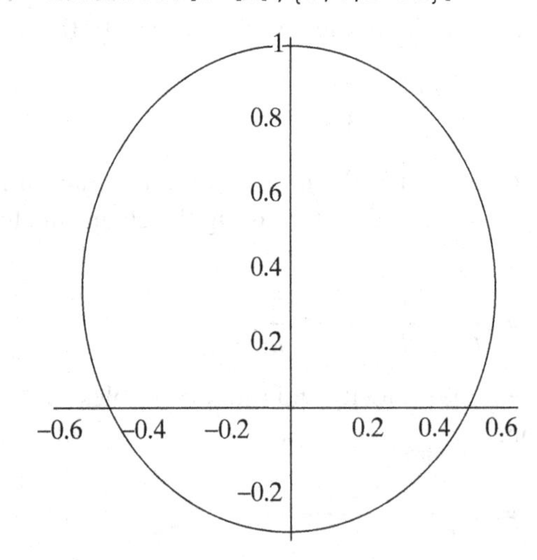


```
In[6]:= r[t_] = 1/(2 -Sin[t]])
```

(b)
```
Out[6]=     1
        ---------
        2 -Sin[t]
```

Figure 1.13. Example 1.4.

After taking the square root of both sides, the variables can be separated. Consulting a table of integrals yields the following result:

$$\theta + C_3 = \arcsin\left(\frac{GMr - C_1^2}{r\sqrt{G^2M^2 + C_1^2C_2}}\right)$$

or

$$r = \frac{A}{B - \sin(\theta + C_3)}.$$

The easiest way to see that this relationship between variables describes an ellipse is to choose values for A, B, and C_3, such as 1, 2, and zero, and computer plot the result as shown in Figure 1.13. The result is an ellipse with foci at (0,0) and (0,2/3). The more rigorous approach to showing that the above solution is indeed an ellipse is, for example, to transform the x, y equation for an ellipse, $(x/a)^2 + (y/b)^2 = 1$, into polar coordinates ρ, ϕ. That is, where ρ, ϕ are measured from the center of the ellipse, use the transformation $x = \rho\cos\phi$ and $y = \rho\sin\phi$ to obtain

$$\frac{a^2}{\rho^2} = 1 + \frac{e^2}{b^2}\sin^2\phi.$$

To convert this equation into one only involving the r, θ of the above analysis, which have their origin at one focus of the ellipse, write, from a diagram of the ellipse, $\rho\cos\phi = e + r\cos\theta$ and $\rho\sin\phi = r\sin\theta$, where e, the eccentricity of the ellipse is related to the major axis lengths as $e^2 = a^2 - b^2$. After some algebra, the result is

$$r = \frac{-b/e}{(a/e) - \cos\theta}.$$

This is the same result as above when the phase angle in the above sine function is taken to be $\pi/2$, a rotation of 90°.

2 Mechanical Vibrations: Practice Using the Lagrange Equations

2.1 Introduction

The focus of this textbook is on the vibrations of engineering structures, not mechanisms. However, this chapter focuses on pendulums as representative of mechanisms. Pendulums are rarely a part of an engineering structure.[1] However, because the motions of pendulums are familiar to everyone, they do provide a comparatively simple means for both visualizing and explaining some basic aspects of more general vibratory systems. Pendulums also provide an opportunity to consolidate the lessons on dynamics set forth in the first chapter without the complication of dealing with flexible structures. As an aside, pendulums also provide a relatively simple introduction to the quite challenging topic of nonlinear vibrations. Thus, despite their limited relevance to engineering structures in general, this introductory study of structural dynamics begins with the study of the back-and-forth motion of pendulums.

The *static equilibrium position* (SEP) of any dynamical system is the deflected position of that system in response to all the applied static loads and their support reactions, if any. A stable pendulum system is defined as any system that, when displaced from its static equilibrium position, tends to return to that SEP as a result of the presence of a gravitational force field or other force field. An example of body forces other than gravitational forces that stabilize a structural system is the centrifugal force field acting on a rotating helicopter blade. Consider a rotating helicopter blade that is unrestrained at its outer end and has a single, horizontal hinge line that is located at the end of the blade adjacent to the blade's vertical axis of rotation. The hinge allows the blade to flap up and down something like a bird wing. For a viewer rotating with the blade, the undisturbed blade will have a near-horizontal SEP resulting from the balance between the blade's distributed weight forces, lift forces, and centrifugal forces. If a disturbance causes the rotating blade to flap upward, the outwardly directed centrifugal forces acting along the length of the blade produce a

[1] There are exceptions. Pendulums have been used for the gravity gradient stabilization of orbiting satellites and also as vibration dampers for structures and rotary machinery. As shown shortly, helicopter blades can also be classified as pendulums.

moment about the horizontal hinge axis that causes the blade to rotate downward. Similarly, if the blade undergoes a disturbance that initially causes the blade to rotate downward from its near-horizontal SEP, then the centrifugal forces along the blade immediately produce a moment that causes the blade to move upward. Thus, in either the presence or absence of aerodynamic forces and a gravitation field, a rotating helicopter blade is a stable pendulum because there is an associated centrifugal force field that always tends to return the helicopter blade to its SEP after a disturbance moves the blade away from the SEP.

An unstable pendulum system tends to move away from its SEP under the action of whatever body forces are inherent to that system. An example of an unstable pendulum could be a sharpened pencil standing on its point in a normal gravitational field. The (unstable) static equilibrium position is the vertical line upward from the pencil point tip. Stability itself is a secondary issue for this textbook because all but one of the example applications of structural dynamics analysis to be discussed in this chapter are of unconditionally stable systems (within the limits of the mathematical model). The one example problem that is the exception is discussed in Section 2.10 of this chapter (followed by Exercise 2.11). There is also a brief overview of stability presented in Endnote (1).

A mechanical system can merit the description of being a pendulum system even though there are significant external forces other than restorative or destabilizing forces acting on the system. All that is necessary for a system to be classified as a pendulum system is that the field forces make a significant (big enough to be included in the descriptive equations) contribution to either stabilizing or destabilizing the motion of the mechanical system.

2.2 Techniques of Analysis for Pendulum Systems

One basic choice for structural dynamics analyses is between force methods of analysis and energy methods of analysis. Force methods are grounded on Newton's second and third laws. Those two laws involve the vector quantities of force, displacement, velocity, and acceleration. Recall that energy, in all its various forms, is the capacity for doing work. All work/energy expressions have the advantage of being scalar quantities. One form of an energy method of analysis is the writing of conservation of energy equations. This chapter points out both the usefulness and the danger of energy conservation equations. Another energy approach is to write the Lagrange equations of motion, which can be viewed as a restatement of Newton's second law in terms of energy quantities. One objective of this chapter is to convince the reader that, relative to Newton's force equations of motion, the Lagrange equations are generally the much more convenient analytical tool, particularly if there is any complexity to the system being studied.

As a first explanatory analysis, consider the simple mechanical system of Figure 2.1(a), which is restricted to move in the plane of the paper. In this simplified mathematical model, the rod that connects the bob mass m to the pivot is modeled as massless. That is, all the mass of the system is concentrated in the bob at the bottom of the rod. (This apparently crude type of modeling is justified later.) In this case, the

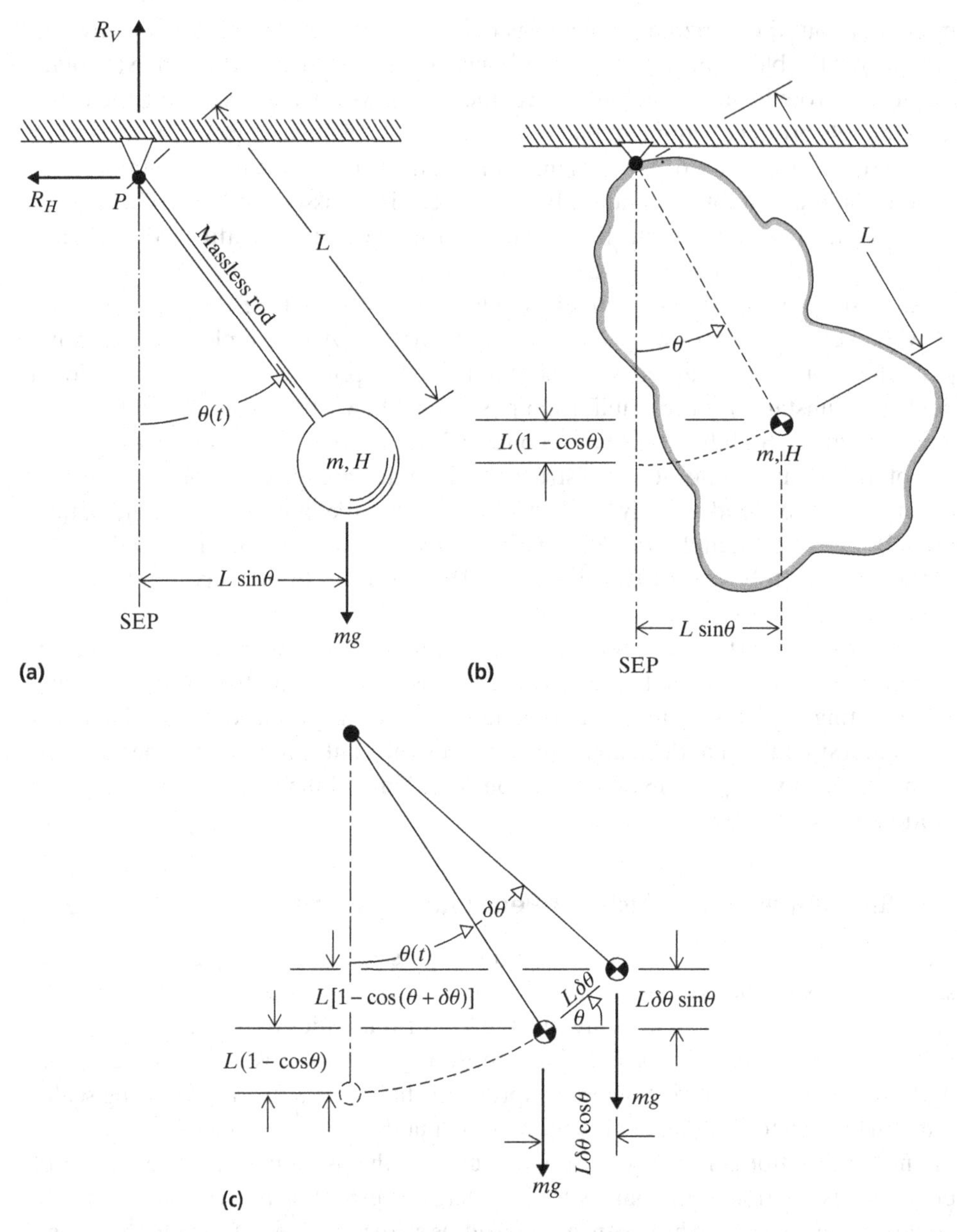

Figure 2.1. (a) Displaced system. The single generalized coordinate θ fully locates the position of every mass particle in the system. (b) A general shape pendulum system whose equation of motion is exactly the same as the system in Figure 2.1(a). (c) The system's actual displacement augmented by a positive virtual displacement (shown as greatly exaggerated) causing the gravitational force to do negative (ficticious) work.

bob mass is described by two parameters, m and H. The quantity m is just the total mass of the bob, whereas H is the mass moment of inertia of the bob about an axis perpendicular to the plane of the paper and passing through the center of mass of the bob.

When discussing the dynamics of masses, the word *system* always needs precise definition. Recall that Newton's second law, and thus all other formulas describing the motion of mass that can be derived from Newton's second law, apply only to a fixed, identifiable, quantity of mass. Through familiar usage, the closed boundaries of almost all mass systems under discussion will be self-evident and not require comment. However, the present recognition of the importance of precisely defining what does and does not belong to the closed system under discussion (and thus which forces are internal and which forces are external) prompts the explicit definition of this system to be the rigid, massless rod connecting the pivot point to the bob, and the bob itself.[2] In simple terms, the system under study is everything that moves.

Figure 2.1(a) shows the system deflected to the reader's right. The position of the rigid bar is measured by the time-varying value of the angle θ. The implication of the single arrowhead in the diagram's dimensioning for θ is that θ has positive values when the rigid bar is deflected to the right of the vertical SEP and negative values when the rigid bar is rotated to the left of the SEP. Figure 2.1(a) also shows that every mass particle of the rigid bob is precisely located by this angular measure, θ. The reverse is also true: the location of the mass particles of the rigid bob defines the value of θ. Since in a structural dynamics analysis the generalized coordinates of a mechanical system unambiguously locate the instantaneous positions of all of the mass of the system, and because all the mass of this system is concentrated in the bob, θ is the sole generalized coordinate of what then is this single degree of freedom system. (Recall that a generalized coordinate is often referred to as a degree of freedom (DOF); and the unique number of DOF required to fully describe all the possible motions of a system is a inherent characteristic of the dynamical system.) In mathematical terms, $q_1 = \theta$.

The mechanical system of Figure 2.1(a) is a stable pendulum system because, regardless of whether the system is deflected to the left or the right, the gravitational force on the mass produces a moment about the pivot point that tends to cause the system to rotate back to its below-the-pivot, vertical SEP. It is purposeful that Figure 2.1(a), which is intended to guide the following analyses, shows the pendulum system displaced from its vertical position rather than just hanging vertically. It is strongly suggested that all pendulum analyses begin with a diagram of all the components of the pendulum system displaced in positive directions from the SEP. Without a sketch of the system in a general displaced configuration, the analyst risks overlooking forces and moments associated with such a general displacement or generalized coordinates necessary to describe the general motion.

In this simple pendulum system, every mass particle rotates in the plane of the paper about the fixed axis seen end-on as the pivot point, point P. Thus it is possible to use the corollary of Newton's second law that states that the sum of the torques about any fixed axis (M_{FA}) is equal to the mass moment of inertia about that fixed axis (H_{FA}) multiplied by the second time derivative of the angular rotation about the

[2] In fluid dynamics, the application of Newton's second law to a specific "control" volume in the flow field is really the repeated application of Newton's second law at different times to the series of masses that occupy that volume at those time points.

fixed axis, here called $\ddot{\theta}$. This equation is Eq. (1.5a).[3] Therefore all that is necessary to complete this fixed axis differential equation of motion is to write the mathematical descriptions of the torque and the moment of inertia, both about the fixed axis running through the pivot point. These torque and moment of inertia descriptions must be in terms of the given quantities. Using the parallel axis theorem for mass moments of inertia developed in the first chapter, and noting that the external force reaction at the pivot point has no moment arm about the fixed axis, leads to

$$H_{FA} = H + mL^2 \qquad M_{FA} = -mgL\sin\theta$$

$$\therefore (H + mL^2)\ddot{\theta} = -mgL\sin\theta,$$

where mg is the only external force producing a moment about the fixed axis. The last equation is called the equation of motion. Equations of motion are usefully written in the form where all the terms involving the unknown quantities (in this case, the one time function, θ) and their derivatives are placed on the left-hand side of the differential equation. Hence the equation of motion should be presented as

$$(H + mL^2)\ddot{\theta} + mgL\sin\theta = 0. \tag{2.1}$$

Solutions to equations of motion are discussed later in this chapter in order to first focus on writing such equations of motion.

The use of the fixed axis (FA) version of Newton's second law was quite efficient in this very simple case where a fixed axis exists. However, many mechanical, or even pendulum, systems have more complicated motions than simply that of a rotation about a fixed axis. Thus for the sake of preparing for more challenging circumstances, the writing of the equation of motion for the above simple pendulum system is now repeated using the more general, planar rotational form of Newton's second law. This more general form is that for the CG case, which is Eq. (1.5b)

$$H_{CG}\ddot{\theta} = M_{CG}. \tag{2.2}$$

In this system, the center of mass is at the center of the bob. The rotation of the bob is the same as the rotation of the pendulum. Thus, for this example problem, the left-hand side of the above equation is simply $H\ddot{\theta}$. The moment about the center of mass is positive in the same counterclockwise direction that θ is positive. Thus

$$H\ddot{\theta} = M_{CG} = R_H L\cos\theta - R_V L\sin\theta,$$

where the weight force has no moment arm. The horizontal and vertical reactions, R_H and R_V (called the forces of constraint), that appear in the above expression for the moments about the center of mass are determined by the use of the two equations $f_{hor} = ma_{hor}$ and $f_{vrt} = ma_{vrt}$. The respective horizontal and vertical accelerations of the center of mass are determined by twice differentiating with respect to time the horizontal and vertical deflections of the center of mass. The one point that bears particular attention is that the reaction R_H was drawn positive to the left, whereas, because of the positive direction of θ, the horizontal deflection, velocity,

[3] If the bob rotated independently of the pendulum arm, then a second generalized coordinate would be required, say $\phi(t)$, and Eq. (1.4b) would have to be used.

and acceleration of the center of mass are positive to the right. Thus, a negative sign, in this case placed on the displacement, is necessary to adjust for the opposite directions. Hence, after careful use of the chain rule for ordinary derivatives

$$R_H = m\frac{d^2}{dt^2}(-L\sin\theta)$$

$$\text{or} \quad R_H = -mL\ddot{\theta}\cos\theta + mL\dot{\theta}^2\sin\theta$$

$$\text{and} \quad R_V - mg = m\frac{d^2}{dt^2}(L - L\cos\theta)$$

$$\text{or} \quad R_V = mg + mL\ddot{\theta}\sin\theta + mL\dot{\theta}^2\cos\theta$$

$$\text{so} \quad M_{CG} = -mgL\sin\theta - mL^2\ddot{\theta}.$$

Substitution of this last moment equation into Eq. (2.2) again yields the proper equation of motion, Eq. (2.1). This second analysis is clearly more complicated, typically so, than the first analysis that took note of the fixed axis of the motion.

Before proceeding to the explanation of the last of the three major methods of analysis presented for this simple pendulum problem, consider Figure 2.1(b). This figure shows the deflected position of a mass m of general shape that is pinned to a rigid support so as to rotate in the plane of the paper. The center of mass is a distance L below the pivot point. The mass moment of inertia of this mass of general shape, at the mass center, about an axis perpendicular to the plane of the paper, is H. Clearly there is no analytical difference between this dynamical system and that of Figure 2.1(b) because all the quantities that are relevant to the equations of motion for these two systems are the same. Hence the simplified mathematical model used in Figure 2.1(a) is justified.

The previous two analyses are written in terms of forces and moments, the quantities directly associated with Newton's second law and its immediate corollaries. Although it is not evident in this simple example, analyses based directly on Newton's laws are seldom as simple as energy methods in dynamical analyses of structural systems. Thus it is necessary to master energy methods in general and practice the use of the Lagrange equations of motion in particular. The most convenient way of writing the Lagrange equations for the dynamical analysis of structures is

$$\frac{d}{dt}\left(\frac{\partial T}{\partial \dot{q}_i}\right) - \frac{\partial T}{\partial q_i} + \frac{\partial (U+V)}{\partial q_i} = Q_i, \tag{2.3}$$

where, again, q_i is the ith generalized coordinate of the mechanical or structural system whose motion is being described, T is the system's total kinetic energy, U is the strain energy of the elastic portion of the system, V is any other significant potential energy possessed by the system, and Q_i is the net generalized force corresponding to the ith generalized coordinate representing those aspects of the system that are not, or cannot, be included in the potential functions U, V. If there are n generalized coordinates, then a Lagrange equation must be written for each of the n DOF.

After identifying the system DOF, an analysis employing the Lagrange equation begins with writing the mathematical expression for the kinetic energy, T, in terms of (i) the time derivatives of the DOF (called the generalized velocities) and perhaps (ii) the DOF themselves and perhaps (iii) time itself. That is, in mathematical terms,

$T = T(\dot{q}_i, q_i, t)$. In the case of the system of Figure 2.1(a), where of course θ is the sole DOF, from Eq. (1.17)

$$T = \tfrac{1}{2}m\, v_{CG}{}^2 + \tfrac{1}{2}H_{CG}\, \dot{\theta}^2_{CG}.$$

The quantity v_{CG} is the total rectilinear velocity of the center of mass and, of course, θ_{CG} is the rotation at the center of mass. That is,

$$\theta_{CG} = \theta_{FA} = \theta$$

$$\text{and} \quad v^2 = v^2_{hor} + v^2_{vrt} = \left[\frac{d}{dt} L \sin\theta\right]^2 + \left[\frac{d}{dt} L(1 - \cos\theta)\right]^2$$

$$= L^2\dot{\theta}^2\cos^2\theta + L^2\dot{\theta}^2\sin^2\theta = L^2\dot{\theta}^2$$

$$\text{thus} \quad T = \tfrac{1}{2}(mL^2 + H)\dot{\theta}^2 = \tfrac{1}{2}H_{FA}\dot{\theta}^2.$$

Of course, the total rectilinear velocity of the center of mass in this simple case is nothing more than the tangential velocity of the center of mass at the center of the bob. Thus this simple result for the total rectilinear velocity could have been written immediately.

The next step is to write the potential energy expressions. In this case, because the rod and mass of Figure 2.1(a) or the amorphous mass of Figure 2.1(b) are modeled as being rigid, they do not store work because of elastic deformations. Thus the strain energy, U, of this system is zero. There is, of course, a potential energy, V, that results from the gravitational field. The magnitude of the gravitational potential energy is simply the negative of the work done by the gravitational force, mg, as the bob mass m moves along its circular arc from an arbitrary (but convenient) datum to another arbitrary position. Let the datum be at $\theta = 0$ (the SEP) and let the arbitrary position of the bob mass be specified by a positive value of the coordinate theta. Then, from Figure 2.1(b), the vertical rise of the bob mass from the datum is $L(1 - \cos\theta)$. Since the gravitational force is directed downward, and the displacement of the force is upward, the work done by the gravitational force when it moves from the datum position to the arbitrary position is a negative quantity. Again, because any potential energy is the negative of the corresponding work quantity, the gravitational potential energy in this case is $V = +mgL(1 - \cos\theta)$.

The last item to be discussed in this application of the Lagrange equation to this single-DOF system is the generalized force Q_θ. Generalized forces are always obtained by calculating the virtual work done by all the external forces acting on the system that are *not* accounted for by any of the potential energy expressions. (The inertia forces are always accounted for by the kinetic energy expression.) Note (i) the two support forces at the pivot do not move as a result of the virtual displacement $\delta\theta$, so they do no virtual work; (ii) the gravitational force has been fully incorporated into the gravitational potential energy; and (iii) there are no other forces to consider. Thus the conclusion is that the one generalized force, Q_θ, for this system is zero.

It is worth reiterating that the above V, Q assignments are not the only possible choice. The analyst could equally well choose a zero value to the potential energy and leave to the generalized force expression the means for incorporating the gravitational force into the analysis. To see how this is done, view Figure 2.1(c). This figure shows the entire system moved through a positive, virtual, increment in the generalized coordinate, $\delta\theta$. (If there were other generalized coordinates, they would be incremented separately, in turn.) As with all such virtual displacements, the variation in θ, $\delta\theta$, is positive in the same sense that θ is positive. Recall that a real force doesn't change its magnitude or direction in response to a virtual displacement. Then this virtual displacement $\delta\theta$ causes the gravitational force to rise vertically and thus do virtual work of the amount

$$\delta W = -mg\{L[1 - \cos(\theta + \delta\theta)] - L(1 - \cos\theta)\}$$
$$\text{where} \quad \cos(\theta + \delta\theta) = \cos\theta - \delta\theta \sin\theta$$

because $\delta\theta$ is very small. Thus

$$\delta W = mgL\delta\theta \sin\theta.$$

A simpler approach to again obtain the same result would have been to just use the component of the weight force in the tangential direction of the virtual displacement. The next step towards determining the generalized force is to compare this particular value for the virtual work to the virtual work definition of generalized forces. As Eq. (1.12a), the generalized force definition, makes clear, the generalized forces corresponding to each generalized coordinate are merely the coefficients of the variations on that generalized coordinate in that general virtual work expression:

$$\delta W \equiv \sum_{i=1}^{n} Q_i \delta q_i = Q_1 \delta q_1 \equiv Q_\theta \delta\theta$$
$$\text{here} \quad \delta W = -mgL\delta\theta \sin\theta$$
$$\text{thus} \quad Q_\theta = -mgL \sin\theta.$$

Substitution of the above expressions for T, $(U = 0)$, and V or Q_θ (not both because they both account for the same gravitational force), into the Lagrange equation for the DOF θ, leads to the same equation of motion previously obtained, Eq. (2.1). In summary, Eq. (2.1) is most easily obtained by the fixed axis formula. However, the fixed axis formula is quite limited in its applicability. Among the two general approaches, Newton's and the Lagrange equations, the Lagrange equation is no more work than Newton's equations and, with familiarity, will generally be much less work.

2.3 Example Problems

Pendulums can have all sorts of appearances. The following example problems, and the exercises at the end of the chapter, illustrate some of their many varieties of form. In all cases, the gravitational vector is downward and the systems under study are without any form of energy dissipative friction such as that between two wet or dry surfaces sliding on each other. The second example problem is a caution that not everything that looks like a pendulum is really a pendulum.

EXAMPLE 2.1 Write the equation of motion for the mechanical system of Figure 2.2(a), where the system is pictured in its undeflected position. Again the

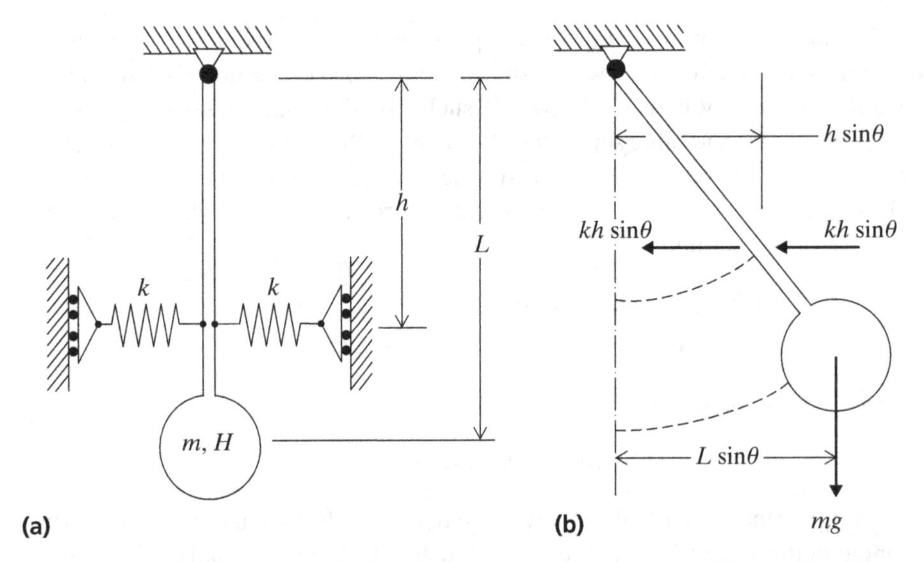

Figure 2.2. Example 2.3: Simple pendulum with elastic supports. (a) Undefected system. (b) Deflected system.

pendulum arm and the bob are modeled as (relatively) rigid, and the only flexible elements of the system are represented by the coiled springs.

COMMENT Note that when elastic springs are present, as is the case here, it is always to be understood unless otherwise specified, that (i) the springs are massless, (ii) the springs embody linear force–displacement relationships (i.e., force equals stiffness factor k times displacement), and (iii) the springs retain their original line-of-force orientation. In other words, with respect to the third item, after a deflection of this pendulum arm to the left or the right, both outer supports of each spring move so that each spring remains horizontal. The reason for this convention is that if the position of the outer spring supports were fixed, and the deflection of the spring became more than several percent of the length of the spring, then the force in each spring would begin to vary nonlinearly with respect to the horizontal deflection at each inner spring connection. Since, in structural analyses, springs are often mathematical representations for other, more complicated structural elements or collections of elements, this is a nonlinearity not worth including. Thus spring lengths are pieces of information that are never needed for a linear analysis.

SOLUTION The first step in the analysis is to redraw the system in its deflected position. Note again that the translational and rotational position of all the mass in the system (the bob) is fully specified by the single DOF θ. Thus θ is q_1. After identifying the generalized coordinate(s), the Lagrange equation quantities T, U, V, and Q_θ are the next order of business. The addition of (massless) elastic springs does nothing to alter the kinetic energy of this system from that of the system of Figures 2.1(a)–(c). Thus, again

$$T = \tfrac{1}{2}mL^2\dot{\theta}^2 + \tfrac{1}{2}H\dot{\theta}^2.$$

The difference between this system and the system of Figures 2.1(a)–(c) is the presence of the linearly elastic springs that store elastic strain energy, U. As discussed in greater detail in Section 3.4, the strain energy expression for any single linear spring, where u_1, u_2 are the end deflections of the spring in the direction of the spring axis, is

$$U = \tfrac{1}{2}\lfloor u_1 \quad u_2 \rfloor \begin{bmatrix} +k & -k \\ -k & +k \end{bmatrix} \begin{Bmatrix} u_1 \\ u_2 \end{Bmatrix},$$

where the above square matrix is the spring finite element stiffness matrix. Multiplying out the triple matrix product produces the general result for a spring that $U = \tfrac{1}{2}k(u_1 - u_2)^2$, where the difference between the two spring end displacements is, of course, the spring stretch. Another way of looking at this is to picture the linear force–deflection plot for the elastic spring. Since the deflection $u_2 - u_1$ increases proportionally to the applied force $k(u_2 - u_1)$, the work done by the applied force on the spring is the triangular area under the straight line plot, which is $\tfrac{1}{2} * (u_2 - u_1) * k(u_2 - u_1)$. This work is stored in the spring as the strain energy of the spring. Note in passing that the one-half factor is often an indicator of linear elasticity.

Since, for the left spring, $u_1 = 0, u_2 = h\sin\theta$ and for the right spring $u_1 = h\sin\theta, u_2 = 0$, then, for both springs together, $U = kh^2\sin^2\theta$. The above general expressions for the strain energy for a spring is used frequently in this textbook. Finally, the V, Q trade-off is the same as in the first discussed pendulum problem. Substitution of T, U, V, and Q into the Lagrange equation, and differentiating accordingly, yields, after a small bit of trigonometry, the equation of motion

$$(H + mL^2)\ddot{\theta} + mgL\sin\theta + kh^2\sin 2\theta = 0. \tag{2.4}$$

Note again that the only addition to the analysis of the previous problem is the addition of moments because of the elastic springs. ★

EXAMPLE 2.2 Consider the rod and bob system of Figure 2.3, which in this case is drawn deflected upward through an angle θ from its horizontal static equilibrium position. This is *not* a pendulum system even though, with the deflection in this upward direction, the gravitational force tends to return the system to its undeflected position. This system fails being a pendulum system because, when the system is rotated downward, the gravitational field force does not tend to return the system to its undeflected

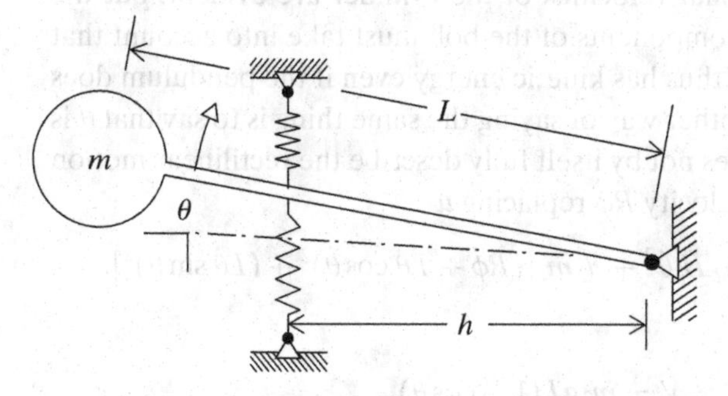

Figure 2.3. Example 2.2: A vibratory system that appears to be a pendulum but is not.

position but rather tends to move the system further from its undeflected position. Regardless of the direction of the system displacement, the gravitational field (or other force field) must always move the system back toward (or away from) the static equilibrium position for the system to be classified as a pendulum system. The writing of the equations of motion for systems like this is postponed until later. ★

EXAMPLE 2.3 Figure 2.4(a) shows an upper cylinder that rolls on a flat surface without slipping. The upper cylinder's center of mass is at its geometric center. The upper cylinder supports a free-swinging pendulum that in turn supports an external, time-varying, horizontal force that is applied at the pendulum center of mass. Write the equations of motion for this system.

SOLUTION It is a straightforward but lengthy procedure to (i) write five Newtonian force and moment equations and (ii) to use three of those five equations to eliminate the two internal reaction force components between the rolling cylinder and the pendulum and the rolling friction force between the cylinder and the flat surface. The result is one equation of motion for each of the two rigid masses. However, the simpler analysis presented below is based on the Lagrange equations. Again, the first step in a Lagrange analysis is to choose the generalized coordinates. With the no-slip condition between the cylinder and the flat surface, the distance u that the cylinder rolls from the datum is related to the rotation angle of the cylinder ϕ by the equation $R\phi = u$. Thus u and ϕ are not independent variables, and only one of the two can be a generalized coordinate. (Recall that all system DOF must be wholly independent quantities.) Arbitrarily choose ϕ as the generalized coordinate that precisely locates the position of the rolling cylinder. With the rolling cylinder located thusly, the mass of the pendulum bob is located by use of the additional measure θ. Hence θ is the second and final generalized coordinate for this pendulum system. Again, ϕ locates the position of all the mass of the rolling cylinder, whereas ϕ and θ together locate the position of all the mass of the pendulum. Thus all of the mass of the system is located unambiguously by these two DOF. Note that the presence or absence of forces and moments has no bearing on the choice of the generalized coordinates.

Now it is a matter of going through the T, U, V, Q checklist. The kinetic energy, T, is never more than the rectilinear and rotational kinetic energy of each system mass. The rectilinear and rotational velocities of the cylinder are evident, but the horizontal and vertical velocity components of the bob must take into account that bob moves with the cylinder and thus has kinetic energy even if the pendulum does not swing. See Figure 2.4(b). Another way of saying the same thing is to say that θ is a relative coordinate in that it does not by itself fully describe the rectilinear motion of the bob. Therefore, with the velocity $R\phi$ replacing u

$$T = \tfrac{1}{2} m_1 (R\dot{\phi})^2 + \tfrac{1}{2} H_1 \dot{\phi}^2 + \tfrac{1}{2} H_2 \dot{\theta}^2 + \tfrac{1}{2} m_2 [(R\dot{\phi} + L\dot{\theta}\cos\theta)^2 + (L\dot{\theta}\sin\theta)^2].$$

Furthermore,

$$U = 0 \qquad V = m_2 g L(1 - \cos\theta).$$

The only part of the problem not yet accounted for is the externally applied force of arbitrary magnitude, $F(t)$. Since no potential is available for such a nonconservative

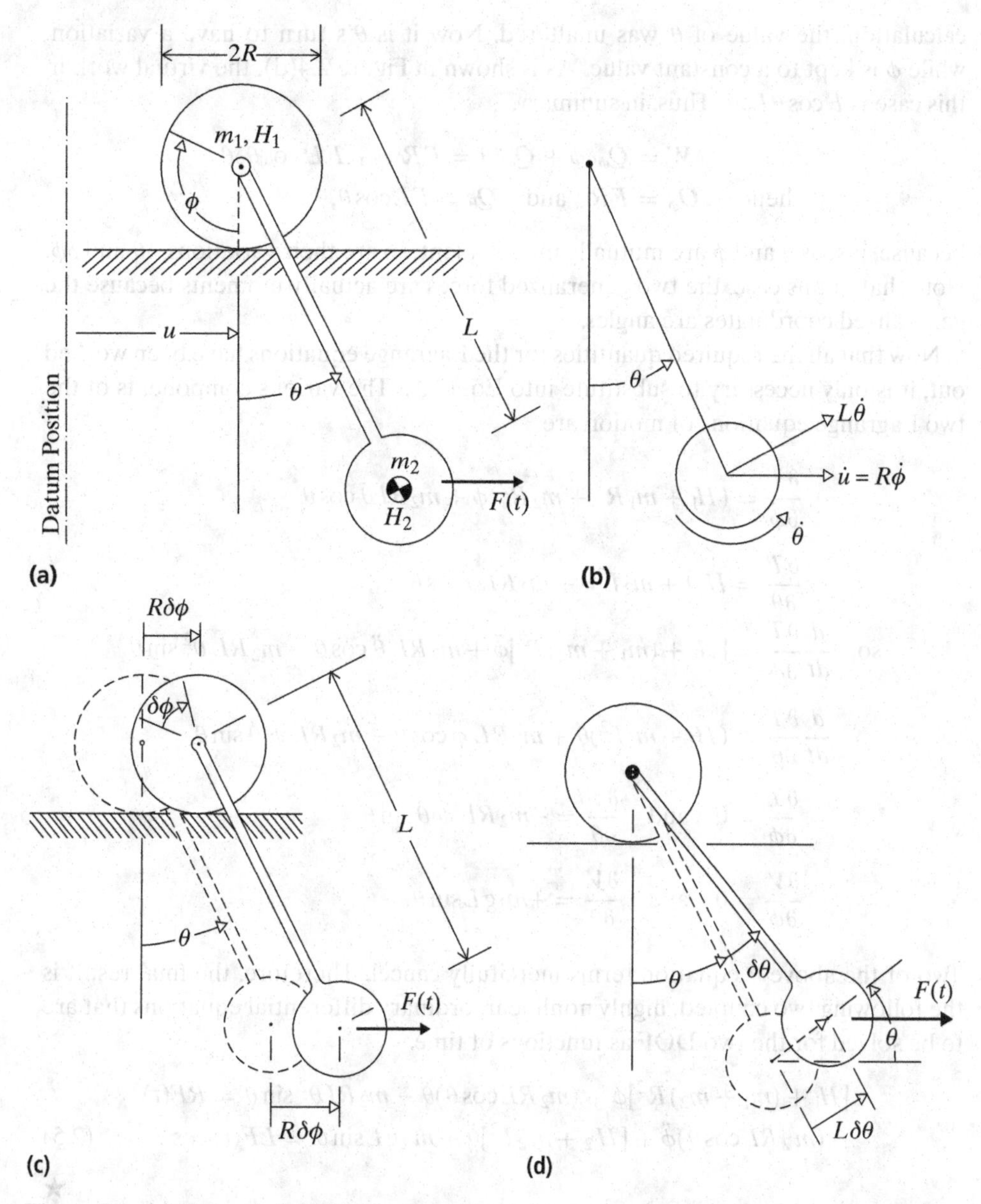

Figure 2.4. (a) Example 2.3: Deflected position of two-mass pendulum system. (b) Velocity diagram for the second mass. (c) Exaggerated first virtual displacement, $\delta\phi$. (d) Exaggerated second virtual displacement, $\delta\theta$.

force, it has to be included in the generalized force terms of the Lagrange equations. As always, the calculation of the generalized forces proceeds from the introduction of positively directed virtual deflections, one at a time, in each of the system DOF for the purpose of calculating the resulting virtual work. Start with a virtual deflection $\delta\phi$. As can be seen from Figure 2.4(c), $\delta\phi$ results in a horizontal (virtual) movement of the applied force of magnitude $R\delta\phi$. Since virtual work is always a matter of real forces and moments moving through, respectively, virtual translations and rotations, this external force does virtual work of magnitude $\delta W = FR\delta\phi$. Note that in this calculation, the value of θ was unaltered. Now it is θ's turn to have a variation, while ϕ is kept to a constant value. As is shown in Figure 2.4(d), the virtual work in this case is $F\cos\theta L\delta\theta$. Thus, in summary,

$$\delta W = Q_\phi \delta\phi + Q_\theta \delta\theta = FR\delta\phi + FL\cos\theta\delta\theta$$
$$\text{hence} \quad Q_\phi = FR \quad \text{and} \quad Q_\theta = FL\cos\theta,$$

because, just as θ and ϕ are mutually independent, so are their variations, $\delta\theta$ and $\delta\phi$. Note that in this case, the two generalized forces are actually moments because the generalized coordinates are angles.

Now that all the required quantities for the Lagrange equations have been worked out, it is only necessary to substitute into Eq. (2.3). The various components of the two Lagrange equations of motion are

$$\frac{\partial T}{\partial \dot\phi} = (H_1 + m_1R^2 + m_2R^2)\dot\phi + m_2RL\dot\theta\cos\theta$$

$$\frac{\partial T}{\partial \dot\theta} = H_2\dot\theta + m_2L^2\dot\theta + m_2RL\dot\phi\cos\theta$$

$$\text{so} \quad \frac{d}{dt}\frac{\partial T}{\partial \dot\phi} = [H_1 + (m_1+m_2)R^2]\ddot\phi + m_2RL\,\ddot\theta\cos\theta - m_2RL\,\dot\theta^2\sin\theta$$

$$\frac{d}{dt}\frac{\partial T}{\partial \dot\theta} = (H_2 + m_2L^2)\ddot\theta + m_2RL\,\ddot\phi\cos\theta - m_2RL\,\dot\phi\,\dot\theta\sin\theta$$

$$\frac{\partial T}{\partial \phi} = 0 \quad \text{and} \quad \frac{\partial T}{\partial \theta} = -m_2RL\,\dot\phi\,\dot\theta\sin\theta$$

$$\frac{\partial V}{\partial \phi} = 0 \quad \text{and} \quad \frac{\partial V}{\partial \theta} = +m_2gL\sin\theta.$$

Two of the above θ equation terms mercifully cancel. Therefore, the final result is the following two coupled, highly nonlinear, ordinary differential equations that are to be solved for the two DOF as functions of time.

$$\begin{aligned}
&[H_1 + (m_1+m_2)R^2]\ddot\phi + (m_2RL\cos\theta)\ddot\theta - m_2RL\dot\theta^2\sin\theta = RF(t)\\
&(m_2RL\cos\theta)\ddot\phi + [H_2 + m_2L^2]\ddot\theta + m_2gL\sin\theta = LF(t)\cos\theta.
\end{aligned} \tag{2.5}$$

★

EXAMPLE 2.4 Write the equation of motion for the smaller cylinder that rolls without slipping on the inside of the static, larger circular cylinder, as shown in end view in Figure 2.5. The smaller cylinder has mass m, and its center of mass is at its geometric center. The smaller cylinder also has a mass moment of inertia H about its center of mass.

SOLUTION The no-slip constraint on the system means, from the geometry of the two cylindrical surfaces, that $R\theta = r\phi$. Thus the position of the rolling cylinder is completely specified by the use of either of these two angular measures, ϕ or θ. Arbitrarily choose θ as the single DOF for the rolling cylinder. Now it is just a matter of going

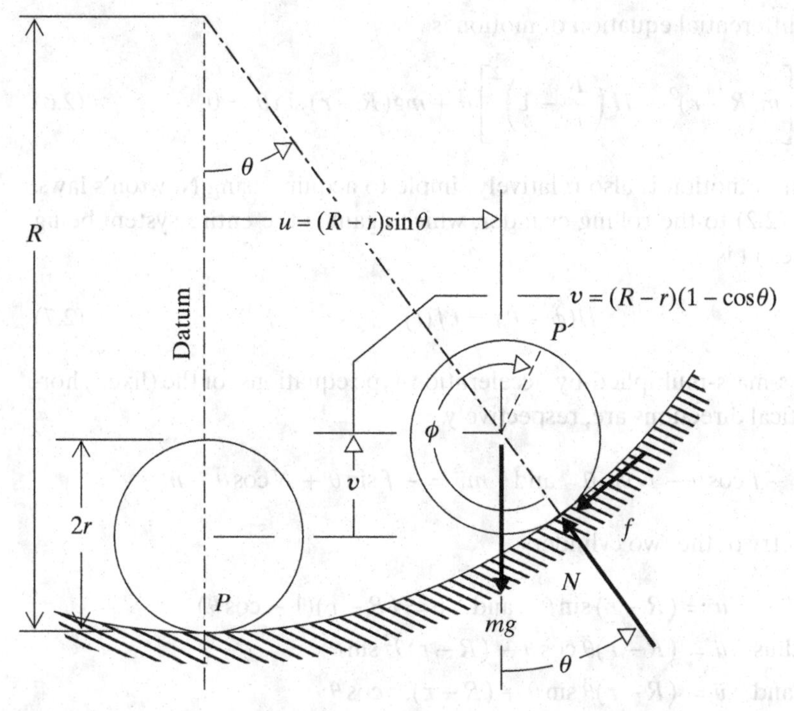

Figure 2.5. Example 2.4: A circular cylinder of radius r rolling inside another of radius R.

through the T, U, V, Q checklist for the Lagrange equation. However, the kinetic energy expression does require some explanation. The total rectilinear velocity (i.e., the square root of the squares of the two orthogonal velocity components) of the rolling cylinder is again the tangential velocity, $(R - r)\dot{\theta}$. The total angular velocity of the rolling cylinder (the reason for this example) is not simply the time derivative of ϕ, but rather the time derivative of ϕ less θ. The reason for this is that, from Newton's laws on which the Lagrange equation is based, the angular velocity must be measured with respect to a fixed axis (e.g., a vertical axes), not with respect to a rotating axis such as that between the center of the larger cylinder and the point of contact between the two cylinders.

Another way of determining the correct rotational velocity of the smaller cylinder is first to recall that kinetic energy depends only on the instantaneous mass velocities (i.e., on the instantaneous displacements whose time derivatives yield the velocities) and not the path taken to achieve those velocities or displacements. Second, note that the arbitrary deflected position of the mobile cylinder, as shown in Figure 2.5, can be achieved by taking the path where the mobile cylinder translates through space, from the SEP datum at the bottom of the larger cylinder, up and to the right, and then rotates into its final position. Note that after the two translations, the vertical line between the cylinder center and the original contact point P rotates only through the angle $\phi - \theta$ to reach point P′. Thus

$$T = \tfrac{1}{2}m[(R-r)\dot{\theta}]^2 + \tfrac{1}{2}H(\dot{\phi} - \dot{\theta})^2 = \tfrac{1}{2}\left[m(R-r)^2 + H\left(\frac{R}{r} - 1\right)^2\right]\dot{\theta}^2$$

$$V = mg(R-r)(1-\cos\theta) \qquad U = 0 \qquad Q = 0.$$

Therefore the differential equation of motion is

$$\left[m(R-r)^2 + H\left(\frac{R}{r}-1\right)^2\right]\ddot{\theta} + mg(R-r)\sin\theta = 0. \tag{2.6}$$

This equation of motion is also relatively simple to acquire using Newton's laws. First apply Eq. (2.2) to the rolling cylinder, which again is the entire system being analyzed. The result is

$$H(\ddot{\phi} - \ddot{\theta}) = rf(t). \tag{2.7}$$

The force-equals-mass-multiplied-by-acceleration-type equations for the (fixed) horizontal and vertical directions are, respectively,

$$m\ddot{u} = -f\cos\theta - N\sin\theta \quad \text{and} \quad m\ddot{v} = -f\sin\theta + N\cos\theta - mg.$$

From the geometry of the two cylinders

$$\begin{aligned} u &= (R-r)\sin\theta \quad \text{and} \quad v = (R-r)(1-\cos\theta) \\ \text{thus} \quad \ddot{u} &= (R-r)\ddot{\theta}\cos\theta - (R-r)\dot{\theta}^2\sin\theta \\ \text{and} \quad \ddot{v} &= (R-r)\ddot{\theta}\sin\theta + (R-r)\dot{\theta}^2\cos\theta. \end{aligned}$$

Substituting these accelerations into the force equations yields

$$\begin{aligned} m(R-r)\ddot{\theta}\cos\theta - m(R-r)\dot{\theta}^2\sin\theta &= -f\cos\theta - N\sin\theta \\ m(R-r)\ddot{\theta}\sin\theta + m(R-r)\dot{\theta}^2\cos\theta &= -f\sin\theta + N\cos\theta - mg. \end{aligned}$$

Inspection of these two equations indicates that they can be greatly simplified by multiplying the first of the two by $\cos\theta$, multiplying the second by $\sin\theta$, and adding the two resulting equations to obtain

$$m(R-r)\ddot{\theta} = -f - mg\sin\theta.$$

Substituting this last equation into the moment equation, Eq. (2.7), yields

$$\left[H\left(\frac{R}{r}-1\right) + mr(R-r)\right]\ddot{\theta} + mgr\sin\theta = 0.$$

This equation can be seen to be identical to Eq. (2.6), the previously derived equation of motion, when it is multiplied by the quantity $[(R-r)/r]$. An important aspect of this Newtonian derivation of the equation of motion is that this second derivation reinforces the conclusion that the friction force, which enforces the no-slip condition at the cylinder boundary, does not appear in the final form of the equation of motion. This conclusion fully establishes the fact that this friction force, which is a constraint or reaction force, does no work on the rolling cylinder. Therefore this friction force of constraint, unlike other friction forces, is not an energy dissipative force, and it truly has no place in the energy expressions of the Lagrange equation. ★

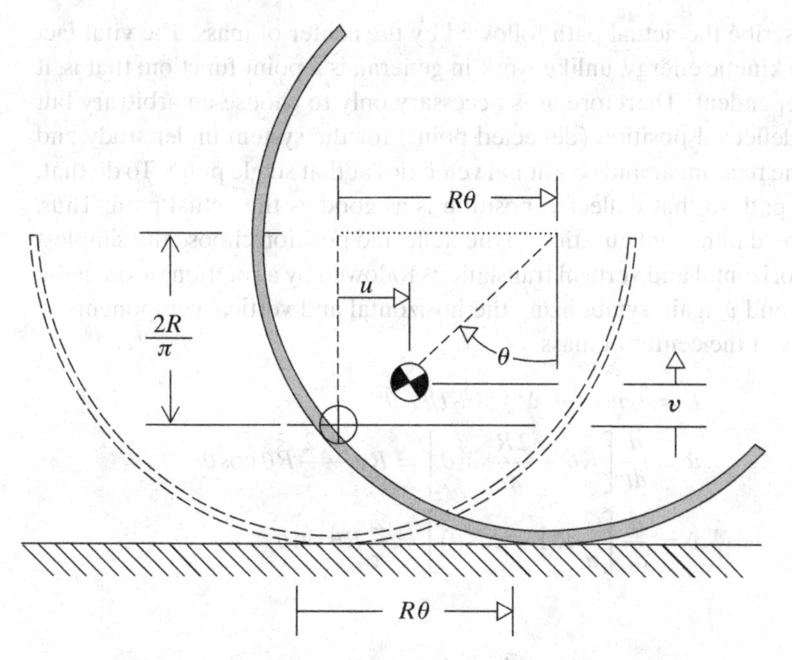

Figure 2.6. Example 2.5: Rocking half cylinder rolling without slipping.

EXAMPLE 2.5 Write the equation of motion for the rigid, hollow, half cylinder of thin ($t \ll R$) material shown in end view both in its datum and displaced configurations in Figure 2.6. The half cylinder rocks back and forth without slipping on its supporting flat surface. The reader is invited to confirm that the center of mass of the half cylinder is located a distance $2R/\pi$ below the full cylinder center, and that $H_{cg} = mR^2[1 - (4/\pi^2)]$, where m is the total mass of the half cylinder.

SOLUTION This is a pendulum problem because regardless of which of the two directions the cylinder is displaced, left or right, the gravitational field causes the half cylinder to roll back toward its datum, its static equilibrium position, where its plane of symmetry is vertical. Note that the geometry of the half cylinder is such that the full cylinder center is always a distance R above the flat surface. Thus, as is shown in the figure, the half cylinder center point always moves entirely horizontally a distance equal to that moved by the contact point between the half cylinder and the flat surface.

Since the half cylinder rolls without slipping, the angle of rotation, θ, can be used to describe both the rotation of the half cylinder and the translation of its center of mass. Thus θ is chosen as the single generalized coordinate for the rocking half cylinder. With no half cylinder deformations and no external forces other than the gravitational force that is to be accounted for by the potential energy V, both U and Q are zero. Therefore the equation of motion is developed by writing the expressions for T and V. The total kinetic energy is the kinetic energy of translation and the kinetic energy of rotation. Since there is no fixed axis about which this system rotates, the analysis must focus on the motion of the system center of mass. A simple formula for the actual motion of the center of mass is not apparent. (The path in the plane of the paper that the center of mass actually traces out is a curtate cycloid.) Fortunately, it is not at all necessary to describe the actual path followed by the center of mass. The vital fact is again that the kinetic energy, unlike work in general, is a point function; that is, it is never path dependent. Therefore, it is necessary only to choose an arbitrary but representative deflected position (deflected point) for the system under study and then calculate the rectilinear and rotational velocities at that single point. To do that, *any* convenient path to that deflected position is as good as the actual path. Thus, to travel from the datum configuration to the deflected position choose the simplest possible path: horizontal and vertical translations followed by a rotation through the angle θ. With $\dot{u}$ and $\dot{v}$ again symbolizing the horizontal and vertical components of the total velocity at the center of mass

$$T = \tfrac{1}{2}m(\dot{u}^2 + \dot{v}^2) + \tfrac{1}{2}H_{CG}\dot{\theta}^2$$

$$\dot{u} = \frac{d}{dt}\left[R\theta - \frac{2R}{\pi}\sin\theta\right] = R\dot{\theta} - \frac{2}{\pi}R\dot{\theta}\cos\theta$$

$$\dot{v} = \frac{d}{dt}\left[\frac{2}{\pi}R(1-\cos\theta)\right] = \frac{2}{\pi}R\dot{\theta}\sin\theta.$$

So

$$\dot{u}^2 + \dot{v}^2 = R^2\dot{\theta}^2\left(1 + \frac{4}{\pi^2} - \frac{4}{\pi}\cos\theta\right)$$

$$\text{and} \quad T = \tfrac{1}{2}mR^2\dot{\theta}^2(2 - \frac{4}{\pi}\cos\theta)$$

$$\text{and} \quad V = mg\frac{2}{\pi}R(1-\cos\theta).$$

Now it is simply a matter of substituting into the Lagrange equation and simplifying the result. The necessary derivatives, and the simplified form of the equation of motion in standard form, are

$$\frac{\partial T}{\partial\dot{\theta}} = 2mR^2\dot{\theta}\left(1 - \frac{2}{\pi}\cos\theta\right)$$

$$\frac{d}{dt}\frac{\partial T}{\partial\dot{\theta}} = 2mR^2\ddot{\theta}\left(1 - \frac{2}{\pi}\cos\theta\right) + 2mR^2\dot{\theta}\left(\frac{2}{\pi}\dot{\theta}\sin\theta\right)$$

$$-\frac{\partial T}{\partial\theta} = -mR^2\dot{\theta}^2\left(\frac{2}{\pi}\sin\theta\right)$$

$$\text{and} \quad \frac{\partial V}{\partial\theta} = \frac{2}{\pi}mgR\sin\theta$$

$$\text{thus} \quad \left(1 - \frac{2}{\pi}\cos\theta\right)\ddot{\theta} + \frac{\dot{\theta}^2}{\pi}\sin\theta + \frac{g}{\pi R}\sin\theta = 0. \tag{2.8}$$

★

EXAMPLE 2.6 Write the equation of motion for the "double pendulum" shown in Figure 2.7. The new feature of this example problem is that the two pendulum system is being driven by an enforced base motion, $h(t)$, which has a known time history, rather than being driven by an applied load.

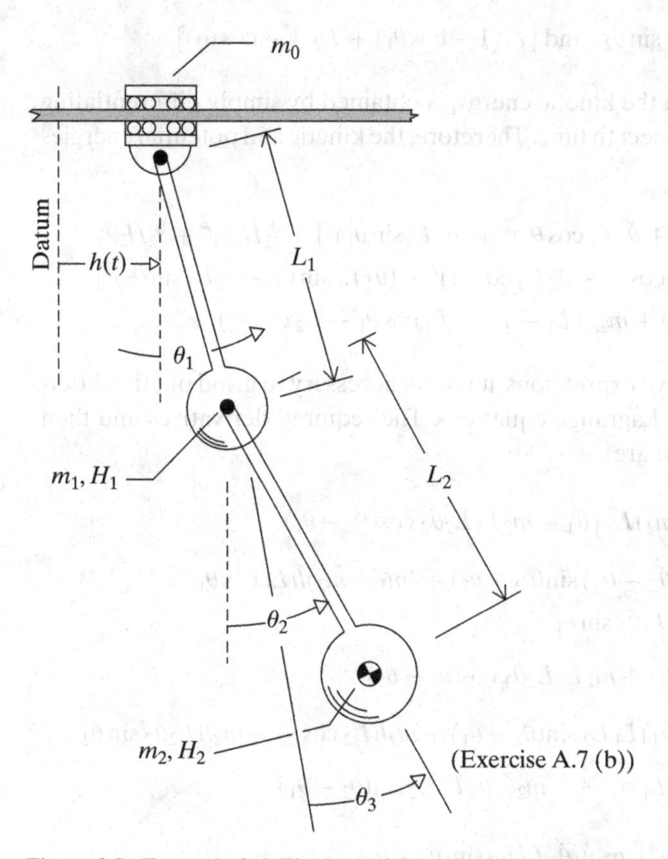

Figure 2.7. Example 2.6: The base-driven double pendulum shown in its deflected position.

SOLUTION The first step is defining the system and selecting the generalized coordinates. The first question is whether the mass of the moving support, m_0, should be included within the system boundaries. It is not evident that it is not necessary or useful to include the moving support within the system. Therefore m_0 will be included in the system so that later it can be seen that it has no part in the equations of motion. The position of the moving support mass, m_0, is fully described by the known quantity $h(t)$. Thus no generalized coordinate is required for the moving support. One generalized coordinate is required for each of the two pendulums. Choose θ_1 and θ_2, as indicated in Figure 2.7, to be the required DOF. [See Exercise 2.7(b) for another choice of DOF involving θ_3.] Since each pendulum arm is rigid, the strain energy, U, is zero. Since the gravitational forces are the only explicit, external forces, and because they are conveniently described by the gravitational potential function, the two generalized forces Q_1, Q_2 are also zero. Thus it is only necessary to write expressions for the kinetic and potential energies.

The combination of the mass values, the horizontal distances, and the vertical distances each of the three masses move are

(i) m_0, $h(t)$, and 0;

(ii) m_1, $(h + L_1 \sin\theta_1)$, and $L_1(1 - \cos\theta_1)$; and

(iii) m_2, $(h + L_1 \sin\theta_1 + L_2 \sin\theta_2)$ and $[L_1(1 - \cos\theta_1) + L_2(1 - \cos\theta_2)]$.

The velocities, and hence the kinetic energy, is obtained by simply differentiating the above distances with respect to time. Therefore, the kinetic and potential energies are

$$\begin{aligned} T &= \tfrac{1}{2}m_0\dot{h}^2 + \tfrac{1}{2}m_1[(\dot{h} + \dot{\theta}_1 L_1 \cos\theta_1)^2 + (\dot{\theta}_1 L_1 \sin\theta_1)^2] + \tfrac{1}{2}H_1\dot{\theta}_1{}^2 + \tfrac{1}{2}H_2\dot{\theta}_2^2 \\ &\quad + \tfrac{1}{2}m_2[(\dot{h} + \dot{\theta}_1 L_1 \cos\theta_1 + \dot{\theta}_2 L_2 \cos\theta_2)^2 + (\dot{\theta}_1 L_1 \sin\theta_1 + \dot{\theta}_2 L_2 \sin\theta_2)^2] \\ V &= m_1 g L_1(1 - \cos\theta_1) + m_2 g(L_1 + L_2 - L_1\cos\theta_1 - L_2\cos\theta_2). \end{aligned}$$

After simplifying the above expressions, it is only necessary to grind out the differentiations called for by the Lagrange equations. The required derivatives and then the two equations of motion are

$$\begin{aligned} \frac{d}{dt}\frac{\partial T}{\partial\dot{\theta}_1} &= \left[H_1 + (m_1 + m_2)L_1^2\right]\ddot{\theta}_1 + m_2 L_1 L_2\ddot{\theta}_2\cos(\theta_2 - \theta_1) \\ &\quad - m_2 L_1 L_2\dot{\theta}_2(\dot{\theta}_2 - \dot{\theta}_1)\sin(\theta_2 - \theta_1) + (m_1 + m_2)\ddot{h} L_1 \cos\theta_1 \\ &\quad - (m_1 + m_2)\dot{h} L_1\dot{\theta}_1 \sin\theta_1 \\ \frac{d}{dt}\frac{\partial T}{\partial\dot{\theta}_2} &= \left(H_2 + m_2 L_2^2\right)\ddot{\theta}_2 + m_2 L_1 L_2\ddot{\theta}_1\cos(\theta_2 - \theta_1) \\ &\quad - m_2\dot{\theta}_1(\dot{\theta}_2 - \dot{\theta}_1)L_1 L_2\sin(\theta_2 - \theta_1) + m_2\ddot{h} L_2\cos\theta_2 - m_2\dot{h} L_2\dot{\theta}_2\sin\theta_2 \\ -\frac{\partial T}{\partial\theta_1} &= (m_1 + m_2)\dot{h}\dot{\theta}_1 L_1\sin\theta_1 - m_2\dot{\theta}_1\dot{\theta}_2 L_1 L_2\sin(\theta_2 - \theta_1) \\ -\frac{\partial T}{\partial\theta_2} &= m_2\dot{h}\dot{\theta}_2 L_2\sin\theta_2 + m_2\dot{\theta}_1\dot{\theta}_2 L_1 L_2\sin(\theta_2 - \theta_1) \\ \frac{\partial V}{\partial\theta_1} &= (m_1 + m_2)L_1 g\sin\theta_1 \quad \text{and} \quad \frac{\partial V}{\partial\theta_2} = m_2 L_2 g\sin\theta_2 \end{aligned}$$

$$\begin{aligned} \text{thus} \quad & \left[H_1 + (m_1 + m_2)L_1^2\right]\ddot{\theta}_1 + m_2 L_1 L_2\ddot{\theta}_2\cos(\theta_2 - \theta_1) \\ & \quad - m_2 L_1 L_2\dot{\theta}_2^2\sin(\theta_2 - \theta_1) + (m_1 + m_2)g L_1\sin\theta_1 \\ & = -(m_1 + m_2)\ddot{h} L_1\cos\theta_1 \\ \text{and} \quad & (H_2 + m_2 L_2^2)\ddot{\theta}_2 + m_2 L_1 L_2\ddot{\theta}_1\cos(\theta_2 - \theta_1) \\ & \quad - m_2\dot{\theta}_1^2 L_1 L_2\sin(\theta_1 - \theta_2) + m_2 g L_2\sin\theta_2 \\ & = -m_2\ddot{h} L_2\cos\theta_2. \end{aligned} \tag{2.9}$$

Note that, as mentioned above, the mass of the support does not enter into the equations of motion. Only the second time derivative of the enforced displacement, $h(t)$, the motion that drives the system, enters into the equations of motion in combination with the pendulum masses in the form of equivalent externally applied moments, that is, as mass–acceleration–distance terms.

EXAMPLE 2.7 For the case of small angular excursions from the vertical, write the equation of motion for the single spherical pendulum; that is, a simple pendulum supported by a universal joint such that the pendulum arm can rotate freely about both a horizontal axis in the plane of the paper and another such axis perpendicular

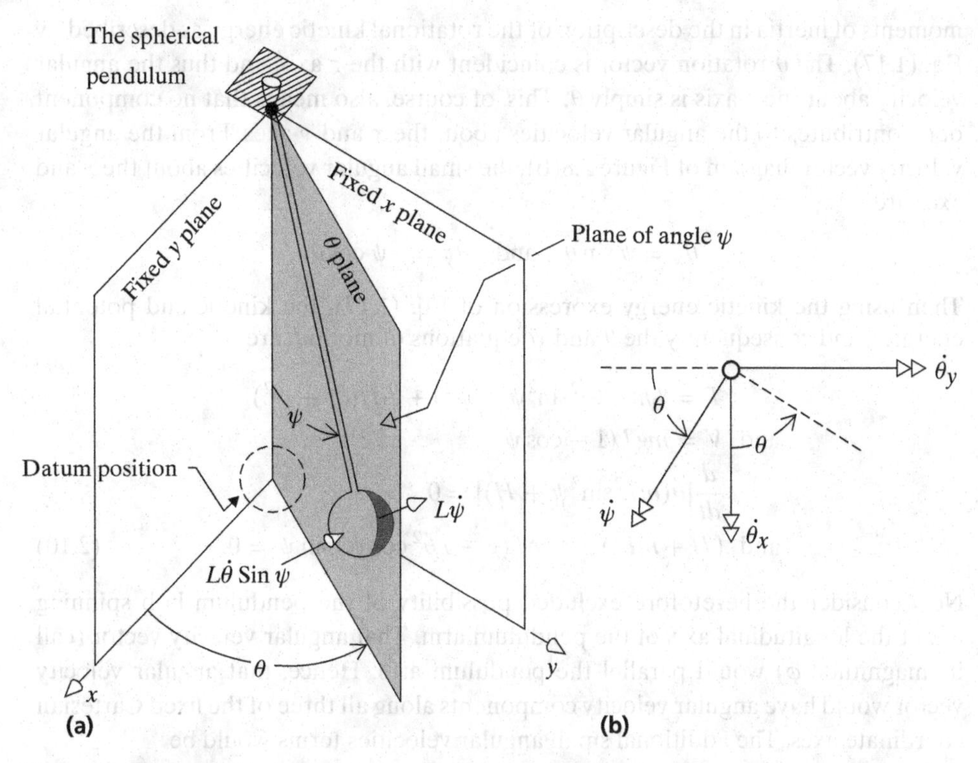

Figure 2.8. Example 2.7: (a) Displaced pendulum in three-space whose center of mass position is defined by θ and ψ. (b) Top view of angular velocity vectors.

to the plane of the paper. However, the joint is such that the bob mass cannot rotate about the longitudinal axis of the pendulum arm. See Figure 2.8(a). The pendulum arm length is L, and the bob mass parameters are simply m, and $H_{xx} = H_{yy} = H_{zz} = H$, while all mass products of inertia are zero. There are neither externally forced motions or externally applied forces or moments.

SOLUTION The bob center of mass moves on a spherical surface. Since the spherical radius coordinate has the fixed value L, the bob center of mass can be located at any point in time by the use of just the azimuth angle θ and the coordinate ψ. Hence, as shown in Figure 2.8, choose as the two generalized coordinates θ and ψ.[4] (If the prohibited spin about the pendulum arm axis were possible, then a third angular DOF would be required.) The two tangential velocity vectors corresponding to the θ and ψ DOF are $L\dot{\theta}\sin\psi$ (a velocity vector perpendicular to the "revolving door" plane of the pendulum arm) and $L\dot{\psi}$ (a vector within the "revolving door" plane).[5] These two velocity components are orthogonal to each other. Thus the sum of their squares is the square of the magnitude of the total rectilinear velocity vector. Having developed the quantities necessary to describe the translational portion of the kinetic energy, it is now necessary to turn to the fixed axis angular velocities to be used with the mass

[4] The "small" (right-hand rule) rotation angles being added vectorially are ψ and θ.
[5] Review the results of Exercise 1.6(b).

moments of inertia in the description of the rotational kinetic energy as described by Eq. (1.17). The $\dot{\theta}$ rotation vector is coincident with the z axis, and thus the angular velocity about the z axis is simply $\dot{\theta}$. This, of course, also means that no component of $\dot{\theta}$ contributes to the angular velocities about the x and y axes. From the angular velocity vector diagram of Figure 2.8(b), the small angular velocities about the x and axes are

$$\dot{\theta}_x = \dot{\psi} \sin\theta \quad \text{and} \quad \dot{\theta}_y = -\dot{\psi} \cos\theta.$$

Then using the kinetic energy expression of Eq. (1.17), the kinetic and potential energies, and consequently the θ and ψ equations of motion, are

$$\begin{aligned} T &= \tfrac{1}{2}mL^2(\dot{\theta}^2\sin^2\psi + \dot{\psi}^2) + \tfrac{1}{2}H(\dot{\theta}^2 + \dot{\psi}^2) \\ \text{and}\quad V &= mgL(1 - \cos\psi) \\ &\frac{d}{dt}[\dot{\theta}(mL^2\sin^2\psi + H)] = 0 \\ \text{and}\quad &(H + mL^2)\ddot{\psi} + mL(g - L\dot{\theta}^2\cos\psi)\sin\psi = 0. \end{aligned} \tag{2.10}$$

Now consider the heretofore excluded possibility of the pendulum bob spinning about the longitudinal axis of the pendulum arm. That angular velocity vector (call its magnitude $\dot{\phi}$) would parallel the pendulum arm. Hence, that angular velocity vector would have angular velocity components along all three of the fixed Cartesian coordinate axes. The additional small angular velocities terms would be

$$\dot{\theta}_x = -\dot{\phi}\cos\theta\sin\psi \qquad \dot{\theta}_y = -\dot{\phi}\sin\theta\sin\psi \qquad \dot{\theta}_z = \dot{\phi}\cos\psi.$$

★

2.4 Interpreting Solutions to Pendulum Equations

The previous section provides a variety of pendulum problems. More challenging pendulum vibration problems can be found in the exercises. A mastery of those example problems provides a firm basis for writing the equations of motion for any mechanical system (as opposed to a structural system). An inspection of the equations of vibratory motion associated with these example problems is sufficient to draw the conclusion that they are generally nonlinear ordinary differential equations. One important aspect of the (economical) practice of engineering is avoiding the rigors of solving nonlinear differential equations when it is possible to gain useful information from the corresponding linear forms of the same equations. The most direct way to avoid the nonlinearities associated with pendulum problems is to limit the range of applicability of the pendulum differential equations by limiting the magnitudes of the generalized coordinates. When the rotational angles are "small," the nonlinear sine functions of those rotation angles can be replaced by the linear functions that are simply the rotation angles themselves. That is, when θ is "small," sin θ can be approximated by θ. Similarly, the cosine of a small angle can be approximated by the value one after, not before,[6] the equation(s) of motion are obtained from the general form of the Lagrange equations. Of course, the validity of these two

[6] Before doing the differentiation required by the Lagrange equation, when the DOF θ is small, the cosine function can replaced by the first *two* terms of its series expansion, $1 - \tfrac{1}{2}\theta^2$.

trigonometric approximations depends on the accuracy required for the analysis. When the argument of these two functions is as large as two-tenths of a radian (a bit over 11°), then the largest error is in the cosine function, and that error is less than 2%. That is well within the usually accepted range of "engineering error." This assertion on the magnitude of the error of that approximation can be checked either by writing out the first few terms of Taylor series expansions for the sine and cosine functions or simply by use of a hand calculator. Thus, when the simple pendulum of Section 2.2 swings through the not insignificant, two-sided arc of 23° or less, the equation of motion of that pendulum, Eq. (2.1), can be approximated by replacing $\sin\theta$ by θ itself to get the following linear differential equation, which is in error by less than 1%:

$$(H + mL^2)\ddot{\theta} + mgL\theta = 0. \tag{2.11}$$

This linear differential equation (linear because θ and its derivatives appear only to the first power) is much easier to solve than the equation with the sine function. A brief discussion of an analytical (as opposed to numerical) solution of the original differential equation that includes the nonlinear sine function is discussed in Endnote (1). Numerical solutions to nonlinear ordinary differential equations are discussed in Chapter 9.

The purpose of this section is to discuss the various features of the mathematical solution to this typical, linear, vibratory equation of motion, Eq. (2.11). At this point in the subject development, the solution to this linear differential equation, although simple to derive, is only stated. The stated solution forms can be substantiated easily by direct substitution into Eq. (2.11). This is a matter of organizational economy. The full derivation of the solution to a slightly more inclusive linear differential equation is postponed to Chapter 5. As for the solution itself, first note that the solution to the above second-order, linear ordinary differential equation must have two constants of integration. The solution for the angular position of the pendulum arm at any point in time can be written in any one of the following three entirely equivalent mathematical forms

$$\theta(t) = C_1 \sin\omega t + C_2 \cos\omega t$$

$$\text{or} \quad \theta(t) = C_0 \sin(\omega t + \psi_1) = C_0 \cos(\omega t - \psi_2)$$

$$\text{where} \quad \omega = \sqrt{\frac{mgL}{H + mL^2}}$$

$$\text{and} \quad C_1 = C_0 \cos\psi_1 = C_0 \sin\psi_2 \quad C_2 = C_0 \sin\psi_1 = C_0 \cos\psi_2$$

$$\text{or} \quad C_0 = \sqrt{C_1^2 + C_2^2} \quad \text{and} \quad \tan\psi_1 = \frac{C_2}{C_1} = \cot\psi_2,$$

where the constants of integration in the first-solution form are C_1, and C_2, in the second-solution form they are C_0, and ψ_1, and in the third-solution form they are C_0 and ψ_2.

Note that the parameter ω is, in this case as it is to be in all cases, the square root of the ratio of the coefficient of the deflection term (θ) over the coefficient of the acceleration term ($\ddot{\theta}$). Focus on the second-solution form $C_0 \sin(\omega t + \psi_1)$, or the similar third-solution form $C_0 \cos(\omega t + \psi_2)$. The term C_0 is called the (total) *amplitude* of the motion because it describes how far the pendulum swings to one side or the other. The parameters C_1 and C_2 of the first-solution form are only partial amplitudes, as is seen clearly on the last of the above equation lines. The parameters ψ_1 and ψ_2 are called *phase angles*. Phase angles have only one effect. Phase angles shift the point where the sine or cosine function crosses the horizontal axis of the independent variable, which in this case is time. For example, if the first of these phase angles were to have the value $\pi/2$, then the sine function with that phase angle would be shifted to the left along the horizontal axis to the point where it would become a cosine function. Similarly, if the second phase angle had the value $\pi/2$, then the cosine function would become a sine function.[7]

Since θ is only a function of time, the various pairs of constants of integration (C_1, C_2), (C_0, ψ_1), and (C_0, ψ_2) depend on the values of the initial conditions. Let the symbols for the initial conditions be as follows:

$$\text{initial deflection} \quad \Theta_0 = \theta(0)$$
$$\text{and initial velocity} \quad \dot{\Theta}_0 = \dot{\theta}(0).$$

Setting time equal to zero in (i) the first of the above three solution forms for θ as a function of time, the one with the constants of integration C_1 and C_2; and (ii) the angular velocity equation, which is determined by differentiating this same angular deflection equation, yields

$$C_1 = \frac{\dot{\Theta}_0}{\omega} \quad \text{and} \quad C_2 = \Theta_0.$$

Therefore the first, and perhaps simplest, solution form becomes

$$\theta(t) = \frac{\dot{\Theta}_0}{\omega} \sin \omega t + \Theta_0 \cos \omega t. \tag{2.12}$$

The other two ways of writing the small deflection solution in terms of the initial conditions are easily obtained by using the relationships among the various constants of integration C_1, C_2, C_0, ψ_1, ψ_2, set forth above. Consider now either the single sine function or the single cosine function form of the solution. Either of these solution forms shows that for any combination of initial conditions, a plot of the back-and-forth motion of the pendulum arm is a sinusoidal plot. (Recall the cosine function is the same as the sine function except for a ninety degree phase shift.) As will be seen, sinusoidal motion is very common for vibrating systems in the (approximate) absence of energy dissipating friction or nonlinearities. It is so common that there is a special name for this type of motion. Such a motion is called *harmonic motion*, and it is so important that it is worth examining in detail. Figure 2.9(a) shows two different time histories for a simple pendulum where both amplitudes[8] are less than, say, 0.2 rad (11°) for the sake of the validity of the linear differential equation and its solution. The solid line time history in Figure 2.9(a) is for the case where the initial

[7] A simple basis for these statements can be the only trigonometric identities that need to be memorized: $\sin(\alpha \pm \beta) = \sin\alpha\cos\beta \pm \cos\alpha\sin\beta$ and $\cos(\alpha \pm \beta) = \cos\alpha\cos\beta \mp \sin\alpha\sin\beta$.

[8] Recall that as defined above, the "amplitude" is the extent of the motion past the static equilibrium position (where θ equals zero), on either side of the horizontal axis.

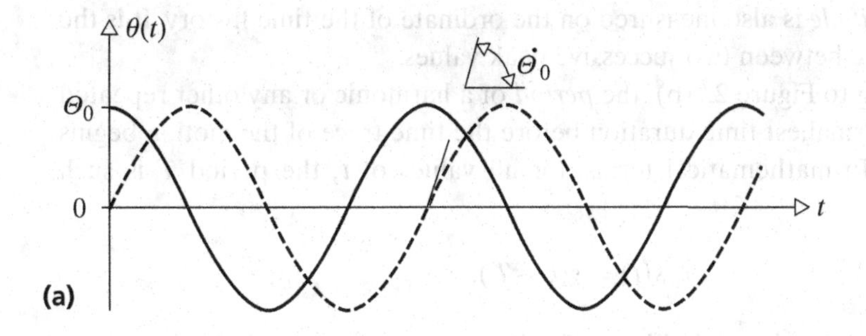

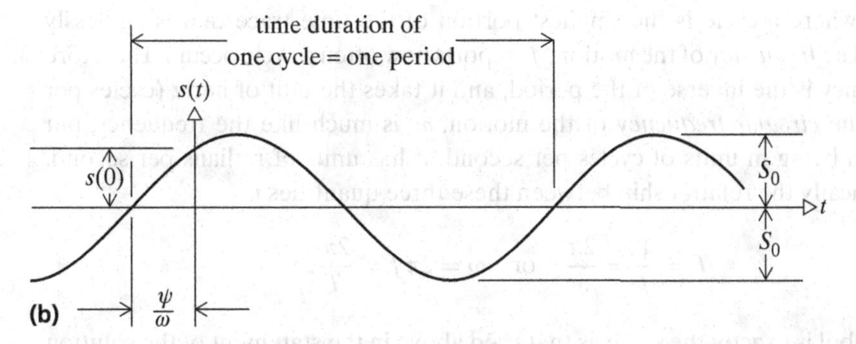

Figure 2.9. Graphs for the linearized (undamped) vibratory equation of motion, Eq. (2.11). (a) A time history (solid line) corresponding to an initial displacement, Θ_0, and another (dashed line) for an initial velocity, $\dot{\Theta}_0$.
(b) A plot of a general harmonic response $s(t) = S_0 \sin(\omega t + \psi)$ used to identify the amplitude S_0, the period $2\pi/\omega$, and the phase angle ψ.

(i.e., starting) conditions for the pendulum are an initial displacement of magnitude, Θ_0, and no initial angular velocity. In other words, it is the case where the pendulum arm is rotated through the angle Θ_0 and then released without a shove. From the above mathematics, this response history is a cosine function. The dashed line is the time history when the pendulum starts from its static equilibrium position, that is, the downward vertical position, and then is given a shove (i.e., an initial velocity is imparted to the pendulum without there being an initial deflection). From the above mathematics, this time history is a sine function. Both these time histories correspond closely to our observation of the vibrations of pendulums and vibrating elastic structures (play with a weight suspended from a rubber band). This correspondence between this mathematical description and the observed time history of an actual vibration is close but not exact. In a physical system, the amplitudes will sooner or later noticeably decrease. This decrease is said to be becasue of energy dissipating "damping forces." The discussion of damping forces is postponed to Chapter 5.

Figure 2.9(b) is another harmonic time history, $s(t)$, which is labeled so as to define general terms that are used hereafter. The motion in this case can be mathematically described as either a sine function with a phase angle or a cosine function with a different phase angle. Referring to this diagram, the following definitions are universal. As above, the amplitude of the motion is the vertical distance on the plot from the undeflected position, $s(t) = 0$, to the peak value where $s(t)$ has its maximum positive or its maximum negative value. The amplitude is always reported as a positive value.

The *double amplitude* is also measured on the ordinate of the time history. It is the positive difference between two successive peak values.

Again referring to Figure 2.9(b), the *period* of a harmonic or any other repeated motion, T, is the smallest time duration before the time trace of the motion begins repeating itself. In mathematical terms, for all values of t, the period T is such that

$$s(t) = s(t + T).$$

For vibratory motions, the period has units of seconds, or, more precisely, seconds per cycle, where a cycle is the smallest portion of the time trace that is endlessly repeated. The *frequency* of the motion, f, reports how often a cycle occurs. Therefore the frequency is the inverse of the period, and it takes the unit of hertz (cycles per second). The *circular frequency* of the motion, ω, is much like the frequency, but rather than being in units of cycles per second, it has units of radians per second. Mathematically the relationship between these three quantities is

$$T = \frac{1}{f} = \frac{2\pi}{\omega} \quad \text{or} \quad \omega = 2\pi f = \frac{2\pi}{T}.$$

This ω symbol is exactly the same as that used above in the statement of the solution for the displacement of the pendulum arm because $C_0 \sin(\omega T + \psi_1) = C_0 \sin(\psi_1)$. Clearly the word *frequency* can have two meanings: cycles per second and radians per second. In a mathematical analysis, the word *frequency* generally means circular frequency because it is more convenient to write an ω rather than $2\pi f$ in such solutions as that of Eq. (2.12), which typifies all harmonic solutions. However, when discussing data, such as experimental data, the term *frequency* usually means hertz (cycles per second). Of course it is important that the units of frequency either be evident from a mathematical expression or be explicitly stated. Since this is an analysis textbook rather than a technical report, *frequency* from this point onward means circular frequency, that is, ω, in radians per second. Finally, Figure 2.9(b) can also be used to review the meaning of that essential quantity called the phase angle of the motion. Again, in the typical mathematical expression for a harmonic motion, $s(t) = S_0 \sin(\omega t + \psi)$, the phase angle, ψ, is the part of the argument of the sine function that does not contain time as a factor. Figure 2.9(b) shows the typical shift in the zero crossing of the sine form of the function $s(t)$ as the result of a positive phase angle.

It will be seen that when the vibratory system is free of externally applied contact loads or imposed base motions, the values of the periods and frequencies of the motion depend only on the inherent characteristics of the vibratory system: the mass, geometric, and elastic parameters of the system. Such periods and frequencies are called *natural periods* and *natural frequencies*. It will be shown that the number of natural frequencies or natural periods is equal to the number of DOF the system possesses.

It is worth repeating that when there is a single DOF, and hence a single equation of motion, the natural (circular) frequency of such a system is always the square root of the coefficient of the deflection term (the zeroth time derivative term) divided by the coefficient of the acceleration term. That is, for the generic, linear, one DOF equation of motion (the dynamic force equilibrium equation)

$$m\,\ddot{u}(t) + k\,u(t) = F(t),$$

where (m, k) are constants, no matter how complicated, the single natural frequency of this system is

$$\omega_1 = \sqrt{\frac{k}{m}}.$$

No such simple relation exists when there are more than one DOF. As above, from this point onward, to distinguish natural frequencies from frequencies associated with applied loads and enforced motions, the natural frequencies are denoted with positive integers as subscripts.

Concluding this introductory comment on natural frequencies and natural periods, Ref. [2.1] provides a conservation of energy solution for simple pendulum amplitudes that are not limited to being small. That nonlinear solution shows that the value of the simple pendulum natural frequency decreases (and the value of the natural period increases) as the pendulum swing becomes larger. In other words, the natural frequency and period are not constant for large deflections but depend on the magnitude of the deflection. The mathematics of the large amplitude solution for a simple pendulum is presented in Endnote (1). In the case of structures (as opposed to mechanisms such as pendulums) the geometry of the deflection usually causes the structure to become stiffer (bigger k) and causes the natural frequencies to increase with increasing deflection. However, if the deflections pass the elastic limit into the plastic region, then the structure can become considerably less stiff over the peak parts of the amplitude, and the natural frequencies will now decrease with larger amplitudes. Note that in these circumstances of changing geometry or plasticity, the vibratory motion will no longer be sinusoidal. However, the motion will still be back and forth and thus close enough to being harmonic that all the terms defined above will still be relevant. In this textbook, the emphasis is on linear elastic systems where the geometry is not appreciably altered by the deflections, and the materials remain elastic.

2.5 Linearizing Differential Equations for Small Deflections

The previous section showed that when the amplitude of the simple pendulum of Section 2.2 is limited to being small enough to allow the replacement of the sine function by its argument, the equation of motion of the simple pendulum has a harmonic solution. Again, a harmonic solution function is either a sine or cosine function. The differential equations of the more complicated pendulums of Section 2.3 can also be linearized on the same basis. Whenever the amplitude of the motion is small enough to allow the sine of the unknown deflection function to be replaced by the unknown function itself, the cosine of that function can be replaced by 1.0 with only slightly less accuracy. Thus, for example, the linearized (i.e., small deflection) form of Eq. (2.4) is

$$(H + mL^2)\ddot{\theta} + (mgL + 2kh^2)\theta = 0.$$

Next in complexity with respect to demonstrating the linearization of pendulum equations of motion are Eqs. (2.5). These are the two nonlinear equations of motion for the problem where a simple pendulum is suspended from a cylinder rolling on a flat surface. The third term of the first of these two equations of motion contains the frequently occurring product of $\dot{\theta}^2$ and the sine of θ. This term can be discarded as very small compared to, for example, the second term in the same equation, the one containing $\ddot{\theta}$ whenever the applied force $F(t)$ results in near harmonic motion for the angles ϕ and θ at amplitudes less than, say, two-tenths of a radian. The reason this third term is much smaller than the second term in these circumstances is

$$\begin{aligned} &\theta(t) \approx \Theta \sin \omega t \quad \text{and} \quad \sin\theta \approx \theta \leq \Theta \\ \text{therefore} \quad &\ddot{\theta}(t) \approx \omega^2 \Theta \sin \omega t \leq \omega^2 \Theta \\ \text{and similarly} \quad &\dot{\theta}^2 \sin\theta \leq \omega^2 \Theta^2 \Theta = \omega^2 \Theta^3. \end{aligned}$$

Thus the third term is proportional to the third power of the small amplitude (a value less than 0.2), whereas the second term of this equation of motion is only proportional to the first power of the small amplitude. Hence the third term, with the amplitude being limited to 0.2 rad, is at most, 4% of the second term of the same equation. Thus the third term can be eliminated without loss of engineering accuracy. Therefore the linear forms of Eqs. (2.5) are

$$\begin{aligned} [H_1 + (m_1 + m_2)R^2]\ddot{\phi} + m_2 RL\ddot{\theta} &= RF(t) \\ m_2 RL\ddot{\phi} + (H_2 + m_2 L^2)\ddot{\theta} + mgL\theta &= LF(t). \end{aligned}$$

The linearization of the remainder of the example problem equations of motion duplicates the above-discussed steps. Thus such linearizations are left to the exercises. As an aside, note that, from the first of the above two linearized equations, when the externally applied force $F(t)$ is zero, the two angular accelerations become directly proportional to each other. Even without the applied force being zero, equation compactness can be achieved by using the first of the above equations to eliminate $\ddot{\phi}$ in the second equation. Then there is but one equation in terms of one unknown function.

As a repeated comment on the process of linearization, note that the above discussion focused on carrying out the process of linearization after obtaining the nonlinear differential equation. Often the linearization can be done more conveniently at an earlier stage, with the sines of small angles being replaced by the small angles. However, it may or may not be possible to replace the cosines of small angles by 1: As an example of a case where it is not correct to so linearize the cosine function, recall that the potential energy of many pendulums is proportional to $(1 - \cos\theta)$. This factor would then be zero in those circumstances, and the gravitational effect would be completely lost from the analysis. In such cases, the cosine of the small angle needs to be replaced by the first two terms in its Taylor series expansion, which are $1 - \theta^2/2$.

2.6 Summary

Pendulum problems begin the discussion of vibratory systems because the back-and-forth swinging motion of the pendulum arm, the system vibration, is familiar to all readers. Therefore, the parameters of the mathematical solution to any linear or nonlinear pendulum differential equation (period, frequency, amplitude, phase angle, initial deflection, etc.) are quantities the reader can readily visualize. Pendulums of different configurations also offer practice in writing system equations of motion that do not require a knowledge of structural mechanics. The writing of these pendulum equations of motion provides the reader with (i) a review of the Lagrange equations of motion and (ii) a reminder of the limitations of present analytical techniques for solving nonlinear differential equations and hence the importance of the numerical techniques to be discussed in Chapter 9. Further opportunities for practicing the application of the Lagrange equations are found in the exercises at the end of the chapter.

Occasionally, toward the end of chapters, there are sections marked by a double asterisk (**). These sections deal with topics that are more complicated than are appropriate for a one-semester course. The next two sections fit this mold. Subjects off the main path followed by this book are relegated to endnotes. This chapter has an endnote that continues the discussion of stability questions in the context of the general, discrete system vibratory equations.

2.7 **Conservation of Energy versus the Lagrange Equations**

The Lagrange equations are the preferred means for describing the dynamics of mechanisms and structures. Generally, for pendulums and structures alike, the Lagrange equations, just like Newton's laws, involve accelerations, velocities, and deflections. Conversely, the conservation of energy equation involves only velocities and deflections. Therefore it may seem that in the circumstances of no frictional (i.e., no energy dissipative) forces, or no externally applied forces, that the conservation of energy equation offers (i) the advantage of greater simplicity and (ii) the advantage of having, in effect, eliminated the accelerations and thereby accomplished one of the two required integrations over time that are needed to solve the Lagrange equations. That is, a conservation of energy equation requires only one integration, whereas a Lagrange equation requires two. This is more than a matter of less required effort. It is more a matter that first-order differential equations are generally much easier to solve than second-order differential equations, which are other than linear equations with constant coefficients. Indeed, the conservation of energy equation does offer these important advantages, and this limited method should be part of an analyst's repertoire. Again, the advantage of having, in effect, accomplished one time integration is especially important in the case of nonlinear equations of motion. However, a certain amount of caution is necessary in the application of the conservation of energy equation. It is necessary to point out that the conservation of energy equation may not produce the correct answer when there is an enforced motion any more than it would when there is an externally applied force. The following example problem illustrates this statement.

EXAMPLE 2.8 A thin, stiff, uniform bar of constant cross section A, length L, mass density ρ, and total mass $m = \rho AL$ is supported by a hinge at its top end. The rod rotates about a vertical axis with a constant angular velocity ω_0, as shown in Figure 2.10. In Figure 2.10 the axis of the hinge is seen in end view as point P. This hinge axis, of course, rotates with the rod and is always perpendicular to the plane defined by the rod and the rotation axis. Since the elastic deformations of the rod as a result of the actions of the gravitational and centrifugal forces are much smaller than the overall motion of the rod, and because the focus here is on describing the overall motion of the rod, the rod is modeled as rigid. The tasks to be completed are (i) determine θ_0, a constant angle that defines the inclined, static equilibrium position of the rotating rod, and (ii) derive the equation of motion for small flapping oscillations, $\theta_1(t)$, about the static equilibrium position of part (i). The quantities $\theta_1(t)$ and $\theta(t) = \theta_0 + \theta_1(t)$ are illustrated in Figure 2.10(b). (Note that if the rod were a uniform helicopter blade, it would also be necessary to include lift forces and, as a result, the static equilibrium angle could be above the horizontal.)

SOLUTION (a) Using Newton's equations first, the static equilibrium position is determined by the zero sum of (i) the moment about the hinge caused by the weight of the rod and (ii) the moment about the hinge caused by centrifugal forces acting on the rod. The gravitational moment about point P is simply obtained by multiplying the vertically directed gravitational force, mg, by the horizontal distance from the axis of rotation to the center of mass, $\frac{1}{2}L\sin\theta_0$. To check this result, and, more importantly, to lay a foundation for the calculation of the moment as a result of the centrifugal forces, repeat the calculation of the gravitational moment as a sum (integral) of differential moments. That is, consider a differential length of the uniform bar, dx, as shown in Figure 2.10. The force acting on this differential length is proportional to the force acting on the entire bar, $(mg)(dx/L)$. Combining this vertically acting force with its horizontal moment arm yields

$$M_P = \int_0^L (x\sin\theta_0)\left(\frac{mg}{L}dx\right) = \frac{mg}{L}\sin\theta_0 \int_0^L x\,dx = \tfrac{1}{2}mgL\sin\theta_0,$$

which, of course, is the same as the answer previously obtained.

Since any differential element of the bar moves in a circle lying on a horizontal plane, there are centrifugal accelerations directed away from the axis of rotation. Note that it is incorrect to place the total centrifugal force at the center of mass because the magnitude of the centrifugal force is not constant along the length of the rod as is the weight per unit length. Therefore it is properly cautious to calculate the moment created by the centrifugal accelerations by again considering a length dx of the uniform rod. The centrifugal moment about the hinge axis produced by the mass of this differential portion of the rod is the vertical (moment arm) distance to this differential length, $x\cos\theta_0$, multiplied by the horizontal centrifugal force. The centrifugal force acting on this differential sized length of the bar is equal to the differential mass multiplied by the local centrifugal acceleration; that is, $dF_c = (\rho A dx)(r\omega_0^2)$, where $r = x\sin\theta_0$. As with the gravitational moment, the total centrifugal moment is obtained by integrating the product of all these differential

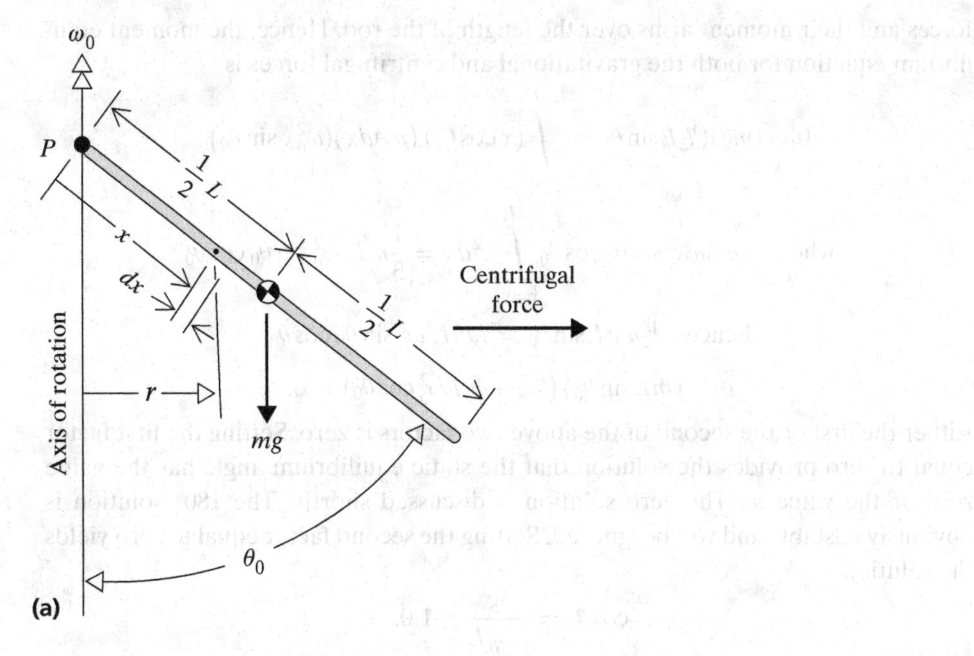

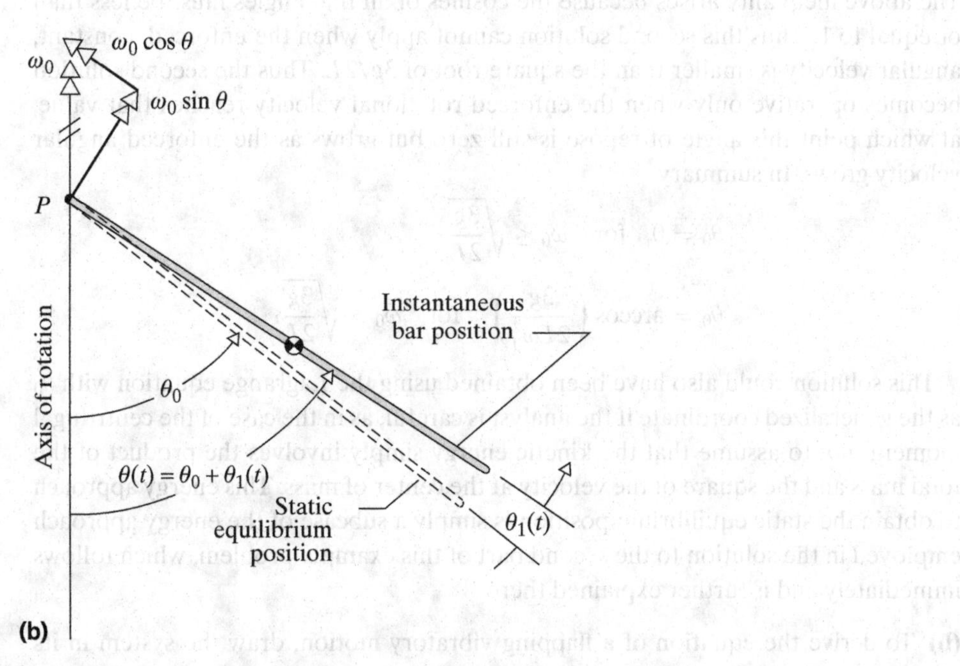

Figure 2.10. Example 2.8: (a) The static equilibrium position, θ_0, is determined by the balance between the weight force and the centrifugal force. (b) An illustration of the up (and down) vibratory motion of the bar as it rotates about a vertical axis.

forces and their moment arms over the length of the rod. Hence, the moment equilibrium equation for both the gravitational and centrifugal forces is

$$0 = (mg)(\tfrac{1}{2}L\sin\theta_0) - \int_{x=0}^{L} (x\cos\theta_0)\,(\rho A dx)(\omega_0^2 x\sin\theta_0)$$

$$\text{where}\quad \rho A\omega_0{}^2 \sin\theta_0\cos\theta_0 \int_{x=0}^{L} x^2 dx = \frac{1}{3}mL^2\omega_0^2\sin\theta_0\cos\theta_0$$

$$\text{hence}\quad \tfrac{1}{2}mgL\sin\theta_0 = \tfrac{1}{3}mL^2\omega_0^2\sin\theta_0\cos\theta_0$$

$$\text{or}\quad (mL\sin\theta_0)\left(\tfrac{1}{2}g - \tfrac{1}{3}L\omega_0^2\cos\theta_0\right) = 0.$$

Either the first or the second of the above two factors is zero. Setting the first factor equal to zero provides the solution that the static equilibrium angle has the value zero or the value π. The zero solution is discussed shortly. The 180° solution is obviously unstable and will be ignored. Setting the second factor equal to zero yields the solution

$$\cos\theta_0 = \frac{3g}{2\omega_0^2 L} \leq 1.0.$$

The above inequality arises because the cosines of all real angles must be less than or equal to 1. Thus this second solution cannot apply when the enforced, constant, angular velocity is smaller than the square root of $3g/2L$. Thus the second solution becomes operative only when the enforced rotational velocity reaches that value, at which point this angle of repose is still zero but grows as the enforced angular velocity grows. In summary

$$\theta_0 = 0 \quad \text{for} \quad \omega_0 \leq \sqrt{\frac{3g}{2L}}$$

$$\theta_0 = \arccos\left(\frac{3g}{2L\omega_0^2}\right) \quad \text{for} \quad \omega_0 > \sqrt{\frac{3g}{2L}}.$$

This solution could also have been obtained using the Lagrange equation with θ_0 as the generalized coordinate if the analyst is careful, as in the case of the centrifugal moment, *not* to assume that the kinetic energy simply involves the product of the total mass and the square of the velocity at the center of mass. This energy approach to obtain the static equilibrium position is simply a subcase of the energy approach employed in the solution to the second part of this example problem, which follows immediately and is further explained there.

(b) To derive the equation of a flapping vibratory motion, draw the system in its displaced configuration as shown in Figure 2.10(b) where a nonzero static equilibrium position is assumed for the sake of generality. Since the static equilibrium position is now known, and because the angular velocity about the vertical axis is a known constant, only the time-varying, vibratory position of the rod is unknown. Just like locating the static equilibrium position, only a single angular coordinate is required to locate the flapping rod in its vertical plane. Let that DOF be

$$\theta(t) = \theta_0 + \theta_1(t), \tag{2.13}$$

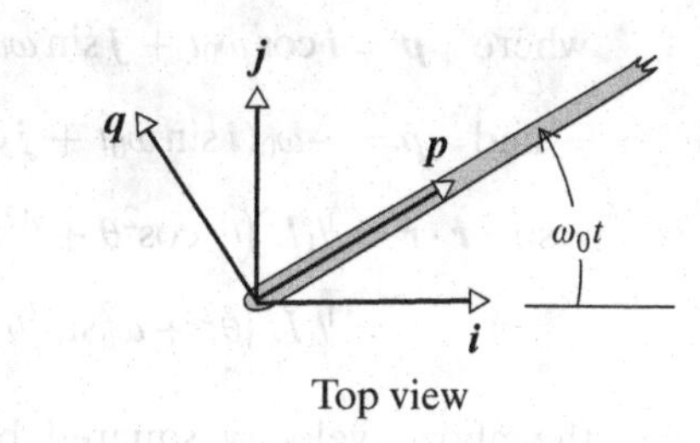

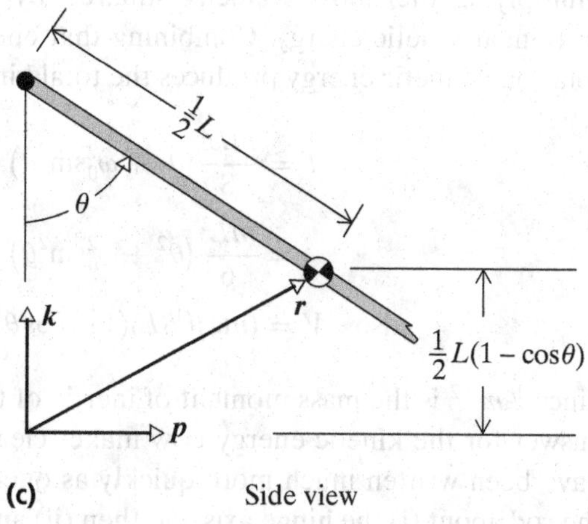

Figure 2.10. Example 2.8 (*continued*): (c) Location and orientation of selected vectors.

where, again, these quantities are defined in Figure 2.10(b).

The rotational kinetic energy of the rod is approximated by noting that the mass moment of inertia of the thin rod about its own longitudinal axis is negligible. If it were not negligible, the corresponding angular velocity from Figure 2.10(b) is $\omega_0 \cos\theta$. From the same figure, rotation of the rod about the axis lying in the plane of the paper, which is perpendicular to the rod's longitudinal axis, is $\omega_0 \sin\theta$. The corresponding mass moment of inertia at the center of mass can be calculated to be $mL^2/12$. The angular velocity about the third orthogonal axis, the one perpendicular to the plane of the paper, is simply $\dot{\theta}$, and the value of its corresponding mass moment of inertia at the center of mass is also $mL^2/12$. That takes care of the three orthogonal components of the rotational kinetic energy. Now the two nonzero components of the angular velocity can used to write the rotational kinetic energy as $T_{rot} = \frac{1}{2}(mL^2/12)(\dot{\theta}^2 + \omega_0^2 \sin^2\theta)$.

This system is sufficiently complicated that it is wise, when seeking the rectilinear component of the kinetic energy, to proceed from first principles. For this purpose, set up two sets of orthogonal unit vectors. Let $\boldsymbol{p}$ and $\boldsymbol{q}$ be horizontal unit vectors that originate at the center of mass of the rod when the rod is vertical. Let these two vectors rotate about the axis of rotation such that $\boldsymbol{p}$ is always in the plane defined by the rod and the axis of rotation, as shown in Figures 2.10(c). Let $\boldsymbol{i}$, $\boldsymbol{j}$, and $\boldsymbol{k}$ be the usual Cartesian unit vectors fixed in space, and let them have the same origin. Now, to calculate the rectilinear velocity of the center of mass of the rod, introduce the position vector $\boldsymbol{r}(t)$, whose time derivative is the velocity at the center of mass. Then

$$\boldsymbol{r}(t) = \boldsymbol{p}[\tfrac{1}{2}L\sin\theta] + \boldsymbol{k}[\tfrac{1}{2}L(1 - \cos\theta)]$$
$$\dot{\boldsymbol{r}}(t) = \dot{\boldsymbol{p}}[\tfrac{1}{2}L\sin\theta] + \boldsymbol{p}[\tfrac{1}{2}L\dot{\theta}\cos\theta] + \boldsymbol{k}[\tfrac{1}{2}L\dot{\theta}\sin\theta]$$

$$\text{where} \quad \boldsymbol{p} = \boldsymbol{i}\cos\omega_0 t + \boldsymbol{j}\sin\omega_0 t$$

$$\text{and} \quad \dot{\boldsymbol{p}} = -\omega_0(\boldsymbol{i}\sin\omega_0 t + \boldsymbol{j}\cos\omega_0 t) = +\boldsymbol{q}\omega_0$$

$$\text{so} \quad \dot{\boldsymbol{r}}\cdot\dot{\boldsymbol{r}} = {}^1\!/\!_4 L^2(\dot{\theta}^2\cos^2\theta + \dot{\theta}^2\sin^2\theta + \omega_0^2\sin^2\theta)$$

$$= {}^1\!/\!_4 L^2(\dot{\theta}^2 + \omega_0^2\sin^2\theta).$$

Multiplying the above velocity squared by one-half of the mass provides the rectilinear kinetic energy. Combining that energy with the previously determined rotational kinetic energy produces the total kinetic energy, which is

$$T = \frac{mL^2}{8}(\dot{\theta}^2 + \omega_0^2\sin^2\theta) + \frac{mL^2}{24}(\dot{\theta}^2 + \omega_0^2\sin^2\theta)$$

$$= \frac{mL^2}{6}(\dot{\theta}^2 + \omega_0^2\sin^2\theta)$$

$$\text{also} \quad V = (mg)({}^1\!/\!_2 L)(1 - \cos\theta).$$

Since ${}^1\!/\!_3 mL^2$ is the mass moment of inertia of the rod about the rod end, the above answer for the kinetic energy now makes clear that the total kinetic energy could have been written much more quickly as one-half the mass moment of inertial of the rod about (i) the hinge axis and then (ii) an axis perpendicular to the hinge axis; each multiplied by its respective angular velocity.

Before proceeding to the Lagrange equation, calculate the virtual work to be sure that it is zero as it would seem at first glance. Since the inertial, including the centrifugal, and gravitational forces are included in the kinetic and potential energies, the only external forces not accounted for are those that produce the moment that maintains the constant angular velocity about the vertical axis of rotation. The vector for this moment is orthogonal to the vector[9] for the virtual rotation $\delta\theta$. Therefore the virtual work is indeed zero.

Substituting the above kinetic and potential energies into the Lagrange equation yields the equation of motion

$$\frac{mL^2}{3}\ddot{\theta} - \frac{mL^2\omega_0^2}{3}\sin\theta\cos\theta + \frac{mgL}{2}\sin\theta = 0. \tag{2.14}$$

Note that this equation cannot be linearized immediately because θ is not necessarily a small angle. Theta need not be a small angle because it includes the perhaps large angle of the static equilibrium position, θ_0. Therefore, with the intent of obtaining a linear equation so that a first estimate of the small vibratory deflection, $\theta_1(t)$, natural frequency, or natural period is possible, substitute Eq. (2.13) into Eq. (2.14). With $\theta_1(t)$ regarded as being small in the usual sense of the sine of this angle being approximately equal to this angle, and its cosine being approximately equal to 1

$$\sin(\theta_0 + \theta_1) = \sin\theta_0 + \theta_1\cos\theta_0 \quad \text{and} \quad \cos(\theta_0 + \theta_1) = \cos\theta_0 - \theta_1\sin\theta_0.$$

[9] From Ref. [2.3], small (10° or less for engineering accuracy) angular rotations can be treated as vector quantities, and virtual rotations are very small.

After discarding the square of θ_1, and simplifying, the equation of motion in terms of θ_1 is as follows:

$$\ddot{\theta}_1(t) + \left(\frac{3g}{2L}\cos\theta_0 - \omega_0^2\cos 2\theta_0\right)\theta_1(t) = \sin\theta_0\left(\omega_0^2\cos\theta_0 - \frac{3g}{2L}\right) = 0.$$

Again, the right-hand side is zero because of the solution for the static equilibrium angle: $\cos\theta_0 = 3g/2L\omega_0^2$. The coefficient of the above zeroth derivative term, which is the square of the small vibratory deflection natural frequency, can be better understood to be a positive quantity by rewriting it as follows:

$$\begin{aligned}\omega_1^2 &= \frac{3g}{2L}\cos\theta_0 - \omega_0^2\cos 2\theta_0\\ &= \frac{3g}{2L}\cos\theta_0 - \omega_0^2\cos^2\theta_0 + \omega_0^2\sin^2\theta_0\\ &= \omega_0^2\sin^2\theta_0.\end{aligned}$$

Returning to the main purpose of this example, now write the conservation of energy equation, $(d/dt)(T+V) = 0$. This equation, at first glance, would seem to be valid because the only immediately evident external force is the gravitational force, and that is included in the potential energy term. From above

$$T + V = \frac{mL^2}{6}(\dot{\theta}^2 + \omega_0^2\sin^2\theta) + \frac{mgL}{2}(1 - \cos\theta) = \text{const.}$$

$$\frac{d}{dt}(T+V) = \dot{\theta}\left[\frac{mL^2}{3}(\ddot{\theta} + \omega_0^2\sin\theta\cos\theta) + \frac{mgL}{2}\sin\theta\right] = 0$$

$$\text{or}\quad \frac{mL^2}{3}\ddot{\theta} + \frac{mL^2\omega_0^2}{3}\sin\theta\cos\theta + \frac{mgL}{2}\sin\theta = 0. \tag{2.15}$$

A comparison of Eqs. (2.14) and (2.15) shows there is a sign difference between the two equations with respect to the second term of these equations. The conservation of energy equation is wrong because the energy is not constant. To see why the energy of the system is not conserved, consider the rod at two different positions in its up and down flapping vibration (i) where the rod has swung up to the maximum value of θ and (ii) where the rod has swung down to the minimum value of θ. At both positions the kinetic energy associated with the motion in the plane of the rod and the vertical axis of rotation is zero. This is so because at the instants that the rod occupies those positions, the rod has come to a halt in that plane in order to change the direction of its swing in that plane. In the upper inclined position, the kinetic energy of rotation about the vertical axis is greater than that at the lower inclined position because, from the above, the angular velocity is $\omega_0\sin\theta$ and $\sin\theta$ increases as θ increases when θ is below a value of 90°, as it must be here. The translational kinetic energy is also greater at the upper position because the center of mass radius of the tangential velocity is also larger. The higher elevation of the center of mass at the upper position also means that the potential energy is also greater at the upper position. Hence the total energy is greater at the upper rod position than at the lower rod position, and thus the energy is not constant. The energy levels change because the constant angular velocity ω_0 is only possible by means of a time-varying moment

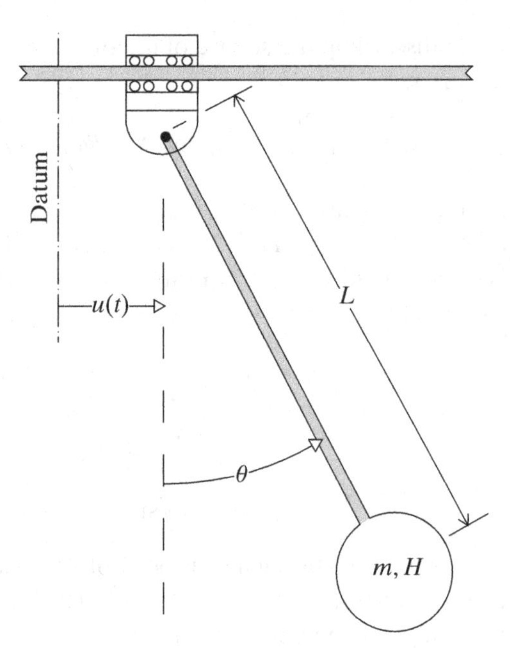

Figure 2.11. Horizontally base excited pendulum.

acting about the axis of rotation that pushes the system along when the rod swings above the rod static equilibrium position and slows down the system when the rod swings below its static equilibrium position. Thus this moment is always adding work (energy) to, or deleting work (energy) from, this pendulum system. ★

2.8 **Nasty Equations of Motion**

In certain circumstances, even when the motion of a single, simple-appearing pendulum system is restricted to small deflections, there may not be the slightest hope of even an approximate analytical solution. That is, the equation of motion will require a numerical solution as discussed in Chapter 9. For the sake of a stark contrast to a later example that illustrates this point, first consider the support motion activated pendulum of Figure 2.11. This one-DOF system, with its horizontally moving support is little different from the previously studied pendulum systems. The derivation of its equation of motion is straightforward. With θ as the single generalized coordinate

$$\begin{aligned} T &= \tfrac{1}{2}H\dot{\theta}^2 + \tfrac{1}{2}m[(\dot{u} + L\dot{\theta}\cos\theta)^2 + (L\dot{\theta}\sin\theta)^2] \\ &= \tfrac{1}{2}H\dot{\theta}^2 + \tfrac{1}{2}m[\dot{u}^2 + 2\dot{u}\dot{\theta}L\cos\theta + L^2\dot{\theta}^2] \end{aligned}$$

$$\text{and} \quad V = mgL(1 - \cos\theta)$$

$$\text{therefore} \quad (H + mL^2)\ddot{\theta} + mgL\sin\theta = -m\ddot{u}L\cos\theta.$$

This equation of motion is easily linearized when the base motion is harmonic with a magnitude such that θ is always small enough that $\sin\theta$ can be replaced by θ, and $\cos\theta$ can be replaced by 1.

To obtain an equation of motion for a similar, simple-appearing system that requires a numerical solution, again consider the pendulum system of Figure 2.11. This time, however, let the rail on which the pendulum support moves be oriented vertically in such a way that there is no interference between the vertical rail supporting the pinned end of the pendulum and the pendulum itself swinging in the plane of the paper. Let the vertical base motion be described by the time function $v(t)$, where, as per usual, $v(0) = \theta(0) = 0$ is the datum point for the system. Then

$$T = \tfrac{1}{2}H\dot{\theta}^2 + \tfrac{1}{2}m\left(\left[\frac{d}{dt}(L\sin\theta)\right]^2 + \left\{\frac{d}{dt}[v + L(1 - \cos\theta)]\right\}^2\right)$$

$$= \tfrac{1}{2}H\dot{\theta}^2 + \tfrac{1}{2}m(\dot{v}^2 + 2\dot{v}\dot{\theta}L\sin\theta + L^2\dot{\theta}^2)$$

and $\quad V = mg[v + L(1 - \cos\theta)]$

therefore $\quad (H + mL^2)\ddot{\theta} + mL(g + \ddot{v})\sin\theta = 0.$

To better contrast these results, let θ be restricted to small deflections in the usual sense and let the previous, horizontal base motion and the present vertical base motion be harmonic; that is, let

$$u(t) = u_0 \sin\omega_0 t \quad \text{and} \quad v(t) = v_0 \sin\omega_0 t,$$

where u_0 and v_0 are the small amplitudes of the base motion. Then the respective equations of motion corresponding to these two differently directed excitations are

$$(H + mL^2)\ddot{\theta} + mgL\theta = -mL\omega_0^2 u_0 \sin\omega_0 t$$

$$(H + mL^2)\ddot{\theta} + mL\left(g - \omega_0^2 v_0 \sin\omega_0 t\right)\theta = 0.$$

The first of these two equations of motion is a linear, ordinary differential equation with constant coefficients. As shown in Chapter 5, this equation of motion is easily solved. The second equation of motion is also a linear, ordinary differential equation. However, in the case of the second equation, its θ coefficient is not a constant. Since this coefficient varies with time in a nonpolynomial manner, this equation can be expected to require a numerical solution.

One thing can be learned immediately from the second of the above equations, the one corresponding to the vertical excitation. To understand the point to be made, first consider two similar, easily solved, linear, ordinary differential equations with constant coefficients

$$(1)\colon \ddot{q}(t) + \omega_1^2 q(t) = 0 \qquad (2)\colon \ddot{q}(t) - \omega_1^2 q(t) = 0.$$

Their respective solutions can be written as

$$(1)\colon q(t) = C_1 \sin\omega_1 t + C_2 \cos\omega_1 t \qquad (2)\colon q(t) = C_1 \sinh\omega_1 t + C_2 \cosh\omega_1 t.$$

Thus the change in sign for the coefficient of the deflection term changes the solution from one that is strictly bounded in the first case to one that is unbounded in the second case. Although the first motion is called vibratory, the second motion is termed *unstable*. Returning to the pendulum with the vertical excitation, the coefficient of the θ term becomes negative, and the motion unstable, whenever

$$\omega_0^2 v_0 > g.$$

The next section further explores questions of system stability.

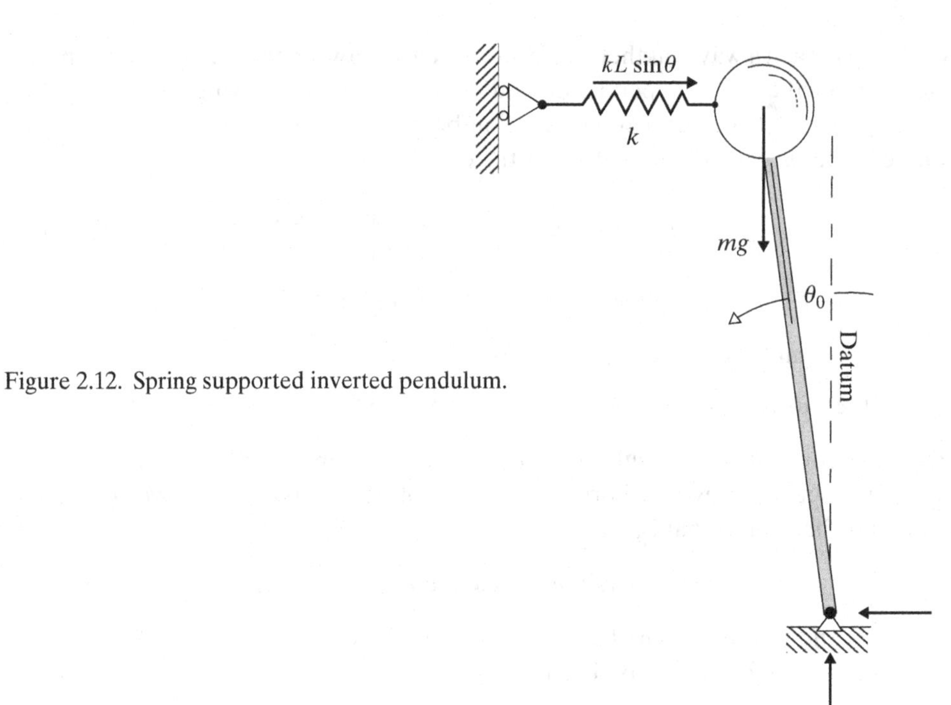

Figure 2.12. Spring supported inverted pendulum.

2.9 **Stability of Vibratory Systems**

As shown in the third section of Chapter 4, it is important to clearly identify the system's static equilibrium position. Consider the inverted pendulum system of Figure 2.12. It is necessary to consider the possibility of more than one static equilibrium position for this pendulum. Let the static equilibrium position be measured by the rotational angle θ_0, which is an unknown constant to be determined. Setting the sum of the moments about the base equal to zero yields the static equilibrium result

$$0 = (kL\sin\theta_0)(L\cos\theta_0) - mg(L\sin\theta_0)$$
$$\text{or} \quad \sin\theta_0(kL\cos\theta_0 - mg) = 0$$

so either

$$\sin\theta_0 = 0, \text{ which implies } \quad \theta_0 = 0$$

or

$$\cos\theta_0 = \frac{mg}{kL}, \text{ which implies } \quad kL \geq mg.$$

Thus, when the weight force is less than kL, there are two possible static equilibrium positions. The rod is in static equilibrium when it is vertical ($\theta_0 = 0$), and it also is in static equilibrium when it is at the positive or negative angle described by θ_0 equal to the arccosine of mg/kL.

The following *static* stability analysis shows that only the vertical position, of the two positions, is stable. With θ_0 as the generalized coordinate, the total potential energy of this pendulum system is

$$U + V = \tfrac{1}{2}k(L\sin\theta_0)^2 - mgL(1 - \cos\theta_0).$$

With the kinetic energy being zero as it is in all static cases, the Lagrange equation of motion is the equation of static equilibrium. That equation is

$$\frac{\partial(U+V)}{\partial\theta_0} = 0 = kL^2 \sin\theta_0 \cos\theta_0 - mgL\sin\theta_0.$$

for the same equilibrium state results as above. It can be shown that, just as the first derivative of the potential energy yields the equilibrium states, the stability nature of the equilibrium states can be determined in an energy conservative case like this by examining the second partial derivative of the potential energy. See Ref. [2.2]. Whenever this second partial derivative is positive, the system is stable. Whenever this second partial derivative is negative, the system is unstable. The second partial derivative has the form

$$\frac{\partial^2(U+V)}{\partial\theta_0^2} = kL^2(\cos^2\theta_0 - \sin^2\theta_0) - mgL\cos\theta_0.$$

Substituting θ_0 equals zero, its value for the first static equilibrium state, shows that equilibrium state is stable whenever kL is greater than mg, and it is unstable when the inequality is reversed. Substituting the value of θ_0 for the second equilibrium state, where again $\cos\theta_0 = mg/kL$ yields

$$\frac{1}{kL^2}\frac{\partial^2(U+V)}{\partial\theta_0^2} = (\cos^2\theta_0 - \sin^2\theta_0) - (\cos\theta_0)\cos\theta_0 = -\sin^2\theta_0 < 0$$

for all values of θ_0. Therefore this second equilibrium state, where the inverted pendulum system is in an off-vertical position, is always unstable. That is, the pendulum system will remain in its off-vertical static equilibrium position before falling down for approximately the same period of time a pencil balanced on its sharpened point remains in that static equilibrium position. Hence, the conclusion relating to the vibrations of this system that this pendulum system can vibrate only about its vertical equilibrium position and only then if $kL > mg$. With θ_0 equal to zero, and $\theta(t)$ the time-varying position of the pendulum arm measured from the vertical, the linearized equation of motion and the system natural frequency of vibration is

$$mL^2\ddot{\theta}(t) + (kL^2 - mgL)\theta(t) = 0$$

$$\text{where} \quad \omega_1 = \sqrt{\frac{k}{m} - \frac{g}{L}}.$$

This solution also shows that there cannot be a real first natural frequency for this one DOF system, unless $kL > mg$. Indeed, this identical result for the limitation on the system parameters can be viewed as now having been obtained by a "dynamical analysis" as opposed to the previous "energy approach." This indicates another use for dynamic analyses.

To be a bit more general in the discussion, accept that in later chapters it is shown that the force free vibration of any *one*-DOF mechanical system is described by the following linear ordinary differential equation:

$$\overline{m}\,\ddot{u}(t) + \overline{c}\dot{u}(t) + \overline{k}u(t) = 0.$$

This differential equation of motion with its three constant coefficients can be rewritten in terms of just two system parameters (i.e., two ratios) by simply dividing this equation by the leading coefficient. In terms of the system natural frequency ω_1, and the system damping factor (the significance of which is to be discussed in Chapter 5) ζ_1, the system equation of motion becomes

$$\ddot{u} + (\bar{c}/\bar{m})\dot{u} + (\bar{k}/\bar{m})u(t) = 0$$

$$\text{or} \quad \ddot{u} + 2\zeta_1\omega_1\dot{u} + \omega_1^2 u(t) = 0$$

$$\text{where} \quad \omega_1 = \sqrt{\frac{\bar{k}}{\bar{m}}} \quad \text{and} \quad \zeta_1 = \frac{\bar{c}}{2\sqrt{\bar{k}\bar{m}}}. \tag{2.16}$$

It will be seen that whenever the damping factor ζ_1 has a value less than 1 (the common situation for structures whereby vibrations are possible), the solution to the system differential equation of motion can be written as

$$u(t) = C_0 \exp(-\zeta_1\omega_1 t) \cos\left(\omega_1 t\sqrt{1-\zeta_1^2} - \psi\right),$$

where C_0 and ψ are the two constants of integration for the second-order differential equation. Of course, as discussed previously, these two constants of integration are related to the initial deflection and the initial velocity of the one-DOF system. Since it is always to be understood that the initial conditions are at time zero, the time variable t has only positive values. First consider the case where the system parameters, the natural frequency and the damping factor, are both positive. Inspection of the above mathematical solution for the motion of the one-DOF system, $u(t)$, again shows that the cosine function part of the solution describes a back-and-forth, constant amplitude vibratory motion. The new element, the negative exponential function part of the solution modifies the constant amplitude part of the motion, steadily decreasing it as time progresses. This decrease in vibratory amplitudes with increasing time is exactly what is observed with wholly stable physical systems undergoing force free (i.e., "natural") vibrations.

To focus in the simplest possible way on the effect of negative values of the parameters on the single degree of freedom system, begin by letting the effective damping coefficient, $\bar{c}$, and hence the damping factor, ζ, temporarily be zero. Also, without any loss of generality, let the "effective mass," $\bar{m}$, always be a positive number. If $\bar{k}$ (and $\bar{c}$) are also positive quantities, then the situation is a stable vibration as mentioned in the preceding paragraph. If, however, the coefficient $\bar{k}$ is a negative quantity, as it could be for the inverted pendulum system discussed in this section, then the natural frequency of vibration is an imaginary number and as such has no physical meaning as a vibration frequency. In this case, what happens to the mathematical solution for the motion of the mass, Eq. (2.16), is that the cosine of the imaginary number (ignoring the phase angle ψ for simplicity)[10] becomes a hyperbolic cosine. The hyperbolic cosine, with an argument proportional to time, describes a motion of the mass that is not vibratory, but just as described above, a motion that continuously and steadily (i.e., monotonically) increases in magnitude as time increases. Such a system where the mass moves with ever increasing magnitude in a single direction

[10] For a brief discussion of why it is justified to ignore the damping factor and the phase angle in this argument, see the footnotes attached to Endnote (2).

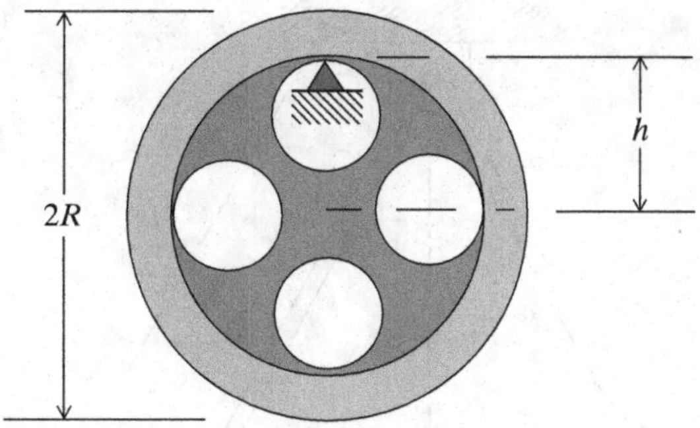

Figure 2.13. Exercise 2.1: Measuring a mass moment of inertia.

is called a *statically unstable*, or a system undergoing *divergence*. A marble placed at the top of the smooth outer surface of a bowling ball typifies this situation. Again, with respect to the inverted pendulum problem, if $mg > kL$, then the pendulum is statically unstable or divergent.

Returning to the full $\overline{m}, \overline{c}, \overline{k}$ system, this time let the effective damping coefficient, $\overline{c}$, and thus the damping factor, ζ, be the sole negative parameter. The above solution for the motion of the mass, Eq. (2.16) is still valid, and the only change in the form of the solution is that the exponential function now has a positive argument proportional to time rather than a negative argument. That is, the cosine function is unaltered, but the function that modifies the amplitudes of the cosine function, the exponential function, undergoes a change from being a "squeezer" to being an "expander." Thus the mathematics reveals that the motion of the mass is vibratory as a result of the cosine function, but the amplitudes of the vibration steadily increase. Such a system is called *dynamically unstable*, and the system is also said to be experiencing *flutter*. What is true for single degree of freedom systems regarding divergence and flutter is also basically true for multidegree of freedom systems. However, the explanation offered here of how this is so requires more advanced mathematical skills, and for that reason the explanation is relegated to Endnote (2) and left for more advanced readers. An application of these stability concepts to an important engineering problem is discussed in Section 11 of Chapter 7.

REFERENCES

2.1 Symon, K. R., *Mechanics*, 2nd Ed., Addison–Wesley, Reading, MA, 1960.

2.2 Dym, C. L., and I. H. Shames, *Solid Mechanics, A Variational Approach*, McGraw-Hill, New York, 1973.

2.3 Donaldson, B. K., *Analysis of Aircraft Structures: An Introduction*, McGraw-Hill, New York, 1993. (ed.)

2.4 Hodgman, C. D. (ed.), *C. R. C. Standard Mathematical Tables*, 12th ed. Chemical Rubber, Cleveland, 1959.

CHAPTER 2 EXERCISES (answers in Appendix I)

2.1 **(a)** As shown in Figure 2.13, a symmetric, but irregularly shaped, circular object is suspended a distance h above its geometric center. The symmetry of the object and

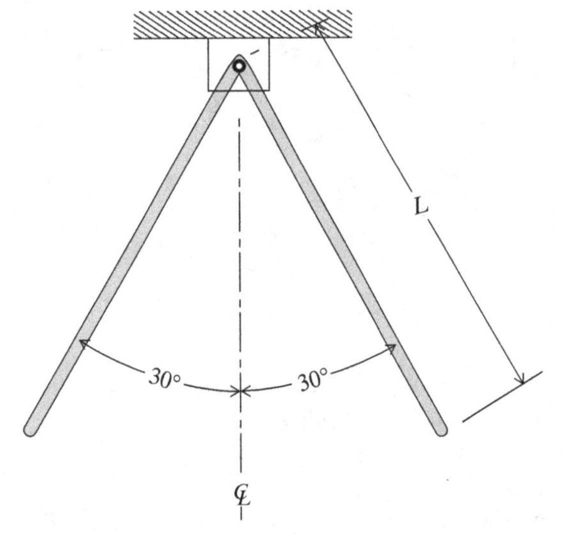

Figure 2.14. Exercise 2.3: A V pendulum.

the homogeneity of the material gives assurance that the geometric center is also the center of mass. The circular object has a weight w, an outer radius R, and an average thickness P and its astrological sign is Leo. When set to swinging in the plane of the paper from its support point, through small angular amplitudes, its observed period of vibratory motion is T. What is the magnitude of the object's mass moment of inertia about an axis through the center of mass and perpendicular to the plane of the paper?

(b) Conceptually design an experiment using a pendulum for the accurate determination of the acceleration of gravity. Hint: One approach is to consider using a long object of imperfect geometry that can be physically suspended from two different points along its length where each suspension point is a different, far distance from the object's center of mass. The two suspensions can provide two pieces of information that allow the calculation of the mass moment of inertia and then the acceleration of gravity.

2.2 **(a)** Linearize Eq. (2.6), and write the expression for the small deflection natural (circular) frequency and natural period.

(b) Linearize Eq. (2.8), and provide the solution for the small deflection natural (circular) frequency.

(c) Linearize Eqs. (2.9), and provide the solution for the (small deflection) natural period.

2.3 **(a)** Write the small deflection equation of motion, and determine the period of vibration for the inverted V pendulum shown in Figure 2.14. The pendulum swings only within the plane of the paper and is thus a single degree of freedom system. For this part of the exercise, omit the effect of the mass moments of inertia of each arm by only taking into account the mass of each pendulum arm, which is M.

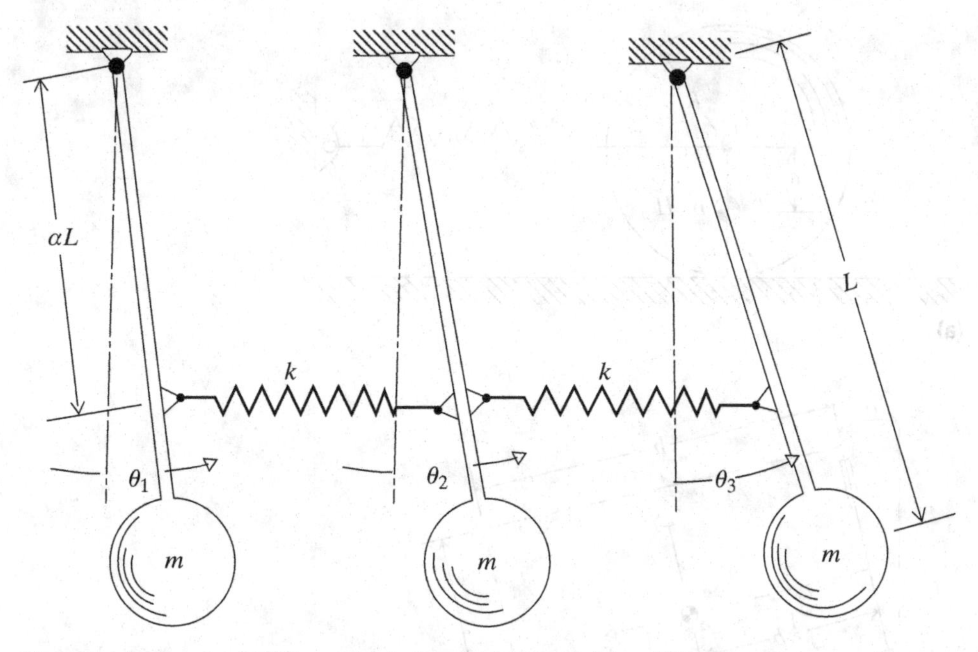

Figure 2.15. Exercise 2.4: Three elastically coupled pendulums. (This vibratory system is examined again in Chapter 6.)

(b) Improve the answer for part (a) by now including the effect of the mass moments of inertia of each of the pendulum arms. Let the thicknesses of the two arms be much smaller than the lengths. Thus the mass moment of inertia of each arm about an axis perpendicular to the plane of the paper, and passing through the center of mass of the arm, is $ML^2/12$. (Can you verify this moment of inertia value?)

(c) Note that the long thin rod has a mass moment of inertia about its center of mass that is much larger than that of a compact body with the same mass. In this context, what was the error associated with ignoring the mass moments of inertia of the two bars as indicated by your answers to parts (a) and (b)?

2.4 By separately grouping the acceleration and deflection terms for each of the three equations of motion, write the (small deflection, linear) matrix equation of motion that describes the motions of the three pendulums that are connected as shown in Figure 2.15. These pendulums swing only within the plane of the paper. To simplify the algebra, let $g/L = \alpha^2(k/m)$, where α is the nondimensional constant that locates the spring connections. Let the spring constant k be sufficiently "soft" that the spring forces are of the same order of magnitude as the gravitational forces. Note that the mass moment of inertia of each pendulum bob about its own center of mass is negligible in this case.

2.5 (a) Consider a circular cylinder of mass m shown in Figure 2.16(a). The cylinder rolls back and forth, without slipping, because its center of mass is located a distance a below the geometric axis of the cylinder. If H is the mass moment of inertia of the cylinder about the center of mass, what then is the small deflection vibratory period

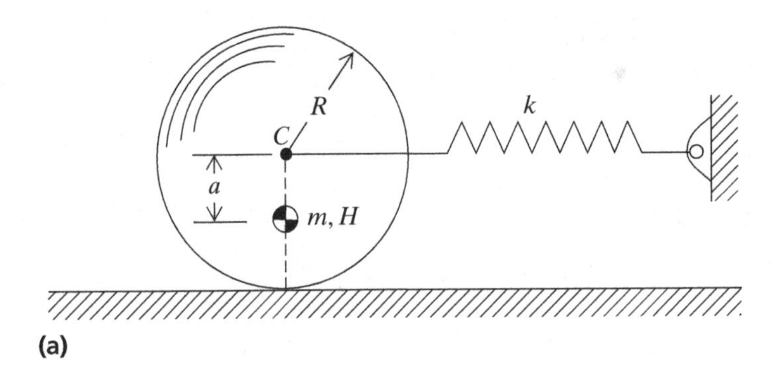

(a)

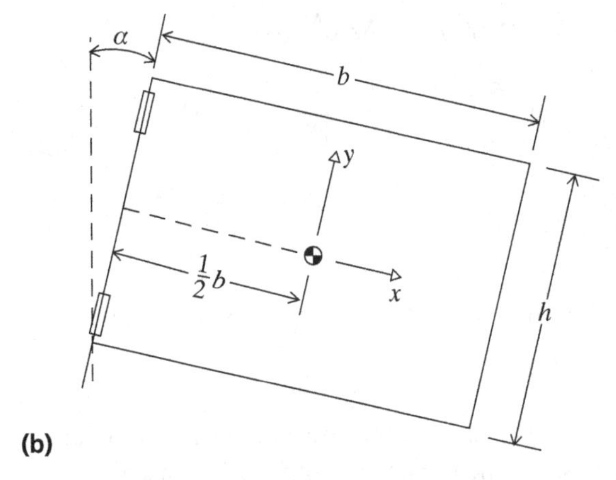

(b)

Figure 2.16(a). Exercise 2.5(a). (b) Exercise 2.5(b): The swinging gate pendulum.

for the cylinder? Note that the horizontal, linear spring is attached to the geometric center of the circular cylinder.

(b) The (rigid) thin, garden gate of Figure 2.16(b) has a uniformly distributed mass. The gate's total mass $m = \rho bht_0$, where ρ is the constant mass density and t_0 is the constant gate thickness. The gate's hinge line is off vertical by an angle α. Determine the natural frequency of its vibration as it swings about its hinge axis. Hint: Verify that with $t_o^2 \ll h^2, b^2$, the gate's mass moment of inertia about an axis parallel to the hinge line at the center of mass is

$$H_{CG} = \frac{\rho h}{12}\left(bt_0^3 + b^3 t_0\right) \approx \rho hbt_0\left(\frac{b^2}{12}\right) = \frac{mb^2}{12}.$$

(c) Write the equations of motion for the spring-pendulum system shown in Figure 2.16(c) using the DOF of Figure 2.16(d).

(d) Linearize the equations of motion of part (c) above by setting $L + v \approx L$; $\dot{\theta}^2 \approx 0$; $v\dot{\theta} \approx 0$; $\cos\theta \approx 1$; and $\sin\theta \approx \theta$.

2.6 **(a)** Not all pendulums look like the usual pendulum situation of a solid object moving in a gravitational field. Consider the mercury manometer tube shown in Figure 2.17. Its inner cross-sectional area is A, and the total, U-shaped, fluid column

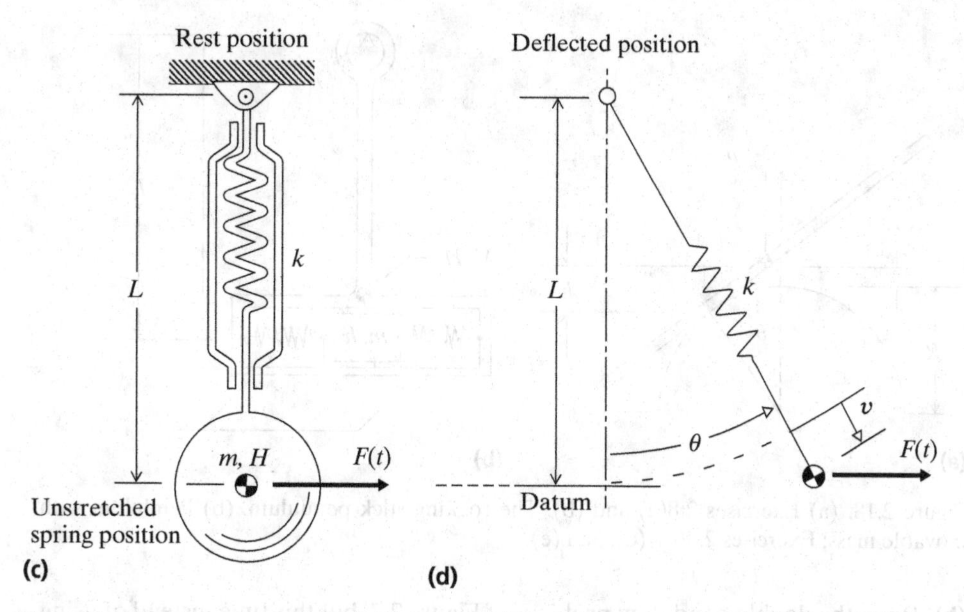

Figure 2.16(c). Exercise 2.5(c). (d) Exercise 2.5(c).

length is L. Both halves of the tube are open to the same atmospheric pressure, and thus the static equilibrium position of the mercury is that where both fluid columns have the same height. The incompressible mercury (mass density ρ) can be set to sloshing back and forth in the tube. Ignoring friction and other small effects, determine the period of this sloshing vibration by writing the equation of motion of the mercury column.

(b) If one of the two tops of the above manometer were now connected to a pressure tap that caused one fluid column to be higher than the other, would this disparity in column heights affect the vibratory period?

2.7 (a) Write the equations of motion for the pendulum system of Figure 2.4, which is now modified by changing the surface on which the roller travels from a flat surface to that of a circular cylinder like the support surface of Figure 2.5. Let the support surface have a radius R and the roller have a radius r.

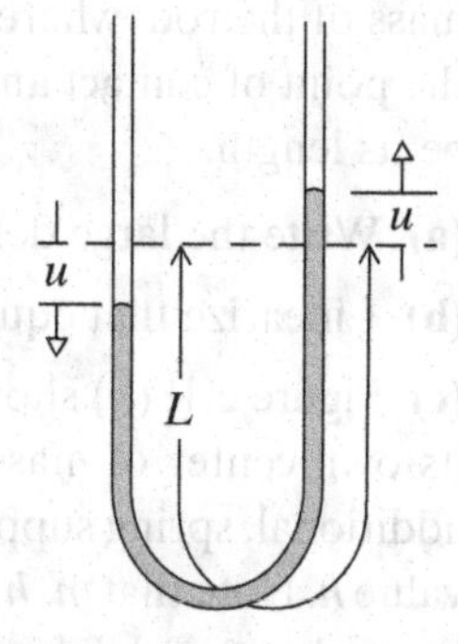

Figure 2.17. Exercise 2.6: Sloshing manometer tube with equal pressure openings.

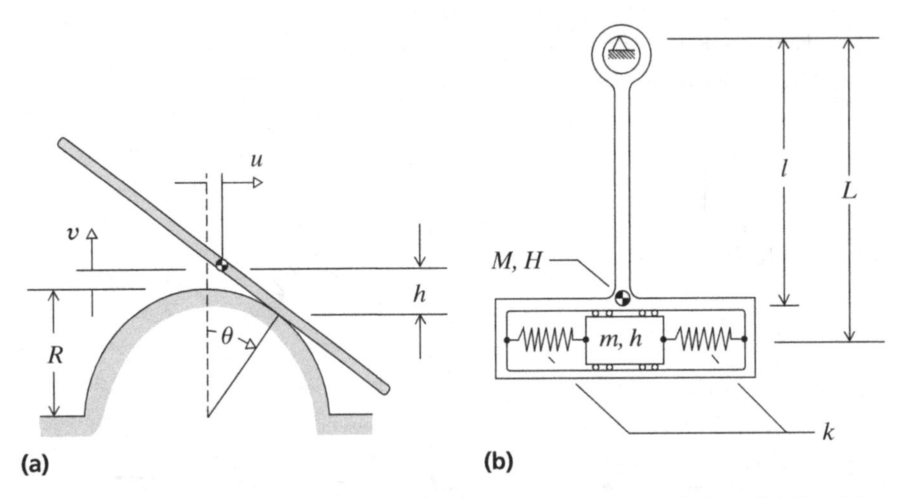

Figure 2.18. (a) Exercises 2.8(a) and (b): The rocking stick pendulum. (b) Pendulum with movable mass; Exercises 2.8(c), (d), and (e).

(b) Redo the double pendulum problem of Figure 2.7, but this time instead of using the previously identified (absolute) generalized coordinate θ_2 as the second DOF, use the generalized coordinate θ_3, which is defined as the rotation of the lower pendulum arm relative to the rotation of the upper pendulum arm. In other words, θ_3 is θ_2 minus θ_1.

(c) Show that the equation of motion for the pendulum system of Example 2.1 is unaltered even if the two springs have equal preloads. Note that with only slightly unequal preloads, the pendulum would just move slightly off-vertical so as to equalize the forces in the springs; that is, the system would move to a slightly nonvertical static equilibrium position from which it would vibrate as before.

For the eager

2.8 Figure 2.18(a) shows a thin rod rocking (without slipping) back and forth on a circular cylindrical surface. The axis of the rod is perpendicular to the axis of the cylinder, and the single generalized coordinate θ is the angle measured from the vertical through the center of the cylindrical surface to the point of contact with the rod. Note that u, v are the horizontal and vertical distances traversed by the center of mass of the rod, whereas ℓ and h are the horizontal and vertical distances between the point of contact and the rod center of mass. Let m be the mass of the rod, and L be its length.

(a) Write the large deflection equation of motion.

(b) Linearize that equation to obtain the small deflection natural frequency.

(c) Figure 2.18(b) shows a pendulum of mass M and mass moment of inertia about its own center of mass of magnitude H. Enclosed within the pendulum bob is an additional, spring supported, mass m with its CG mass moment of inertia having the value h. Note that m, h are not part of M, H. The two centers of mass are respectively located at ℓ and L from the fulcrum, and these CGs move only in the plane of the

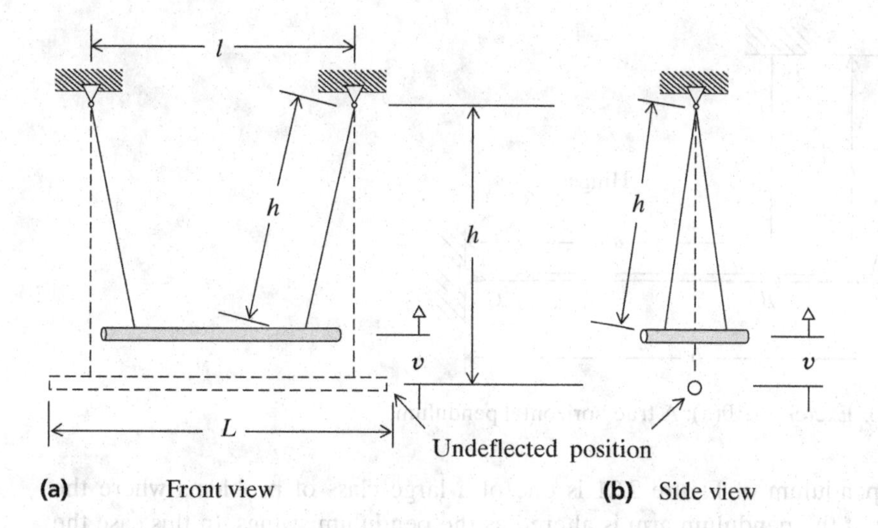

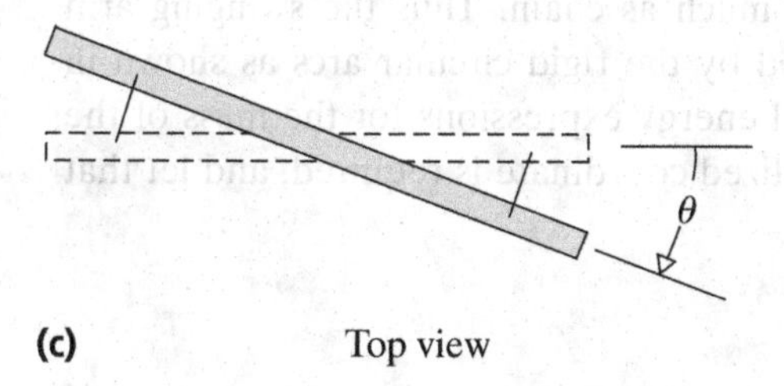

Figure 2.19. Exercise 2.9: The twisting trapeze pendulum.

paper. Modeling the pendulum system as frictionless and without external excitation, write the large rotation equations of motion for this system, and then cast the linear forms of these equations in matrix form. As per usual, the springs are to be modeled as massless, and the static equilibrium position of the mass m is located on the axis of the pendulum arm.

(d) How would the task of part (c) change if the two springs were infinitely stiff?

(e) How would the mass m move if the two springs had zero stiffness?

2.9 Figure 2.19 shows the geometry of a trapeze and three views of the trapeze undergoing vibratory motion as the trapeze twists about its axis of symmetry. (Be sure to note that the trapeze does not swing in or out of the paper in such a manner that the bar remains parallel to its original position.) Write the *small* deflection expressions for the kinetic and potential energies in terms of a single generalized coordinate of your choice.

2.10 **(a)** Figure 2.20 illustrates a "horizontal" pendulum problem. The horizontal pendulum arm of length b is rigid and massless. At the wall, the arm is supported by a universal joint that allows the pendulum to swing in and out of the plane of the paper. At a distance a from the wall, the rod is supported by a massless string of length h. The string remains taut and unstretched at all times. Write the small deflection equation of motion and determine the small deflection natural frequency.

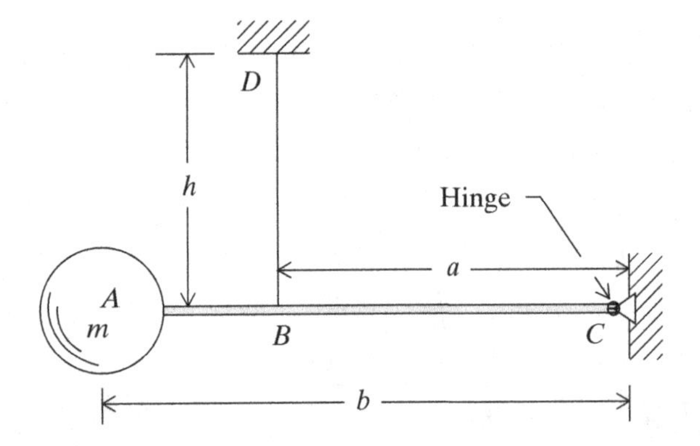

Figure 2.20. Exercise 2.10(a): A true horizontal pendulum.

(b) The pendulum of Figure 2.21 is one of a large class of problems where the geometry of the pendulum arm is altered as the pendulum swings. In this case the arm is axially rigid, but laterally flexible, much as chain. Thus the swinging arm conforms to the shape constraints imposed by the rigid circular arcs as shown in the sketch. Write the kinetic and potential energy expressions for the mass of the pendulum bob. Note that only one generalized coordinate is required, and let that DOF be the θ indicated on the figure.

For the especially eager

2.11 Consider the inverted pendulum problem of Figure 2.12 when the rectilinear spring attached to the bob is replaced by a rotational spring (i.e., a linear spring that generates a moment to oppose a rotation, as per $M = K\theta$, just a rectilinear spring generates a force to oppose a translation, as per $F = ku$). Let the rotational spring be attached to the base of the rod. Using any method:

(a) Determine all possible static equilibrium positions.

(b) Determine the stability of those SEPs,

(c) Determine the natural frequency of small vibrations about all stable SEPs.

2.12 Reconsider the one DOF system of Figure 2.5 where now the concave support surface has a radius that varies with θ; that is, $R = R(\theta)$, where $R(\theta)$ is a very smooth, everywhere concave, function of θ, and $R'(0) = 0$. $R(\theta)$ is symmetric about $\theta = 0$.

(a) Show that in this case the relationship between the two angular velocities, $\dot{\phi}$ and $\dot{\theta}$, is

$$r\dot{\phi} = \frac{d}{dt}\int_0^\theta \left[R^2 + \left(\frac{dR}{d\theta}\right)^2\right]^{1/2} d\theta = \dot{\theta}\left[R^2 + \left(\frac{dR}{d\theta}\right)^2\right]^{1/2}.$$

(b) In terms of the two rotational angles ϕ and θ, write the kinetic energy and potential energy expressions for the rolling cylinder. (This is a challenging task.)

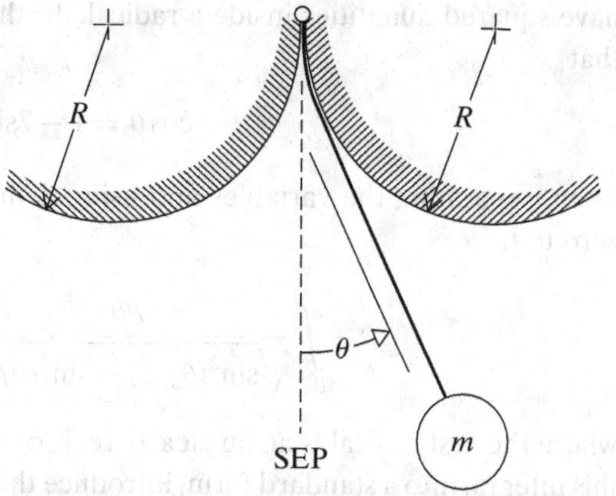

Figure 2.21. Exercise 2.10(c): A pendulum arm that is inextensible but wholly flexible laterally and of length L.

ENDNOTE (1): THE LARGE-DEFLECTION, SIMPLE PENDULUM SOLUTION

Consider the nonlinear differential equation for the large deflections of a simple pendulum of arm length L

$$\ddot{\theta} + \frac{g}{L}\sin\theta = 0.$$

The $\ddot{\theta}$ term can be rewritten as follows:

$$\frac{d\dot{\theta}}{dt} = \frac{d\dot{\theta}}{d\theta}\frac{d\theta}{dt} = \dot{\theta}\frac{d\dot{\theta}}{d\theta}.$$

Substituting the above product for $\ddot{\theta}$ and then multiplying through by $d\theta$, the integration of the differential equation becomes

$$\int \dot{\theta}\, d\dot{\theta} = -\frac{g}{L}\int \sin\theta\, d\theta + C_1.$$

Carrying out the indefinite integration leads to

$$\frac{1}{2}\dot{\theta}^2 = \frac{g}{L}\cos\theta + C_1.$$

Now choose the initial conditions for this problem to be an initial deflection of θ_0 and a zero initial velocity. Thus, at time zero

$$0 = \frac{g}{L}\cos\theta_0 + C_1 \quad \text{or} \quad C_1 = -\frac{g}{L}\cos\theta_0.$$

Hence, after substitution for the constant of integration, taking the square root, and rearranging the result, where, in general, $\theta_0 \geq \theta$,

$$\frac{d\theta}{dt} = \pm\sqrt{\frac{2g}{L}}\sqrt{\cos\theta - \cos\theta_0}.$$

Note that because the radical involving the cosines is real, and the time rate of change of θ is negative because in the first half cycle of the pendulum vibration the value of θ must be decreasing after the pendulum is released from its initial position θ_0. Hence the negative sign must be chosen for the square root. Generally, it is better to have squared quantities inside a radical. To this end use the trigonometric identity that

$$\cos\theta = 1 - 2\sin^2\frac{\theta}{2}.$$

Then, separating the variables and using definite integration over the time period zero to t,

$$-\int_{\theta_0}^{\theta} \frac{d\theta}{\sqrt{\sin^2(\theta_0/2) - \sin^2(\theta/2)}} = 2\sqrt{\frac{g}{L}}\int_0^t dt,$$

where the first radical is again clearly real for all values of θ_0 less than $\pm\pi/2$. To put this integral into a standard form, introduce the notation $\sin(\theta_0/2) = 1/k$, and factor out that quantity from the first radical. Then select as the upper limit of integration the point in time where the pendulum swings from its initial deflection to the vertical, which, in terms of time, is one quarter of a period, that is, select $t = T/4$. Reversing the limits of integration on the left-hand side and carrying out the integration on the right-hand side yields

$$\int_0^{\theta_0} \frac{d(\theta/2)}{\sqrt{1 - k^2\sin^2(\theta/2)}} = \frac{T}{4}\sqrt{\frac{g}{L}}\,\sin\left(\frac{\theta_0}{2}\right).$$

The left-hand side integral is known as an "incomplete elliptic integral of the first kind." Tabulated values of these integrals can be found, for example, in Ref. [2.4]. This integral can be evaluated here by use of a binomial expansion, in which case this integral becomes

$$\int_0^{\theta_0}\left[1 + \frac{k^2}{2}\sin^2\left(\frac{\theta}{2}\right) + \frac{3k^4}{8}\sin^4\left(\frac{\theta}{2}\right) + \frac{5k^6}{16}\sin^6\left(\frac{\theta}{2}\right) + \cdots\right] d\left(\frac{\theta}{2}\right).$$

The nth term in the above series is

$$\frac{(2n-3)!k^{2m-2}\sin^{2m-2}(\theta/2)}{2^{m-3}(m-1)!(m-2)!}.$$

The above integral can be evaluated once a value of θ_0 is chosen. If, for example, the initial deflection is chosen to be the convenient value of 90°, then the above series of integrals can be easily evaluated using beta functions. With any choice of θ_0, it is clear that the natural period T will not be a constant but depend on θ_0 by depending on k.

ENDNOTE (2): DIVERGENCE AND FLUTTER IN MULTIDEGREE OF FREEDOM, FORCE FREE SYSTEMS

The last paragraph of this chapter states without proof that multidegree of freedom structural systems free of externally applied, time-varying forces and moments have roughly the same divergence and flutter responses that single degree of freedom systems have. The following is the justification for that statement. As is seen in Chapter 5, the general "force free," n-DOF, linear equation of motion can be written in matrix form as

$$[m]\{\ddot{u}\} + [c]\{\dot{u}\} + [k]\{u\} = \{0\}, \tag{2.17}$$

where $[m]$, $[c]$, and $[k]$ are, respectively, the system mass (or inertia), damping, and stiffness matrices, all of which are of size $N \times N$. The next three chapters also demonstrate that these three matrices can be made to be symmetric matrices for almost all structures, and therefore they are assumed to be symmetric in the discussion that follows. The $N \times 1$ column matrix $\{u\} = \{u(t)\}$ is the matrix of generalized coordinates that locates the time-varying positions of the system masses, and, of course, each dot represents one differentiation with respect to time. An example problem at the end of this endnote partially illustrates the above statements.

Although not necessarily the best choice of a coordinate transformation, the topic development begins with the following conceptually simple coordinate transformation between the original (physically based) generalized coordinates $u(t)$, and a new set of generalized coordinates, $q(t)$, where[11]

$$\{u(t)\} = [m]^{-1/2}\{q(t)\}.$$

The above inverse square root of the mass matrix, a matrix of constants, is easily obtained if the mass matrix is a diagonal matrix, as is often the case for simple structures. If the mass matrix is not a diagonal matrix, then its inverse square root is obtained from the eigenvalues of the mass matrix as explained in the last section of Chapter 9. Substituting this coordinate transformation and premultiplying both sides of the equation of motion by the same inverse square root of the mass matrix leads to

$$\left[m^{-1/2} m^1 m^{-1/2}\right]\{\ddot{q}\} + \left[m^{-1/2} c\, m^{-1/2}\right]\{\dot{q}\} + \left[m^{-1/2} k\, m^{-1/2}\right]\{q\} = \{0\}.$$

From the definition of the inverse square root of a matrix, the first coefficient matrix is the identity matrix. Therefore, the above equation can be rewritten using the following symbols where the definitions of the new matrix symbols are obvious

$$\{\ddot{q}\} + 2[\zeta\omega]\{\dot{q}\} + [\omega^2]\{q\} = \{0\}.$$

It is important to note that both of the new constant coefficient matrices of the above equation are equal to their own matrix transposes and thus are also symmetric matrices. As discussed in Chapter 9, the matrix solution to the above second-order modified matrix equation of motion can be written as

$$\begin{aligned} \{q(t)\} &= \exp(-[\zeta\omega]t)\left(\sin([\omega_d]t)\{C_1\} + \cos([\omega_d]t)\{C_2\}\right) \\ \text{where} \quad [\omega_d]^2 &\equiv [\omega]^2 - [\zeta\omega]^2 \end{aligned} \tag{2.18}$$

[11] From Chapter 5, a more efficient transformation is one based on a Cholesky decomposition of the mass matrix. However, the transformation based on the square root of the mass matrix requires less explanation at this point.

and where $\{C_1\}$ and $\{C_2\}$ are $n \times 1$ column matrices of constants of integration that, of course, are related to the system's $2N$ initial conditions. Again, an explanation of the mathematics of functions of matrices can also be found in Ref. [2.3]. This solution statement for $\{q\}$ can be validated by substitution into the corresponding matrix equation of motion. Only the validity of the first (the sine) portion of the solution need be demonstrated below

$$\{\dot{q}\} = -[\zeta\omega]\{q\} + [\omega_d]\exp(-[\zeta\omega]t)\cos([\omega_d]t)\{C_1\}$$
$$\{\ddot{q}\} = +[\zeta\omega]^2\{q\} - 2[\zeta\omega][\omega_d]\exp(-[\zeta\omega]t)\cos([\omega_d]t)\{C_1\}$$
$$- [\omega_d]^2\exp(-[\zeta\omega]t)\sin([\omega_d]t)\{C_1\}$$

$$\text{so} \quad \{\ddot{q}\} + 2[\zeta\omega]\{\dot{q}\} + [\omega]^2\{q\}$$
$$= [\zeta\omega]^2\{q\} - [\omega_d]^2\{q\} - 2[\zeta\omega]^2\{q\} + [\omega]^2\{q\} \equiv \{0\} \quad Q.E.D.$$

Following the same discussion pattern that is followed for the single-DOF system, first, temporarily, let the damping matrix, $[c]$, be null,[12] and consider only the cosine portion of the solution[13] of Eq. (2.18). Then, in these much reduced, but still indicative, circumstances, $[\omega_d]^2 = [\omega]^2$. Next decompose the $[\omega^2]$ matrix into the product of its eigenvectors and eigenvalues; that is, write $[\omega]^2 = [\Phi][\Lambda][\Phi]^t$, where $[\Lambda]$ is the diagonal matrix of the n eigenvalues. If any of these eigenvalues is negative, then in the eigenvalue–eigenvector decomposition of $[\omega] = [\Phi][\Lambda]^{\frac{1}{2}}[\Phi]^t$, the middle, diagonal matrix of eigenvalue square roots will contain an imaginary number. In that case, the cosine of that imaginary number will be a real hyperbolic cosine that will grow without bound as time increases. Thus in the number of degrees of freedom for which there are negative eigenvalues for the $[\omega^2]$ matrix, there will be that same number of static instabilities (also called divergences). Hence, for a structural system to be statically stable, all the eigenvalues of the $[\omega^2]$, matrix, and thus all the eigenvalues of the stiffness matrix, have to be positive. Another way of saying the same thing is to say that either the $[\omega^2]$ matrix or the stiffness matrix has to be "positive definite." (If one these two "equivalent" matrices is positive definite, so too is the other.)

Similarly, if any of the n eigenvalues of the matrix $[\zeta\omega]$ happens to be negative, then the exponential function will have a positive argument. Thus as time steadily increases, so too will the values of those exponential functions associated with the negative eigenvalues. Since these exponential functions serve as both an upper and lower bound for the amplitudes of the vibration, this means, of course, that the amplitudes of the vibration increase without bound. Therefore there will be as many dynamical instabilities as there are negative eigenvalues of the $[\zeta\omega]$ matrix. Conversely, for the system to be dynamically stable and not flutter, the $[\zeta\omega]$ matrix, or, equivalently, the damping matrix, must be positive definite.

[12] Positive damping (such as any form of friction) is stabilizing, so setting the theoretical value of the damping to zero does not falsely expand the stability boundary.

[13] This is equivalent to having initial displacements but not initial velocities. The justification for this step is that the general characteristics of the system's linear behavior are unaltered by the nature of the system initial conditions.

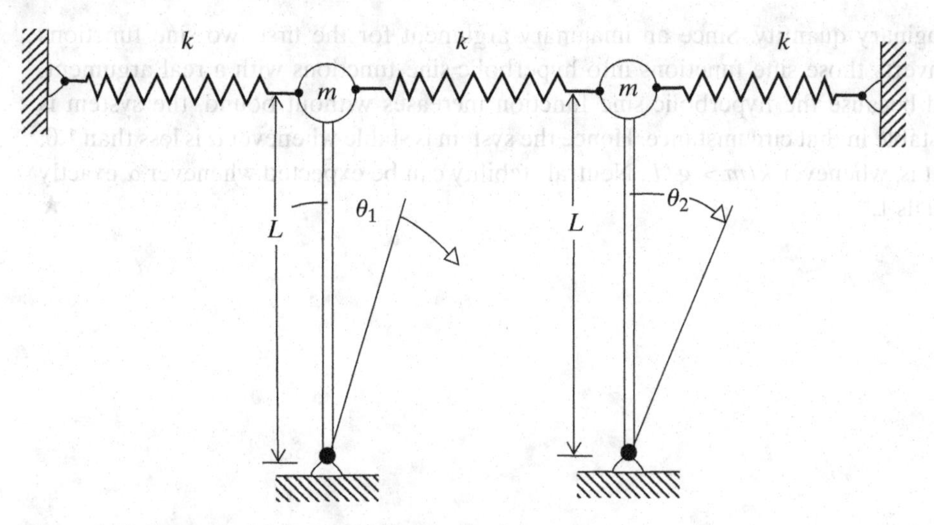

Figure 2.22. Example 2.9: Inverted double pendulum system.

EXAMPLE 2.9 Consider the double inverted pendulum of Figure 2.22. The reader can check that the matrix equation of motion is

$$\begin{bmatrix} 1 & 0 \\ 0 & 1 \end{bmatrix} \begin{Bmatrix} \theta_1 \\ \theta_2 \end{Bmatrix} + \begin{bmatrix} \left(2\frac{k}{m} - \frac{g}{L}\right) & -\frac{k}{m} \\ -\frac{k}{m} & \left(2\frac{k}{m} - \frac{g}{L}\right) \end{bmatrix} \begin{Bmatrix} \theta_1 \\ \theta_2 \end{Bmatrix} = \begin{Bmatrix} 0 \\ 0 \end{Bmatrix}.$$

Since $[\backslash m \backslash]$ in this simple case is the identity matrix $[\,I\,]$, which is equal to its own inverse square root, no transformation of coordinates is necessary to obtain a solution to this matrix equation of motion. As was done in the triple pendulum problem of Figure 2.15 to simplify the algebra, let α be a nondimensional proportionality factor such that

$$\frac{g}{L} = \alpha \frac{k}{m}.$$

Using the techniques discussed in Chapter 6, the solutions that describe the motion of this double inverted pendulum problem are

$$\theta_1(t) = C_1 \sin(\omega_1 t + \psi_1) + C_2 \sin(\omega_2 t + \psi_2)$$
$$\theta_2(t) = C_1 \sin(\omega_1 t + \psi_1) - C_2 \sin(\omega_2 t + \psi_2),$$

where C_1, C_2, ψ_1, and ψ_2 are the four initial condition-dependent constants of integration that are to be expected for the solution of any two second-order ordinary differential quations,

$$\omega_1^2 = (1 - \alpha)\frac{k}{m} \quad \text{and} \quad \omega_2^2 = (3 - \alpha)\frac{k}{m}.$$

Given the above information, determine the value for α that constitutes the boundary point between stability and instability for this mass system.

SOLUTION From examining the solutions for the two natural frequencies of vibration, it should be clear that if α is greater than 1.0, the first natural frequency will be an imaginary quantity. Since an imaginary argument for the first two sine functions converts those sine functions into hyperbolic sine functions with a real argument, and because the hyperbolic sine function increases without bound, the system is unstable in that circumstance. Hence the system is stable whenever α is less than 1.0; that is, whenever $k/m > g/L$. Neutral stability can be expected whenever α exactly equals 1.0. ★

3 Review of the Basics of the Finite Element Method for Simple Elements

3.1 Introduction

The first two chapters provided the basics of small-rotation dynamics with applications to rigid bodies or near rigid bodies. The purpose of this chapter is to provide an introduction to the finite element method (FEM) of structural analysis to the limited extent necessary for the use of this textbook. That is, the present discussion of the FEM is limited mostly to one-dimensional structural elements. Specifically, all the example problems and exercises deal only with linear, beam finite elements and linear, spring finite elements. Neither of these elements require the finite series sophistication of two- or three-dimensional elements such as plate or solid finite elements. A very brief introduction to multidimensional finite elements is presented in Endnote (1). If the reader is already familiar with the FEM, then this chapter can be skipped, particularly because the next chapter provides ample further review of this topic. Since the finite element method is so extensively used for, and so particularly suited to, structural dynamics analyses,[1] no other method of structural analysis is used for the calculations presented in this textbook. For the sake of instruction, the use of the FEM in this textbook is oriented to hand calculations rather than the use of one of the many available and routinely used commercial software programs that all do essentially the same things and differ only in style. Thus the reader should be able to gain insight into what all such FEM programs need to do.

In this chapter only, the point of view is that of static analyses, that is, the FEM is considered in relation to loads that do not vary with time. This approach is both possible and useful because the Lagrange equations represent a summation of different types of forces to form what is sometimes referred to as equations of "dynamic equilibrium." Specifically, the Lagrange equations sum the (i) generalized inertia forces, which are the $(d/dt)(\partial T/\partial \dot{q}) - \partial T/\partial q$ mass–acceleration terms; (ii) internal generalized elastic forces, which are the $\partial U/\partial q$ stiffness–deflection terms; and (iii) the externally applied forces and the internal dissipative forces, which are collectively

[1] In engineering, the initial development of the FEM is credited to the leadership of the structural dynamics group of the Boeing Co. in Seattle, Washington. See Ref. [3.1]

the $\partial V/\partial q$ and the Q terms. This mathematical separation of types of forces in the Lagrange equations allows the separate consideration of the deflection dependent $\partial U/\partial q$ elastic forces regardless of whether the applied loads are time invariant.

3.2 Generalized Coordinates for Deformable Bodies

The definition of generalized coordinates, or degrees of freedom (DOF), as quantities that determine the system's deflected position, was originally introduced in Chapter 1 in relation to rigid bodies. The same definition for generalized coordinates applies to flexible bodies. However, unless certain restrictions are implemented, the flexibility of the body can require the unnecessary use of huge numbers of DOF to merely approximate the deflected position of even the simplest of flexible bodies. This difficulty concerning the required numbers of generalized coordinates can be illustrated by considering a uniform, long beam, of length L, which, for the sake of simplicity, has a doubly symmetric cross section. Let the loci of beam cross-section centroids and the x axis coincide. Let the y and z axes be the cross-sectional axes of symmetry. See Figure 3.1(a). Let the beam be subjected to an arbitrary, y-direction (distributed) force loading per unit of beam length, $f_y(x)$. Then the deflections of the beam centroidal axis in the positive y direction, call those deflections $v(x)$, are related to the arbitrary applied distributed loading by the differential equation,[2] $EI_{zz}v''''(x) = f_y(x)$. The arbitrary force input, $f_y(x)$, over the span length produces an arbitrary deflection output over the span length. To see that a very large number of generalized coordinates is required to approximate such an arbitrary deflection function, $v(x)$, even when that function has continuous first derivatives as all beam deflection functions must, consider the following argument in relation to this beam of length L. One generalized coordinate is needed to specify $v(L/2)$, the vertical deflection of the beam axis at $x = L/2$. Two additional, entirely independent, generalized coordinates are needed to fix the deflections at the quarter points, $x = L/4$ and $x = 3L/4$. Four additional generalized coordinates are needed to fix the deflections at the remaining eighth points $x = L/8, 3L/8, 5L/8,$ and $7L/8$. Then eight additional DOF are needed to specify the deflections at the remaining sixteenth points, and so on. Hence the above-mentioned apparent requirement for large numbers of DOF when the body is flexible. Is there a limit, no matter how large, on this number of DOF needed to fully specify the position of the beam of length L? This is the same as asking the following question. Can the deflection of an intermediate point on the beam axis be determined from knowing the deflections on either side of this intermediate point? That the answer is "no" can be understood by fixing the deflections at the adjacent points and introducing an additional concentrated force of arbitrary magnitude at that intermediate point. This additional concentrated force can cause the intermediate point's original deflection to move up or down from what it would be without that arbitrary force. It does not matter if the additional deflection relative to the adjacent points is quite small. It suffices that the center point deflection is thus

[2] The quantity E is Young's modulus, I_{zz} is the cross section's area moment of inertia about the z axis. A simple way to derive this equation is to twice differentiate the following familiar beam bending equation $EI_{zz}v''(x) = M_z(x)$, where the derivative of the internal bending moment is the internal shear force and the derivative of the shear force is the external loading per unit length.

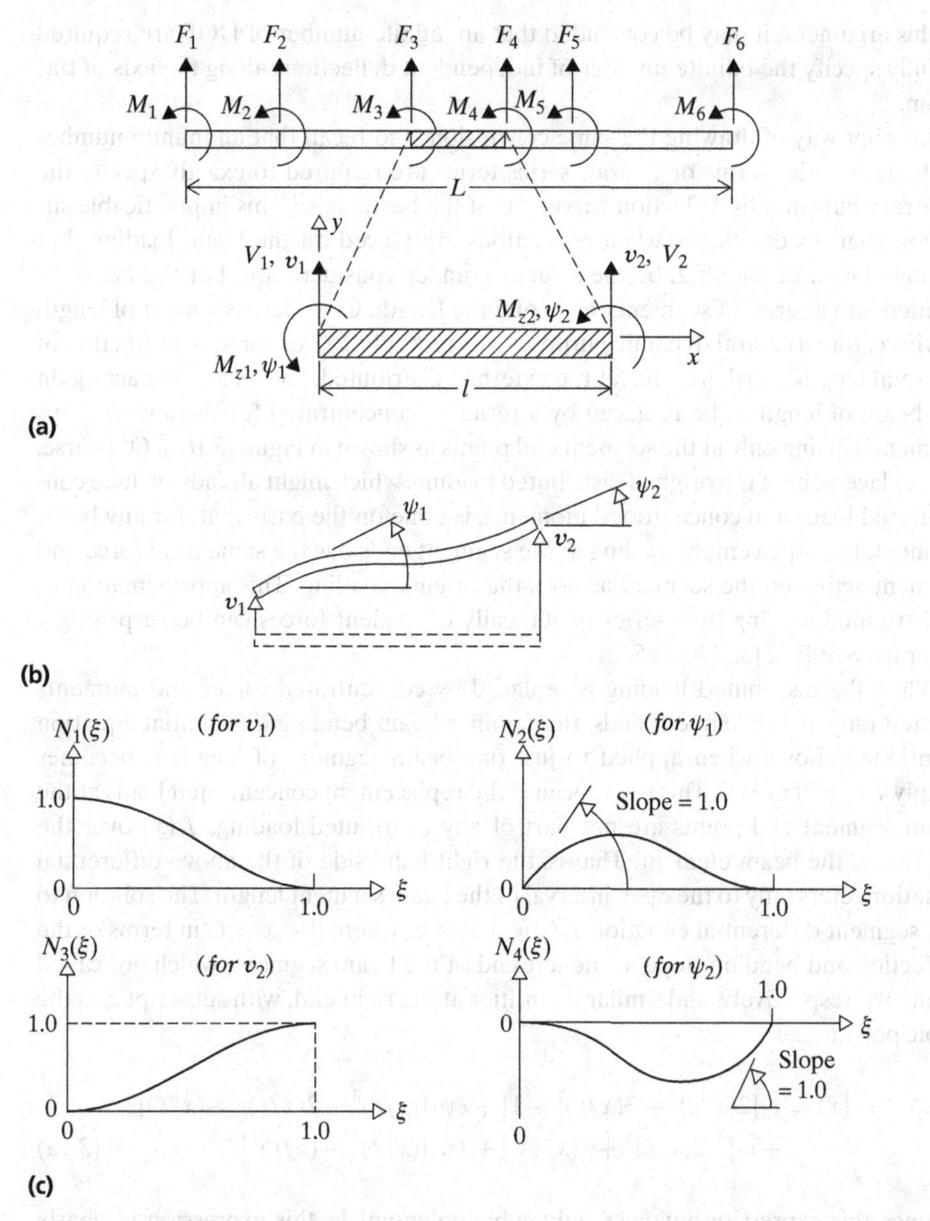

Figure 3.1. Finite element modeling for beam bending. (a) Beam partitioned into several beam elements with the external loading represented by a series of discrete forces and moments along the beam length and: (i) the (internal) shearing forces and bending moments acting at the single element ends; (ii) the generalized coordinates at the single element ends (nodes). (b) A beam element whose deflected position is "controlled" by the four generalized coordinates. (c) Plots of the four shape functions associated with the beam element generalized coordinates showing a unit value associated with the corresponding generalized coordinate and zero deflection or slope at the location of the other three generalized coordinates.

independent of the two adjacent point deflections. Therefore, it may be concluded that two generalized coordinates defining the deflections at the two adjacent points are not sufficient to determine the deflection at the center point when the magnitude of the additional concentrated force at the center point is arbitrary. Thus by extension of this argument, it may be concluded that an infinite number of DOF are required to fully specify the infinite number of independent deflections along the axis of this beam.

Another way of drawing the same conclusion is to recall that an infinite number of Taylor's series terms or Fourier series terms are required to exactly specify the arbitrary but smooth deflection curve $v(x)$of the beam axis.[3] This impracticable situation changes drastically when restrictions are placed on the beam loading. Let a single beam of length L be the structure under consideration. Let the beam be divided into a series of segments, each of finite length. Consider a segment of length ℓ, where, for structural dynamic purposes, ℓ is typically $L/8$ or some such fraction of the total length. Furthermore, let the external, distributed, lateral loading acting on the beam of length L be replaced by a series of concentrated forces (and perhaps moments) acting only at the segment end points as shown in Figure 3.1(a). Of course, the replacement of the original distributed loading, which might already include concentrated loads and concentrated moments, is done on the basis that, for any beam segment, the replacement loading at the segment ends has the same total force and moment acting on the segment as does the original loading. This approximation of a distributed loading by a series of statically equivalent forces can be surprisingly accurate. See Ref. [3.2], p. 685.

When the distributed loading is replaced by concentrated forces and moments located only at the segment ends, the familiar beam bending differential equation mentioned above, when applied to just one beam segment of length ℓ, becomes simply $EI_{zz}v''''(x) = 0$. This is so because the replacement concentrated loads at the beam segment end points are not part of any distributed loading, $f_y(x)$,over the interior of the beam element. That is, the right-hand side of the above differential equation refers only to the open interval of the beam segment length. The solution to this segment differential equation $EI_{zz}v''''(x) = 0$, where $0 < x < \ell$, in terms of the deflection and bending slope at the left end of the beam segment, which are called v_1 and ψ_1, respectively, and similar quantities at the right end, with subscript 2, is the cubic polynomial

$$
\begin{aligned}
v(x) = {} & v_1[2(x/\ell)^3 - 3(x/\ell)^2 + 1] + \ell\psi_1[(x/\ell)^3 - 2(x/\ell)^2 + (x/\ell)] \\
& + v_2[-2(x/\ell)^3 + 3(x/\ell)^2] + \ell\psi_2[(x/\ell)^3 - (x/\ell)^2]. \qquad (3.1a)
\end{aligned}
$$

Since this expression involves only cubic polynomials, this expression is clearly a solution for the above fourth-order beam bending differential equation. Furthermore, it contains the required four unknown constants of integration, which again are the deflection boundary conditions v_1, ψ_1, v_2, and ψ_2. From this solution it is apparent that the deflection function $v(x)$, for all values of x, is controlled entirely by the four quantities v_1, ψ_1, v_2, and ψ_2. The spatial variable x merely identifies the point on the segment axis whose deflection is determined by the values of v_1, ψ_1, v_2, and ψ_2. Hence, those four end-point deflections are the beam-bending degrees of freedom for the beam segment without interior loading. As is shown in Figure 3.1(b), bending

[3] In this case the Fourier coefficients would be the generalized coordinates.

a flexible ruler will confirm the complete physical control that these four quantities have on the deflected shape of the entire ruler.

As an aside, it is useful to know the following terminology. In mathematical terms, the cubic polynomial terms associated with each of the generalized coordinates are called cubic *splines*. In FEM terms, each of these polynomials is called a *shape function*. This latter name comes from the fact that these cubic polynomials individually describe a deflection shape that corresponds to a unit value of their associated DOF and zero values for the other three DOF. This is illustrated in Figure 3.1(c), where $x/\ell = \xi$. Since four is a manageable number of degrees of freedom for a beam segment bending in a single plane, this segmentation scheme and load limitation approach is adopted for beam bending and all other types of beam deflections, as well as the deflections of plate segments, shell segments, solid segments, and so on. Furthermore, if the beam bending deflections vary with time, so too do the generalized coordinates. That is, for the vibrating beam bending segment

$$v(x,t) = v_1(t)[2(x/\ell)^3 - 3(x/\ell)^2 + 1] + \ell\psi_1(t)[(x/\ell)^3 - 2(x/\ell)^2 + (x/\ell)] + v_2(t)[-2(x/\ell)^3 + 3(x/\ell)^2] + \ell\psi_2(t)[(x/\ell)^3 - (x/\ell)^2]. \tag{3.1b}$$

In summary, generalized coordinates, or DOF, for flexible systems describe the deflected shape of the system in exactly the same fashion as they do for rigid systems. However, to make the number of DOF manageable, the actual loading on the system is approximated by a series of concentrated forces and concentrated moments. The selected geometric points that are the locations of these concentrated forces and moments in or on the flexible body are called the system *nodes*, and each of the finite sized structural segments between the nodes is called a *finite element*. The term *finite* distinguishes these elements from the differential sized elements that are a basis of a calculus based analysis and, hence, a classical structural analysis.

3.3 Element and Global Stiffness Matrices

To make use of finite element modeling in static or dynamic structural analyses, it is necessary to develop the means to describe the force–deflection relationship for each type of finite element. After an appropriate deflection function expression has been obtained in terms of suitable generalized coordinates and shape functions, such as Eq. (3.1), the most straightforward way of approaching this task is to enforce element equilibrium by use of the principle of virtual work ($\delta W_{ex} = \delta U$) and use the element material equations.[4] Equally efficient is the use of the principle of the minimum value of the total potential energy. Since some readers may be vague in their understanding of these closely related principles, and since this textbook employs only beam or still simpler finite elements, then the derivation of the force–deflection equations for the beam bending finite element that follows forgoes the use of those sophisticated

[4] Since a discussion of the complications of plasticity is unnecessary to explaining structural dynamics, all structures herein are considered to be elastic. The principle of virtual work is included within Hamilton's principle, but the previous derivation of Hamilton's principle did not include essential details of the principle of virtual work.

general principles.[5] Instead, just the previously detailed concept of virtual work is used.

The internal bending moments and shear forces at the nodes of the above beam bending finite element can be determined by using Eq. (3.1b) and the familiar strength of materials equations $M_z(x) = EI_{zz}v''(x)$ and $V_y(x) = EI_{zz}v'''(x)$ at both element end points. Thus after twice and thrice differentiating the Eq. (3.1b) polynomial solution for the lateral deflection $v(x)$, setting $x = 0$ or $x = \ell$ to localize the bending moment and shearing force results to segment ends 1 and 2, respectively, and arranging the answers in matrix form, the result is

$$\begin{Bmatrix} V_{y1} \\ M_{z1} \\ V_{y2} \\ M_{z2} \end{Bmatrix} = \frac{EI_{zz}}{L^3} \begin{bmatrix} 12 & 6L & -12 & 6L \\ 6L & 4L^2 & -6L & 2L^2 \\ -12 & -6L & 12 & -6L \\ 6L & 2L^2 & -6L & 4L^2 \end{bmatrix} \begin{Bmatrix} v_1 \\ \psi_1 \\ v_2 \\ \psi_2 \end{Bmatrix}.$$

The brief form of this matrix relationship is written as $\{Q\} = [k_e]\{q\}$, where the square matrix, including the scalar factor, is called[6] the beam *element stiffness matrix* for bending in a single plane. Of course, if the beam were also bent in the x, z plane, there would be another such fourth-order algebraic matrix equation involving the nodal shear forces and bending moments of that plane, as well as the nodal deflections of that plane. (Those deflections are written here as $w_1, \theta_1, w_2,$ and θ_2.) Furthermore, as discussed in Section 3.4, there is also a 2×2 stiffness matrix for beam torsion, with twisting DOF ϕ_1 and ϕ_2, and another for beam extension with DOF u_1 and u_2. Hence the full beam element force–deflection matrix equation involves a 12×12 element stiffness matrix where there are eight DOF for bending in two orthogonal planes, two DOF for twisting, and two DOF for extension.

The next step is to proceed from the above force–deflection description of one segment of a beam, that is, a single beam finite element, to a description of the force–deflection relations for an entire beam or a large frame or grid structure composed of many beams. To accomplish this step, first note that the external virtual work of the shear forces and bending moments acting on the one beam segment, call it the jth beam segment, is simply

$$\delta W_{ex}^{(j)} = V_{y1}^{(j)}\delta v_1^{(j)} + M_{z1}^{(j)}\delta\psi_1^{(j)} + V_{y2}^{(j)}\delta v_2^{(j)} + M_{z2}^{(j)}\delta\psi_2^{(j)}$$

$$= \delta\lfloor v_1 \quad \psi_1 \quad v_2 \quad \psi_2\rfloor^{(j)} \begin{Bmatrix} V_{y1} \\ M_{z1} \\ V_{y2} \\ M_{z2} \end{Bmatrix}^{(j)} = \lfloor\delta q\rfloor^{(j)}\{Q\}^{(j)}.$$

Remember that here the delta operator acts on only the deflection quantities.

Recall that, by definition, an elastic body is one in which all external work done on the body is stored in the body as recoverable internal strain energy, U. Since the

[5] The differential equation solution approach used here to determine a beam stiffness matrix will not suffice for, for example, plate, shell, and solid finite elements.

[6] In general terms, a stiffness coefficient or factor is the opposite of a flexibility coefficient or factor. A stiffness factor k generally relates force and deflection as $F = ku$, whereas a flexibility coefficient c generally relates these quantities as $u = cF$.

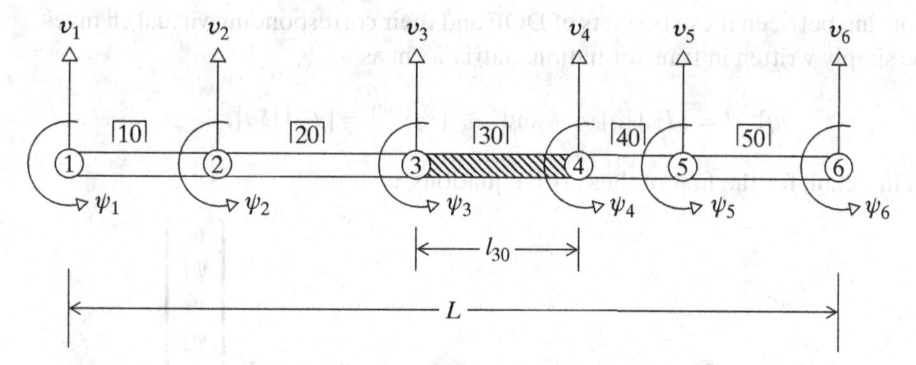

Figure 3.2. Example beam global (or system) generalized coordinates.

total internal energy of any body is equal to the sum of the internal energy of each of its parts, the virtual strain energy of the entire elastic beam or frame consists of the sum of the virtual strain energy of each of the N beam segments. That is

$$\delta U = \sum_{j=1}^{N} \delta U^{(j)} = \sum_{j=1}^{N} \lfloor \delta q \rfloor^{(j)} \{Q^{(in)}\}^{(j)} = \sum_{j=1}^{N} \lfloor \delta q \rfloor^{(j)} [k_j] \{q\}^{(j)}, \tag{3.2}$$

where the above individual finite element force–deflection relationship $\{Q\}^{(e)} = [k_e]\{q\}^{(e)}$ has been used. To make Eq. (3.2) useful, it is necessary to transition from the various sets of generalized coordinates for the individual finite elements, many of which overlap with those of adjacent finite elements, to the (unique) generalized coordinates of the overall structure. This latter set of DOF are called the *system* or *global* generalized coordinates. Like the element DOF, the global coordinates, defined below, are also located at the nodes of the structure but may or may not be in the same direction as the *element* or *local* coordinates. In this textbook for the sake of simplicity for hand calculations, the few beam example problems that require rotations of any kind only involve right-angle rotations. Thus the topic of rotations becomes unimportant to the present explanation. In commercial finite element analysis programs, the required rotations of DOF for actual structures are automatically calculated from the input geometry.

The idea that defines the global DOF is that they completely specify the displacements of the entire structure by specifying the deflections of all of the individual finite elements. For the beam of Figure 3.1, the 12 selected global DOF, v_1 through ψ_6, are shown in Figure 3.2. Note that in Figure 3.2 the nodes are simply labeled 1, 2, 3, . . . , whereas the finite elements are numbered 10, 20, 30, ... Since these selected system DOF accomplish the task of fully specifying the deflections of the five finite elements that constitute this single beam structure, they are a valid set of system/global generalized coordinates.

To illustrate the relation between the global generalized coordinates, $\{q\}$, and each of the local coordinate sets, $\{q\}^{(j)}$, consider beam finite element 30. The element or local DOF of beam element 30 are $v_1^{(30)}$, $\psi_1^{(30)}$, $v_2^{(30)}$, and $\psi_2^{(30)}$. In global terms, these same coordinates are v_3, ψ_3, v_4, and ψ_4, without any superscripts. Thus the relationship between these two sets of DOF and their corresponding virtual changes can be simply written in transformation matrix form as

$$\{q\}^{(30)} = [T_{30}]\{q\} \qquad \text{and} \qquad \{\delta q\}^{(30)} = [T_{30}]\{\delta q\},$$

which in detail, for the first of these two equations, is

$$\begin{Bmatrix} v_1 \\ \psi_1 \\ v_2 \\ \psi_2 \end{Bmatrix}^{(30)} = \begin{bmatrix} 0 & 0 & 0 & 0 & 1 & 0 & 0 & 0 & 0 & 0 & 0 & 0 \\ 0 & 0 & 0 & 0 & 0 & 1 & 0 & 0 & 0 & 0 & 0 & 0 \\ 0 & 0 & 0 & 0 & 0 & 0 & 1 & 0 & 0 & 0 & 0 & 0 \\ 0 & 0 & 0 & 0 & 0 & 0 & 0 & 1 & 0 & 0 & 0 & 0 \end{bmatrix} \begin{Bmatrix} v_1 \\ \psi_1 \\ v_2 \\ \psi_2 \\ v_3 \\ \psi_3 \\ v_4 \\ \psi_4 \\ v_5 \\ \psi_5 \\ v_6 \\ \psi_6 \end{Bmatrix}.$$

This type of coordinate transformation equation can be used to advantage in the above virtual work expression, Eq. (3.2). Substituting for the element DOF vectors, while recalling that transposing a matrix product requires reversing the product order, yields

$$\begin{aligned} \delta U &= \sum_{j=1}^{5} \lfloor \delta q \rfloor [T_j]^t [k_j][T_j]\{q\} = \sum_{j=1}^{5} \lfloor \delta q \rfloor [K_j]\{q\} \\ &= \lfloor \delta q \rfloor \left(\sum_{j=1}^{5} [K_j] \right) \{q\}, \end{aligned}$$

where each $[K_j]$ is, in this case, a 12×12 matrix. The sum of those five 12×12 matrices is another 12×12 matrix. Call that sum simply $[K]$, the system stiffness matrix. Therefore the above equation for the internal strain energy of the entire structure can be written as

$$\delta U = \lfloor \delta q \rfloor [K]\{q\}. \tag{3.3}$$

Note that each $[K_j]$ is a symmetric matrix because the corresponding $[k_j]$ is symmetric. Since the sum of symmetric matrices is symmetric, so too is $[K]$. The symmetry of the structural stiffness matrix is a reflection of Maxwell's reciprocity theorem [3.1,3.2].

Even when the structural system being analyzed requires just hundreds of global DOF, the above transformation matrices $[T_j]$, and the expanded element stiffness matrices $[K_j]$, would consume enormous computer storage. Fortunately, the actual creation of those transformation matrices is entirely unnecessary. To see how the structural stiffness matrix can be created without resort to transformation matrices, return to the above illustration for beam element 30. In more concise *sub*matrix form, the above expanded form of the element stiffness matrix, $[K_{30}]$, can be rewritten as

$$[K_{30}] = [T_{30}]^t[k_{30}][T_{30}] = \begin{bmatrix} 0 \\ I \\ 0 \end{bmatrix} [k_{30}][0 \quad I \quad 0] = \begin{bmatrix} 0 & 0 & 0 \\ 0 & k_{30} & 0 \\ 0 & 0 & 0 \end{bmatrix},$$

where each of the submatrices is 4×4. If the same thing is done for finite element 10, the result is

$$[K_{10}] = [T_{10}]^t[k_{10}][T_{10}] = \begin{bmatrix} I \\ 0 \\ 0 \end{bmatrix} [k_{10}][I \quad 0 \quad 0] = \begin{bmatrix} k_{10} & 0 & 0 \\ 0 & 0 & 0 \\ 0 & 0 & 0 \end{bmatrix}$$

and therefore

$$[K_{10} + K_{30}] = \begin{bmatrix} k_{10} & 0 & 0 \\ 0 & k_{30} & 0 \\ 0 & 0 & 0 \end{bmatrix}.$$

Hence, from these examples, the following important conclusion. The structural stiffness matrix can be built entry by entry by simply positioning each individual element stiffness matrix entry in the global stiffness matrix at the row and column position that corresponds to the associated global degrees of freedom. Hence, rather than use any of those unwieldy coordinate transformations, the system stiffness matrix is always constructed in this direct superposition fashion. In the case of the single beam being discussed, the final matrix sum $[K_{10} + K_{20} + K_{30} + K_{40} + K_{50}] = [K]$ is one where $[K_{20}]$ overlaps $[K_{10}]$ and $[K_{30}]$ because the global DOF for those elements overlap, and indeed all the elements overlap in a sequential case. To clearly see this type of superposition result in a simple form, let the EI and ℓ of each of the five finite elements of the single beam be the same. Then placing each of the element stiffness matrix entries in their proper position in the global stiffness matrix, and adding superimposed terms, leads to

$$[K]\{q\} = \frac{EI}{\ell^3} \begin{bmatrix} 12 & 6\ell & -12 & 6\ell & 0 & 0 & 0 & 0 & 0 & 0 & 0 & 0 \\ 6\ell & 4\ell^2 & -6\ell & 2\ell^2 & 0 & 0 & 0 & 0 & 0 & 0 & 0 & 0 \\ -12 & -6\ell & 24 & 0 & -12 & 6\ell & 0 & 0 & 0 & 0 & 0 & 0 \\ 6\ell & 2\ell^2 & 0 & 8\ell^2 & -6\ell & 2\ell^2 & 0 & 0 & 0 & 0 & 0 & 0 \\ 0 & 0 & -12 & -6\ell & 24 & 0 & -12 & 6\ell & 0 & 0 & 0 & 0 \\ 0 & 0 & 6\ell & 2\ell^2 & 0 & 8\ell^2 & -6\ell & 2\ell^2 & 0 & 0 & 0 & 0 \\ 0 & 0 & 0 & 0 & -12 & -6\ell & 24 & 0 & -12 & 6\ell & 0 & 0 \\ 0 & 0 & 0 & 0 & 6\ell & 2\ell^2 & 0 & 8\ell^2 & -6\ell & 2\ell^2 & 0 & 0 \\ 0 & 0 & 0 & 0 & 0 & 0 & -12 & -6\ell & 24 & 0 & -12 & 6\ell \\ 0 & 0 & 0 & 0 & 0 & 0 & 6\ell & 2\ell^2 & 0 & 8\ell^2 & -6\ell & 2\ell^2 \\ 0 & 0 & 0 & 0 & 0 & 0 & 0 & 0 & -12 & -6\ell & 12 & -6\ell \\ 0 & 0 & 0 & 0 & 0 & 0 & 0 & 0 & 6\ell & 2\ell^2 & -6\ell & 4\ell^2 \end{bmatrix} \begin{Bmatrix} v_1 \\ \psi_1 \\ v_2 \\ \psi_2 \\ v_3 \\ \psi_3 \\ v_4 \\ \psi_4 \\ v_5 \\ \psi_5 \\ v_6 \\ \psi_6 \end{Bmatrix}.$$

The process of positioning the element stiffness matrix entries in the global stiffness matrix is called *assembling* the global stiffness matrix. The above single-beam example produces the typical result that the nonzero matrix entries are clustered around

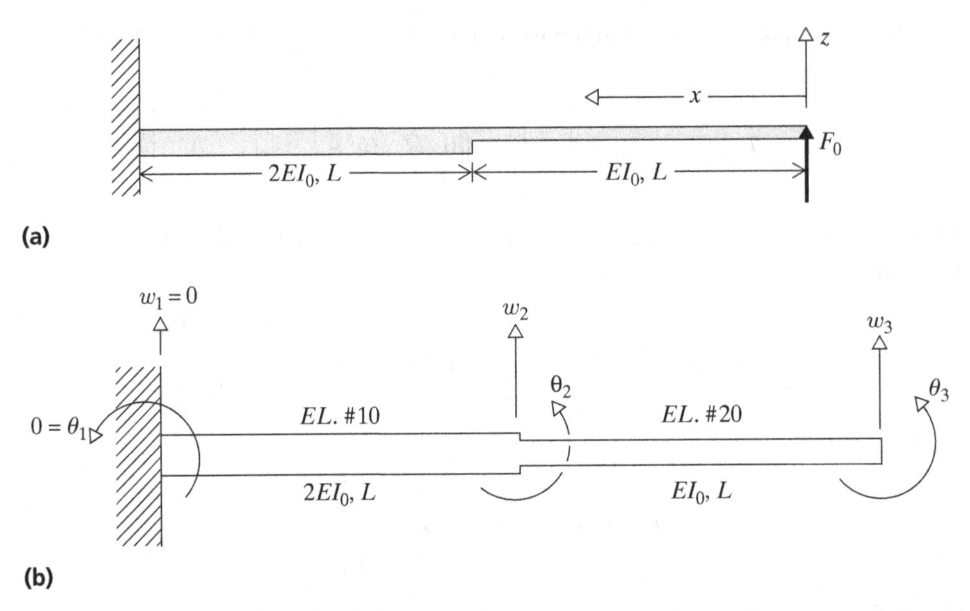

Figure 3.3. Example 3.1: (a) Analysis model of nonuniform, cantilevered beam with tip load, F_0. (b) Global generalized coordinates for the corresponding two-element FEM model.

the matrix's main diagonal. Such matrices are called *banded*. All large structural systems have stiffness matrices that exhibit this characteristic, at least in part.

Beyond the assembling of the global stiffness matrix, there are two other aspects of the finite element method that need attention even in a static load case. These two items are the creation of the global load vector and the use of the system boundary conditions (BCs). These aspects of the FEM are examined in the following example problem.

EXAMPLE 3.1 Consider the tip-loaded, nonuniform, cantilevered beam shown in Figure 3.3(a). As always, let the y and z axes be the principal axes of the beam cross sections. Write the matrix equation $\{Q\} = [K]\{q\}$, whose solution is the selected DOF for this structural system.

SOLUTION The finite element model for this structure that achieves engineering accuracy depends on the loading. In this case of just a static tip force F_0, the finite element model need not be more complicated than the two-element model shown in Figure 3.3(b), which accounts for the piecewise change in the beam cross section at midspan. However, if there were a static external loading per unit length of the beam structure, then at least six, possibly eight, beam finite elements would be required for engineering accuracy for such a structure rather than just the two finite elements used here. Along with the process of choosing the number of finite elements is the process of choosing an appropriate set of global DOF. Again, because this example problem is simply a case of beam bending in the plane of the paper under the action of a single, concentrated applied load, the six-system DOF shown in Figure 3.3(b) are wholly sufficient for these two finite elements. Note that the selected directions of these global DOF reflect those of the element DOF. That is, there is a global degree of freedom concurrent with each element DOF and vice versa. However, a global DOF can correspond to more than one element DOF as is seen in the case of global DOF w_2 and θ_2. Again, whereas the element DOF are unchanging for a particular type of element, the number of system DOF changes with the refinement of the mathematical model chosen to represent the structure.

The first step toward writing the requested matrix equation $\{Q\} = [K]\{q\}$ is to determine the system stiffness matrix $[K]$. To this end, the precise correspondence of the element DOF to the global DOF is simply established by inspecting the global FEM model of Figure 3.3(b) so as to determine which global DOF appear where the element DOF would be in the (4×1) element DOF vector. Hence in terms of the global DOF, the element 10 the virtual strain energy, or its negative, the internal virtual work, is

$$\delta U_{10} = \lfloor \delta q \rfloor [k_{10}]\{q\}$$

$$= \lfloor \delta w_1 \quad \delta\theta_1 \quad \delta w_2 \quad \delta\theta_2 \rfloor \frac{2EI_0}{L^3} \begin{bmatrix} 12 & 6L & -12 & 6L \\ 6L & 4L^2 & -6L & 2L^2 \\ -12 & -6L & 12 & -6L \\ 6L & 2L^2 & -6L & 4L^2 \end{bmatrix} \begin{Bmatrix} w_1 \\ \theta_1 \\ w_2 \\ \theta_2 \end{Bmatrix}.$$

In the hand calculations that follow, it is convenient to keep just EI_0/L^3 as a factor for all the stiffness matrices in anticipation of combining those matrices. Therefore, insert the above factor 2 into the element stiffness matrix so as to obtain

$$\delta U_{10} = \lfloor \delta w_1 \quad \delta\theta_1 \quad \delta w_2 \quad \delta\theta_2 \rfloor \frac{EI_0}{L^3} \begin{bmatrix} 24 & 12L & -24 & 12L \\ 12L & 8L^2 & -12L & 4L^2 \\ -24 & -12L & 24 & -12L \\ 12L & 4L^2 & -12L & 8L^2 \end{bmatrix} \begin{Bmatrix} w_1 \\ \theta_1 \\ w_2 \\ \theta_2 \end{Bmatrix}.$$

As for element 20, also in terms of the system DOF

$$\delta U_{20} = \lfloor \delta w_2 \quad \delta\theta_2 \quad \delta w_3 \quad \delta\theta_3 \rfloor \frac{EI_0}{L^3} \begin{bmatrix} 12 & 6L & -12 & 6L \\ 6L & 4L^2 & -6L & 2L^2 \\ -12 & -6L & 12 & -6L \\ 6L & 2L^2 & -6L & 4L^2 \end{bmatrix} \begin{Bmatrix} w_2 \\ \theta_2 \\ w_3 \\ \theta_3 \end{Bmatrix}.$$

The next step can be either to assemble the system stiffness matrix or to apply the system BCs. In a commercial FEM program, the BCs would be applied as the last step before the solution to the resulting force–deflection simultaneous equations. This is so because BCs generally cannot be accurately estimated in many cases. Therefore the analyst may wish to examine the results of several BC choices. Then it becomes more efficient to do the BCs last. However, in hand calculations such as these, there is less bookkeeping if the BCs are done first and then the global stiffness matrix is assembled. In this example, the BCs, which are always stated in terms of the global DOF, are

$$w_1 = \theta_1 = 0.$$

These boundary conditions have no effect on the second element. For the first element, substitute the above boundary conditions into each of the two DOF vectors in the element 10 strain energy expression. The multiplication of the two zeros of the two DOF vectors ($[\delta q]$ and $\{q\}$) effectively eliminate the first two rows and first two columns of that stiffness matrix. Discarding these rows and columns, the virtual strain energy of the first beam element can be rewritten more concisely as

$$\delta U_{10} = \lfloor \delta w_2 \quad \delta\theta_2 \rfloor \frac{EI_0}{L^3} \begin{bmatrix} 24 & -12L \\ -12L & 8L^2 \end{bmatrix} \begin{Bmatrix} w_2 \\ \theta_2 \end{Bmatrix}.$$

All the preparation necessary to obtain the global stiffness matrix is now complete. It only remains to superimpose the stiffness matrices for the two elements. For example, the stiffness term in the w_2 row and the θ_2 column of the stiffness matrix for element number 10 is added to the corresponding term of the stiffness matrix of element number 20, and so on. The final result, the global stiffness matrix and displacement vector, is

$$[K]\{q\} = \frac{EI_0}{L^3} \begin{bmatrix} 36 & -6L & -12 & 6L \\ -6L & 12L^2 & -6L & 2L^2 \\ -12 & -6L & 12 & -6L \\ 6L & 2L^2 & -6L & 4L^2 \end{bmatrix} \begin{Bmatrix} w_2 \\ \theta_2 \\ w_3 \\ \theta_3 \end{Bmatrix}.$$

Note that, just like the element stiffness matrices, the global stiffness matrix has to be symmetric, and all diagonal elements must be positive numbers. These two requirements can serve as weak checks on the correct hand assembly of the global stiffness matrix.

The next step is to create the load vector. This vector of generalized forces is obtained by writing the expression for the external virtual work, which is

$$\delta W_{ex} = 0\,\delta w_2 + 0\,\delta\theta_2 + F_0\,\delta w_3 + 0\,\delta\theta_3 = \lfloor \delta w_2 \quad \delta\theta_2 \quad \delta w_3 \quad \delta\theta_3 \rfloor \begin{Bmatrix} 0 \\ 0 \\ F_0 \\ 0 \end{Bmatrix}.$$

The above four-term description of the external virtual work is the most convenient form for this particular simple problem statement. Since the deflections at the clamped beam end have fixed zero values, they do not have a variation. Nevertheless, the force and moment reactions at the clamped beam end could have been included in this expression for the external virtual work for the sake of considering other boundary conditions at a later time.

Now that expressions have been developed above for the strain energy and the external virtual work, the next step is to combine the two on the basis $\delta W_{ex} = \delta U$. Again, this vital link can be viewed either as (i) one way of writing the principle of virtual work or (ii) in less sophisticated terms, as a variation on the equation of conservation of energy for the elastic system.[7] This identity $\delta W_{ex} = \delta U$ is detailed as $\lfloor \delta q \rfloor \{Q\} = \lfloor \delta q \rfloor [K]\{q\}$, or $\lfloor \delta q \rfloor (\{Q\} - [K]\{q\}) = 0$. Since the vector $\lfloor \delta q \rfloor$ of this two-term product is entirely arbitrary, the quantity in parentheses must be the source of the zero result. Therefore $\{Q\} = [K]\{q\}$. This last result now can be the object of all static FEM analyses without further discussion of virtual energies. In this example

[7] The principle of virtual work applies to energy dissipative systems as well.

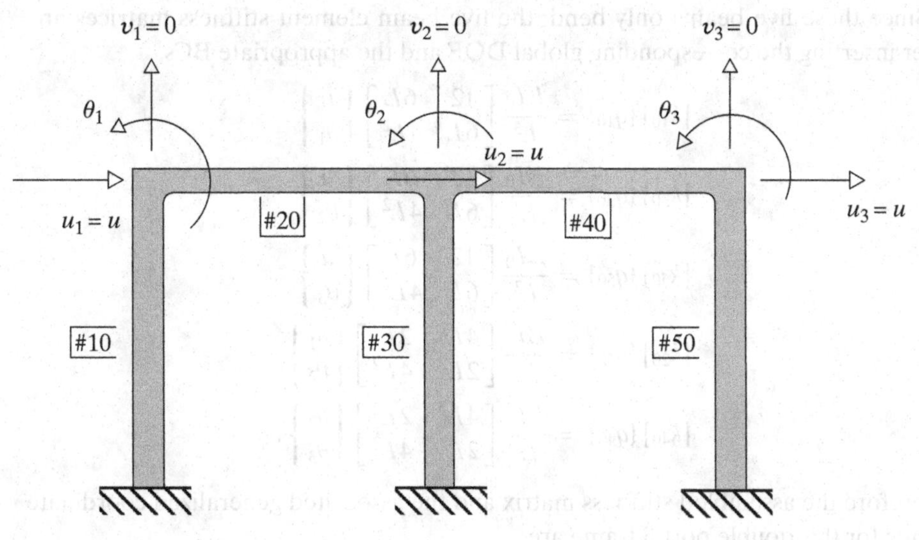

Figure 3.4. Example 3.2.

problem, the equation $\{Q\} = [K]\{q\}$ is a set of four simultaneous equations, in matrix form, that is now solved for the four entries in the DOF vector. In detail, this matrix equation is

$$\begin{Bmatrix} 0 \\ 0 \\ F_0 \\ 0 \end{Bmatrix} = \frac{EI_0}{L^3} \begin{bmatrix} 36 & -6L & -12 & 6L \\ -6L & 12L^2 & -6L & 2L^2 \\ -12 & -6L & 12 & -6L \\ 6L & 2L^2 & -6L & 4L^2 \end{bmatrix} \begin{Bmatrix} w_2 \\ \theta_2 \\ w_3 \\ \theta_3 \end{Bmatrix}.$$

★

EXAMPLE 3.2 Write the FEM matrix equation for the two-bay, one-story portal frame FEM model shown in Figure 3.4. Each of the five beam elements has the same length, L, and the same bending stiffness coefficient, EI_0. Let the external loading be forces (not shown) of magnitude F_0 acting to the left at each of the two upper corners of the frame where the DOF u_1 and u_3 are shown.

SOLUTION Planar beam frames are structures that bend only within their plane. Recall that long beams are much stiffer in stretching than bending.[8] Thus, as is customary for the linear analysis of beam frames, all beams in the frame are modeled to be rigid with respect to stretching. Thus, in this model, the stretching DOF v_1, v_2, v_3 are set equal to zero, and the nonzero translational bending DOF all have the same value; that is, $u_1 = u_2 = u_3 = u$. The three rotational (θ) DOF are unaffected by this constraint on axial rigidity. The use of the axial rigidity approximation not only greatly simplifies the problem, but it also helps avoid matrix ill-conditioning in large structures as a result of mixing the numerically small bending stiffnesses with the numerically, very large, axial stiffnesses, in the same stiffness matrix.

[8] For example, it is much easier to bend a common 12-in. ruler 1 in. than stretch the ruler 1/100 of an inch.

Since these five beams only bend, the five beam element stiffness matrices are, after inserting the corresponding global DOF and the appropriate BCs,

$$[k_{10}]\{q_{10}\} = \frac{EI_0}{L^3}\begin{bmatrix} 12 & 6L \\ 6L & 4L^2 \end{bmatrix}\begin{Bmatrix} u \\ \theta_1 \end{Bmatrix}$$

$$[k_{30}]\{q_{30}\} = \frac{EI_0}{L^3}\begin{bmatrix} 12 & 6L \\ 6L & 4L^2 \end{bmatrix}\begin{Bmatrix} u \\ \theta_2 \end{Bmatrix}$$

$$[k_{50}]\{q_{50}\} = \frac{EI_0}{L^3}\begin{bmatrix} 12 & 6L \\ 6L & 4L^2 \end{bmatrix}\begin{Bmatrix} u \\ \theta_3 \end{Bmatrix}$$

$$[k_{20}]\{q_{20}\} = \frac{EI_0}{L^3}\begin{bmatrix} 4L^2 & 2L^2 \\ 2L^2 & 4L^2 \end{bmatrix}\begin{Bmatrix} \theta_1 \\ \theta_2 \end{Bmatrix}$$

$$[k_{40}]\{q_{40}\} = \frac{EI_0}{L^3}\begin{bmatrix} 4L^2 & 2L^2 \\ 2L^2 & 4L^2 \end{bmatrix}\begin{Bmatrix} \theta_2 \\ \theta_3 \end{Bmatrix}.$$

Therefore the assembled stiffness matrix and the associated generalized coordinate vector for this double portal frame are

$$[K]\{q\} = \frac{EI_0}{L_3}\begin{bmatrix} 36 & 6L & 6L & 6L \\ 6L & 8L^2 & 2L^2 & 0 \\ 6L & 2L^2 & 12L^2 & 2L^2 \\ 6L & 0 & 2L^2 & 8L^2 \end{bmatrix}\begin{Bmatrix} u \\ \theta_1 \\ \theta_2 \\ \theta_3 \end{Bmatrix}.$$

The virtual work expression for the two leftward-acting, external forces located at the upper frame corners is

$$\delta W_{ex} = (-F_0)\delta u_1 + (-F_0)\delta u_3 = -2F_0\delta u.$$

Therefore the final force–deflection matrix equation is

$$\begin{Bmatrix} -2F_0 \\ 0 \\ 0 \\ 0 \end{Bmatrix} = \frac{EI_0}{L_3}\begin{bmatrix} 36 & 6L & 6L & 6L \\ 6L & 8L^2 & 2L^2 & 0 \\ 6L & 2L^2 & 12L^2 & 2L^2 \\ 6L & 0 & 2L^2 & 8L^2 \end{bmatrix}\begin{Bmatrix} u \\ \theta_1 \\ \theta_2 \\ \theta_3 \end{Bmatrix}.$$

Again, all the quantities in the generalized force vector are known, while all the quantities in the DOF vector are unknown. The solution of these four simultaneous equations for the unknown DOF permits, via the element stiffness matrices, the analyst to determine both the element deflections, the support reactions, and the internal bending moments and shearing forces and their corresponding stresses. ★

3.4 More Beam Element Stiffness Matrices

The previous section focuses on elastic bending of a beam axis in a single principal plane of the beam cross section. This section introduces beam twisting and beam extension about and along the same beam axis. Again, the creation of the associated element stiffness matrices is entirely a matter of familiar static force analysis. This unsophisticated approach does not compromise the usefulness of these stiffness

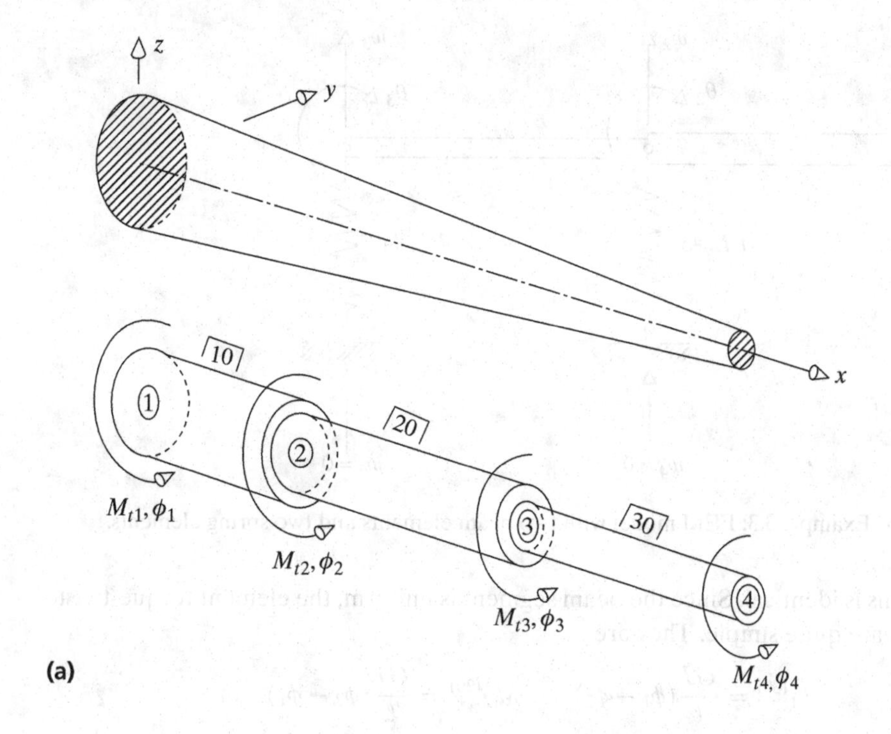

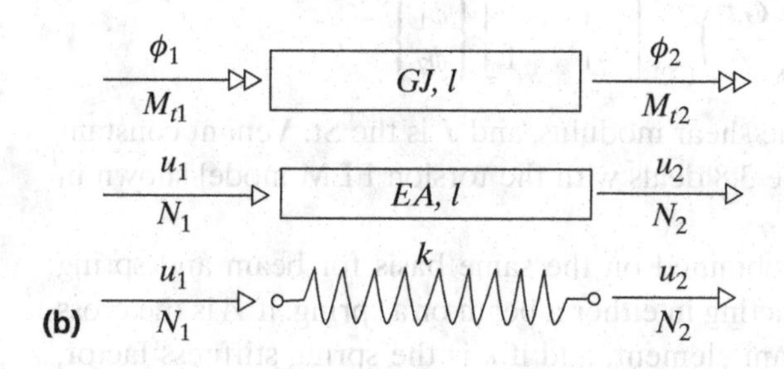

Figure 3.5. (a) Beam element modeling. (b) Sign conventions for beam element twisting and extension, and spring extension.

matrices in dynamic analyses at all. This is so because, as mentioned above, and as is seen in subsequent chapters, when the Lagrange equations are applied to a vibrating structure, the forces that are proportional to deflections, the elastic forces, represented by the description $[K]\{q\}$, are entirely separate from the forces that are dependent on velocities or accelerations.

Consider the twisting of a nonuniform beam subjected to a torque distributed over the length of the beam. Let the beam geometry be approximated by a series of just three uniform beam elements as illustrated in Figure 3.5(a). Let the entire distributed torsional loading acting on the beam be replaced by equivalent concentrated torques, M_t, at the four nodes. Let beam twisting finite element number 30, as shown in the first part of Figure 3.5(b), represent a typical such element. Note that, just as is the case for all elements, the sign convention for the forcelike quantities and for the

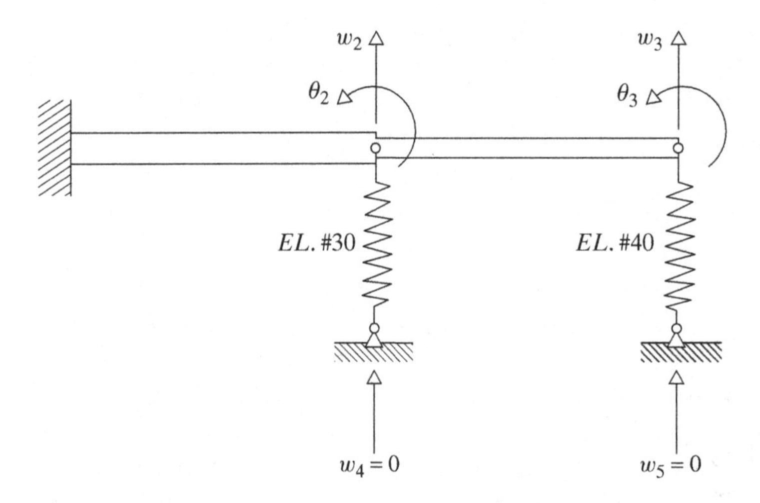

Figure 3.6. Example 3.3: FEM model with two beam elements and two spring elements.

deflections is identical. Since the beam segment is uniform, the element torque-twist relations are quite simple. They are

$$M_{t1}{}^{(30)} = \frac{GJ}{\ell}(\phi_1 - \phi_2) \qquad M_{t2}{}^{(30)} = \frac{GJ}{\ell}(\phi_2 - \phi_1) \quad \text{or}$$

$$\begin{Bmatrix} M_{t1} \\ M_{t2} \end{Bmatrix}^{(30)} = \left(\frac{GJ}{\ell}\right)_{(30)} \begin{bmatrix} 1 & -1 \\ -1 & 1 \end{bmatrix} \begin{Bmatrix} \phi_1 \\ \phi_2 \end{Bmatrix}^{(30)},$$

where ϕ is the beam twist, G is the shear modulus, and J is the St. Venant constant for uniform torsion [3.1]. Exercise 3.8 deals with the torsion FEM model shown in Figure 3.5.

Quite similar expressions are obtained on the same basis for beam and spring extension. If N is the axial force acting in either a beam or a spring, if A is the cross sectional area of the uniform beam element, and if k is the spring stiffness factor, then the beam extension and spring element stiffness relations are, on the same basis as above

$$\begin{Bmatrix} N_1 \\ N_2 \end{Bmatrix}^{(e)} = \left(\frac{EA}{\ell}\right)_{(e)} \begin{bmatrix} 1 & -1 \\ -1 & 1 \end{bmatrix} \begin{Bmatrix} u_1 \\ u_2 \end{Bmatrix}^{(e)}$$

$$\text{and} \quad \begin{Bmatrix} N_1 \\ N_2 \end{Bmatrix}^{(e)} = k_e \begin{bmatrix} 1 & -1 \\ -1 & 1 \end{bmatrix} \begin{Bmatrix} u_1 \\ u_2 \end{Bmatrix}^{(e)}.$$

Again, assembling element stiffness matrices such as these is illustrated in the following examples and later chapters.

EXAMPLE 3.3 Modify the two beam element structure of Figure 3.3, Example 3.1, by adding a vertically oriented, rectilinear spring with spring constant $5EI_0/L^3$ at the structure's midspan, and another such spring with spring constant $7EI_0/L^3$ at the free end of the structure, as shown in Figure 3.6. Then write the applied force–deflection matrix equation.

SOLUTION These two additions make the structure stiffer, that is, able to bear a larger load for the same deflection. This increase in the physical stiffness of the structure is, of course, reflected in the enhancement of the global stiffness matrix entries. Remember that these springs are just another pair of structural elements.[9] In this case, their boundary conditions, strain energies, and associated 1×1 (i.e., scalar) stiffness matrices, are

$$w_4 = w_5 = 0$$

$$\delta U_3 = \delta w_2 \frac{5EI_0}{L^3} w_2 \qquad \delta U_4 = \delta w_3 \frac{7EI_0}{L^3} w_3 .$$

Assembling the stiffness matrices of these additional elements in exactly the same way, that is, adding corresponding 1×1 matrix terms as determined by the global DOF (augmenting the (1,1) and (3,3) entries by 5 and 7, respectively) produces the new global stiffness matrix equation

$$\begin{Bmatrix} 0 \\ 0 \\ +F_0 \\ 0 \end{Bmatrix} = \frac{EI_0}{L^3} \begin{bmatrix} 41 & -6L & -12 & 6L \\ -6L & 12L^2 & -6L & 2L^2 \\ -12 & -6L & 19 & -6L \\ 6L & 2L^2 & -6L & 4L^2 \end{bmatrix} \begin{Bmatrix} w_2 \\ \theta_2 \\ w_3 \\ \theta_3 \end{Bmatrix}.$$

★

EXAMPLE 3.4 Using one beam finite element for each beam member, write the FEM force–deflection matrix equation for the two beam structure and loading sketched in Figure 3.7(a). For ease of hand assembly, let $GJ = \frac{1}{2}EI$.

SOLUTION The two beam structure shown in Figure 3.7(a) is called a beam "grid" because the loading causes the planar structure to deflect out of its plane. Specifically, the externally applied moment M_1 causes beam 10 to bend and beam 20 to twist, and M_2 causes beam 10 to twist and beam 20 to bend. Therefore, bending and twisting element stiffness matrices are needed for both beams.[10] Thus adopt the system DOF shown in Figure 3.7(b). Note that, for example, all ϕ vectors are directed in the same direction. Although not important in such a small problem, this uniformity of direction for all like named terms is convenient for interpreting computer solutions for large problems. The price worth paying for this uniformity is that ϕ is a twisting DOF for beam 10, whereas it is a bending slope DOF for beam 20. Otherwise, the selected global DOF are much like the various element DOF.

The element stiffness matrix for beam 10, after combining the bending and the twisting components and using the given relationship $GJ/L = (\frac{1}{2}L^2)(EI/L^3)$ and

[9] Structural spring elements are sometimes called "point elements" or "zero dimensional elements" because there is no spring length associated with the stiffness description.

[10] A commercial FEM program starts by automatically using all 12 beam DOF and then during the solution process deleting those DOF that are not load activated such as, in this example, the nodal translations in the y direction.

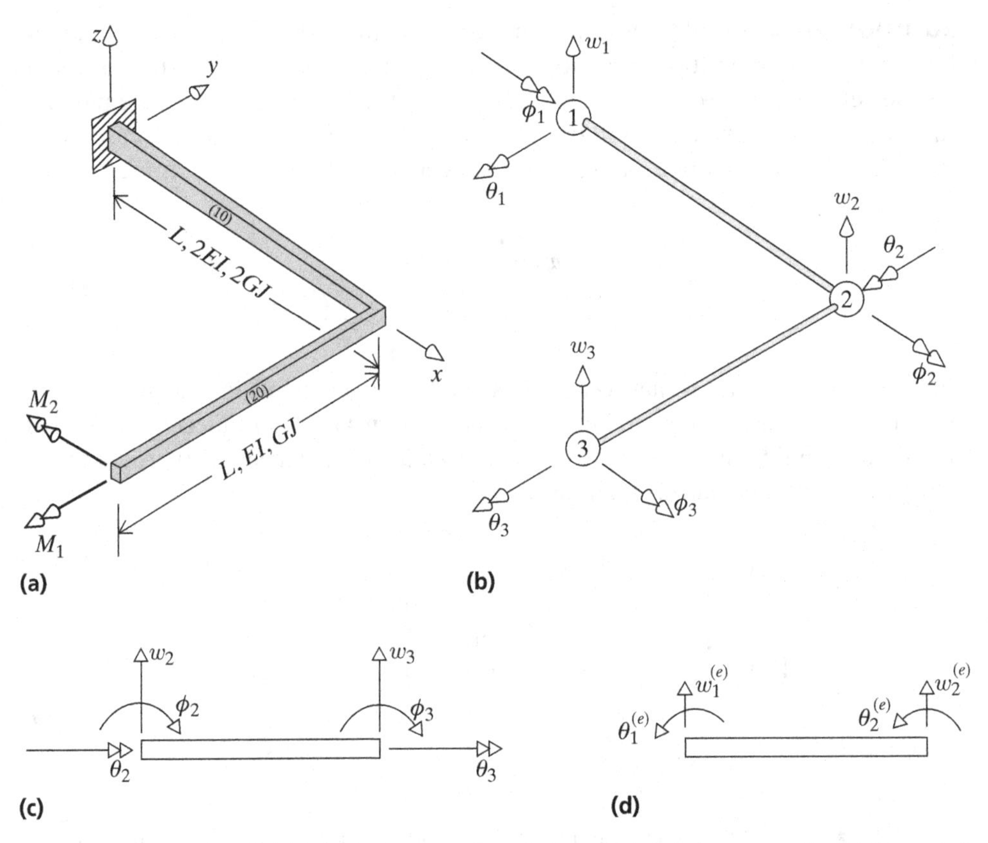

Figure 3.7. Example 3.4: (a) Simple beam grid analysis model. (b) Corresponding FEM model. (c) Beam 20 global coordinates. (d) Standard element coordinates for the x, z plane.

noting that the positive directions of the chosen global DOF are aligned with the positive directions of the element DOF

$$[k_{10}]\{q\} = \frac{EI}{L^3}\begin{bmatrix} 24 & -12L & 0 \\ -12L & 8L^2 & 0 \\ 0 & 0 & L^2 \end{bmatrix}\begin{Bmatrix} w_2 \\ \theta_2 \\ \phi_2 \end{Bmatrix}.$$

Beam 20 requires some care. As shown in Figure 3.7(c), when the lower numbered node of beam element 20 is placed on the left-hand side as per usual, the selected system beam bending slope DOF ϕ_2 and ϕ_3 are directed oppositely to the standard element beam bending slopes ψ of Figure 3.1(a), which is repeated for the x, z plane in Figure 3.7(d). To most easily explain the process for sorting out this circumstance where global DOF are opposite to element DOF, it is convenient to revert back to the expression for the beam virtual strain energy. For bending alone, the virtual strain energy expression for beam 20 in terms of the global DOF is

$$\delta U_{20} = \lfloor \delta w_2 \quad -\delta\phi_2 \quad \delta w_3 \quad -\delta\phi_3 \rfloor \frac{EI}{L^3}\begin{bmatrix} 12 & 6L & -12 & 6L \\ 6L & 4L^2 & -6L & 2L^2 \\ -12 & -6L & 12 & -6L \\ 6L & 2L^2 & -6L & 4L^2 \end{bmatrix}\begin{Bmatrix} w_2 \\ -\phi_2 \\ w_3 \\ -\phi_3 \end{Bmatrix}.$$

What is most convenient is to transfer the minus signs from the DOF vectors to their factors within the stiffness matrix. In this manner, the second and fourth rows and the second and fourth columns of the element stiffness matrix are multiplied by −1. This results in a checker board of sign changes that looks like

$$\begin{matrix} + & - & + & - \\ - & + & - & + \\ + & - & + & - \\ - & + & - & + \end{matrix}$$

Note that because there are no sign changes on the main diagonal all elements on that diagonal remain positive. The result is

$$\delta U_{20} = \lfloor \delta q \rfloor [k_{20}] \{q\} = \delta \lfloor w_2 \quad \phi_2 \quad w_3 \quad \phi_3 \rfloor \frac{EI}{L^3} \begin{bmatrix} 12 & -6L & -12 & -6L \\ -6L & 4L^2 & 6L & 2L^2 \\ -12 & 6L & 12 & 6L \\ -6L & 2L^2 & 6L & 4L^2 \end{bmatrix} \begin{Bmatrix} w_2 \\ \phi_2 \\ w_3 \\ \phi_3 \end{Bmatrix}.$$

Note that instead of having the minus signs on the third row and column, they now appear on the first row and column.

It is still necessary to add the twisting terms. Since the global twisting DOF are aligned with the element twisting DOF, it is necessary only to make the transition, as before, from GJ/L to $(EI/L^3)(\frac{1}{2}L^2)$. This twisting matrix equation by itself is then

$$[k_{20}]\{q\} = \frac{EI}{L^3} \begin{bmatrix} \frac{1}{2}L^2 & -\frac{1}{2}L^2 \\ -\frac{1}{2}L^2 & \frac{1}{2}L^2 \end{bmatrix} \begin{Bmatrix} \theta_2 \\ \theta_3 \end{Bmatrix}.$$

The external virtual work is

$$\delta W_{ex} = M_1 \delta\theta_3 - M_2 \delta\phi_3.$$

Now that the element matrices in terms of the global DOF, and the virtual work expression have been determined, the final system force–deflection matrix equation can be written. Again, the system stiffness matrix is obtained by mapping the above element matrices on the system matrix according to the row and column of the global DOF that multiply each element stiffness matrix entry. The final result is

$$\begin{Bmatrix} 0 \\ 0 \\ 0 \\ 0 \\ +M_1 \\ -M_2 \end{Bmatrix} = \frac{EI}{L^3} \begin{bmatrix} 36 & -12L & -6L & -12 & 0 & -6L \\ -12L & 8.5L^2 & 0 & 0 & -0.5L^2 & 0 \\ -6L & 0 & 5L^2 & 6L & 0 & 2L^2 \\ -12 & 0 & 6L & 12 & 0 & 6L \\ 0 & -0.5L^2 & 0 & 0 & 0.5L^2 & 0 \\ -6l & 0 & 2L^2 & 6L & 0 & 4L^2 \end{bmatrix} \begin{Bmatrix} w_2 \\ \theta_2 \\ \phi_2 \\ w_3 \\ \theta_3 \\ \phi_3 \end{Bmatrix}.$$

★

EXAMPLE 3.5 Set up the FEM force–deflection matrix equation for the two beam grid shown in Figure 3.8(a) using the DOF shown in Figure 3.8(b). Again let $GJ = \frac{1}{2}EI$.

SOLUTION Here the challenge is dealing with beams of different lengths. Specifically, for beam 10, everywhere the element stiffness matrix template requires a length ℓ

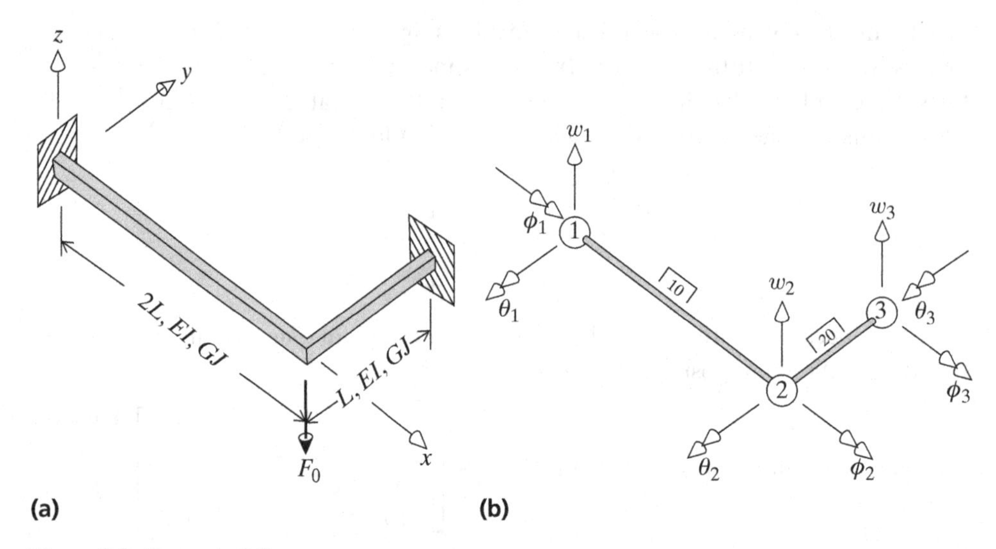

Figure 3.8. Example 3.5.

(i.e., both within the matrix and in its initial factor), the value $2L$ needs to be used. Therefore, using as always the same EI/L^3 initial factor for ease of hand assembly, the beam 10 stiffness matrix, after consolidating all factors, is

$$[k_{10}]\{q\} = \frac{EI}{L^3}\begin{bmatrix} 1.5 & -1.5L & 0 \\ -1.5L & 2L^2 & 0 \\ 0 & 0 & \tfrac{1}{4}L^2 \end{bmatrix}\begin{Bmatrix} w_2 \\ \theta_2 \\ \phi_2 \end{Bmatrix}.$$

Note that, in agreement with everyday experience, the longer the beam the lesser the stiffness.

In the case of this beam 20, the global bending DOF are all aligned in the positive directions of the element bending DOF, but the global twisting DOF are opposite to the positive directions of the element twisting DOF. Hence the twisting strain expression for beam 20 has the form

$$\delta U_{20} = \lfloor -\delta\theta_2 \quad -\delta\theta_3 \rfloor \frac{\tfrac{1}{2}EI}{L^3}\begin{bmatrix} L^2 & -L^2 \\ -L^2 & L^2 \end{bmatrix}\begin{Bmatrix} -\theta_2 \\ -\theta_3 \end{Bmatrix}.$$

Redeploying the negative signs from both DOF vectors into the square matrix leaves the square matrix unchanged. This is always the case for 2×2 matrices. Hence the complete beam 20 element stiffness matrix, after application of the right end BCs is

$$[k_{20}]\{q\} = \frac{EI}{L^3}\begin{bmatrix} 12 & 0 & 6L \\ 0 & \tfrac{1}{2}L^2 & 0 \\ 6L & 0 & 4L^2 \end{bmatrix}\begin{Bmatrix} w_2 \\ \theta_2 \\ \phi_2 \end{Bmatrix}.$$

Therefore the system force–deflection matrix equation to be solved simultaneously for the three unknown DOF is

$$\begin{Bmatrix} -F_0 \\ 0 \\ 0 \end{Bmatrix} = \frac{EI}{L^3}\begin{bmatrix} 13.5 & -1.5L & 6L \\ -1.5L & 2.5L^2 & 0 \\ 6L & 0 & 4.25L^2 \end{bmatrix}\begin{Bmatrix} w_2 \\ \theta_2 \\ \phi_2 \end{Bmatrix}.$$

★

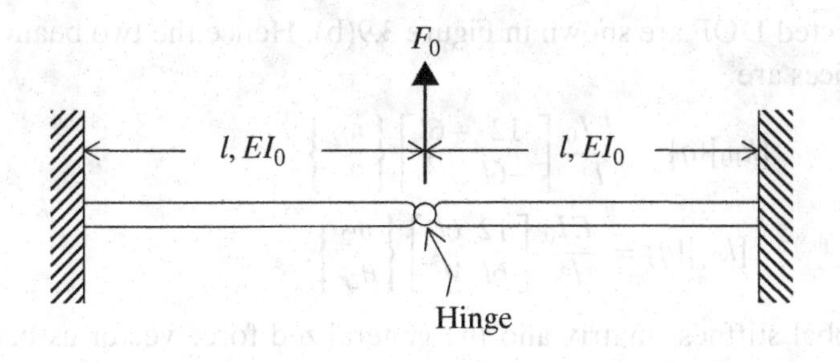

(a)

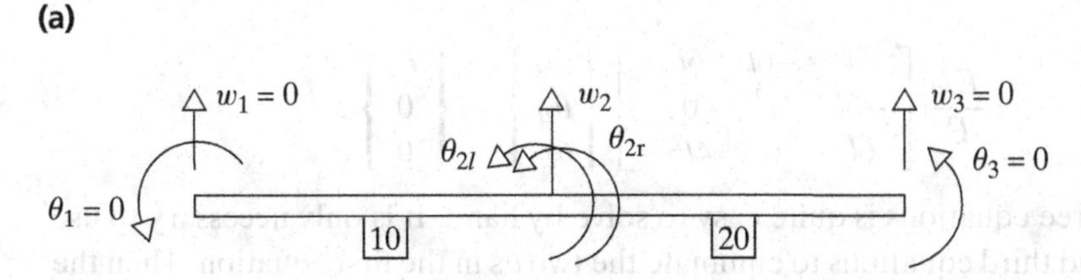

(b)

Figure 3.9. Example 3.7: (a) Analysis model. (b) FEM model.

EXAMPLE 3.6 Once again modify the beam structure of the previous example problem by adding a vertically oriented spring. Let this spring be attached either above or below node 2 and have a stiffness factor of $k = 5EI/L^3$.

SOLUTION The addition of the spring further enhances the stiffness (i.e., reduces the deflections) of the structure at node 2. All that is necessary to complete the alteration of the above system matrix equation is to add a 5 to the (1,1) entry in the stiffness matrix. The result is

$$\begin{Bmatrix} -F_0 \\ 0 \\ 0 \end{Bmatrix} = \frac{EI}{L^3} \begin{bmatrix} 18.5 & -1.5L & 6L \\ -1.5L & 2.5L^2 & 0 \\ 6L & 0 & 4.25L^2 \end{bmatrix} \begin{Bmatrix} w_2 \\ \theta_2 \\ \phi_2 \end{Bmatrix}. \quad \bigstar$$

Up to this point, all connections between beams are modeled as "rigid" connections, meaning that there is complete continuity of deflections and of bending slopes and of twists. Not all beam connections fit this description. If a connection has little rotational stiffness, then the engineering model might well represent the connection as a hinge. The next example problem shows the simple way in which the finite element method deals with such a hinge connection. Note that a hinge maintains continuity of deflections but not continuity of bending slopes. Thus there is only one deflection DOF at a hinge, but different bending slope DOF on each side of the hinge.

EXAMPLE 3.7 Determine the upward deflection at the hinge of the simple two beam structure shown in Figure 3.9(a).

SOLUTION The selected DOF are shown in Figure 3.9(b). Hence the two beam element stiffness matrices are

$$[k_{10}]\{q\} = \frac{EI_0}{l^3}\begin{bmatrix} 12 & -6l \\ -6l & 4l^2 \end{bmatrix}\begin{Bmatrix} w_2 \\ \theta_{2l} \end{Bmatrix}$$

$$[k_{20}]\{q\} = \frac{EI_0}{l^3}\begin{bmatrix} 12 & 6l \\ 6l & 4l^2 \end{bmatrix}\begin{Bmatrix} w_2 \\ \theta_{2r} \end{Bmatrix}.$$

Assembling the global stiffness matrix and the generalized force vector as before leads to

$$\frac{EI_0}{l^3}\begin{bmatrix} 24 & -6l & 6l \\ -6l & 4l^2 & 0 \\ 6l & 0 & 4l^2 \end{bmatrix}\begin{Bmatrix} w_2 \\ \theta_{2l} \\ \theta_{2r} \end{Bmatrix} = \begin{Bmatrix} F_0 \\ 0 \\ 0 \end{Bmatrix}.$$

This set of three equations is quite easy to solve by hand. It is only necessary to use the second and third equations to eliminate the two θs in the first equation. Then the solution result is

$$w_2 = \frac{F_0 l^3}{6EI_0}.$$

This solution can easily be checked on the basis that the tip stiffness of a cantilevered beam is $3EI_0/\ell^3$. Since there are two such cantilevered beams in this structure, their combined stiffness produces the above result. ★

The previous example problems were limited to structures that were either beams or beam frames, or, separately, beam grids. That is, all the previous problems were two dimensional. In addition, the loadings were limited to applied forces and moments. One of the important benefits provided by the finite element method is the ease with which it deals with three-dimensional structures and support motions. (The FEM also handles temperature changes with ease, but the slow pace of most temperature changes makes this topic of no interest for this textbook.) The following example is a simple illustration of such a capability relative to support motions.

EXAMPLE 3.8 Consider the three-beam grid/frame shown in Figure 3.10. The structural support at the right end has slowly shifted position since original construction so that the total shift is described by the two horizontal vector components, u_0 and v_0. This support motion deforms the structure. Therefore it produces the same effect as that produced by forces and moments. Using the FEM, write the force–displacement relationship for this structure. Use the usual approximation that the axial stiffnesses of the beams are infinite relative to the bending stiffnesses. The square cross sections are the same for all three beams, and thus the stiffness coefficient in each bending plane is simply EI_0. The torsional stiffness factor $GJ = \frac{1}{2}EI_0$.

SOLUTION The first task is to select the generalized coordinates for the structure. At each beam end there are, in general, three translational DOF and three rotational DOF for a total of six. Since the left-hand support is fixed, no DOF are required there. At the right-hand support, just the DOF corresponding to u_0 and v_0 are required. At the corners of the structure, the original twelve DOF are reduced by the zero

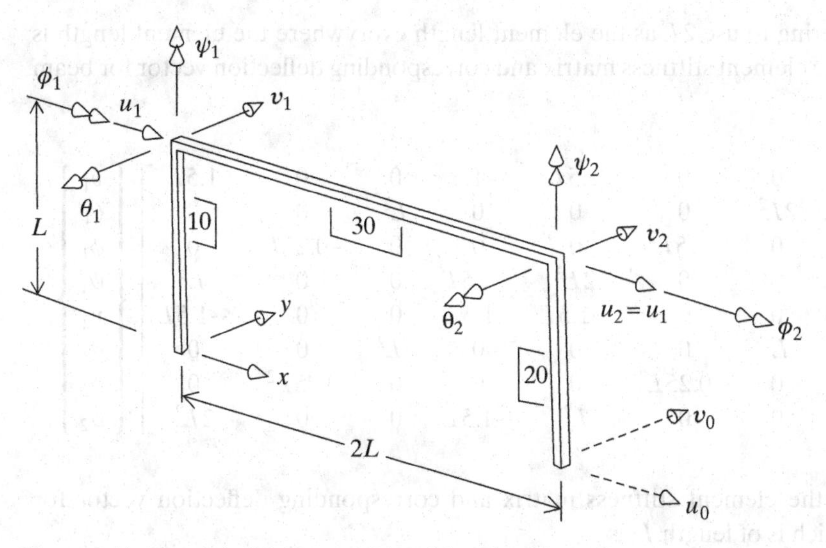

Figure 3.10. Example 3.8: Beam grid and frame subjected to a support movement.

axial deflection constraint. That is, because there can be no upward deflection at the corners, and because u_1 must equal u_2, these twelve DOF are reduced to nine. The nine DOF for the corners of the structure are illustrated in the figure.

Now it is a matter of writing the element stiffness matrices for each of the three beams labeled 10, 20, and 30. Consider beam-column 10. Note that for small deflections, the bending in the x, z plane is wholly independent of bending in the y, z plane, and those two sets of deflections are independent of the twisting of this beam. Thus the total element stiffness matrix for beam 10 is created by the superposition of the bending stiffness matrices for the two orthogonal planes along with the twisting stiffness matrix. To deal with any differences between the element DOF and the system DOF with regard to their respective positive directions, it is suggested that the reader make a separate sketch of each element of the given structure for each bending plane. Remember to keep the lower numbered node at the left end of the beam element.

When those three stiffness matrices are combined with the deflections at the left support set to zero, the result for beam-column 10 is as follows, where the beam element length is L:

$$[k_{10}]\{q\} = \frac{EI_0}{L^3}\begin{bmatrix} 12 & 0 & 6L & 0 & 0 \\ 0 & 12 & 0 & 6L & 0 \\ 6L & 0 & 4L^2 & 0 & 0 \\ 0 & 6L & 0 & 4L^2 & 0 \\ 0 & 0 & 0 & 0 & 0.5L^2 \end{bmatrix}\begin{Bmatrix} u_1 \\ v_1 \\ \theta_1 \\ \phi_1 \\ \psi_1 \end{Bmatrix}.$$

Note that because all the off-diagonal terms in the fifth row and fifth column are zero, there is indeed no interaction between the twisting DOF, ϕ, and these other four DOF. Furthermore, again because of the zero off-diagonal terms, this small deflection matrix requires that the deflection and bending slope in one plane have no interaction with those of the other plane.

Remembering to use $2L$ as the element length everywhere the element length is called for, the element stiffness matrix and corresponding deflection vector for beam 30 is

$$\frac{EI_0}{L^3}\begin{bmatrix} 1.5 & 0 & 0 & 1.5L & -1.5 & 0 & 0 & 1.5L \\ 0 & 2L^2 & 0 & 0 & 0 & L^2 & 0 & 0 \\ 0 & 0 & 0.25L^2 & 0 & 0 & 0 & -0.25L^2 & 0 \\ 1.5L & 0 & 0 & 2L^2 & -1.5L & 0 & 0 & L^2 \\ -1.5 & 0 & 0 & -1.5L & 1.5 & 0 & 0 & -1.5L \\ 0 & L^2 & 0 & 0 & 0 & 2L^2 & 0 & 0 \\ 0 & 0 & -0.25L^2 & 0 & 0 & 0 & 0.25L^2 & 0 \\ 1.5L & 0 & 0 & L^2 & -1.5L & 0 & 0 & 2L^2 \end{bmatrix}\begin{Bmatrix} v_1 \\ \theta_1 \\ \phi_1 \\ \psi_1 \\ v_2 \\ \theta_2 \\ \phi_2 \\ \psi_2 \end{Bmatrix}.$$

Similarly, the element stiffness matrix and corresponding deflection vector for beam 20, which is of length L, is

$$[k_{30}]\{q\} = \frac{EI_0}{L^3}\begin{bmatrix} 12 & 0 & 6L & 0 & 0 & -12 & 0 \\ 0 & 12 & 0 & 6L & 0 & 0 & -12 \\ 6L & 0 & 4L^2 & 0 & 0 & -6L & 0 \\ 0 & 6L & 0 & 4L^2 & 0 & 0 & -6L \\ 0 & 0 & 0 & 0 & 0.5L^2 & 0 & 0 \\ -12 & 0 & -6L & 0 & 0 & 12 & 0 \\ 0 & -12 & 0 & -6L & 0 & 0 & 12 \end{bmatrix}\begin{Bmatrix} u_1 \\ v_2 \\ \theta_2 \\ \phi_2 \\ \psi_2 \\ u_0 \\ v_0 \end{Bmatrix}.$$

The last two rows corresponding to the known values of the support deflections would provide the horizontal reactions at the right-hand support once the deflections are calculated. Since there is no present interest in those two reactions, discard those last two rows. Then, using the rules of matrix multiplication, subdivide this 5×7 matrix into a 5×5 matrix that involves only the unknown DOF, and two 5×1 matrices that involve only the known support movements

$$\frac{EI_0}{L^3}\begin{bmatrix} 12 & 0 & 6L & 0 & 0 \\ 0 & 12 & 0 & 6L & 0 \\ 6L & 0 & 4L^2 & 0 & 0 \\ 0 & 6L & 0 & 4L^2 & 0 \\ 0 & 0 & 0 & 0 & 0.5L^2 \end{bmatrix}\begin{Bmatrix} u_1 \\ v_2 \\ \theta_2 \\ \phi_2 \\ \psi_2 \end{Bmatrix} - \frac{EI_0 u_0}{L^3}\begin{Bmatrix} 12 \\ 0 \\ 6L \\ 0 \\ 0 \end{Bmatrix} - \frac{EI_0 v_0}{L^3}\begin{Bmatrix} 0 \\ 12 \\ 0 \\ 6L \\ 0 \end{Bmatrix},$$

The above 5×5 matrix is the element stiffness matrix for beam-column 20. As should be expected, it is exactly the same matrix as that for beam-column 10. The 5×1 matrices will soon be parts of the equivalent load vector as described below.

At this point the global stiffness matrix can be assembled from the above three element stiffness matrices. Note that because there are no externally applied loads directly corresponding to the nine retained DOF, the original external load vector is all zeros. Therefore, the equation involving the assembled stiffness matrix is

$$\begin{Bmatrix} 0 \\ 0 \\ 0 \\ 0 \\ 0 \\ 0 \\ 0 \\ 0 \\ 0 \end{Bmatrix} = \frac{EI}{L^3} \begin{bmatrix} 24 & 0 & 6L & 0 & 0 & 0 & 6L & 0 & 0 \\ 0 & 13.5 & 0 & 6L & 1.5L & -1.5 & 0 & 0 & 1.5L \\ 6L & 0 & 6L^2 & 0 & 0 & 0 & L^2 & 0 & 0 \\ 0 & 6L & 0 & 4.25L^2 & 0 & 0 & 0 & -0.25L^2 & 0 \\ 0 & 1.5L & 0 & 0 & 2.5L^2 & -1.5L & 0 & 0 & L^2 \\ 0 & -1.5 & 0 & 0 & -1.5L & 13.5 & 0 & 6L & -1.5L \\ 6L & 0 & L^2 & 0 & 0 & 0 & 6L^2 & 0 & 0 \\ 0 & 0 & 0 & -0.25L^2 & 0 & 6L & 0 & 4.25L^2 & 0 \\ 0 & 1.5L & 0 & 0 & L^2 & -1.5L & 0 & 0 & 2.5L^2 \end{bmatrix} \begin{Bmatrix} u_1 \\ v_1 \\ \theta_1 \\ \phi_1 \\ \psi_1 \\ v_2 \\ \theta_2 \\ \phi_2 \\ \psi_2 \end{Bmatrix} - \frac{EI_0}{L^3} \begin{Bmatrix} 12u_0 \\ 0 \\ 0 \\ 0 \\ 0 \\ 12v_0 \\ 6Lu_0 \\ 6Lv_0 \\ 0 \end{Bmatrix}.$$

Transposing the last vector on the right-hand side to the left-hand side completes the process of obtaining the $\{Q\} = [K]\{q\}$ equation. In this way it is clear that the new left-hand side vector is the "equivalent" or "effective" generalized force vector. Note that the units of each effective generalized force vector entry are those of force for the rows corresponding to a deflection, and the units are those of a moment for the rows corresponding to a rotation. ★

The next two chapters provide more example problems illustrating the use of FEM concepts. Therefore, any further practice needed in choosing DOF, creating system stiffness matrices from element stiffness matrices, and creating applied load vectors can be left to those examples found in subsequent chapters and, of course, to the exercises found in this and those chapters.

3.5 Summary

The dynamic equations central to this textbook are the Lagrange equations that were derived from Hamilton's principle. The static form of Hamilton's principle is obtained by setting the kinetic energy equal to zero and having the sole remaining quantity in the integrand, the virtual work, be independent of the time variable t. The result is that the total virtual work, which is the sum of the external virtual work, δW_{ex}, plus the internal virtual work, δW_{in}, is zero. This is called the principle of virtual work. When the internal forces are energy conservative, as they are for perfectly elastic bodies, then it is convenient to introduce a potential function for the elastic forces. This potential function, the negative of the work done by the internal elastic forces, is symbolized as U and is called the strain energy. Therefore, in terms of the strain energy, the principle of virtual work can be written in the form where the virtual work of the external loads is equal to the virtual change in the strain energy potential; that is, $\delta W_{ex} = \delta U$.

The linear finite element method of static structural analysis is most conveniently based on the principle of virtual work, and matrix algebra. The external virtual work, which is always calculated directly from the loads applied to the structural system, can be used to define quantities called the generalized forces, Q, according to the basic formulation $\delta W_{ex} = \lfloor \delta q \rfloor \{Q_{ex}\}$. This is just a reflection of the basic concept that forces moving through displacements, real or virtual, do work, real or virtual. As for the other part of the above statement of the principle of virtual work, the virtual strain energy is $\delta U = -\delta W_{in} = +\lfloor \delta q \rfloor \{Q_{elas}\} = \lfloor \delta q \rfloor [K]\{q\}$, where these internal elastic generalized forces are the previously discussed internal moments and shear forces at the finite element nodes. From the rules that govern the variational operator

$$\text{since} \quad \delta U = \lfloor \delta q \rfloor [K]\{q\}, \quad \text{then} \quad U = \frac{1}{2}\lfloor q \rfloor [K]\{q\}.$$

It is later shown that when this form of the strain energy is differentiated with respect to each of the generalized coordinates in turn, and the result cast in matrix form, the result is again that the generalized elastic forces acting on the finite element are $[K]\{q\}$.

Again, to focus on the dynamics of structures, this textbook almost always uses only the simplest of structural elements to create illustrative finite element models. Hence, only the small-displacement stiffness matrices of springs, bars, and beams are employed in the example problems offered here. Even then, because the choice is often sufficient for illustrative purposes, the corresponding structural elements are usually confined to being parts of planar structures. The analyst can be comfortable with these structural modeling simplifications only when he or she knows how to deal with the complications that are being avoided. The next several paragraphs discuss three slight complications that are generally avoided in subsequent chapters for the sake of focusing on the dynamics of the structures. In the order discussed, these slight complications are beams with nonsymmetrical cross sections, beams with shear flexibility, and, finally, finite elements arbitrarily oriented in three-dimensional space. The endnotes further extend the introduction to the FEM by briefly discussing two sophisticated aspects of the finite element method. Endnote (1) discusses the simplest of the two-dimensional elements, a thin rectangular slab. Note that this finite element can bear inplane tensile, compressive (but without buckling effects), and shearing loadings. As such, this sort of finite element could easily have been included as a wall or bulkhead in many of the elastic frame models discussed in this textbook. Endnote (2) serves two purposes. By looking at curved beam elements, Endnote (2) illustrates another approach to obtaining finite element stiffness matrices, and it also makes evident that even some one-dimensional elements can be complicated.

Of the three one-dimensional elements cited above as the elements that are used here to illustrate the various aspects of structural dynamics, (i.e., the spring, bar, and beam elements), the beam element is the most complicated. Beam bending has been, and will be, treated here on the basis that the y and z axes are the principal axes of the beam cross section, and shearing deformations can be completely neglected. However, commercial, digital computer, finite element programs make no such restrictions. First, commercial finite element programs generally allow the input of an area product of inertia I_{yz}, which is a measure of the lack of symmetry of the beam cross-section geometry with respect to the y and z axes. If there is just one axis of symmetry for the beam cross section, and either of the y or z axes is coincident with that axis of symmetry, then the product of inertia is zero. See Ref. [3.2], p. 250 ff. An angle iron, where the centoidal axes, the z and y axes, parallel the web and flange respectively, would be a case where the product of inertia would not be zero, but would be negative or positive depending on which direction, left or right, the flange extends. Because of the desire for simplicity in modeling the beam connections, the beam boundary conditions, it is usually not advantageous to rotate the convenient horizontal and vertical cross-sectional axes with a nonzero product of inertia to principal axes where the product of inertial is zero. Hence the need to know the formulation of the beam bending stiffness matrix when the product of inertia, I_{yz}, is not zero. The derivation proceeds from the virtual strain energy expression, Hooke's law, and engineering beam theory, also known as the Bernoulli–Euler hypothesis

$$\delta U \equiv \iiint \sigma_{xx}\delta\epsilon_{xx} d(\text{vol.})$$

$$\text{where} \quad \sigma_{xx} = -E[y\,v''(x) + z\,w''(x)]$$

$$\text{and} \quad \delta\epsilon_{xx} = -y\,\delta v''(x) - z\,\delta w''(x),$$

where, again $v(x)$ is the beam axis deflection in the y coordinate direction, and w is the same in the z coordinate direction. Substituting the stress and virtual strain expressions into the strain energy integral, and then carrying out the integration over the volume by first integrating over the area, where

$$I_{yy} = \iint z^2 dA \qquad I_{zz} = \iint y^2 dA \qquad I_{yz} = \iint yz\,dA$$

and using the same cubic spline shape functions set out in Eq. (3.1), leads to the 8×8 beam bending stiffness matrix below. For the sake of conciseness, let the 4×4 beam bending matrix without its EI/ℓ^3 coefficient be symbolized, at its first entry, as $[12\ 6\ell \ldots]$. Then the 8×8, small deflection, bending stiffness matrix for arbitrary orientations of the y and z axes, along with the corresponding vector of generalized coordinates, is

$$[k]\{q\} = \begin{bmatrix} \dfrac{EI_{zz}}{\ell^3}[12\ \ 6\ell\ldots] & \dfrac{EI_{yz}}{\ell^3}[12\ \ 6\ell\ldots] \\ \dfrac{EI_{yz}}{\ell^3}[12\ \ 6\ell\ldots] & \dfrac{EI_{yy}}{\ell^3}[12\ \ 6\ell\ldots] \end{bmatrix} \begin{Bmatrix} v_1 \\ \psi_1 \\ v_2 \\ \psi_2 \\ w_1 \\ \theta_1 \\ w_2 \\ \theta_2 \end{Bmatrix},$$

where again ψ is the bending slope deflection in the x, y plane, and θ is the same in the x, z plane.

Commercial finite element programs also allow for shearing as well as bending flexibility. As can be demonstrated from those few available theory of elasticity solutions for statically, laterally, loaded beams with rectangular cross sections, shearing deflections in long beams are negligible. However, a case can be made for the inclusion of shearing flexibility or shearing stiffness in vibratory analysis. The first reason for including shearing deformations is that, as is shown in detail in Figure 8.1, when a beam vibrates at frequencies that are multiples of what will be called its first natural frequency, the "bending length" of the long beam can be a fraction of the original length, making the long beam into a series of short beams. For that deflection pattern, the shearing deflection in those short beams generally will still be secondary to the bending deflection but now can be somewhat significant. That is, it is possible that the percentage of the total deflection attributable to shearing can exceed the standard allowance for engineering approximations for a given load, usually 7% or so. The second reason for including shearing deformations is that by so doing, the beam finite element is rendered more flexible, less stiff.[11] This is of value because the beam bending element stiffnesses described by the above beam stiffness matrices overestimate the actual beam stiffness. To understand this point, consider a simply supported beam subjected to a uniform lateral load per unit of beam length. The deflection solution for the beam is a quartic polynomial. A lateral loading that varied linearly over the length of the beam would result in a deflection solution that was a fifth-order polynomial. More complicated loadings would result in higher order polynomials or analytical functions whose power series expansions would involve an infinite number of distinct polynomial terms. These facts are in contrast to the above development of the beam bending stiffness matrix that depended on cubic polynomials over segments of the total beam length as approximations to actual deflection. Often times, for large, complex structures, the analyst selected element length is the same as the beam length. This situation results in a cubic polynomial approximating a higher order polynomial. This situation can be usefully viewed as a case where the coefficients of the polynomial terms above the third order in the expression representing the actual deflection pattern for the beam are forced to be zero by the finite element approximation. Requiring those deflection coefficients to be zero places constraints on the beam deflection just the same as adding another support to the simply supported beam under discussion. The added support also constrains the deflection, in this case the total deflection, to be zero at the added support position. It should be clear that the added constraint would lessen the overall deflections of the beam. Thus the added support makes the beam stiffer or less flexible. The addition of shearing deflections to the beam bending model partly offsets this increased stiffness.

All the previous developments involving beam bending deflections have been based on the Bernoulli–Euler approximation for beam bending, often called engineering beam theory. This approximation excludes shearing strains by requiring the cross sections to always remain perpendicular to the beam axis. On the basis of that excellent approximation, other than rigid body rotations, all beam element rotations, or more precisely, changes in those rotations (called curvatures), are directly related to bending moments by such familiar formulas as $EIw''(x) = M(x)$.

The simplest approach to including shearing deflections (i.e., shearing rotations due to shearing forces) in the beam element is to focus on the beam axis of a single element. The element nodes, with their w_1 and w_2 DOF, lie at each end of this axis. The shearing rotations can be separated from the bending rotations by simply requiring the element's beam axis to rotate from its original horizontal orientation while remaining straight (no curvatures). At the same time require that the cross-sectional areas retain their vertical orientation much like the individual cards of a stack of playing cards remain parallel to the tabletop as the stack is shifted from being straight to having an overhang. This geometry means that there is now other than

[11] Again, flexibility is the inverse of stiffness. For example, EA/ℓ is a stiffness factor, whereas ℓ/EA is a flexibility factor.

a 90° angle between each of the beam cross sections and the straight beam axis. By definition, this small change in the original 90° angle is the shearing strain. Therefore, in mathematical terms, the shearing strain is

$$\gamma = \frac{1}{\ell}(w_2 - w_1) = \frac{1}{\ell}\lfloor -1 \quad +1 \rfloor \begin{Bmatrix} w_1 \\ w_2 \end{Bmatrix}.$$

Using Hooke's law

$$\tau \equiv \sigma_{sh} = \frac{G}{\ell}\lfloor -1 \quad +1 \rfloor \begin{Bmatrix} w_1 \\ w_2 \end{Bmatrix}.$$

Hence the virtual strain energy expression is

$$\delta U = \iiint \tau \, \delta\gamma \, d(\text{vol.})$$
$$= \lfloor \delta w_1 \quad \delta w_2 \rfloor \iiint \frac{1}{\ell} \begin{Bmatrix} -1 \\ +1 \end{Bmatrix} \frac{G}{\ell} \lfloor -1 \quad +1 \rfloor dA \, dx \begin{Bmatrix} w_1 \\ w_2 \end{Bmatrix},$$

where the DOF vectors have been factored out of the volume integral because they do not vary with x, y, or z. Carrying out the integration yields

$$\delta U = \lfloor \delta w_1 \quad \delta w_2 \rfloor \frac{GA}{\ell} \begin{bmatrix} +1 & -1 \\ -1 & +1 \end{bmatrix} \begin{Bmatrix} w_1 \\ w_2 \end{Bmatrix}. \tag{3.4}$$

The above argument based on the approximate shearing strain at the beam axis is a simplification that ignores the fact that the shearing stress varies over the beam cross-sectional area and, therefore, so too does the shearing strain. That is, the shearing strain over the entire beam cross section cannot be simply determined, as above, from the use of the lateral translation DOF w_1 and w_2 at the beam axis. To make a long story short, from Ref. [3.2], p. 540, one possible adjustment is to apply a correction factor γ to the cross-sectional area A. Thus the expression used is

$$\delta U = \lfloor \delta w_1 \quad \delta w_2 \rfloor \frac{G(\gamma A)}{\ell} \begin{bmatrix} +1 & -1 \\ -1 & +1 \end{bmatrix} \begin{Bmatrix} w_1 \\ w_2 \end{Bmatrix}.$$

This nondimensional correction factor, γ, is generally close to a value of 1. For a rectangular cross section, it has the value 5/6. One general expression for thin cross–sections, from the same reference, is

$$\frac{1}{\gamma} = \frac{A}{I_{yy}^2} \int_{s=0}^{s=h} \frac{[Q_y(s)]^2}{t(s)} ds,$$

where A is the total cross-sectional area; s is an arc length coordinate that follows the centerline of the segments of the thin cross section; h is the maximum value of the arc length coordinate, the sum of all the lenths of all the flanges and webs; t is the local thickness, and Q is the same first moment of partial area that the reader first encountered in his or her elementary strength of materials course in the formula for beam shearing stress: $\tau = VQ/(It)$. Many commercial finite element programs provide an alternative and much simpler approach to entering the effective cross-sectional area for shear. To understand this alternative, consider a beam with a wide flange cross section, often called an H cross section. Consider the verticals of the H

to be the beam flanges, and the horizontal cross bar of the H to be the beam web. Let the y axis be coincident with the centerline of the web, and the z axis parallel to the flanges. For shearing forces acting in the y and z directions, the respective *effective* area inputs for the shearing stiffness matrix of Eq. (3.4) would then be the two quantities A_y and A_z. The former quantity, A_y, would be only the area of the flanges, or five-sixths thereof, whereas the latter quantity would be only the area of the web, or five-sixths thereof. These choices are made on the basis of the strength of materials solutions for shear stresses in thin beams. In those solutions, the dominant shear stresses are always aligned with the centerline of the thin flange or thin web.

The above shearing stiffness matrix generally has much larger values than the bending stiffness[12] matrix. To make a comparison that supports inclusion of the shearing stiffness terms, consider the δw_1 and w_1 entries of the bending and shearing stiffness matrices. The bending matrix term is $12EI/\ell^3$, whereas that for the shearing matrix is, ignoring the γ term near 1, GA/ℓ. To make a direct comparison, multiply the term GA/ℓ top and bottom, by ℓ^2. Then the ratio of the shearing stiffness term to the bending stiffness term is $GA\ell^2$ to $12EI$. If this ratio is applied to a long beam, the $GA\ell^2$ term would overwhelm the moment of inertia factor. Thus the beam would be much, much stiffer in shearing than bending. In such a case the beam would properly be modeled as having infinite shearing stiffness that is the same as saying that the shearing flexibility would not be included in the analysis. However, when the long beam is divided into several beam elements, the length of one of these finite elements is sometimes such that this disparity is only, say, 10:1 or 20:1. Then with E being roughly two and a half times larger than G, and taking into account the factor 12, it can be seen that in this somewhat extreme case the bending stiffness is greater than the shearing stiffness, the reverse of the normal situation. Then the shearing flexibility is greater than the bending flexibility and should be included. Therefore when using a commercial program, expect to find an opportunity for data input that includes the shear modulus even when no beam twisting is expected in the structure being modeled and data input for the cross-sectional areas (A, A_y, A_z) even when there is no significant beam extension is expected. However, again, for the sake of focusing on the dynamics of structures, this textbook will ignore beam axial and beam shear deflections, and all beam area products of inertia will be zero.

The third ordinary complication avoided in this textbook is arbitrary orientations of structural elements in three-dimensional space. Rather than use a beam for discussion purposes, the topic can be best illustrated using a bar element. The bar element sign convention and stiffness matrix equation in terms of the bar's two axial displacements u_1 and u_2 exactly mimic those of the spring where the spring stiffness factor k is replaced by the bar stiffness factor (EA/L). Note that because bars of pin-jointed trusses usually have constant cross sections and thus are constant force elements, there is no reason to subdivide a bar into more than one element. Thus L is almost always the full bar length. Hence, in the first of the two equations below, just like

[12] Recall that stiffness in the inverse of flexibility, and the greater the flexibility, the greater the deflections.

for the spring element, where again N is the bar axial force and u is the bar axial deflection:

$$\begin{Bmatrix} N_1 \\ N_2 \end{Bmatrix} = \frac{EA}{L} \begin{bmatrix} +1 & -1 \\ -1 & +1 \end{bmatrix} \begin{Bmatrix} u_1 \\ u_2 \end{Bmatrix} = \frac{EA}{L} \begin{bmatrix} +1 & 0 & -1 & 0 \\ 0 & 0 & 0 & 0 \\ -1 & 0 & +1 & 0 \\ 0 & 0 & 0 & 0 \end{bmatrix} \begin{Bmatrix} u_1 \\ v_1 \\ u_2 \\ v_2 \end{Bmatrix}.$$

The latter of the above two matrix equations is just the same as the first matrix equation but for the addition of generalized coordinates that describe deflections perpendicular to the bar axis. Note, for small displacements, there are no (shear) forces associated with those lateral, v, deflections, and the lateral deflections do not affect the axial forces.[13] The above 4×4 planar bar stiffness equation is not very useful in terms of the two deflections along the bar axis and the two deflections perpendicular to the bar axis unless all bars are either horizontal or vertical. What is needed is the bar stiffness matrix for a bar oriented in a plane at an arbitrary angle, or oriented in space by a set of angles, but written in terms of global DOF that parallel the selected Cartesian coordinate system for the planar or spatial truss. Such global DOF greatly simplify system assembly. To achieve such element stiffness equations, it is just a matter of writing rotation equations for the bar oriented DOF in terms of the horizontal and vertical system DOF. First for a planar truss bar element, where the element DOF are distinguished by an overbar, the global DOF are unmodified, and the angle β is measured counterclockwise from the (usually horizontal) direction of the global u degrees of freedom. That matrix rotation equation is

$$\begin{Bmatrix} \overline{u_1} \\ \overline{v_1} \\ \overline{u_2} \\ \overline{v_2} \end{Bmatrix} = \begin{bmatrix} \cos\beta & \sin\beta & 0 & 0 \\ -\sin\beta & \cos\beta & 0 & 0 \\ 0 & 0 & \cos\beta & \sin\beta \\ 0 & 0 & -\sin\beta & \cos\beta \end{bmatrix} \begin{Bmatrix} u_1 \\ v_1 \\ u_2 \\ v_2 \end{Bmatrix}.$$

Designate this 4×4 (planar) rotation matrix as $[R]$. Then, to obtain the bar element stiffness matrix in terms of the global DOF, substitute the above rotation equation into the virtual strain energy form of the above 4×4 element stiffness matrix equation. That is, write

$$\delta U_{el} = \lfloor \delta\overline{q} \rfloor [\overline{k_{el}}] \{\overline{q}\} = \lfloor \delta q \rfloor [R]^T [\overline{k_{el}}] [R] \{q\} = \lfloor \delta q \rfloor [k_{el}] \{q\}.$$

Therefore the element stiffness matrix equation for the arbitrarily oriented bar element in two-space is

$$\begin{Bmatrix} X_1 \\ Y_1 \\ X_2 \\ Y_2 \end{Bmatrix} = \frac{EA}{L} \begin{bmatrix} \cos^2\beta & \cos\beta\sin\beta & -\cos^2\beta & -\cos\beta\sin\beta \\ \cos\beta\sin\beta & \sin^2\beta & -\cos\beta\sin\beta & -\sin^2\beta \\ -\cos^2\beta & -\cos\beta\sin\beta & \cos^2\beta & \cos\beta\sin\beta \\ -\cos\beta\sin\beta & -\sin^2\beta & \cos\beta\sin\beta & \sin^2\beta \end{bmatrix} \begin{Bmatrix} u_1 \\ v_1 \\ u_2 \\ v_2 \end{Bmatrix},$$

[13] Although, as is stated by the given matrix equation, the bar axial force N due to the two stretching DOF is $EA(\Delta u/L)$, and is zero for the two lateral deflections, a more precise value for the axial force due to difference in the lateral deflections is easily worked out to be $N = 1/2EA(\Delta v/L)^2$. This latter term, because of the squaring, is usually quite small and hence negligible.

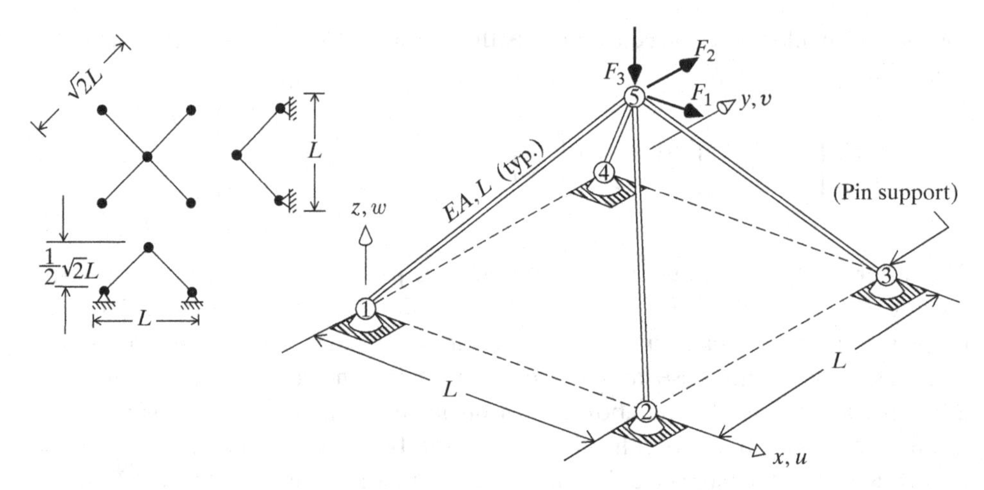

Figure 3.11. Example 3.9: Three-dimensional truss model whose geometry outlines a pyramid.

where X and Y are the bar force components in the u and v directions. The following example problem extends the above result to bars oriented in three-dimensional space and demonstrates the use of the subsequent 6×6 stiffness matrices.

EXAMPLE 3.9 Consider the four-bar, five-node, pin-jointed spatial truss, shown in Figure 3.11. The joint at the base of each of the four bars is fixed against translation. Each of the four faces of the pyramid whose outline is formed by the bars is an equilateral triangle. Hence each bar length is L. Each bar has a stiffness coefficient of EA. Begin the analysis by presuming that bar buckling is not a concern, and then write the global $\{Q\} = [K]\{q\}$ equation for this structure. After the forces in the bar are calculated, then the assumption of no buckling can be verified or adjusted. The element stiffness matrix for a bar in three-space in terms of the two sets of u, v, and w global DOF, where $\overline{x}$ is directed along the axis of the bar in question, is

$$[k_{el}]\{q\} = \frac{EA}{L}\begin{bmatrix} +C & -C \\ -C & +C \end{bmatrix}\begin{Bmatrix} u_1 \\ v_1 \\ w_1 \\ u_2 \\ v_2 \\ w_2 \end{Bmatrix},$$

where, for example, $(\bar{x}, z)$ represents the angle between the bar axis and the z axis, and

$$[C] = \begin{bmatrix} \cos^2(\overline{x}, x) & \cos(\overline{x}, x)\cos(\overline{x}, y) & \cos(\overline{x}, x)\cos(\overline{x}, z) \\ \cos(\overline{x}, x)\cos(\overline{x}, y) & \cos^2(\overline{x}, y) & \cos(\overline{x}, y)\cos(\overline{x}, z) \\ \cos(\overline{x}, x)\cos(\overline{x}, z) & \cos(\overline{x}, y)\cos(\overline{x}, z) & \cos^2(\overline{x}, z) \end{bmatrix}.$$

As was done in the two-dimensional case, this bar element stiffness matrix, using u, v, w DOF paralleling the convenient Cartesian coordinate axes x, y, z while the bar element itself has an arbitrary orientation in space, can be derived on the basis of relating, on the one hand, the three DOF that are oriented along the axis of the bar

($\overline{u}$) and perpendicular to the axis of the bar ($\overline{v}$, $\overline{w}$), with its known element stiffness matrix, to, on the other hand, the above convenient DOF u, v, w. The basis of the relationship at either of the bar element's two nodes is the fact that the total deflection at the node is the same vector in either the bar-oriented Cartesian coordinate system or the convenient Cartesian coordinate system. That is,

$$\overline{u}\overline{\boldsymbol{i}} + \overline{v}\overline{\boldsymbol{j}} + \overline{w}\overline{\boldsymbol{k}} = u\boldsymbol{i} + v\boldsymbol{j} + w\boldsymbol{k}.$$

Now the transformation matrix relating the two sets of DOF can be constructed by taking the dot product on both sides of the above equation with first $\overline{\boldsymbol{i}}$, then $\overline{\boldsymbol{j}}$, and then $\overline{\boldsymbol{k}}$. The result is

$$\overline{u} = u\cos(\overline{x}, x) + u\cos(\overline{x}, y) + w\cos(\overline{x}, z)$$

and so forth, or, for either node

$$\begin{Bmatrix} \overline{u} \\ \overline{v} \\ \overline{w} \end{Bmatrix} = \begin{bmatrix} \cos(\overline{x}, x) & \cos(\overline{x}, y) & \cos(\overline{x}, z) \\ \cos(\overline{y}, x) & \cos(\overline{y}, y) & \cos(\overline{y}, z) \\ \cos(\overline{z}, x) & \cos(\overline{z}, y) & \cos(\overline{z}, z) \end{bmatrix} \begin{Bmatrix} u \\ v \\ w \end{Bmatrix},$$

or, for both nodes

$$\{\overline{u}\} = [T_{bar}]\{u\} \quad \text{and} \quad \lfloor \overline{u} \rfloor = \lfloor u \rfloor [T_{bar}]^t.$$

The above coordinate transformation can now be applied to the bar's virtual strain energy expression

$$\delta U_{bar} = \lfloor \delta\overline{u} \rfloor [\overline{k}_{el}] \{\overline{u}\} = \lfloor \delta u \rfloor [T_{bar}]^t [\overline{k}_{el}][T_{bar}]\{u\},$$

where

$$[\overline{k}_{el}]\{\overline{u}\} = \begin{bmatrix} +1 & 0 & 0 & -1 & 0 & 0 \\ 0 & 0 & 0 & 0 & 0 & 0 \\ 0 & 0 & 0 & 0 & 0 & 0 \\ -1 & 0 & 0 & +1 & 0 & 0 \\ 0 & 0 & 0 & 0 & 0 & 0 \\ 0 & 0 & 0 & 0 & 0 & 0 \end{bmatrix} \begin{Bmatrix} \overline{u}_1 \\ \overline{v}_1 \\ \overline{w}_1 \\ \overline{u}_2 \\ \overline{v}_2 \\ \overline{w}_2 \end{Bmatrix}$$

and therefore $[k_{el}] = [T_{bar}]^t[\overline{k}_{el}][T_{bar}]$ as shown above.

Return now to the solution of the pyramid truss problem. To be clear on the geometry of the pyramid, for example, the cosines of the angles between the bar 15 axis and the x, y, and z directions are, respectively, $\cos(60°) = ½$, $\cos(60°) = ½$, $\cos(45°) = \sqrt{2}/2$. Note that these angles are not the same for all four bars.

SOLUTION The direction cosines for element 25 are $\cos(120°) = -½$, $\cos(60°) = ½$, $\cos(45°) = \sqrt{2}/2$. The direction cosines for element 35 are $\cos(120°) = -½$, $\cos(120°) = -½$, $\cos(45°) = \sqrt{2}/2$. The direction cosines for element 45 are $\cos(60°) = ½$, $\cos(120°) = -½$, $\cos(45°) = \sqrt{2}/2$. Hence the element stiffness matrices for the four bars in the global coordinate system are

$$[k_{15}]\{q\} = \frac{EA}{4L}\begin{bmatrix} 1 & 1 & \sqrt{2} \\ 1 & 1 & \sqrt{2} \\ \sqrt{2} & \sqrt{2} & 2 \end{bmatrix}\begin{Bmatrix} u_5 \\ v_5 \\ w_5 \end{Bmatrix}$$

$$[k_{25}]\{q\} = \frac{EA}{4L}\begin{bmatrix} 1 & -1 & -\sqrt{2} \\ -1 & 1 & \sqrt{2} \\ -\sqrt{2} & \sqrt{2} & 2 \end{bmatrix}\begin{Bmatrix} u_5 \\ v_5 \\ w_5 \end{Bmatrix}$$

$$[k_{35}]\{q\} = \frac{EA}{4L}\begin{bmatrix} 1 & 1 & -\sqrt{2} \\ 1 & 1 & -\sqrt{2} \\ -\sqrt{2} & -\sqrt{2} & 2 \end{bmatrix}\begin{Bmatrix} u_5 \\ v_5 \\ w_5 \end{Bmatrix}$$

$$[k_{45}]\{q\} = \frac{EA}{4L}\begin{bmatrix} 1 & -1 & \sqrt{2} \\ -1 & 1 & -\sqrt{2} \\ \sqrt{2} & -\sqrt{2} & 2 \end{bmatrix}\begin{Bmatrix} u_5 \\ v_5 \\ w_5 \end{Bmatrix}.$$

Hence the assembled global stiffness matrix equation is

$$\{Q\} = \begin{Bmatrix} F_1 \\ F_2 \\ -F_3 \end{Bmatrix} = [K]\{q\} = \frac{EA}{L}\begin{bmatrix} 1 & 0 & 0 \\ 0 & 1 & 0 \\ 0 & 0 & 2 \end{bmatrix}\begin{Bmatrix} u_5 \\ v_5 \\ w_5 \end{Bmatrix},$$

which is very easily solved for the joint 5 deflections in the global coordinate system. Then by use of the element stiffness matrix equation and the above global deflections, the components of the bar forces can be obtained. Then use of the direction cosines provides the bar axial forces. For example, from the above solution

$$u_5 = \frac{F_1 L}{EA} \quad v_5 = \frac{F_2 L}{EA} \quad w_5 = -\frac{F_3 L}{2EA}$$

the three components of the axial force in bar 15 are

$$\begin{Bmatrix} N_x \\ N_y \\ N_z \end{Bmatrix}^{(15)} = \frac{EA}{4L}\begin{bmatrix} 1 & 1 & \sqrt{2} \\ 1 & 1 & \sqrt{2} \\ \sqrt{2} & \sqrt{2} & 2 \end{bmatrix}\begin{Bmatrix} u_5 \\ v_5 \\ w_5 \end{Bmatrix} = \frac{1}{4}\begin{bmatrix} 1 & 1 & -\frac{\sqrt{2}}{2} \\ 1 & 1 & -\frac{\sqrt{2}}{2} \\ \sqrt{2} & \sqrt{2} & -1 \end{bmatrix}\begin{Bmatrix} F_1 \\ F_2 \\ F_3 \end{Bmatrix}.$$

Then, the axial force in bar 15 is

$$N = N_x \cos(\overline{x}, x) + N_y \cos(\overline{x}, y) + N_z \cos(\overline{x}, z) = \frac{1}{2}N_x + \frac{1}{2}N_y + \frac{\sqrt{2}}{2}N_z.$$

Once the bar forces are obtained, they can be compared to the Euler or elastoplastic buckling load for the bars. ★

The above example shows how simple three-dimensional structures are dealt with using hand calculations. Machine calculations, as might be expected, usually follow a different order. In most software, the analyst inputs the nodal locations throughout the structure, and the element types (e.g., beams, membranes, rigid elements, etc.) that connect those system nodes. Then the software calculates whatever element dimensions, such as bar length, that are required. So far, with one exception, all the example problem connections between beam structural elements have been "rigid" connections that provide complete continuity of all deflections. The exercises further consider other possibilities such as a hinged connections. Software generally, as a default option, identifies the system DOF at each node as the three deflections and three rotations at each node paralleling an analyst selected spatial coordinate system provided as an option by the software. These DOF may then be increased or reduced in number by the analyst by changing the rigid connection to, say, a hinged connection or by selecting only certain types of elements such as bars whereby the rotational DOF become superfluous. That is, the DOF are ascertained from a knowledge of the types of element DOF associated with each type of structural element that connects at that node and by whatever coding is required to alter a rigid connection. The analyst further inputs whatever additional mass, material, and geometric information is necessary to fully identify the structure under study. Then, with the further specification of the loads and support conditions, the computer numerically processes that information to obtain the desired solution often in terms of pages and pages, or screens and screens, of data printout and drawings. Since the quantity of solution data is often so large, the solution data may even have to be animated on the monitor screen so as to assist in its comprehension.

REFERENCES

3.1 Turner, M. J., R. W. Clough, H. C. Martin, and L. J. Topp, "Stiffness and deflection analysis of complex structures," *J. Aeronaut. Sci.* vol. 23, 9, 1956.

3.2 Donaldson, B. K., *Analysis of Aircraft Structures: An Introduction*, McGraw-Hill, New York, 1993.

CHAPTER 3 EXERCISES

Note: All beam loadings in this chapter, and all subsequent chapters, are to be understood to be acting within a plane of symmetry of the beam cross section.

3.1 **(a)** Consider a single, uniform, simply supported beam whose length is L and whose stiffness coefficient is EI. The beam is loaded by a single, concentrated, lateral force of magnitude F_0 placed at the center of the beam span. Using just one beam bending finite element for the left half of the beam, and another for the right half-span, determine the lateral deflection at midspan.

(b) Redo the previous problem, but this time let the boundary conditions be clamped supports at both beam ends. Note the percentage change in deflection between these two extremes of beam boundary condition modeling.

3.2 **(a)** As a variation on the above problem, consider the beam-spring structure shown in Figure 3.12(a), where the sole loading is an externally applied moment of magnitude M_0 at midspan. Using the DOF shown in Figure 3.12(b), write the full seven-DOF $\{Q\} = [K]\{q\}$ equation for this structural system and loading. Just use the generic symbol R to represent the support reactions. Then apply the boundary conditions to obtain the two DOF matrix equation to be solved for the unknown DOF.

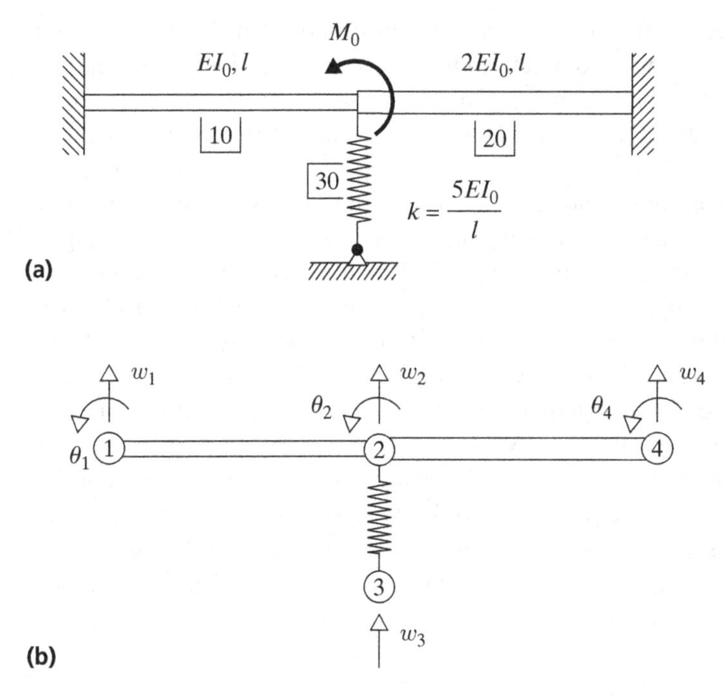

Figure 3.12. Exercise 3.2(a): (a) Math model. (b) FEM model.

(b) Consider the cantilevered beam structure that is shown in Figure 3.13. The structure is loaded by a tip force whose three components are F_1, F_2, and F_3. The beam of length 2ℓ is labeled beam 12 on the basis of its end nodes. Similarly, the upper beam of length ℓ is labeled beam 23. That figure also displays the corresponding finite element model and all the required DOF. Write the $\{Q\} = [K]\{q\}$ equation that can be solved for the unknown 10 DOF.

3.3 All eight beams shown in Figure 3.14 have the same length and have the same doubly symmetric cross section. Specifically, the four beam-columns of the structure shown in Figure 3.14 have the same bending stiffness coefficient, EI, about the x axis as the y axis. The torsional stiffness coefficient for all eight beams is $GJ = \frac{1}{2}EI$. The applied loading (not shown) consists of two sets of loads. The first set of equal magnitude loads are x-direction forces F_1 placed at the upper corners labeled 1 and 3. The second set of equal magnitude loads, which act along with the first set, are two y-direction forces F_2 placed at corners 1 and 2. For small deflections (as always), write the smallest size matrix equation $\{Q\} = [K]\{q\}$ that can be used to determine the deflections of this structure under this loading.

3.4 Consider the fixed-end-supported, three-beam-segment, two-spring structure loaded by two moment components M_1 and M_2 at the right-hand corner, and a lateral force F_1 at the left corner, as shown in Figure 3.15. If the numerical value of the torsional stiffness coefficient GJ for each beam segment is half that of the bending stiffness coefficient EI_0, write the final matrix equation $\{Q\} = [K]\{q\}$ that can be solved for the unknown DOF.

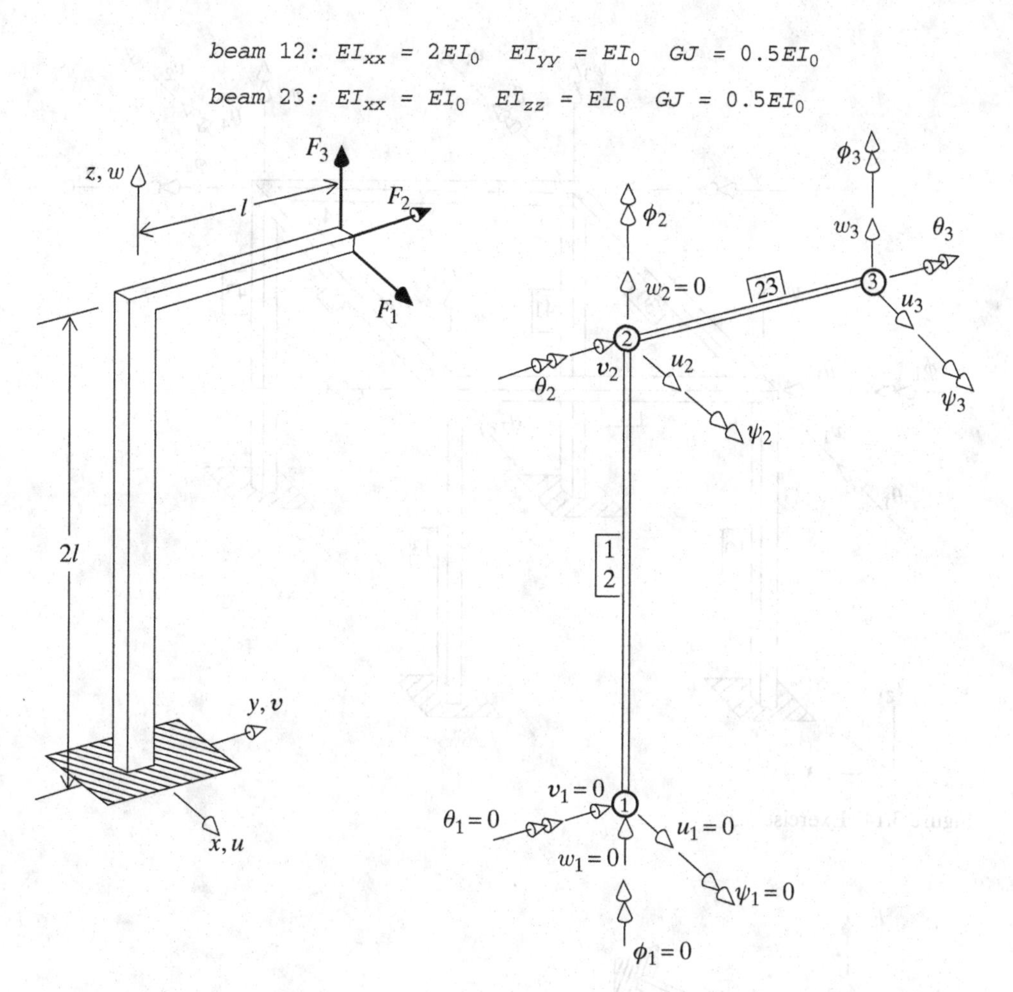

Figure 3.13. Exercise 3.2(b).

3.5 Consider the structural and FEM models of the small, two-dimensional, frame shown in Figure 3.16. The beams have been modeled as inextensible. Hence the only DOF required are as shown. Note that the base of the left-hand support is subjected to an enforced horizontal motion labeled $u_0(t)$. Let this horizontal enforced motion, $u_0(t)$ occur slowly so that no significant amount of kinetic energy is engendered. Therefore, for this small deflection analysis, only the maximum value of $u_0(t)$ is significant, and as a result, this is simply a static analysis. Write the matrix equation to be solved for the unknown global DOF.

3.6 Occasionally the stiffnesses of components of mechanisms are modeled as combinations of springs in series and springs in parallel. The above-described FEM allows the quick mathematical description of the overall stiffness properties of such a system without reference any rules about springs in parallel or series. To illustrate this statement, consider the three-DOF spring system of Figure 3.17, where horizontal motion only is possible. The system is shown in its displaced configuration. Show that

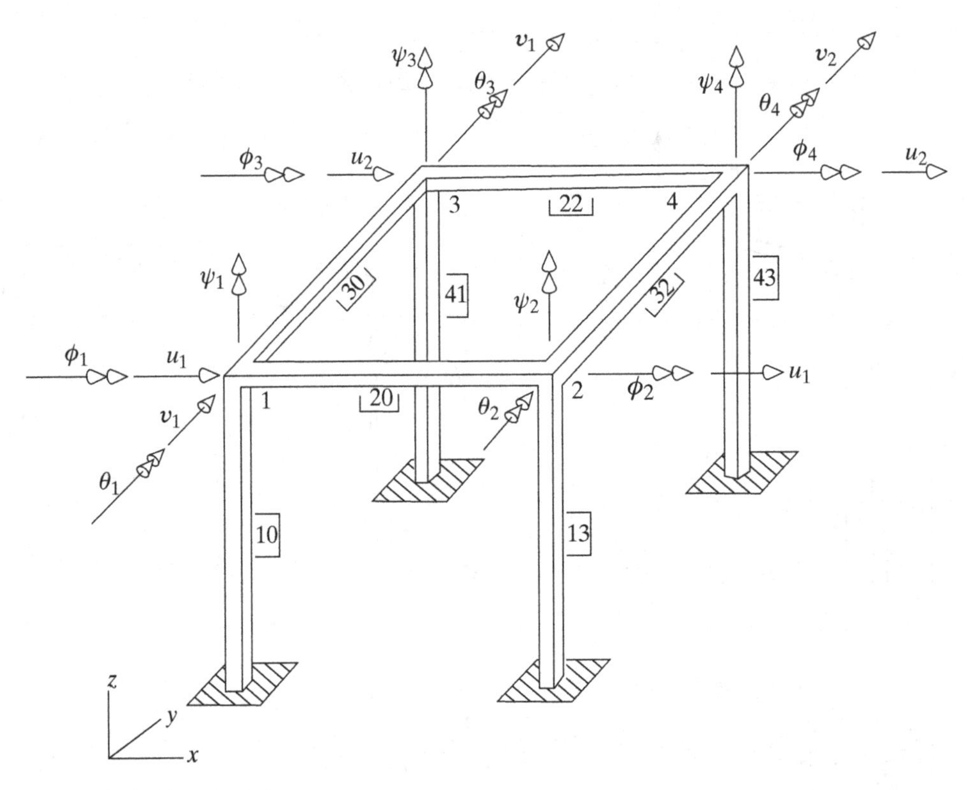

Figure 3.14. Exercise 3.3.

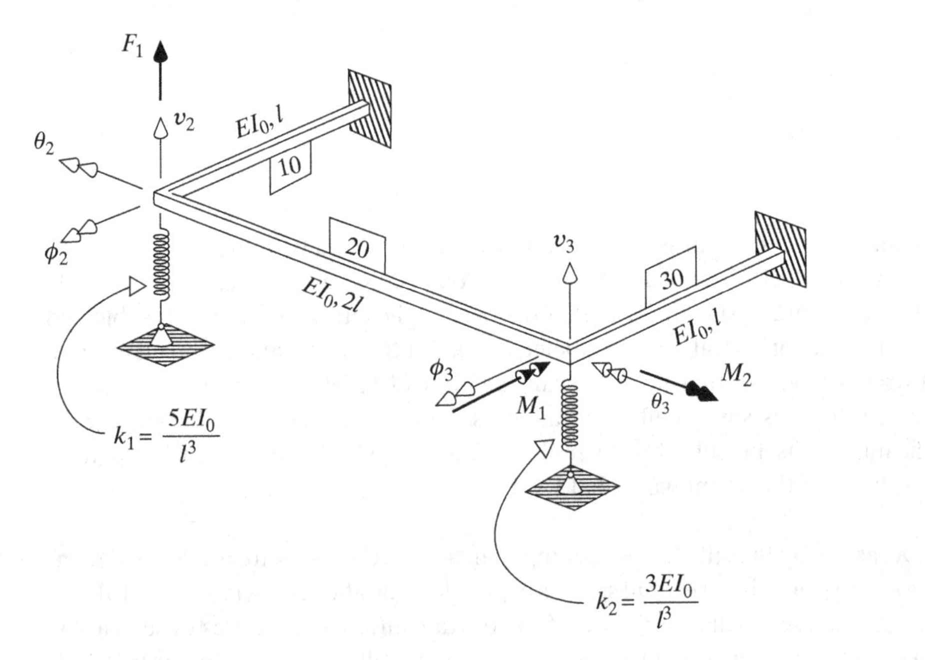

Figure 3.15. Exercise 3.4.

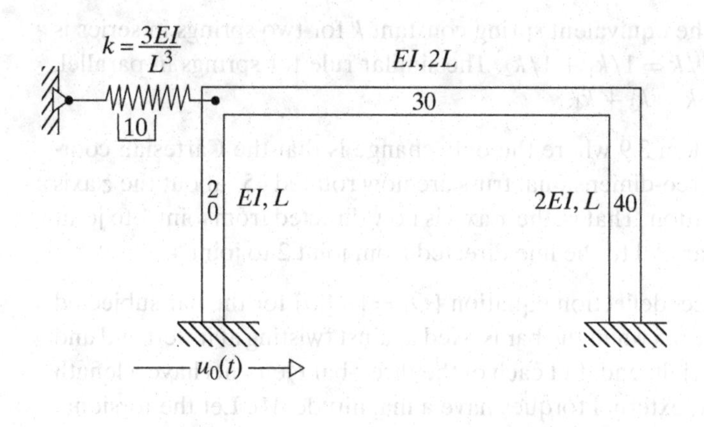

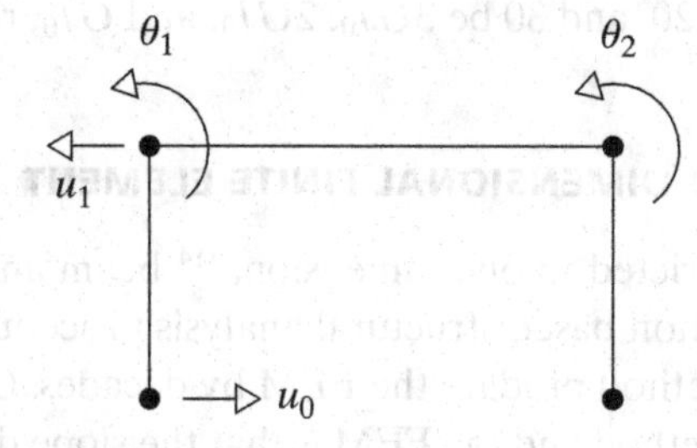

Figure 3.16. Exercise 3.5.

the matrix force–displacement equation is

$$\begin{Bmatrix} 0 \\ F_1 \\ F_2 \end{Bmatrix} = \begin{bmatrix} k_1 + k_2 & 0 & -k_2 \\ 0 & (k_3 + k_4 + k_5) & -k_5 \\ -k_2 & -k_5 & (k_2 + k_5) \end{bmatrix} \begin{Bmatrix} u_1 \\ u_2 \\ u_3 \end{Bmatrix}.$$

Note that the lack of a structural connection between node 1 and the "cart" that is the equivalent of node 2 is represented by the zero matrix entries in positions (1,2) and (2,1) of the above 3×3 stiffness matrix. A small number of simultaneous, algebraic equations such as the above are easily and accurately solved by any number of digital computer programs or even by hand. However, the hand calculation would be even simpler if the two springs with stiffnesses k_1 and k_2 were combined by means of the rule for springs in series, thus eliminating the u_1 DOF, leaving only two simultaneous

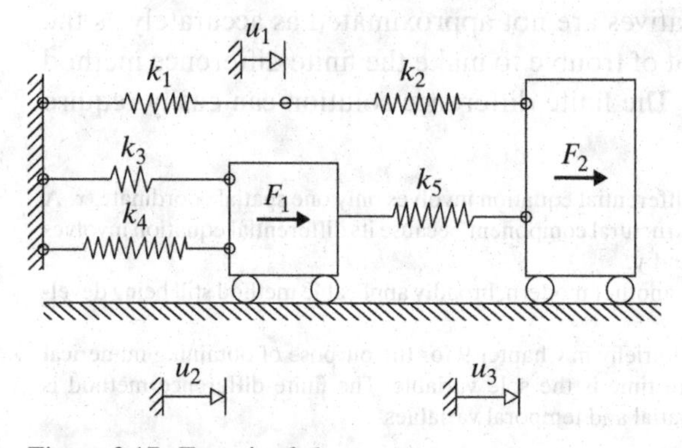

Figure 3.17. Exercise 3.6.

equations to be solved. The equivalent spring constant k for two springs in series is easily determined to be $1/k = 1/k_1 + 1/k_2$. The similar rule for springs in parallel, as are k_3 and k_4, is simply $k = k_3 + k_4$.

3.7 Redo Example Problem 3.9 where the only change is that the Cartesian coordinates of the four-bar, three-dimensional truss are now rotated 45° about the z axis from their original orientation. That is, the x axis is now directed from joint 1 to joint 3, and the y axis is now parallel to the line directed from joint 2 to joint 4.

3.8 Write the matrix force–deflection equation $\{Q\} = [K]\{q\}$ for the bar subjected to torques shown in Figure 3.5 when the bar is fixed against twisting at its left end and free from constraint at its right end. Let each of the three-bar elements have a length L, and each of the applied external torques have a magnitude M_0. Let the torsional stiffness coefficients of bars 10, 20, and 30 be $3GJ_0$, $2GJ_0$, and GJ_0, respectively.

ENDNOTE (1): A SIMPLE TWO-DIMENSIONAL FINITE ELEMENT

The finite element method restricted to one-dimensional[14] beam and bar elements has a long history. Other deflection based structural analysis procedures such as the widely used slope-deflection method predate the FEM by decades. One connection between the slope-deflection method and the FEM is that the slope deflection equations are just part of the beam finite element stiffness matrix equation. The FEM became distinct and widely accepted as the most useful, general structural analysis tool only when it was extended to two-dimensional problems.[15] Two-dimensional problems are approachable by other methods of analysis. Before the development of the FEM, the two most popular, general methods of solving multidimensional structural problems were the *Rayleigh–Ritz* (or just *Ritz*) method (RRM) and the *finite difference method*.[16] Both methods can be applied to either energy formulations or differential equation formulations of a structural problem. A major drawback of the elegant Rayleigh–Ritz method is that for each distint problem, the analyst is required to select a set of suitable functions for an (approximate) finite series solution. These functions might be difficult to find in some circumstances and always require careful consideration. In any event, because of the necessity of evaluating many integrals, the Rayleigh–Ritz method was not easily automated by use of software and thus not easily applicable to large problems in two or three dimensions. The finite difference method is easily coded and broadly applicable, but if applied to differential equations, the higher partial derivatives are not approximated as accurately as the lower derivatives, and it can be a lot of trouble to make the finite difference method adaptable to arbitrary geometries. The finite difference solution can easily require

[14] A beam is one dimensional because its differential equation involves only one spatial coordinate, x. A plate, for example, is a two-dimensional structural component because its differential equation involves two independent spatial coordinates, x and y.

[15] The boundary element method (BEM) is another modern, broadly applicable method still being developed around the world.

[16] The finite difference method is discussed briefly in Chapter 9 for the purpose of obtaining numerical solutions to differential equations where time is the sole variable. The finite difference method is generally applicable to problems with spatial and temporal variables.

a solution of more equations than the FEM. Thus the FEM was a major advance because (i) relative to the RRM, its selected functions[17] are always the same for a given type of element and thus do not require any effort of selection on the part of the analyst and (ii) all its steps are easily programmed. Relative to the finite difference method, the FEM adapts more easily to geometric and changing material complexities.

The finite element method, as used today, falls in the category of a deflection-based,[18] finite series solution. Any complete[19] series can be used as a basis for FEM structural element, but because unadorned polynomial terms are so efficiently calculated, compared to, say, sine terms, they are universially preferred even though they lack other desirable properties such as orthogonality. The choice of polynomial series terms to represent deflections generally would not be a good choice except when they are applied to just a small portion of the total structure. For example, if the chosen polynomial were limited to being just a linear polynomial, such a choice normally would lead to a very poor approximation for elastic deflections over an entire range of the structure. If, however, a straight-line approximation is applied to just a small portion of the structure over and over again, then, for example, like an n-sided, regular polygon, it can closely approximate a smooth circle. Recall that this type of approximation of a complicated deflection function over a extensive domain by a piecewise series of simpler functions over subdomains is what was done in the beginning of this chapter for a one-dimensional analysis when a cubic polynomial was applied to eight or more segments of a beam.

The plane stress problem is here chosen to illustrare, in the simplest possible fashion, the extension of the FEM to multidimensional problems. Recall that a plane stress problem is one where there exists a direction z such that all stress components in that direction are zero. That is, there is a z direction for the given structure such that

$$\sigma_{xz} = \sigma_{yz} = \sigma_{zz} = 0.$$

Such circumstances, at least approximately, are commonplace when a structure, or a portion of a structure, has a geometry where one dimension is much smaller than the other two dimensions. The web or flange of a wide flange beam are examples where portions of the structure meet this geometric circumstance and the above three stresses are much less than the other three stresses.

One circumstance that exactly meets the requirements of the above plane stress definition is that where the entire external loading acting on a flat, thin structure is limited to acting only in the plane of the thin structure and is constant over the thickness. Such a representative plane stress problem is sketched in Figure 3.18(a).

[17] In this sense of using a set of selected functions to build a series solution, the finite element method is a specialized form of the Rayleigh–Ritz method

[18] A force-based finite element method was extensively developed but proved to be inferior to the deflection based FEM.

[19] A complete series, like a Fourier series with terms $\sin(n\pi x/L)$ and $\cos(m\pi x/L)$, for all positive interger value of n is one that can be used to represent any piecewise continuous function. If just one of those series terms is omitted, say $\sin(8\pi x/L)$, then the remaining infinite series is no longer complete because all the remaining series terms cannot represent (i.e., sum to) the missing function $\sin(8\pi x/L)$.

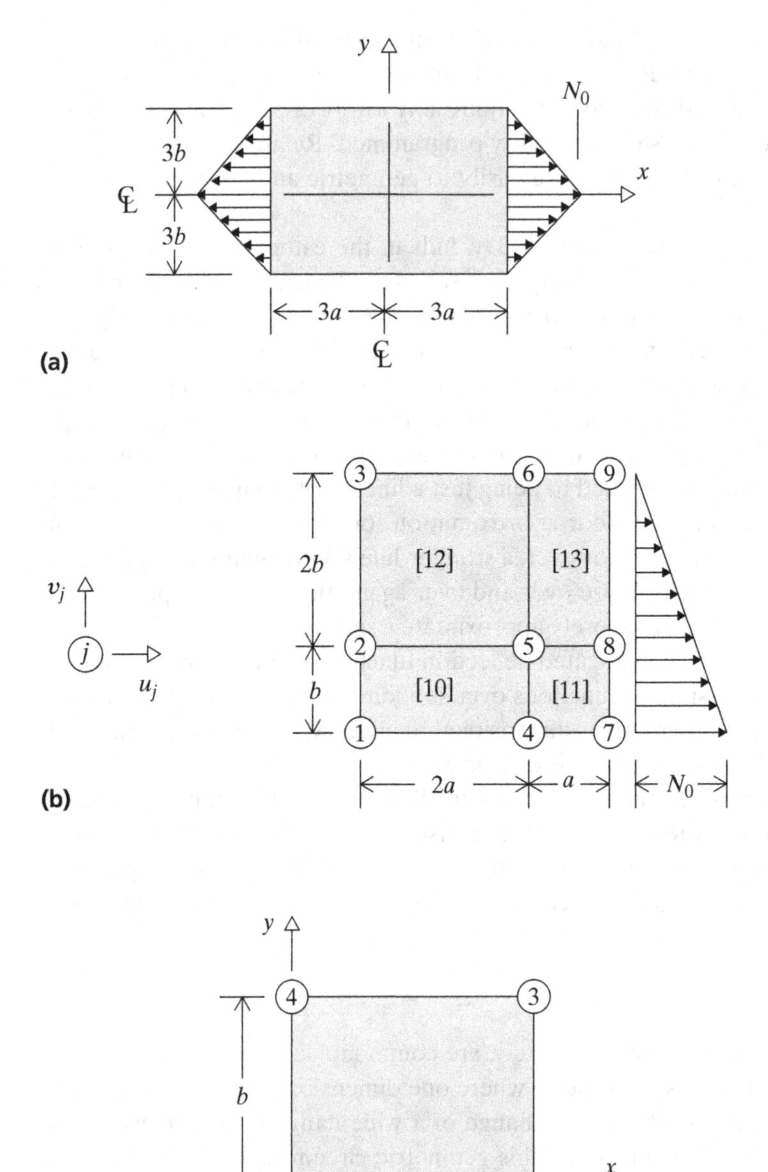

Figure 3.18. (a) Slab loaded at its ends. (b) Crude finite element model of one-quarter of the slab. (c) Simple, rectangular, four-node, plane stress finite element.

This specific thin slab has a constant thickness h in the z direction. For the sake of simplicity, let the slab be homogeneous, isotropic, and linearly elastic. These choices will allow the later use of the simplest form of Hooke's law throughout the slab. Any solution for the $u(x, y)$ and $v(x, y)$ deflections and the three nonzero stresses σ_{xx}, σ_{xy}, and σ_{yy} in this slab begins with noting the double symmetry of the geometry and loading. This double symmetry allows the analyst to drastically reduce the number of selected nodes by focusing on just one quadrant of the slab, say, the upper right quadrant. Figure 3.18(b) shows the upper right quadrant very crudely modeled with just four rectangular, plane stress, finite elements. The elements are selected to be smaller where the stresses and deflections are expected to be greater. Again, the specific purpose of this endnote is to explain how such a rectangular, plane stress, finite element is developed. That development serves as an introduction to the development of all other multidimensional finite elements.

The multidimensional finite element is usually based on representing the solution for the internal deflections as a finite portion of a complete infinite series. Again, the finite series that is almost always preferred is an unadorned polynomial series even though such a series lacks such desirable characteristics as the mutual orthogonality of the series terms. A polynomial series does have the great advantage that the values of its terms are very quickly calculated by digital computer. That is, there are far fewer steps to the calculation of xy^4 than to the calculation of $\sin(x)$. Thus, for this plane stress problem, let the approximate solution for the in-plane[20] deflection field be initially written as follows

$$u(x, y) = a_0 + a_1x + a_2y + a_3x^2 + a_4xy + a_5y^2 + a_6x^3 + \cdots$$
$$v(x, y) = b_0 + b_1x + b_2y + b_3x^2 + b_4xy + b_5y^2 + b_6x^3 + \cdots,$$

where the coefficients a_j and b_j are to be determined. The first question to be answered is how many of these terms should be included in the finite series approximation for the deflection field. Consider Figure 3.18(c), which shows just one of these rectangular, plane stress finite elements. Note that, as per usual for such elements, nodes have been placed at the corners of this finite element. As indicated in Figure 3.18(b), the nodal deflections of the slab as a whole are the global DOF of the structure. Since, for any one element, there are only four nodal deflections in, say, the x direction, and because these four values of the nodal deflections can uniquely determine only the values of four of the coefficients a_j, then the number of terms to be used in each of the above series solutions is four.

The next question concerns which four terms are best. Again, using the $u(x, y)$ series to represent either of the two deflection series, note that without the a_0 term, there can be no representation of a rigid body motion[21] for this finite element. Thus that term must be selected because, otherwise, a rigid body motion would result in a contradictory strain. The a_1 term is vital because without it, there can be no constant normal strain in the x direction, ϵ_{xx}. This constant strain possibility is, of course, necessary to accomodate a constant loading case. The coefficient a_2 is required for a constant shearing strain. The fourth and last term must be one of the three quadratic terms because the lower the order of the polynomial, the more basic the corresponding motion. The best choice for the fourth term is the a_4 term. The primary reason for this choice is that along any specific edge of the rectangular finite element, both the u and v deflections will vary linearly, and those linearly varying edge deflections

[20] Deflections in the z direction are of no importance.

[21] A rigid body motion is one where all parts of the element have exactly the same translational motion. This means that the deflection within the finite element is independent of both x and y. Such a motion is possible only when all the a_j but a_0 are equal to zero.

are completely determined by the values of the two sets of nodal deflections at the ends of the outer edge. Therefore, both deflections will be continuous from one finite element to an adjacent finite element because the adjacent element would share the same two nodes. Such deflection continuity is a requirement for the application of the principle of virtual work to the entire structure. Hence, the selected approximating series are

$$u(x, y) = a_0 + a_1 x + a_2 y + a_4 xy$$
$$v(x, y) = b_0 + b_1 x + b_2 y + b_4 xy.$$

This may seem to be too simplistic a series representation for a deflection field, and it certainly would be if it were applied over any significant portion of the slab. However, if, as argued above, this deflection approximation is applied over only a small portion of the slab, then, as the many-sided, regular polygon closely approximates the circle, many of these elements will closely approximate the true deflection field. Even so, at the expense of a greater number of calculations, this displacement field approximation could be improved by placing more nodes on the boundaries of, or even inside, the finite element. For example, additional nodes could be placed at each of the four edge midpoints for a total of eight nodes. Then eight terms could be retained in each deflection series, making those series complete through the quadratic terms, plus two cubic terms x^2y and xy^2. Since a quadratic polynomial has three coeffiicients, the quadratic form of the displacements at each of the element edges would be fully defined by the three nodes along that edge, and therefore the adjacent element that shares those same three nodes would have fully continuous deflections with the other rectangular element.

The next step for the four node plane stress element is to determine the values of the a_j coefficients in terms of the element nodal deflections. In this case, this is easily done from the continuity of the nodal and internal deflections. For example, from Figure 3.18(c), where $x_1 = 0$, $x_2 = a$, $y_1 = 0$, and $y_2 = b$,

$$u_1 = a_0 \qquad u_2 = a_0 + a_1 a \qquad u_3 = a_0 + a_1 a + a_3 b + a_4 ab \qquad u_4 = a_0 + a_2 b.$$

Solving for the a_j coefficients in terms of the nodal deflections, and then substituting back into the finite series approximations for the deflections u and v, allows those two deflections to be written in the following form, where again the polynomial functions $N_j(x, y)$ are called *shape functions*:

$$u(x, y) = \sum_{j=1}^{4} u_j N_j(x, y) \qquad v(x, y) = \sum_{j=1}^{4} v_j N_j(x, y)$$

$$\text{where} \quad N_1 = \left(1 - \frac{x}{a}\right)\left(1 - \frac{y}{b}\right) \qquad N_2 = \left(\frac{x}{a}\right)\left(1 - \frac{y}{b}\right)$$
$$N_3 = \left(\frac{x}{a}\right)\left(\frac{y}{b}\right) \qquad N_4 = \left(1 - \frac{x}{a}\right)\left(\frac{y}{b}\right). \tag{3.5}$$

Note again that each nondimensional shape function has the property that the function has the value 1.0 at the location of its corresponding node, and the value zero at all other nodes. These shape functions, and many other shape functions, can be easily deduced directly from that set of characteristics.

Once the element shape functions have been determined, the remainder of the preparation for the use of the principle of virtual work is routine. To see where the discussion is going, recall that in this case where three of the stresses are zero, the principle of virtual work can be stated as

$$\delta W_{ex} = \delta U = h \iint [\sigma_{xx}\delta\epsilon_{xx} + \sigma_{yy}\delta\epsilon_{yy} + \sigma_{xy}\delta\gamma_{xy}]dx\,dy.$$

Therefore the necessary tasks ahead are (i) to determine the strains from the above deflections, (ii) to take the variations of those strains, and (iii) to determine the stresses from the strains using Hooke's law. The strains, arranged in matrix form, are

$$\begin{Bmatrix} \epsilon_{xx} \\ \epsilon_{yy} \\ \gamma_{xy} \end{Bmatrix} = \begin{Bmatrix} \dfrac{\partial u}{\partial x} \\ \dfrac{\partial v}{\partial y} \\ \dfrac{\partial u}{\partial y} + \dfrac{\partial v}{\partial x} \end{Bmatrix} = [B]\{q\},$$

where the details of the $[B]$ and $\{q\}$ matrices in the latter matrix equation are

$$\begin{Bmatrix} \epsilon_{xx} \\ \epsilon_{yy} \\ \gamma_{xy} \end{Bmatrix} = \left[\begin{matrix} -\frac{1}{a}\left(1-\frac{y}{b}\right) & \frac{1}{a}\left(1-\frac{y}{b}\right) & \frac{1}{a}\frac{y}{b} & -\frac{1}{a}\frac{y}{b} \\ 0 & 0 & 0 & 0 \\ -\frac{1}{b}\left(1-\frac{x}{a}\right) & -\frac{1}{b}\frac{x}{a} & \frac{1}{b}\frac{x}{a} & \frac{1}{b}\left(1-\frac{x}{a}\right) \end{matrix} \right.$$

$$\left. \begin{matrix} 0 & 0 & 0 & 0 \\ -\frac{1}{b}\left(1-\frac{x}{a}\right) & -\frac{1}{b}\frac{x}{a} & \frac{1}{b}\frac{x}{a} & \frac{1}{b}\left(1-\frac{x}{a}\right) \\ -\frac{1}{a}\left(1-\frac{y}{b}\right) & \frac{1}{a}\left(1-\frac{y}{b}\right) & \frac{1}{a}\frac{y}{b} & -\frac{1}{a}\frac{y}{b} \end{matrix} \right] \begin{Bmatrix} u_1 \\ u_2 \\ u_3 \\ u_4 \\ v_1 \\ v_2 \\ v_3 \\ v_4 \end{Bmatrix}.$$

The virtual strains are simply $\{\delta\epsilon\} = [B]\{\delta q\}$. In the absence of temperature changes, Hooke's law can be written as $\{\sigma\} = [D]\{\epsilon\}$, which in detail is

$$\begin{Bmatrix} \sigma_{xx} \\ \sigma_{yy} \\ \sigma_{xy} \end{Bmatrix} = \frac{E}{1-\nu^2} \begin{bmatrix} 1 & \nu & 0 \\ \nu & 1 & 0 \\ 0 & 0 & \frac{(1-\nu)}{2} \end{bmatrix} \begin{Bmatrix} \epsilon_{xx} \\ \epsilon_{yy} \\ \gamma_{xy} \end{Bmatrix}.$$

Thus the virtual strain energy can be written as

$$\delta U = h\lfloor \delta q \rfloor \left(\int_0^a \int_0^b [B]^T[D][B]dx\,dy \right) \{q\} = \lfloor \delta q \rfloor [k]\{q\}.$$

Therefore the derivation of the 8×8, symmetric, element stiffness matrix is just a lengthy matter of carrying out the matrix products and integrations enclosed in paranthesis. For example, the $(\delta u_1, u_1)$ entry of the 8×8 stiffness matrix is

$$k_{11} = \frac{Eh}{1-\nu^2} \int_0^a \int_0^b \left[\frac{1}{a^2} \left(1 - \frac{y}{b}\right)^2 + \frac{1-\nu}{2b^2} \left(1 - \frac{x}{a}\right)^2 \right] dx\, dy$$
$$= \frac{Eh}{3(1-\nu^2)} \left(\frac{b}{a} + \frac{1-\nu}{2} \frac{a}{b} \right).$$

The remaining 63 element stiffness matrix entries, 63 because of the use of matrix symmetry as a calculation check on the 36 different entries, are left to the reader as exercises. Note that the element aspect ratio a/b and its inverse appear separately in this entry of the element stiffness matrix. Experience has shown that this and other rectangular elements are more accurate when that aspect ratio is close to 1 and less than 2.

Returning to the specific problem of Figure 3.18(a), the above plane stress element stiffness matrices are assembled into a global stiffness matrix in exactly the same way as was done for beam elements, which was on the basis of the correspondences between the element DOF and the global DOF. Once assembled, the boundary conditions resulting from the symmetry are

$$u_1 = u_2 = u_3 = 0 \qquad v_1 = v_4 = v_7 = 0.$$

These zero boundary conditions eliminate six columns of the global stiffness matrix and cause six rows to be set aside. This process reduces the original 18×18 global stiffness matrix to a 12×12 matrix.

The generalized force vector, $\{Q\}$, for this problem is worth attention. Specifically, the task is to determine the entries in the generalized force vector for this case where the only externally applied loading on the upper-right quadrant of the slab is the loading per unit length that is linearly distributed over the right edge of the quadrant. As ever, the generalized force vector can be computed from the virtual work expression, which for the right edge of the quadrant can be written as $\delta W = \int N(y)\delta u(3a, y)dy$. This integral form needs to be applied to elements 11 and 13. For element 11, from Eq. (3.5), where the local x coordinate has the value a, and the local and global y coordinates coincide, the virtual deflection for the right side of element 11 in terms of the global DOF is simply

$$\delta u(a, y) = \delta u_7 \left(1 - \frac{y}{b}\right) + \delta u_8 \frac{y}{b}.$$

For element 13, again using Eq. (3.5) but replacing the local y with the global $y - b$ and the local b with the global measure $2b$,

$$\delta u(3a, y) = \delta u_8 \left(1 - \frac{y-b}{2b}\right) + \delta u_9 \frac{y-b}{2b} = \frac{1}{2}\left(3 - \frac{y}{b}\right)\delta u_8 - \frac{1}{2}\left(1 - \frac{y}{b}\right)\delta u_9.$$

Note that, for example, the coefficient of δu_8 is zero at the location of the u_9 DOF where $y = 3b$, and the coefficient of δu_9 is zero at $y = b$, where the u_8 DOF is located. The edge loading is easily described in terms of the global y as $N_0[1 - y/(3b)]$.

Therefore, utilizing the general expression $\delta W = \lfloor \delta q \rfloor \{Q\}$, the virtual external work for element 11 can be written as

$$\delta W_{ex}^{(11)} = \int_{y=0}^{b} N_0 \left(1 - \frac{y}{3b}\right) \delta u(y) dy = \int_{y=0}^{b} N_0 \left(1 - \frac{y}{3b}\right) \lfloor \delta u_7 \quad \delta u_8 \rfloor \begin{Bmatrix} 1 - \dfrac{y}{b} \\ \dfrac{y}{b} \end{Bmatrix} dy$$

$$\text{hence} \begin{Bmatrix} Q_7 \\ Q_8 \end{Bmatrix} = N_0 b \begin{Bmatrix} \dfrac{8}{18} \\ \dfrac{7}{18} \end{Bmatrix}.$$

Similarly for element 13

$$\delta W_{ex}^{(13)} = \int_{y=0}^{b} N_0 \left(1 - \frac{y}{3b}\right) \lfloor \delta u_8 \quad \delta u_9 \rfloor \begin{Bmatrix} \dfrac{3}{2} - \dfrac{y}{2b} \\ -\dfrac{1}{2} + \dfrac{y}{2b} \end{Bmatrix} dy = \lfloor \delta u_8 \quad \delta u_9 \rfloor \; N_0 b \begin{Bmatrix} \dfrac{4}{9} \\ \dfrac{2}{9} \end{Bmatrix}.$$

Hence the only nonzero entries in the 12×1 generalized force vector are the following total values corresponding to the u_7, u_8, and u_9 DOF:

$$\begin{aligned} \{Q\}^T &= \frac{N_0 b}{18} \lfloor 0 \quad . \quad . \quad . \quad 0 \quad 8 \quad 15 \quad 4 \quad 0 \quad . \quad . \quad . \quad 0 \rfloor \\ &= N_0 b \lfloor 0 \quad . \quad . \quad . \quad 0 \quad 0.444 \quad 0.833 \quad 0.222 \quad 0 \quad . \quad . \quad . \quad 0 \rfloor. \end{aligned}$$

Note the total force sums to $(3/2)N_0$ as it should. This was a lengthy process. An alternative to the above procedure is to simply group the load per unit length over the length around each node. That is, for example, for DOF u_9, take the distributed loading $N_0[1 - y/(3b)]$ over half the depth of the element, b, to get $Q_9 = (1/2)(N_0/3)(b) = (1/6)N_0 b$. If this apportioning process is completed, the approximate generalized force vector is

$$\{Q\}^T = N_0 b \lfloor 0 \quad . \quad . \quad . \quad 0 \quad 0.458 \quad 0.875 \quad 0.167 \quad 0 \quad . \quad . \quad . \quad 0 \rfloor.$$

Again the total force by this second method adds to $(3/2)N_0$, as it should, but there is a 25% error in the smallest entry, a 5% error in the largest entry, and a 3% error in the middle entry. These errors would be even less if the finite element grid were not so crude as to have only two elements along the loaded edge. If the original element grid involved a larger, and hence a more reasonable, number of rectangular elements then simple minded grouping of the load around the nodes would be more satisfactory, whereas the hand calculation of the "exact" generalized force vector would be much more lengthy and thus unreasonable.

In summary, multidimensional finite element stiffness matrices are generally based on the approximation of the deflection field as finite polynomial series. The coefficients of those series are then related to the nodal deflections, which are the element DOF. These two steps can often be skipped by writing the expressions for the internal element deflections directly in terms of the product of the element DOF and their corresponding shape functions. This is possible because of the characteristic of all shape functions that they have the value of 1 at the location of their corresponding DOF and the value zero at the locations of all the other DOF. Once the shape functions have been determined, and interelement deflection continuity assured, then it is a straightforward process of (i) differentiating the deflection expressions to obtain the strains; (ii) using a mathematical description of the material behavior to obtain the stresses from the strains; and then (iii) substituting the stresses and the virtual strains into the virtual strain energy integral, which is the more complicated part of the principle of virtual work. Evaluating that integral identifies the element stiffness matrix in the form $\lfloor \delta q \rfloor [k_{el}]\{q\}$. Then it is a matter of assembling the system stiffness marix $[K]$, and determing the values of the generalized force vector $\{Q\}$ from the system external virtual work expression. The final step of equating $\{Q\}$ to $[K]\{q\}$ enforces equilibrium throughout the system.

It must be noted that the above-discussed rectangular plane stress element is limited to modeling rectangular slabs, or, as they are often called, rectangular membranes. More general planform geometries with straight edges require triangular elements. Curved edge geometries can be modeled using what are called isoparametric elements, or elements created with one or more curved boundaries.

ENDNOTE (2): THE CURVED BEAM FINITE ELEMENT

Again, the small deflection finite elements used in this textbook are limited to the simplest types because they are solely for instructional purposes on the topic of structural dynamics. For example, when the straight beam element experiences all possible types of beam displacements, it has only twelve DOF. When the straight beam area product of inertia is zero, then the four lateral deflection and bending slope DOF for bending in one principal plane do not interact with the four DOF for bending in the other principal plane.[22] Furthermore, those eight bending DOF are uncoupled with both the two DOF for beam twisting and the two DOF for beam extension. As is done extensively in this textbook, this extensive uncoupling, allows the hand work selection of only those smaller portions of the total 12×12 beam stiffness matrix that correspond to the loading and subsequent expected motions. The stiffness matrices of more sophisticated finite elements generally do not possess such extensive uncoupling, even when the element is one dimensional.[23] Perhaps the simplest element beyond the straight beam element with respect to uncoupling is the beam element that is curved in a circular arc in only one plane. The analyst has two choices when modeling a curved beam structure. As one possibility, the analyst can represent a curved beam as a series of small, rotated, straight beam segments much like a high-order, regular polygon approximates a circle. As a second possibility, the analyst can use a fewer number of curved beam elements. The first point of this endnote is to show that the latter are much more complicated than straight beam elements. The second point of this endnote is to

[22] Being "uncoupled" is the mathematical term corresponding to a lack of interaction in the physical world.

[23] An element is "one dimensional" when, in the equation that describes its elastic behavior, there is but a single independent spatial variable. A straight or curved beam is such a one-dimensional element because the beam deflections are solely functions of the length along the beam axis, say, x or s. A spring is a zero-dimensional element, whereas a plate is a two-dimensional element because the lateral deflections of the plate midplane are functions of two coordinates, say, x and y.

demonstrate another approach to determining stiffness matrices for one-dimensional elements.

This endnote begins, but because of the complexity does not complete, the derivation of the element stiffness matrix for a beam curved in a single plane, where the constant radius of curvature is 10 or more times larger than the depth of the beam. This is a common circumstance, for example, in aircaft and submarines, and therefore one of some importance. The importance of the above qualification about the large radius of curvature (small curvature) relative to the depth of the beam is that this qualification allows the use of straight beam strength of materials theory to describe the curved beam segment.

The method that is used here for setting up the large radius curved beam stiffness matrix is one where an older method of structural analysis is applied to the structural element to first determine the element's flexibility characteristics rather than directly ascertain the element's stiffness characteristics. Then the flexibility equations are inverted to get the stiffness equations that form the element stiffness matrix. For the sake of clarity and confidence, this procedure is now illustrated by being first applied to a straight beam bending finite element. To that end, consider the usual four-DOF beam bending finite element as is shown in Figure 3.1. Begin the analysis by temporarily clamping the beam element at node 2. The nodal loads at node 1 are, of course, the shear force and moment V_1 and M_1. The selected method of analysis in this illustrative case is that of writing and integrating the familiar beam bending differential equation

$$EIv''(x) = V_1 x - M_1 \rightarrow EIv'(x) = \tfrac{1}{2}V_1 x^2 - M_1 x + C_1$$

$$\rightarrow EIv(x) = \tfrac{1}{6}V_1 x^3 - \tfrac{1}{2}M_1 x^2 + C_1 x + C_2.$$

Applying the BCs that $v(L) = v'(L) = 0$ yields the result

$$C_1 = -\tfrac{1}{2}V_1 L^2 + M_1 L \qquad C_2 = \tfrac{1}{3}V_1 L^3 - \tfrac{1}{2}M_1 L^2.$$

Inserting these solutions into the expressions for the element lateral deflection and bending slope yields

$$EIv'(x) = -\tfrac{1}{2}V_1(L^2 - x^2) + M_1(L - x)$$

$$EIv(x) = \tfrac{1}{6}V_1(2L^3 - 3Lx + x^3) - \tfrac{1}{2}M_1(L^2 - 2Lx + x^2).$$

By setting $x = 0$, the following matrix form of the relationships between the generalized displacements and generalized forces at node 1 are obtained immediately

$$\begin{Bmatrix} v_1 \\ \psi_1 \end{Bmatrix} = \frac{L^3}{EI}\begin{bmatrix} 1/3 & -1/(2L) \\ -1/(2L) & 1/L^2 \end{bmatrix}\begin{Bmatrix} V_1 \\ M_1 \end{Bmatrix}.$$

Inverting this matrix yields

$$\begin{Bmatrix} V_1 \\ M_1 \end{Bmatrix} = \frac{EI}{L^3}\begin{bmatrix} 12 & 6L \\ 6L & 4L^2 \end{bmatrix}\begin{Bmatrix} v_1 \\ \psi_1 \end{Bmatrix},$$

which is the familiar first 2×2 submatrix of the beam bending stiffness matrix. Similarly, by clamping the beam element at node 1, and considering the generalized deflections and loads at node 2, the second of the 2×2 submatrices on the stiffness matrix diagonal is obtained as

$$\begin{Bmatrix} V_2 \\ M_2 \end{Bmatrix} = \frac{EI}{L^3} \begin{bmatrix} 12 & -6L \\ -6L & 4L^2 \end{bmatrix} \begin{Bmatrix} v_2 \\ \psi_2 \end{Bmatrix}.$$

The off-diagonal 2×2 submatrices are obtained by writing the two equilibrium equations for the beam element. These equations are

$$V_1 = -V_2 \qquad M_1 = -V_2 L - M_2.$$

Casting these equations into matrix form and then substituting the above flexibility submatrix yields

$$\begin{Bmatrix} V_1 \\ M_1 \end{Bmatrix} = -\begin{bmatrix} 1 & 0 \\ L & 1 \end{bmatrix} \begin{Bmatrix} V_2 \\ M_2 \end{Bmatrix} = -\begin{bmatrix} 1 & 0 \\ L & 1 \end{bmatrix} \frac{EI}{L^3} \begin{bmatrix} 12 & -6L \\ -6L & 4L^2 \end{bmatrix} \begin{Bmatrix} v_2 \\ \psi_2 \end{Bmatrix}$$

$$\text{or} \quad \begin{Bmatrix} V_1 \\ M_1 \end{Bmatrix} = \frac{EI}{L^3} \begin{bmatrix} -12 & 6L \\ -6L & 2L^2 \end{bmatrix} \begin{Bmatrix} v_2 \\ \psi_2 \end{Bmatrix}.$$

Rewriting the equilibrium equations as

$$V_2 = -V_1 \qquad M_2 = +V_1 L - M_1$$

and then following the procedure immediately above provides the other off-diagonal 2×2 submatrix. Of course, combining the four submatrices completes the rederivation of the straight beam finite element stiffness matrix for bending in one plane.

Return now to the primary concern of this endnote. The geometry and generalized coordinates of a curved beam finite element are shown in Figures 3.19(a) and 3.19(b). Note again that this beam element, like most curved beams, is curved only in a single plane, which in this case is designated the x, z plane. Parts (a) and (b) of Figure 3.19 each show, as discussed below, the six DOF that interact among themselves, but not with the six DOF of the other figure. For simplicity, the beam product of inertia is taken to be zero. The equilibrium equations that will be written shortly for this beam segment will show the corresponding interrelationships among the various types of element generalized forces. That is, these equilibrium equations will support the idea that, for example, the six generalized coordinates of Figure 3.19(a) are coupled among themselves. In terms of generalized forces and referring to Figure 3.19(a), a nonzero value for a shearing force at the node 2 end, such as V_3, will produce not only a shearing force at the node 1 end, V_1, but it will also produce both a bending moment and a twisting moment at node 1, such as M_1 and M_3, respectively. This interaction can be more easily visualized when (i) the total arc angle β is 90°; (ii) the beam is cantilevered at node 1, and (iii) loaded by the V_3 shear force at node 2 in the direction of the DOF v_2. Similarly, twisting the beam at end 2 bends the beam at end 1, and bending the beam at end 2 twists the beam at end 1, and so on. Therefore the twisting and out-of-plane bending motions are inseparable, that is, coupled.

Figure 3.19(b) shows the DOF associated with beam extension and bending within the plane of curvature. These two types of motion are also coupled. To understand this coupling, let the angle β be 180°, making the beam resemble and inverted U, and again let the beam be cantilevered at node 1. Then, for example, if at node 2 there is a nonzero but small value for w_2, then the beam is bent in its plane as the mouth of

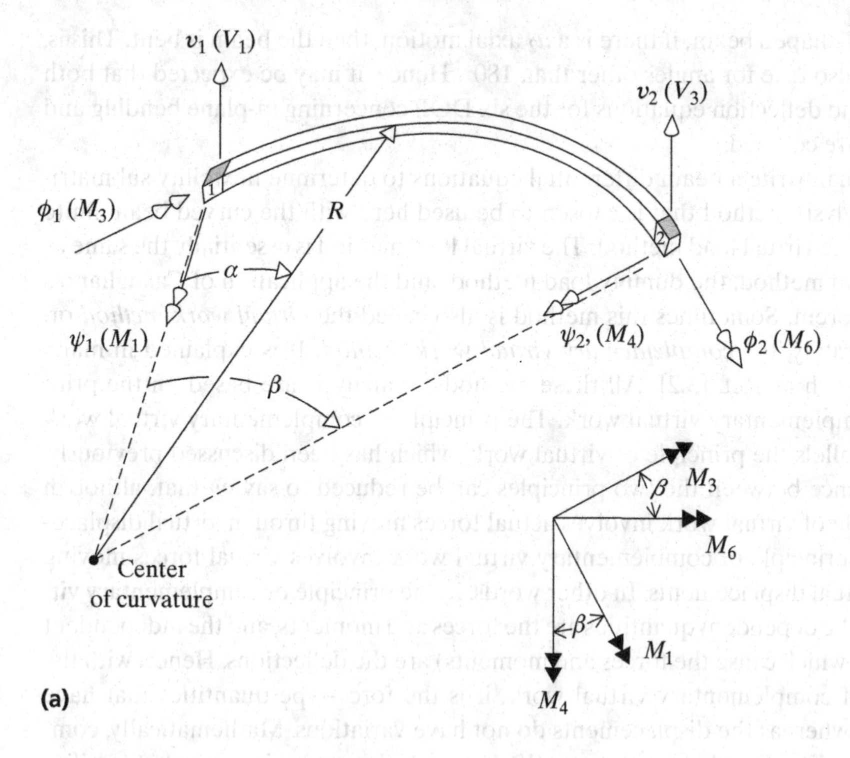

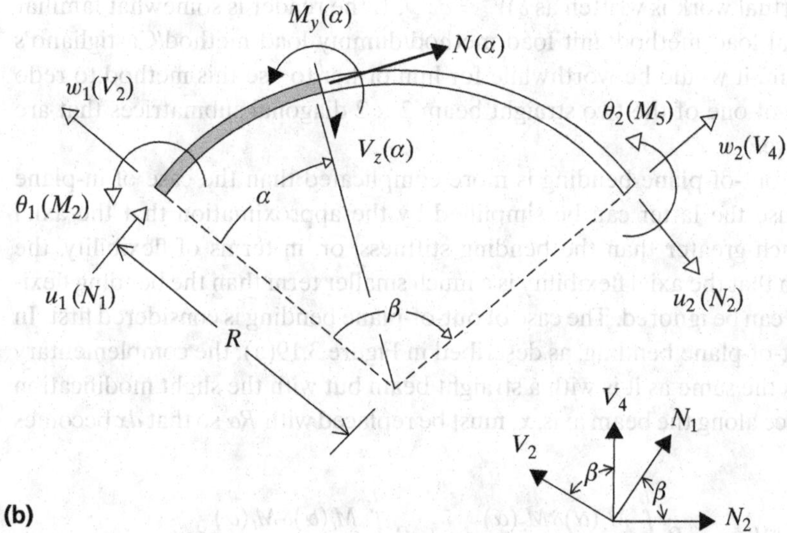

Figure 3.19. (a) Curved beam loaded and deflected out of its plane. (b) Plan view of curved beam bending in its own plane. The generalized forces corresponding to the generalized coordinates are shown in parentheses.

the U opens up.[24] However, the beam is also stretched by the w_2 motion, or in force terms, by the V_4 shear force that is associated with the w_2 motion, because that force is in the same direction as the beam axis at the top of the inverted U. Similarly, for

[24] Bending can be viewed simply as a change in curvature.

this same U-shaped beam, if there is a u_2 axial motion, then the beam is bent. This is, of course, also true for angles other than 180°. Hence, it may be expected that both the force and deflection equations for the six DOF concerning in-plane bending and extension are coupled.

Rather than write a beam differential equations to determne flexibility submatrices, the analysis method that is chosen to be used here with the curved beam finite element is the virtual load method. The virtual load method is essentially the same as the unit load method, the dummy load method, and the application of Castigliano's second theorem. Sometimes this method is also called the *virtual work method* or, more accurately, the *complementary virtual work method.* It is explained in many textbooks such as Ref. [3.2]. All these methods of analysis are based on the principle of complementary virtual work. The principle of complementary virtual work closely parallels the principle of virtual work, which has been discussed previously. The difference between the two principles can be reduced to saying that, although the principle of virtual work involves actual forces moving through virtual displacements, the principle of complementary virtual work involves virtual forces moving through actual displacements. In other words, in the principle of complementary virtual work, the dependent quantities are the forces and moments, and the independent quantities (which cause the forces and moments) are the deflections. Hence, with the principle of complementary virtual work, it is the force-type quantities that have variations, whereas the displacements do not have variations. Mathematically, complementary virtual work is written as $\delta W^* = q\delta Q$. If the reader is somewhat familiar with the virtual load method/unit load method/dummy load method/Castigliano's second theorem, it would be worthwhile for him or her to use this method to redo the derivation of one of the two straight beam 2×2 diagonal submatrices that are rederived above.

The case of out-of-plane bending is more complicated than the case of in-plane bending because the latter can be simplified by the approximation that the axial stiffness is much greater than the bending stiffness, or, in terms of flexibility, the approximation that the axial flexibility is a much smaller term than the bending flexibility and thus can be ignored. The case of out-of-plane bending is considered first. In the case of out-of-plane bending, as described in Figure 3.19(a), the complementary virtual work is the same as it is with a straight beam but with the slight modification that the distance along the beam axis, x, must be replaced with $R\alpha$ so that dx becomes $Rd\alpha$. Thus

$$\delta W^*_{ex} = R\int_0^L \frac{M_z(\alpha)\delta M_z(\alpha)}{EI} d\alpha + R\int_0^L \frac{M_t(\alpha)\delta M_t(\alpha)}{GJ} d\alpha,$$

where the internal complementary strain energy, on the above right-hand side, includes contributions from beam twisting as well as beam bending, which as explained above, are inseparable for a curved beam loaded out of its plane. The effects of beam shearing could also be included by simply adding a third integral, but to focus on the more important aspects of the problem, shearing effects are ignored.

Since the beam cross-section area product of inertia has been specified to be zero, the overall stiffness matrix for bending out of the plane of curvature is a 6×6

submatrix of the total beam 12×12 beam stiffness matrix. Minicking the procedure followed above for the straight beam, the values of this 6×6 submatrix are determined by obtaining four 3×3 submatrices, starting with the 3×3 submatrices that lie on the main diagonal of the 6×6 submatrix. To obtain the first of the diagonal 3×3 submatices, clamp the beam element at node 2, and require the left end of the curved beam element to displace through arbitrary but positive values of the three node 1 generalized coordinates: the DOF v_1, ψ_1, and ϕ_1. In this case of an arbitrary deflection at node 1 while node 2 is clamped, there are six unknown nodal forces. There are the three out-of-plane nodal loads, V_1, M_1, and M_3, that accomplish the arbitrary displacements at node 1 and the three corresponding reactions at node 2. Again, Figure 3.19(a) shows the loads at node 1 and the corresponding reactions at node 2. Since there are six unknown force-type quantities and but three equilibrium equations, there are three redundant loads. It is immaterial which three of the six loads are selected as the redundant reactions. The choice made here is that the three reactions at end 1 are selected as the redundant reactions. Then the equilibrium equations that determine the values of the reactions at end 2, once the values of the reactions at end 1 are discovered, are

$$V_3 = -V_1 \qquad M_4 = V_1 R \sin\beta + M_1 \cos\beta - M_3 \sin\beta$$
$$M_6 = V_1 R(1 - \cos\beta) + M_1 \sin\beta + M_3 \cos\beta.$$

The actual load system bending and twisting moment expressions can also be obtained from Figure 3.19(a). The results, which closely mimic those above, are

$$M_z(\alpha) = V_1 R \sin\alpha + M_1 \cos\alpha - M_3 \sin\alpha$$
$$M_t(\alpha) = V_1 R(1 - \cos\alpha) + M_1 \sin\alpha + M_3 \cos\alpha.$$

To solve for the three redundant force quantities, three independent virtual load systems are required. These three virtual load systems, all of which are required to be in equilibrium, are illustrated in Figure 3.20(a). Note that there is no complementary virtual work done at node 2 because there are no actual deflections there now that that end is temporarily fixed. Therefore, when the above virtual internal bending and twisting moments are inserted into the complementary virtual work equations, the result is the following three equations to be initially solved for the three DOF

$$v_1 \delta V_1 = \frac{R^2}{EI_{zz}} \int_{\alpha=0}^{\beta} [V_1 R \sin\alpha + M_1 \cos\alpha - M_3 \sin\alpha][\delta V_1 \sin\alpha] d\alpha$$
$$+ \frac{R^2}{GJ} \int_{\alpha=0}^{\beta} [V_1 R(1 - \cos\alpha) + M_1 \sin\alpha + M_3 \cos\alpha][\delta V_1 (1 - \cos\alpha)] d\alpha$$

$$\psi_1 \delta M_1 = \frac{R}{EI_{zz}} \int_{\alpha=0}^{\beta} [V_1 R \sin\alpha + M_1 \cos\alpha - M_3 \sin\alpha][+\delta M_1 \cos\alpha] d\alpha$$
$$+ \frac{R}{GJ} \int_{\alpha=0}^{\beta} [V_1 R(1 - \cos\alpha) + M_1 \sin\alpha + M_3 \cos\alpha][\delta M_1 \sin\alpha] d\alpha$$

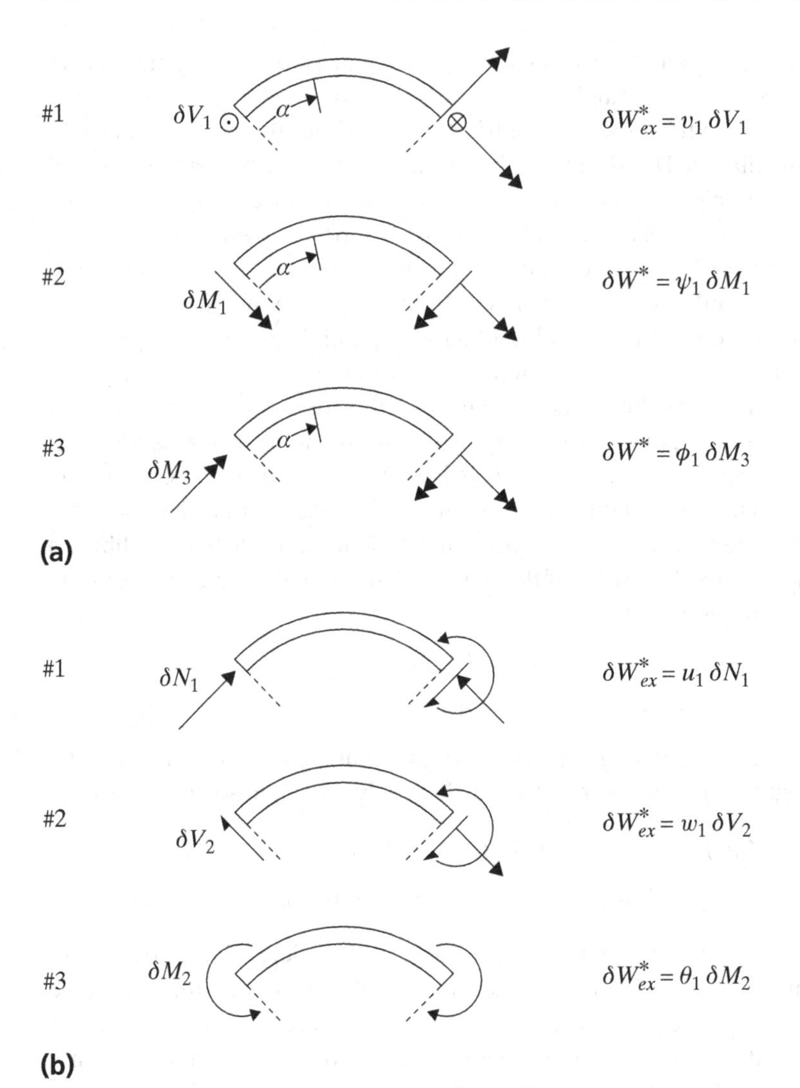

Figure 3.20. (a) The three virtual load systems needed to determine the actual displacements at node 1 of the curved beam element for out-of-plane bending and twisting. (b) The three virtual load systems needed to determine the actual displacements at node 1 for in-plane bending and extension.

and

$$\phi_1 \delta M_3 = \frac{R}{EI_{zz}} \int_{\alpha=0}^{\beta} [V_1 R \sin\alpha + M_1 \cos\alpha - M_3 \sin\alpha][-\delta M_3 \sin\alpha] d\alpha$$

$$+ \frac{R}{GJ} \int_{\alpha=0}^{\beta} [V_1 R(1 - \cos\alpha) + M_1 \sin\alpha + M_3 \cos\alpha][\delta M_3 \cos\alpha] d\alpha.$$

Again, in each of the above equations the arbitrary values of the virtual loads cancel. Now it is a matter of evaluating the above six integrals, which break down to five different integrals over α. These five different integrals and their evaluations are

$$\int_0^\beta \sin\alpha d\alpha = 1 - \cos\beta \qquad \int_0^\beta \cos\alpha d\alpha = \sin\beta \qquad \int_0^\beta \sin\alpha\cos\alpha d\alpha = \frac{1}{2}\sin^2\beta$$

$$\int_0^\beta \sin^2\alpha d\alpha = \frac{1}{2}\left(\beta - \frac{1}{2}\sin 2\beta\right) \qquad \int_0^\beta \cos^2\alpha d\alpha = \frac{1}{2}\left(\beta + \frac{1}{2}\sin 2\beta\right).$$

Therefore the equations for the first three generalized coordinates in terms of the nodal loads are

$$\begin{aligned}
v_1 = {} & \frac{R^2}{2EI_{zz}}\left[V_1 R\left(\beta - \frac{1}{2}\sin 2\beta\right) + M_1\sin^2\beta - M_3\left(\beta - \frac{1}{2}\sin^2\beta\right)\right] \\
& + \frac{R^2}{GJ}\left[V_1 R\left(3\beta - 4\sin\beta + \frac{1}{2}\sin 2\beta\right) + 2M_1\left(1 - \cos\beta - \frac{1}{2}\sin^2\beta\right)\right. \\
& \left. + M_3\left(2\sin\beta - \beta - \frac{1}{2}\sin 2\beta\right)\right] \\
\psi_1 = {} & \frac{R}{2EI_{zz}}\left[-V_1 R\sin^2\beta + M_1\left(\beta + \frac{1}{2}\sin 2\beta\right) - M_3\sin^2\beta\right] \\
& + \frac{R}{2GJ}\left[V_1 R\left(2 - 2\cos\beta - 2\sin^2\beta\right) + M_1\left(\beta - \frac{1}{2}\sin 2\beta\right) + M_3\sin^2\beta\right] \\
\phi_1 = {} & \frac{R}{2EI_{zz}}\left[-V_1 R\left(-\beta + \frac{1}{2}\sin 2\beta\right) - M_1\sin^2\beta + M_3\left(\beta - \frac{1}{2}\sin 2\beta\right)\right] \\
& + \frac{R}{2GJ}\left[-V_1 R\left(2\sin\beta - \beta - \frac{1}{2}\sin 2\beta\right) + M_1\sin^2\beta + M_3\left(\beta + \frac{1}{2}\sin 2\beta\right)\right].
\end{aligned}$$

The next step is to cast these three equations into the matrix form

$$\begin{Bmatrix} v_1 \\ \psi_1 \\ \phi_1 \end{Bmatrix} = \begin{bmatrix} \frac{R^3}{2EI_{zz}}\left(\beta - \frac{1}{2}\sin 2\beta\right) + \frac{R^3}{2GJ}\left(3\beta - 4\sin\beta + \frac{1}{2}\sin 2\beta\right) & \cdot & \cdot \\ \cdot & \cdot & \cdot \\ \cdot & \cdot & \cdot \end{bmatrix} \begin{Bmatrix} V_1 \\ M_1 \\ M_3 \end{Bmatrix}, \tag{3.3}$$

where dots have been used to indicate the other eight entries in the submatrix to save space. The coefficient matrix in the above equation is, of course a flexibility matrix. Obtaining the inverse of this flexibility submatirx completes the task of determining the first of the four 3×3 stiffness submatrices for out-of-plane bending. Obtaining the algebraic form of the inverse matix is easily done using a digital computer program such as Mathematica. Unfortunately the printout of the solution is four pages in length and is not reproduced here. This fact does not render this approach impractical because when this stiffness matrix component is required for a specific calculation, it is necessary only to first write the entries of the flexibility matrix as numerical values and then, if necessary, invert that numerically valued flexibility matrix.

The second of the main diagonal 3×3 stiffness matrices is obtained from the corresponding equations for the three DOF v_2, ψ_2, and ϕ_2. These equations can be obtained by reversing the above procedure, that is, by clamping node 1 and requiring three arbtirary deflections at node 2. Since there is nothing new about that process, the details and results are omitted. To obtain, say, the upper-right off-diagonal stiffness submatrix from its corresponding flexibility matrix, the same equilibrium equations stated above can be used:

$$V_3 = -V_1 \qquad M_4 = V_1 R \sin\beta + M_1 \cos\beta - M_3 \sin\beta$$
$$M_6 = V_1 R(1 - \cos\beta) + M_1 \sin\beta + M_3 \cos\beta,$$

which in matrix form are

$$\begin{Bmatrix} V_3 \\ M_4 \\ M_6 \end{Bmatrix} = \begin{bmatrix} -1 & 0 & 0 \\ R\sin\beta & \cos\beta & -\sin\beta \\ R(1-\cos\beta) & \sin\beta & \cos\beta \end{bmatrix} \begin{Bmatrix} V_1 \\ M_1 \\ M_3 \end{Bmatrix}.$$

Now it is a matter of substituting the inverse of Eq. (3.3) for the right-hand vector. Once this off-diagonal 3×3 stiffness submatrix is thus obtained, the other such three by three off-diagonal stiffness matrix is just the transpose of this one because of the symmetry of the overall stiffness matrix.

Now for the deflections in the plane of curvature. The process for determining the flexibility matrix, and then its inverse, for deflections within the plane of the curved beam element, is, of course, much the same as that above. The values of the three DOF on the left-hand side of the beam element are calculated in terms of the six generalized element forces shown in Figure 3.19(b). This is accomplished by having three real deflections at the left-hand side corresponding to the three DOF at element node 1, while the curved beam element is cantilevered at node 2. In this case the equality between the complementary external virtual work and the complementary strain energy is

$$\delta W_{ex}^{*} = \int_0^L \frac{M_y\, \delta M_y}{EI} dx + \int_0^L \frac{N\, \delta N}{EA} dx,$$

where, again, dx is replaced by $R\,d\alpha$, and L is equal to $R\beta$. Again, shearing deflections are ignored. From Figure 3.19(b), the equilibrium equations for determining the element generalized forces at node 2 in terms of those at node 1, are

$$N_2 = -N_1 \cos\beta + V_2 \sin\beta \qquad V_4 = -N_1 \sin\beta - V_2 \cos\beta$$
$$M_5 = N_1 R(1 - \cos\beta) + V_2 R \sin\beta - M_2.$$

The signs on these equations can be checked by simply choosing convenient, fixed values of β. Mimicking the above equations, the moment and axial force at a typical position along the curved beam, that is, at α, are

$$M_y(\alpha) = N_1 R(1 - \cos\alpha) + V_2 R \sin\alpha - M_2 \qquad N(\alpha) = -N_1 \cos\alpha + V_2 \sin\alpha.$$

The three virtual load systems required for this three redundant load system are shown in Figure 3.20(b). The values of the virtual axial forces and the virtual bending moments for each of these virtual load systems (VLS) are

$$VLS1:\quad \delta N(\alpha) = -\delta N_1 \cos\alpha \quad \delta M(\alpha) = R\,\delta N_1(1-\cos\alpha)$$

$$VLS2:\quad \delta N(\alpha) = \delta V_2 \sin\alpha \quad \delta M(\alpha) = R\,\delta V_2 \sin\alpha$$

$$VLS3:\quad \delta N(\alpha) = 0 \quad \delta M(\alpha) = -\delta M_2.$$

The next step is to substitute the above actual loads and, in turn, the above virtual loads into the equality between the complementary external virtual work and the complementary strain energy. The result for the first virtual load system is

$$u_1\delta N_1 = \frac{R}{EA}\int_0^\beta [-N_1\cos\alpha + V_2\sin\alpha][-\delta N_1\cos\alpha]d\alpha$$

$$+\frac{R}{EI}\int_0^\beta [N_1R(1-\cos\alpha) + V_2R\sin\alpha - M_2][\delta N_1 R(1-\cos\alpha)]d\alpha.$$

For the second virtual load system

$$w_1\delta V_2 = \frac{R}{EA}\int_0^\beta [-N_1\cos\alpha + V_2\sin\alpha][\delta V_2\sin\alpha]d\alpha$$

$$+\frac{R}{EI}\int_0^\beta [N_1R(1-\cos\alpha) + V_2R\sin\alpha - M_2][\delta V_2 R\sin\alpha]d\alpha.$$

For the third virtual load system

$$\theta_1\delta M_2 = \frac{R}{EA}\int_0^\beta [-N_1\cos\alpha + V_2\sin\alpha][0]d\alpha$$

$$+\frac{R}{EI}\int_0^\beta [N_1R(1-\cos\alpha) + V_2R\sin\alpha - M_2][-\delta M_2]d\alpha.$$

These three equations simplify as

$$u_1 = N_1\left[\frac{(\beta + \tfrac{1}{2}\sin 2\beta)}{2EA} + \frac{R^2}{2EI_{yy}}(\beta - \tfrac{1}{2}\sin 2\beta)\right] - M_2\left[\frac{R}{EI_{yy}}\sin\beta\right]$$

$$-V_2\left[\frac{\sin^2\beta}{2EA} + \frac{R^2}{2EI_{yy}}\sin^2\beta\right]$$

$$w_1 = -N_1\left[\frac{\sin^2\beta}{2EA} + \frac{R^2}{EI_{yy}}(\sin^2\beta - 2 + 2\cos\beta)\right] - M_2\left[\frac{R}{EI_{yy}}(1-\cos\beta)\right]$$

$$+ V_2\left[\frac{\beta - \tfrac{1}{2}\sin 2\beta}{2EA} + \frac{R^2}{EI_{yy}}(\beta - \tfrac{1}{2}\sin 2\beta)\right]$$

$$\theta_1 = -N_1\left[\frac{R^2}{EI_{yy}}(\beta - \sin\beta)\right] + M_2\left[\frac{R}{EI_{yy}}\beta\right] - V_2\left[\frac{R^2}{EI_{yy}}(1 - \cos\beta)\right].$$

As before, these three equations can be cast in matrix form and the flexibility coefficient matrix can be inverted to obtain the corresponding stiffness submatrix. The inversion is possible because the beam is clamped at node 2, and thus no singularity inducing rigid body motion is possible.

As in the case of out-of-plane bending, looking through the back of the paper that bears the diagram for the generalized forces and generalized coordinates, so that node 2 appears to the left and node 1 appears on the right, the above equations can be adjusted so as to obtain the equations for u_2, w_2, and θ_2 in terms of N_2, V_2, and M_2. Of course, there are necessary sign changes on the slope and moment, and the axial displacement and axial force. The result is

$$u_2 = N_2\left[\frac{(\beta + \tfrac{1}{2}\sin 2\beta)}{2EA} + \frac{R^2}{2EI_{yy}}(\beta - \tfrac{1}{2}\sin 2\beta)\right] - M_5\left[\frac{R}{EI_{yy}}\sin\beta\right]$$

$$+ V_4\left[\frac{\sin^2\beta}{2EA} + \frac{R^2}{2EI_{yy}}\sin^2\beta\right]$$

$$w_2 = N_2\left[\frac{\sin^2\beta}{2EA} + \frac{R^2}{EI_{yy}}(\sin^2\beta - 2 + 2\cos\beta)\right] + M_5\left[\frac{R}{EI_{yy}}(1 - \cos\beta)\right]$$

$$+ V_4\left[\frac{\beta - \tfrac{1}{2}\sin 2\beta}{2EA} + \frac{R^2}{EI_{yy}}(\beta - \tfrac{1}{2}\sin 2\beta)\right]$$

$$\theta_2 = -N_2\left[\frac{R^2}{EI_{yy}}(\beta - \sin\beta)\right] + M_5\left[\frac{R}{EI_{yy}}\beta\right] + V_4\left[\frac{R^2}{EI_{yy}}(1 - \cos\beta)\right].$$

Now, as above, it is necessary to cast these three equations into matrix form and invert the flexibility coefficient matrix for the element generalized forces N_2, V_4, and M_5 in terms of the above three DOF. Then, using the previously determined equilibrium equations that provide the element generalized forces (N_2, V_4, and M_5) in terms of (N_1, V_2, and M_3), the quantities (N_1, V_2, and M_3) can also be written in terms of the above three DOF.

In the case of the three equations for the previous three DOF, v_1, ψ_1, and ϕ_1, there were no simplifications possible because GJ and EI are of the same order of magnitude. It is often the case that $R^2EA \gg EI$. When that is the case, all of the above terms with EA in the denominator can be discarded as too small to be of concern. That greatly simplifies the inversion of these equations and makes the in-plane curved beam more suitable for use in those cases where out of plane motions are effectively constrained.

4 FEM Equations of Motion for Elastic Systems

4.1 Introduction

Structural dynamic analyses usually begin with the preparation of data input to a commercially available finite element method (FEM) digital computer program with the capability of processing time-varying loads. The geometric, material, mass, and applied load data required are the same for all such programs and, for that matter, all such hand calculations. Generally unseen by the analyst, the finite element program takes the above input data and creates the computer's equivalent of the structure's equations of motion. The purpose of this and the next chapter is to teach what such FEM programs routinely do in the way of writing equations of motion for actual elastic structures by developing in the reader the ability to write such equations of motion by hand. Using FEM models of simple structures, this chapter concentrates on the two indispensable aspects of all structural system equations of motion: (i) the inertia forces in the form of a mass or inertia matrix multiplied by the second time derivatives of the generalized coordinates (called the generalized accelerations), $-[m]\{\ddot{q}\}$; and (ii) the linearly elastic restoring forces in the form of a stiffness or elastic matrix multiplied by the generalized coordinates, $-[k]\{q\}$. The negative sign is part of the elastic force description because, as will be seen, these are the forces acting on the system masses, which are equal and opposite to the forces acting on the system elastic elements. The negative sign on the inertia forces arises from rewriting Newton's second law as $\boldsymbol{F} - m\boldsymbol{a} = 0$ and treating the quantity $-m\boldsymbol{a}$, the inertia force, as just another force to be summed with the sum of the externally applied forces, $\boldsymbol{F}$.

Even though a vibration can occur in the absence of externally applied loads represented mathematically by the time-varying generalized force vector, $\{Q(t)\}$, the task of describing this vector is also addressed in this chapter by means of further practice in writing virtual work statements. A detailed discussion of the final component of the general linear form of the equations of motion, the energy dissipative forces, $-[c]\{\dot{q}\}$, is postponed to the next chapter. Later chapters deal with the processes used to solve the equations of motion.

By way of review, recall from Chapter 2 that stable pendulums are defined as mechanisms where gravity, or another acceleration field, always contributes a restoring force that causes the pendulum to swing back toward its static equilibrium position regardless of the direction of the deflection. Structures differ from pendulums in that, although gravity almost always loads a structure, there is at least one deflection direction where the gravity field does not supply forces that tend to return the structure to its static equilibrium position. The primary restoring forces for structures are their own internal elastic forces that result from the deformations of the structure. This conclusion can be reached by considering an initially straight, long, thin ruler. Let one end of the ruler (as a cantilevered beam) be held firmly at the edge of a table and the other, unsupported, end be deflected downward (an initial condition) and released. The ruler quickly moves back up to its originally straight configuration and then beyond that straight configuration to an upward deflection that closely resembles the initial downward deflection. The ruler then moves downward and then upward, and so on. In other words the ruler vibrates after release from its initial conditions. The forces that continually move the ruler up or down relative to its original, straight configuration cannot be external contact forces, because there are none. Nor can the restoring forces be because of the gravity field because the gravity field acts only downward. The only other possibility is the internal forces that result from the deformations of the ruler. Unless the deformations are large enough that the material is stressed beyond its yield point, those internal forces are entirely elastic forces and moments that can be closely approximated by familiar formulas from strength of materials.

Elastic forces were first encountered in this textbook in a particularly simple form in Chapter 2. Those were the forces that resulted from the inclusion of linearly elastic springs in some of the pendulum systems of that chapter. Although coiled spring elements are seldom found in actual structures, they are sometimes used in structural analyses as convenient mathematical representations of other, generally complicated, elastic bodies that the analyst does not want to model in detail. Therefore, for this reason and because of their simplicity, spring elements are again used in the hand calculations of this chapter. However, to give the example problems a greater resemblance to real structures, rod and beam structural elements are emphasized. Since plate elements, plain stress elements, and so on, are commonly included in FEM analysis in exactly the same way as spring, rod, and beam elements are included, there is no need to include these types of linear finite elements in the example problems that follow. These more complicated elements are mentioned only to remind the reader that they too could be included in the example problems in exactly the same way that the beam and spring elements are included, but at the expense of more computational bookkeeping.

4.2 Structural Dynamic Modeling

Every engineering structure has both inherent mass and elastic characteristics. The elastic modeling of a structure for a dynamic analysis is not different in style from that for a static analysis. Indeed, present computer speeds and storage capacities have grown so fast and large that it is now sometimes practical to use a very highly detailed static analysis model of a structure in a dynamic analysis. However, because the solution process for the dynamic problem is more complicated than that for a static analysis, the analyst is often prompted to select an elastic model for the dynamic analysis that is significantly simpler (fewer DOF) than that for the static analysis.

The primary difference between a static and dynamic model of a structure is the mass modeling. In a static analysis, the mass, by means of gravitational acceleration, provides only a set of, usually minor, externally applied static forces. In a dynamic analysis the inertia forces associated with the mass can be critical loads. To understand the necessity of mass modeling, consider the uniform, simply supported beam of length $8L$ shown in Figure 4.1(a). In addition to the weight loading, let that beam be subjected to a uniform, external loading per unit length symbolized as f_0. The force per unit length because of the weight of the beam is $\rho g A$, where ρ is the beam's mass density, g is the acceleration of gravity, and A is the beam's cross-sectional area. The usual sketch for a static analysis of this beam looks something like that shown in Figure 4.1(b). Then, for example, the static deflection at midspan can be calculated any number of ways with the result

$$w(4L) = -\frac{160}{3}\,\frac{(f_0 + \rho g A)L^4}{EI}.$$

Now let the f_0 portion of the external lateral loading vary with time in a rapid manner and now be symbolized as $f(t)$. Consider whether the solution for the midspan deflection is now

$$w(4L, t) \stackrel{?}{=} -\frac{160}{3}\,\frac{[f(t) + \rho g A]L^4}{EI}.$$

This answer is incorrect because it does not account for the additional loading resulting from the accelerations experienced by the mass of the beam all along the beam length as the beam moves. In other words, in a dynamic analysis, the motion of the mass induces a set of inertia loads that depend on the beam deflections (via accelerations) as well as the magnitude of the masses. Since these inertia loads in turn cause or modify the beam deflections, there exists an interaction of masses and deflections. Algebraic equations, such as the above, can never reflect such a feedback loop. Only differential equations, where the various load components can be written as proportional to deflections and the time derivatives of deflections, can describe the deflection–load feedback mechanism.

For the purposes of creating a mathematical model of the simple beam structure shown in Figure 4.1(a), one that is suitable for a dynamic analysis, the mass characteristics of the structure can be modeled in either of two distinct ways. The first form of mass description is simply to describe the mass of any particular structural element as being distributed as it actually is distributed throughout the structural element. This distributed mass approach is, of course, in good agreement with the reality of any individual structural element. However, when this approach is coupled with the FEM to form the "consistent mass" approach, as discussed in Endnote (4) of this chapter, the result is more computational expense without any increase in accuracy. When the distributed mass approach is coupled with the differential equation method of analysis, as discussed in Chapter 8, then closed-form, analytical solutions are generally feasible only when a beam structure consists only of one or two structural

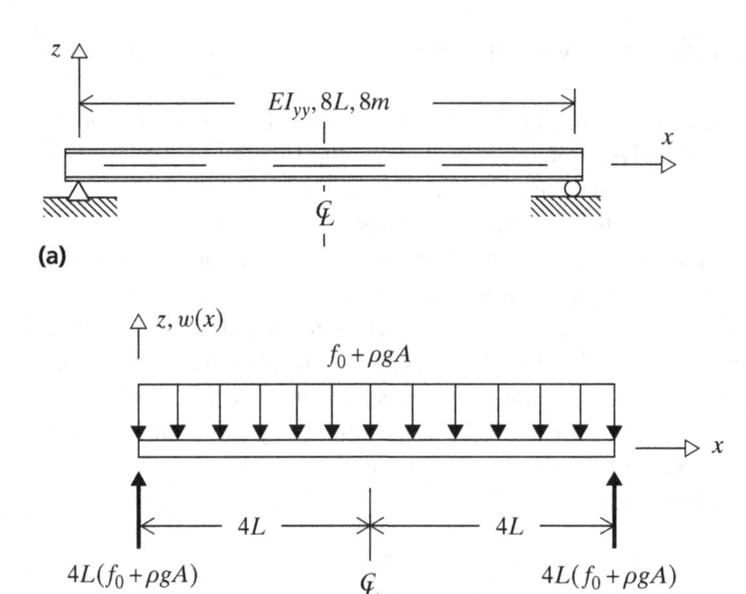

(b)

L L L L
m m m m

(c)

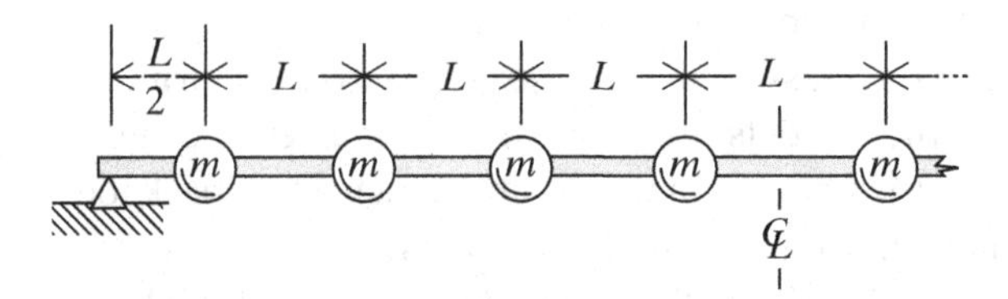

(d)

Figure 4.1. Beam mass modeling schemes. (a) Typical uniform beam of total length $8L$. (b) Simply supported beam subjected to gravity and uniformly applied load. (c) A possible mass modeling scheme. (d) An alternative mass modeling scheme.

elements, the mass density is constant, and the geometry is uniform.[1] Even for such simple structures as single plates, such restrictive conditions are necessary for open-form analytical solutions. Thus the use of distributed mass modeling is of only secondary importance.

The generally far more useful approach to the mass modeling of engineering structures is discrete mass modeling. Discrete mass modeling for vibrating structures is the

[1] There are some exceptions. See Chapter 8 for more on the benefits and difficulties of distributed mass modeling.

process by which the distributed mass of the real structure is collected by the analyst in the form of rigid masses at discrete, that is, isolated, points within the geometry of the structure. Mathematically, this mass discretization approach is equivalent to the reduction of each partial differential equation[2] of the distributed mass model into a series of ordinary differential equations. As an example of this type of mass modeling, consider again the simply supported, uniform beam of length $8L$, total mass $8m$, that is shown in Figure 4.1(a). Let the y and z axes be the principal[3] axes of the cross section, and let the beam-bending stiffness coefficient for the x, z plane be EI. Let this beam undergo only an up and down bending motion within the plane of the paper. To that end, let the loci of the beam cross section's shear centers coincide with the loci of centroids so that the beam is not twisting as it undergoes small bending deflections. The first step in the discrete mass modeling procedure is to concentrate all the uniformly distributed beam mass along the loci of the beam cross-section centroids to temporarily form a line of mass. This step is consistent with the strength of materials theory of beam bending in which the three-dimensional beam stress resultants, such as internal forces and moments, are concentrated along the axis that is the loci of the cross-section centroids. The second step is to gather up the mass distributed along the loci of centroids, at a relatively small number of selected centroidal points along the beam span, and thus form a necklace of discrete mass beads. The collection of the individual discrete masses of each structural element forms the model of the mass distribution for the entire structure.

In this beam example, a suitable mass modeling could be either of the two schemes shown in Figures 4.1(c) and (d). In both of these approximations, the beam is divided into eight equal lengths. In Figure 4.1(c), the first discretization scheme, the uniform mass in each segment of length L is totaled, and half of the total is placed at the end points of each segment. Thus at each interior end point, two segment masses of magnitude $m/2$ come together from adjacent beam segments to form one mass of magnitude m as shown in the figure. Now consider the half of the segment mass that is sent to a point where there is a rigid support. Since this mass of magnitude $m/2$ cannot move vertically as the beam bends (the only time-varying deflection under consideration), this end mass cannot develop any kinetic energy. Since the only energy a rigid mass can possibly develop is kinetic energy, this $m/2$ mass and the other such mass at the other support make no contribution to the energy terms of the Lagrange equations of motion for the beam. Thus all such nonmoving masses can be omitted from the beam dynamic model.[4]

The kinetic energy associated with the beam mass moments of inertia about a centroidal y axis, and the corresponding beam bending slope angular velocities, is also omitted from the mass model because this form of the beam kinetic energy proves to be quite small relative to the kinetic energy associated with the up-and-down motion

[2] A partial differential equation is required to describe the beam motion because there are two independent variables, distance, x, and time, t.

[3] The beam cross-sectional area moments of inertia about the principal axes are the maximum and minimum area moments of inertia, and the beam cross-sectional product of inertia is zero. Thus a beam deflection in the plane of the paper can occur without a concurrent deflection out of the plane of the paper.

[4] This loss of mass, usually quite small, must be accounted for when summing the total mass for a structural model as a check on the completeness of the mass modeling of the entire structure.

of the discrete beam masses. However, although the mass moments of inertia of the beam structural elements are not important, there is always the possibility that nonstructural masses attached to the elastic structure contribute substantial mass moments of inertia to the rigid masses of the mass model. These additional mass terms can be modeled as either integral to the discrete mass or, as is usually the case for computer-based models, as additional masses located at their own center of mass and connected to the structural mass by means of a short, and therefore rigid, connection. Both of these techniques are employed in example problems of this chapter.

In the second mass scheme, shown in Figure 4.1(d), the structural mass of each of the eight beam segments is collected at the midpoint of its segment. The elastic elements that connect the discrete, rigid masses in both these mass models are the same type of beam finite elements, but in this second case, there are nine beam elements (two with span lengths of $L/2$) connecting the eight masses, whereas in the first scheme there are only eight beam elements connecting seven masses. Endnote (1) presents results of calculations for the natural frequencies of this beam that is mathematically modeled using these two competing mass discretization schemes. Note again that: (i) the natural frequencies are numbered according to their increasing magnitude, (ii) the number of calculated natural frequencies cannot be greater than the number of DOF, and (iii) the higher numbered of the calculated natural frequencies for a given mathematical model are generally of poorer accuracy. The two tables of Endnote (1) show that the computer-calculated natural frequencies for the case where the beam segment mass is lumped at the segment ends and those frequencies for the case where the mass is lumped at the segment center are the same for equal numbers of beam segments. Moreover, the lower numbered half of either set of calculated frequencies agree quite well with what are called the "exact" natural frequency solutions that are calculated in Chapter 8. These exact solution results are strength of materials solutions obtained from solving a differential equation that models the beam as a mass and stiffness continuum. The Chapter 8 simply supported beam solution for the nth natural frequency and the nth deflection pattern, $w_n(x)$, is

$$\omega_n = \frac{n^2\pi^2}{64}\sqrt{\frac{EI}{mL^3}} \quad \text{and} \quad w_n(x) = \sin\frac{n\pi x}{L}.$$

A brief explanation of why these two models, with their different total number of masses, give the same natural frequency results is offered in Endnote (2). When the mass properties are discretized from an infinite number of points to a small number of points, the discrete model solution natural frequency results are, of course, only approximate. There is one thing about the two beam models of Figures 4.1(c) and (d) that is misleading. In these two models there are approximately equal numbers of elastic elements and rigid masses (mass elements). In almost all finite element models of engineering structures, the number of mass elements is much less than the number of elastic elements. That is, the description of the mass properties of a structural system does not have to be nearly as refined as the description of the elastic properties to obtain the same accuracy of solution. Since the number of equations to be solved depends on the number of masses, it is desirable to use as few masses as reasonably possible.

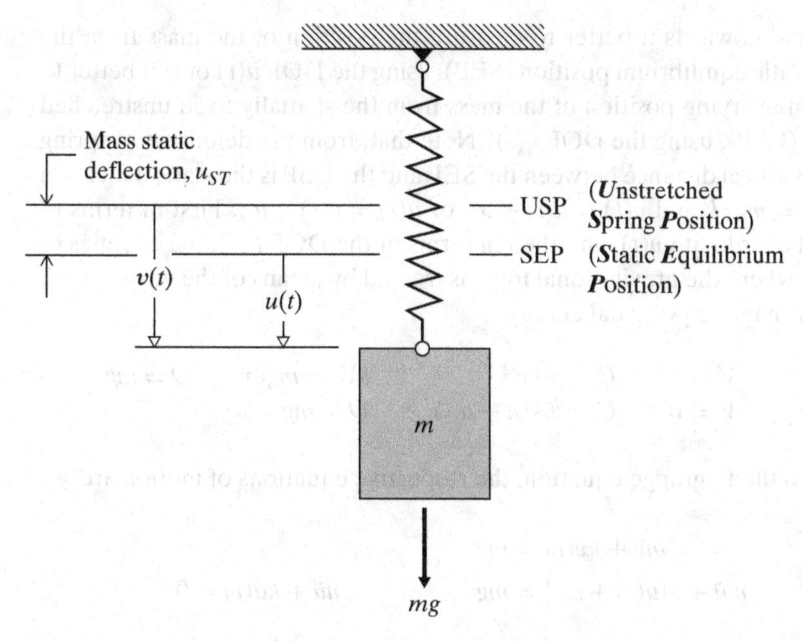

Figure 4.2. Simplest possible structural dynamic model: one spring element and one mass element.

In summary, the modeling of structures for the purposes of dynamic analyses almost always begins with (i) an elastic model that, with today's computer capacity, is sometimes as refined as it would be for a general static loading; and (ii) a far less refined mass model consisting of discrete masses that, when summed, roughly equal the mass of the structure.[5] The discrete masses are commonly and accurately, if inelegantly, referred to as "lumped" masses.

4.3 Isolating Dynamic from Static Loads

The previous section explained that structural dynamic models generally consist of discrete, rigid masses and a variety of elastic elements ranging from the simplest elements, springs and beams, to the most complicated. Structural dynamic models may also contain rigid elements other than lumped masses when one part of the structure is very much stiffer than other parts. The use of rigid elements helps avoid numerical ill-conditioning of the stiffness matrix. To illustrate the next important point, the first lumped mass structure to be analyzed has the simplest possible dynamic model: one lumped mass (with no flexibility) and one elastic spring (with no mass). See Figure 4.2. Since the spring only stretches or contracts, this drawing is meant to imply that the only motion possible for the single mass is an up-and-down motion. Since there is only one motion for the mass, this is a one-DOF system. However, the question arises as to which is the better choice for the zero value position of the single generalized coordinate that locates the vertical deflection of the mass as

[5] Again, one check on a computer generated mass model is to sum the mass properties and compare them with design or experimental results.

it vibrates up and down. Is it better to measure the position of the mass from the spatially fixed static equilibrium position (SEP), using the DOF $u(t)$ or is it better to measure the time-varying position of the mass from the spatially fixed unstretched spring position (USP), using the DOF $v(t)$? Note that, from the definition of spring stiffness, k, the vertical distance between the SEP and the USP is the static deflection of the mass $u_{st} = mg/k$ and $v(t) = u(t) + u_{st}$ or $u(t) = v(t) - u_{st}$. First in terms of the generalized coordinate $v(t)$, and then in terms of the DOF $u(t)$, the energies of this system are, where the gravitational force is treated by means of the virtual work equation rather than the potential energy,

$$T = \tfrac{1}{2}m\dot{v}^2 \qquad V = 0 \qquad U = \tfrac{1}{2}kv^2 \qquad \delta W = mg\delta v \Rightarrow Q = mg$$
$$T = \tfrac{1}{2}m\dot{u}^2 \qquad V = 0 \qquad U = \tfrac{1}{2}k(u + u_{st})^2 \qquad Q = mg.$$

Substituting into the Lagrange equation, the respective equations of motion are

$$m\ddot{v} + kv(t) = mg$$
$$\text{and} \qquad m\ddot{u} + k[u(t) + u_{st}] = mg \qquad \text{or} \qquad m\ddot{u} + ku(t) = 0.$$

The reader is invited to rederive these equations using the potential energy expression for the gravitational force rather than the virtual work expression. Clearly the equation in terms of the DOF measured from the SEP produces a slightly simpler equation by virtue of the fact that the static, gravitational force is offset in the equation of motion by the equal and opposite force in the spring that results from the static deflection of the spring.

The slight advantage obtained in this simplest of all structural dynamic models by using the static equilibrium position as the datum for the DOF $u(t)$ can be a substantial advantage in more complicated cases. To demonstrate this advantage that the general linear matrix equation of motion can be written without the least attention to *any* static (i.e., time-invariant) loads, consider the general, small-deflection matrix differential equation that includes both static loads Q_{st} and dynamic loads $Q(t)$ and their corresponding deflections. Such a division of loads into these two categories is permissible because the basic viewpoint is that load systems are independent of each other. Furthermore, these two categories are mutually exclusive and exhaustive. As is soon to be seen, this general lumped mass equation for small deflections is

$$[m]\{\ddot{q}\} + [c]\{\dot{q}\} + [k]\{q\} = \{Q_{st}\} + \{Q(t)\},$$

where, as discussed in detail in the next chapter, the square matrix $[c]$ and the generalized velocity vector are mostly used to describe energy dissipative forces. This matrix equation is a linear equation. Therefore, the total deflection vector $\{q(t)\}$ is the superposition of the static and dynamic deflections, that is, $\{q(t)\} = \{q_{st}\} + \{q_{dy}(t)\}$. The static deflections are specifically the solution to the static load equation $[k]\{q_{st}\} = \{Q_{st}\}$, which is the static equilibrium equation. This total deflection equation $\{q(t)\} = \{q_{st}\} + \{q_{dy}(t)\}$ means that the static deflections, $\{q_{st}\}$, are as always measured from the unloaded position of the structure (the USP), and the dynamic deflections, $\{q_{dy}(t)\}$, are measured from the static deflection position (the SEP). Substitution into the above matrix equation of motion of the total deflection in terms of its static and dynamic components leads to

$$[m]\{\ddot{q}_{dy}\} + [c]\{\dot{q}_{dy}\} + [k]\{q_{dy}\} + [k]\{q_{st}\} = \{Q_{dy}\} + \{Q_{st}\}$$

$$\text{or} \quad [m]\{\ddot{q}_{dy}\} + [c]\{\dot{q}_{dy}\} + [k]\{q_{dy}\} = \{Q_{dy}\},$$

where the second of these two equations results from subtracting the static equilibrium equation from the first equation. Thus the final form of the matrix equation of motion involves only the dynamic loads and their corresponding deflections. Therefore, all the static loads, regardless of their origins, have no place in the final form of the linear equation of motion. Thus, for example, the dynamic analysis of an aircraft need not, and thus would not, include the steady-state lift, thrust, or drag forces and moments or the gravitational forces. The dynamic analysis of a large building would not include steady-state winds or fixed gravitational loads. In summary, the effect of this important superposition conclusion is that, for all subsequent linear analyses:

(i) All generalized coordinates for the motion of the structure are to be measured from the deflected position of the structure that is produced by all the static loads acting on the structural system. That position is called the static equilibrium position (SEP).
(ii) Since the internal forces that result from the internal stresses produced by the static deflections, and the externally applied static loads cancel each other, that is, $-[k]\{q_{st}\} + \{Q_{st}\} = \{0\}$, neither of these types of forces is to be part of the equations of motion.

Remember that the simplicity obtained from totally ignoring all static loads and their effects depends on the total deflections remaining sufficiently small that no nonlinear effects become significant. Once the deflections become finite, that is, become large enough to have a significant effect on either the magnitudes or the lines of action of the internal and external loads relative to the geometry of the structure, the superposition of the static and dynamic deflections is no longer valid, and the generalized coordinates need to originate at the zero strain state.

There is one point of possible confusion that needs to be addressed. Following the above conclusion, gravity forces are now to be completely ignored. Why were they not also ignored in the analysis of pendulums where they are obviously essential to the analysis? The answer to this question is that in the case of all pendulums, the gravity field does not produce static generalized forces. Consider any of the pendulum problems of Chapter 2. In each case the static gravity force, mg, acting on each bob or center of mass has a time-varying moment arm. Hence, the static force and the time-varying moment arm produce a time-varying moment that is the generalized force for the system. However, for structures undergoing small displacements, the moment arms of the gravity forces are either constant or too small to be significant.

4.4 Finite Element Equations of Motion for Structures

Again, almost all structural dynamics analyses today are carried out using a large or a small commercial, finite element, digital computer program. When such programs and a suitable computer are available, it is quite unusual for it to be possible to do a cheaper, quicker, or more accurate structural analysis using some method other than the FEM. Before getting into the details of the use of finite element method in a dynamic analysis, by means of example problems, an overview of the matrix structural dynamics analysis is presented.

The basis for the structural equations of motion is the Lagrange equations. Thus it is necessary to discuss the matrix forms of the kinetic energy, potential energy, and strain energy. As is soon illustrated, a nice feature of elastic structures undergoing small vibrations is that the kinetic energy always depends exclusively on the generalized velocities of the discrete masses; that is, on the time derivatives of the DOF, and not on the DOF themselves or time, t, explicitly. Thus, for elastic structures, all the derivatives of the kinetic energy with respect to the generalized coordinates, the second terms of the Lagrange equations, are always zero. Hence, as will be seen, for the small vibrations of structures, the kinetic energy of the entire structure can always be written as

$$T(\dot{q}_1, \dot{q}_2, \ldots, \dot{q}_n) = \tfrac{1}{2}\lfloor \dot{q} \rfloor [M] \{\dot{q}\}.$$

This expression is, of course, just another manifestation of the familiar kinetic energy form of one-half the mass multiplied by the velocity squared. When developed from a kinetic energy expression as indicated above, the system mass matrix $[M]$ always is a symmetric matrix. As mentioned in Chapter 1, the mass matrix is also a positive definite matrix because the system kinetic energy $(\tfrac{1}{2}\Sigma m_i \dot{\boldsymbol{r}}_i \cdot \dot{\boldsymbol{r}}_i)$ is never a negative quantity. See Exercise 4.17 for a proof of these assertions regarding the mass matrix. As for the potential energy, the separation of dynamic from static (generalized) forces, as discussed previously, means that the gravitational potential energy of structures, V, can always be set to zero.

The strain energy, U, from the finite element method theory for any single, linearly elastic structural element, such as a small deflection beam element, has form parallel to that of the kinetic energy, which is

$$U(q_1, q_2, \ldots, q_n) = \tfrac{1}{2}\lfloor q \rfloor^{(e)} [k_e] \{q\}^{(e)},$$

where the element stiffness matrix $[k_e]$ is symmetric. As was demonstrated in the previous chapter and will be again demonstrated shortly, a stiffness matrix for a structure composed of a collection of connected linearly elastic structural elements, symbolized as $[K]$, is developed from such symmetric element stiffness matrices $[k_e]$ by a summation of all such element strain energies to obtain the system strain energy. That summation guarantees that all such resulting system (or global) stiffness matrices $[K]$ are also symmetric. The symmetry characteristic for both the element and global stiffness matrices follows from Maxwell's reciprocity theorem.[6] The positive definiteness of the global stiffness matrix $[K]$ is a direct result of the elastic strain energy, as a form of stored work, always being a positive quantity for any nonzero strain state. However, neither individual element stiffness matrices nor the summation of

[6] The basis of Maxwell's theorem is that the work done on an elastic body by a first set of loads moving through deflections caused by a second, independent load set is equal to the work done by the second load set moving through the deflections caused by the first load set. This leads to the conclusion that flexibility matrices, and hence stiffness matrices, are symmetric.

element stiffness matrices are positive definite until after the application of deflection boundary or other conditions that prohibit or remove rigid body motion.

The individual Lagrange equations for each structural DOF can be ordered according to the ordering of their DOF and then stacked one upon another to form the matrix equation. Collectively these equations with the DOF vector $\{q\}$ can be compactly symbolized as

$$\frac{d}{dt}\frac{\partial T}{\partial\{\dot{q}\}} + \frac{\partial U}{\partial\{q\}} = \{Q\}, \tag{4.1}$$

where, as always, $\{Q\}$ is the vector of generalized forces that are determined from the virtual work expression for the time-varying forces and moments. The above partial derivatives with respect to column vectors are *not* a new adventure in partial differentiation. They are only a means of indicating a succession of individual partial derivatives with respect to the individual elements of the column vector. When the resulting equations are then stacked one on top of the other, they can be recodified in the matrix form shown below. There is a reason, other than conciseness, for adopting this column vector notation for derivatives. That reason is that this notation gives the appearance that when (i) the triple matrix product expressions for the kinetic energy and the strain energy of the are formally substituted into Eq. (4.1), (ii) the usual product rule for derivatives is observed with the understanding that a derivative with respect to a row matrix is the same as a derivative with respect to a column matrix, and (iii) transposes are taken of separate double products involving the symmetric mass and symmetric stiffness matrices, the apparent result is

$$[M]\{\ddot{q}\} + [K]\{q\} = \{Q\}. \tag{4.2}$$

This "result," Eq. (4.2), for the partial differentiation in Eq. (4.1) can be verified by the much more rigorous reasoning presented in Endnote (3). Of course, the part the finite element method plays directly is that of the construction of the system or global stiffness matrix, $[K]$, from the element stiffness matrices, $[k_e]$, for each of the various elements of the total structure. The creation of the stiffness matrix is generally, by far, the most time-consuming part of the structural dynamics analysis preceding the solution of the matrix equation of motion. Thus nearly everything else in the analysis preparation should be focused on simplifying the preparation of, and reducing the size of, the stiffness matrix.

One possible way of reducing the size of the stiffness matrix, and thus the size of the problem, is now illustrated by considering again the simply supported beam of Figure 4.1(c). Let there not be any externally applied loads, nor any initial conditions specified. As is discussed at greater length later, all that can be done with an unloaded structure without specified initial conditions is to calculate the natural frequencies and the deflections associated with those natural frequencies. Since this single-beam structure is symmetric about its center, the mode shapes (the deflection pattern associated with each natural frequency) of this lumped mass model are either symmetric or antisymmetric about the beam center.[7] This knowledge allows the division of this

[7] Recall the statement that the nth mode shape for the distributed mass model beam of Figure 4.1(a) is $v_n(x) = \sin(n\pi x/8L)$. Thus odd values of n yield symmetric mode shapes, whereas even values of n produce antisymmetric mode shapes about the beam center.

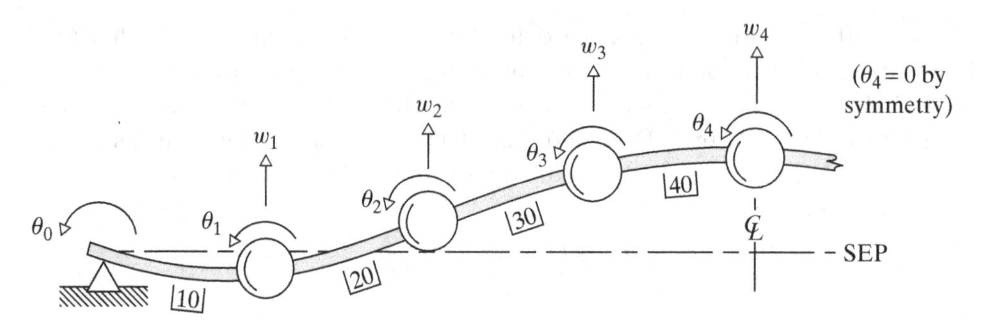

Figure 4.3. Exaggerated, general, symmetric beam deflection pattern. Since the static (and dynamic and combined) deflections are small, the beam static equilibrium position can be taken to be the straight line between support points.

analysis into two smaller analyses: one tailored to the symmetric mode shapes and frequencies and the second tailored to the antisymmetric mode shapes and frequencies. In either case, it is necessary only to focus on, say, the left half of the beam. This is, of course, just one example of an approach that greatly reduces the number of DOF, and thus the size of the stiffness matrix, to be considered.

To carry out the symmetric analysis of this beam, choose the mass modeling of Figure 4.1(c), which is more convenient for this hand analysis than the beam mass model of Figure 4.1(d), because there are only four, identical beam elements with which to contend. Each beam element has a lateral deflection DOF and a bending slope DOF at each of its two ends. Each of these four beams has the element stiffness matrix

$$\text{for beam bending:} \quad [k_e]\{q_e\} = \frac{EI}{L^3}\begin{bmatrix} 12 & 6L & -12 & 6L \\ 6L & 4L^2 & -6L & 2L^2 \\ -12 & -6L & 12 & -6L \\ 6L & 2L^2 & -6L & 4L^2 \end{bmatrix}\begin{Bmatrix} w_1 \\ \theta_1 \\ w_2 \\ \theta_2 \end{Bmatrix}. \qquad (e)$$

Of course, each pair of global DOF are located also at the structural element connection points, called the nodes of the structural system. See Figure 4.3. In terms of the global degrees of freedom, the element stiffness matrices are as follows. For beam element 10, where the first listed global lateral deflection DOF, w_0, must be zero because of the simple support BC at the left end of the beam element

$$[k_{10}]\{q_{10}\} = \frac{EI}{L^3}\begin{bmatrix} 4L^2 & -6L & 2L^2 \\ -6L & 12 & -6L \\ 2L^2 & -6L & 4L^2 \end{bmatrix}\begin{Bmatrix} \theta_0 \\ w_1 \\ \theta_1 \end{Bmatrix}.$$

For beam element 20, where none of the global DOF are zero,

$$[k_{20}]\{q_{20}\} = \frac{EI}{L^3}\begin{bmatrix} 12 & 6L & -12 & 6L \\ 6L & 4L^2 & -6L & 2L^2 \\ -12 & -6L & 12 & -6L \\ 6L & 2L^2 & -6L & 4L^2 \end{bmatrix}\begin{Bmatrix} w_1 \\ \theta_1 \\ w_2 \\ \theta_2 \end{Bmatrix}.$$

For beam 30, which is just like beam element 20,

$$[k_{30}]\{q_{30}\} = \frac{EI}{L^3}\begin{bmatrix} 12 & 6L & -12 & 6L \\ 6L & 4L^2 & -6L & 2L^2 \\ -12 & -6L & 12 & -6L \\ 6L & 2L^2 & -6L & 4L^2 \end{bmatrix}\begin{Bmatrix} w_2 \\ \theta_2 \\ w_3 \\ \theta_3 \end{Bmatrix}.$$

For beam element 40, where θ_4 must be zero because of the symmetry of the complete beam deflection pattern in this analysis,

$$[k_{40}]\{q_{40}\} = \frac{EI}{L^3}\begin{bmatrix} 12 & 6L & -12 \\ 6L & 4L^2 & -6L \\ -12 & -6L & 12 \end{bmatrix}\begin{Bmatrix} w_3 \\ \theta_3 \\ w_4 \end{Bmatrix}.$$

Assembling (i.e., superimposing) the four-element beam stiffness matrices to create the global stiffness matrix is accomplished by placing all the element stiffness matrix terms into the global matrix in row and column positions that correspond to their respective global DOF and summing where necessary. For example, in the third row of $[K]$ corresponding to θ_1, and the second column corresponding to w_1, add $-6L$, $+6L$, 0, 0 from elements 10, 20, 30, and 40, respectively, to get the global stiffness matrix entry of zero. This summation process can be accomplished quickly when it is done one element stiffness matrix at a time. In this example the assembly process produces the following global stiffness matrix and deflection vector for the eight global DOF of the left half of the complete beam

$$[K]\{q\} = \frac{EI}{L^3}\begin{bmatrix} 4L^2 & -6L & 2L^2 & 0 & 0 & 0 & 0 & 0 \\ -6L & 24 & 0 & -12 & 6L & 0 & 0 & 0 \\ 2L^2 & 0 & 8L^2 & -6L & 2L^2 & 0 & 0 & 0 \\ 0 & -12 & -6L & 24 & 0 & -12 & 6L & 0 \\ 0 & 6L & 2L^2 & 0 & 8L^2 & -6L & 2L^2 & 0 \\ 0 & 0 & 0 & -12 & -6L & 24 & 0 & -12 \\ 0 & 0 & 0 & 6L & 2L^2 & 0 & 8L^2 & -6L \\ 0 & 0 & 0 & 0 & 0 & -12 & -6L & 12 \end{bmatrix}\begin{Bmatrix} \theta_0 \\ w_1 \\ \theta_1 \\ w_2 \\ \theta_2 \\ w_3 \\ \theta_3 \\ w_4 \end{Bmatrix}.$$

The global stiffness matrix and global degree of freedom matrix for the antisymmetric analysis is assembled from the same matrices detailed above for beam elements 10, 20, and 30. The only difference between the symmetric and antisymmetric analyses is in the matrices for beam element 40. For the antisymmetric case, $w_4 = 0$, and θ_4 is not zero. The remainder of the detailing of the antisymmetric form of the beam element 40 stiffness matrix, and the assembly of the antisymmetric global stiffness matrix, are left to the reader as an exercise.

It is quite possible to combine the separate symmetric and antisymmetric analyses into one analysis. The price to be paid for that action is the number of global degrees of freedom for the combined analysis is fully twice that for either previous analysis, and the stiffness and mass matrices have four times as many entries. This would roughly mean four times the computational effort when solving Eq. (4.2). That choice still may be the more convenient choice for a small problem such as this example problem. However, for the analysis of a structure with many more elements and elements of greater complexity than a few beam elements, it is likely that the analyst would want to take the time to sort out the symmetric modal solutions from the antisymmetric solutions when possible.

The mass matrix for the symmetric beam analysis is determined from the kinetic energy expression. After, for the sake of exposition, temporarily assigning moment of inertias, H, to each full discrete mass, the kinetic energy for the half beam of Figure 4.3 is

$$T = \tfrac{1}{2}m\left(\dot{w}_1^2 + \dot{w}_2^2 + \dot{w}_3^2 + \tfrac{1}{2}\dot{w}_4^2\right) + \tfrac{1}{2}H\left(\tfrac{1}{2}\dot{\theta}_0^2 + \dot{\theta}_1^2 + \dot{\theta}_2^2 + \dot{\theta}_3^2\right).$$

Thus, the first part of Eq. (4.2) for the symmetric deflection analysis, $[m]\{\ddot{q}\}$, is

$$\begin{bmatrix} \tfrac{1}{2}H & & & & & & & \\ & m & & & & & & \\ & & H & & & & & \\ & & & m & & & & \\ & & & & H & & & \\ & & & & & m & & \\ & & & & & & H & \\ & & & & & & & \tfrac{1}{2}m \end{bmatrix} \begin{Bmatrix} \ddot{\theta}_0 \\ \ddot{w}_1 \\ \ddot{\theta}_1 \\ \ddot{w}_2 \\ \ddot{\theta}_2 \\ \ddot{w}_3 \\ \ddot{\theta}_3 \\ \ddot{w}_4 \end{Bmatrix}.$$

The presence of the one-half factor in the first entry in the mass matrix is the result of having only half of the distributed mass of beam element 10 assigned to the discrete mass at the left support. The one-half of the last entry in the mass matrix is the result of the beam's center mass being located on the axis of symmetry. This central location dictates that half of that mass be included in the description of the left half of the beam, whereas the other half would be part of the model for the right half of the beam.

The moments of inertia for each mass, H, have not been mentioned in connection with this beam model previous to this description of the kinetic energy. Recall that the analyst rarely bothers to make moment of inertia estimates for structural mass. Usually, only when the beam or other structural element directly supports nonstructural masses that move with the elastic element are mass moment of inertia estimates included in the analysis. The nonstructural masses could be, for example, machinery, in the case of heavy structures, or electronics, in the case of light structures. If the nonstructural masses cannot be modeled as rigid, then at least a crude finite element model would be necessary for these items, making them part of the structural system. If the nonstructural masses are partially filled tanks of liquids, then the problem is more complicated, and the reader is referred to the technical journals.

Although, it is seldom worthwhile to estimate the moment of inertia terms for the beam itself, there is an elegant way to make such estimates about which the reader should be aware. As mentioned previously, the mass moments of inertia of the beam segments themselves can be estimated by the use of the *consistent mass matrix* approach. This approach is based on the use in the beam kinetic energy integral of the same cubic spline finite element shape functions that are used in the strain energy integral to develop the element stiffness matrices. See Endnote (4) for a discussion of this element mass matrix.

If there are no nonstructural masses to enter into the mass matrix, and, if the routine decision is made not to undertake the expense of machine calculations for the mass moments of inertia terms of consistent mass matrices, then the mass matrix and acceleration vector for the symmetric analysis of the beam of Figure 4.3 can be rearranged to have the form

$$m\begin{bmatrix} 1 &&&&&&& \\ & 1 &&&&&& \\ && 1 &&&&& \\ &&& \tfrac{1}{2} &&&& \\ &&&& 0 &&& \\ &&&&& 0 && \\ &&&&&& 0 & \\ &&&&&&& 0 \end{bmatrix}\begin{Bmatrix} \ddot{w}_1 \\ \ddot{w}_2 \\ \ddot{w}_3 \\ \ddot{w}_4 \\ \ddot{\theta}_0 \\ \ddot{\theta}_1 \\ \ddot{\theta}_2 \\ \ddot{\theta}_3 \end{Bmatrix}.$$

Any rearrangement of the rows and columns of the mass matrix and the rows of the acceleration vector requires exactly the same rearrangement for rows and columns of the stiffness matrix and the deflection vector. This rearrangement where null submatrices are created in the mass matrix allows for a beneficial reduction in the size of the matrix equations of motion. To be a bit more general, introduce an arbitrary vector of externally applied time-varying generalized loads $\{Q(t)\}$ and rewrite the complete form of the reordered equations of motion as

$$\begin{bmatrix} M & 0 \\ 0 & 0 \end{bmatrix}\begin{Bmatrix} \ddot{w} \\ \ddot{\theta} \end{Bmatrix} + \begin{bmatrix} K_{ww} & K_{w\theta} \\ K_{\theta w} & K_{\theta\theta} \end{bmatrix}\begin{Bmatrix} w \\ \theta \end{Bmatrix} = \begin{Bmatrix} Q_w \\ Q_\theta \end{Bmatrix}.$$

When the matrix products of the submatrices are written out, the result is

$$[M]\{\ddot{w}\} + [K_{ww}]\{w\} + [K_{w\theta}]\{\theta\} = \{Q_w\}$$
$$[K_{\theta w}]\{w\} + [K_{\theta\theta}]\{\theta\} = \{Q_\theta\}.$$

Using a matrix inverse, the second of these two matrix equations can be solved for $\{\theta\}$, and that result substituted into the first of these two equations. The result is

$$[M]\{\ddot{w}\} + \left[K_{ww} - K_{w\theta}K_{\theta\theta}^{-1}K_{\theta w}\right]\{w\} = \{Q_w\} - \left\{K_{w\theta}K_{\theta\theta}^{-1}Q_\theta\right\}. \tag{4.3}$$

This process is illustrated in Example 6.7.

The size of this equation of motion is only the size of $\{w\}$, which is always approximately half the size of the original vector of unknown deflections. In addition, by means of a judicious choice on where to locate the number of lumped masses necessary for the accuracy desired, this process can be used to eliminate as much as three-quarters or more of the original number of DOF. There is, of course, a price to be paid for this important advantage. It is necessary to use software that causes the computer to either invert the submatrix $[K_{\theta\theta}]$ by whatever means and carry out three matrix multiplications or follow the much more economical Guyan reduction procedure discussed briefly in Ref. [4.1]. Especially with the choice of the Guyan reduction procedure, it is worth this computational price in almost all circumstances. Note that because $[K_{\theta\theta}]$ is a square submatrix of a positive definite matrix, it is always invertible. Also note that the new stiffness matrix $[K_{ww} - K_{w\theta}K_{\theta\theta}^{-1}K_{\theta w}]$ is symmetric because $[K_{w\theta}]$ is the transpose of $[K_{\theta w}]$. The importance of the symmetry of the mass and stiffness matrices is explained when solutions to the matrix equations of motion are sought.

The foregoing discussion of the details of a simple beam problem was made overly long by excursions into the use of symmetry, the reduction of the number of DOF, and modeling for moments of inertia. The writing of the finite element equations of motion is normally done automatically by software after the creation of the finite element model. Even the creation of the finite element model is now quite often computer assisted using software referred to as "preprocessors." Postprocessors display the analysis results graphically in easily understood formats. The current software trend is to integrate pre- and postprocessors with all sorts of analysis and design packages. To understand what the analysis program is doing for the analyst, the following section offers example problems.

4.5 Finite Element Example Problems

The following example problems are intended to emphasize the creation of the mass matrices and the applied load vectors of the finite element matrix equations of motion. The stiffness matrices are also provided for those who would like to challenge themselves on the assembly of those matrices. The first two sets of problems are planar grid problems, whereas the last pair of related problems are planar frame problems. The planar beam grid and beam frame example problems are limited to beams with zero and 90° angles of orientation. This restricted geometry is sufficient for explanatory purposes. The reason that beams oriented at odd angles in three-space are avoided here is because such beams, like all finite elements at odd angles, require the introduction of a rotation matrix transformation between the element DOF and the system DOF of the form $\{q\}^{(e)} = [R_e]\{q\}$. As explained in the latter part of Section 3.5, the beam rotation matrix $[R]$ is a matrix of direction cosines that identifies the spatial direction of the beam axis. Thus the element stiffness matrices in terms of system coordinates become $[R_e]^t[k_e][R_e]$. Although this requirement for rotation matrices for beams with odd orientations makes hand calculations unnecessarily complicated, that is not the case for bars. Hence, one planar truss example problem is provided.

The example problems are mostly presented in sets of three, where the first problem of the set involves simpler mass modeling and externally applied, time-varying loads. The middle example extends the first problem by slightly complicating the mass modeling. The last of the related example problems deals with the same structure as the middle problem, but this time the structure is subjected to a time-varying base motion, also called a support motion. Base motions, which result in equivalent applied loads, are particularly important because they are so common. Sources of support motions range from earthquakes to rough roads, runways, and seas.

EXAMPLE 4.1 Write the matrix equation of motion for the small beam grid of Figure 4.4(a) for lateral (bending and twisting) motions out of the plane of the beam grid. The relatively very large beam axial stiffnesses prevent any significant motion in the plane of the grid. At each of the two internal nodes, that is, at the internal beam junctures, there is a large nonstructural mass that is modeled as a rigid body.

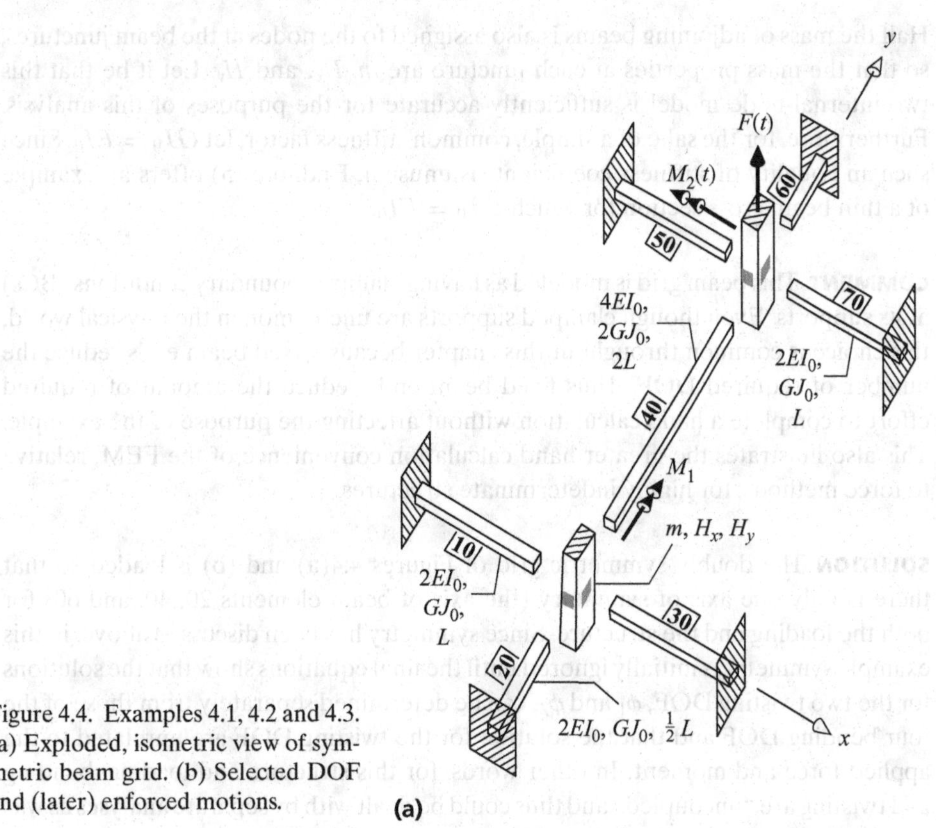

Figure 4.4. Examples 4.1, 4.2 and 4.3. (a) Exploded, isometric view of symmetric beam grid. (b) Selected DOF and (later) enforced motions.

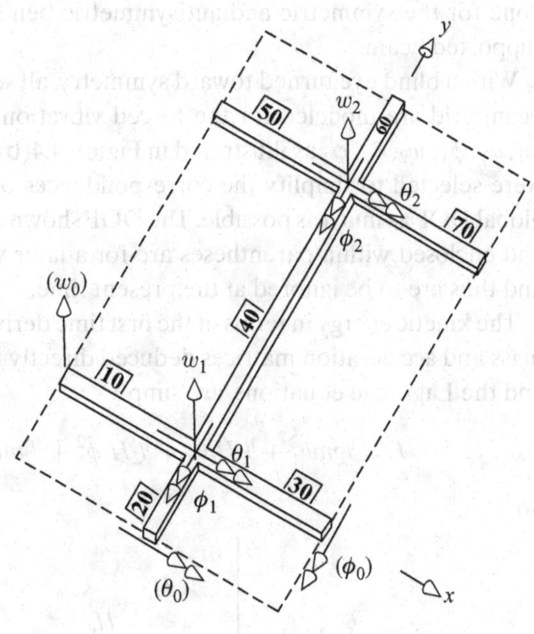

Half the mass of adjoining beams is also assigned to the nodes at the beam junctures, so that the mass properties at each juncture are m, H_x, and H_y. Let it be that this two-internal-node model is sufficiently accurate for the purposes of this analysis. Furthermore, for the sake of a simple, common stiffness factor, let $GJ_0 = EI_0$. Since such an equality of stiffness coefficients is unusual, Endnote (5) offers an example of a thin beam cross section for which $GJ_0 = EI_0$.

COMMENT This beam grid is modeled as having clamped boundary conditions (BCs) at its supports. Even though clamped supports are uncommon in the physical world, this choice is common throughout this chapter because fixed beam ends reduce the number of required DOF. Thus fixed beam ends reduce the amount of required effort to complete a hand calculation without affecting the purpose of the example. This also illustrates the greater hand calculation convenience of the FEM, relative to force methods, for highly indeterminate structures.

SOLUTION The doubly symmetric grid of Figures 4.4(a) and (b) is loaded so that there is only one axis of symmetry (the axis of beam elements 20, 40, and 60) for both the loading and the structure. Since symmetry has been discussed above, in this example symmetry is initially ignored until the final equations show that the solutions for the two twisting DOF, ϕ_1 and ϕ_2, can be determined separately from those of the four bending DOF and that the solution for the twisting DOF are unrelated to the applied force and moment. In other words, for this structure and loading, bending and twisting are "uncoupled" and thus could be dealt with by separate analyses as was done for the symmetric and antisymmetric bending of the above-discussed simply supported beam.

With a blind eye turned toward symmetry, all seven beams and both masses of the beam grid are modeled for the forced vibration analysis. Choose the global DOF $w_1, \theta_1, \phi_1, w_2, \theta_2, \phi_2$ as illustrated in Figure 4.4(b). The directions of the global DOF were selected to simplify the correspondences between the element DOF and the global DOF as much as possible. The DOF shown in Figure 4.4(b) with zero subscripts and enclosed within parentheses are for a later variation on this example problem and thus are to be ignored at the present time.

The kinetic energy in terms of the first time derivatives of the six DOF, and then the mass and acceleration matrices deduced directly from the kinetic energy expression and the Lagrange equations, are simply

$$T = \tfrac{1}{2}m\dot{w}_1^2 + \tfrac{1}{2}H_x\dot{\theta}_1^2 + \tfrac{1}{2}H_y\dot{\phi}_1^2 + \tfrac{1}{2}m\dot{w}_2^2 + \tfrac{1}{2}H_x\dot{\theta}_2^2 + \tfrac{1}{2}H_y\dot{\phi}_2^2.$$

So

$$[m]\{\ddot{w}\} = \begin{bmatrix} m & & & & & \\ & H_x & & & & \\ & & H_y & & & \\ & & & m & & \\ & & & & H_x & \\ & & & & & H_y \end{bmatrix} \begin{Bmatrix} \ddot{w}_1 \\ \ddot{\theta}_1 \\ \ddot{\phi}_1 \\ \ddot{w}_2 \\ \ddot{\theta}_2 \\ \ddot{\phi}_2 \end{Bmatrix}.$$

As per usual the stiffness matrix requires a bit more effort for a hand calculation. Consider beam element 10. As is the case for all subsequent beam elements, the task here is to adapt the standard forms of the beam bending and twisting stiffness matrices and their deflection vectors to the circumstances of beam 10. To this end, the first thing to notice is that in response to the applied loading, M_2, beam 10, like some other beam elements of this structure, twists, as well as bends, as the grid vibrates out of its plane. Therefore combine the standard element beam bending and twisting stiffness matrices into one beam 10 element stiffness matrix. The beam 10 rows and columns are arranged to conform with the ordering of the nonzero global DOF at the juncture of beam elements 10, 20, and 30. In this text, the usual ordering of the global DOF for any node of a planar structure is (i) rectilinear deflections by alphabetical order, (ii) bending slopes, and (iii) angles of twist. In beam 10, the element bending slope DOF correspond to the global twisting DOF, ϕ_1, and the element twisting DOF correspond to the global bending slope DOF, θ_1. Thus the bending slope and twisting rows and columns in the standard beam element stiffness matrix need to be interchanged so as to correspond to the order of the DOF in the global stiffness matrix. The standard beam element stiffness matrix also needs to be adjusted so that all three DOF at the left end of the beam element are zero, and the bending stiffness factor is $2EI_0$ rather than simply EI_0. When these adjustments have been made, then the element stiffness matrix and its corresponding global DOF vector are

$$[k_{10}]\{w_{10}\} = \frac{EI_0}{L^3}\begin{bmatrix} 24 & & -12L \\ & L^2 & \\ -12L & & 8L^2 \end{bmatrix}\begin{Bmatrix} w_1 \\ \theta_1 \\ \phi_1 \end{Bmatrix}.$$

The element stiffness equation for beam element 50 is exactly the same but for subscripts two replacing subscripts one in the deflection vector. The same sort of change in subscripts holds true for beam elements 30 and 70. The stiffness matrices for element 30 and the remaining beam elements are

$$[k_{30}]\{w_{30}\} = \frac{EI_0}{L^3}\begin{bmatrix} 24 & & 12L \\ & L^2 & \\ 12L & & 8L^2 \end{bmatrix}\begin{Bmatrix} w_1 \\ \theta_1 \\ \phi_1 \end{Bmatrix}$$

$$[k_{20}]\{w_{20}\} = \frac{EI_0}{L^3}\begin{bmatrix} 192 & -48L & \\ -48L & 16L^2 & \\ & & 2L^2 \end{bmatrix}\begin{Bmatrix} w_1 \\ \theta_1 \\ \phi_1 \end{Bmatrix}$$

$$[k_{60}]\{w_{60}\} = \frac{EI_0}{L^3}\begin{bmatrix} 192 & 48L & \\ 48L & 16L^2 & \\ & & 2L^2 \end{bmatrix}\begin{Bmatrix} w_2 \\ \theta_2 \\ \phi_2 \end{Bmatrix}$$

$$[k_{40}]\{w_{40}\} = \frac{EI_0}{L^3}\begin{bmatrix} 6 & 6L & & -6 & 6L & \\ 6L & 8L^2 & & -6L & 4L^2 & \\ & & L^2 & & & -L^2 \\ -6 & -6L & & 6 & -6L & \\ 6L & 4L^2 & & -6L & 8L^2 & \\ & & -L^2 & & & L^2 \end{bmatrix}\begin{Bmatrix} w_1 \\ \theta_1 \\ \phi_1 \\ w_2 \\ \theta_2 \\ \phi_2 \end{Bmatrix}.$$

When all seven element stiffness matrices are assembled into the global stiffness matrix, the result is

$$[K]\{w\} = \frac{EI_0}{L^3}\begin{bmatrix} 246 & -42L & & -6 & 6L & \\ -42L & 26L^2 & & -6L & 4L^2 & \\ & & 19L^2 & & & -L^2 \\ -6 & -6L & & 246 & 42L & \\ 6L & 4L^2 & & 42L & 26L^2 & \\ & & -L^2 & & & 19L^2 \end{bmatrix}\begin{Bmatrix} w_1 \\ \theta_1 \\ \phi_1 \\ w_2 \\ \theta_2 \\ \phi_2 \end{Bmatrix}.$$

As mentioned above, an inspection of the mass and stiffness matrices shows a lack of matrix coupling[8] between the w's and θ's, on one hand, and the ϕ's, on the other hand. Therefore it is mathematically clear that physically the mass of this beam grid can twist about the y axis without deflecting laterally (i.e., moving in the z direction) or twisting about the x axis. This ability of the mass of this symmetric beam grid to twist without having to have concurrent lateral deflections should be studied until it is evident. Typically, this lack of coupling means the 6×6 matrix equation could be separated into a 4×4 matrix equation and a 2×2 matrix equation, which would reduce the solution effort. Such reductions would be important in a larger problem.

The last step in the preparation of the finite element matrix equations of motion is writing the expression for the virtual work to obtain the generalized forces acting on the grid. In this case the virtual work expression is particularly simple:

$$\delta W = 0\delta w_1 + 0\delta\theta_1 - M_2\delta\phi_1 + F(t)\delta w_2 - M_1(t)\delta\theta_2 + 0\delta\phi_2 = \lfloor Q \rfloor\{\delta w\}$$

$$\text{so} \quad \lfloor Q \rfloor = \lfloor 0 \quad 0 \quad -M_2 \quad +F(t) \quad -M_1(t) \quad 0 \rfloor.$$

Putting the three above parts of the equation of motion together provides the complete matrix equation $[M]\{\ddot{w}\} + [K]\{w\} = \{Q\}$. ★

Note the dashed line in Figure 4.4(b) that encloses the beam grid. Consider for the moment that this dashed line is the edge of a thin, flat plate. Then the beam grid becomes a reinforcement for this plate. The point is that this variation on the problem does not require additional DOF to account for the deflections of the six plate bending elements that are now part of the elastic model, as long as they, like the beams, have clamped BCs at their outer edges. The inclusion of the plate elements

[8] "Coupling" terms within a matrix are those off-diagonal matrix entries that mathematically connect one type of displacement to another. For example, through the mechanism of the (1,2) term of the above grid stiffness matrix, the y-direction bending slope term θ_1 causes an internal, lateral force at interior node 1. That force, in turn, causes a vertical deflection w_1. Thus the (1,2), and the similar (2,1), entries couple θ_1 and w_1 so that one cannot occur without the occurrence of the other.

would require only the additional superposition of the plate element stiffness entries (i.e., more stiffness) in the global stiffness matrix and the addition of parts of the plate masses to the mass terms of the mass matrix.

EXAMPLE 4.2 Redo the previous example of Figure 4.4 where this time the centers of mass of both of the discrete masses at the beam junctures are a distance h above the plane of the grid rather than located at the origins of the generalized coordinates in the plane of the grid.

SOLUTION Since the mass centers are now higher relative to the plane of the beam grid, they have more rotational inertia about the nodes. Thus it can be expected that the natural frequencies of the beam grid are now lower than what they were in Example 4.5 because (i) the stiffness of the structure has not been altered and (ii), as touched upon lightly in Chapter 2, the natural frequencies depend on the square root of the ratio of stiffness to mass terms.

Again, the stiffness matrix is wholly unaffected by the higher placement of the lumped mass centers of mass. In the case of these three applied loads, the virtual work is also unchanged. The only thing that changes is the kinetic energy expression. The kinetic energy has to be augmented by the horizontal velocity components at the centers of mass resulting from the rotations at the nodes where the DOF are located. The new expression for the kinetic energy and the new mass matrix are

$$T = \tfrac{1}{2} m\dot{w}_1^2 + \tfrac{1}{2}(H_x + mh^2)\dot{\theta}_1^2 + \tfrac{1}{2}(H_y + mh^2)\dot{\phi}_1^2 + \tfrac{1}{2} m\dot{w}_2^2 + \tfrac{1}{2}(H_x + mh^2)\dot{\theta}_2^2 + \tfrac{1}{2}(H_y + mh^2)\dot{\phi}_2^2$$

so

$$[m]\{\ddot{w}\} = \begin{bmatrix} m & & & & & \\ & H_x + mh^2 & & & & \\ & & H_y + mh^2 & & & \\ & & & m & & \\ & & & & H_x + mh^2 & \\ & & & & & H_y + mh^2 \end{bmatrix} \begin{Bmatrix} \ddot{w}_1 \\ \ddot{\theta}_1 \\ \ddot{\phi}_1 \\ \ddot{w}_2 \\ \ddot{\theta}_2 \\ \ddot{\phi}_2 \end{Bmatrix}.$$

★

EXAMPLE 4.3 Redo the beam grid problem of Example 4.1, where this time, rather than externally applied forces and moments driving the structure, there are the three following simultaneous base motions illustrated in Figure 4.4(b): (i) the base support for beam element 10 is moving up and down with a time-varying motion $w_0(t)$, (ii) the base support of beam element 20 is rotating about an axis parallel to the y axis with a time variation $\theta_0(t)$, and (iii) the base support of beam element 30 is rotating about the x axis with a motion $\phi_0(t)$. The other boundary conditions remain unchanged. The positive directions of these known, enforced motions, as illustrated, match the positive directions of the global DOF.

SOLUTION The substitution of enforced base motions for applied forces and moments has no effect whatever on the system mass and stiffness matrices. They and their associated generalized acceleration and deflection DOF vectors are wholly unchanged. The only changes take place in the generalized force vector. There is a new generalized force vector $\{Q\}$ because the applied base motions cause accelerations in the mass of the structure, and acceleration multiplied by mass is the same as force. The mathematically discrete masses of the analysis model are set in motion through the agency of the elastic beams that connect the discrete masses to the moving supports. The generalized force vector, which is now a set of what are called equivalent forces and moments, reflects this transmission of the support motion to the masses by means of the stiffnesses of the structural elements connecting the moving supports to the discrete masses.

The calculation of the magnitudes of these equivalent forces and moments is a conceptually straightforward procedure using the element stiffness matrices. That is, the procedure is simple if a software package keeps track of the sign conventions. Consider beam element 10 as pictured in Figures 4.4(a) and (b). The bending and torsional stiffness matrix for beam element 10, in terms of the global DOF and a nonzero vertical motion at the left end, is

$$[k_{10}]\{w\} = \frac{EI_0}{L^3}\begin{bmatrix} 24 & -24 & 0 & 12L \\ -24 & 24 & 0 & -12L \\ 0 & 0 & L^2 & 0 \\ 12L & -12L & 0 & 8L^2 \end{bmatrix}\begin{Bmatrix} w_0 \\ w_1 \\ \theta_1 \\ \phi_1 \end{Bmatrix} = \begin{Bmatrix} \text{Force at wall} \\ \text{Force, rt. end} \\ \text{Torque, rt. end} \\ \text{Moment, rt. end} \end{Bmatrix}.$$

Here, of course, the quantity $w_0(t)$ is not an unknown deflection as are the DOF $w_1(t)$, $\phi_1(t)$ and $\theta_1(t)$. That is, $w_0(t)$ is already a known quantity, as opposed to the DOF, which are to be determined in the solution phase of the analysis. Hence, to facilitate the solution process, the known motion $w_0(t)$ needs to be separated from the unknown deflections in the DOF vector. Separate $w_0(t)$ from the unknown DOF in the following manner:

$$\begin{Bmatrix} \text{Force, rt. end} \\ \text{Mom't, rt. end} \\ \text{Torque, rt. end} \end{Bmatrix} = \frac{EI_0}{L^3}\begin{Bmatrix} -24 \\ 0 \\ 12L \end{Bmatrix} w_0(t) + \frac{EI_0}{L^3}\begin{bmatrix} 24 & 0 & -12L \\ 0 & L^2 & 0 \\ -12L & 0 & 8L^2 \end{bmatrix}\begin{Bmatrix} w_1 \\ \theta_1 \\ \phi_1 \end{Bmatrix},$$

where the first row (the first equation) of the previous matrix equation has been set aside as not being useful at this point in the analysis. It is not useful now because that equation involves not only the unknown DOF but also the unknown reaction at the wall and thus is no help in determining the solution for the unknown DOF. However, this discarded equation can be used to calculate that reaction at the left end of beam 10 after the DOFs have been determined.

Note that the second right-hand side part of the above equation is exactly the same contribution that beam element 10 made previously to the global stiffness matrix as shown in the previous problem. Thus, as stated above, this process does not lead to any changes in the global stiffness matrix. Note also that the first part of the right-hand side, the quantities associated with the equivalent load vector do have units of force or moment as appropriate to that row.

The corresponding equivalent load terms for beam element 20 are obtained in exactly the same way as those for beam element 10. In the original 6×6 beam bending and twisting element stiffness matrix, the (zero-valued) element bending slope DOF at the previously fully clamped end is replaced by the known, now nonzero, deflection function $\theta_0(t)$. Then the $\theta_0(t)$ stiffness factors in the rows corresponding to the retained DOF are separated from the rest of the element stiffness matrix with the result

$$\begin{Bmatrix} \text{Force, rt. end} \\ \text{Mom't rt. end} \\ \text{Torque, rt. end} \end{Bmatrix} = \frac{EI_0}{L^3}\begin{Bmatrix} -48L \\ 8L^2 \\ 0 \end{Bmatrix}\theta_0(t) + \frac{EI_0}{L^3}\begin{bmatrix} 192 & -48L & 0 \\ 48L & 16L^2 & 0 \\ 0 & 0 & 2L^2 \end{bmatrix}\begin{Bmatrix} w_1 \\ \theta_1 \\ \phi_1 \end{Bmatrix}.$$

The enforced motion at the right end of beam 30 is also a time-varying rotation. However, the mathematics of this case require special care because the enforced motion $\phi_0(t)$, although a "twisting" motion for the grid as a whole, is an enforced bending slope for beam element 30. That is, in usual FEM terms, the element DOF relation to the global DOF is the relation $\theta_2^{(30)} = \phi_0(t)$. Making this substitution and substituting the other global DOF and BCs for the element DOF so that all references are global, and reordering, yields

$$\begin{Bmatrix} \text{Force, left end} \\ \text{Torque, left end} \\ \text{Mom't, left end} \end{Bmatrix} = \frac{EI_0}{L^3}\begin{Bmatrix} 12L \\ 0 \\ 4L^2 \end{Bmatrix}\phi_0(t) + \frac{EI_0}{L^3}\begin{bmatrix} 24 & 0 & 12L \\ 0 & L^2 & 0 \\ 12L & 0 & 8L^2 \end{bmatrix}\begin{Bmatrix} w_1 \\ \theta_1 \\ \phi_1 \end{Bmatrix}.$$

When the three components of the equivalent force vector are transposed to the right-hand side of Eq. (4.2) and combined, the result is

$$\{Q\} = \frac{EI_0}{L^3}\begin{Bmatrix} 24w_0 + 48L\theta_0 - 12L\phi_0 \\ -8L^2\theta_0 \\ -12Lw_0 - 4L^2\phi_0 \\ 0 \\ 0 \\ 0 \end{Bmatrix}.$$

A partial check on this answer can be had by referring to Figure 4.4(b) and visualizing that when the DOF at node 1 are required to have zero values, w_0 causes a positive (upward) force to act at node 1, and, in the global sense, a negative twisting moment at node 1. Thus w_0 entries should appear only in the first and third rows of the generalized force vector, with the signs mentioned, as they do. Similarly, θ_0 causes a positive force, and, in the global sense, a negative bending moment at node 1. Therefore this θ_0 term should appear with those signs in the first and second rows of the generalized force vector, as it does. Similarly ϕ_0 causes both a negative force and a negative twisting moment (torque) at node 1, and hence this term should appear in the first and third rows with those signs, as it does. ★

Note that if there were both enforced motions and applied loads acting on a structure, then, if the total deflections are small, superposition is valid. That is, obtaining the matrix equation of motion for both types of loading is simply a matter of putting both the applied load vector and the equivalent load vector on the right-hand side of Eq. (4.2) and then combining the two vectors into one.

EXAMPLE 4.4 Using the FEM, write the equations of motion for the cantilevered beam of Figure 4.5(a) in matrix form.

COMMENT Note that in addition to the usual elastic beam elements, the cantilevered beam model of Figure 4.5(a) also includes two rigid elements that connect the lumped masses to the nodes of the elastic elements. The mass of these rigid elements, as well as the mass of the other structural components, is included in the lumped masses. This rigid element type of modeling can be expected to be an option of any commercial FEM program when the center of the lumped mass associated with the node point is specified as being located away from the node point of the elastic elements. In this example the offset distance is only in the y direction. It could just as well also be in the x and z directions.

The use of rigid elements of any kind is desirable whenever one part of a structure is much stiffer than other parts. Specifically, the use of rigid elements simplifies the system model and avoids the matrix ill-conditioning that is possible when stiffness matrix entries vary by a few orders of magnitude. In structural dynamics terms, because the smaller stiffness terms are associated with the largest deflections, they are more important than the larger stiffness terms. Eliminating those larger stiffness terms by means of rigid elements avoids the larger stiffness terms obscuring the lesser stiffness terms. In this case, the rigid element modeling is wholly appropriate if $e \ll L$ because the shorter the span length of an elastic element, the stiffer the element. An example of an actual engineering structure that fits this circumstance is an aircraft wing such as that of a glider or long-range aircraft. Modeling such a wing as a beam, it is appropriate to note that the span length is much larger than the chord length (planform width), which in turn is much larger than the wing thickness. Therefore it is possible to model, at least for a preliminary analysis, the wing as rigid across the short chord length and thickness, but flexible over the long span length.

Rigid elements other than bar elements, and partially elastic, partially rigid elements are also common. For example, a later set of example problems again illustrates the use of beam elements that are flexible in bending but rigid with respect to axial deflections. The viewpoint that rigid elements are nothing more than a set of algebraic relationships between DOF, relationships sometimes called multipoint constraints, becomes more evident in those examples.

SOLUTION The first question to be answered is whether the system DOF should be centered at the nodes of the elastic elements or at the centers of mass. Recall that the description of the elastic properties of the structural system is more complicated than the description of the system inertial properties. Placing the global DOF only at the nodes of the elastic elements[9] simplifies the description of the elastic forces. See Figure 4.5(b). The components of the velocities at the lumped masses, both translational and rotational, necessary to write the kinetic energy expression, can be determined

[9] Occasionally it is not possible to directly match global DOF to element DOF, say, at a beam element end. See Example 4.12 for an illustration of this situation and its remedies.

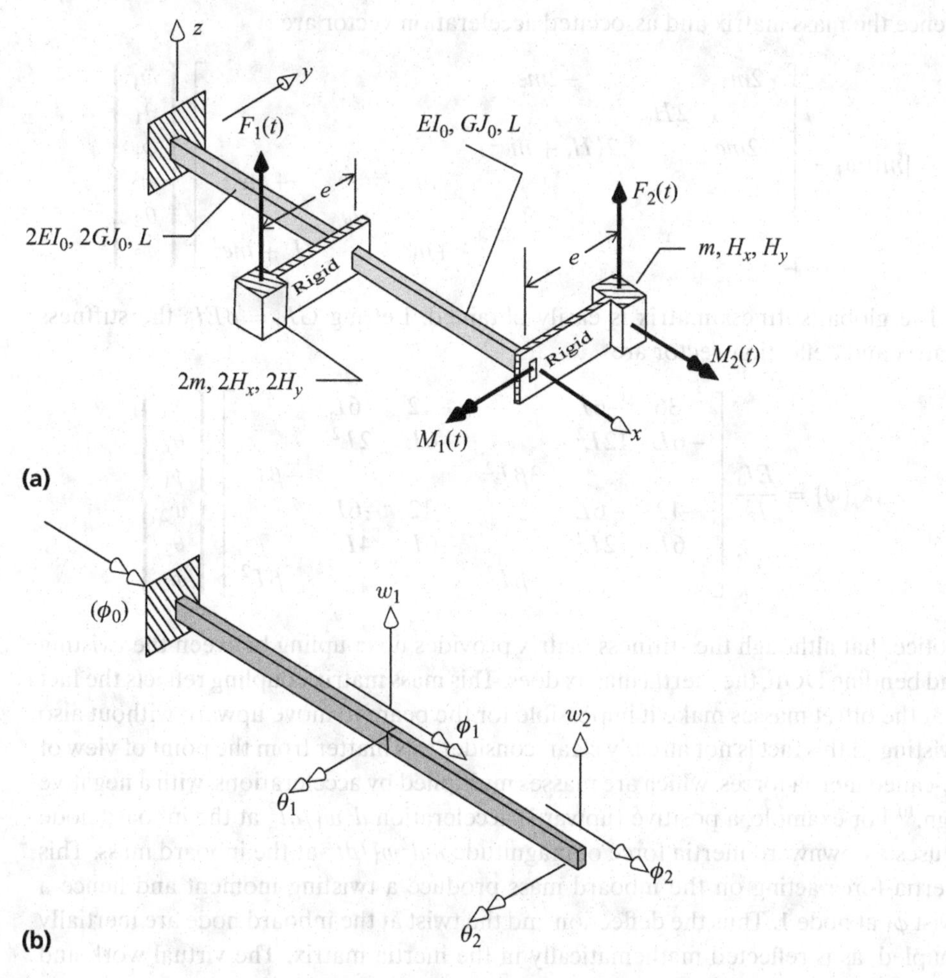

Figure 4.5. Examples 4.4, 4.5, and 4.6: (a) Beam grid featuring rigid elements. (b) Selected degrees of freedom plus an enforced motion for Example 4.6.

by simply taking, one at a time, each of the DOF and visualizing the affected lumped mass moving in response to that DOF while all the other DOF have zero values. (This is much the same approach taken to write the virtual work expressions.) For example, with all the other DOF being zero, a positive velocity dw_1/dt at the inboard node produces an upward component of velocity dw_1/dt at the mass of magnitude $2m$. The rotational velocity $d\theta_1/dt$ does not produce a vertical motion at that same mass, but the rotational velocity $d\phi_1/dt$ does produce a downward velocity component of magnitude $e\,d\phi_1/dt$. Hence the total vertical velocity at the inboard mass is $(\dot{w}_1 - e\dot{\phi}_1)$. Thus the total kinetic energy expression, which is not often much more complicated than this, is

$$T = \tfrac{1}{2}(2m)(\dot{w}_1 - e\dot{\phi}_1)^2 + \tfrac{1}{2}(2H_y)\,\dot{\theta}_1^2 + \tfrac{1}{2}(2H_x)\,\dot{\phi}_1^2$$
$$+ \tfrac{1}{2}m(\dot{w}_2 + e\dot{\phi}_2)^2 + \tfrac{1}{2}H_y\,\dot{\theta}_2^2 + \tfrac{1}{2}H_x\,\dot{\phi}_2^2.$$

Hence the mass matrix and associated acceleration vector are

$$[m]\{\ddot{w}\} = \begin{bmatrix} 2m & & -2me & & & \\ & 2H_y & & & & \\ -2me & & 2(H_x+me^2) & & & \\ & & & m & & +me \\ & & & & H_y & \\ & & & +me & & H_x+me^2 \end{bmatrix} \begin{Bmatrix} \ddot{w}_1 \\ \ddot{\theta}_1 \\ \ddot{\phi}_1 \\ \ddot{w}_2 \\ \ddot{\theta}_2 \\ \ddot{\phi}_2 \end{Bmatrix}.$$

The global stiffness matrix is easily obtained. Letting $GJ_0 = \beta EI_0$ the stiffness matrix and deflection vector are

$$[k]\{w\} = \frac{EI_0}{L^3}\begin{bmatrix} 36 & -6L & & -12 & 6L & \\ -6L & 12L^2 & & -6L & 2L^2 & \\ & & 3\beta L^2 & & & -\beta L^2 \\ -12 & -6L & & 12 & -6L & \\ 6L & 2L^2 & & -6L & 4L^2 & \\ & & -\beta L^2 & & & \beta L^2 \end{bmatrix} \begin{Bmatrix} w_1 \\ \theta_1 \\ \phi_1 \\ w_2 \\ \theta_2 \\ \phi_2 \end{Bmatrix}.$$

Notice that although the stiffness matrix provides no coupling between the twisting and bending DOF, the inertia matrix does. This mass matrix coupling reflects the fact that the offset masses make it impossible for the beam to move upward without also twisting. If this fact is not already clear, consider this matter from the point of view of so-called inertia forces, which are masses multiplied by accelerations, with a negative sign.[10] For example, a positive (upward) acceleration d^2w_1/dt^2 at the inboard node causes a downward inertia force of magnitude md^2w_1/dt^2 at the inboard mass. This inertia force acting on the inboard mass produce a twisting moment and hence a twist ϕ_1 at node 1. Thus the deflection and the twist at the inboard node are inertially coupled, as is reflected mathematically in the inertia matrix. The virtual work and applied load vectors are

$$\delta W = F_1\,\delta w_1 + 0\,\delta\theta_1 - eF_1\,\delta\phi_1 + F_2\,\delta w_2 + M_1\,\delta\theta_2 + (M_2 + eF_2)\,\delta\phi_2$$

$$\text{so} \quad \lfloor Q \rfloor = \lfloor F_1 \quad 0 \quad -eF_1 \quad F_2 \quad M_1 \quad (M_2 + eF_2) \rfloor.$$

Again, the equations of motion for the cantilevered beam system are completed when the above specified matrices are placed in the standard form of Eq. (4.2), $[M]\{\ddot{q}\} + [K]\{q\} = \{Q\}$.

In review, the equations of motion are developed by using the FEM element stiffness matrices to provide the necessary system stiffness matrix and, separately, the kinematics of the masses to provide the necessary mass matrix. The virtual work calculation provides the external generalized force vector. ★

[10] Inertia forces are, for example, the forces people experience acting on their bodies during a carnival ride or a ride on a high-speed elevator in a tall building. When the elevator starts (accelerates) upward, a person's body is pushed toward the elevator floor by those quite real forces. Thus all inertia forces act opposite to the direction of the associated acceleration, a. This fact is accounted for by use of a negative sign. That is, an inertia force acting on a mass m equals $-ma$.

EXAMPLE 4.5 Repeat the above example problem where the one outboard mass of magnitude m is now offset not only a distance e in the positive y direction but also a distance ℓ in the positive x direction.

SOLUTION The only changes necessary are changes in the mass matrix and the applied force vector. The stiffness matrix does not change because the DOF are the same, and the two beam elements have not been altered. As a result of the change in the location of the outboard mass, the new kinetic energy expression is

$$\begin{aligned} T &= \tfrac{1}{2}(2m)(\dot{w}_1 - e\dot{\phi}_1)^2 + \tfrac{1}{2}(2H_y)\,\dot{\theta}_1^2 + \tfrac{1}{2}(2H_x)\,\dot{\phi}_1^2 \\ &\quad + \tfrac{1}{2}m(\dot{w}_2 + e\dot{\phi}_2 + \ell\dot{\theta}_2)^2 + \tfrac{1}{2}H_y\,\dot{\theta}_2^2 + \tfrac{1}{2}H_x\,\dot{\phi}_2^2. \end{aligned}$$

The new mass matrix and the (same) acceleration vector are

$$[m]\{\ddot{w}\} = \begin{bmatrix} 2m & & -2me & & & \\ & 2H_y & & & & \\ -2me & & 2(H_x + me^2) & & & \\ & & & m & m\ell & me \\ & & & m\ell & (H_y + m\ell^2) & me\ell \\ & & & me & me\ell & (H_x + me^2) \end{bmatrix} \begin{Bmatrix} \ddot{w}_1 \\ \ddot{\theta}_1 \\ \ddot{\phi}_1 \\ \ddot{w}_2 \\ \ddot{\theta}_2 \\ \ddot{\phi}_2 \end{Bmatrix}.$$

The virtual work and the applied load vector have only one change

$$\delta W = F_1\,\delta w_1 - eF_1\,\delta\phi_1 + F_2\,\delta w_2 + (M_1 + \ell F_2)\,\delta\theta_2 + (M_2 + eF_2)\,\delta\phi_2$$

so $\lfloor Q \rfloor = \lfloor F_1 \quad 0 \quad -eF_1 \quad F_2 \quad (M_1 + \ell F_2) \quad (M_2 + eF_2) \rfloor.$ ★

EXAMPLE 4.6 State the matrix equation of motion for the previous problem when the following changes are made: (i) remove all the applied loads F_1 through M_2 and (ii) now drive the beam by means of an enforced twisting motion $\phi_0(t)$ at the wall support.

SOLUTION The mass matrix is unaltered because the kinetic energy is controlled entirely by the time derivatives of the DOF, and the DOF are the same symbolic values regardless of whether the structural system is driven by applied loads or by support motions. The stiffness matrix is also the same because the elastic elements and their DOF are unchanged. However, it is necessary to reexamine the element stiffness matrix for the inboard beam element to obtain the new equivalent applied load vector. After inserting the zero BCs, which eliminates the first two columns, and therefore setting aside the first two rows, the remaining terms are

$$[k_{10}]\{w\} = \frac{EI_0}{L^3} \begin{bmatrix} 2\beta L^2 & & & -2\beta L^2 \\ & 24 & -12L & \\ & -12L & 8L^2 & \\ -2\beta L^2 & & & 2\beta L^2 \end{bmatrix} \begin{Bmatrix} \phi_0 \\ w_1 \\ \theta_1 \\ \phi_1 \end{Bmatrix}.$$

Setting aside the first of these four rows because the associated full equation involves the unknown torque reaction at the wall, the equivalent load vector for the entire

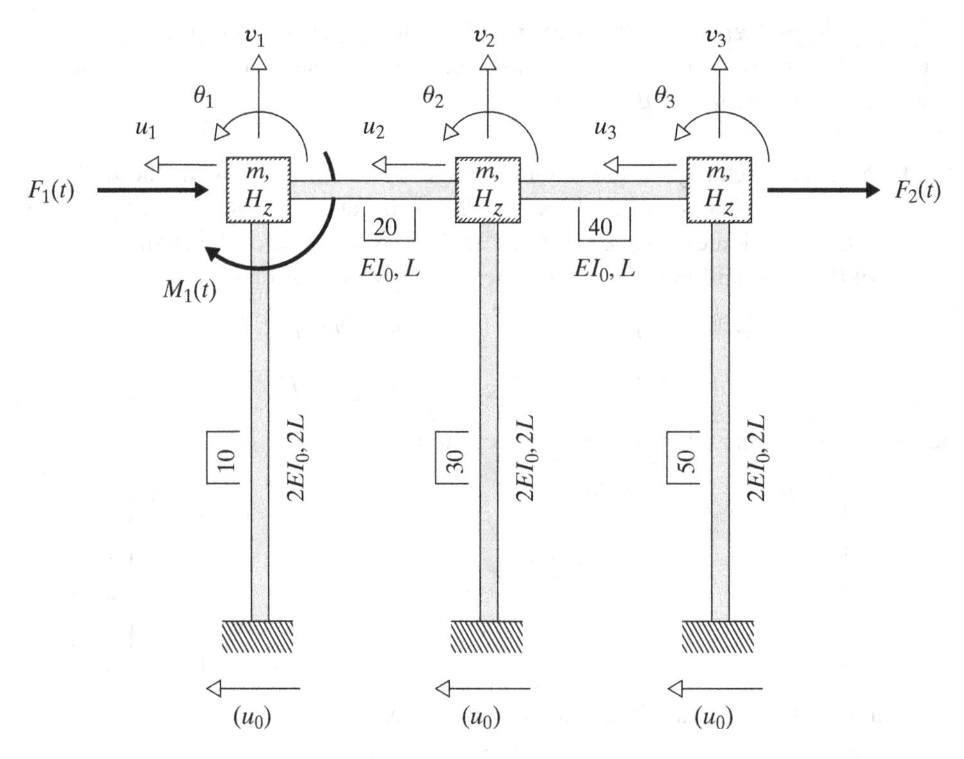

Figure 4.6. Examples 4.7 and 4.8: Two bay planar frame with mass modeling and applied loading.

system, from the above (4,1) entry, is, after transposing to the right-hand side of the equation to better fit the page

$$\lfloor Q \rfloor = \frac{2\beta EI_0\, \phi_0(t)}{L} \lfloor 0 \quad 0 \quad 1 \quad 0 \quad 0 \quad 0 \rfloor.$$

★

EXAMPLE 4.7 Write the matrix equation of motion for the five-beam-element planar frame of Figure 4.6 where, as per usual, the axial stiffness of each beam is much greater than its bending stiffness, that is, $EA_0/L \gg EI_0/L^3$. The applied loading is the set of three external loads F_1, F_2, and M_1, as shown. (The indicated base motion, $u_0(t)$, is reserved for the next example problem.) Again, to reduce the number of required generalized coordinates for this hand calculation, let the supports at the system base be fixed supports.

SOLUTION Again begin the hand calculation by choosing the global DOF to achieve the best fit with the element DOF. This means, for example, that for the beam element 10 right-hand vertical deflection DOF to coincide with the global DOF at node 1, the global DOF u_1 must be positive to the left. If u_1 were instead to be positive to the right, then minus signs would have to be introduced into the beam element 10 stiffness matrix. In a machine calculation, the software automatically accounts for any such bothersome sign differences.

Now, because each of the beams is ever so much stiffer axially than it is laterally the axial stiffness of each beam is approximated as being infinite. This partly rigid beam idealization is always accomplished by putting constraints on the appropriate system generalized coordinates. In this case, the constraints are

$$v_1 = v_2 = v_3 = 0 \quad \text{and} \quad u_1 = u_2 = u_3 = u.$$

There are no constraints on the rotations. Thus the kinetic energy and the mass matrix and acceleration vector are

$$T = \frac{1}{2}(m+m+m)\,\dot{u}^2 + \frac{1}{2}H_z\dot{\theta}_1^2 + \frac{1}{2}H_z\dot{\theta}_2^2 + \frac{1}{2}H_z\dot{\theta}_3^2$$

$$[M]\{\ddot{u}\} = \begin{bmatrix} 3m & & & \\ & H_z & & \\ & & H_z & \\ & & & H_z \end{bmatrix} \begin{Bmatrix} \ddot{u} \\ \ddot{\theta}_1 \\ \ddot{\theta}_2 \\ \ddot{\theta}_3 \end{Bmatrix}.$$

The element stiffness matrices are the same for beam elements 10, 30, and 50. Those for beam elements 20 and 40 are the same. The stiffness matrices for the beam elements are

$$[k_{10}]\{u_{10}\} = \frac{EI_0}{L^3}\begin{bmatrix} 3 & -3L \\ -3L & 4L^2 \end{bmatrix}\begin{Bmatrix} u \\ \theta_1 \end{Bmatrix}$$

$$[k_{30}]\{u_{30}\} = \frac{EI_0}{L^3}\begin{bmatrix} 3 & -3L \\ -3L & 4L^2 \end{bmatrix}\begin{Bmatrix} u \\ \theta_2 \end{Bmatrix}$$

$$[k_{50}]\{u_{50}\} = \frac{EI_0}{L^3}\begin{bmatrix} 3 & -3L \\ -3L & 4L^2 \end{bmatrix}\begin{Bmatrix} u \\ \theta_3 \end{Bmatrix}$$

and

$$[k_{20}]\{u_{20}\} = \frac{EI_0}{L^3}\begin{bmatrix} 4L^2 & 2L^2 \\ 2L^2 & 4L^2 \end{bmatrix}\begin{Bmatrix} \theta_1 \\ \theta_2 \end{Bmatrix}$$

$$[k_{40}]\{u_{40}\} = \frac{EI_0}{L^3}\begin{bmatrix} 4L^2 & 2L^2 \\ 2L^2 & 4L^2 \end{bmatrix}\begin{Bmatrix} \theta_2 \\ \theta_3 \end{Bmatrix}.$$

The assembled structural stiffness matrix is

$$[K]\{u\} = \frac{EI_0}{L^3}\begin{bmatrix} +9 & -3L & -3L & -3L \\ -3L & 8L^2 & 2L^2 & 0 \\ -3L & 2L^2 & 12L^2 & 2L^2 \\ -3L & 0 & 2L^2 & 8L^2 \end{bmatrix}\begin{Bmatrix} u \\ \theta_1 \\ \theta_2 \\ \theta_3 \end{Bmatrix}.$$

After writing the virtual work expression, the applied load vector is

$$\{Q\} = \begin{Bmatrix} -F_1 - F_2 \\ -M_1 \\ 0 \\ 0 \end{Bmatrix}.$$

Putting these matrices into the standard Lagrange equation form of $[M]\{\ddot{q}\} + [K]\{q\} = \{Q\}$ completes the example. ★

EXAMPLE 4.8 Redo the above problem after removing the applied forces and moment and applying the same enforced base motion $u_0(t)$ simultaneously at all three supports, as indicated in Figure 4.6.

COMMENT It is always implied, in these problems where beam-columns can be in compression, that the compressive loads produced by the motion and weight of the masses are below the beam-column buckling loads. Otherwise, the stiffness of a perfect column with respect to lateral bending loads is drastically lessened at or above the buckling load. In those circumstances, the problem becomes nonlinear.

SOLUTION Once again, neither the mass matrix nor the system stiffness matrix is affected by the change in the loading. All that needs to be done is to determine the equivalent load vector from the beam element stiffness matrices for elements 10, 30, and 50. After, for each of these elements, inserting the zero bending slope at the base, and setting aside the top two rows of the original 4 × 4 element stiffness matrix, the equivalent generalized force vector (with units that check) is

$$\lfloor Q \rfloor = \frac{EI_0\, u_0(t)}{L^3} \lfloor (3+3+3) \quad -3L \quad -3L \quad -3L \rfloor. \qquad \bigstar$$

4.6 Summary

Since almost all modern engineering structures have, at the least, complicated geometries, almost all structural dynamic analyses today are carried out by means of digital computer software that encodes the FEM. This chapter begins the process of clarifying what is accomplished by FEM software by here building the matrix equations of motion by hand for small problems using relatively simple structural elements. All structural dynamics software employing the FEM, and all such hand calculations, must, at a minimum, (i) identify the system (global) DOF, (ii) create the system mass matrix, (iii) assemble a global stiffness matrix from the individual element stiffness matrices, and (iv) assemble the applied load vector from a virtual work calculation and/or assemble a support motion equivalent load vector from element stiffness matrices. Note that temperature changes are usually sufficiently slow to develop that they fall within the static load category. Therefore, thermal effects are ignored in this textbook. However, there are cases where thermal effects occur sufficiently quickly (relative to the first natural period) or sufficiently affect the system stiffness that they must be considered. For example, thermal deformations can be part of the dynamic deformations of a spacecraft whose rotations produce cyclic heating and cooling. For example, the Hubble space telescope, when first launched, experienced a thermo-servo-elastic instability. If thermal expansion results in large compressive loads (compared to buckling loads), then the stiffnesses of the affected structure may be lessened to the point where such changes need incorporation into the dynamic analysis.

Return to the above four listed tasks performed by FEM software for dynamic analyses. The hand calculations of this textbook are intended to produce an understanding of those software steps. Illustrations of the first, third, and fourth tasks began in the previous chapter. For hand calculations, selection of the system DOF requires an understanding of the major motions of the structural system. Again, commercial FEM software usually approaches the DOF selection by presuming that at each node there are three unknown, orthogonal, translational DOF and three unknown, orthogonal, rotational DOF, and complete continuity of translational and rotational connections from each structural element to all other structural elements connected to that node. As was discussed in the previous chapter, if a DOF is inappropriate, or a connection at a node is not "rigid," as would be the case for a connection modeled as a hinge, then the software would require data input that identifies which DOF are to be removed or made not continuous for which structural elements.

EXAMPLE 4.9 Reconsider the one-beam and two-column frame/grid structure of Example 3.8 shown in Figure 3.10, which shows the applicable DOF for this system. The stiffness matrix for that structure was developed in that example. Let the mass supported by this structure be modeled as shown in Figure 4.7(a), where the 1 stands for m_1, and so on. Using the given DOF, arranged in the same order as that used for the deflection vector, write the mass matrix for this structure. For the readers convenience, Figure 4.7(b) repeats a sketch of the applicable DOF.

COMMENT The mass modeling illustrated in Figure 4.7(a) is more common than that where values for mass moments of inertia and mass products of inertia are estimated for a given system node. Here there are no such mass moments of inertia associated with any of the four masses because, for example, the first and third masses and their small, rigid interconnections represent two representative parts of a single, more complicated (nearly rigid) mass and similarly for the second and fourth masses. This approach has the advantage that the software estimates the mass moments and products of inertia to be associated with the node. For study purposes, note that this example is not as complicated as the general case where there are rigid bar extensions in three orthogonal coordinate directions, and a full set of three angular DOF. This latter case is addressed in Exercise 4.13, and should be carefully studied as an application of the material of Figure 1.7.

SOLUTION The motions of masses three and four are determined by the translations and rotations at masses one and two. The kinetic energy expression can be written as follows:

$$T = \tfrac{1}{2}\{m_1(\dot{u}_1^2 + \dot{v}_1^2) + m_3[(\dot{u}_1 - e_y\dot{\psi}_1)^2 + (\dot{v}_1 + e_x\dot{\psi}_1)^2 + (e_x\dot{\theta}_1 + e_y\dot{\phi}_1)^2]$$
$$+ m_2(\dot{u}_1^2 + \dot{v}_2^2) + m_4[(\dot{u}_1 - e_z\dot{\theta}_2)^2 + (\dot{v}_2 - e_z\dot{\phi}_2)^2]\}.$$

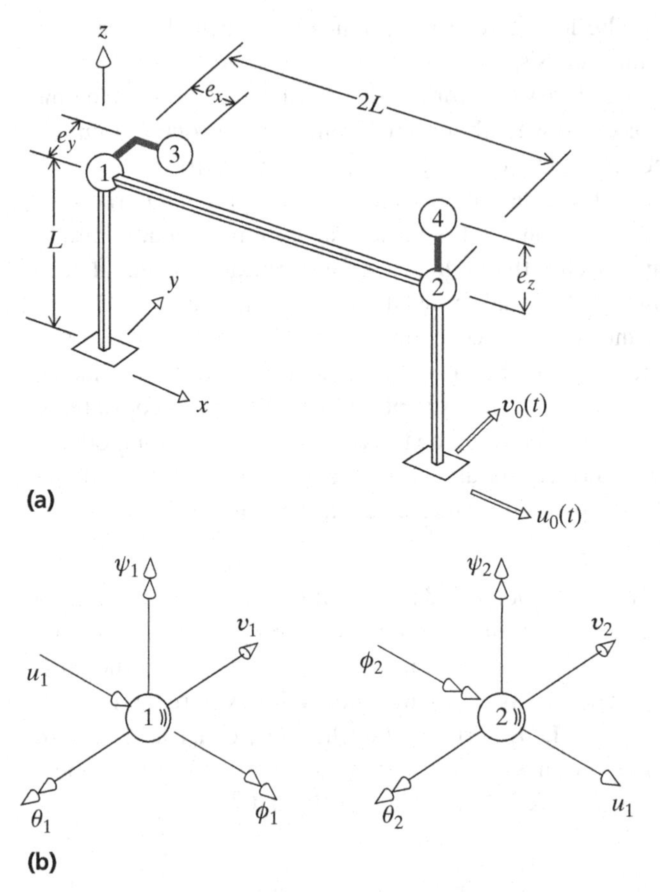

Figure 4.7. Example 4.9. (a) Structure and mass modeling (b) Generalized coordinates.

After factoring the above kinetic expression into the form $1/2\lfloor \dot{q} \rfloor [m]\{\dot{q}\}$, the above kinetic energy expression leads via the Lagrange equations to the following mass matrix and acceleration vector:

$$\begin{bmatrix} m_0 & 0 & 0 & 0 & -m_3 e_y & 0 & -m_4 e_z & 0 & 0 \\ 0 & m_1+m_3 & 0 & 0 & m_3 e_x & 0 & 0 & 0 & 0 \\ 0 & 0 & m_3 e_x^2 & m_3 e_x e_y & 0 & 0 & 0 & 0 & 0 \\ 0 & 0 & m_3 e_x e_y & m_3 e_y^2 & 0 & 0 & 0 & 0 & 0 \\ -m_3 e_y & m_3 e_x & 0 & 0 & m_3\left(e_x^2+e_y^2\right) & 0 & 0 & 0 & 0 \\ 0 & 0 & 0 & 0 & 0 & m_2+m_4 & 0 & -m_4 e_z & 0 \\ -m_4 e_z & 0 & 0 & 0 & 0 & 0 & m_4 e_z^2 & 0 & 0 \\ 0 & 0 & 0 & 0 & 0 & -m_4 e_z & 0 & m_4 e_z^2 & 0 \\ 0 & 0 & 0 & 0 & 0 & 0 & 0 & 0 & 0 \end{bmatrix} \begin{Bmatrix} \ddot{u}_1 \\ \ddot{v}_1 \\ \ddot{\theta}_1 \\ \ddot{\phi}_1 \\ \ddot{\psi}_1 \\ \ddot{v}_2 \\ \ddot{\theta}_2 \\ \ddot{\phi}_2 \\ \ddot{\psi}_2 \end{Bmatrix},$$

where $\quad m_0 = m_1 + m_2 + m_3 + m_4.$ ★

The previous example problems emphasized beam bending. Figures 4.8 and 4.9 illustrate other types of elastic systems eminently suited to FEM description.

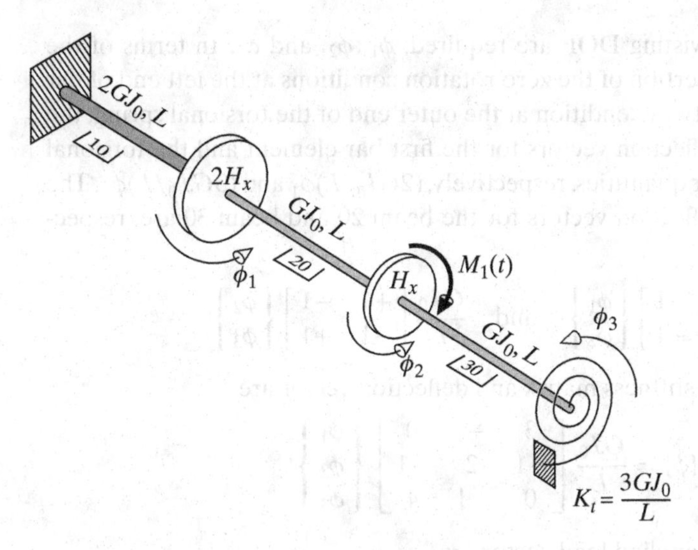

Figure 4.8. Examples 4.10a and 4.10b.

Regardless of the nature of the structural system under study, this chapter has developed two strategies and two facts regarding structural dynamic analyses that should always be kept in mind. They are, again:

1. Ignore all static loads (including linear prestressing) and all static deflections as long as the total deflections are small.

2. Use generalized coordinates associated with the elastic elements because the elastic elements are more complicated than the mass elements.

3. Any mass matrix is always symmetric and positive definite.

4. Any system stiffness matrix compiled from elastic elements stiffness matrices also is always symmetric, and after the application of BCs that prohibit rigid body motion, the stiffness matrix is positive definite. (If rigid body motion is possible, the stiffness matrix is merely nonnegative definite, and, as will be seen, there will be as many zero valued natural frequencies as there are possible rigid body motions.)

EXAMPLE 4.10A Write the matrix equation of motion for the torsional system of Figure 4.8. The spiral symbol at the near end of the bar system represents a linearly elastic torsional spring, that is, a perfectly elastic spring that resists the twisting of the beam with a twisting moment directly proportional to the twist at its point of connection to the bar. The proportionality factor between the torque and the rotation is the spring constant K_t. The element stiffness matrix for a torsional spring is of the same form as that for a translational spring. With ϕ_1 being the twist at one end of the torsional spring and ϕ_2 being the twist at the other end (often zero), the torsional spring element stiffness matrix is

$$\{M_t\} = [k_e]\{\phi_e\} \quad \text{or} \quad \begin{Bmatrix} M_{t1} \\ M_{t2} \end{Bmatrix}^{(e)} = K_t \begin{bmatrix} +1 & -1 \\ -1 & +1 \end{bmatrix} \begin{Bmatrix} \phi_1 \\ \phi_2 \end{Bmatrix}^{(e)}.$$

SOLUTION Only three twisting DOF are required: ϕ_1, ϕ_2, and ϕ_3. In terms of the global DOF, after the insertion of the zero rotation conditions at the left end of bar element 10 and the zero twist condition at the outer end of the torsional spring, the stiffness matrices and deflection vectors for the first bar element and the torsional spring reduce to the scalar quantities, respectively, $(2GJ_0/L)\phi_1$ and $(3GJ_0/L)\phi_3$. The stiffness matrices and deflection vectors for the beam 20 and beam 30 are, respectively,

$$\frac{GJ_0}{L}\begin{bmatrix} +1 & -1 \\ -1 & +1 \end{bmatrix}\begin{Bmatrix} \phi_1 \\ \phi_2 \end{Bmatrix} \quad \text{and} \quad \frac{GJ_0}{L}\begin{bmatrix} +1 & -1 \\ -1 & +1 \end{bmatrix}\begin{Bmatrix} \phi_2 \\ \phi_3 \end{Bmatrix}.$$

Therefore the assembled stiffness matrix and deflection vector are

$$[K]\{\phi\} = \frac{GJ_0}{L}\begin{bmatrix} 3 & -1 & 0 \\ -1 & 2 & -1 \\ 0 & -1 & 4 \end{bmatrix}\begin{Bmatrix} \phi_1 \\ \phi_2 \\ \phi_3 \end{Bmatrix}.$$

The mass matrix and the applied load vector are

$$[M] = H_x\begin{bmatrix} 2 & & \\ & 1 & \\ & & 0 \end{bmatrix} \quad \text{and} \quad \{Q\} = \begin{Bmatrix} 0 \\ -M_1(t) \\ 0 \end{Bmatrix}.$$

Substitution into $[M]\{\ddot{q}\} + [K]\{q\} = \{Q\}$ completes the setup of the matrix equation of motion. ★

EXAMPLE 4.10B Redo the above example problem where now, rather than a clamped boundary at the left end of bar 10, there is an enforced, rotational base motion $\Phi(t)$.

SOLUTION The mass matrix is unaltered because there is no change where the two rotational masses are located. The previous applied load vector is unchanged, but now it must be augmented by the effects of the base motion. The process of determining the equivalent applied loads created by base motions begins with the element stiffness matrices of each element connected to a moving support. Here, to illustrate a very slightly different style of analysis, the full element stiffness for beam element 10 is assembled into the global stiffness matrix. As a result, $[K]$ has been increased in size to account for the additional (known) DOF-type motion at the left support. The expanded global stiffness matrix is

$$\{M_t\} = [K]\{\phi\} = \frac{GJ_0}{L}\begin{bmatrix} 2 & -2 & 0 & 0 \\ -2 & 3 & -1 & 0 \\ 0 & -1 & 2 & -1 \\ 0 & 0 & -1 & 4 \end{bmatrix}\begin{Bmatrix} \Phi \\ \phi_1 \\ \phi_2 \\ \phi_3 \end{Bmatrix},$$

where $\{M_t(t)\}$ is the 4×1 vector of external moments at the system nodes. The top entry of this vector is the moment at the base support. Rewriting the right-hand side form of this moment vector by reducing it in row size so that it corresponds only to

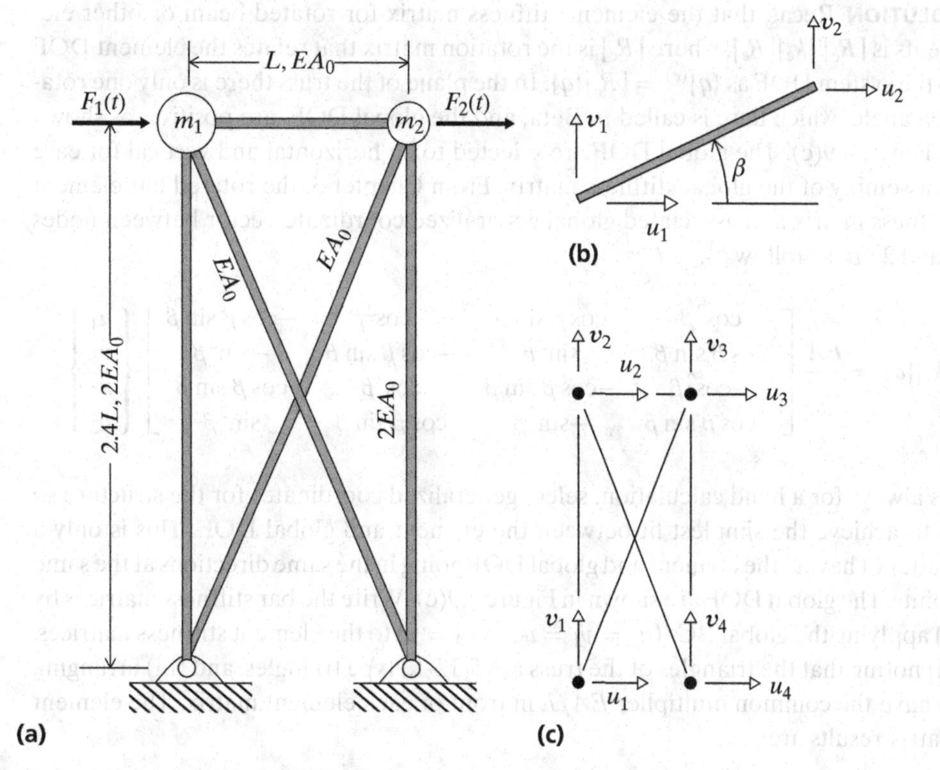

Figure 4.9. Example 4.11: (a) Pin-jointed truss and applied loads. (b) Bar element DOF and orientation angle β. (c) Selected global DOF.

the system DOF (in effect, setting aside the top row) and separating out the factors of the time function $\Phi(t)$, leads to the three row result

$$\{M_t\} = \frac{GJ_0}{L}\begin{Bmatrix}-2\\0\\0\end{Bmatrix}\Phi(t) + \frac{GJ_0}{L}\begin{bmatrix}3 & -1 & 0\\-1 & 2 & -1\\0 & -1 & 4\end{bmatrix}\begin{Bmatrix}\phi_1\\\phi_2\\\phi_3\end{Bmatrix}.$$

Therefore the complete matrix equation of motion is

$$\begin{bmatrix}2 & & \\ & 1 & \\ & & 0\end{bmatrix}\begin{Bmatrix}\ddot{\phi}_1\\\ddot{\phi}_2\\\ddot{\phi}_3\end{Bmatrix} + \frac{GJ_0}{H_x L}\begin{bmatrix}3 & -1 & 0\\-1 & 2 & -1\\0 & -1 & 4\end{bmatrix}\begin{Bmatrix}\phi_1\\\phi_2\\\phi_3\end{Bmatrix} = \frac{GJ_0\Phi(t)}{H_x L}\begin{Bmatrix}2\\0\\0\end{Bmatrix} - \frac{M_1(t)}{H_x}\begin{Bmatrix}0\\1\\0\end{Bmatrix}.$$

★

EXAMPLE 4.11 Write the matrix equations of motion for the planar, five-bar, pin-jointed truss, loaded as shown in Figure 4.9(a).

COMMENT Since the deformations in this truss structure are axial deformations, the bar axial deformations cannot be neglected relative to other types of deformations because there are no other types of deformation in this mathematical model. This lack of other types of deformation is a result of approximating the truss joints as being pinned, which means the joints cannot develop bending moments.

SOLUTION Recall that the element stiffness matrix for rotated beam or other elements is $[R_e]^t[k_e][R_e]$, where $[R_e]$ is the rotation matrix that relates the element DOF to the system DOF as $\{q\}^{(e)} = [R_e]\{q\}$. In the plane of the truss there is only one rotation angle, which here is called β. Beta, and the global DOF, are positive, as shown in Figure 4.9(c). The global DOF are selected to be horizontal and vertical for ease of assembly of the global stiffness matrix. From Chapter 3, the rotated bar element stiffness matrix and associated global generalized coordinate vector between nodes 1 and 2 are as follows[11]:

$$[k_e]\{q\} = \frac{EA}{L}\begin{bmatrix} \cos^2\beta & \cos\beta\sin\beta & -\cos^2\beta & -\cos\beta\sin\beta \\ \cos\beta\sin\beta & \sin^2\beta & -\cos\beta\sin\beta & -\sin^2\beta \\ -\cos^2\beta & -\cos\beta\sin\beta & \cos^2\beta & \cos\beta\sin\beta \\ -\cos\beta\sin\beta & -\sin^2\beta & \cos\beta\sin\beta & \sin^2\beta \end{bmatrix}\begin{Bmatrix} u_1 \\ v_1 \\ u_2 \\ v_2 \end{Bmatrix}.$$

As always for a hand calculation, select generalized coordinates for the structure so as to achieve the simplest fit between the element and global DOF. This is only a matter of having the element and global DOF going in the same directions at the same points. The global DOF are shown in Figure 4.9(c). Write the bar stiffness matrices by (i) applying the global BCs ($u_1 = v_1 = u_4 = v_4 = 0$) to the element stiffness matrices, (ii) noting that the triangles of the truss are 5-12-13 type triangles, and (iii) arranging to have the common multiplier EA/L in front of each element matrix. The element matrix results are

$$[k_{1,2}]\{q\} = \frac{2EA}{2.4L}[1]\{v_2\} = \frac{EA}{L}[0.83333]\{v_2\}$$

$$[k_{3,4}]\{q\} = \frac{EA}{L}[0.83333]\{v_3\}$$

$$[k_{2,3}]\{q\} = \frac{EA}{L}\begin{bmatrix} +1.0 & -1.0 \\ -1.0 & +1.0 \end{bmatrix}\begin{Bmatrix} u_2 \\ u_3 \end{Bmatrix}$$

$$[k_{1,3}]\{q\} = \frac{EA}{L}\begin{bmatrix} 0.05690 & 0.13655 \\ 0.13655 & 0.32772 \end{bmatrix}\begin{Bmatrix} u_3 \\ v_3 \end{Bmatrix}$$

$$[k_{2,4}]\{q\} = \frac{EA}{L}\begin{bmatrix} +0.05690 & -0.13655 \\ -0.13655 & +0.32772 \end{bmatrix}\begin{Bmatrix} u_2 \\ v_2 \end{Bmatrix}.$$

The number of significant figures used in the above matrices is wholly sufficient for numerical accuracy for a problem of this very small size. However, more significant figures may well be necessary for larger problems to avoid computational inaccuracies in the processing required by the matrix equation solution process. However, any inaccuracy in the description of the angles in the actual construction is treated by

[11] Setting β equal to zero causes the bar element to be horizontal and the element and global DOF to coincide. The 4×4 element stiffness matrix then has two rows and two columns of only zeros, whereas the other two rows and two columns contain the easily derived, original 2×2 bar stiffness matrix, $[k_e]$, introduced in Section 3.4. Thus it is clear that the 4×4 element stiffness matrix centered in the product $[R_e]^t[k_e][R_e]$ is just the original 2×2 matrix expanded by rows and columns of zeros.

altering the βs. Assemble the above five element matrices to obtain the following global stiffness matrix:

$$[K]\{q\} = \frac{EA}{L}\begin{bmatrix} 1.0569 & -0.13655 & -1.0000 & \\ -0.13655 & 1.1610 & & \\ -1.0000 & & 1.0569 & 0.13655 \\ & & 0.13655 & 1.1610 \end{bmatrix}\begin{Bmatrix} u_2 \\ v_2 \\ u_3 \\ v_3 \end{Bmatrix}.$$

For the mass matrix, the kinetic energy is simply

$$T = \frac{1}{2}m_1\left(\dot{u}_2^2 + \dot{v}_2^2\right) + \frac{1}{2}m_2\left(\dot{u}_3^2 + \dot{v}_3^2\right),$$

which can be put into matrix form as

$$T = \frac{1}{2}\lfloor \dot{u}_2 \quad \dot{v}_2 \quad \dot{u}_3 \quad \dot{v}_3 \rfloor \begin{bmatrix} m_1 & & & \\ & m_1 & & \\ & & m_2 & \\ & & & m_2 \end{bmatrix}\begin{Bmatrix} \dot{u}_2 \\ \dot{v}_2 \\ \dot{u}_3 \\ \dot{v}_3 \end{Bmatrix}.$$

The generalized force vector, as always, is obtained from the virtual work expression $\delta W = F_1\delta u_2 + F_2\delta u_3$. Therefore,

$$\{Q\}^T = \lfloor F_1 \quad 0 \quad F_2 \quad 0 \rfloor,$$

Now it is simply a matter of placing the above mass matrix, stiffness matrix, generalized coordinate vector, and generalized force vector into $[M]\{\ddot{q}\} + [K]\{q\} = \{Q\}$.

★

4.7 **Offset Elastic Elements**

The final example problem explores the matter of beam element DOF (or other types of element DOF) offset from the global generalized coordinates. The direct way of dealing with offsets is relating the local element coordinates to the global DOF by writing a coordinate transformation matrix. As will be seen, this is exactly analogous to the above example problem where a rotation transformation matrix was introduced. Again, the local stiffness matrix ends up being sandwiched between the transpose of this coordinate transformation matrix and the coordinate transformation matrix itself. Thus, in the example discussed below, which is an analysis of the two-beam structure shown in Figure 4.9, the use of a coordinate transformation matrix from the local beam coordinates of the short beam to the global DOF (the DOF of the long beam) provides a short beam element stiffness matrix that is immediately ready to be assembled into the global stiffness matrix. Of course, commercial software accomplishes these steps without any effort on the part of the analyst.

EXAMPLE 4.12 The structure shown in Figure 4.10 is capable of moving up, down, and sideways; that is, in both the z and the y directions. The very large axial stiffnesses of the beams prevents motion in the x direction and also prevents rotation about the z axis. Therefore, use the indicated global coordinates $\lfloor q \rfloor = \lfloor w \quad \theta \quad \phi \quad v \rfloor$ to write the small-deflection, matrix equation of motion for this clamped, two-beam, one-rigid-mass grid-frame structure. For the short beam, let $EI_{yy} = EI_{zz} = EI$, the (area)

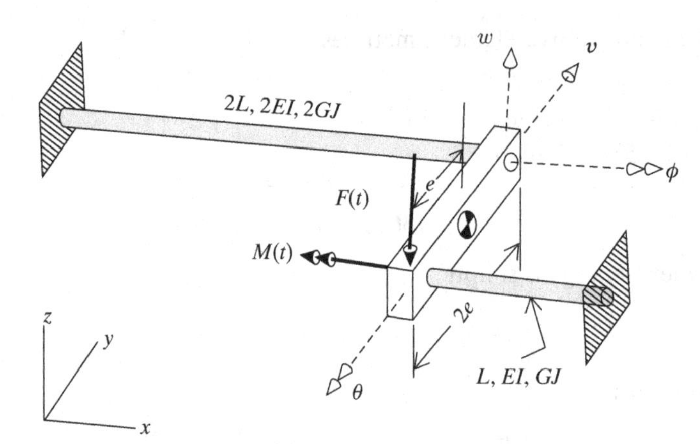

Figure 4.10. Example 4.12.

product of inertia $EI_{yz} = 0$, and let $GJ = \tfrac{1}{2}EI$. Let the corresponding bending and twisting stiffness coefficients for the long beam be twice those values. The rigid bar of length $2e$ has a mass m, and mass moments of inertia $H_x = H_y = \tfrac{1}{2}me^2$.

SOLUTION The kinetic energy of the rigid mass is $T = \tfrac{1}{2}m[\dot{v}^2 + (\dot{w} - e\dot{\phi})^2] + \tfrac{1}{4}me^2[\dot{\theta}^2 + \dot{\phi}^2]$. The system mass matrix and the stiffness matrix for the long beam are

$$[M] = \begin{bmatrix} m & & -me & \\ & \frac{me^2}{2} & & \\ -me & & \frac{3me^2}{2} & \\ & & & m \end{bmatrix} \qquad [k_1]\{q\} = \frac{EI}{L^3}\begin{bmatrix} 3 & -3L & & \\ -3L & 4L^2 & & \\ & & \frac{L^2}{4} & \\ & & & 3 \end{bmatrix}\begin{Bmatrix} w \\ \theta \\ \phi \\ v \end{Bmatrix}.$$

Since for the short beam the element DOF can be described in terms of the global DOF as

$$\lfloor q_{short} \rfloor = \lfloor (w - 2e\phi) \quad \theta \quad \phi \quad v \rfloor,$$

then

$$[k_2] = \frac{EI}{L^3}\begin{bmatrix} 1 & 0 & 0 & 0 \\ 0 & 1 & 0 & 0 \\ -2e & 0 & 1 & 0 \\ 0 & 0 & 0 & 1 \end{bmatrix}\begin{bmatrix} 12 & 6L & 0 & 0 \\ 6L & 4L^2 & 0 & 0 \\ 0 & 0 & \frac{L^2}{2} & 0 \\ 0 & 0 & 0 & 12 \end{bmatrix}\begin{bmatrix} 1 & 0 & -2e & 0 \\ 0 & 1 & 0 & 0 \\ 0 & 0 & 1 & 0 \\ 0 & 0 & 0 & 1 \end{bmatrix}$$

thus

$$[k_2] = \frac{EI}{L^3}\begin{bmatrix} 12 & 6L & -24e & 0 \\ 6L & 4L^2 & -12eL & 0 \\ -24e & -12eL & (\tfrac{1}{2}L^2 + 48e^2) & 0 \\ 0 & 0 & 0 & 12 \end{bmatrix}$$

and

$$[K] = [k_1] + [k_2].$$

The virtual work leads to $\lfloor Q \rfloor = \lfloor -F(t) \quad 0 \quad (2eF(t) - M(t)) \quad 0 \rfloor$. Now all the components of $[M]\{\ddot{q}\} + [K]\{q\} = \{Q\}$ are in place. ★

REFERENCES

4.1 Donaldson, B. K., *Analysis of Aircraft Structures: An Introduction*, McGraw-Hill, New York, 1993.

4.2 Shames, I. H., and C. L. Dym, *Energy and Finite Element Methods in Structural Mechanics*, Hemisphere, Washington, D.C., 1985.

4.3 Temple, G., and W. G. Bickley, *Rayleigh's Principle, and Its Applications to Engineering*, Dover, New York, 1956.

4.4 Meirovitch, L., *Analytical Methods in Vibrations*, Macmillan, New York, 1967.

CHAPTER 4 EXERCISES

4.1 **(a)** Reconsider the beam of Figure 4.5. At the free end of that two-beam-element beam, add a third massless beam element of length L and double stiffness values exactly like that of the inboard beam. Like that previously inboard beam element, let the new beam element also be clamped at its outer end so that the new three-element beam model is clamped at both its outer ends. Write the new stiffness matrix.

(b) If the lumped masses of the structure pictured in Figure 4.5 are unaltered by the addition of the third beam element as in part (a), does the mass matrix change? Does the applied load vector change?

(c) What changes in the analysis would be necessary if the BC at the base of the frame of Figure 4.6 were changed from those of fixed beam ends to simply supported beam ends?

4.2 **(a)** The one-beam, three-spring system shown in Figure 4.11(a) is undergoing small, force free vibrations. ("Force free vibrations" means that the system was set into motion at time zero and, in the absence of energy dissipative forces, has been vibrating ever since.) Each of the two masses rotate and move vertically within the plane of the paper as indicated by the labeled DOF. Write the matrix equation of motion for this system.

(b) As above, but now the vibratory motion is forced (caused) by a base motion $w_0(t)$, positive up, at the left-hand support, a counterclockwise bending moment $M_0(t)$ acting on the smaller of the two masses, and a downward force $F_0(t)$ acting on the larger mass.

(c) The up-and-down displacement of the fixed support for the massless, cantilevered beam shown in Figure 4.11(b) is described by the time function $v(t)$. Let the beam-supported mass be connected at its center of mass to the cantilevered beam tip by a frictionless pin. Let the mass start its motion, $u(t)$, at time zero without either an angular deflection or an angular velocity. Note that from the concept of conservation of angular momentum (which is Eq. (1.3b) in the absence of a moment acting on the mass), the stipulated frictionless pin connection means that the mass continues to

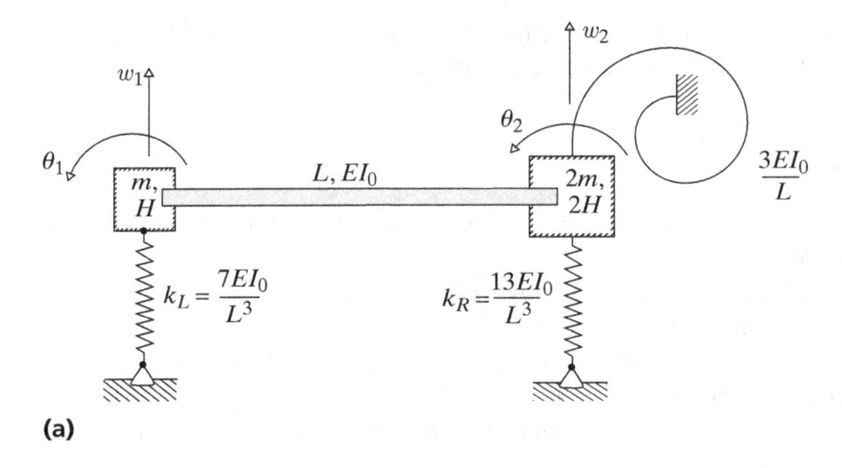

(a)

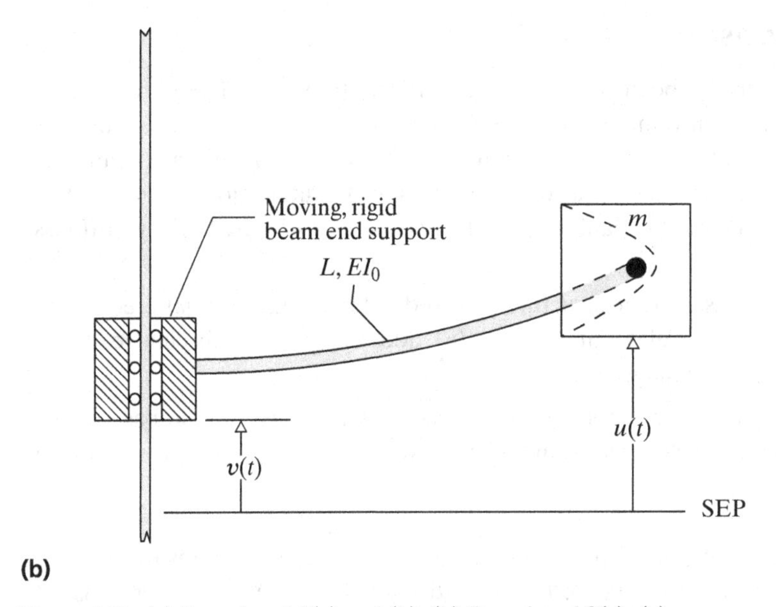

(b)

Figure 4.11. (a) Exercises 4.2(a) and (b). (b) Exercises 4.2(c)–(e).

have a zero rotation as the mass moves up and down. Thus this is a system that requires only the single translational DOF, $u(t)$. Use the finite element method beam element stiffness matrix to write the small deflection equation of motion for this system in terms of $u(t)$.

(d) Write the equation of motion for the system in part (c) in terms of the DOF $w(t)$, which is the deflection of the beam tip relative to the beam support. Show that either equation of motion provides the same expression for the natural frequency for this single-DOF system. (Since the natural frequency of any system is independent of any applied loads or enforced motions, in this case it can be interpreted to be the vibratory frequency when $v(t) = 0$.)

(e) Would the equation of motion for the system of part (c) change if the rail guiding the rigid beam support were horizontal rather than vertical as shown, and the beam supporting the mass were vertical rather than horizontal?

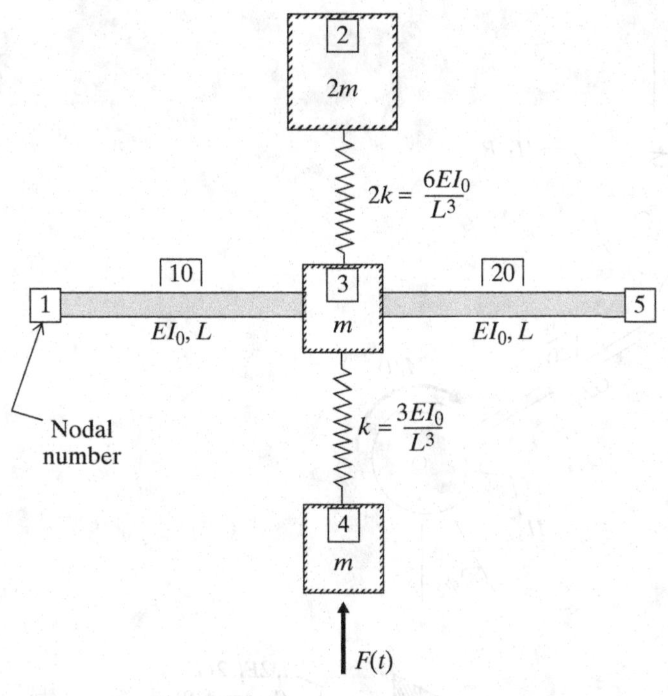

Figure 4.12. Exercise 4.3

4.3 **(a)** The three-mass-element, two-beam-element, two-spring-element structural model of Figure 4.12 is one where the structural and nonstructural mass has been lumped as shown. Note that the system is subjected to an applied force acting at node 4. If the beam ends at nodes 1 and 5 are fixed, then write the matrix equation for up-and-down motion within the plane of the paper.

(b) As above, but this time there are simple supports at nodes 1 and 5.

(c) As in part (a), where the far beam ends are clamped, but this time there is a frictionless hinge at the point (node 3) where the central, lumped mass is attached to the two beam and two spring ends. In this case, the central mass does not rotate.

(d) As in part (a), but this time the applied force is removed, and the system is now driven by means of a known vertical motion $w_1(t)$, positive up, occurring at node 1.

(e) As above, but now let the enforced motion be $\theta_5(t)$.

4.4 **(a)** Write the stiffness matrix for the antisymmetric vibrations of the beam of Figure 4.3.

(b) Is there any difference between the mass matrices for the symmetric and antisymmetric matrix equations of motion?

4.5 The example analyses of the beam grid of Figure 4.4 concerned the out-of-plane bending and torsion of that structural system. Now consider the same structure as a beam frame, where now the vibratory motion occurs entirely within the x, y plane. Change the original applied loads to a single moment, M_3, whose vector axis parallels the z axis and that is located at the juncture of beam elements 40, 50, 60, and 70. That is, replace $F(t)$ with $M_3(t)$ and delete $M_1(t)$ and $M_2(t)$. Let the mass moments of inertia about the z axis for both lumped masses be H_z, and let the bending stiffness in

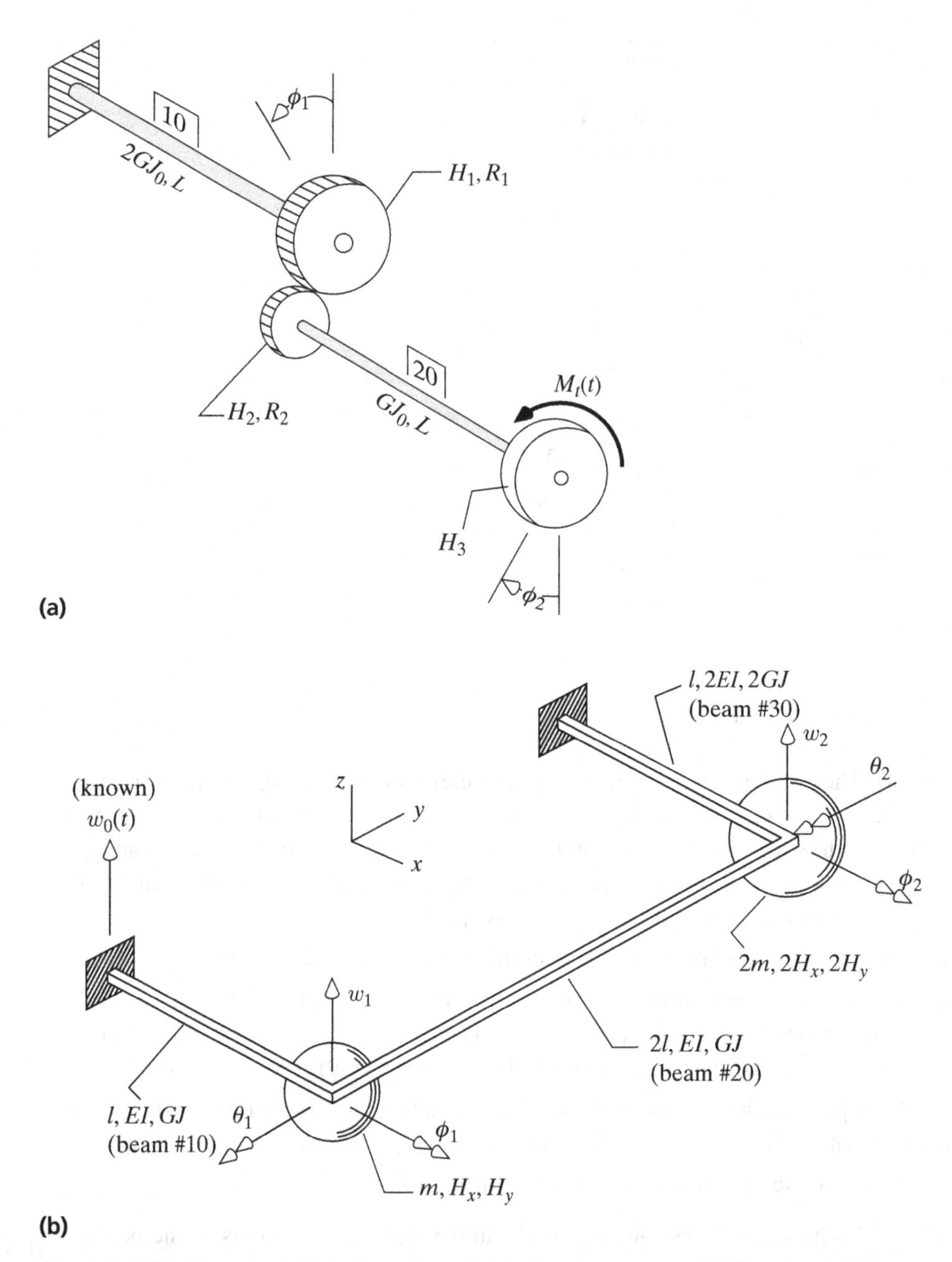

Figure 4.13(a). Exercise 4.9(a): Geared torsional vibration system. (b) Exercise 4.9(b).

the plane be the same as it is out of the plane. Write the matrix equations of motion for this system. Treat the axial (i.e., longitudinal) stiffness of the beam elements as infinite.

4.6 **(a)** Redo Example Problem 4.6 (Figure 4.5) where now the enforced motion at the wall is only $w_0(t)$.

(b) Redo Example Problem 4.6 (Figure 4.5) where now the enforced motion at the wall is only $\theta_0(t)$.

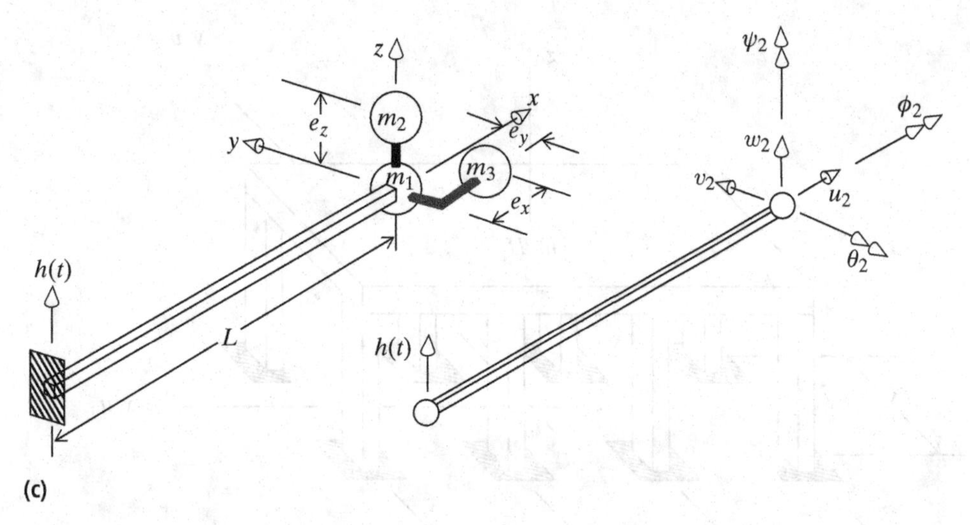

Figure 4.13(c). Exercise 4.9(c): Contilevered beam with mass modeling and selected global DOF.

(c) Redo Example Problem 4.8 where now the enforced base motion is only a rotation about the z axis at the right-hand base support.

4.7 **(a)** Reconsider the beam of Figure 4.1(a). Are the natural frequencies of this beam increased or decreased when additional nonstructural mass is distributed along the length of the beam?

(b) As above, are the natural frequencies of the original beam increased or decreased if the beam cross section is rearranged so as to increase the beam cross-sectional area moment of inertia I_{yy} without altering the beam mass per unit of beam length? That is, increase I_{yy} without altering A, the beam cross-sectional area.

4.8 **(a)** Consider any simply supported, uniform beam. Is the first natural frequency increased or decreased if all the mass of the beam is concentrated at the center of the beam? (Only an answer based on the reader's intuition is expected at this point.)

(b) As above, but this time in accordance with the discrete mass modeling techniques discussed in this chapter, the entire distributed mass is replaced by half the total mass lumped at midspan.

4.9 **(a)** Write the equations of motion in symmetric matrix form for the torsional vibration of the geared system shown in Figure 4.13(a).

(b) Write the equations of motion in symmetric matrix form for the two-mass beam grid shown in Figure 4.13(b). As per usual, treat all beam elements as inextensible and be assured that all motions other than those indicated by the labeled DOF are prevented by unseen constraints. Let $GJ = EI$, and be sure to note that the system is being driven by a known, vertical motion, $w_0(t)$, at the left-hand support.

(c) The uniform, cantilevered beam shown in Figure 4.13(c) supports a three part tip mass and is being driven by a support motion $h(t)$ as shown. As pointed out previously, it is generally more advantageous to select generalized coordinates that

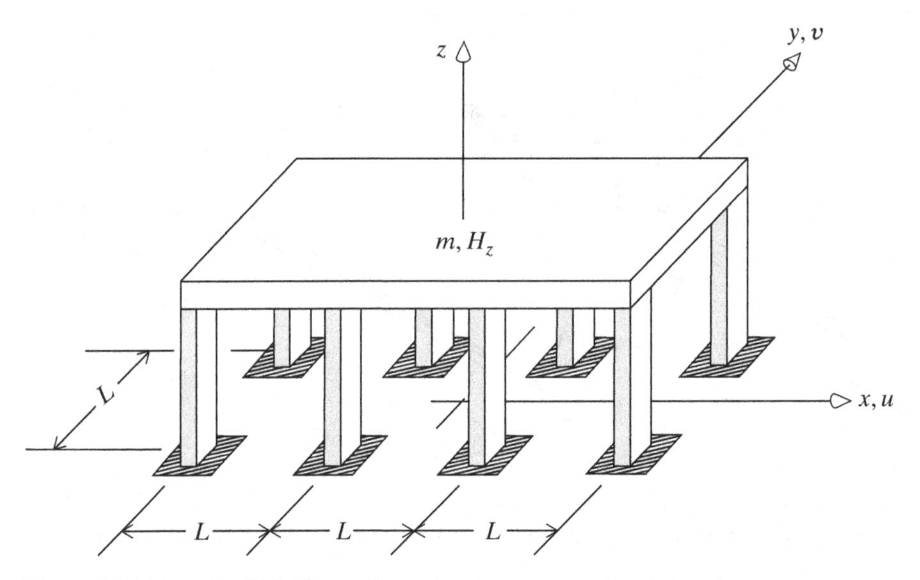

Figure 4.14. Exercise 4.10: Three-dimensional structure with eight columns.

simplify the writing of the stiffness matrix while allowing complications in the mass matrix. Therefore, using as appropriate, the standard DOF indicated in the sketch in the order $\{v_2 \quad \psi_2 \quad w_2 \quad \theta_2 \quad \phi_2\}$, write the stiffness matrix for this beam. To that end, as per usual, model the beam as having no shearing deformations, and note that here

$$EI_{yy} = 3EI_0, \quad EI_{zz} = 2EI_0, \quad GJ = EI_0, \quad EA = \infty,$$

where, for example, I_{yy} is the area moment of inertia about the y axis. Hint: If you are not sure a motion will occur, include that DOF to be safe. If the motion doesn't occur, the associated DOF will have a zero solution. Determine the equivalent generalized force vector, $\{Q_{eff}\}$.

(d) Write the mass matrix for the same single, cantilevered beam of part (c). Note again that in lieu of estimating mass moments of inertia, the mass is modeled in three interconnected parts as shown. Since $L \gg e$, each of the three short extensions e_j is to be considered rigid. Use the same DOF employed in part (c).

4.10 The structure of Figure 4.14 is a roof slab supported by eight identical beam-columns that are fixed at their bases and rigidly attached to the roof. The roof is sufficiently stiff in its own plane to be treated with respect to motion in the x, y plane as a single rigid mass of lateral dimensions L *by* $3L$. The bending stiffness coefficients of the columns are EI_0 about the y axis, and $2EI_0$ about the x axis. Crudely model the individual beam torsional stiffness coefficients, GJ, as negligible.

(a) What is the natural frequency of this structure for a side-sway motion in the x direction?

(b) As above, but for a side-sway motion in the y direction.

(c) As above, but for a rotation about the z axis.

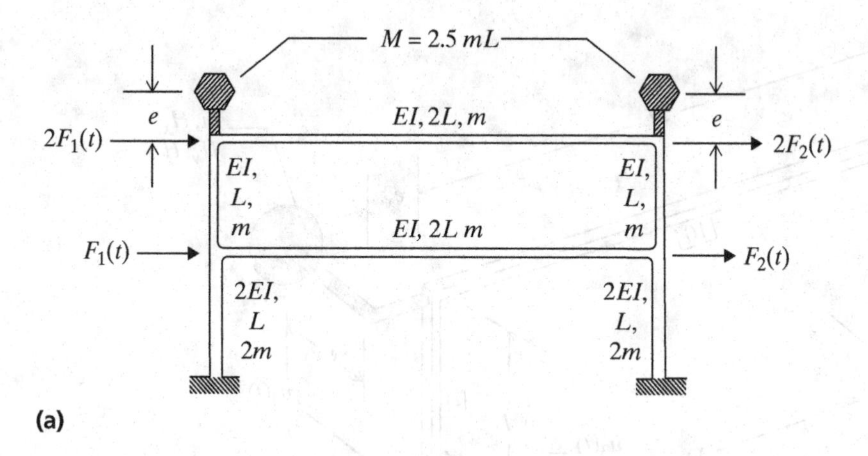

(a)

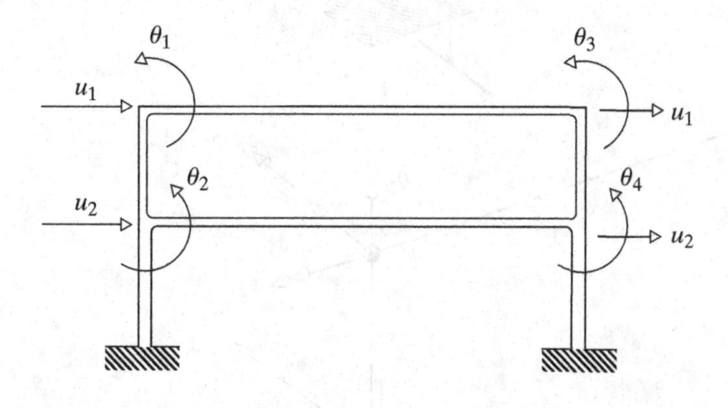

(b)

Figure 4.15. (a) Exercise 4.11: Two-story frame with mass modeling. (b) Exercise 4.11: degrees of freedom.

(d) When the beam-columns are modeled as axially rigid as well as the roof being modeled as rigid, are there natural frequencies for this model other than the three considered above?

4.11 Consider the two-story, clamped frame shown in Figure 4.15(a). The beams that comprise the frame have, as indicated in the sketch, uniformly distributed masses of magnitudes m or $2m$, where m is in units of mass per unit of beam length. The frame also supports a rigid mass of magnitude $2.5mL$ at the top of each beam-column on a rigid, vertical extension of magnitude e above the roof beam centerlines. These rigid, vertical extensions are rigidly connected to the frame corners. As per usual, treat all the beams as inextensible. Thus choose the two lateral DOF and four rotational DOF shown in Figure 4.15(b).

(a) Convert the distributed masses of the beams to discrete masses. For the sake of simplicity, lump mass only at the four beam joints. Do not make any provision for mass moments of inertia associated with the mass of the beams.

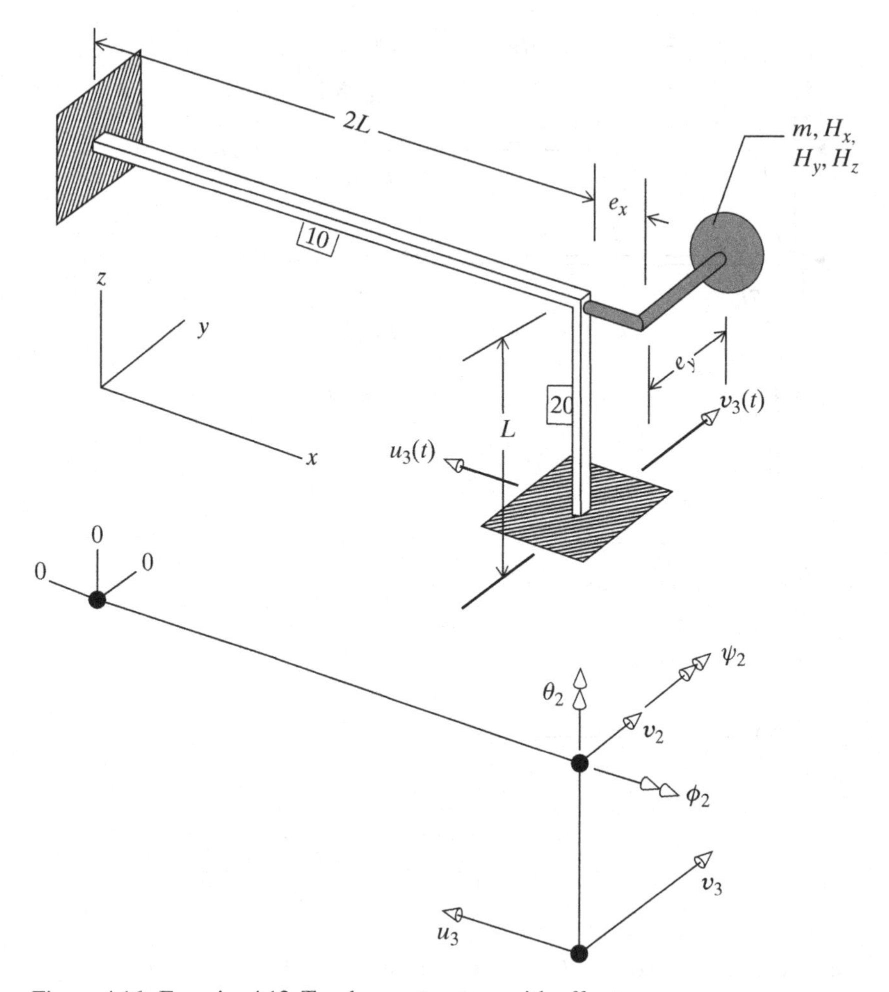

Figure 4.16. Exercise 4.12. Two beam structure with off-set mass.

(b) Write the mass matrix for this frame.

(c) Write the stiffness matrix for this frame.

(d) Write the applied load vector for this frame.

4.12 Consider the two-beam-element, grid-frame, shown in Figure 4.16. Note the coordinate system and the required four nonzero, DOF and two enforced base motions. Understand that beam 10 is being forced to bend about its y axis by the enforced base motion $u_3(t)$. Beam 10 is also being forced to bend about its z axis and twist about its x axis by the $v_3(t)$ enforced base motion. The two excitation functions, $u_3(t)\cdot$ and $v_3(t)$ can be considered the x and y components of a single base motion. There is only one mass and it is offset in the x and y directions from node 2 by two rigid extensions of lengths e_x and e_y. The data for the beam stiffness coefficients are

$$\text{beam 10:} \quad EI_{yy} = 5EI_0 \quad EI_{zz} = 3EI_0 \quad GJ = EI_0$$
$$\text{beam 20:} \quad EI_{xx} = EI_0 \quad EI_{yy} = 2EI_0 \quad GJ = EI_0.$$

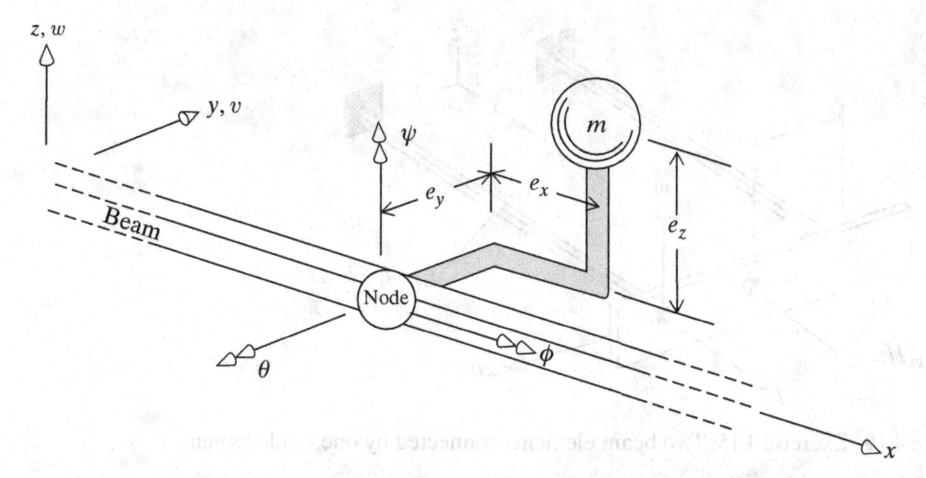

Figure 4.17. Exercise 4.13. Mass with three offsets from its associated node.

(a) Write the kinetic energy expression and state the mass matrix when the deflection vector entries are ordered as follows: v_2, θ_2, ϕ_2, and ψ_2.

(b) Write the stiffness matrix and equivalent force vector for this structure using the same deflection vector.

4.13 Consider the lumped mass offset by rigid extensions from the beam node shown in Figure 4.17. Write the kinetic energy expression for that mass m. All six of the possible DOF at the node are active except for the axial displacement u. Use the kinetic energy expression to create the mass matrix and acceleration vector for this node.

4.14 Write the Lagrange equations of motion for the three-bar truss supporting the mass m and the applied force $F(t)$, as shown in Figure 4.18.

For the eager

4.15 Write the free vibration equations of motion in symmetric matrix form for the cantilevered beam grid shown in Figure 4.19. Let the distance between the two beam

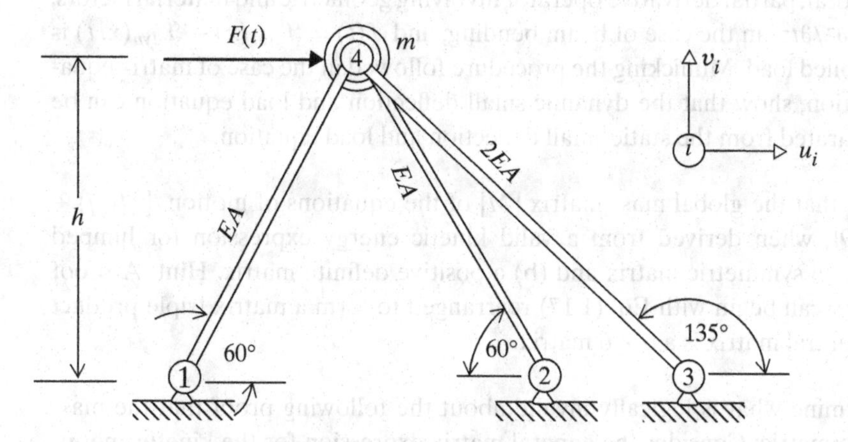

Figure 4.18. Exercise 4.14: Planar truss.

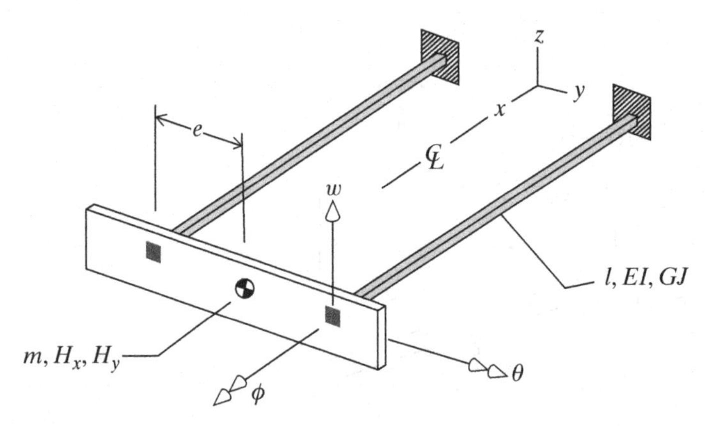

Figure 4.19. Exercise 4.15: Two beam elements connected by one rigid element.

tips be $2e$, and model the structural element connecting the beam tips as rigid. Let $GJ = EI$ for the beam elements of length ℓ. Be sure to understand that only the three indicated DOF are necessary to describe this structural system because the motion of the right-hand beam tip controls the motion of the left-hand beam tip through the constraints imposed by the rigid connection between the two beam tips.

4.16 Chapter 8, which deals with continuous structural elements (for example, beams, plates, etc., modeled with distributed mass), shows that the general description of the (small) motion for such a system is one or more linear partial differential equations of the form

$$\mathcal{P}[w(x,t)] + \mathcal{H}[w(x,t)] = \mathcal{L}(t),$$

where t is time; x represents spatial variable(s) in one-, two-, or three-space; $w(x,t) = w_{stat}(x) + w_{dyn}(x,t)$ is the unknown total deflection; $\mathcal{P}$ is a linear, even-order, spatial, partial derivative operator involving geometric and material parameters, such as $(\partial^2/\partial x^2)[EI(x)(\partial^2/\partial x^2)]$ in the case of beam bending; $\mathcal{H}$ is a linear, even-order, temporal, partial derivative operator involving geometric and material factors, such as $\rho A(\partial^2/\partial t^2)$ in the case of beam bending; and $\mathcal{L}(t) = \mathcal{L}_{stat}(x) + \mathcal{L}_{dyn}(x,t)$ is the total applied load. Mimicking the procedure followed in the case of matrix equations of motion, show that the dynamic small deflection and load equation can be entirely separated from the static small deflection and load equation.

4.17 Prove that the global mass matrix $[M]$ of the equations of motion, $[M]\{\ddot{q}\} + [K]\{q\} = \{Q\}$, when derived from a valid kinetic energy expression for lumped masses, is (a) a symmetric matrix and (b) a positive definite matrix. Hint: A proof for symmetry can begin with Eq. (1.17) rearranged to form a matrix triple product where the central matrix is a 6×6 matrix.

4.18 Determine what is logically invalid about the following proof that the mass matrix is symmetric: Consider the general matrix expression for the kinetic energy.

Since $2T = \lfloor \dot{q} \rfloor [M] \{\dot{q}\}$ is a scalar, which is a 1×1 matrix, its transpose is equal to itself:

$$\therefore \quad 2T = \lfloor \dot{q} \rfloor [M] \{\dot{q}\} = \lfloor \dot{q} \rfloor [M]^T \{\dot{q}\}$$
$$\therefore \quad \lfloor \dot{q} \rfloor \left([M] - [M]^T\right) \{\dot{q}\} = 0.$$

Since the velocity vector is entirely arbitrary, the false conclusion is that the mass matrix equals its own transpose and hence is symmetric.

ENDNOTE (1): MASS REFINEMENT NATURAL FREQUENCY RESULTS

The Abaqus digital computer program is a large, commercially available, finite element analysis program. In units of hertz (cycles per second, that is, f not ω), the following natural frequency results were obtained using Abaqus for the simply supported beam model of Figure 4.1(c). For the purposes of these calculations, $m =$ 1.0 lb.-sec^2/in., *total beam span length* $= 8L = 80.0$ in., $E = 10{,}000{,}000$ psi, and the square beam cross section is 2 in. $\times$ 2 in. Note again that natural frequencies are always numbered and subscripted so that the lowest value is the first natural frequency, the second lowest is the second natural frequency, and so on. This numbering system generally corresponds to the importance of those natural frequencies. The mass modeling used for the first table numerical results placed half the beam segment mass at the segment ends.

I. Natural frequencies for mass at segment ends

Freq. no.	Two segments/ one mass	Four segments/ three masses	Eight segments/ seven masses	Exact strength of material solution
1	2.8135	2.833	2.834	2.834
2		11.254	11.333	11.336
3		23.895	25.460	25.506
4			45.016	45.345

Notes: 1. The one mass of the one-mass case has magnitude $4m$ and is located at $4L = 40$ in. from the left-hand beam end.

2. Each mass of the three mass case is $2m$, and they are located at 20, 40, and 60 in. of the span.
3. Each mass of the seven mass case is m, and they are located at 10-in. intervals.
4. Since the two-segment, one-mass case is a one-DOF system, there is only one natural frequency and similarly there are only three natural frequencies for the three-mass case, and so on.
5. Symmetry was not used in this analysis.

The frequency result for a one-beam-segment, one-mass case (a one-DOF system), where the part of the beam mass is discretized at the beam center, can easily be calculated by hand because the discrete stiffness k at the simply supported beam center is $48EI/(8L)^3$. If, for example, the entire beam mass $8m$ is placed at the beam center, the estimate of the first natural frequency, the square root of $k/(8m)$ is 1.99 Hz, a very poor estimate. However, if only half the beam mass, $4m$, is placed at the beam center, as is done using the discretization process of the first table, then the result is the much more accurate value of 2.813 Hz. Further, notice that this one-mass case, and all the other discrete mass cases, do not include any estimates for the beam mass moment of inertia. Be advised that the strength of materials "exact" solution also does not include any effect of the beam mass moments of inertia. Thus it is not correct to use this data to support the correct conclusion that those beam mass moments of inertia are unimportant.

For the sake of comparisons, the mass modeling for this second table placed the lumped mass at the center of the beam segment.

II. Natural frequencies for mass at segment centers

Freq. no.	Two segments/ one mass	Four segments/ three masses	Eight segments/ eight masses	Exact strength of material solution
1	2.8135	2.833	2.834	2.834
2	7.758	11.254	11.333	11.336
3		23.895	25.460	25.506
4		31.831	45.016	45.345

Notes: 1. Each mass of the two mass case has magnitude $4m$, and they are located at $2L = 20$ in. and $6L = 60$ in.
2. Each mass of the four mass case is $2m$, and they are located at 10, 30, 50, and at 70 in.
3. Each mass of the eight mass case is m, and they are located at 10-in. intervals with the left-most mass located at a distance of 5 in. from the left end.

The conclusion is that there is no difference between the two approaches to creating a discrete mass model with respect to the accuracy of the calculated results. (The reason why this is so is discussed in the next endnote.) However, placing the discretized mass at the beam segment ends sometimes leads to a slightly simpler model per beam segment, as is the case here, in that there are one fewer masses and one fewer beam finite elements with that approach relative to placing the mass at the beam segment center.

As an aside, note that the lower the number of the natural frequency, the better the accuracy of the FEM calculation.

ENDNOTE (2): THE RAYLEIGH QUOTIENT

The two purposes for this endnote are (i) to first introduce the Rayleigh quotient, which is a special application of the Rayleigh–Ritz or Ritz method of analysis; and (ii) to explain why the two different mass models of Figures 4.1(c) and (d) produce the same finite element solutions for their first four natural frequencies. Today, the frequency of use of the Rayleigh quotient is much like that of the Rayleigh–Ritz method (RRM), which is seldom used. In the wake of the development of the computer based FEM (which can be viewed as a subcase of the RRM), the usefulness today of the Rayleigh–Ritz method is mostly confined to providing qualitative insights rather than quantitative answers. That is the reason why this once-important topic is now discussed in an endnote.

As for the second purpose of this endnote, recall that the segment centered beam mass model has eight equal masses of magnitude m, whereas the segment end beam mass model has only seven masses of magnitude m, each of which is located halfway between the positions of two of the masses of the eight-mass model. That is, the models are different enough to raise the question as to why they produce the same natural frequency results. Since an extensive introduction to the Rayleigh quotient is necessary before proceeding further, the purpose of explaining the equivalence of the two mass models is not realized until the last two paragraphs.

As for the Rayleigh quotient, consider the simply supported beam of Figure 4.1(a) as it undergoes bending vibrations within the plane of the paper. As the figure indicates, there are no external loads acting on the beam system, so the beam's vibration is termed *natural* or *(force) free*. Let the vibration be solely the result of an initial displacement. Furthermore, as has been done in all the previous analyses, allow the approximation that all energy dissipative forces, which are always present in any real system, are absent here. In this circumstance, the beam is said to be "undamped." (As is seen later, this is a good approximation over a short period of time.) Then, the total energy of the beam system, the sum of the kinetic and internal strain energies, is a positive constant. That is, $\mathscr{E} = T + U = \text{const.} > 0$, where neither T nor U is ever a negative quantity.[12] (Recall, that because this is a linear structural dynamics problem, the static gravitational forces and their potentials can be ignored when calculating the purely dynamic results such as natural frequencies and natural vibration deflection shapes.) At the time the beam starts its vibration from its initial deflection shape, and at each later point in time when each point along the axis of the beam reaches its peak deflection position, before reversing direction in its back-and-forth motion, the velocity of every point on the beam axis is zero.[13] Hence at this point in time the kinetic energy is necessarily zero. Since the total energy is a positive constant, this must also be the point in the vibration cycle when the positive valued strain energy is a maximum. Therefore, $\mathscr{E} = U_{max}$. There is also another point in time when the entire beam passes through its horizontal static equilibrium position. At that point in time, because there are no deflections beyond the SEP, the dynamic portion of the beam strain energy is zero. Thus, as above, it is also true that $\mathscr{E} = T_{max}$. This specialized form of the conservation of energy equation, $T_{max} = U_{max}$, can be used to estimate the natural frequencies of structures such as the vibrating beams of Figures 4.1(c) and (d).

To estimate the value of the small-deflection, first natural frequency for the two different beam mass models under discussion, it is necessary only to estimate (i.e., carefully guess) the distribution of vibratory amplitudes of the masses (called the *mode shape*) of each beam as it vibrates at its first natural frequency. To explain, a mode shape is a set of the relative amplitude values of the various mass components of the structure. Mode shapes are often arranged as simply the ratios of deflection

[12] The fact that T is "positive definite," the mathematical way of saying never negative, stems from the basic notion that any motion imparts (positive) kinetic energy to a mass. This positive definiteness is reflected in the general kinetic energy equation, Eq. (1.17). When the body axes are chosen to be principal axes that cause the matrix $[H]$ to be a diagonal matrix, it is clear that T always depends on the squares of velocities. Ref. [4.2], Chapter 1 exercises, establishes the positive definiteness of the strain energy.

[13] Verification that undamped motion is such that all points along the beam length reach their maximum deflections at the same time, and together pass through the static equilibrium position at the same time, must wait until Chapter 5.

amplitudes of each mass element to the maximum deflection amplitude of the structure as the structure undergoes a natural frequency vibration. The estimation of the relative amplitudes is at the heart of the Rayleigh quotient. These estimates of the true deflection shapes are sometimes called *assumed modes*. See Refs. [4.3, 4.4]. When there is only one assumed mode, the analysis method is called the Rayleigh quotient or Rayleigh method. Ritz extended the Rayleigh method by using more than one assumed mode.

Confidence in estimating the required deflection patterns of force free vibrating structures with simple geometries and isotropic materials comes quickly with experience gained by observing mode shape solutions obtained by other methods. If no guide in the form of previous experience is available, a satisfactory estimate of the vibratory amplitudes associated with the first natural frequency for even a very complicated structure is any set of amplitudes proportional to the structure's static deflections due to gravity. See also Ref. [4.1]. The reason that the Rayleigh quotient once held considerable importance is that even if the guess for the deflection pattern for the vibrating structure is only a rough approximation to the true amplitude shape, the calculated result for the natural frequency is much more accurate than that of the selected deflection amplitude estimate. It is worth repeating that even a somewhat crude deflection estimate produces a satisfactory approximation for the natural frequency [4.3].

To use the Rayleigh quotient for the estimation of the first natural frequencies of the lumped mass beams of Figures 4.1(c) and (d), guess that the approximate, overall deflection shape of both beams vibrating at their first natural frequencies is the function $\sin(\pi x/8L)$. This is an easy approximation to make for the lumped-mass beam models because, as noted previously, that is the "exact" strength of materials solution for the similar, uniformly distributed mass beam model. Thus, where the constant A is a small[14] but arbitrary amplitude (not an area), the vertical deflection of any point along the beam can be written as the continuous function

$$w(x,t) = A\sin(\pi x/8L)\sin\omega t.$$

Then the velocity and curvature of any point along the beam are, respectively,

$$(d/dt)w(x,t) = A\omega\sin(\pi x/8L)\cos\omega t.$$

$$\text{and}\quad (d^2/dx^2)w(x,t) = -A(\pi/8L)^2\sin(\pi x/8L)\sin\omega t.$$

On the way to writing $T_{max} = U_{max}$, the general equations for the beam kinetic energy and elastic strain energy are [4.1,4.3]

$$T = \tfrac{1}{2}\int_{\ell} m(x)[\dot{w}(x,t)]^2 dx$$

$$\text{and}\quad U = \tfrac{1}{2}\int_{\ell} EI(x)[w''(x,t)]^2 dx$$

14 "Small" so as to avoid nonlinear elastic effects.

$$\text{where here}\quad m(x) = \sum_{i=1}^{n} m_i \delta(x - x_i)$$

$$\text{and}\quad EI(x) = EI = \text{const.}, \tag{4.4}$$

where ℓ is the total beam length ($8L$ in this case), $m_i = m$, x_i is the lengthwise coordinate location of the lumped mass, and the upper summation index n is 7 for the beam model of Figure 4.1(c) and 8 for the beam model of Figure 4.1(d). The Dirac delta function, $\delta(x - x_i)$, of the mass summation expression is explained in detail in Chapter 7. For the moment, just consider that roughly it produces a unit value in the evaluation of the integral whenever $x = x_i$, and zero elsewhere.

At the respective maximum values of the kinetic and strain energies, their associated sine and cosine time functions have their maximum values, 1. Therefore,

$$\begin{aligned}
T_{max} &= \tfrac{1}{2}\sum_{i=1}^{n} A^2 m_i \omega^2 \int_0^{8L} \delta(x - x_i)\sin^2\left(\frac{\pi x}{8L}\right) dx \\
&= \tfrac{1}{2} A^2 m_i \omega^2 \sum_{i=1}^{n} \sin^2\left(\frac{\pi x_i}{8L}\right) \\
U_{max} &= \tfrac{1}{2} EI\left(\frac{\pi}{8L}\right)^4 A^2 \int_0^{8L} \sin^2\left(\frac{\pi x}{8L}\right) dx \\
&= \tfrac{1}{4} EI\left(\frac{\pi}{8L}\right)^4 A^2 \int_0^{8L} \left[1 - \cos\left(\frac{\pi x}{4L}\right)\right] dx \\
&= EI\left(\frac{\pi}{8L}\right)^4 A^2 (2L).
\end{aligned}$$

With the deflection amplitudes canceling, $T_{max} = U_{max}$ produces the result

$$\omega^2 = \frac{\pi^4 EI}{1024\, m_i L^3} \frac{1}{\gamma}$$

$$\text{where}\quad \gamma = \sum_{i=1}^{n} \sin^2\left(\frac{\pi x_i}{8L}\right).$$

It should be clear that the use of the Rayleigh method is simpler when the mass distribution is modeled as being distributed.

From the above solution for the square of the natural frequency, it is clear that the only difference between RRM analysis for the beam of Figure 4.1(c) and that of Figure 4.1(d) lies in the computation of γ. The simplest way to show that γ is the same for both beam models is to make a hand calculation for both cases. Using the symmetry of the sine function about the beam center makes each calculation consist only of summing four terms. The required values of x_i for the beam of Figure 4.1(c) are simply L, $2L$, $3L$, (each term to be counted twice) and (to be counted just once) $4L$. The x_i values for the beam of Figure 4.1(d) (each term to be counted twice) are $L/2$, $3L/2$, $5L/2$, and $7L/2$. The result for each sum is the same value 4.0. Hence the RRM estimate for the first natural frequency is the same for both forms of mass modeling.

ENDNOTE (3): THE MATRIX FORM OF THE LAGRANGE EQUATIONS

The matrix form of the Lagrange equations presented in Eq. (4.2) can be justified as follows. Start with the demonstrated forms of the kinetic energy and the strain energy,

$$T = \tfrac{1}{2}\lfloor \dot{q} \rfloor [M]\{\dot{q}\} \qquad U = \tfrac{1}{2}\lfloor q \rfloor [K]\{q\},$$

where, again, this form of the kinetic energy comes directly from the previously derived equation $T = \tfrac{1}{2}\sum m_j \dot{\boldsymbol{r}}_j \bullet \dot{\boldsymbol{r}}_j$ and a representation of the position vector as a function of the generalized coordinates. The above form of the strain energy was here obtained from a virtual work expression for the elastic forces acting on the discrete masses. The variation of these two quantities, because of the symmetry of the mass and stiffness matrices, is

$$\delta T = \lfloor \delta\dot{q} \rfloor [M]\{\dot{q}\} \qquad \delta U = \lfloor \delta q \rfloor [K]\{q\}.$$

Using the chain rule for differentiation, the variation of the kinetic energy and the variation of the strain energy can also be represented as

$$\delta T = \Sigma \frac{\partial T}{\partial \dot{q}_j}\delta\dot{q}_j = \lfloor \delta\dot{q} \rfloor \left\{\frac{\partial T}{\partial \dot{q}_j}\right\} \qquad \delta U = \Sigma \frac{\partial U}{\partial q_j}\delta q_j = \lfloor \delta q \rfloor \left\{\frac{\partial U}{\partial q_j}\right\}.$$

Subtract the corresponding quantities of the above two sets of equations and obtain

$$0 = \lfloor \delta\dot{q} \rfloor \left([M]\{\dot{q}\} - \left\{\frac{\partial T}{\partial \dot{q}_j}\right\}\right) \qquad 0 = \lfloor \delta q \rfloor \left([K]\{q\} - \left\{\frac{\partial U}{\partial q_j}\right\}\right).$$

Now the argument is made that in each of the above two equations, the row matrix of the varied quantity is entirely arbitrary with respect to the quaintity within the parantheses. That is, the terms of either row matrix can be chosen arbitrarily. Whatever the choices made for the row matrices, the matrix product is zero. The only way that this is possible is that the quantities within the parantheses must be zero. Thus the conclusion

$$[M]\{\dot{q}\} = \left\{\frac{\partial T}{\partial \dot{q}_j}\right\} \qquad [K]\{q\} = \left\{\frac{\partial U}{\partial q_j}\right\}.$$

Insertion of the right-hand side vectors into the Lagrange equation, and noting that the mass matrix is a matrix of constants, yields Eq. (4.2).

ENDNOTE (4): THE CONSISTENT MASS MATRIX

The FEM consistent mass matrix for the distributed mass of a single, uniform beam element that is bending in the x, z plane, is derived, as are all mass matrices, from the kinetic energy expression for the beam element. Without accounting for rotary inertia effects for the beam cross section, the beam bending kinetic energy expression is

$$T = \tfrac{1}{2}\int_0^{\ell} \rho(x)A(x)[\dot{w}(x,t)]^2 dx. \tag{4.5}$$

Let the area and mass density be constants. Then use the same four beam bending shape functions, $N_1(x)$ through $N_4(x)$ (the four cubic splines) for the element lateral deflection that are used to obtain the beam bending element stiffness matrix (the reason for the characterization of the mass matrix as "consistent" with the stiffness matrix). That is, use the velocity distribution description

$$\dot{w}(x,t) = \lfloor N_1(x) \quad N_2(x) \quad N_3(x) \quad N_4(x) \rfloor \begin{Bmatrix} \dot{w}_1 \\ \dot{\theta}_1 \\ \dot{w}_2 \\ \dot{\theta}_2 \end{Bmatrix},$$

where

$$\begin{aligned} N_1(x) &= 2(x/\ell)^3 - 3(x/\ell)^2 + 1 \\ N_2(x) &= \ell[(x/\ell)^3 - 2(x/\ell)^2 + (x/\ell)] \\ N_3(x) &= -2(x/\ell)^3 + 3(x/\ell)^2 \\ N_4(x) &= \ell[(x/\ell)^3 - (x/\ell)^2], \end{aligned}$$

in the kinetic energy integral. Then evaluate the integral, and cast the result in the usual matrix form of $T = \frac{1}{2}\{q_e\}^t[m_e]\{q_e\}$. Insertion of this kinetic energy expression into the Lagrange equations leads directly to the result

$$[m_e]\{\ddot{q}_e\} = \frac{\rho A \ell}{420} \begin{bmatrix} 156 & 22\ell & 54 & -13\ell \\ 22\ell & 4\ell^2 & 13\ell & -3\ell^2 \\ 54 & 13\ell & 156 & -22\ell \\ -13\ell & -3\ell^2 & -22\ell & 4\ell^2 \end{bmatrix} \begin{Bmatrix} \ddot{w}_1 \\ \ddot{\theta}_1 \\ \ddot{w}_2 \\ \ddot{\theta}_2 \end{Bmatrix}. \qquad (e)$$

This result has the apparent advantage that it provides an estimate of the rotary inertia effects associated with the element nodal rotations that result from the distribution of the mass along the beam axis. However, studies have shown that this advantage does not seem to lead to increased numerical accuracy. Moreover, this consistent mass matrix requires many more machine calculations than the unrefined mass matrix previously discussed, which is, for the sake of comparison, when the element mass is lumped at the element ends,

$$[m_e]\{\ddot{q}_e\} = \frac{\rho A \ell}{420} \begin{bmatrix} 210 & 0 & 0 & 0 \\ 0 & 0 & 0 & 0 \\ 0 & 0 & 210 & 0 \\ 0 & 0 & 0 & 0 \end{bmatrix} \begin{Bmatrix} \ddot{w}_1 \\ \ddot{\theta}_1 \\ \ddot{w}_2 \\ \ddot{\theta}_2 \end{Bmatrix}.$$

Hence distributed mass matrices are not often used.

ENDNOTE (5): A BEAM CROSS SECTION WITH EQUAL BENDING AND TWISTING STIFFNESS COEFFICIENTS

The material properties G and E are related by the formula

$$G = \frac{E}{2(1+\nu)},$$

where ν is the Poisson ratio. Therefore, to begin this example design problem in a reasonable manner, choose the Poisson ratio for steel, which is 0.25. Then $G = 0.4E$. Therefore to obtain the equality $GJ = EI$, it is necessary for $J = 2.5I$. A beam cross

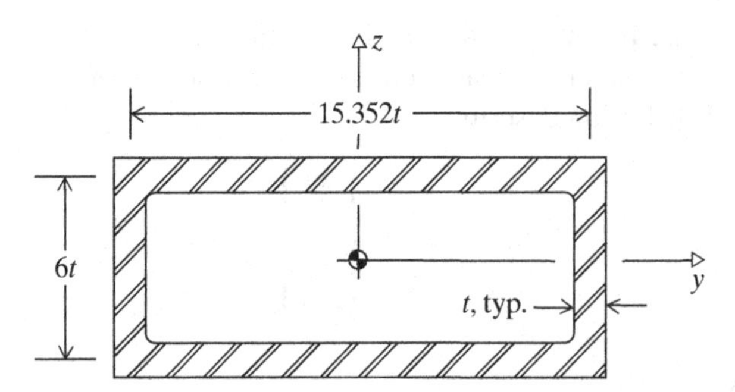

Figure 4.20. Endnote (4): A steel beam cross section for which $GJ = EI_{yy}$.

section with such a relationship between the torsion constant and the area moment of inertia must be relatively stiff in torsion. Hence the easiest way to create such a beam is to look at closed-beam cross sections. A thin circular cross section falls short in that the J/I ratio is only 2.0 rather that the required 2.5. See Ref. [4.1]. Considering thin rectangular cross sections, fixing the thickness as t and the depth at $6t$, and then calculating the required width produces the result shown in Figure 4.20. For this compact cross section, $J = 795t^4$ and $I_{yy} = 318t^4$.

5 Damped Structural Systems

5.1 Introduction

The purpose of this chapter is to introduce damping forces into the structural equations of motion. Simply speaking, damping forces are internal or external friction forces that dissipate the energy of the structural system. Although damping forces are usually much smaller than their companion inertia and elastic forces, they nevertheless can have a significant affect on a vibratory motion, especially after many periods of vibration, or when the system is vibrating at one of certain important frequencies called the system's natural frequencies. This chapter describes various ways of characterizing damping and explains how the damping properties of a vibratory system can be measured. Solutions for the motion of one-DOF systems are presented for force free and certain applied forces to better explain the role that damping plays in structural systems.

5.2 Descriptions of Damping Forces

When an actual, force free, structural system is set in motion by means of initial deflections or initial velocities, or both, any point within the system generally vibrates with amplitudes that are very little different over short time intervals; that is, time intervals lasting typically five or fewer periods of the vibration. Figure 5.1(a) shows the calculated amplitude–time trace of such a vibration where the period T of the vibration is 1 sec and the initial displacement has a unit value. As will soon be seen, the sinusoidal expression that describes the force free motion of a one-DOF undamped system, has to be modified, in this case by an exponential multiplier, when one representative form of system damping is present. The Mathematica instruction is at the top of the graph and is easy to read as plot

$$e^{(-0.01t)} \cos 2\pi t, \qquad 0 \leq t \leq 10.$$

The selected value of the friction measure (the 0.01 of the exponential function) is on the low side for a typical structure but somewhat representative of a new aluminum structure that is well built, firmly supported (or wholly unsupported, such

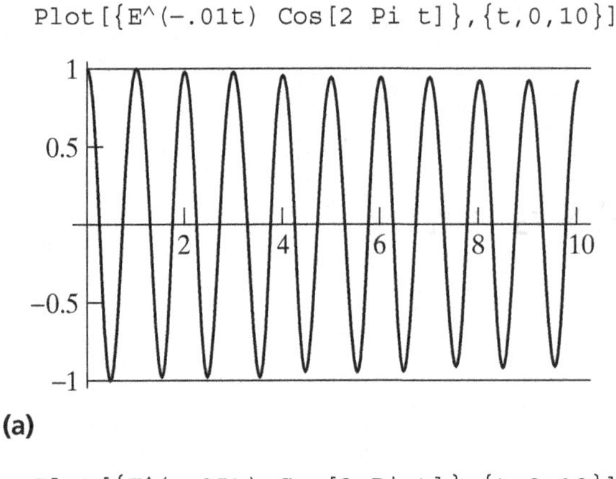

(a)

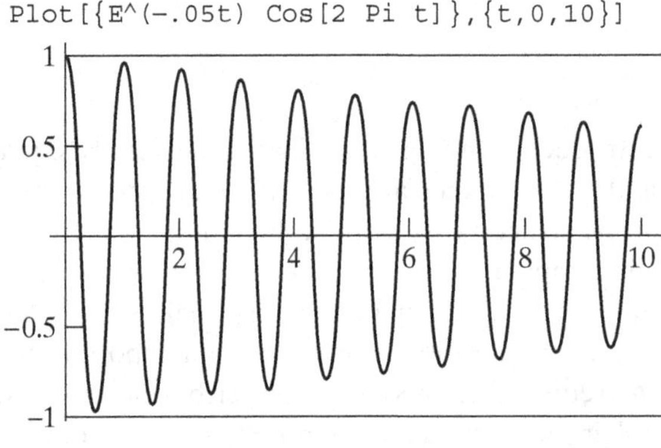

(b)

Figure 5.1. Typical vibratory time histories (from Mathematica).

as an aircraft), and well maintained.[1] Figure 5.1(a) demonstrates for such a structure that the effects of friction can be entirely omitted from the mathematical model of the structure, and still the change in the original value of the amplitude of the vibration during the first few cycles is still within engineering accuracy. If, however, the structure is "loose" (i.e., the rivets or bolts are not tight so that there is a lot of sliding friction or the supports are not wholly firm), then the decay in the vibratory amplitudes can be such as shown in Figure 5.1(b) or even greater. In this latter figure, there is a very noticeable decrease in the vibratory amplitudes over five cycles that well exceeds the limits of engineering accuracy if the effects of the frictional forces are ignored.

The loss of deflection amplitude is also a loss of strain amplitude. For example, from FEM theory, the general, small deflection matrix equation suitable for any structural finite element is $\{\epsilon\} = [B]\{q\}$, where $\{\epsilon\}$ is the vector of engineering strains and $[B]$ is a coefficient matrix of spatial variables. This linear equation shows the unsurprising relationship that the strain magnitudes are directly proportional to deflection (DOF)

[1] From Ref. [5.1], the measure of damping used in the exponential function for a riveted steel structure is 0.03 (and less for a welded steel structure). For aluminum aircraft, it is usually set at 0.02 or a bit more.

magnitudes. The loss of deflection magnitude is a loss of strain energy, because, as before, $U = \frac{1}{2}\{q\}^t[K]\{q\}$. Since, as will be seen, velocity amplitude is nearly proportional to deflection amplitude, there is also always a loss of kinetic energy. The kinetic energy for structural systems undergoing small displacements, from Section 4.4, can always be written as $T = \frac{1}{2}\{\dot{q}\}^t[M]\{\dot{q}\}$. Therefore the decrease in amplitude indicates a decrease in the sum of these two energies, $T + U$, which is the total mechanical energy of the structural system at any time. Dissipated energy is the name applied to the reduction in the total mechanical energy. The dissipative energy can take the form of an increase in heat energy, the work of (local) plastic deformations, sound energy, and so on. From a use of the above strain energy and kinetic energy expressions, the reverse argument can be made that a loss of total mechanical energy can be seen as a cause for the loss of vibratory amplitude. The reason that friction forces always involve an energy loss is simply that, by definition, friction forces always oppose the motion and hence always do negative work. Doing negative work on any system is the same as decreasing the total mechanical energy of that system.

To avoid any possible confusion, note that the so-called friction force associated with an ideally round body rolling on an ideally flat surface, the force that enforces the nonslip rolling condition discussed in Chapter 2, is actually a force of constraint and as such does zero work. The loss of energy (i.e., the source of negative work done on the system) that occurs when an actual wheel, of, say, an automobile, rolls on a flat surface is mostly because of the otherwise round wheel base being flattened by the weight the wheel supports. As a result of this flattening, to maintain its kinetic energy, the wheel requires a constant applied torque, say, from the engine, to overcome the resisting moment that is the product of the weight on the wheel and half the chord length of the flattened arc that is in contact with the surface. The wheel weight also slightly depresses the flat surface. Hence the wheel will always be situated in a slight depression and therefore always rolling uphill.

Returning to the discussion of friction, note that friction is everywhere. Therefore every system in motion has negative work done on it by friction forces. If that lost energy is not offset by positive work done on the system by some external source, the sum of the kinetic and strain energy will continuously decrease. It should be intuitively clear that a continued decrease in the total mechanical energy means that the magnitude of a vibratory motion of the system will decrease and eventually cease. This effect of a lessening of the magnitude of the vibratory motion, because of energy-reducing friction is called *damping*. There are only three general types of damping (friction) forces acting on a structure in motion. They are (i) the internal friction forces in the structural material itself that arise when that material is strained; (ii) the external friction forces that are system boundary contact forces between the structure being studied and another dry, solid body; and (iii) the external friction forces that are system boundary contact forces as the solid object moves within a viscous fluid or is otherwise in contact with a viscous fluid moving relative to all or part of its boundary surface. These latter friction forces are generally called drag forces. The mathematical descriptions of these three types of damping forces have been, of necessity, determined experimentally.

The first type of damping, that due to internal friction, is called *material damping* or *solid damping*. The energy lost through the mechanism of solid damping in structural

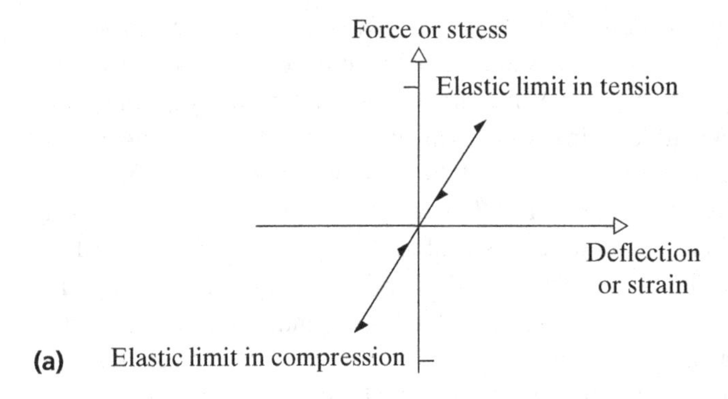

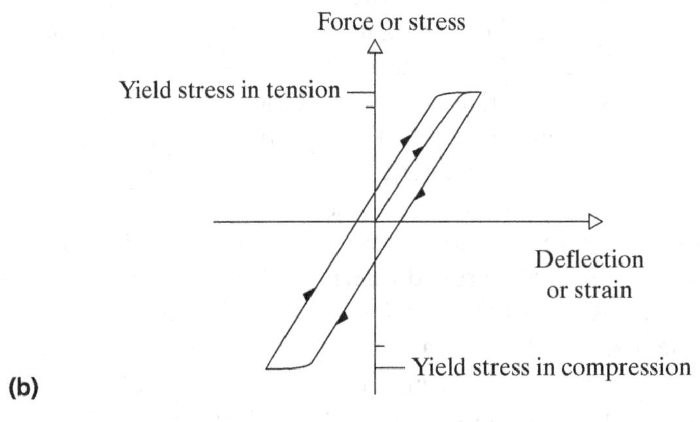

Figure 5.2. Cyclic force–deflection plots for (a) amplitudes below and above the elastic limit (σ_e), and for (b) amplitudes at or slightly above the yield stress (σ_y), where $\sigma_y > \sigma_e$.

materials is largely independent of the frequency of the vibration but is proportional to the square of the strain amplitude. Thus in a structural system undergoing small displacements, where the strain amplitude is proportional to the deflection amplitude, the energy loss is proportional to the square of the deflection amplitude. This type of energy loss can be related to an engineering description of the material that is centered on the material's stress–strain (force–deflection) diagram. Consider a structural material specimen undergoing a constant amplitude vibration. Figure 5.2(a) is a plot of the vibratory linear force–deflection history over many cycles for what is termed an elastic material. A (perfectly) elastic material is an idealization that, by definition, is a material where there are no internal friction forces, and thus no energy is dissipated internally during loading or unloading. Thus, in the absence of other types of damping, too, the repeated loading and unloading curves of the force–deflection (or stress–strain) plot lie on top of each other, and all work[2] done on the material when the material is loaded is wholly recovered during the unloading. Hence the

[2] Since for a force–displacement plot, an increment in the ordinate (force) multiplied by any increment in the abscissa (collinear displacement) is work, the area under a force displacement plot represents work done on the system. Similarly, as discussed in Endnote (1) of Chapter 1, the area under a stress–strain plot is work per unit of material volume.

undamped vibration of a (perfectly) elastic system can maintain constant vibratory amplitudes without the need for an externally applied force to supply energy to the system. Therefore, from an energy viewpoint of an elastic material, the lack of energy loss and the impossibility of energy gain (because there is no external force) during a vibratory cycle also means that when the force or stress returns to its zero value, so too must the deflection or strain. No real solid material is perfectly elastic. All real structural materials possess internal frictional forces. However, on much the same basis that damping effects can be ignored for a few cycles of vibration as discussed above, the even smaller deviations from the single line plot of Figure 5.2(a) for structural materials such as steel, aluminum, fiber-reinforced composites, and so on, can be approximated as nonexistent when the number of loading cycles are few. From Ref. [5.1], the equivalent material damping values for steel and aluminum are just 0.0006 and 0.0002, respectively, which are a lot less than the structural damping value 0.01 used in the demonstration of slow amplitude decay shown in Figure 5.1(a).

Figure 5.2(b) is another stress–strain history for a structural material, but in this case the applied stress clearly exceeds the yield stress, and there are readily apparent inelastic (i.e., plastic) deformations. The resulting full cycle loading and unloading path is called a *hystereses loop*. The area within the hystereses loop, the difference between the loading and unloading lines, represents the large quantity of energy lost per cycle. The lost energy needs to be replaced by work done on the material by an externally applied force for the amplitude of the motion to remain constant as indicated in the figure. This type of severe material damping is also called *hystereses damping*. The stress in Figure 5.2(b) is, for the most part, proportional to the strain in this case where the stress equals or only slightly exceeds the yield stress. Therefore it is possible to argue on this geometric basis for the experimental result that the energy loss per cycle is proportional to the square of the strain amplitude. That is, because the area under both the loading and unloading curves is proportional to the product of stress and strain, so is the difference between the two curves. With stress proportional to strain as cited above, the area of the hystereses curve is thus almost proportional to the strain amplitude squared. The analytical treatment of material damping is postponed to the next section (5.3) in order to first establish some simpler mathematical results.

Now consider the second type of damping force named above, the friction force that arises at the contact surface between two dry solid objects. Specifically, consider a clean, flat block sliding on a clean, flat tabletop. From elementary physics, the *Coulomb* (or dry) friction force between the block and the tabletop is solely proportional to the force normal to the plane of contact, which here is simply the weight of the block. The proportionality constant is called the dynamic Coulomb, or dry, friction coefficient. Actually there are two Coulomb friction coefficients. The static friction coefficient describes the force in the plane of the boundary surface between the block and the tabletop when the block is at rest and the dynamic friction coefficient describes the force between the two when the block is moving on the tabletop. The dynamic friction coefficient is a lesser value than the static friction coefficient. The fact that the Coulomb friction force dissipates energy is evident if the block resting on the tabletop is given an initial velocity. The block soon comes to a complete stop. Its initial kinetic energy is

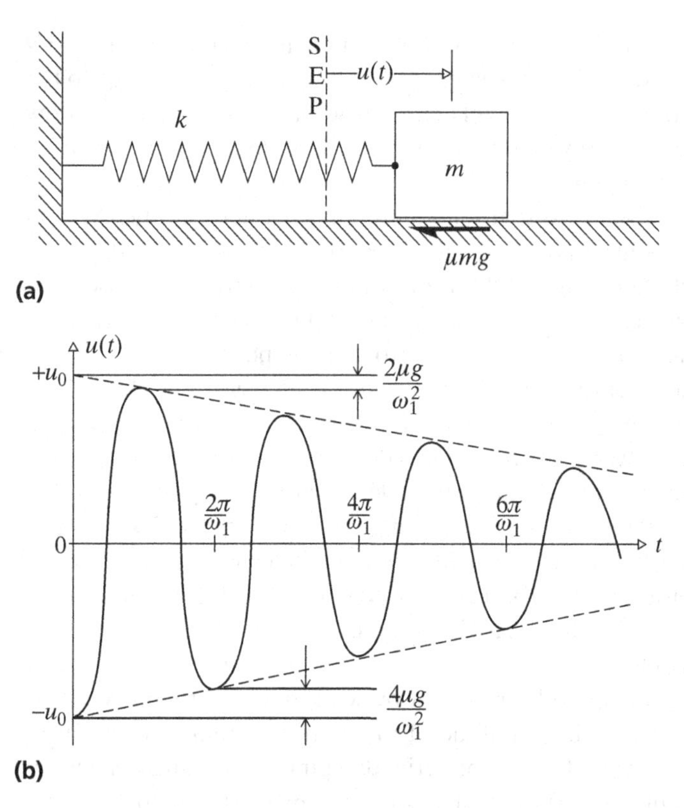

Figure 5.3. (a) Coulomb-damped oscillator. (b) Time history of same after initial deflection u_0 of the left.

totally dissipated into noise, heat, and some polishing of the tabletop and block bottom.

To understand the effect that this type of dry friction force has on a vibration, first recall that the simple spring mass system of Figure 4.2 (called an *oscillator*) is all that is necessary to represent inertial and elastic forces. Thus this one-DOF system is used here as a simple device to investigate different types of damping. Consider the oscillator with Coulomb damping shown in Figure 5.3(a). In the figure, the oscillator is shown moving to the right during its first half cycle of motion. At the time shown in the figure, the mass has a positive horizontal deflection $u(t)$ after starting at a point $u(0) = -u_0$ to the left of the zero deflection point. The zero deflection point is the unstretched spring position that also would be the static equilibrium point of the system if there were no Coulomb friction present in the system. The starting point for this motion, the deflection point $-u_0$, is chosen to be sufficiently to the left of the SEP that the magnitude of the force in the deflected spring, ku_0, a force to the right, is greater than the leftward directed static friction force opposing the start of the motion. Therefore, when the system was released at time zero, the mass immediately starts its motion to the right.

With this problem setup, at the time illustrated in the figure, the system has both a positive deflection and a positive velocity. Since the friction force opposes the motion to the right, the friction force acts toward the left. In general, the friction force always acts in the direction opposite to that of the velocity, $\dot{u}$. To mathematically describe the motion of this sliding block, use the following notation: (i) μ (mu without a subscript) is the dynamic Coulomb friction coefficient, which again is a lesser value than the static Coulomb friction coefficient μ_{stat}; (ii) ω_1 is the natural circular frequency, the square root of the ratio of the stiffness factor, k, to the mass value, m, as discussed in Chapter 2; and (iii) $\mu_{stat}mg$ is the static friction force that must be overcome before motion can be initiated once the mass is stationary. Then from Newton's second law, the equation for the first half cycle of motion, whether u is positive or not, is simply

$$m\ddot{u} = -ku - \mu mg$$

or

$$\ddot{u} + \omega_1^2 u = -\mu g.$$

For initial conditions of $\dot{u}(0) = 0$ and $u(0) = -u_0$, where again it is stipulated that u_0 is such that the initial spring force ku_0 is large enough to overcome the static friction force, the solution to this first half cycle equation of motion for the deflection $u(t)$ is[3]

$$u(t) = -u_0 \cos \omega_1 t - \frac{\mu g}{\omega_1^2}(1 - \cos \omega_1 t).$$

Again, this solution is only valid during the motion to the right. Differentiating this solution to obtain the equation for the velocity, and then setting that velocity equal to zero to determine the time when the mass slows to a stop (i.e., obtaining the time that marks the end of the first half cycle) leads to the equation

$$\left(u_0 - \frac{\mu g}{\omega_1^2}\right) \omega_1 \sin \omega_1 t = 0.$$

Consider the possibility of the factor in parentheses being zero. Reasoning from

$$\mu_{stat}mg > \mu mg$$

and multiplying the factor in parentheses by the nonzero quantity k leads to the result

$$ku_0 - \mu mg > ku_0 - \mu_{stat}\, mg > 0.$$

Therefore, being greater than zero, the factor in parentheses is not zero. Hence, the sine function must produce the zero. This leads to the conclusion that the mass slows to a stop at time $t_s = \pi/\omega_1$. Substitution of this time value into the equation for the deflection shows that when the mass stops, it is located at $u(t = \pi/\omega_1) = +u_0 - 2\,\mu g/\omega_1^2$. This result shows that over the period of one half cycle the amplitude decreases by the amount $2\,\mu g/\omega_1^2$. This result is the same for every half cycle, so the reduction in amplitude for a full cycle is twice that value. Repeating this procedure for each cycle reveals the deflection amplitude time history shown in Figure 5.3(b).

[3] A solution technique for this type of ordinary differential equation is discussed shortly. For the time being, the solution can be validated by substitution into the original differential equation and boundary conditions.

The motion stops when at the end of a half cycle the amplitude of the motion is such that the spring force can no longer overcome the static friction force.

Now consider the first of the two subcases of the third type of friction force, which is the drag force acting on a solid object moving through a viscous gas or a viscous liquid. The motion opposing drag force defies any simple mathematical characterization. The usual mathematical approach is to write that the drag force is equal to a nondimensional drag coefficient, C_d, multiplied by the dynamic pressure (one half the mass density of the fluid, ρ, multiplied by the velocity of the object relative to the fluid squared) multiplied by (typically) a cross-sectional area, S. The difficulty with this formulation is that the drag coefficient varies with the velocity (more precisely, the Reynolds number), although there are large ranges of velocity for which it is reasonably constant. See, for example, Ref. [5.2], p. 17.

Consider again the oscillator of Figure 4.2. Let the oscillator be immersed in a gas or a liquid, and let the drag caused by this fluid be the only source for energy loss. Let most of the velocity range of the mass be such that the drag coefficient is nearly a constant throughout most of that range. Then, because the drag force is then proportional to the square of the velocity of the mass, and it is always directed opposite to the velocity vector, the oscillator equation of motion for the majority of the motion, after division by the mass value, is

$$\ddot{u}(t) + 0.1\,\dot{u}^2 sgn[\dot{u}] + u(t) = 0,$$

where the value of k/m is chosen to be 1.0 and the middle term numerical coefficient of 0.1 is chosen for the value of the ratio $C_d \rho S/2m$. This nonlinear differential equation does not have a known analytical solution. However, Figure 5.4 shows two Runge-Kutta numerical solutions obtained from Mathematica.[4] Both time histories are for a zero initial velocity. The first trace is for an initial deflection of 0.5 and the second trace is for an initial deflection of 1.0. Since the damping force is proportional to the velocity squared, and, as is soon seen, the velocity is nearly proportional to deflection, it is therefore understandable that the amplitudes in the case of the larger initial deflection decrease more rapidly than the amplitudes of the case of the smaller initial deflection. In both cases the decrease in the amplitudes is not linear as it is in the case of Coulomb damping. Rather, the decrease diminishes with time in a manner suggestive of an exponential function; that is, very much as in Figure 5.1.

The second facet of the third type of damping force is *viscous damping*. This is the type of damping that occurs approximately when two lubricated solid objects move relative to each other in such a way that the fluid lubricant between them is mostly sheared. It is also the type of approximate drag force acting on bodies moving through a fluid at low speeds; for example, cylinders and spheres at Reynolds numbers of 0.1 to 4.0 [5.1, p. 16]. The viscous damping force is then closely proportional to the first power of the relative velocity and oppositely directed. The proportionality constant, c, is called the *viscous damping coefficient*. Consider again the oscillator of Figure 4.2. This time let the oscillator be subjected only to viscous damping. The

[4] Mathematica is software for doing mathematics developed by Wolfram Research of Champaign, Illinois. Its Runge-Kutta (see Chapter 9) numerical integration routine was verified over 10 cycles against as exact solution for an oscillator with viscous damping, a form of damping soon to be discussed.

```
NDSolve[{u''[t] + 0.1(u'[t])^2 Sign[u'[t]] +
u[t]==0,u'[0]==0, u[0]==0.5}, u, {t,0,46}]

Out[1]=
      {{u->InterpolatingFunction[{0.,46.},
      <>]}}

In[2]: =
        Plot[u[t]/.%,{t,0,46}]
```

(a)

```
NDSolve[{u''[t] + 0.1(u'[t])^2 Sign[u'[t] +
u[t]==0,u'[0]==0, u[0]==1.0}, u, {t,0,50}]

Out[2] =
      {{u->InterpolatingFunction[{0.,50.},
      <>]}}

In[3]: =
        Plot[u[t]/.%,{t,0,50}]
```

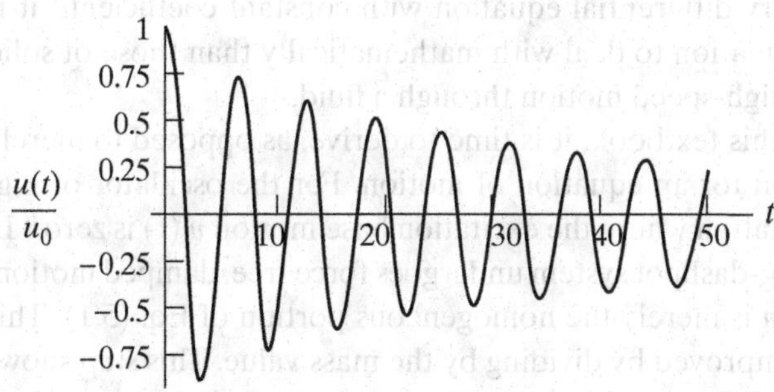

(b)

Figure 5.4. Oscillator response to velocity-squared damping for two different initial deflections ($\omega = 1$) from Mathematica.

standard mathematical model symbol for representing the presence of viscous damping is a massless dashpot.[5] An example of actual dashpots are the "shocks" used to

[5] Energy is dissipated when the piston shears the viscous fluid encased within the cylinder. Work done on a viscous fluid is obviously lost because the fluid returns to its equilibrium condition without storing energy in any form.

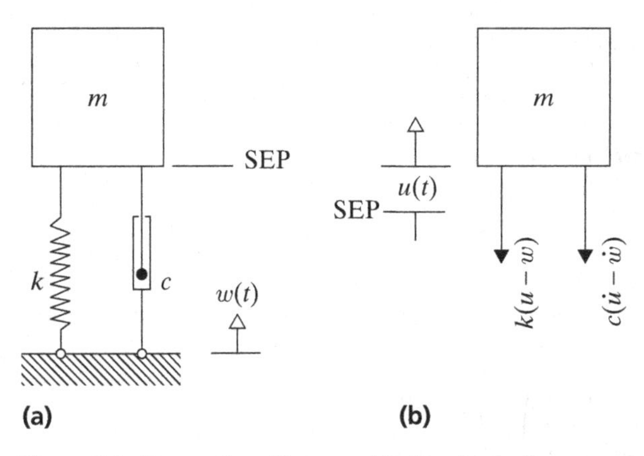

Figure 5.5. Damped oscillator and its free body diagram of external forces acting on the mass: (a) System at rest. (b) Free body diagram of deflected system.

dampen rough road induced automobile vibrations. See Figure 5.5(a). Let the base of this oscillator of this figure undergo a known upward deflection $w(t)$. Then the elastic force (positive upward since u and its time derivatives are positive upward) acting on the mass at its point of contact with the spring is $-k(u - w)$ regardless of whether u is greater than w. Similarly, the force acting on the mass at its point of contact with the dashpot is always the damping coefficient multiplying the relative velocity of the ends of the dashpot, to the first power; that is, the viscous damping force is $-c(\dot{u} - \dot{w})$. See Figure 5.5(b). Therefore from Newton's second law

$$m\ddot{u} = -k(u - w) - c(\dot{u} - \dot{w})$$

$$\text{or} \quad m\ddot{u} + c\dot{u} + ku(t) = c\dot{w} + kw(t). \tag{5.1}$$

Since this is a linear, ordinary differential equation with constant coefficients, it is a much easier differential equation to deal with mathematically than those of solid damping and moderate- to high-speed motion through a fluid.

Now, for the first time in this textbook, it is time to derive, as opposed to merely stating, an analytical solution for an equation of motion. For the oscillator of Figure 5.5, first consider the situation where the excitation base motion $w(t)$ is zero.[6] In this situation, the mass-spring-dashpot system undergoes force free, damped motion. Then the equation of motion is merely the homogeneous portion of Eq. (5.1). This equation statement can be improved by dividing by the mass value. This step shows that the force free, damped motion is dependent only on two system parameters, k/m and c/m. Let k/m be again defined as ω_1^2, which again is the square of the undamped natural frequency of this one-DOF system. Let c/m be defined as $2\zeta_1\omega_1$, where the nondimensional parameter ζ is called the *viscous damping factor*. A quick calculation shows that the algebraic relationship between the damping factor and the damping coefficient is

$$\zeta_1 = \frac{c}{2\sqrt{km}}.$$

[6] Section 5.3 and Chapter 7 discuss general solutions to this type of equation where the equivalent applied force is not zero.

Now that the two parameters, ω and ζ, of the vibratory motion have been defined, let the solution process for the above homogeneous ordinary differential equation begin with the standard (and always successful) trial solution $u(t) = A\exp(rt)$, where A and r are constants whose values are to be discovered. Substitution of this trial solution into the equation of motion leads to

$$\left(r^2 + 2\zeta_1\omega_1 r + \omega_1^2\right) Ae^{rt} = 0.$$

Since the exponential function is never zero even for complex values of r, and because a zero value for A would lead only to a (rejected) trivial solution $u(t) = 0$, both these terms can be canceled from the left-hand side of the above equality. The conclusion is that r must have the two values that satisfy the remaining quadratic equation, called the "auxiliary equation." Those values of r are

$$r_1 = -\zeta_1\omega_1 - \omega_1\sqrt{\zeta_1^2 - 1} \quad \text{and} \quad r_2 = -\zeta_1\omega_1 + \omega_1\sqrt{\zeta_1^2 - 1}.$$

Substitution of these roots into the solution form $u(t) = Ae^{rt}$ leads to the two independent solutions to the homogeneous portion of Eq. (5.1). Hence, the total homogeneous solution is

$$u(t) = \exp(-\zeta_1\omega_1 t)\left[A_1\exp\left(-\omega_1 t\sqrt{\zeta_1^2 - 1}\right) + A_2\exp\left(+\omega_1 t\sqrt{\zeta_1^2 - 1}\right)\right]. \tag{5.2a}$$

The nature of this force free solution clearly depends on the value of the parameter ζ relative to 1.0. First consider the case where ζ is greater than 1. Then $\sqrt{\zeta^2 - 1}$ is a positive number that is less than ζ. Thus the solution that is Eq. (5.2) is the sum of two independent exponential functions each of which decreases monotonically with time. The term with the exponent

$$\left(-\zeta_1 + \sqrt{\zeta_1^2 - 1}\right)\ \omega_1 t$$

decreases more slowly over time than the other term and thus dominates the motion. The important conclusion is that for this range of values for the damping factor, that is when ζ_1 is greater than 1, the motion of the mass described mathematically by the sum of the two exponential terms in Eq. (5.2) is not vibratory. That is, the mass never passes through the static equilibrium position, but merely approaches the SEP as an asymptote. This nonvibratory case where ζ is greater than 1 is called the overdamped case.

Consider the case where ζ equals 1. This borderline case is called the critical damping case. In this case the two quadratic roots are both $r = -\omega_1$. This repeated root case means that the solution, Eq. (5.2), must be modified by the insertion of a t factor for one of the now-identical exponential terms to obtain a second independent solution for the second-order differential equation of motion. Therefore the solution for the case of equal roots for the auxiliary equation is, as can be checked by insertion back into the equation of motion,

$$u(t) = B_1 e^{-\omega_1 t} + B_2 t e^{-\omega_1 t}.$$

Here again both parts indicate that the motion is one of exponential decay rather than a vibratory motion, where the second term dominates. Figures 5.6(a) and (b), which offer response curves for both initial velocities and initial deflections when ζ is 1 or greater, illustrate the important fact that the return to the (near-) static equilibrium position is quickest when the damping factor equals 1. This is an important observation for designing dashpots for vibratory systems like automobiles, large gun carriages, and so on, where it is desirable for the mass to return to its nominal position as quickly as possible.

Now consider the case where ζ is less than 1. This case is called the underdamped case. With ζ less than 1, the quantity within the radical in the solution for the roots r_1 and r_2 is now less than zero. To clearly identify the radical as an imaginary number, rewrite the radical as

$$\sqrt{\zeta_1^2 - 1} = i\sqrt{1 - \zeta_1^2}.$$

Now the rewritten radical itself is always a real number. Using this notation, the underdamped solution for the oscillator motion is

$$u(t) = \exp(-\zeta_1\omega_1 t)\left[\overline{C_1}\exp\left(-i\omega_1 t\sqrt{1-\zeta_1^2}\right) + \overline{C_2}\exp\left(+i\omega_1 t\sqrt{1-\zeta_1^2}\right)\right]. \quad (5.2b)$$

Having the solution in terms of the sum of complex quantities is not convenient. Since $u(t)$ is a real quantity, so too must be the right-hand side of the above equality when taken as a whole. To rewrite the right-hand side in a form that is unmistakably that of a real expression, recall that

$$e^{i\theta} = \cos\theta + i\sin\theta.$$

Using this relation, Eq. (5.2b) can be rewritten as follows

$$u(t) = \exp(-\zeta_1\omega_1 t)\left[(\overline{C_1}+\overline{C_2})\cos\left(\omega_1 t\sqrt{1-\zeta_1^2}\right) + i(\overline{C_2}-\overline{C_1})\sin\left(\omega_1 t\sqrt{1-\zeta_1^2}\right)\right]$$

or

$$u(t) = \exp(-\zeta_1\omega_1 t)\left[C_1\sin\left(\omega_1 t\sqrt{1-\zeta_1^2}\right) + C_2\cos\left(\omega_1 t\sqrt{1-\zeta_1^2}\right)\right], \quad (5.3a)$$

where the constants of integration have been regrouped so that

$$C_1 = i(\overline{C_2} - \overline{C_1}) \quad \text{and} \quad C_2 = (\overline{C_2} + \overline{C_1}).$$

Each term on the right-hand side to the solution for $u(t)$ is now apparently a real quantity for all values of time. To establish this as fact, first note that because ζ, ω, time, and the radical are real quantities, the only possible quantities that could possibly be complex are the two constants C_1, C_2. Since they are constants, their value does not change with time. Therefore look at, for example, the specific times

$$t = \frac{\pi}{2\omega_1\sqrt{1-\zeta_1^2}} \quad \text{and} \quad t = \frac{\pi}{\omega_1\sqrt{1-\zeta_1^2}}.$$

At the first of these two time points, the cosine function has a zero value and the sine function has the value 1. This isolates the C_1 factor. Divide the real left-hand side by all the remaining terms on the right-hand side but for C_1. Since the new left-hand side is the ratio of real numbers, it too must be real. Therefore C_1 must be real.

```
In[50]:= Plot[{(1 + t)* Exp[-t],0.5 * Exp[-2. * t]
        ((1 + 2/Sqrt[3])* Exp[ + t * Sqrt[3]] + (1 - 2/Sqrt[3])
          * Exp[-t * Sqrt[3]]),
      0.5 * Exp[-3.*t]((1 + 3/Sqrt[8])*Exp[+t*Sqrt[8]] +
         (1 - 3/Sqrt[8]) * Exp[-t * Sqrt[8]])},
      {t, 0, 9}, Plotstyle  →{{}, Dashing[{0.01}], Dashing[{0.03}]}]
```

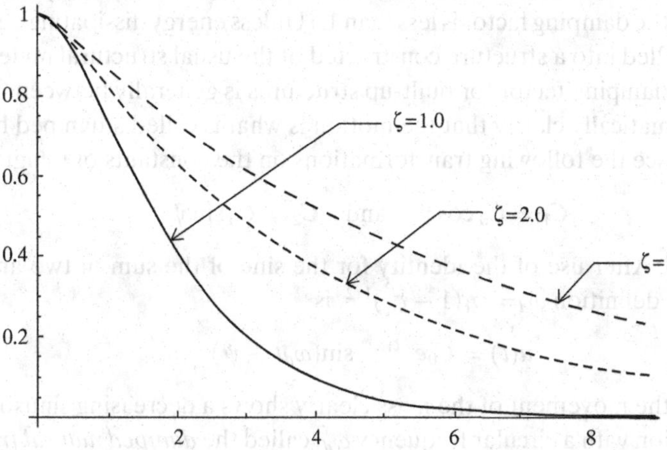

(a)

```
Out[50]= - Graphics -
```

```
In[55]: = Plot[{1/Sqrt[1.1^2 - 1] * Exp[-1.1 * t] * Sinh[t
  * Sqrt[1.1^2 - 1]],
 1/Sqrt[2^2 - 1] * Exp[-2 * t] * Sinh[t * Sqrt[2^2 - 1]],
 1/Sqrt[3^2 - 1] * Exp[-3 * t] * Sinh[t * Sqrt[3^2 - 1]]},
 {t, 0, 10}, Plotstyle → {{}, Dashing[{0.01}], Dashing[{0.03}]}]
```

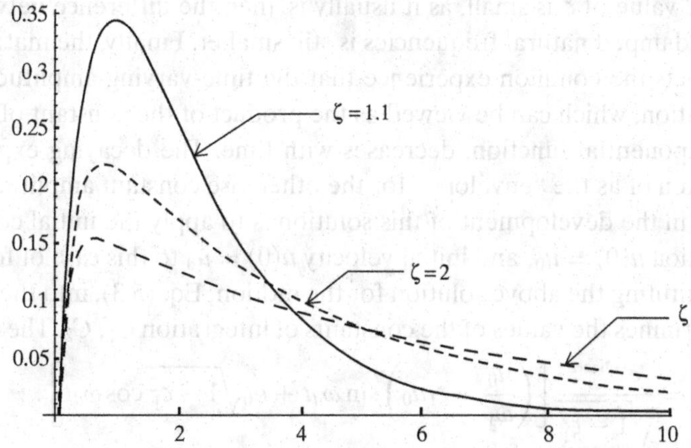

(b)

```
Out[55]= - Graphics -
```

Figure 5.6. (a) Critically damped and overdamped oscillator response to an initial deflection for different values of the damping factor. (b) The response to an initial velocity. All plots for $\omega = 1.0$.

Repeating this argument for the second of the above times shows that C_2 is also real. This means that the original constants of integration, those with the overline, are a complex conjugate pair.

The presence of the sine and cosine functions shows that this deflection solution is the vibratory solution that has been sought and that the viscously damped oscillator vibrates only if the damping factor is less than 1. (Unless energy dissipating devices are purposely installed into a structure constructed of the usual structural materials and fastenings, the damping factor for built-up structures is generally between 0.005 and 0.1.) To mathematically clarify that the motion is what is called a damped harmonic motion, introduce the following transformations on the constants of integration

$$C_1 = C_0 \cos \psi \quad \text{and} \quad C_2 = C_0 \sin \psi$$

into Eq. (5.3a). After use of the identity for the sine of the sum of two angles, the result, with the definition $\omega_d = \omega_1(1 - \zeta_1^2)^{1/2}$, is

$$u(t) = C_0 e^{-\zeta_1 \omega_1 t} \sin(\omega_d t + \psi). \tag{5.3b}$$

This result for the movement of the mass clearly shows a decreasing sinusoidal (i.e., vibratory) motion with a circular frequency ω_d, called the *damped natural frequency*, and a phase angle, ψ, that is just another constant of integration. This conclusion justifies the introductory plots of Figure 5.1. The adjective "natural" is appropriate for the above terms because this result is for a force free vibration. Thus the damped natural frequency is an inherent characteristic of the damped system just as the undamped natural frequency is an inherent characteristic of an undamped system. Note that if there is no damping (ζ equals zero), then the natural frequency is simply ω_1, exactly as before. If the value of ζ is small, as it usually is, then the difference between the damped and undamped natural frequencies is still smaller. Finally, the mathematics again fully reflects the common experience that the time-varying amplitude of the sinusoidal vibration, which can be viewed as the product of the constant of integration and the exponential function, decreases with time. The decaying exponential function is spoken of as the "envelope" for the otherwise constant amplitudes.

The last step in the development of this solution is to apply the initial conditions of initial deflection $u(0) = u_0$, and initial velocity $\dot{u}(0) = \dot{u}_0$ to this case of force free vibration. Substituting the above solution for the motion, Eq. (5.3), into these initial conditions determines the values of the constants of integration C_1, C_2. The result is

$$u(t) = \frac{e^{-\zeta_1 \omega_1 t}}{\sqrt{1 - \zeta_1^2}} \left[\left(\frac{\dot{u}_0}{\omega_1} + \zeta_1 u_0 \right) \sin \omega_d t + u_0 \sqrt{1 - \zeta_1^2} \cos \omega_d t \right]. \tag{5.4}$$

Similarly, the constants of integration for Eq. (5.3b) are

$$C_0 = \frac{1}{\omega_1} \sqrt{\dot{u}_0^2 - 2\zeta_1 \omega_1 \dot{u}_0 u_0 + \omega_1^2 u_0^2} \quad \text{and} \quad \tan \psi = \frac{\omega_d u_0}{\dot{u}_0 + \zeta_1 \omega_1 u_0}.$$

Figures 5.7(a), (b), (c), and (d) show the effects of different initial conditions and different subcritical damping factors.

The last type of damping to be discussed is unlike material damping, or Coulomb damping, or any damping dependent on some power of a relative fluid velocity. This damping is not related to a specific physical phenomena, but it is a representation of a composite of all three forms of damping. Rather than introduce this type of

```
In[1]:= u1[t_, zeta_,idf1t_] = Exp[-zeta*2*Pi*t]
        ((zeta/Sqrt[1-zeta^2])idf1t*Sin[2*Pi*
        Sqrt[1-zeta^2]*t] + idf1t*Cos[2*Pi*
        Sqrt[1-zeta^2]t])
```

Out[1]= $\frac{E^{-2\pi t\ zeta}(\text{idf1t Cos}[2\pi t\sqrt{1-zeta^2}] + \text{idf1t zeta Sin}[2\pi t\ \sqrt{1-zeta^2}])}{\sqrt{1-zeta^2}}$

```
In[4]:= Plot[{u1[t,0.05,1.5],u1[t,0.05,1],
        u1[t,0.05,.5]},{t,0,2.2}]
```

1.5
Large $u(0)$
1
0.5
Small $u(0)$
0.5 1 1.5 2
−0.5
−1

(a) Out[4]= -Graphics-

```
In[11]:= u2[t_,zeta_,ivel_]= Exp[-zeta*2*Pi*t]
         (ivel/(2*Pi*Sqrt[1-zeta^2])) Sin[2*Pi
         *Sqrt[1-zeta^2] t ]
```

Out[11]= $\frac{E^{-2\pi t\ zeta}\ \text{ivel Sin}\left[2\pi t\ \sqrt{1-zeta^2}\right]}{2\pi\sqrt{1-zeta^2}}$

```
In[13]:= Plot[{u2[t,0.05,1],u2[t,0.05,2],
         u2[t,0.05,3]},{t,0,2}]
```

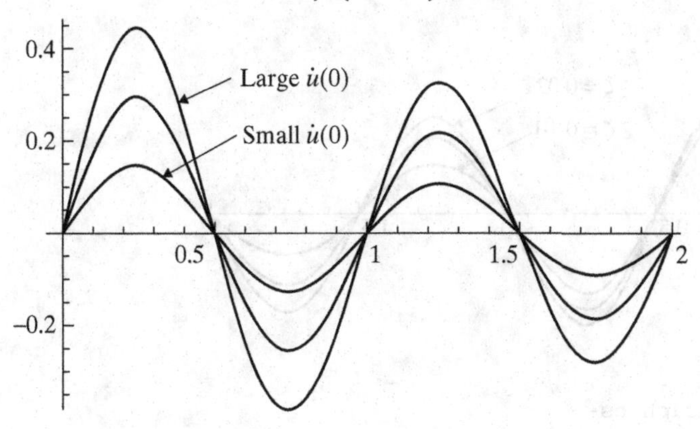

(b) Out[13]= -Graphics-

Figure 5.7. (a) Time history responses of a lightly damped oscillator for large, medium, and small initial deflections (`idf1t`) when the initial velocity is zero. (b) Time history responses of a lightly damped oscillator for large, medium, and small initial velocities (`ivel`) when the initial deflection is zero.

```
In[1]:= u1[t_,zeta_,idflt_] = Exp[-zeta*2*Pi*
        t](zeta/Sqrt[1-zeta^2]) idflt*Sin[2*Pi
        *Sqrt[1-zeta^2]*t] + idflt*Cos[2*Pi*
        Sqrt[1-zeta^2] t ])
```

Out[1]= $E^{-2\pi t\ zeta}\left(\text{idflt Cos}[2\pi\ t\ \sqrt{1-zeta^2}]+\frac{\text{idflt zeta Sin}[2\pi\ t\ \sqrt{1-zeta^2}]}{\sqrt{1-zeta^2}}\right)$

```
In[3]:= Plot[{u1[t,0.01,1],u1[t,0.05,1],u1[t,
        0.1,1,]},{t,0,2.2}]
```

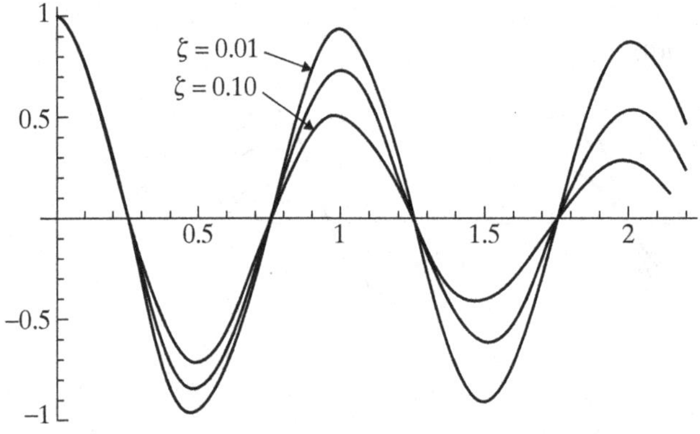

(c) Out[3]= -Graphics-

```
In[5]:= u2[t_,zeta_,ivel_]=
        Exp[-zeta*2*Pi*t](ivel/2*Pi*Sqrt
        [1-zeta^2])) Sin[2*Pi*Sqrt[1-zeta^2]t]
```

Out[5]= $\frac{E^{-2\pi t\ zeta}\ \text{ivelSin}[2\ \pi\ t\ \sqrt{1-zeta^2}]}{2\ \pi\ \sqrt{1-zeta^2}}$

```
In[15]:= Plot[{u2[t,0.02,1],u2[t,0.05,1],u2[t,
         0.1,1]},{t,0,2},PlotRange->{-0.2,0.3}
```

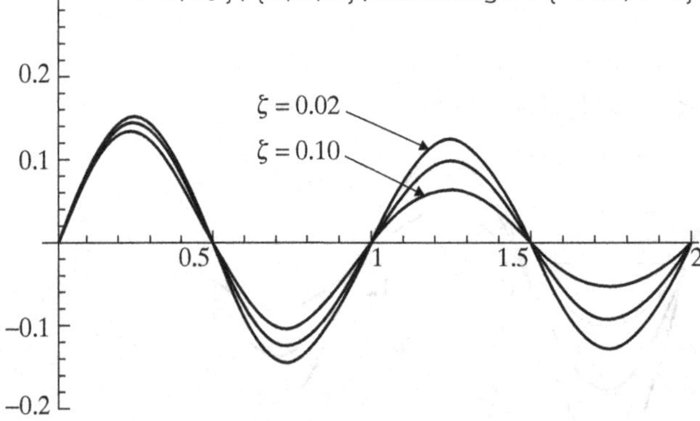

(d) Out[15]= -Graphics-

Figure 5.7. (c) Time history responses for oscillators with zero initial velocity, the same initial deflection, and light damping factors having values 0.01, 0.05, and 0.10. (d) Time history responses for an oscillator having the same initial velocity, zero initial deflection, and damping factors of 0.02, 0.05, and 0.10.

damping with a single-DOF system as was used above, now consider a multidegree of freedom structural system that is undergoing some sort of vibratory motion in response to applied, time-varying applied loads, or equivalent loads. Furthermore, recall the concept of "dynamic equilibrium," which is simply rewriting Newton's second law as $\Sigma \boldsymbol{F} - m\boldsymbol{a} = 0$ and noting that the quantity $-m\boldsymbol{a}$ has the units of force. Since it has the units of force, the $-m\boldsymbol{a}$ term can be labeled the *inertia force* and grouped with all the other forces to allow the original equation to appear as simply $\Sigma \boldsymbol{F} = 0$. Since on its surface, this latter equation looks like the static equilibrium equation, but it actually includes a force that is acceleration dependent, this equation is called the equation of *dynamic equilibrium*. Dynamic equilibrium says that the *vector sum* of all the forces acting on a mass system, including the those forces that are motion dependent like the inertia forces, is zero. From the above viewpoint, rewrite Eq. (4.2), as follows, where again $\{Q(t)\}$ is the vector of applied loads, and the force vector representing all the system damping has been introduced

$$\{Q(t)\} - [m]\{\ddot{q}(t)\} - [k]\{q(t)\} + \{Q_{damp}(t)\} = \{0\}. \tag{5.5}$$

Again, this form of the standard undamped matrix equation of motion can be interpreted as the zero vector sum of the four types of forces acting on the vibratory system: applied, inertial, elastic, and damping, respectively.[7] Consider the damping force vector, $\{Q_{damp}\}$, as another, separate, set of applied forces. Recall that the mathematical linearity of viscous damping provides the only easily solved single-DOF differential equation. Thus the first step toward writing a linear, multiple-DOF, matrix differential equation for a damped system, one that can be readily solved, is accomplished by writing the damping force vector in the coefficient matrix form $-[C]\{\dot{q}\}$, where the generalized velocities are only raised to the first power. This, too, is viscous damping.

As shown in Chapter 7, it turns out that the standard solution process is further greatly simplified if the *damping matrix* $[C]$ can be written as a weighted sum of the mass and stiffness matrix. Although this result is not really what happens in real structures, it is nevertheless not a too unreasonable mathematical description of the damping matrix for the following reasons. Consider, for example, the truss of Figure 4.8 when it is subjected to a dynamic loading. The back-and-forth deflections of the truss may produce dry friction damping at the bolted and riveted joints. The energy dissipated by the Coulomb damping for any joint of the truss will be dependent on, along with the tightness of the joint, the time-varying elastic forces in the bars that cause the slippage at the joints. Thus it is not unreasonable to say the damping force vector is somewhat proportional to $[K]\{q\}$. However, in the same rough way, the aerodynamic and material damping is generally going to occur where the mass is found. Hence, mathematically, it is not unreasonable to say that the damping force vector is somewhat proportional to the mass matrix. Putting these two ideas together, the damping matrix for *proportional damping*, sometimes known as *Rayleigh damping*, is written as

$$[C] = \alpha[M] + \beta[K], \tag{5.6}$$

[7] The minus sign on these elastic forces, for example, results from the fact that these are the elastic forces acting on the masses, whereas $+[k]\{q\}$ are the elastic forces acting on the elastic elements themselves.

where α and β are constants of proportionality with different units. The discussion concerning the utility of proportional damping and the determination of these proportionality constants is postponed to Chapter 7.

Finally, an even more convenient approach to linearizing the damping contribution to the equations of motion is used in later chapters. This approach is not to estimate the damping coefficient matrix at all, not even in the form of proportional damping. Rather, this approach is to wait until the equations of motion have been reduced by means of a transformation (explained in Chapter 7) and then, even for cases where there are hundreds of DOF, estimate a very small number of what will be called *modal* damping factors to be used in the final solution for the motion of the structural system.

5.3 The Response of a Viscously Damped Oscillator to a Harmonic Loading

The previous section detailed the solution for the force free vibratory motion of an oscillator whose damping is modeled as viscous damping. That free vibration solution contains the two constants of integration and is, of course, without reference to any (equivalent) externally applied force. This section continues the discussion of the viscously damped oscillator and the solution for its motion by adding to the osciallator system an externally applied force that varies significantly with time. For present purposes, time-varying forces can be conveniently divided into two categories. The first category contains those forces that vary over many, many periods of the vibration. The second category contains those forces that vary over only several periods of vibration, or an even shorter time interval. The most prominent of the forces in the first category is the harmonic force, which is a force that can be represented mathematically as

$$Q(t) = Q_0 \sin \omega_f t, \tag{5.7}$$

where Q_0 is the constant amplitude of the force, and ω_f is the vibratory frequency of this external force. Other long-acting, time-varying forces are often also periodic and thus can be represented by a Fourier series. For an oscillator with a single discrete mass of magnitude m, with a spring constant k, and an viscous damping coefficient c,[8] subjected to the above applied force, the equation of motion is

$$\begin{aligned} m\ddot{q} + c\dot{q} + kq &= Q_0 \sin \omega_f t \\ \text{or} \quad \ddot{q} + 2\zeta\omega_1\dot{q} + \omega_1^2 q &= \omega_1^2 \frac{Q_0}{k} \sin \omega_f t, \end{aligned} \tag{5.8}$$

where again, ζ the damping factor, equals one-half the ratio of the damping coefficient, c, to the square root of the product of the mass, m, and the spring stiffness, k. The solution to this second-order differential equation is, as usual, obtained in two

[8] The application of a harmonic force to a Coulomb-damped oscillator is examined in the exercises at the end of this chapter.

parts. The first part of the total solution is the complementary solution, which is the solution to the homogeneous equation

$$\ddot{q}(t) + 2\zeta\omega_1\dot{q}(t) + \omega_1^2 q(t) = 0.$$

This equation already has been solved in the previous section. The solution to the above differential equation is Eq. (5.3b), which is, again,

$$q_{comp}(t) = C_0 e^{-\zeta\omega_1 t}\sin(\omega_d t + \psi). \qquad (5.3b)$$

Again, the quantities C_0 and ψ are the two constants of integration that are to be expected to be part of the complementary solution of a second-order ordinary differential equation. The second part of the solution is called the particular solution. It does not contain any constants of integration. It simply is the function that when substituted into the left-hand side of Eq. (5.8) produces the right-hand side term. The simplest procedure for obtaining the particular solution for this equation is to use the differential equation solution technique called the method of undetermined coefficients. This technique calls for a trial solution that includes all the distinct derivatives of the function found on the right-hand side of the original equation. Thus, in this case, the trial solution has the form

$$q_{part}(t) = A_1 \sin\omega_f t + A_2 \cos\omega_f t,$$

where A_1 and A_2 are the coefficients to be determined by substituting this trial solution into the original differential equation and then equating the coefficients of the sine and cosine functions. Those coefficients must be equal because the sine and cosine functions are linearly independent.[9] This is an unnecessarily lengthy process. However, for the sake of comparison between this standard approach that uses only real functions and the more efficient complex algebra approach that is used below, Endnote (1) details the real function approach, which, of course, produces the same end result as that obtained below.

It turns out that, because of presence of the damping force, the solution for the deflection function $q(t)$ is not just in (+) or out (−) of phase with the input force $Q_0 \sin\omega_f t$ but differs by a phase angle less than 180°. To recall the meaning of phase angles, recall that the sine function lags the cosine function by 90° because $\sin\omega_f t = \cos(\omega_f t - \pi/2)$, and the sine function leads or lags the negative of the sine function by 180° because $-\sin\omega_f t = \sin(\omega_f t \pm \pi)$. In other words, the constant term in the argument of a sine or cosine function is the phase angle.

The best way to deal with any phenomenon involving phase angles is to use complex algebra. This amounts to just replacing the real trigonometric function $\sin(\omega_f t)$ with the complex exponential function $\exp(i\omega_f t)$. This replacement is not done on the basis that their functional values are equal, because they are never equal. The replacement is done on the basis that both functions possess the same vital characteristic of

[9] Two functions $f(t)$ and $g(t)$ are *linearly independent* if $c_1 f(t) + c_2 g(t) = 0$ for all values of t only if the constants c_1 and c_2 must be zero for this equality to hold. Note that (functional) independence is different from linear independence. For example, by the above definition, the function $\sin x$ is linearly independent of $\cos x$, but because $\sin x$ can be easily calculated from $\cos x$, these two functions are not (functionally) independent.

repeating themselves, that is, being periodic, with period $2\pi/\omega_f$. Thus either function can be used to mathematically represent a harmonic motion. Hence, the procedure that is used here to obtain the particular solution for the generalized coordinate $q(t)$ is to recast the force input and the trial solution in complex notation as

$$Q(t) = Q_0\, e^{i\omega_f t} \quad \text{and} \quad q_{par}(t) = A_0\, e^{i\omega_f t}.$$

The one complication here is that, whereas Q_0 is a real number, the to-be-determined magnitude of the deflection response, A_0, can be a complex constant. This greater generality for the amplitude of the response allows for the phase angle difference between the force input and the deflection output. With these simply stylistic changes in the representation of the force and the deflection, Eq. (5.8) becomes

$$\ddot{q} + 2\zeta\omega_1\dot{q} + \omega_1^2 q = \omega_1^2 \frac{Q_0}{k} \exp(i\omega_f t).$$

Substituting the complex trial solution $q_{part} = A_0 \exp(i\omega_f t)$ leads to the result

$$\left[(\omega_1^2 - \omega_f^2) + 2i\zeta\omega_1\omega_f\right] A_0 \exp(i\omega_f t) = \omega_1^2 \frac{Q_0}{k} \exp(i\omega_f t).$$

Since the exponential function is never zero, it can be canceled. Dividing by the first natural frequency squared, defining the *frequency ratio* as $\omega_f/\omega_1 = \Omega_1$, and then solving for the nondimensional ratio $A_0 k/Q_0$, leads to

$$\frac{A_0 k}{Q_0} = \frac{1}{(1 - \Omega_1^2) + 2i\zeta\Omega_1}. \tag{5.9a}$$

Both the amplitude $|A_0|$ and the phase angle of the deflection response relative to that of the force input, where the force input has a zero phase angle because Q_0 is a positive real number, can be obtained from this complex expression. For the amplitude, note that $|kA_0/Q_0| = |kA_0|/|Q_0| = k|A_0|/Q_0$, and similarly for the right-hand side. Note further that the amplitude of any complex number is the square root of the sum of the squares of its real and imaginary parts. Then

$$\frac{k|A_0|}{Q_0} = \frac{1}{\sqrt{(1 - \Omega_1^2)^2 + (2\zeta\Omega_1)^2}}. \tag{5.9b}$$

The phase angle between the force input and the deflection output is obtained from the complex form of the quantity kA_0/Q_0. To determine its magnitude, it is necessary to write the right-hand side ratio in the form $a + bi$. This is accomplished by multiplying the numerator and denominator by the complex conjugate of the denominator. That result is

$$\begin{aligned}\frac{kA_0}{Q_0} &= \frac{1}{(1 - \Omega_1^2) + 2i\zeta\Omega_1} * \frac{(1 - \Omega_1^2) - 2i\zeta\Omega_1}{(1 - \Omega_1^2) - 2i\zeta\Omega_1}\\ &= \frac{(1 - \Omega_1^2) - 2i\zeta\Omega_1}{(1 - \Omega_1^2)^2 + (2\zeta\Omega_1)^2}.\end{aligned}$$

A plot of this right-hand side complex number on the complex plane[10] in the form of an amplitude line from the origin to its real and imaginary coordinates shows that, when the frequency ratio is less than 1, the amplitude line lies in the fourth quadrant because its real part is positive, and its imaginary part is negative. When the frequency ratio is greater than 1, this amplitude line lies in the third quadrant. Recall that the positive rotations represented by $\omega_f t$ are counterclockwise as time increases. Thus deduce that the lesser angle between the real number of the input force and the complex number representing the deflection output is a negative angle, or in physical terms, a lag angle. Calling the lag angle ϕ, from the complex plane plot

$$\phi = \arctan\left(\frac{2\zeta\Omega_1}{1-\Omega_1^2}\right). \tag{5.10}$$

It makes sense that the deflection response lags the force input because the system deflection, as a response to the applied (equivalent) force, must first experience the magnitude and direction of the force before it can respond. That is, the deflection cannot anticipate the force. Thus the deflection must lag the force, which is represented mathematically by a negative phase angle.

The above results can reorganized as follows. From Eq. (5.9) and the basic relationships

$$Q(t) = Q_0\, e^{i\omega_f t} \quad \text{and} \quad q_{par}(t) = A_0\, e^{i\omega_f t},$$

the particular solution can be written in complex algebra form as

$$q_{par}(t) = \frac{1/k}{(1-\Omega_1^2) + 2i\zeta\Omega_1} Q_0\, e^{i\omega_f t} \equiv Q_0 H(i\Omega_1)\, e^{i\omega_f t}$$

where the definition of the important new function, $H(i\Omega_1)$, called the *frequency response function*, for this case, is clear. The square root of -1 is part of the argument of the frequency response function only as a reminder that this a complex function. A frequency response function is a dynamic form of a flexibility coefficient.[11] Frequency response functions play important roles in vibratory analyses such as random vibration analyses. Appendix II demonstrates that frequency response functions are essentially Fourier transforms of impulse response functions, which are introduced in Chapter 7 and which form the basis of much of the solution processes discussed in this textbook.

The particular solution can also be formed from the above in terms of real quantities. First, from (i) the fact that any complex number $z = x + iy$ can be written in its polar form $\rho\exp(i\theta)$, and thus $A_0 = |A_0|\exp(-i\phi)$; (ii) from Eq. (5.9b)

$$|A_0| = \frac{Q_0/k}{\sqrt{(1-\Omega_1^2)^2 + (2\zeta\Omega_1)^2}}$$

[10] The standard representation of the coordinate axes of the complex plane is one where the real axis is the horizontal axis and the imaginary axis is the vertical axis. Of course these axes cross at 0 and $0i$.

[11] See Endnote (2) of Chapter 3 for a discussion of flexibility coefficients and flexibility matrices, which, in general terms, are the inverse of stiffness matrices.

and (iii) $q_{par}(t) = A_0 \exp(i\omega_f t)$; the result is, where ϕ is defined in Eq. (5.10),

$$q_{par}(t) = \frac{Q_0/k}{\sqrt{\left(1 - \Omega_1^2\right)^2 + (2\zeta\Omega_1)^2}} \exp(i\omega_f t - i\phi).$$

Now the conversion to a real form is the reverse of what was done originally, which then was simply replacing the sine function with the complex exponential function. Hence, reversing that procedure, the particular solution in wholly real terms is

$$q_{par}(t) = \frac{Q_0/k}{\sqrt{\left(1 - \Omega_1^2\right)^2 + (2\zeta\Omega_1)^2}} \sin(\omega_f t - \phi). \tag{5.11}$$

Substitution of Eq. (5.11) into the original differential equation, Eq. (5.8b) demonstrates the validity of this expression for the particular solution and that of the complex algebra process that produced this expression. When making that check substitution, realize that from Eq. (5.10) that for the lag angle ϕ

$$\sin\phi = \frac{2\zeta\Omega_1}{\sqrt{\left(1 - \Omega_1^2\right)^2 + (2\zeta\Omega_1)^2}} \quad \text{and} \quad \cos\phi = \frac{1 - \Omega_1^2}{\sqrt{\left(1 - \Omega_1^2\right)^2 + (2\zeta\Omega_1)^2}}.$$

The total solution for the response is, of course, the sum of the complementary solution, Eq. (5.3b), and the particular solution, Eq. (5.11), which is

$$q(t) = C_0\, e^{-\zeta\omega_1 t} \sin(\omega_d t + \psi) + \frac{Q_0/k}{\sqrt{\left(1 - \Omega_1^2\right)^2 + (2\zeta\Omega_1)^2}} \sin(\omega_f t - \phi). \tag{5.12}$$

Again, this is the total one-DOF viscously damped deflection response to a constant magnitude harmonic input force $Q_0 \sin\omega_f t$. The constants of integration are, again, determined by the initial conditions. Be sure to understand that when an applied force is present, the process of determining the constants of integration must include the particular solution.

Note again that because the negative exponential factor, $-\zeta\omega_1 t = -2\pi\zeta(t/T_1)$, the first part of this deflection response fades away to very near zero after just a few periods. However, the second part of the solution maintains a constant amplitude for as long as the applied force persists. Hence, the first part of the solution is called the *transient solution*,[12] whereas the second part of the solution is called the *steady-state solution*. The peak displacement, and hence the peak internal force, may occur in the first few cycles. This matter is investigated in Chapter 7. However, for long-term effects, such as fatigue considerations, or for any circumstances beyond a few cycles, only the steady-state solution is of interest. Therefore, for the remainder of this section the focus is on the steady-state solution.

There is an important conclusion to be drawn from the above steady-state solution for the magnitude of the deflection response and a point worth noting about the phase angle of the deflection response relative to the input force. Examine the nondimensional solution for the amplitude of the deflection response, Eq. (5.9b), or the similar deflection amplitude function that is part of Eq. (5.12). The deflection

[12] In the case of a damping factor equal to 0.08, which is representative of some small steel structures, after just nine cycles, the damped amplitude solution of Eq. (5.12) is only 1% of its initial value.

```
Plot[{h[w,0.05],h[w,0.1],h[w,0.15],h[w,0.7]},
{w,0,4}]-> Automatic, AspectRatio->1]
```

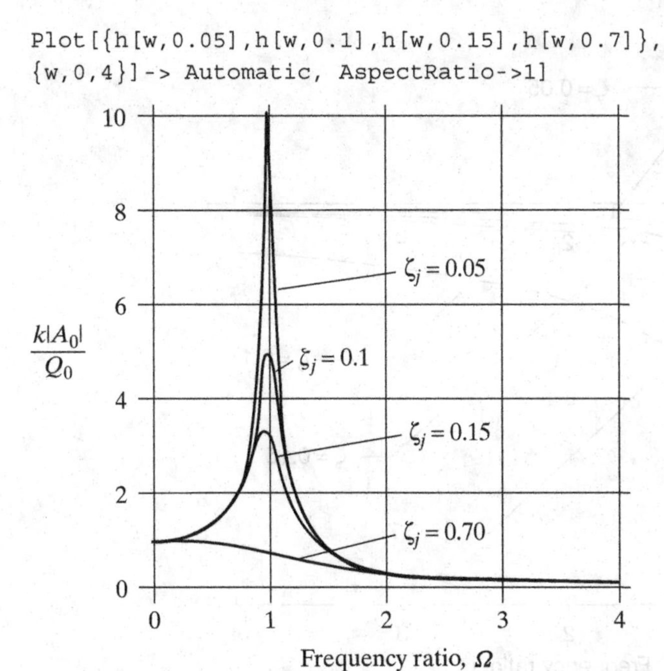

Figure 5.8. Nondimensional amplitude of the steady-state harmonic response as a function of the frequency ratio Ω and the damping ratio ζ as a plot parameter.

amplitude clearly depends on two parameters, the frequency ratio, Ω_1, and the viscous damping factor, ζ. A plot of this amplitude response, using the frequency ratio as the abscissa and the damping factor as the parameter of the family of curves, is shown in Figure 5.8. When the frequency ratio is zero, the applied force is a static force, and the applied force is entirely opposed by the spring force. As the frequency ratio increases to a value close to 1, the deflection response increases rapidly to a sharp peak when the damping factor is small, as it often is. At this point, as shown in Figure 5.9(a), the phase angle of the deflection response, and hence the spring force that is proportional to the deflection response, is 90°. Hence the spring force no longer opposes the applied force because the spring force is perpendicular to the applied force. As illustrated in Figure 5.9(b), a force polygon with the correct phase angles would show that the spring force now only opposes the inertial force, whereas only the damping force opposes the applied force. Therefore it is clear that when the frequency ratio is 1, the damping force is quite important.

Large deflections mean large internal (spring) forces, which means that the integrity of the structure is threatened. This phenomenon of a dynamic response many times the magnitude of a corresponding static response, when the forcing frequency nears or equals the natural frequency, is called *resonance*. Despite the fact that, in this case, the peak response occurs slightly before $\omega_f = \omega_1$, resonance is defined to occur at a frequency ratio of 1. The ratio of the resonant response to the static response is called the *magnification factor*. Even though the above conclusions were drawn on the basis on just a one-DOF structure, in the next two chapters it will be demonstrated that an N degree of freedom structure has N natural

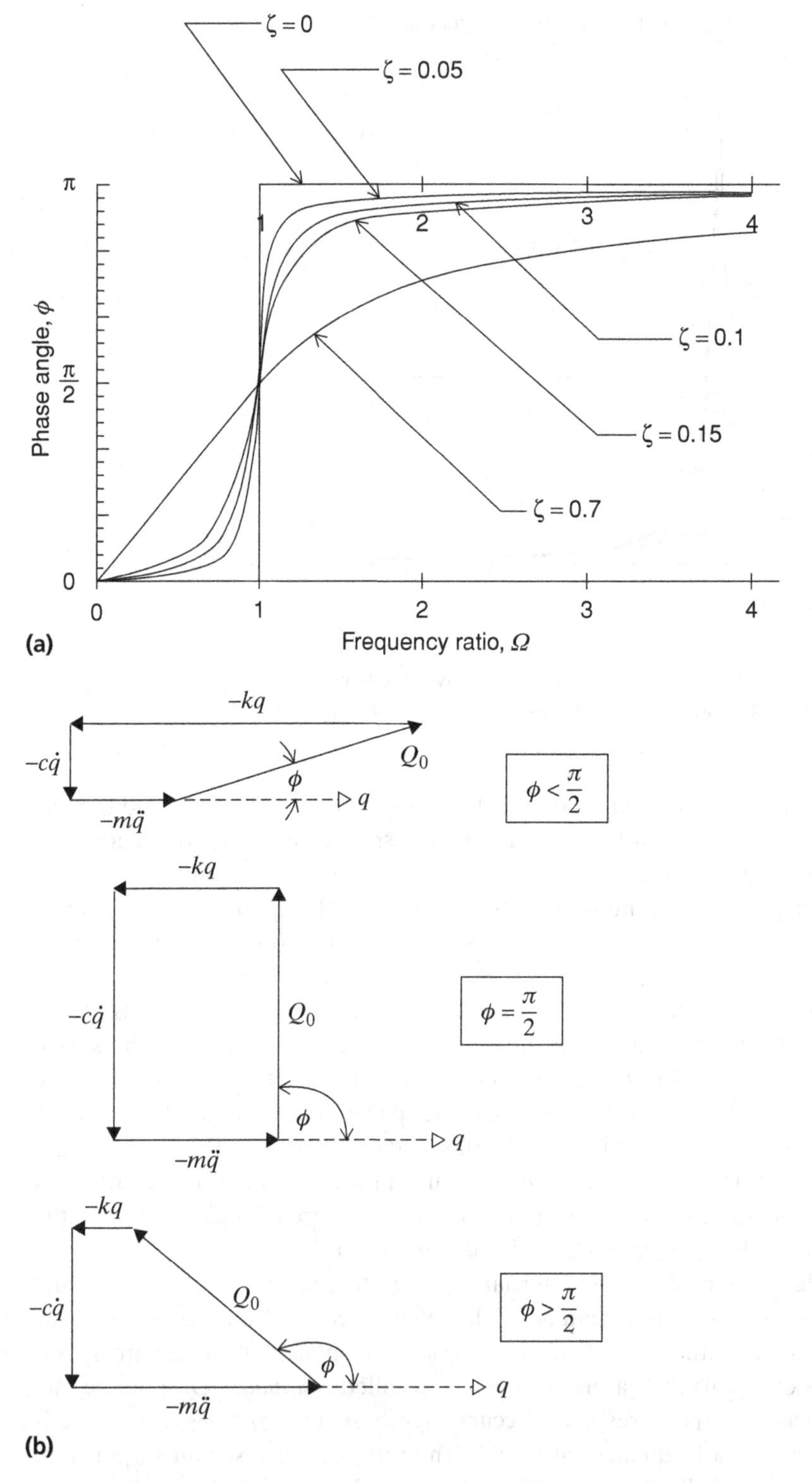

Figure 5.9. (a) Variation of lag angle with frequency ratio and damping factor. (b) Phase relationships among the applied, spring, damping, and inertia forces for harmonic motion for frequency ratio values less than one-half, equal to one, and equal to one and a half.

frequencies, and resonance is possible for each of those natural frequencies. However, the increased damping associated with the higher numbered natural frequencies means that only the first several frequencies are of interest. Nevertheless, obviously, resonance is a major concern whenever machinery with rotating (even slightly unbalanced) parts are attached to a structure.

Returning to Figure 5.9(a), the plot that shows how the lag angle varies with respect to the frequency ratio and the damping factor. For typically low values of the damping factor, the phase angle at resonance changes rapidly from near zero to near 180°. The benefit that can be had from this rapid change in phase angle is that it can be used to more precisely locate a resonant peak, where the phase angle is 90°, than is possible just looking at an amplitude response, which for higher numbered natural frequencies, can be somewhat broad.

There is another similar, complex algebra oscillator solutions worth noting. Consider the case where the oscillator is driven by a base motion $h(t) = h_0 \sin \omega_f t$. The equation of motion for the generalized coordinate $u(t)$ is

$$m\ddot{u}(t) + c\dot{u}(t) + ku(t) = c\dot{h}(t) + kh(t)$$
$$\text{or} \quad \ddot{u} + 2\zeta_1\omega_1\dot{u} + \omega_1^2 u = 2\zeta_1\omega_1\dot{h} + \omega_1^2 h.$$

The task is to determine the amplitude and the phase angle of the response $u(t)$ in terms of the frequency ratio, $\Omega_1 = \omega_f/\omega_1$, and the damping factor, ζ. Again, complex algebra is used to obtain the solution for $|u_0/h_0|$, where here u_0, as opposed to its absolute value, is a complex quantity. Of course the input amplitude h_0 is a real number. Thus the first step of the solution process is to represent the enforced deflection input $h(t) = h_0 \sin \omega_f t$ by its associated complex form, which is $h_0 \exp(i\omega_f t)$, and to similarly represent the response as $u_0 \exp(i\omega_f t)$. Substituting these two complex forms into the equation of motion leads to

$$\left[(\omega_1^2 - \omega_f^2) + 2i\zeta_1\omega_1\omega_f\right] u_0 = \left[\omega_1^2 + 2i\zeta_1\omega_1\omega_f\right] \; h_0.$$

Dividing by the first natural frequency squared and solving for the nondimensional response u_0/h_0 leads to

$$\frac{u_0}{h_0} = \frac{1 + 2i\zeta_1\Omega_1}{(1 - \Omega_1^2) + 2i\zeta_1\Omega_1} \quad \text{and} \quad u(t) = h_0 H(i\Omega_1)\, e^{i\omega_f t}.$$

Clearly the input amplitude, in this case h_0, identifies the type of frequency response function. Since $|u_0/h_0| = |u_0|/|h_0| = |u_0|/h_0$, similarly, for the right-hand side,

$$\frac{|u_0|}{h_0} = \frac{\sqrt{1 + (2\zeta_1\Omega)^2}}{\sqrt{(1 - \Omega^2)^2 + (2\zeta_1\Omega)^2}}.$$

Note that here, too, there is a large amplitude at a unit value of the frequency ratio whenever, typically, ζ is small. So, again, define resonance as occurring when the frequency ratio is 1. At resonance, the amplitude ratio in this case is quite close to the previous high value of $1/(2\zeta_1)$ when the damping ratio is small. Thus it matters little if the input is a harmonic force or a harmonic base motion.

The phase angle, which is again a lag angle, is obtained from the above complex form of the quantity u_0/h_0. Again, to determine its magnitude, it is necessary to write the above deflection ratio in the form $a + bi$. Again this is accomplished by multiplying the numerator and denominator by the complex conjugate of the denominator. That result is

$$\frac{u_0}{h_0} = \frac{1 + 2i\zeta_1\Omega}{(1 - \Omega^2) + 2i\zeta_1\Omega} * \frac{(1 - \Omega^2) - 2i\zeta_1\Omega}{(1 - \Omega^2) - 2i\zeta_1\Omega}$$
$$= \frac{[(1 - \Omega^2) + (2\zeta_1\Omega)^2] - 2i\zeta_1\Omega^3}{(1 - \Omega^2)^2 + (2\zeta_1\Omega)^2}.$$

Therefore, in this case the phase angle ϕ is equal to

$$\phi = \arctan\left[\frac{2\zeta_1\Omega^3}{(1 - \Omega^2) + (2\zeta_1\Omega)^2}\right]$$

and it is a lag angle because the imaginary component of the complex deflection response is negative.

As is seen above, a lightly damped oscillator has a large resonant peak response of a magnitude close to $1/2\zeta$ relative to the static deflection produced by a static form of the same magnitude. It is possible that the resonating oscillator could fail at such a large deflection magnitude. Then, if the applied load, or a base motion, were to start at a zero frequency and increase to some constant operating frequency greater than the oscillator's natural frequency, the question arises as to the feasibility of actually reaching that higher operating frequency. This question, called the start-up problem, can be addressed by investigating the response of the oscillator in its corresponding most severe situation. To this end, let the input force be a harmonic force whose forcing frequency is the same as the oscillator's natural frequency. To further pursue a worst-case scenario as well as to simplify the problem, let the damping be zero. Then the equation of motion can be written as

$$\ddot{u}(t) + \omega_1^2 u(t) = \omega_1^2 \frac{F_0}{k} \sin \omega_1 t.$$

Since there is no damping, there is no need to resort to the use of complex algebra. The complementary solution is now well known to the reader as

$$u(t)_{comp} = C_1 \sin \omega_1 t + C_2 \cos \omega_1 t.$$

Note that the particular solution, for this situation where the input (driving) frequency is ω_1, cannot be a constant multiplied by $\sin \omega_1 t$ or $\cos \omega_1 t$ because those functions, parts of the complementary solution, will always yield zero when substituted into the left-hand side of the differential equation. Thus the simplest method for finding a particular solution, the method of undetermined coefficients fails in this circumstance. The particular solution for this linear equation is obtainable by use of the always successful mathematical method for linear equations called *variation of parameters*. Rather than follow that lengthy procedure, it is easier just to confirm the solution given below by substituting back into the given differential equation. The sum of the complementary and particular solutions is

$$u(t) = C_1 \sin \omega_1 t + C_2 \cos \omega_1 t - \frac{1}{2}\omega_1 t \frac{F_0}{k} \cos \omega_1 t.$$

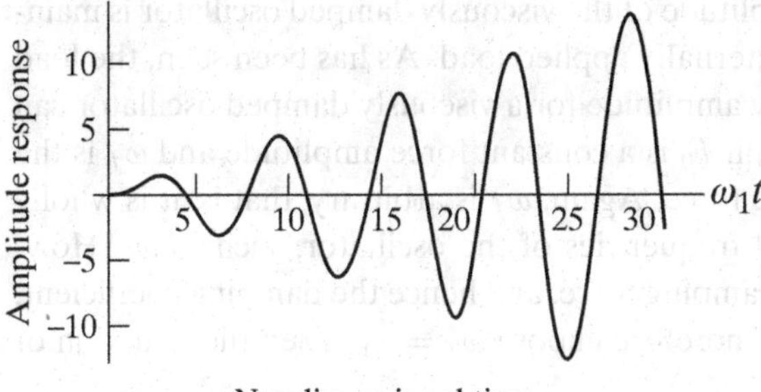

Figure 5.10. The increase in the amplitude of the motion of a one-DOF system mass when both the initial deflection and the initial velocity are zero, and the forcing frequency coincides with the natural frequency.

Initial conditions of zero deflection and zero velocity for the generalized coordinate $u(t)$ yield $C_2 = 0$ and $C_1 = {}^1\!/\!_2 F_0$. Therefore the solution incorporating these zero initial conditions is

$$u(t) = -{}^1\!/\!_2 F_0\,(\omega_1 t \cos \omega_1 t - \sin \omega_1 t)\,.$$

A plot of this undamped response is shown in Figure 5.10. Clearly, even with the forcing frequency "dwelling" at the natural frequency, the very large deflection response of the undamped system is reached only after many periods of time. Therefore it may be concluded that after a start-up from zero frequency, a fairly rapid increase in the rotator *rpm* to reach the operating frequency value causes a rapid traversal of a natural frequency, and this circumstance does not cause any difficulties whatever because the response amplitude does not have sufficient time to build.

5.4 Equivalent Viscous Damping

An important conclusion to be drawn from the above calculations of oscillator amplitude and frequency response for Coulomb, aerodynamic or hydrodynamic, and underdamped viscous damping, as evidenced from the previous figures of this chapter, is that the shape of the oscillator response is similar in all these cases. The one-DOF responses are either sinusoidal with an exponentially decreasing amplitude or, in appearance, very close to that. Not only is the form of the amplitude envelope similar, but the previous calculations show also that the damped frequency of vibration is close in value to the undamped natural frequency. Thus, from its similar response vis-à-vis the other damping cases, and from its simplicity because of linearity, it is apparent that viscous damping, although not common in itself, can be conveniently used to represent all common forms of structural system damping in all such circumstances. This is the standard approach to the mathematical representation of damping. That is, the standard procedure is to use an equivalent viscous damping coefficient for all types of actual damping, singularly or in combination. The precise equivalence between viscous damping and another type of damping is, quite reasonably, based on equal amounts of energy dissipated per cycle of a constant amplitude vibration.

Consider the case where the amplitude of the viscously damped oscillator is maintained at a constant value by an externally applied load. As has been seen, the load that maintains a constant vibratory amplitude for a viscously damped oscillator can be written as $F_0 \sin \omega_f t$, where again F_0 is a constant force amplitude and ω_f is the forcing frequency of this harmonic force. Again, ω_f is arbitrary, that is, it is wholly independent of any of the natural frequencies of the oscillator, such as ω_1. However, as seen in Figure 5.9(b), the damping force, and hence the damping coefficient, are most important at resonance. Therefore choose $\omega_f = \omega_1$. Then the equation of motion for the oscillator is

$$m\ddot{u} + c\dot{u} + ku(t) = F_0 \sin \omega_1 t. \tag{5.8'}$$

As has been seen, the solution for $u(t)$ consists of a transient part and a steady-state part. Since the transient part of the motion can be very short lived, only the steady-state portion of the motion, $u(t) = A\sin(\omega_1 t - \phi_1)$, will be considered. With that simplification, it is a straightforward matter to calculate the work done by the viscous damping force per cycle, which is the object of this discussion. Using the standard definition of work, and integrating over the time interval of one cycle,

$$W_{cycle} = \int_0^T (-c\dot{u})du = -c \int_0^{2\pi/\omega_1} \dot{u}\frac{du}{dt}dt.$$

The velocity terms in the above integrand can be written as

$$\dot{u}(t) = \omega_1 A\cos(\omega_1 t - \phi_1).$$

Therefore the work done by the viscous force each cycle is

$$\begin{aligned} W_{cycle} &= -c\omega_1^2 A^2 \int_0^{2\pi/\omega_1} \cos^2(\omega_1 t - \phi_1)dt \\ &= -\tfrac{1}{2}c\omega_1^2 A^2 \int_0^{2\pi/\omega_1} [1 + \cos 2(\omega_1 t - \phi_1)]dt \\ &= -\pi c\omega_1 A^2, \end{aligned}$$

where the second term of the integrand produces a zero result. The negative sign attached to the value for the work done per cycle means, of course, that the work done by the viscous force is dissipative. This loss of energy is entirely offset by the energy added to the system by the applied force because it turns out that the inertial force and the elastic force do no work on the mass of the oscillator. In other words, this viscous force–applied force energy balance is necessary to maintain the stipulated constant amplitude of forced vibration.

With the above result in hand, it is now possible to return to the idea of obtaining equivalent viscous damping coefficients for "other" types of damping on the basis of equal work dissipated per cycle at the maximum amplitude response. Let the work dissipated per cycle by some other type of damping be $-W_{other}$. Then equating this value for the energy loss to that for viscous damping, where the viscous damping coefficient is now replaced by an equivalent viscous damping coefficient c_{eq}, leads to

$$c_{eq} = \frac{W_{other}}{\pi \omega_1 A^2}.$$

The above result was obtained for an oscillator, which, of course, is a single-DOF system. Multidegree of freedom systems have many natural frequencies. Therefore, the choice is usually made that there is an equivalent viscous damping coefficient, and hence an equivalent viscous damping factor, associated with each natural frequency. Indeed that is the choice made in Chapter 7.

It is now convenient to complete the characterization of the solid or material or hystereses damping case. Recall that within the elastic range, the energy lost from hystereses damping is independent of frequency and proportional to the deflection amplitude, A, squared. Thus mathematically, $W_{mat} = -\alpha A^2$, where α is the constant of proportionality. Equating this energy loss to that for equivalent viscous damping leads to the result $c_{eq} = \alpha/\pi\omega$. Substituting this result into the viscous damping equation, the pseudo material damping equation, with an arbitrary applied force $F(t)$, becomes

$$\ddot{u} + \frac{\alpha}{\pi\omega}\dot{u} + \omega^2 u(t) = F(t).$$

Again it is quite useful to introduce complex algebra into the representation of the deflection or force because of the ease with which complex algebra represents phase angles. Expect the above response $u(t)$ to be very nearly harmonic. Then, for example, if the deflection is represented by the function $\sin\omega_1 t$, then the velocity is represented by the product $\omega_1 \cos\omega_1 t$. The cosine leads the sine by a phase angle of 90°. In the terms of complex algebra, a phase angle advance of 90° is accomplished by a multiplication by $i = (-1)^{1/2}$. Therefore the cosine function of the velocity term can be replaced by the factor $i\omega$ multiplied by the sine function, which is associated with the deflection function. Therefore, the differential equation for the material damped oscillator can be, and is, rewritten as

$$\ddot{u} + \left(k + \frac{i\alpha}{\pi}\right) u(t) = \ddot{u} + k(1 + i\gamma)\, u(t) = F(t), \tag{5.13}$$

where γ is called the *material damping factor* and $k(1 + i\gamma)$ is called the *complex stiffness* or, occasionally, the complex damping. The steady-state solution for the above differential equation is developed in Example 5.5.

The above transition from solid damping into equivalent viscous damping is a bit misleading in that, in other damping type cases, the translation is not quite so simple. For example, consider a velocity-squared-type damping. Approximating the deflection response as $A\sin\omega t$, the damping force is proportional to $A^2\cos^2\omega t$. The integration of the damping force over the distance of one cycle produces a energy loss proportional to A^3. Equating that result with the equivalent viscous damping result leads to an equivalent viscous damping coefficient dependent on the amplitude of the response, A, which would generally be unknown. Thus an iteration would generally be necessary to obtain the best estimate of the equivalent viscous damping coefficient or equivalent viscous damping factor. Fortunately, in many cases, a somewhat rough estimate is all that is necessary to account for the effects of damping in many circumstances. Finally, note that proportional damping, discussed above, is an example of equivalent viscous damping because it leads to damping terms that are proportional to the first powers of the velocities.

5.5 Measuring Damping

More so than mass and stiffness, damping is measured indirectly. Mass is measured by measuring weight, which can be measured by means of balances, calibrated deflection devices, and so on. Stiffness is measured by measuring deflection responses to known loads. Damping is measured by setting the system in motion, postulating equivalent viscous damping, and then measuring the amplitudes of its successive peaks. Consider any two successive peaks. These peaks are separated in time by one damped period, $T_d = 2\pi/\omega_d$. Using Eq. (5.3b) as the representation of this (equivalently) viscous damping response, the ratio of the amplitude of the first peak to the amplitude of the second peak is

$$\frac{u(t)}{u(t+T_d)} \equiv \frac{A_1}{A_2} = \frac{e^{-\zeta\omega t}\sin(\omega_d t + \psi)}{e^{-\zeta\omega(t+T_d)}\sin(\omega_d t + \omega_d T_d + \psi)}.$$

However,

$$\omega_d T_d = 2\pi \quad \text{and because} \quad \sin(\theta + 2\pi) = \sin\theta$$

$$\frac{A_1}{A_2} = e^{+\zeta\omega T_d} = \exp\left(\frac{2\pi\zeta}{\sqrt{1-\zeta^2}}\right) \quad \text{or} \quad \frac{\zeta}{\sqrt{1-\zeta^2}} = \frac{1}{2\pi}\ln\frac{A_1}{A_2}.$$

The natural logarithm of the amplitude ratio is often referred to as the *logarithmic decrement*, often symbolized by δ. The difficulty here is that if the trace of the vibration response is like that pictured in Figure 5.1(a), then the measurement of the ratio A_1/A_2 is quite likely to be inaccurate. To overcome this measurement difficulty, consider n products of such amplitude ratios, each equal to $\exp(\zeta\omega T_d)$. That is, consider the product

$$\frac{A_1}{A_2}\frac{A_2}{A_3}\frac{A_3}{A_4}\cdots\frac{A_n}{A_{n+1}} = \left(e^{\zeta\omega T_d}\right)^n = e^{n\zeta\omega T_d} = \exp\left(\frac{2n\pi\zeta}{\sqrt{1-\zeta^2}}\right)$$

$$\text{thus} \quad \zeta \approx \frac{\zeta}{\sqrt{1-\zeta^2}} = \frac{1}{2n\pi}\ln\frac{A_1}{A_{n+1}}. \tag{5.14}$$

The above approximation is valid only when the damping factor is small; that is, when ζ is less than 0.1. Since the ratio A_1/A_{n+1} can be measured much more accurately than the ratio A_1/A_2, the above formula is a practical means of estimating the equivalent damping factor of a structural system whose vibratory amplitudes have been measured as above.

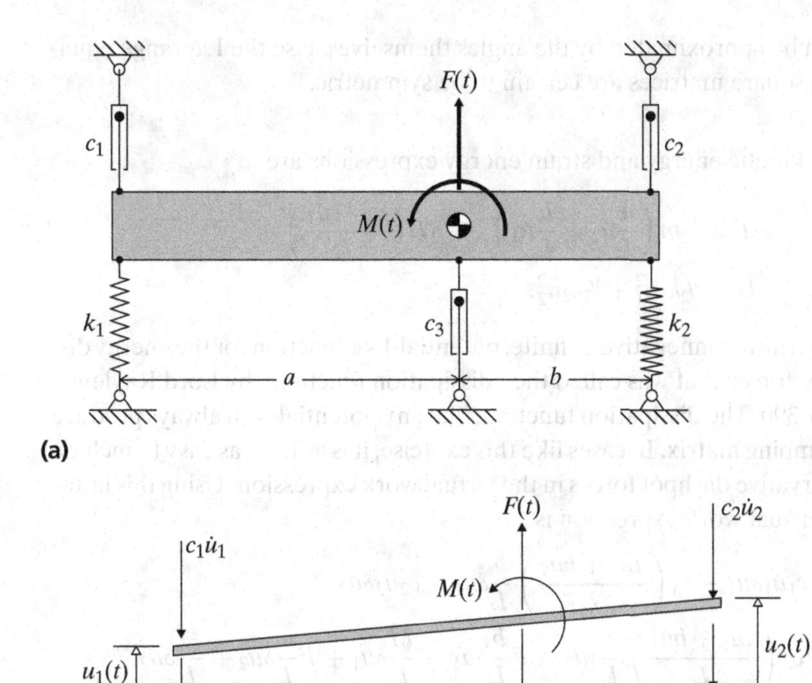

Figure 5.11. (a) Vibratory system of Example 5.1. (b) Free body diagram.

5.6 Example Problems

The following example problem are ones where dashpots are used to explicitly indicate the presence of equivalent viscous damping. These simplified examples could represent those circumstances where the structure or mechanism (or its design) has (or calls for) specific energy dissipating devices, such as the shocks of automobiles, artillery pieces, buildings, the landing gear of flight vehicles, and so on. The dashpots could also be placed in the structural model where the analyst expects there to be significant sources of inherent dry or wet friction to represent that friction. Dashpots could also be part of an analysis that seeks to answer the question whether the dynamic amplitude response of the structure under study is better curtailed by adding more stiffness (and its attendant mass) or adding damping devices or materials (and their attendant cost and mass). However, in the chapter after next the presence of the "dashpots" is implicit rather than explicit.

EXAMPLE 5.1 Consider the rigid bar and elastic spring system shown in Figure 5.11(a). The bar has a mass m and a mass moment of inertia about its CG of magnitude H. Using the two DOF u_1, u_2 shown in Figure 5.11(b), write the matrix equations of motion when (i) the vertical deflections are within the linear range of the springs and (ii) the rigid bar rotations are sufficiently small that the sines or tangents of all angles can be approximated by the angles themselves. Use the Lagrange equations so that all square matrices are certain to be symmetric.

SOLUTION The kinetic energy and strain energy expressions are

$$T = \tfrac{1}{2}m\left(\frac{a}{L}\dot{u}_2 + \frac{b}{L}\dot{u}_1\right)^2 + \tfrac{1}{2}H\left(\frac{\dot{u}_2 - \dot{u}_1}{L}\right)^2$$
$$U = \tfrac{1}{2}k_1 u_1^2 + \tfrac{1}{2}k_2 u_2^2.$$

It is possible to write a nonnegative definite, potential-like function for the energy dissipative dashpot forces that was called the "dissipation function" by Lord Rayleigh. See Ref. [5.3], p. 390. The dissipation function, like any potential, will always produce a symmetric damping matrix. In cases like this exercise, it is as least as easy to include these nonconservative dashpot forces in the virtual work expression. Using this latter approach, the virtual work expression is

$$\delta W = -c_1\dot{u}_1\delta u_1 - c_3\left(\frac{a\dot{u}_2 + b\dot{u}_1}{L}\right)\frac{b}{L}\delta u_1 - c_2\dot{u}_2\delta u_2$$
$$- c_3\left(\frac{a\dot{u}_2 + b\dot{u}_1}{L}\right)\frac{a}{L}\delta u_2 + F\frac{b}{L}\delta u_1 - \frac{M}{L}\delta u_1 + F\frac{a}{L}\delta u_2 + \frac{M}{L}\delta u_2.$$

Substitution into the Lagrange equations, factoring into matrix form, and multiplying through by L^2 leads to

$$\begin{bmatrix} H+mb^2 & -H+mab \\ -H+mab & H+ma^2 \end{bmatrix}\begin{Bmatrix} \ddot{u}_1 \\ \ddot{u}_2 \end{Bmatrix} + \begin{bmatrix} c_1L^2 + c_3b^2 & c_3ab \\ c_3ab & c_2L^2 + c_3a^2 \end{bmatrix}\begin{Bmatrix} \dot{u}_1 \\ \dot{u}_2 \end{Bmatrix}$$
$$+ \begin{bmatrix} k_1L^2 & 0 \\ 0 & k_2L^2 \end{bmatrix}\begin{Bmatrix} u_1 \\ u_2 \end{Bmatrix} = L\begin{Bmatrix} Fb - M \\ Fa + M \end{Bmatrix}.$$

Notice that the damping matrix $[C]$ is also symmetric. In the absence of gyroscopic forces, the velocity coefficient matrix is always symmetric. ★

EXAMPLE 5.2 The mass center of mass of the circular object of Figure 5.12 is at its geometric center. The mass moment of inertia of the circular object about the geometric center is H. Write the matrix differential equation of motion for this rigid body-spring-dashpot system on the basis that all energy dissipating friction in the system is modeled by the dashpots shown in the sketch. Use only the generalized coordinates $u(t)$ and $\theta(t)$ shown in the sketch. Note that the system is being driven by an enforced motion $h(t)$ at the far left of the system.

SOLUTION Without loss of generality, let the relative magnitudes of the two-DOF u, θ be such that the springs, and therefore the dashpots, are in tension. (The reader can confirm that any other choice concerning the relative magnitudes of the DOF will produce the same equations of motion.) The expressions for the kinetic energy, strain energy, and the virtual work are as follows:

$$T = \tfrac{1}{2}m\dot{u}^2 + \tfrac{1}{2}H\dot{\theta}^2 \quad \text{and} \quad U = \tfrac{1}{2}k_1(u-h)^2 + \tfrac{1}{2}k_2(R\theta - u)^2$$
$$\delta W = +c_1(r\dot{\theta} - \dot{u})\delta u - c_1(r\dot{\theta} - \dot{u})r\delta\theta - c_2(R\dot{\theta})R\delta\theta.$$

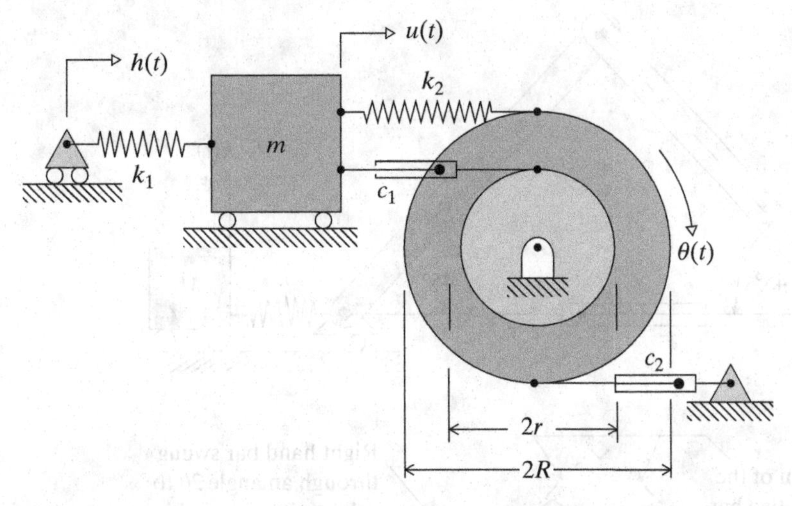

Figure 5.12. Damped mechanical system of Example 5.2.

Substituting into the Lagrange equations, and organizing the various terms into matrix form, leads to

$$\begin{bmatrix} m & 0 \\ 0 & H \end{bmatrix} \begin{Bmatrix} \ddot{u} \\ \ddot{\theta} \end{Bmatrix} + \begin{bmatrix} c_1 & -rc_1 \\ -rc_1 & (r^2c_1 + R^2c_2) \end{bmatrix} \begin{Bmatrix} \dot{u} \\ \dot{\theta} \end{Bmatrix}$$
$$+ \begin{bmatrix} (k_1 + k_2) & -Rk_2 \\ -Rk_2 & R^2k_2 \end{bmatrix} \begin{Bmatrix} u \\ \theta \end{Bmatrix} = \begin{Bmatrix} k_1 h(t) \\ 0 \end{Bmatrix}.$$

Notice that all square matrices are symmetric. ★

EXAMPLE 5.3 Consider the pin-jointed mechanism sketched in Figure 5.13(a). Each of the two uniform bars is a rigid link. Each bar has a length $2L$, a mass m, and a mass moment of inertia about its center of mass, H. The translating mass M is undergoing a forced horizontal motion $h(t)$ that results in only a small rotation $\theta(t)$ (positive clockwise), of the left-hand link from its static equilibrium position at 45°. Write the differential equation of motion, and from that equation, determine the expressions for the undamped natural frequency of this mechanism and its equivalent viscous damping factor.

SOLUTION This example serves as a connection to the geometric reasoning of Chapter 2, and a review of the fact that only mass whose motion needs to be described by one or more generalized coordinates needs to be included within the system mathematical model. As with pendulums, the expression for the kinetic energy, which is a point function, is obtained by choosing a simply understood path for the center of masses from their static equilibrium position at time zero, to their displaced position at time t. The lateral velocity of the left-hand bar center of mass is simply obtained. In particular, there is no axial velocity component for this center of mass, just the lateral component. The velocity components for the right-hand bar center of mass can be ascertained one step at a time. First let the right-hand bar be rigidly fixed to the left-hand bar as the latter rotates clockwise through an angle θ from the original 45°

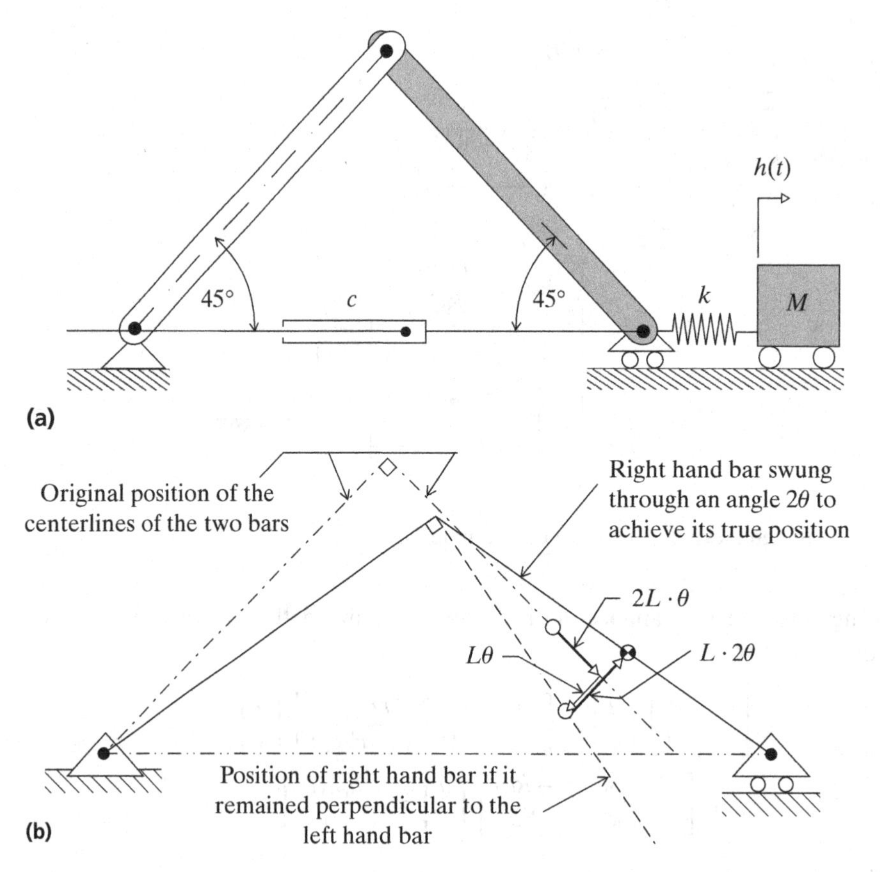

Figure 5.13. (a) Example 5.3: System. (b) Geometry of the deflection of the *CG* of the right-hand bar.

orientation. As shown in Figure 5.13(b), in this step the right-hand bar also rotates clockwise through an angle θ. As a result of this motion, the right-hand bar center of mass moves along its original longitudinal axis a distance $2L\theta$, and it also moves a lateral distance $L\theta$ downward and to the left. In the second step, swing the right-hand bar counterclockwise until the right-hand end of the right-hand bar is once again on a horizontal line from the left-hand hinge. Since the two bars always form an isosceles triangle, this upswing is through an angle 2θ. With this step the center of mass moves laterally upward through the distance $L(2\theta)$. Therefore the net lateral movement of this center of mass is only $L\theta$ upward. Hence, the kinetic and strain energy expressions, and the virtual work expression, are

$$T = \tfrac{1}{2}m(L\dot{\theta})^2 + \tfrac{1}{2}m[(2L\dot{\theta})^2 + (L\dot{\theta})^2] + 2(\tfrac{1}{2}H\dot{\theta}^2) + \tfrac{1}{2}M\dot{h}^2$$
$$U = \tfrac{1}{2}k(h - 2\sqrt{2}L\theta)^2 \quad \text{and} \quad \delta W = -c(2\sqrt{2}L\dot{\theta})2\sqrt{2}L\,\delta\theta.$$

Substitution into the Lagrange equation, and division by 2, leads to

$$(H + 3mL^2)\ddot{\theta} + 4cL^2\dot{\theta} + 4kL^2\theta = \sqrt{2}kLh(t)$$
$$\omega_1^2 = \frac{4kL^2}{H + 3mL^2} \quad \text{and} \quad \zeta = \frac{cL}{\sqrt{k(H + 3mL^2)}}.$$

★

5.7 Harmonic Excitation of Multidegree of Freedom Systems

It is appropriate to consider briefly, but only briefly, a steady-state solution to the N-DOF, equivalent viscous damping equations put forth at the beginning of the previous section, Eqs. (5.8a). It is not often that these equations are ever written in this form because determining the entries to the damping matrix is not often possible with any precision. However, for those rare occasions when such an equation can be written in a reliable fashion, it is possible, on the basis of a lot of computation, to directly obtain a limited solution for the steady-state deflection amplitudes for a fixed value of the forcing frequency, and, as later shown, a solution for arbitrary values of the forcing frequency. A full explanation of the latter result must await the material of Chapter 7.

If the applied loading were to be described by $\{Q(t)\} = \{Q_0\}\exp(i\omega_f t)$, where all Q_0 entries are real numbers, then it would mean that all the applied forces would be in phase with each other, if positive quantities, or 180° out of phase, if negative. However, it is an easy matter when there are N loads to allow for phase differences between the applied loads as well as amplitude differences. Let the amplitudes and phasing of the applied loading vector $\{Q(t)\}$ be described by means of the vector of complex quantities $\{Q^{(r)} + i\,Q^{(i)}\}$ so that $\{Q(t)\} = \{Q^{(r)} + i\,Q^{(i)}\} \times \exp(i\omega_f t)$. Since the input has complex amplitudes, it is to be expected that the output, the steady-state structural response,[13] as described by the time-varying generalized coordinates, also requires a complex algebra description. Therefore write $\{q(t)\} = \{q^{(r)} + iq^{(i)}\}\exp(i\omega_f t)$. Now substitute these complex algebra representations for the force and deflection into the N-DOF equation. Note that the vector of velocities $\{\dot{q}(t)\} = i\omega_f\{q^{(r)} + iq^{(i)}\}\exp(i\omega_f t)$ and then $\{\ddot{q}(t)\} = -\omega_f{}^2\{q^{(r)}+iq^{(i)}\}\exp(i\omega_f t)$. Both sides of the resulting equation involve complex quantities. For the equality to hold, as it does, the real terms on the right-hand side must equal the real terms on the left-hand side. The imaginary terms on the two sides of the equality sign must also be equal. Thus, after canceling the common factor that is the exponential term, the original n-sized matrix equation becomes the following set of $2N$ linear, simultaneous algebraic equations in the $2N$ unknowns $\{q^{(r)}\}$ and $\{q^{(i)}\}$

$$
\begin{aligned}
\left[K - \omega_f^2 M\right]\{q^{(r)}\} + \omega_f[C]\{q^{(i)}\} &= \{Q^{(r)}\} \\
-\omega_f[C]\{q^{(r)}\} + \left[K - \omega_f^2 M\right]\{q^{(i)}\} &= \{Q^{(i)}\}.
\end{aligned}
$$

For a fixed value of the forcing frequency, Gaussian elimination or the Gauss-Seidel method, depending on the magnitude of N, is often the best approach to solving for the complex components of the deflection vector. However, a solution for an arbitrary value of the forcing frequency cannot be readily obtained from the above equations. However, as mentioned above, the steady-state deflection amplitudes can be obtained for arbitrary values of the forcing frequency using the transformation approach outlined in Chapter 9. In that manner, creating a matrix of frequency response functions is possible. The details of this procedure are set forth in an endnote of that chapter.

[13] It is possible to obtain a complementary solution in terms of these matrices and thus demonstrate that in the N-DOF case, too, the complementary solution is transient. See Chapter 9.

5.8 Summary

From an energy viewpoint, an undamped structural vibration is an interchange between kinetic energy and strain energy where the total mechanical energy, $T + U$, is constant. For an undamped system, all the deflections are either in or out of phase. That is, the deflections have either positive or negative values. When the undamped system passes through it static equilibrium position, the dynamic strain energy is zero, but the velocities, and hence the kinetic energy, are a maximum. When the system reaches its maximum deflections, the strain energy is a maximum, but because the velocities are changing direction, the velocities and the kinetic energy are zero. When friction is present, as it is in all real structures, some of the mechanical energy is continually dissipated, that is, channeled into unrecoverable forms of energy such as local plastic deformations and heat. The dissipated energy is viewed as the result of the work done by forces, called damping forces, that oppose the various system motions, relative and absolute. For example, damping forces act on any mechanical system that undergoes elastic or plastic deformations because of internal friction between the crystals or grains of the material as they move relative to each other. Damping forces also act on any mechanical system that is in contact with, and moving relative to, any fluid or other solid. Most mathematical descriptions of actual damping forces lead to differential equations that are cumbersome to solve analytically if they can be solved at all in that manner. On the basis that the general response of a structure to any one form of damping differs little from that of viscous damping, at least when the amplitudes of the motion are large, the engineering approach to an analytical description of damping forces is to merge all the actual damping forces acting on the system into a single equivalent viscous damping force. Viscous damping forces are chosen to represent all types of damping forces because they are proportional to the first power of the velocity of the mass, and thus they lead to linear differential equations whose analytical solutions are easy to obtain. The basis for the equivalence between the equivalent viscous damping force and all the actual damping forces is an equal amount of energy dissipated per cycle at the appropriate system natural frequency.

When employed, the dashpot symbol represents the location and orientation of a specific equivalent viscous damping force in a structural system mathematical model. The magnitude of the dashpot damping force is the product of the relative velocity of the dashpot ends and the associated equivalent viscous damping coefficient, $c_{eq} = c$. In other words, dashpots in association with their relative velocities represent damping forces in exactly the same manner that springs in association with their relative deflections represent generic elastic forces. If there are no known significant, localized sources of damping, then it is better that dashpots not be an explicit part of the system mathematical model. There are three other options for modeling without dashpots. The first of these options lacking dashpots is to introduce general system equivalent viscous damping force vector (i.e., an $N \times 1$ matrix) as being proportional to a combination of the inertial and elastic forces. This option is called using proportional damping. The coefficients of proportionality are best estimated after the transformation introduced in the chapter after next is accomplished. Thus this approach is

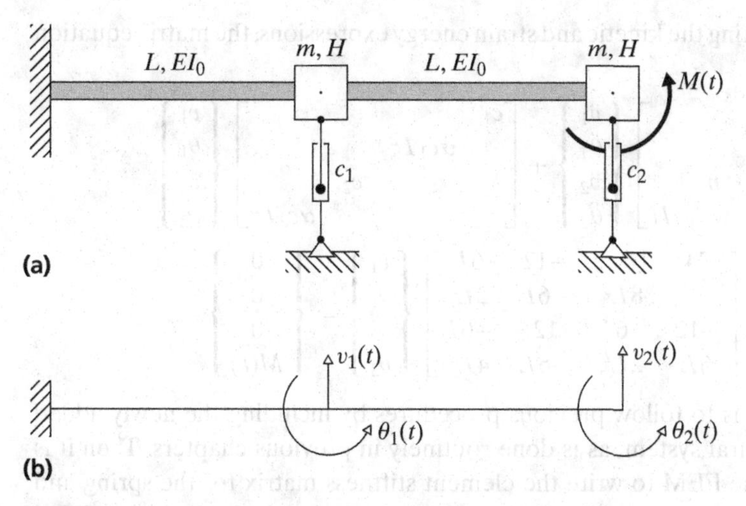

Figure 5.14. Example 5.4: (a) System. (b) Degrees of freedom.

equivalent to the second option, where equivalent viscous damping can be introduced by means of what are called modal damping factors, which are explained in the chapter after next. The third option is to simply ignore damping forces altogether, and thus greatly simplify the analysis, while realizing what are the general energy dissipative effects of damping forces on the motion of the system and its components. For example, realize that because of damping, any maximum vibratory amplitude for an actual system will be less than that calculated for an undamped system. Chapter 7 further examines the role of damping with regard to limiting vibratory amplitudes.

EXAMPLE 5.4 **(a)** Write the matrix equations of motion for the lumped mass, cantilevered, elastic beam model shown in Figure 5.14(a). Not shown on the sketch, but nevertheless present at each of the two masses, is a rotary damper that produces a moment opposing the rotation. The magnitude of this damping moment is $\alpha c_1 L^2 \dot{\theta}_1$ at the first mass, and $\alpha c_2 L^2 \dot{\theta}_2$ at the second mass, where α is a small, nondimensional parameter.

(b) Make the following modifications to the above structure and loading (i) below the beam, add a vertically oriented spring of stiffness $k_1 = 7EI/L^3$ that is attached at its top to the midspan mass and at its bottom to a rigid support that is undergoing a known, positive upward, vertical motion $w(t)$, (ii) let the colocated dashpot with damping coefficient c_1 also have its lower attachment point undergo the same known upward vertical motion $w(t)$, and (iii) remove both of the rotary dampers. Then write the modified matrix equations of motion.

SOLUTION **(a)** Using the DOF shown in Figure 5.14(b), the virtual work expression can be written as

$$\delta W = -c_1 \dot{v}_1 \delta v_1 - c_2 \dot{v}_2 \delta v_2 - \alpha c_1 L^2 \dot{\theta}_1 \delta\theta_1 - \alpha c_2 L^2 \dot{\theta}_2 \delta\theta_2 + M(t)\delta\theta_2.$$

Therefore, after writing the kinetic and strain energy expressions, the matrix equation of motion is

$$\begin{bmatrix} m & & & \\ & H & & \\ & & m & \\ & & & H \end{bmatrix} \begin{Bmatrix} \ddot{v}_1 \\ \ddot{\theta}_1 \\ \ddot{v}_2 \\ \ddot{\theta}_2 \end{Bmatrix} + \begin{bmatrix} c_1 & & & \\ & \alpha c_1 L^2 & & \\ & & c_2 & \\ & & & \alpha c_2 L^2 \end{bmatrix} \begin{Bmatrix} \dot{v}_1 \\ \dot{\theta}_1 \\ \dot{v}_2 \\ \dot{\theta}_2 \end{Bmatrix}$$

$$+ \frac{EI}{L^3} \begin{bmatrix} 24 & & -12 & 6L \\ & 8L^2 & -6L & 2L^2 \\ -12 & -6L & 12 & -6L \\ 6L & 2L^2 & -6L & 4L^2 \end{bmatrix} \begin{Bmatrix} v_1 \\ \theta_1 \\ v_2 \\ \theta_2 \end{Bmatrix} = \begin{Bmatrix} 0 \\ 0 \\ 0 \\ M(t) \end{Bmatrix}.$$

(b) The first option is to follow previous procedures by including the newly added spring in the structural system, as is done routinely in previous chapters. Then it is a matter of using the FEM to write the element stiffness matrix for the spring and replacing the top and bottom spring element DOFs by, respectively, the beam global DOF and the enforced motion. However, because the spring structural element is such a simple element, an alternative approach to the equations of motion is offered. In this alternate approach, do not include the spring in the structural system. Then, to begin with, the mass and stiffness matrices are exactly as they are in part (a). The new external force acting on the system from the spring is $k_1(v_1 - w)$, downward, regardless of whether $v_1(t)$ is greater than $w(t)$. Similarly, from the colocated dashpot, the force transmitted to the structural system is $c_1(\dot{v}_1 - \dot{w})$, also acting downward, again irrespective of whether the top of the dashpot moves upward more or less than the bottom of the dashpot. Then the virtual work expression for the modified structure becomes

$$\delta W = -c_1(\dot{v}_1 - \dot{w})\delta v_1 - k_1(v_1 - w)\delta v_1 - c_2\dot{v}_2\delta v_2 + M\delta\theta_2.$$

Now the above $-c_1\dot{v}_1$ and the $-c_2\dot{v}_2$ terms are transposed to the left-hand side (as positive terms) to form the 4×4 damping matrix as before, but now there are zero entries for the terms corresponding to the angular velocities. The remaining known equivalent force $+c_1\dot{w}(t)$ becomes part of the known generalized force vector $\{Q\}$. The spring term in the δW expression is similarly broken into two parts. The known quantity $+(7EI/L^3)w(t)$ becomes part of the known input, $\{Q\}$. The unknown DOF-dependent term $-(7EI/L^3)v_1(t)$ is transposed to the left-hand side and incorporated into the only place where it can be included, which is in the stiffness matrix. Therefore the end result is

$$\begin{bmatrix} m & & & \\ & H & & \\ & & m & \\ & & & H \end{bmatrix} \begin{Bmatrix} \ddot{v}_1 \\ \ddot{\theta}_1 \\ \ddot{v}_2 \\ \ddot{\theta}_2 \end{Bmatrix} + \begin{bmatrix} c_1 & & & \\ & 0 & & \\ & & c_2 & \\ & & & 0 \end{bmatrix} \begin{Bmatrix} \dot{v}_1 \\ \dot{\theta}_1 \\ \dot{v}_2 \\ \dot{\theta}_2 \end{Bmatrix}$$

$$+ \frac{EI}{L^3} \begin{bmatrix} 31 & & -12 & 6L \\ & 8L^2 & -6L & 2L^2 \\ -12 & -6L & 12 & -6L \\ 6L & 2L^2 & -6L & 4L^2 \end{bmatrix} \begin{Bmatrix} v_1 \\ \theta_1 \\ v_2 \\ \theta_2 \end{Bmatrix} = \begin{Bmatrix} c_1\dot{w} + \dfrac{7EI}{L^3}w \\ 0 \\ 0 \\ M(t) \end{Bmatrix}.$$

★

EXAMPLE 5.5 Determine the steady-state response of a one-DOF (m, k) oscillator to a harmonic input $F(t) = F_0 \sin \omega_f t$ when the damping is material damping. That is, find the steady-state solution to the equation of motion

$$m\ddot{u}(t) + k(1 + i\gamma)u(t) = F_0 \sin \omega_f t.$$

SOLUTION The first step is to divide by m and recast the applied load in complex form as

$$\ddot{u}(t) + \omega_1^2(1 + i\gamma)u(t) = \omega_1^2 \frac{F_0}{k} e^{i\omega_f t}.$$

The solution for the motion of the mass is now written in complex form, too, as $u(t) = A_0 \exp(i\omega_f t)$, where A_0 is a complex quantity or, alternately, writing $u(t) = B_0 \exp(i\omega_f t - \phi)$, where B_0 is a real quantity and the phase difference, ϕ, appears explicitly in the exponential function rather than being hidden in the complex amplitude A_0. Substituting the first form, where

$$\ddot{u}(t) = -\omega_f^2 A_0 e^{i\omega_f t},$$

canceling the common exponential term, dividing by the natural frequency squared, and factoring leads to the following form for the nondimensionalized response amplitude:

$$\frac{A_0 k}{F_0} = \frac{1}{(1 - \Omega_1^2) + i\gamma}.$$

The amplitude of the response is the absolute value of this complex number, which is

$$\frac{k|A_0|}{F_0} = \frac{1}{\sqrt{(1 - \Omega_1^2)^2 + \gamma^2}}.$$

This solution closely parallels that for viscous damping in that at a frequency ratio value of 1, the amplitude of the deflection response is $1/\gamma$, a large number because the material damping factor is a small quantity corresponding to twice the equivalent viscous damping factor. In other words, there is a resonance phenomena with material damping just like there is for viscous damping.

The solution for the phase angle is obtained from the separated real and imaginary parts of the above complex solution. Therefore, multiply both the numerator and the denominator of the above complex solution by the conjugate of the denominator. Then

$$\frac{A_0 k}{F_0} = \frac{(1 - \Omega_1^2) - i\gamma}{(1 - \Omega_1^2)^2 + \gamma^2}.$$

Since the imaginary part of this complex number is negative, the phase angle in the complex plane is in the fourth or third quadrant of the complex plane and thus is a lag angle of magnitude

$$\tan \phi = \frac{\gamma}{1 - \Omega_1^2}.$$

Again, the phase angle is a small angle at low values of the frequency ratio, sharply rises to 90° at resonance, and then slowly approaches 180° as the frequency ratio becomes large. ★

As a final comment, the equations of motion for base motion problems can be written so that the equivalent force vectors either involve (i) displacements and velocities as Eq. (5.1) and the above example problems illustrate or (ii) accelerations. To illustrate the latter case, again consider the oscillator of Figure 5.5 and the resulting equation of motion

$$m\ddot{u}(t) + c\dot{u}(t) + ku(t) = c\dot{w}(t) + kw(t). \tag{5.1}$$

This writing of the equation of motion is appropriate if the input record is provided in terms of measured displacements over time. In that case the velocities of the small damping force can be carefully approximated as the slopes of the displacement record. If, however, the original input record is in terms of accelerations, two integrations would be required to obtain the displacement record. The numerical errors associated with those integrations may well make that approach impractical. Nevertheless an acceleration record is often the basis for input, as it would be for an earthquake. The way to use the acceleration record directly is to use generalized coordinates that describe the relative motion of masses. For example, for the oscillator of Eq. (5.1), rewrite that equation of motion as

$$m\ddot{u} + c(\dot{u} - \dot{h}) + k(u - h) = 0.$$

Now add and subtract the quantity $-m\ddot{h}$ and define the generalized coordinate $q = u - h$, which is the measure of the motion of the mass relative to the support. Then the equation of motion becomes

$$m\ddot{q} + c\dot{q} + kq = -m\ddot{h}.$$

Since the elastic force on the mass is $-kq(t)$, the force on the elastic element, the spring, is $+kq(t)$. Hence a solution for $q(t)$ is useful because the force borne by the elastic element, the object of the structural analysis, is thereby determined. What is true for the oscillator is also true for more elaborate structural systems. Exercises 5.7(b) and (c) examine the same building model using first generalized coordinates that are referenced to fixed coordinate axes and then generalized coordinates that describe increments in deflection.

REFERENCES

5.1 Steidel, R. F., Jr., *An Introduction to Mechanical Vibrations*, Wiley, New York, 1971.

5.2 Schlichting, H., *Boundary Layer Theory*, 4th ed., McGraw-Hill, New York, 1960.

5.3 Meirovitch, L., *Analytical Methods in Vibrations*, Macmillan, New York, 1971.

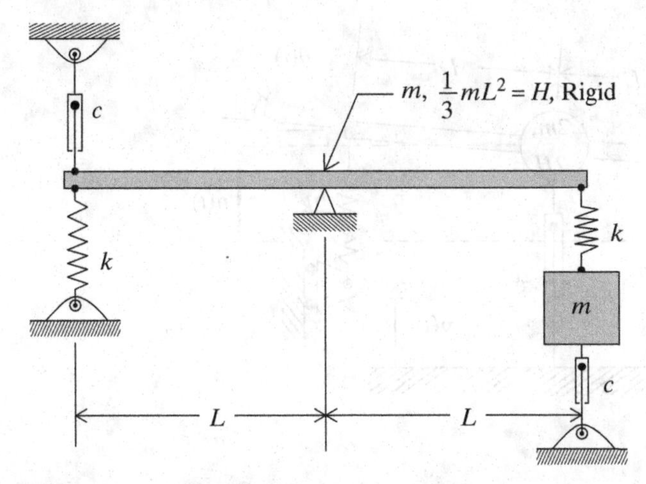

Figure 5.15. Exercise 5.4(a).

CHAPTER 5 EXERCISES

5.1 Determine the transformation for the constants of integration C_1, C_2 so that Eq. (5.3a) can be written in terms of a cosine function rather than a sine function.

5.2 **(a)** Redo Example 5.1 using the Lagrange equations and the following two generalized coordinates: $u(t)$ measured upward at the center of mass and $\theta(t)$ measured counterclockwise about the center of mass. Obtain the matrix equations of motion.

(b) Redo part (a), but this time use Newton's equations of motion to write the matrix equations of motion.

(c) Redo part (a), but this time use as DOF $u_1(t) \equiv v(t)$, the upward deflection of the left-hand end of the system, and $\theta(t)$ the counterclockwise rotation about the left-hand end of the system.

(d) Redo part (c), but this time use Newton's equations of motion to write the matrix equations of motion.

5.3 Write the equation for the equivalent viscous damping coefficient for the following types of damping forces.

(a) Coulomb damping. (Also see Exercises 5.8 and 5.9.)

(b) A damping force for which the energy dissipated per cycle is independent of frequency and proportional to the amplitude of the vibration raised to the 1.75 power. (This represents a fluid drag force acting on a cylinder for a Reynolds number range from 4 to 1000.)

5.4 Write the matrix differential equations of motion for the rigid body-spring-dashpot system shown in:

(a) Figure 5.15, for small rotations of the bar where the mass is uniformly distributed over the rigid bar length.

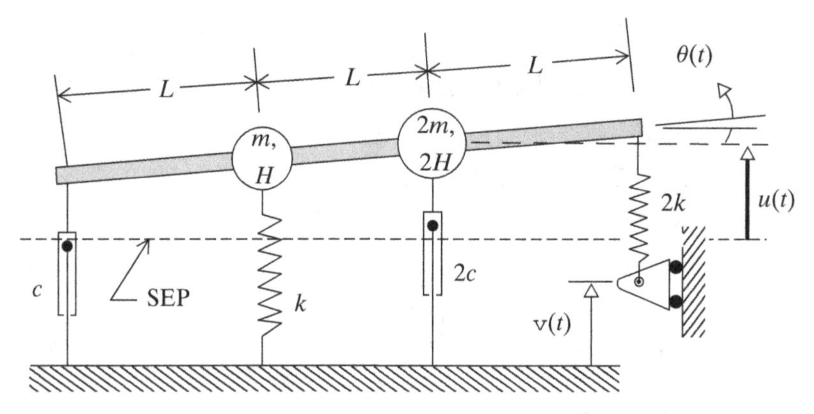

Figure 5.16. Exercise 5.4(b).

(b) Figure 5.16, for small rotations. Here all the mass is lumped at two locations. Note that the DOF are $u(t), \theta(t)$, while $v(t)$ is the enforced motion. In order to simplify the algebra, let $H = mL^2/6$.

(c) Figure 5.17. Use the DOF $u_1(t), u_2(t)$ that are measured from fixed points.

(d) Figure 5.17, but now use the DOF $u_1(t)$ as above and the DOF $u_3(t)$ that is the relative motion between m_1 and m_2.

(e) Figure 5.18, where the left bar is shown rotated through a small angle θ, the single DOF for the system. The rotational spring with stiffness coefficient K that connects the two uniform, rigid bars of length L and mass m near their mutual hinge point provides a moment resisting their relative rotation equal to K multiplied by the angle of their relative rotation. Let $H = mL^2/12$. Furthermore, linearize the equation of motion to obtain the damped natural frequency.

(f) Figure 5.15, but this time let each bar segment be flexible with a stiffness coefficient $EI_0 = const.$

(g) Figure 5.16, but this time let each bar segment be flexible with a stiffness coefficient $EI_0 = const.$

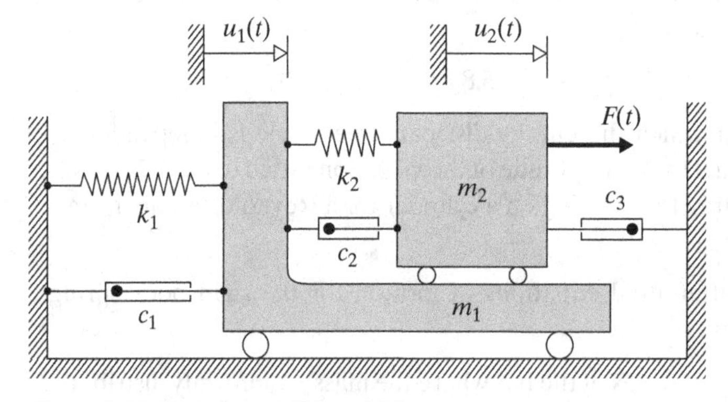

Figure 5.17. Exercises 5.4(c) and (d).

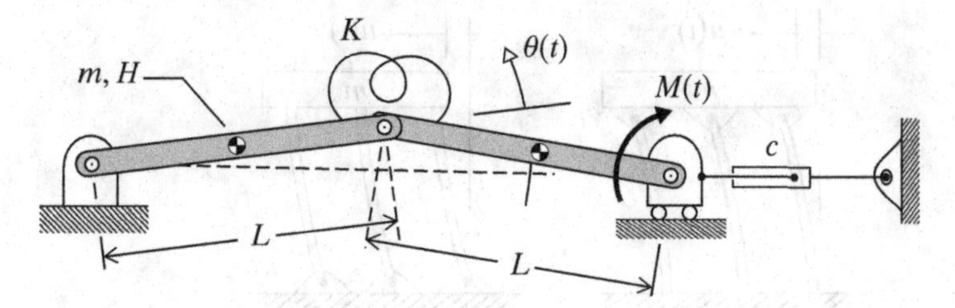

Figure 5.18. Exercise 5.4(e).

(h) Figure 5.19. Use only symmetric matrices to write the equations of motion in matrix form. Note that the system is being driven by an applied moment at the base of the rod.

5.5 The pen trace on a strip chart shows that the amplitudes of a certain vibratory mode decreased to one-third the initial value after 10 cycles. What is the (equivalent) viscous damping factor for this vibration?

5.6 Determine the damped natural frequency of the following flexible, single-DOF systems shown in their deflected state if the (equivalent viscous) damping factor for each indicated motion is 0.2. All beams have a stiffness coefficient EI in each plane of motion and a length L. In some cases, use of the FEM beam stiffness matrix is helpful and in other cases the use of a force (flexibility) method of analysis such as

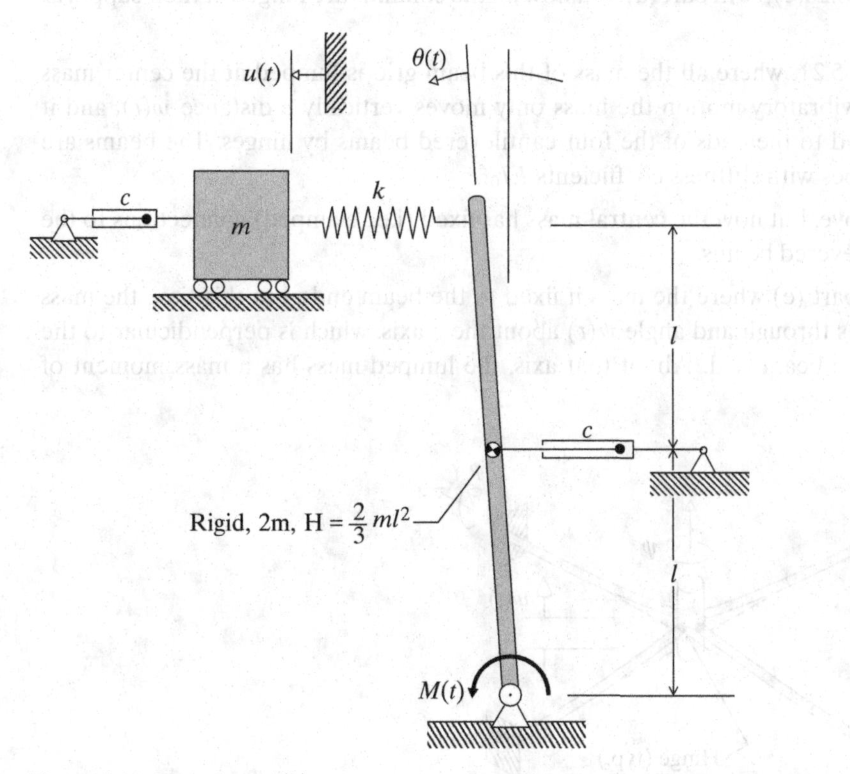

Figure 5.19. Exercise 5.4(h).

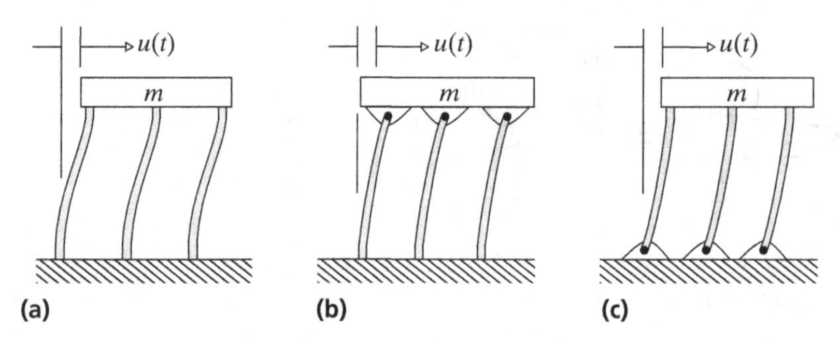

Figure 5.20. Exercise 5.6(a)–(c).

the unit/virtual load method (or Castigliano) or a beam differential equation method $[EIw''(x) = M(x)]$ is more advantageous for determining the stiffness of the structure at the single mass location.

(a) Figure 5.20(a), where the three columns are fixed at their supports and at the rigid top of the structure. All the mass of the structure, m, is concentrated at the rigid top of the structure.

(b) Figure 5.20(b), as above, but now there are hinges between each of the column tops and the rigid top of the structure.

(c) Figure 5.20(c), as in part (a), but now all the columns are hinged at their supports.

(d) Figure 5.21, where all the mass of this beam grid is lumped at the center mass m. In this vibratory motion the mass only moves vertically a distance $w(t)$, and it is connected to the ends of the four cantilevered beams by hinges. The beams are circular pipes with stiffness coefficients EI_0.

(e) As above, but now the central mass has fixed (i.e., clamped) connections to the four cantilevered beams.

(f) As in part (e) where the mass if fixed to the beam ends, but this time the mass only rotates through and angle $\psi(t)$ about the z axis, which is perpendicular to the plane of the beam grid. About that axis, the lumped mass has a mass moment of inertia H_z.

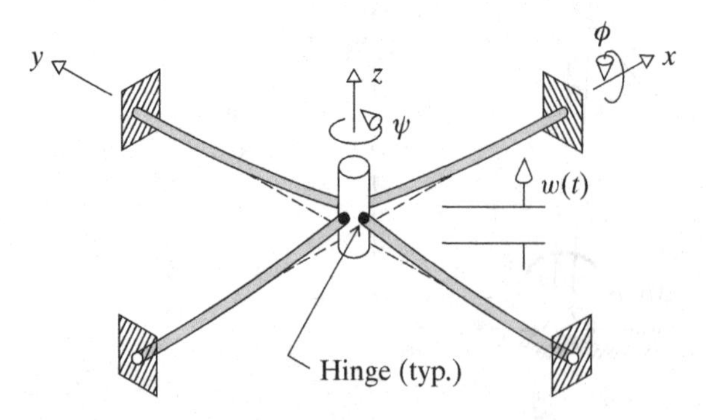

Figure 5.21. Exercises 5.6(d)–(h).

Figure 5.22. (a) Exercise 5.7(a).

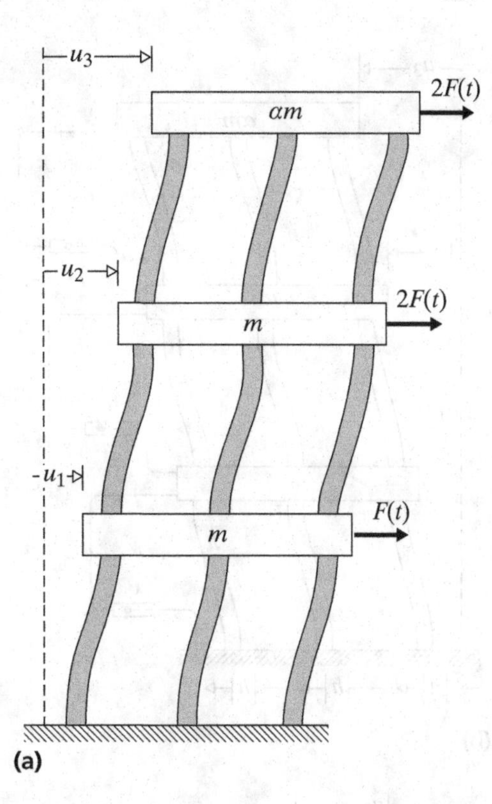

(g) As in part (e), but this time the mass rotates through an angle $\phi(t)$ about the longitudinal axis of one of the beams, say the x axis. The mass moment of inertia about this axis is H_x. Be sure to understand that this motion involves the twisting of two beams and the bending of the other two beams. Let the torsional stiffness coefficient $GJ_0 = \alpha EI_0$.

(h) As in part (g), but this time the mass rotates about an axis that lies at a 45° angle to the x and y axes. This means that all four beams simultaneously bend and twist. Let $H_y = H_x$.

5.7 (a) Write the matrix equations of motion for the force free vibration of the simplified three-story planar building model shown in Figure 5.22(a). The building sways back and forth in the plane of the paper. Each clamped–domped beam column has a length L and a bending stiffness coefficient EI. Let the damping in the three-DOF structure be modeled as proportional to the strain energy. That is, let the damping matrix entries be the same as the nondimensional entries of the stiffness matrix (after division by m and removal of the greatest common factor $36EI/mL^3$). Then let the common factor for the damping matrix be $12\zeta(EI/mL^3)^2$. Thus this choice means that there is a single damping parameter ζ for the entire structure.

(b) As above, but now let the damping in the structure be modeled using the dashpots shown in Figure 5.22(b). Each dashpot has the same equivalent viscous damping coefficient, c. Furthermore, now remove the applied loads $F(t)$, but let the base support have a known horizontal motion described by the time function $h(t)$. With the value of α being 2, write the matrix equations of motion using generalized

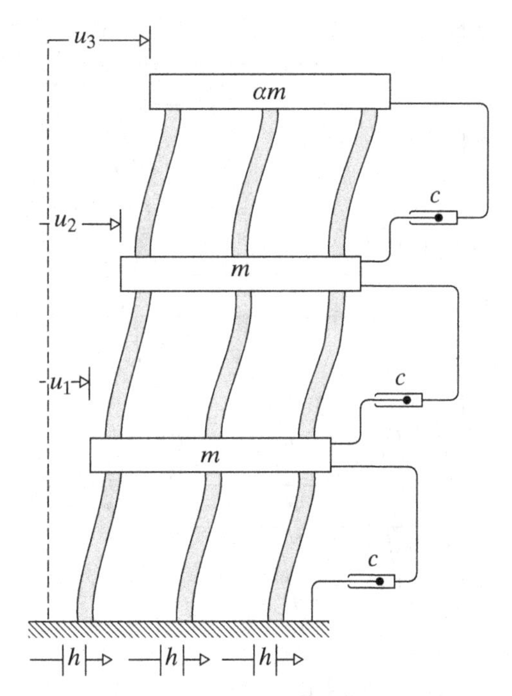

(b)

Figure 5.22. (b) Exercise 5.7(b). (c) Exercise 5.7(c).

(c)

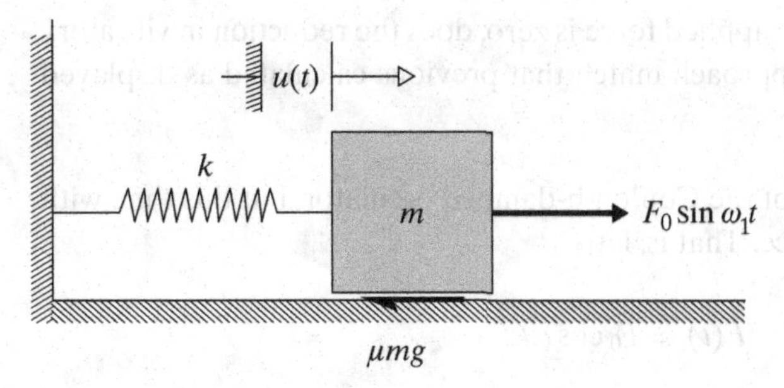

Figure 5.23. Exercise 5.8.

coordinates, as above, that measure the motion of the mass relative to a fixed vertical axis.

(c) Repeat part (b), but this time use generalized coordinates that measure the position of the mass relative to the mass (or support) below the mass under consideration. See Figure 5.22(c). This formulation may be more convenient if the accelerations of the input base motion are the known data rather than the deflection time history of the base motion.

For the eager

5.8 Consider the Coulomb-damped oscillator shown in Figure 5.23. This case differs from that previously discussed within the chapter in that there is an applied harmonic load. The frequency of this applied force is the same as the system natural frequency. At time zero the mass m is released with zero initial velocity from a position $u(0) = 0$, which is located a distance ℓ to the left of the unstretched spring position. Let ℓ be such that the initial spring force, $+k\ell$, is greater than the static friction force. Thus the mass starts in motion to the right immediately even though the applied force has a zero magnitude at time zero.

(a) With ω_1, the oscillator natural frequency, being the square root of the ratio k/m, and u_{stat} being F_0/k, $u_g = mg/k = g/\omega_1^2$, verify that the differential equation of the motion is

$$\ddot{u}(t) + \omega_1^2 u(t) = \omega_1^2(-\mu u_g) + u_{stat}\omega_1^2 \sin \omega t.$$

(b) Verify that the solution to this equation of motion is

$$u(t) = C_1 \sin \omega_1 t + C_2 \cos \omega_1 t - \tfrac{1}{2}\, \omega_1 t\; u_{stat} \cos \omega_1 t - \mu u_g.$$

(c) Verify that the application of the initial conditions $u(0) = 0$, $\dot{u}(0) = 0$ leads to the result

$$u(t) = \tfrac{1}{2} u_{stat}[\sin \omega_1 t - \omega_1 t \cos \omega_1 t] - \frac{\mu g}{\omega_1^2}(1 - \cos \omega_1 t).$$

(d) Determine the distance traveled over the time period of the first half cycle (the time period of applicability of the above differential equation) by finding the time value at which the velocity goes to zero. To do this, let $\pi F_0/2 < \mu mg$, a "weak" force.

(e) When the magnitude of the applied force is zero, does the reduction in vibratory amplitude calculated by this approach match that previous calculated as displayed in Figure 5.3?

5.9 Redo the above problem of the Coulomb-damped oscillator, but this time with a phase shift in the applied force. That is, let

$$F(t) = F_0 \cos \omega t.$$

Note that the solution to the differential equation of motion is the same but for $-t \cos \omega t$ replaced by $+t \sin \omega t$. That is, (a) obtain the general form of the solution for an initial deflection of a distance ℓ to the left of the unstretched spring position where $u(0) = 0$, and zero initial velocity.

(b) Determine when the velocity becomes zero if the magnitude of the applied force F_0 is selected to be $4\mu mg$.

5.10 A horizontal linear spring of stiffness k is connected at its right end to a block mass m resting on a dry surface where the Coulomb friction coefficient between the block and the surface is

$$\mu_{stat} = \frac{kd_0}{mg} = \frac{5}{4}\mu_{dyn}.$$

The left end of the spring is connected to a vehicle that reaches a constant velocity V_0 moving to the left before the vehicle travels the distance d_0. At time zero, the vehicle passes the distance d_0, causing the stretch in the spring to produce a spring force that overcomes the static friction force, and the block begins moving to the left.

(a) Write and solve the equation of motion for the block for this first movement.

(b) Determine an explicit solution for the time t_{stop} at which the block no longer moves to the left.

(c) If $(\omega_1^2 d_0 - \mu_{dyn} g)/\omega_1 V_0 = 1.0$, what happens after the block first comes to a stop?

5.11 If the applied harmonic force acting on an oscillator has a forcing frequency very close to, but not equal to, the natural frequency (i.e., $F(t) = F_0 \sin(\omega_1 \pm \Delta\omega)t$, where $\Delta\omega \ll \omega_1$), what then is the oscillator deflection response? Hint: Write $\sin(\omega_1 \pm \Delta\omega)t$ as $\sin \omega_1 t \cos \Delta\omega t \pm \cos \omega_1 t \sin \Delta\omega t$.

ENDNOTE (1): A REAL FUNCTION SOLUTION TO A HARMONIC INPUT

The equation of motion for the steady-state response of a one-DOF system in the case of a harmonic load can be solved either (i) by use of the complex number representations of the force and displacement as is done in this chapter or (ii) by use of the real functions $\sin \omega_f t$ and $\cos \omega_f t$. This endnote offers the latter approach and demonstrates that such a solution procedure is much more time consuming than the use of the complex algebra representation. To this end, return to Eq. (5.8), which is, again,

$$\ddot{q} + 2\zeta\omega_1\dot{q} + \omega_1^2 q = \frac{F_0}{k}\omega_1^2 \sin\omega_f t.$$

In other words, the present task is to use mathematical procedure called the method of undetermined coefficients to solve for the particular (steady-state) solution of the above linear, ordinary differential equation with constant coefficients. Recall that the complementary (transient) solution is already been determined to be

$$q(t)_{compl} = \exp(-\zeta\omega_1 t)[C_1 \sin\omega_d t + C_2 \cos\omega_d t].$$

The first question to be asked is whether the method of undetermined coefficients is suitable to this equation. This approach is viable only if the known forcing function on the right-hand side of the equation has a finite number of distinct derivatives. This is indeed true for this equation because the sine function's only other distinct derivative is the cosine function. Therefore, when the forcing frequency, ω_f, is different from the natural frequency, ω_1, the trial particular solution can be written as

$$\begin{aligned} q(t) &= A\sin\omega_f t + B\cos\omega_f t \\ \text{where}\quad \dot{q}(t) &= \omega_f A\cos\omega_f t - \omega_f B\sin\omega_f t \\ \ddot{q}(t) &= -\omega_f^2 A\sin\omega_f t - \omega_f^2 B\cos\omega_f t, \end{aligned}$$

where the constants A and B are to be determined. Substituting the above trial solution and its derivatives into the differential equation leads to

$$\begin{aligned} &\sin\omega_f t\left[-\omega_f^2 A - 2\zeta\omega_1\omega_f B + \omega_1^2 A\right] + \cos\omega_f t\left[-\omega_f^2 B + 2\zeta\omega_1\omega_f A + \omega_1^2 B\right] \\ &= \frac{F_0}{k}\omega_1^2 \sin\omega_f t. \end{aligned}$$

Since the sine function and the cosine function are linearly independent, the coefficients of the sine and cosine functions can be equated to each other. Specifically

$$\left(\omega_1^2 - \omega_f^2\right)A - 2\zeta\omega_1\omega_f B = \frac{F_0}{k}\omega_1^2$$

$$2\zeta\omega_1\omega_f A + \left(\omega_1^2 - \omega_f^2\right)B = 0.$$

For a better comparison with the result obtained using complex algebra, in the above two simultaneous equations for the unknown constants A and B divide through by the square of the natural frequency to obtain

$$\begin{aligned} \left(1 - \Omega_1^2\right) A - 2\zeta\Omega_1 B &= \frac{F_0}{k} \\ 2\zeta\Omega_1 A + \left(1 - \Omega_1^2\right) B &= 0. \end{aligned}$$

The solutions for the unknown constants are easily obtained as

$$\frac{Ak}{F_0} = \frac{\left(1 - \Omega_1^2\right)}{\left(1 - \Omega_1^2\right)^2 + (2\zeta\Omega_1)^2} \qquad \frac{kB}{F_0} = \frac{-2\zeta\Omega_1}{\left(1 - \Omega_1^2\right)^2 + (2\zeta\Omega_1)^2}.$$

Substituting these solutions into the original expression for the particular solution, and changing the form of the solution to the standard phase angle form; that is, writing

$$q(t)_{part} = A\sin\omega_f t + B\cos\omega_f t = C\sin(\omega_f t + \psi),$$

$$\text{where} \quad C = \sqrt{A^2 + B^2} \quad \text{and} \quad \tan\psi_j = \frac{B}{A}.$$

leads to the same result obtained using complex algebra, namely

$$C = \frac{F_0/k}{\sqrt{\left(1 - \Omega_1^2\right)^2 + (2\zeta\Omega_1)^2}} \quad \text{and} \quad \tan\psi = -\frac{2\zeta\Omega_1}{1 - \Omega_1^2}.$$

6 Natural Frequencies and Mode Shapes

6.1 Introduction

This chapter begins the mathematical process of solving the linear, matrix differential equation that describes the small deflection, vibratory motion of a structure subjected to an arbitrary, time-varying loading. In this textbook, the loading is limited to being one whose magnitude is known for any time value, as opposed to being a loading whose magnitude can only be described in probabalistic terms. Using the concept of equivalent viscous damping, that matrix equation is

$$[M]\{\ddot{q}(t)\} + [C]\{\dot{q}(t)\} + [K]\{q(t)\} = \{Q(t)\}. \quad (6.1)$$

The overall form of the solution process closely parallels that for a single-DOF dynamical system. First a complementary solution, which contains the constants of integration, is obtained for the unloaded and undamped structural system. Then the response to the applied loading, with or without the effect of damping, is separately obtained. These two solutions can be combined to provide the complete description of the system motion. If required, the values of the constants of integration are then, and only then, determined by the system initial conditions.

This chapter concentrates on the first of the above two steps of the solution process, which is the step dealing with the unloaded and undamped structure. This first step has an important physical basis. To understand that physical basis, consider any lightly damped, multidegree of freedom structure. If such a structure, unloaded with respect to time-varying loads, is set in motion by the application of an arbitrary set of initial conditions, then the resulting vibration will generally appear to be without any discernable pattern. Indeed, the motion of the force free structural system appear to be quite confusing.[1] Specifically, no parts of the system will appear to be moving in or out of phase with any other part of the structural system. Furthermore, in general, no part of the system will display a motion characterized by a constant frequency of motion and near constant amplitude as was the solution and reality for the

[1] The word *chaotic* was avoided only because that word now has a well-established technical meaning that generally refers to dynamic systems whose trajectories are difficult to predict because of nonlinearities and a high susceptibility to large changes in response to small changes in system parameters, including initial conditions.

deflections of the lightly damped single-DOF system. This circumstance is noteworthy because engineers always seek to identify patterns because patterns are very helpful for fostering understanding. Thus it is fortunate that the seeming lack of any pattern to the free vibration of a multidegree of freedom structural system, whose motion was started by use of arbitrary initial conditions, is more apparent than real. As can be demonstrated with actual, small structures, *if* the initial conditions are properly chosen, the entire structure can be made to vibrate in an orderly manner at a single vibratory frequency with near constant amplitudes. That is, there is a vibratory motion where every mass particle in the structure moves either in or out of phase with all other mass particles. To be more specific, in this lightly damped structure, all mass particles will reach there maximum amplitudes at the same time, and then all mass particles will pass through the static equilibrium configuration at their maximum velocities at the same time. Using a C-clamp and a stiff support to reduce foundation damping, cantilevering a long ruler at one end and deflecting and releasing the other end would provide an illustration of this phenomenon.

The elimination of damping from the mathematical treatment of the structural system reduces the unloaded structural model to just its inherent mass and elastic properties. As will soon be seen, such an undamped, multidegree of freedom system mathematical model reflects the above described physical reality by having a solution for a vibration at a single frequency. All such frequencies are called the *natural frequencies*, and the associated patterns of vibratory amplitudes are called the *natural mode shapes*. The natural frequencies and mode shapes are very important when characterizing the inertial and elastic properties of a structure. In particular, the set of the lower numbered natural frequencies and the associated set of mode shapes of a structure prove to be the most important information set available to an engineer for gaining a physical insight into the behavior of that structure when it is subject to time-varying loads.

This chapter concentrates on developing (i) the precise meaning of the phrases *natural frequency* and *mode shape*, in those cases where there is more than one; (ii) the number of, and the mathematical nature of, those natural frequencies and mode shapes; and (iii) various means for calculating their values. Recall that Chapter 2 first explained that for any single-DOF vibratory system, such as a pendulum system, subjected to neither base motions nor externally applied contact loads (i.e., for a force free vibration), the natural frequency f_1 and natural circular frequency ω_1 of the back-and-forth motion are determined by measuring the elapsed time between any two points in the time record of the deflection DOF where, together, the deflection and the velocity repeat themselves. That elapsed time between repetition points is called the period of the vibratory motion, T_1. Again, the quantitative relationships between the circular frequency, the frequency, and the period are, respectively,

$$\omega_1 = 2\pi f_1 = \frac{2\pi}{T_1}.$$

Recall that for the single-DOF vibratory system, typified by a single pendulum, or an oscillator, undergoing small deflections, the (circular) natural frequency squared of the sinusoidal motion is always the ratio of the coefficient of the deflection term to the coefficient of the acceleration term. However, for the matrix equation of motion associated with a multidegree of freedom structural system, instead of a single deflection term and its single scalar stiffness coefficient, there is a deflection vector and its associated stiffness matrix. Instead of a single acceleration term and its mass coefficient, there is an acceleration vector and its associated mass matrix. There is no ratio of matrices corresponding to the important k/m ratio of the single-DOF system. Indeed $[m]^{-1}[k]$ is generally quite different from $[k][m]^{-1}$, or $[m]^{-1/2}[k][m]^{-1/2}$, and it is not clear which of these nondiagonal matrices would be preferred, or if they have any interpretation. Thus the multidegree of freedom problem poses a new challenge.

6.2 Natural Frequencies by the Determinant Method

As before, the only parts of Eq. (6.1) essential to a vibration are elastic and inertial forces. Consider the matrix equation $[M]\{\ddot{q}\} + [K]\{q\} = \{0\}$. The solution for this $[M]$, $[K]$ system equation starts with the conjecture that there is some function of time, $f(t)$, that is common to all the generalized coordinates of an N-DOF system. In the single-DOF case, that function of time is $\sin(\omega_1 t + \psi)$. Such a possibility would produce the "orderly" vibration mentioned in the first section of this chapter. To investigate this possibility, write the deflection vector as $\{q\} = \{A\} f(t)$, where $\{A\}$ is a $N \times 1$ vector of unknown amplitudes (i.e., unknown constants). Substituting this trial solution into the undamped, force free vibration equation yields

$$[M]\{A\}\ddot{f}(t) + [K]\{A\} f(t) = \{0\}. \tag{6.2a}$$

Conceptually premultiplying this matrix equation by the row vector $\{A\}^t$, and rearranging the scalar result yields

$$\frac{\ddot{f}}{f} = -\frac{\lfloor A \rfloor [K]\{A\}}{\lfloor A \rfloor [M]\{A\}}. \tag{6.2b}$$

Since the triple matrix product in the denominator is equal to twice the maximum value of the kinetic energy, this triple matrix product is a positive constant, which is the same as saying that the mass matrix is positive definite.[2] Similarly, the triple matrix product of the numerator is twice the maximum value of the strain energy. Thus the numerator is either zero (in the case of an structure unconstrained by supports) or positive (in the case of supports that constrain rigid body motion). The stiffness matrix is termed *nonnegative definite*. Therefore the right-hand side ratio of Eq. (6.2) is either zero or, because of the negative sign on the right-hand side, a negative real number. Following the symbol choice used for the corresponding single-DOF equation of Chapter 2, call that negative real number $-\omega^2$. Be sure to note that this means that ω^2 itself is always a positive number.

Now consider in turn these two possibilities for Eq. (6.2b), which is a differential equation involving the time function $f(t)$. When the right-hand side of Eq. (6.2b) is

[2] However, when there are zeros on the main diagonal, the numerator and the mass matrix may be only "nonnegative definite" rather than positive definite. However, the elimination of the DOF associated with those zero diagonal mass entries, as discussed in Chapter 3, produces a positive definite numerator and mass matrix.

zero, this differential equation for the time function, for all the DOF amplitudes, is simply

$$\ddot{f}(t) = 0 \quad \text{so that} \quad f(t) = a_1 t + a_2.$$

This solution for the time function, which corresponds to a uniform rigid body motion, is clearly just a reflection of Newton's first law for this unloaded structural system. Since this uniform motion is wholly unimportant to a study of vibratory motions and their associated stresses, it receives no further attention. When the right-hand side of the differential equation for $f(t)$ is equal to the negative of ω^2, the resulting ordinary differential equation is

$$\ddot{f}(t) + \omega^2 f(t) = 0.$$

From the previous chapter, this undamped equation has a solution that can be written as either

$$\begin{aligned} f(t) &= C_0 \sin(\omega t + \psi) \\ \text{or} \quad f(t) &= A_0 \sin \omega t + B_0 \cos \omega t, \end{aligned} \tag{6.3a}$$

where either form contains two constants of integration. Therefore, just like the single-DOF m, k system, the only possible orderly, nontrivial, time history for an undamped, unforced, multidegree of freedom, $[M]$, $[K]$ vibratory system is one that involves a sine function and a cosine function or their combination. Note, moreover, that when substituting the above solution for $f(t)$ into the original trial solution $\{q(t)\} = \{A\} f(t)$, the above unknown constants of integration C_0, or A_0 and B_0, in the above solution for $f(t)$, are simply absorbed into the vector of unknown amplitude constants. That is, for example, after substituting the second of the above solutions for $f(t)$ into the trial solution $\{q\} = \{A\} f(t)$ yields the following partial solution for the DOF vector:

$$\{q(t)\} = \{A\} \sin \omega t + \{B\} \cos \omega t, \tag{6.3b}$$

where $\{A\}$ and $\{B\}$ are the two different column vectors of unknown amplitudes associated with the two different linearly independent solutions, $\sin \omega t$ and $\cos \omega t$. The above equation has $2N$ constants of integration. This is the proper number because the original matrix differential equation represents N individual second-order differential equations. If the alternate phase angle form of the solution were chosen here, the equation corresponding to Eq. (6.3b) would require a different phase angle be associated with each amplitude. The more convenient form of these equivalent solutions will vary from use to use.

The partial solutions represented by the sine and cosine functions that involve a yet to be determined frequency or frequencies, ω, can be further explored. The substitution of that partial solution, Eq. (6.3b), into $[M]\{\ddot{q}\} + [K]\{q\} = \{0\}$, yields the result

$$-\omega^2[M]\{A\} \sin \omega t - \omega^2[M]\{B\} \cos \omega t + [K]\{A\} \sin \omega t + [K]\{B\} \cos \omega t = \{0\}$$

or

$$\left(-\omega^2[M]\{A\} + [K]\{A\}\right) \sin \omega t + \left(-\omega^2[M]\{B\} + [K]\{B\}\right) \cos \omega t = \{0\}.$$

Since the sine function and the cosine function are linearly independent, each of the above expressions in parantheses is individually zero. Further, note that the equation involving $\{B\}$ is exactly the same as the equation involving $\{A\}$. Therefore, it is only necessary to consider, say, the former equation that can be rewritten as

$$[\underline{K} - \omega^2 \underline{M}]\{A\} = \{0\}, \tag{6.4}$$

where the temporary underlines in the above square matrix merely emphasizes the fact that the M, K symbols are the matrices in that mixture of matrix and scalar terms within the matrix brackets.

Consider Eq. (6.4) from the point of view of solving N simultaneous equations for the N unknown amplitudes A_i. The solution of these N homogeneous equations $[K - \omega^2 M]\{A\} = \{0\}$ can be approached from more than one viewpoint. For example, consider multiplying this matrix equation by the inverse of the coefficient matrix so as to obtain

$$\{A\} = [\underline{K} - \omega^2 \underline{M}]^{-1}\{0\} = \{0\}.$$

This result means that all the amplitudes of the motion must always be zero whenever it is possible to construct the inverse of the above coefficient matrix. Zero values for the amplitudes are indeed a solution to Eq. (6.2a). As before, the zero value solution for the amplitudes of motion is called the *trivial solution* because if the amplitudes are zero, then there is no dynamic motion, vibratory or otherwise. Hence the trivial solution is of no interest here. The important thing to note is that the only way the zero solution can be avoided is to insist that this M, K coefficient matrix of the amplitude vector not be invertible, that is to insist that this coefficient matrix be "singular."

The very same conclusion that this coefficient matrix of the amplitude vector must be singular can be reached by solving the amplitude equation, Eq. (6.4), by means of Cramer's rule. By Cramer's rule (see, for example, Ref. [6.1]), the solution for any one of the individual amplitudes A_i can be written as the ratio of two determinants. The denominator is simply the determinant of the coefficient matrix. The numerator is also the determinant of the coefficient matrix, but here the ith column (i.e., the column corresponding to the unknown being determined, A_i) of the numerator is replaced by the right-hand side vector. In this case the right-hand side column is a column of zeroes. This column replacement using the zero right-hand side causes each and every numerator determinant to have a zero value. Therefore, unless the determinant in the denominator is also zero valued, leading to the undefined ratio 0/0, then the only possibility for the amplitudes is the trivial solution. Therefore, again, to avoid the trivial solution, the determinant of the square coefficient matrix is required to be zero, making the matrix of the determinant singular.

There is no difficulty requiring that the coefficient matrix $[K - \omega^2 M]$ be singular, which again is the same as requiring its determinant $|K - \omega^2 M|$ to equal zero. This is so because the unknown ω^2 term appearing in this determinant equation has, hitherto this point, been required only to be a positive real number. This weak constraint, discussed again in Section 6.5, does allow ω^2 to take on whatever value or values are necessary to render the determinant equal to zero. Consider the equation

$|K - \omega^2 M| = 0$. Recall that the expansion of any determinant of size $N \times N$ involves, along with a plus one or minus one factor, a sum of the $N!$ products of N terms composed of one entry of the first column, an entry of the second column in a different row, an entry of the third column in a still different row, and so on, down to the one possible term in the Nth column.[3] Applying this aspect of the expansion rule to the most general form of this determinant equation

$$\begin{vmatrix} K_{1,1} - \omega^2 M_{1,1} & K_{1,2} - \omega^2 M_{1,2} & K_{1,3} - \omega^2 M_{1,3} & \dots \\ K_{2,1} - \omega^2 M_{2,1} & K_{2,2} - \omega^2 M_{2,2} & K_{2,3} - \omega^2 M_{2,3} & \dots \\ K_{3,1} - \omega^2 M_{3,1} & K_{3,2} - \omega^2 M_{3,2} & K_{3,3} - \omega^2 M_{3,3} & \dots \\ \dots & \dots & \dots & \dots \end{vmatrix} = 0$$

leads to the conclusion that the expansion of the Nth order determinant leads, in terms of ω^2, to an Nth order polynomial equation, called the *characteristic equation*. From the theorems of algebra, the next conclusion is that there are N roots (not necessarily distinct) for that polynomial equation; that is, there are no more or less than N solutions for ω^2. Of course, the solutions for ω itself, from the known nonnegative values of ω^2, produces pairs of positive and negative values that are otherwise equal. Since the negative solutions for ω contain no information that is not available from the positive solutions, these negative values are usually ignored.[4] There is a standard convention that must be honored with regard to the positive solutions for ω. These solutions must be ordered (i.e., subscripted) so that the first frequency is the lowest valued frequency solution (called the *fundamental frequency*), and the second frequency is the second lowest value of ω, and so on. Later developments depend on this ordering.

In summary, the above development using a determinant equation to obtain a solution for the natural frequencies of an $[K]$, $[M]$ system shows that an N-DOF system has N natural frequencies. All these natural frequencies are characterized by real, positive numbers that are to be ordered by increasing size. The following three examples illustrate such solutions.

EXAMPLE 6.1 When the triple pendulum of Exercise 2.4 and Figure 2.15 is not subjected to either damping or an external loading, it has the following matrix equation of motion, where $\beta = g/L = \alpha^2 k/m$

$$\begin{bmatrix} 1 & & \\ & 1 & \\ & & 1 \end{bmatrix} \begin{Bmatrix} \ddot{\theta}_1 \\ \ddot{\theta}_2 \\ \ddot{\theta}_3 \end{Bmatrix} + \beta \begin{bmatrix} 2 & -1 & 0 \\ -1 & 3 & -1 \\ 0 & -1 & 2 \end{bmatrix} \begin{Bmatrix} \theta_1 \\ \theta_2 \\ \theta_3 \end{Bmatrix} = \begin{Bmatrix} 0 \\ 0 \\ 0 \end{Bmatrix}.$$

Use the determinant method to calculate the natural frequencies of this there-DOF system.

[3] The necessity of $N!$ products of N terms makes large determinants costly to evaluate, usually much too costly relative to other approaches.

[4] There is the exception that occurs when two-sided Fourier transforms are used for probabilistic estimates for the dynamics of a structure. The use of mathematical transforms is unnecessary for the purposes of this textbook.

SOLUTION The first step toward writing the determinant equation that determines the natural frequencies of the system is to recall that the time history of the free vibratory motion is sinusoidal. That is, because

$$\{\theta\} = \{A\}\sin(\omega t + \phi) \quad \therefore\ \{\ddot{\theta}\} = -\omega^2\{\theta\} = -\omega^2\{A\}\sin(\omega t + \phi).$$

After canceling the $\sin(\omega t + \phi)$ terms that are common to the inertia and elastic forces, the equation of motion becomes

$$-\omega^2\begin{bmatrix} 1 & & \\ & 1 & \\ & & 1 \end{bmatrix}\begin{Bmatrix} A_1 \\ A_2 \\ A_3 \end{Bmatrix} + \beta\begin{bmatrix} 2 & -1 & 0 \\ -1 & 3 & -1 \\ 0 & -1 & 2 \end{bmatrix}\begin{Bmatrix} A_1 \\ A_2 \\ A_3 \end{Bmatrix} = \begin{Bmatrix} 0 \\ 0 \\ 0 \end{Bmatrix}.$$

Combining the left-hand side terms by factoring the amplitude vector, and requiring that the determinant of the square coefficient matrix be zero, leads to

$$\begin{vmatrix} 2\beta - \omega^2 & -\beta & 0 \\ -\beta & 3\beta - \omega^2 & -\beta \\ 0 & -\beta & 2\beta - \omega^2 \end{vmatrix} = 0.$$

The expansion of the above determinant yields the characteristic equation

$$(2\beta - \omega^2)^2(3\beta - \omega^2) - (2\beta - \omega^2)\beta^2 - (2\beta - \omega^2)\beta^2 = 0$$

$$\text{or}\quad (\omega^2 - 2\beta)(\omega^4 - 5\beta\omega^2 + 4\beta^2) = 0.$$

Noticing the unusual circumstance that there is a factor in the above expansion allows a change in the solution process from that of solving for the roots of a cubic polynomial equation to that of solving a quadratic equation and a linear equation. These three roots are immediately evident as

$$\omega_1^2 = \beta \qquad \omega_2^2 = 2\beta \qquad \omega_3^2 = 4\beta$$

$$\text{or}\quad \omega_1 = \sqrt{\beta} \qquad \omega_2 = \sqrt{2\beta} \qquad \omega_3 = 2\sqrt{\beta}.$$

Note again that these roots are ordered by increasing size. Thus substituting these three sets of solutions for frequencies and relative amplitudes into the shorter phase angle form of Eq. (6.3b), the total (undamped) free vibration solution for this simple pendulum system can be written as

$$\begin{Bmatrix} \theta_1(t) \\ \theta_2(t) \\ \theta_3(t) \end{Bmatrix} = \begin{Bmatrix} A_1 \\ A_2 \\ A_3 \end{Bmatrix}^{(1)} \sin(\sqrt{\beta}\, t + \psi_1) + \begin{Bmatrix} A_1 \\ A_2 \\ A_3 \end{Bmatrix}^{(2)} \sin(\sqrt{2\beta}\, t + \psi_2)$$

$$+ \begin{Bmatrix} A_1 \\ A_2 \\ A_3 \end{Bmatrix}^{(3)} \sin(2\sqrt{\beta}\, t + \psi_3).$$

As demonstrated in Example 6.4, the values of six of the nine amplitude factors can be determined from an extension of this calculation discussed in the next section.

In later caluclations, it is demonstrated that three remaining amplitude factors and the three phase angles can be determined by use of the three initial deflections and the three initial velocities. Hence, soon, it will be seen that specific values can be assigned to all presently unspecified quantities. ★

EXAMPLE 6.2 Use the determinant method to calculate the natural frequencies of the three-DOF beam and discrete mass system whose mass and stiffness matrices are as follows:

$$[M] = m\begin{bmatrix} 0.50 & & \\ & 0.50 & \\ & & 1.00 \end{bmatrix} \qquad [K] = k\begin{bmatrix} 9.00 & -3.00 & -2.00 \\ -3.00 & 3.00 & 0 \\ -2.00 & 0 & 2.00 \end{bmatrix}.$$

SOLUTION Immediately write the determinant equation, $|K - \omega^2 M| = 0$, whose solution yields the values of the natural frequencies squared. In this case the characteristic equation, with $\lambda^2 = m\omega^2/k$, is

$$\begin{vmatrix} 9 - 0.5\lambda^2 & -3 & -2 \\ -3 & 3 - 0.5\lambda^2 & 0 \\ -2 & 0 & 2 - \lambda^2 \end{vmatrix} = 0.$$

Expanding the determinant yields the following polynomial equation:

$$24 - 28\lambda^2 + 6.5\lambda^4 - 0.25\lambda^6 = 0$$

$$\text{or} \quad x^3 - 26x^2 + 112x - 96 = 0,$$

where $x = \lambda^2$. In this case it is necessary to solve for the roots of a cubic polynomial. This is a simple matter if appropriate software is available, and it is not at all difficult when a hand-held calculator is available.

There are many, many methods for calculating the roots of a polynomial equation. A straightforward method that possesses considerable generality is the Newton–Raphson method. The Newton–Raphson process is an iterative approach to finding a root r of an equation of the form $F(x) = 0$, meaning that a first guess x_1 for the answer $x = r$ is used as the basis for calculating an improved second guess x_2 for the answer r. The answer r is clearly obtained when the succession of guesses converge, within a specified number of significant figures. The Newton–Raphson method, from Appendix I of Ref. [6.2], is what is called a "tangent method." That is, Newton–Raphson uses $F'(x)$ to determine the value of the next guess. The resulting formula for this process is stated below.

The first guess for the Newton–Raphson iterative approach can be based on a rough sketch of the polynomial or other functional equation. In this case, just by plotting the values of the above polynomial at $x = 0$ and $x = 1$ shows that there is one root slightly beyond $x = 1$. Thus choose the value $x_1 = 1.1$ as a first guess for the root slightly past one. Then the Newton–Raphson iterative procedure can be written in terms of the first guess x_1 and the better, second, estimated value x_2 as

$$x_2 = x_1 - \frac{F(x_1)}{F'(x_1)}.$$

Applying this iterative formula yields the following results:

$$\begin{aligned} x_2 &= 1.1 - \frac{(1.1)^3 - 26(1.1)^2 + 112(1.1) - 96}{3(1.1)^2 - 52(1.1) + 112} \\ &= 1.1 - \frac{-2.929}{58.43} = 1.150 \\ x_3 &= 1.150 - \frac{(1.150)^3 - 26(1.150)^2 + 112(1.150) - 96}{3(1.150)^2 - 51(1.150) + 112} \\ &= 1.150 - \frac{-0.64125}{56.1675} = 1.15114. \end{aligned}$$

Continuing with the above iteration procedure shows that the converged value of this root, to six significant figures (at least three more than is justified by the accuracy of the input or the accuracy of strength of materials beam theory), is $x = 1.15114$. The remaining two roots (4.0 and 20.8489) can be determined by either making another guess for another root and preceding as above or, say, using synthetic division to eliminate this first found root from the original cubic (the reason for the need for additional number of significant figures in the first solution) and thereby obtain a quadratic equation to be solved by use of the quadratic formula for the remaining two roots. Once all three positive, real roots have been determined, their square roots are arranged from lowest to highest. Then the natural frequencies can be presented as

$$\omega_1 = 1.07291\sqrt{\frac{k}{m}} = 1.07\sqrt{\frac{k}{m}} \qquad \omega_2 = 2.00\sqrt{\frac{k}{m}}$$

and

$$\omega_3 = 4.56606\sqrt{\frac{k}{m}} = 4.57\sqrt{\frac{k}{m}}. \qquad \bigstar$$

EXAMPLE 6.3 Determine the natural frequencies of the cantilevered beam structure shown in Figure 3.5 and discussed in Example 3.4. For ease of calculation, adjust the entries in the mass matrix and the entries in the stiffness matrix so that all the entries in any one matrix have the same units. This can be accomplished by, for example, (i) dividing each moment row (rows 2, 3, 5, and 6 of the matrix equations of motion) by the offset distance e; and (ii) in both the acceleration and deflection vectors, replace θ and ϕ, respectively, with $e\theta$ and $e\phi$, which to maintain the equality, means dividing columns 2, 3, 5, and 6 of both the mass and stiffness matrices by the quantity e. Furthermore, to simplify the algebra by having numbers rather than algebraic symbols (the common situation for engineering applications), let $H_x/e^2 = H_y/e^2 = m$, $L/e = 5$, $\beta = 0.2$, and $EI_0/L^3 = k$. This low value of β means that the beam is torsionally weak. With these simplifications, the matrix equation of motion is

$$m\begin{bmatrix} 2 & & -2 & & & \\ & 2 & & & & \\ -2 & & 4 & & & \\ & & & 1 & & 1 \\ & & & & 1 & \\ & & & 1 & & 2 \end{bmatrix}\begin{Bmatrix} \ddot{w}_1 \\ e\ddot{\theta}_1 \\ e\ddot{\phi}_1 \\ \ddot{w}_2 \\ e\ddot{\theta}_2 \\ e\ddot{\phi}_2 \end{Bmatrix}$$

$$+k\begin{bmatrix} 36 & -30 & & -12 & 30 & \\ -30 & 300 & & -30 & 50 & \\ & & 15 & & & -5 \\ -12 & -30 & & 12 & -30 & \\ 30 & 50 & & -30 & 100 & \\ & & -5 & & & 5 \end{bmatrix}\begin{Bmatrix} w_1 \\ e\theta_1 \\ e\phi_1 \\ w_2 \\ e\theta_2 \\ e\phi_2 \end{Bmatrix} = \begin{Bmatrix} 0 \\ 0 \\ 0 \\ 0 \\ 0 \\ 0 \end{Bmatrix}.$$

SOLUTION Again the free vibration case is one of harmonic motion. Thus the acceleration vector is equal to the negative of the displacement vector multiplied by the square of the natural frequency. Making this substitution, and again defining $\lambda^2 = m\omega^2/k$, the determinant that yields the six natural frequencies is

$$\begin{vmatrix} (36-2\lambda^2) & -30 & 2\lambda^2 & -12 & 30 & 0 \\ -30 & (300-2\lambda^2) & 0 & -30 & 50 & 0 \\ 2\lambda^2 & 0 & (15-4\lambda^2) & 0 & 0 & -5 \\ -12 & -30 & 0 & (12-\lambda^2) & -30 & -\lambda^2 \\ 30 & 50 & 0 & -30 & (100-\lambda^2) & 0 \\ 0 & 0 & -5 & -\lambda^2 & 0 & (5-2\lambda^2) \end{vmatrix} = 0.$$

Clearly the expansion of this symmetric, 6×6 determinant by hand is a nightmare. Indeed, with normal mathematical skills, it is virtually impossible to do correctly by hand without substantial effort. (A more practical hand calculation method for a problem this size is discussed later.) Again, with $\lambda^2 = m\omega^2/k$ and the use of the computer program Mathematica [6.3], the expanded determinant equation is

$$8\lambda^{12} - 2,580\lambda^{10} + 232,328\lambda^8 - 5,289,300\lambda^6 + 35,253,800\lambda^4$$
$$- 52,800,000\lambda^2 + 18,000,000 = 0.$$

Before continuing with this polynomial equation, note that counting the sign changes from one coefficient to the next yields six sign changes. Thus Descartes' sign change test for positive roots offers, as expected, the possibility of six positive, λ^2 roots for this polynomial equation. In other words, Descartes' rule allows for a quick (but weak) check on the accuracy of the determinant expansion in that the number of sign changes must equal the number of DOF for the system. The six λ^2 roots of the above polynomial equation, from Mathematica, are 0.489248, 1.42584, 8.15701, 19.6474, 110.271, and 182.51. Thus the six natural frequencies (properly ordered) are

$$\omega_1 = 0.699\sqrt{\frac{k}{m}} \quad \omega_2 = 1.19\sqrt{\frac{k}{m}} \quad \omega_3 = 2.86\sqrt{\frac{k}{m}}$$

$$\omega_4 = 4.43\sqrt{\frac{k}{m}} \quad \omega_5 = 10.5\sqrt{\frac{k}{m}} \quad \omega_6 = 13.5\sqrt{\frac{k}{m}}$$

★

Other natural frequency calculations using the determinant method are discussed in the exercises.

6.3 Mode Shapes by Use of the Determinant Method

There is another important information set that can be obtained from the determinant method. That information is the deflection pattern that is associated with each natural frequency. The deflection pattern associated with the first natural frequency is called the *first mode shape* and that associated with the second natural frequency is called the *second mode shape*, and so on. The term *n*th *mode* is also used. That term refers to two pieces of information, the *n*th natural frequency and the *n*th mode shape. Under very general conditions, as is soon seen, these mode shapes are always distinct (i.e., in mathematical terms the mode shapes are linearly independent vectors), they always are useful for analysis purposes, and they always are useful for vibration testing purposes.

In the case of the determinant method, each mode shape is calculated separately. Return to the general N-DOF free vibration matrix equation, Eq. (6.4).

$$[\underline{K} - \omega^2 \underline{M}]\{A\} = \{0\}.$$

Note again that this matrix equation can be viewed as n simultaneous, linear, algebraic, homogeneous equations in terms of the n unknown amplitudes $\{A\}$. Further recall that (i) there is no solution to these simultaneous equations other than the trivial solution $A_i = 0$ for arbitrary values of ω, and (ii) when the values of ω are those of one of the discrete natural frequencies, then the coefficient matrix for the vector $\{A\}$ is singular. In other words, when substituting any of the natural frequency values back into the coefficient matrix, the determinant of the coefficient matrix takes on a zero value. From the rules that apply to all determinants, a zero value of the determinant implies that one or more rows of that determinant are obtainable as the sum of other rows where each of those other rows is multiplied by some factor. In the case of the $|K - \omega^2 M|$ determinant, there is only one such dependent row when there are no repeated roots to the characteristic equation. In other words, viewed from the point of view that the matrix equation, Eq. (6.4), is a set of n simultaneous equations, it may be concluded that one and only one of those simultaneous equations is dependent on the other $n - 1$ equations. Since that dependent equation is extraneous, it may be eliminated.

As a reminder of exactly what dependent equations are, consider the following three linear, algebraic equations: (i) $2x + 3y - z = 13$, (ii) $4x - 2y + 2z = 6$, and (iii) $2x - 5y + 3z = -7$. Note that the third of these equations is equal to the first subtracted from the second or the first is equal to the third subtracted from the second, and so on. Therefore one of these three equations, the choice is arbitrary, can be selected as the equation that does not contain any information that is not fully present in the other equations. Thus one, any one, of these three algebraic equations can be discarded without the slightest loss of information. What remains is two equations in the three unknowns x, y, and z. Therefore there is no unique solution for the three unknowns. The most that can be done is to determine the values of two of the unknowns, say x and z in terms of the third, y.

In the specific case of the N simultaneous free vibration equations $[K - \omega^2 M]\{A\} = \{0\}$, the one equation (row) to discard when doing hand calculations is usually the most complicated row; that is, usually the row that has the most nonzero entries. The next step is to solve for $N - 1$ unknown amplitudes in terms of an Nth unknown amplitude. To simplify this solution procedure for both hand and machine calculations, choose any one of the A_i terms to have, temporarily, a unit value. Then solve the selected $N - 1$ equations for the remaining $N - 1$ unknowns. There will always be a unique solution for these values when there are no repeated roots[5] to the characteristic equation. Some of the A_i solutions generally will have values greater than 1, whereas others will have values less than 1. The traditional way of presenting the solution data for the amplitudes associated with a particular natural frequency is to divide all the amplitudes by the largest value (positive or negative) in the solution set so that the largest resulting amplitude has the value 1.0. The following example problems illustrate this process.

EXAMPLE 6.4 Calculate the three mode shapes of the triple-pendulum problem of Example 5.1, where again $\beta = g/L = \alpha^2 k/m$.

SOLUTION From that example solution, the amplitude equation for this three-DOF system is

$$\begin{bmatrix} 2\beta - \omega^2 & -\beta & 0 \\ -\beta & (3\beta - \omega^2) & -\beta \\ 0 & -\beta & (2\beta - \omega^2) \end{bmatrix} \begin{Bmatrix} A_1 \\ A_2 \\ A_3 \end{Bmatrix} = \begin{Bmatrix} 0 \\ 0 \\ 0 \end{Bmatrix}.$$

As discussed above, one of these three equations is superfluous. For the sake of convenience, eliminate the middle equation. Now, inserting the previously calculated fundamental natural frequency solution $\omega_1^2 = \beta$ into the first and third equations, obtain the first mode shape by setting $A_3 = 1.0$ and solving those two remaining simultaneous equations that now are

$$(2\beta - \beta)A_1 - \beta A_2 = 0$$
$$-\beta A_2 + (2\beta - \beta)(1.0) = 0.$$

[5] The case of repeated roots is discussed in the next section.

The solutions to these two equations are $A_1 = A_2 = 1.0$. Following the same procedure for the second mode, that is, substituting $\omega_2^2 = 2\beta$ and $A_3 = 1.0$, the matrix equation is

$$\begin{bmatrix} 0 & -\beta & 0 \\ -\beta & \beta & -\beta \\ 0 & -\beta & 0 \end{bmatrix} \begin{Bmatrix} A_1 \\ A_2 \\ 1 \end{Bmatrix} = \begin{Bmatrix} 0 \\ 0 \\ 0 \end{Bmatrix}.$$

The above equations illustrate the unusual circumstance where two of the remaining equations are exactly the same. In such cases, it is of course necessary to eliminate one of the two identical equations to facilitate the solution for all the amplitudes. Throwing away the third equation leads to the result that $A_1 = -1.0$, $A_2 = 0$, and, of course, $A_3 = +1.0$.

For the third mode shape, with the square of the third natural frequency being 4β, after deleting the second equation, the solution for the mode shape is $A_1 = 1.0$, $A_2 = -2.0$, and $A_3 = 1.0$. Recall that the traditional manner of presentation of mode shape data is to have the largest amplitude value take on the value +1.0. Therefore, divide the above third mode shape data by –2. Then the traditional listing of all three mode shapes is

First mode:	1.0	1.0	1.0
Second mode:	–1.0	0.0	1.0
Third mode:	–0.5	1.0	–0.5

Thus the complete solution for the original matrix equation of motion $[M]\{\ddot{q}\} + [K]\{q\} = \{0\}$, in terms of six constants of integration, can be written as

$$\begin{Bmatrix} \theta_1(t) \\ \theta_2(t) \\ \theta_3(t) \end{Bmatrix} = a_1 \begin{Bmatrix} +1 \\ +1 \\ +1 \end{Bmatrix} \sin\left(\sqrt{\beta}t + \psi_1\right) + a_2 \begin{Bmatrix} -1 \\ 0 \\ +1 \end{Bmatrix} \sin\left(\sqrt{2\beta}t + \psi_2\right)$$

$$+ a_3 \begin{Bmatrix} -1/2 \\ +1 \\ -1/2 \end{Bmatrix} \sin\left(2\sqrt{\beta}t + \psi_3\right). \tag{6.5}$$

Figure 6.1(a) is a drawing that attempts to show the physical meaning of the above (relative) amplitude data. In the first mode, all three pendulums swing in unison, at a frequency equal to the first natural frequency, without stretching or squeezing the springs. The all positive numerical values of the amplitudes of the first mode are described in physical terms as being "in phase." That is, at any instant of time, all the pendulums have swung to the right (where the deflections are all positive), or all the pendulums have swung to the left (where all deflections are negative). Since, in this first mode, the springs are not involved in driving the system motion (and are not involved in the total potential energy), the frequency of the vibration is just that of any one of the pendulums vibrating by itself, which is $\sqrt{g/L}$.

In the second mode, the first and third pendulums are out of phase; that is, when one has a positive amplitude, the other has a negative amplitude, and vice versa. The middle amplitude, as stated above, has a zero value at all times. In the third mode, the outer pendulums are in phase, whereas the middle pendulum is out of phase. Note that, other than the support points, in the first mode there are no points (mass

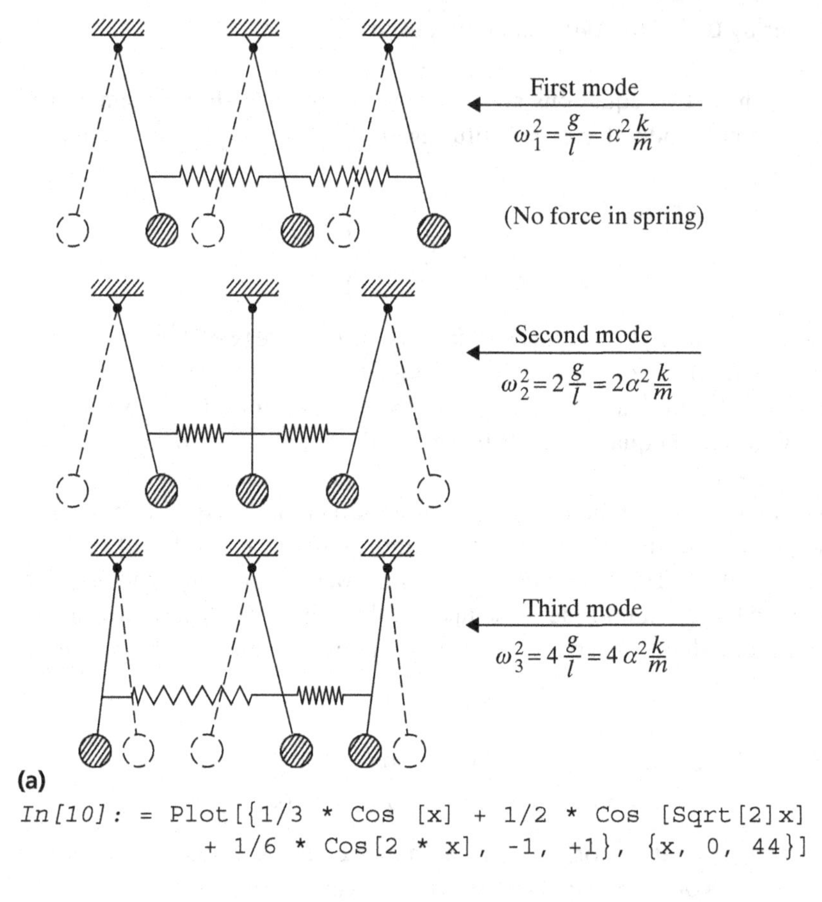


```
In[10]: = Plot[{1/3 * Cos [x] + 1/2 * Cos [Sqrt[2]x]
            + 1/6 * Cos[2 * x], -1, +1}, {x, 0, 44}]
```

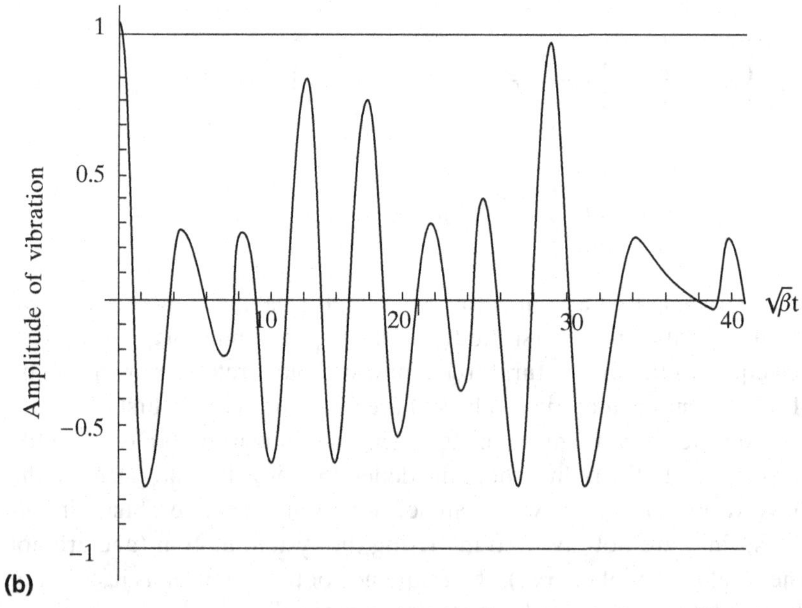


```
Out[10] = -Graphics-
```

Figure 6.1. (a) Model diagrams for the triple-pendulum problem of Examples 6.1 and 6.4. (b) Triple pendulum response for left-most pendulum when only it is given an initial deflection and all pendulums have an initial zero velocity.

points or massless points) in the system that have a zero amplitude. In the second mode, all points along the axis of the second pendulum have a zero amplitude. In terms of left–right motion, the middle pendulum counts as a single point. In the third mode, at the outer one-third points of the spring lengths are the points that have zero amplitude. These points of zero motion are called *node points*. (Be sure to understand that these node points have no relation to the node points of a finite element analysis and clearly distinguish between node and mode.) The above unit increase in the number of node points with the unit increase of the modal identification number is a general occurrence. (The proof of this assertion depends on a later discussion of the mode shapes as general orthogonal vectors.) Finally, recall that these amplitudes are merely the peak values of the harmonic motion of the system as each part of the system vibrates at the natural frequency corresponding to that mode.

Reconsider Eq. (6.5). Note that this equation is a complete description of the motion of the system. It is that because, first of all, it identically satisfies the original *M,K* matrix differential equation, and hence is a solution to that matrix equation. Furthermore, it is a complete solution to the three ordinary, second-order, homogeneous differential equations that comprise the *M,K* matrix differential equation because it contains the correct number of constants of integration, six. Those constants of integration are, of course, a_1, a_2, a_3, ψ_1, ψ_2, and ψ_3. Hence as the complete solution, it fully describes the undamped triple pendulum vibratory motion within the limits of the mathematical model. With the completeness of the solution established, the point can be made that this solution, Eq. (6.5), mathematically validates the earlier observation that when this force free, undamped, system is set in motion by means of arbitrarily chosen initial conditions, the motion appears confusing because it lacks any discernable pattern. Consider, as an example, the following set of initial conditions. Let the left-hand pendulum have a positive initial deflection of magnitude Θ, while the other two pendulums are restrained to have zero initial deflections. Let the initial velocities of all three pendulums be zero. Using the procedure outlined in the next chapter, or simply solving six simultaneous equations, it can be shown that the motion of the left-hand pendulum is

$$\theta_1(t) = \frac{1}{3}\Theta\cos(\sqrt{\beta}t) + \frac{1}{2}\Theta\cos(\sqrt{2\beta}t) + \frac{1}{6}\Theta\cos(2\sqrt{\beta}t).$$

This three-term sum is not like a truncated Fourier cosine series because the frequencies are not integer multiples of the fundamental frequency. The presence of the square root of 2 in the second term guarantees that this sum in not periodic. In fact, because this solution never repeats itself, there is no pattern to this motion discernable by an observer. See Figure 6.1(b). However, if the initial conditions are not chosen arbitrarily but rather are selected to conform with the deflection amplitudes of one of the natural modes, then patterns to the motion are easily observed. For example, as can be shown by either solving simultaneous equations or using the procedure explained in the next chapter, if the selected initial conditions are that all the undamped pendulums are given a positive initial deflection Θ, and a zero initial velocity, then each pendulum will vibrate in phase at a circular frequency equal to $\sqrt{\beta}$. If the initial conditions are such that the left-hand pendulum is given an initial deflection $+\Theta$, while the right-hand pendulum is given an initial deflection of $-\Theta$, then the middle pendulum will remain stationary, while the two outboard pendulums vibrate out of phase at a circular frequency of $\sqrt{2\beta}$. ★

EXAMPLE 6.5 Calculate the first two mode shapes of the three degree of freedom structure described in Example 6.2.

SOLUTION In this case, eliminate the first of the three dependent, free vibration equations $[K - \lambda^2 M]\{q\} = \{0\}$. For the first mode shape (i) substitute the solution for the first nondimensional frequency that is the square of the quantity 1.07291 and (ii) begin by setting the third amplitude equal to the arbitrary value 1.0. (If the first or second amplitude happens to have a larger absolute value, then all amplitudes will be divided by the value of that amplitude so that the largest value in the list of amplitudes will be 1.0.) Hence, for the first mode, the two equations to be solved for the first and second amplitudes of the first mode are

$$-3A_1 + 2.42443A_2 = 0 \quad \text{and} \quad -2A_1 + 0.84886 = 0.$$

The solution to these equations are $A_1 = 0.42443$ and $A_2 = 0.52519$. Thus the first mode shape, or first modal vector, is listed as

$$\lfloor q^{(1)} \rfloor = \lfloor 0.42443 \quad 0.52519 \quad 1.0 \rfloor.$$

This process is repeated for the second mode where λ^2 is the integer value 4. Again guessing that the largest amplitude is the third amplitude with a value 1.0, the equations to be solved are

$$-3A_1 + A_2 = 0 \quad \text{and} \quad -2A_1 = 2,$$

which has the solutions $A_1 = -1$ and $A_2 = -3$. Thus, contrary to the above guess, it turns out that the second amplitude has the largest magnitude. Therefore, to comply with custom, divide all amplitudes by -3 so that the second-mode amplitudes are listed as

$$\lfloor q^{(2)} \rfloor = \lfloor 0.333333 \quad 1.0 \quad -0.333333 \rfloor.$$

★

EXAMPLE 6.6 Using the determinant method, examine the first three mode shapes of the six-DOF, cantilevered beam of Figure 4.5 and Examples 4.4 and 6.3.

SOLUTION Again, from the fact that there are six DOF, and hence five simultaneous equations to be solved, this problem is not well suited to hand calculations. By using a computer solution, after guessing that the last amplitude (the tip twist DOF) in each of the first three modes is the maximum amplitude, the numerical results for the original order of the DOF are

DOF	First mode	Second mode	Third mode
w_1	1.59782	−0.35953	−0.053885
$e\theta_1$	0.564145	−0.10148	−0.072759
$e\phi_1$	0.263477	0.648112	−0.333507
w_2	5.52709	−0.766033	−1.1826
$e\theta_2$	0.901116	−0.0722405	−0.329078
$e\phi_2$	1.0	1.0	1.0

Clearly the first modal amplitudes need to be divided by 5.52709 to normalize the maximum amplitude at 1.0. Similarly, the third mode needs division by −1.1826. The first mode is described as a "first bending mode" simply because the lateral deflection magnitudes w_1, w_2 are significantly greater than the bending slope and twisting amplitudes. Describing the first mode as the first bending mode is not to say that the beam does not twist as it bends up and down. Indeed, the modal solution does show that all three types of deflections do occur simultaneously in what is called a "coupled motion." That modal name only implies that the lateral deflection amplitudes are distinctly more prominent than the bending slope amplitudes and the twisting amplitudes. Similarly, the second mode is identified as a first twisting mode. The qualifier "first" that is applied in the identification of modes one and two is justified on the basis that the amplitudes of the dominant type of motion are in phase (same sign) with each other. That is, there is no nodal point between the amplitudes of the dominant type of motion. Sketches of these three modes are shown in Figure 6.2. The third mode may be described as just a coupled bending-twisting mode. It is also justified, and perhaps more descriptive, to call the third mode the second twisting mode even though the numerical value of the tip lateral deflection amplitude is slightly greater than that of the twist at the beam tip. This is a reasonable description because, for the same size twist at the beam tip, the deflections of the third mode are not nearly as dominant as they are for the first mode, and there is one twisting node between the two twisting DOF. In either case, the names given to the modes are just loosely descriptive and are not that important. ★

6.4 **Repeated Natural Frequencies**

Although possible, it is unusual for the characteristic equation associated with an actual structure to have repeated roots that, of course, lead to repeated natural frequencies. In the case of the planar beam frame and planar beam grid type structures used for illustration in this textbook, repeated roots are only possible if the structural DOF are minimally coupled, or there is at least a two-way symmetry of the structure and its possible motions. (What is meant by a two-way symmetry will be illustrated later in this section.) As an example of a typical beam frame structure where coupling is not minimal and there is only a single axis of symmetry, consider the free vibration

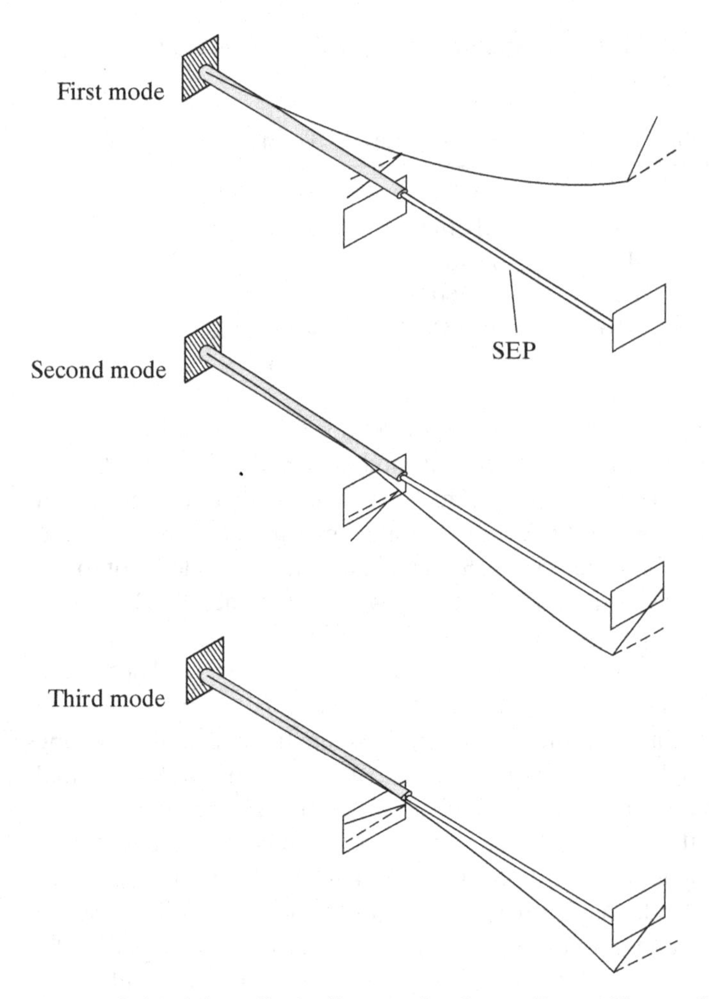

Figure 6.2. Model amplitude diagrams for the cantilevered beam of Figure 4.5.

of the three-story, three-DOF, planar structure shown in Figure 5.22. The upper mass value is αm. Here let the bending stiffness coefficients of the upper columns be βEI. Then, there are no real values of the parameters α, β for which there are repeated roots to this structure's characteristic equation. Again, this is typical of any planar beam frame structure where the beam generalized coordinates are coupled in a normal fashion.

However, it is a simple matter to conceive of this same planar frame extended out of the plane of the paper so as to become a three-dimensional frame structure that has the same appearance and horizontally directed stiffnesses in and out of the plane of the paper (call it the y direction) as it does in the plane of the paper (call it the x direction). Then there will be, in plan view, a square top view and thus a two-way symmetry. Therefore the corresponding side sway frequencies in the x and y directions would be equal. A simpler view of this exact same symmetry where there are equal stiffnesses and masses for two perpendicular motions can be had by

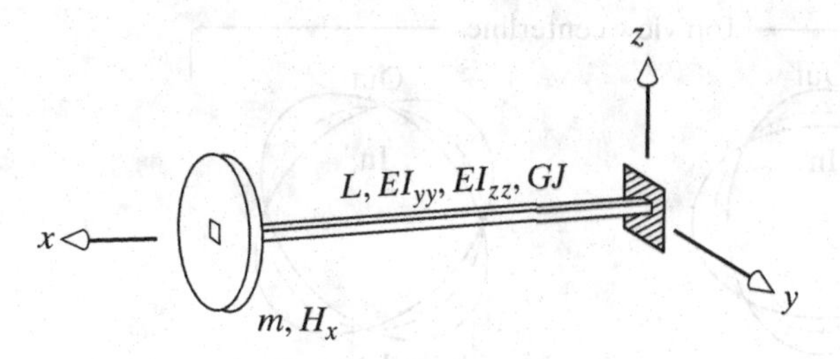

Figure 6.3. Cantilevered beam for discussion of repeated natural frequencies.

considering just a single, cantilevered beam with, say, a square cross section, as shown in Figure 6.3. Although drawn horizontally to save space, let the beam be oriented vertically, and let the beam support a tip mass that is much greater than the mass of the beam. Therefore it would be possible to have a mathematical model where all the system mass is lumped at the beam tip. For simplicity, let the tip mass be connected to the beam by means of a partial ball joint so that the tip mass does not rotate about either the y or z axes, but can twist about the x axis as the beam tip twists. Then, since the axial stiffness is so much greater than the lateral stiffnesses, the three significant system DOF would be the two horizontal, orthogonal beam tip bending deflections and the one twisting rotation. With the cross-sectional centroid and shear centers being coincident, both the mass matrix and stiffness matrix for these three-DOF would be diagonal. The magnitudes of the tip mass, the tip mass (torsional) moment of inertia about the x axis, and the three stiffness coefficients can be arranged so that the respective natural frequencies in bending in the x, z plane, bending in the x, y plane, and twisting about the x axis all have the same value; that is,

$$\omega_{1,2,3} = \sqrt{\frac{3EI_{yy}}{mL^3}} = \sqrt{\frac{3EI_{zz}}{mL^3}} = \sqrt{\frac{GJ}{H_x L}}.$$

Again, the repeated natural frequencies for this cantilevered beam are simple to arrange because the twisting and the two bending motions are physically uncoupled; that is, they do not interact.

Of course, the above cantilevered beam with the same bending natural frequencies in the x, y plane and the x, z plane, and the same natural frequency in twisting about the x axis, is a case of a perfectly built and supported beam with perfectly uniform geometry and material properties. Any actual beam will not be perfect. An interesting phenomena occurs with such a lightly damped, actual beam as a result of its small, unavoidable, imperfections. If, for the above example, the beam tip mass is deflected in the y direction (an initial condition) and released, the beam will begin to vibrate solely in the x, y plane, as would be expected. However, if the vibration persists long enough, the beam will also begin to undergo torsional vibrations and vibrate in the x, z plane as well. In other words, some to the energy of the y-direction vibration will be transferred to a twisting vibration and a z-direction bending vibration and then vice versa. The reasons for this begin with, for example, the fact that the beam tip mass center of gravity cannot be perfectly aligned with the cantilevered beam's

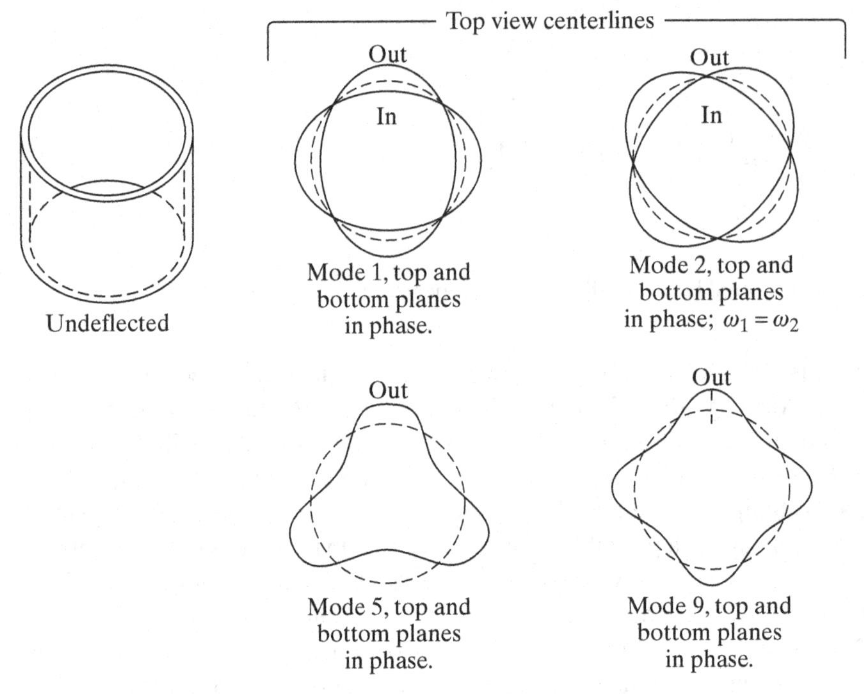

Figure 6.4. Vibratory modes of a short, right circular cylinder. The top view of the undeflected shape is indicated by the dashed line circles.

elastic axis. In general, the mass center will have very small y-direction and z-direction offsets from the elastic axis. In the same fashion as the horizontally offset masses of the beam of Figure 4.5 produce an inertial torque when that beam vibrates laterally, the small imperfection offsets here will eventually produce an inertial torque that stimulates a twisting vibration of the imperfect beam. Then, because of the offsets, the twisting motion produces z-direction inertia forces that cause a z-direction bending of the beam. Another way of saying the above is as follows: because the natural frequency of the initial y-direction bending vibration is nearly the same as the natural frequencies of z-direction bending and twisting, there is a resonance effect where the one vibration drives the other and therefore overcomes the smallness of the imperfections to produce finite amplitudes in horizontal bending and twisting. From an energy viewpoint, because the lightly damped beam is undergoing a force free vibration, the energy content of the beam, kinetic and strain, is slowly diminishing with time. The initial y-direction vibration contained all the system energy. However, through the coupling induced by the imperfections, this total energy content is soon shared by all three types of vibration in constantly changing ratios, until all the energy is dissipated.

Another, far less trivial, case of two-way symmetry is that of a short,[6] unsupported, circular right cylinder as shown in Figure 6.4. The lower vibratory modes of such a

[6] "Short" so as to make the cylinder very stiff in bending along its central axis, which in turn means that the cylinder only bends in its circumferential direction at its lower numbered natural frequencies.

cylinder are dominated by circumferential bending, with greater bending curvatures (for equal amplitudes) resulting in higher strain energies and higher modal numbers. The figure shows top views of the first two mode shapes that are uniform over the height of the cylinder. It is not difficult to see that this pair of vibrations is quite similar to the above discussed case of the three-dimensional frame with the square cross section first used to argue for the existence of equal natural frequencies for different motions. Hence, for the circular cylinder, the values of the first and second natural frequencies are exactly the same. The third and fourth modes also have the same natural frequencies. Their corresponding mode shapes are similar to the first and second modes. In the first two modes, the top and bottom circles deform exactly the same way at the same time, in the third and fourth modes, the top and bottom deflections are out of phase. That is, although, say the north–south axis of the circle at the top end of the cylinder stretches and the top east–west axis contracts, the opposite occurs for the circle at the bottom of the cylinder in the third and fourth modes. The fifth, sixth, seventh, and eighth modes bear the same relationship among themselves as the first, second, third, and fourth modes, but in this group of modes, rather than an oval shape to the modal amplitudes, there is a trefoil shape. Routine numerical calculations for such a cylinder would require, at the minimum, the use of many plate or shell finite elements, particularly in the circumferential direction. Since calculations in this textbook are limited to the use of beam finite elements, such plate and shell calculations are omitted.

A similar circumstance that is more easily grasped is that of square and circular plates possessing two-way symmetry. For an illustrated discussion of simply supported, uniform, square plate vibration solutions, the reader is referred to Ref. [6.4], Section 4.1. The mode shape diagrams of that reference's Figure 4.3 are particularly interesting in that they well illustrate the variety of mode shapes for repeated natural frequencies that can be obtained for square plates. Those plate modal patterns are similar to those of the square 24-beam grid discussed later in this section as the second of two beam grid problems. This second beam grid problem is selected to roughly mimic the mode shapes of a square plate while remaining within the computational constraints adopted in this textbook. The following two example problems illustrate the computation of repeated roots for beam structures.

EXAMPLE 6.7 Consider the z-direction vibrations of the planar, two-story structure shown in Figure 6.5. Let the beam-columns be attached to the rigid masses by frictionless ball joints so that the rigid masses only rotate about the x axis, but not the y axis, or, of course, the z axis. Thus the rigid connections do not transmit a torque between the beams. Therefore one of the cantilevered beams of the beam grid can vibrate through small amplitudes in the x, z plane without exciting a vibration in the other cantilevered beam.

Let each of the four beams be of length L and stiffness coefficient EI. For the inboard rigid mass, with only the DOF w_1, w_2, let the mass and the mass moment of inertia about the x axis be m and $\frac{1}{4}mL^2$, respectively. Let the mass and mass moment of inertia for the outboard (tip) rigid mass be the same quantities multiplied by the factor α. Let the distance between the points of connection to the beams for the rigid

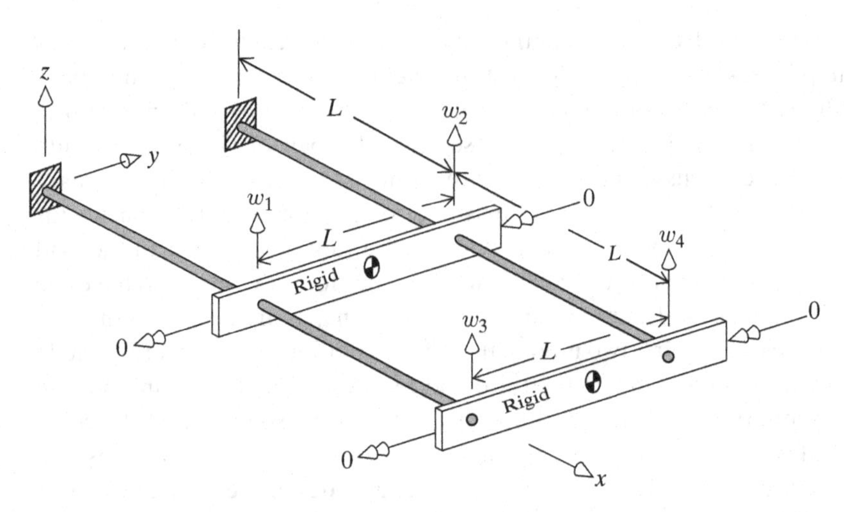

Figure 6.5. Two uncoupled, cantilevered beams constituting one vibratory system with repeated natural frequencies.

masses, that is, the lateral spacing between beams, also be L. Then the kinetic energy of the system is simply

$$T = \frac{1}{2}m\left(\frac{\dot{w}_1 + \dot{w}_2}{2}\right)^2 + \frac{1}{2}\frac{mL^2}{4}\left(\frac{\dot{w}_1 - \dot{w}_2}{L}\right)^2 + \frac{1}{2}\alpha m\left(\frac{\dot{w}_3 + \dot{w}_4}{2}\right)^2 + \frac{1}{2}\frac{\alpha mL^2}{4}\left(\frac{\dot{w}_3 - \dot{w}_4}{L}\right)^2.$$

Therefore the negative of the inertia forces acting on the system masses are

$$[M]\{\ddot{q}\} = \frac{m}{2}\begin{bmatrix} 1 &&&&&&& \\ & 0 &&&&&& \\ && 1 &&&&& \\ &&& 0 &&&& \\ &&&& \alpha &&& \\ &&&&& 0 && \\ &&&&&& \alpha & \\ &&&&&&& 0 \end{bmatrix}\begin{Bmatrix} \ddot{w}_1 \\ \ddot{\theta}_1 \\ \ddot{w}_2 \\ \ddot{\theta}_2 \\ \ddot{w}_3 \\ \ddot{\theta}_3 \\ \ddot{w}_4 \\ \ddot{\theta}_4 \end{Bmatrix}.$$

After assembly of the global stiffness matrix, the negative of the elastic forces acting on the system masses are

$$[K]\{q\} = \frac{EI}{L^3}\begin{bmatrix} 24 &&&& -12 & 6L && \\ & 8L^2 &&& -6L & 2L^2 && \\ && 24 &&&& -12 & 6L \\ &&& 8L^2 &&& -6L & 2L^2 \\ -12 & -6L &&& 12 & -6L && \\ 6L & 2L^2 &&& -6L & 4L^2 && \\ && -12 & -6L &&& 12 & -6L \\ && 6L & 2L^2 &&& -6L & 4L^2 \end{bmatrix}\begin{Bmatrix} w_1 \\ \theta_1 \\ w_2 \\ \theta_2 \\ w_3 \\ \theta_3 \\ w_4 \\ \theta_4 \end{Bmatrix}.$$

The task of solving the equation of motion for the out-of-plane vibrations of this two story planar structure, that is, solving $[M]\{\ddot{q}\} + [K]\{q\} = \{0\}$, can be better understood by first reducing the number of DOF; that is, by eliminating all the DOF for which there are no mass entries, which in this case are the bending slope degrees of freedom. Eliminating all the θs can be accomplished by rearranging the rows and columns of the mass and stiffness matrices and the rows of the acceleration and deflection vectors so that this system equation of motion has the form shown in Eq. (3.3), and below. That is, as discussed previously, rearrange the rows and columns of $[M]\{\ddot{q}\} + [K]\{q\} = \{0\}$ so that this matrix equation can be written in the submatrix form as

$$\begin{bmatrix} M & 0 \\ 0 & 0 \end{bmatrix}\begin{Bmatrix} \ddot{w} \\ \ddot{\theta} \end{Bmatrix} + \begin{bmatrix} K_{ww} & K_{w\theta} \\ K_{w\theta}^T & K_{\theta\theta} \end{bmatrix}\begin{Bmatrix} w \\ \theta \end{Bmatrix} = \begin{Bmatrix} 0 \\ 0 \end{Bmatrix}.$$

Here these various submatrices are

$$[M] = \frac{m}{2}\begin{bmatrix} 1 & & & \\ & 1 & & \\ & & \alpha & \\ & & & \alpha \end{bmatrix} \qquad [K_{ww}] = \frac{EI}{L^3}\begin{bmatrix} 24 & & -12 & \\ & 24 & & -12 \\ -12 & & 12 & \\ & -12 & & 12 \end{bmatrix}$$

$$[K_{w\theta}] = \frac{EI}{L^2}\begin{bmatrix} 0 & & -6 & \\ & 0 & & -6 \\ 6 & & -6 & \\ & 6 & & -6 \end{bmatrix} \qquad [K_{\theta\theta}] = \frac{EI}{L}\begin{bmatrix} 8 & & 2 & \\ & 8 & & 2 \\ 2 & & 4 & \\ & 2 & & 4 \end{bmatrix}$$

After elimination of the θ degrees of freedom, the reduced stiffness matrix, the stiffness matrix for just the lateral deflection DOF is, from Eq. (3.4),

$$[K_r] = \left[K_{ww} - K_{w\theta}K_{\theta\theta}^{-1}K_{w\theta}^t\right].$$

This process is called *static condensation.* The inversion of the theta-theta stiffness submatrix is easily accomplished in this case because there is no mathematical coupling between the odd and even numbered rows and columns. That is, the hand calculation of the inverse is merely a matter of inverting the same 2×2 submatrix twice. Carrying out the Eq. (3.4) calculations leads to the reduced equation of motion in terms of only the lateral deflection DOF, which is

$$\frac{m}{2}\begin{bmatrix} 1 & & & \\ & 1 & & \\ & & \alpha & \\ & & & \alpha \end{bmatrix}\begin{Bmatrix} \ddot{w}_1 \\ \ddot{w}_2 \\ \ddot{w}_3 \\ \ddot{w}_4 \end{Bmatrix} + \frac{EI}{7L^3}\begin{bmatrix} 96 & & -30 & \\ & 96 & & -30 \\ -30 & & 12 & \\ & -30 & & 12 \end{bmatrix}\begin{Bmatrix} w_1 \\ w_2 \\ w_3 \\ w_4 \end{Bmatrix} = \begin{Bmatrix} 0 \\ 0 \\ 0 \\ 0 \end{Bmatrix}.$$

An easy check for the validity of the above stiffness matrix is obtained by using, say, either the unit load method or Castigliano's second theorem to calculate the deflections at the beam center and tip because of z direction, concentrated forces of arbitrary magnitude applied in turn to the beam tip and beam center. These four deflections can be arranged as a 2×2 flexibility matrix for either cantilevered beam of length $2L$. The inverse of this 2×2 flexibility matrix provides the nonzero entries of either the even or odd rows of the above, reduced, stiffness matrix. This flexibility matrix to stiffness matrix check is easy to perform in this case because both of the two cantilevered beams are statically determinate.

Now return to the task of calculating the natural frequencies and mode shapes for this now four-DOF system. Before proceeding to use the determinant method, it is convenient (but not necessary) to assign a numerical value to the parameter α that appears in the mass matrix. Let $\alpha = 0.5$. (Other values of α are considered in the exercises.) Hence let λ be equal to $7mL^3\omega^2/2EI$. Then it is a matter of carefully expanding the 4×4 determinant equation $|K - \lambda M| = 0$. Careful expansion means looking for simplifications that will ease the algebraic burden. One possibility for simplification that suggests itself is letting $p = (96 - \lambda)(12 - 0.5\lambda)$. Then the determinant expansion can be written as $p^2 - 1800p + 810{,}000 = 0$. The roots of this equation are $p = 900{,}900$. Substituting this repeated solution for p into its definition above yields the duplicate equations $\lambda^2 - 120\lambda + 504 = 0$. Therefore, the solutions for λ are 4.3583, 4.3583, 115.64, and 115.64, and the solutions for the natural frequencies are

$$\omega_1 = \omega_2 = 1.116\sqrt{\frac{EI}{mL^3}} \qquad \omega_3 = \omega_4 = 5.748\sqrt{\frac{EI}{mL^3}},$$

where the last two digits in each solution are not to be taken too seriously.

The calculation of the mode shapes is particularly interesting in this repeated root case. Again, because of the lack of mathematical coupling between the even and odd rows of the matrix equation $[K - \lambda M]\{w\} = \{0\}$ to be solved for the mode shapes, it is only necessary to consider the even or odd rows. Choosing, for example, the first and third row, and discarding the third row as redundant, the mode shape equation for the left beam can be written as $(96 - \lambda)w_1 - 30w_3 = 0$. (The mode shape equation for the even number rows, the right-hand beam, is $(96 - \lambda)w_2 - 30w_4 = 0$.) Substituting the first pair of repeated natural frequencies, and then the second pair of repeated natural frequencies, provides the mode shape results

$$\begin{aligned}
\lfloor q^{(1)} \rfloor &= \lfloor 0.3274 \quad 0.000 \quad 1.000 \quad 0.000 \rfloor \\
\lfloor q^{(2)} \rfloor &= \lfloor 0.000 \quad 0.3274 \quad 0.000 \quad 1.000 \rfloor \\
\lfloor q^{(3)} \rfloor &= \lfloor 1.000 \quad 0.000 \quad -0.6547 \quad 0.000 \rfloor \\
\lfloor q^{(4)} \rfloor &= \lfloor 0.000 \quad 1.000 \quad 0.000 \quad -0.6547 \rfloor,
\end{aligned}$$

It is an easy matter to interpret these four mode shapes. In the first and third mode shapes, the left-hand beam is vibrating while the right-hand beam is stationary, and in the second and fourth modes, the reverse is true. These four mode shapes underline the lack of mechanical coupling between the left- and right-hand beams of this structure. Also note that the labeling of the first and second mode shapes could just as well have been reversed, and the same is true for the third and fourth mode shapes. Moreover, the above-listed mode shapes are just one such choice for numerical values. For example, the following choice, which clearly are just combinations of the above mode shapes, also satisfies all the requirements that will soon be discovered to be desirable for all mode shapes

$$\begin{aligned}
\lfloor q^{(1)} \rfloor &= \lfloor \ 0.3274 \quad 0.3274 \quad 1.000 \quad 1.000 \rfloor \\
\lfloor q^{(2)} \rfloor &= \lfloor -0.3274 \quad 0.3274 \quad -1.000 \quad 1.000 \rfloor \\
\lfloor q^{(3)} \rfloor &= \lfloor \ 1.000 \quad 1.000 \quad -0.6547 \quad -0.6547 \rfloor \\
\lfloor q^{(4)} \rfloor &= \lfloor \ 1.000 \quad -1.000 \quad -0.6547 \quad 0.6547 \rfloor.
\end{aligned}$$

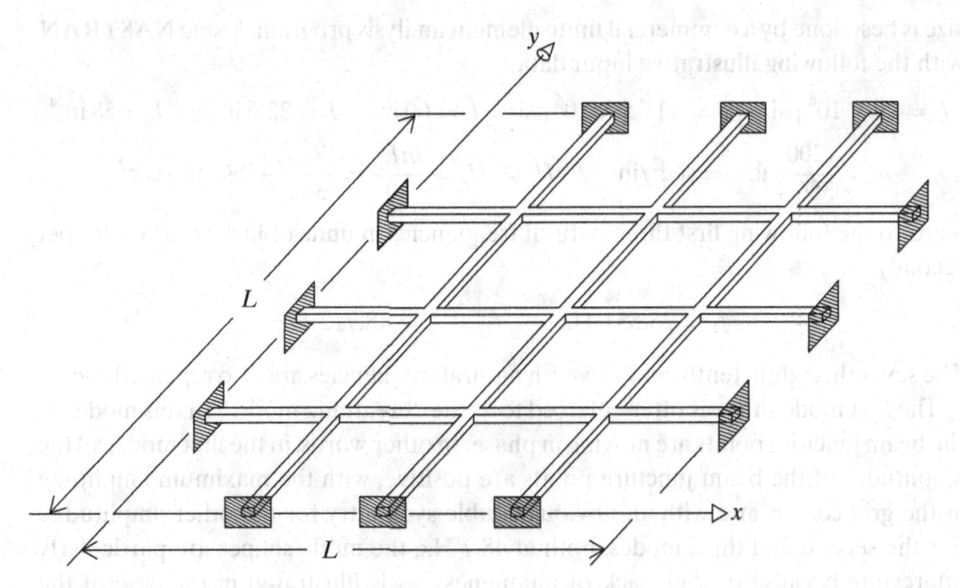

Figure 6.6. Beam grid having repeated natural frequencies.

Here the lack of coupled behavior between the two supporting beams is somewhat disguised as symmetric and antisymmetric mode shapes. What the mathematics of the above mode shapes says is that the two beams are either vibrating together in exactly the same way in the same directions, or they are vibrating in exactly the same way but in opposite directions. ★

In general, the presence of repeated roots cause the determinant of the coefficient matrix in the equation $[K - \omega^2 M]\{A\} = \{0\}$ to have as many dependent rows (to be deleted) as there are repeated roots. For nonrepeated roots, the singularity is only to the first degree (i.e., for an $N \times N$ coefficient matrix, the *rank* of the matrix is $N - 1$), and an arbitrary value can be assigned to only one of the amplitudes. In the above case of the two cantilevered beams (where there are two repeated roots), two rows can be deleted when solving for the mode shapes corresponding to those first two natural frequencies. Thus arbitrary values can be assigned to two of the amplitudes for each of the mode shapes associated with that repeated natural frequency. If the pairs of arbitrary values, viewed as a pair of 2×1 vectors, constitute independent vectors, then the mode shapes will also be independent vectors. The proof that the degree of the singularity is equal to the number of natural frequencies with the same value requires the use of congruence matrix transformations [6.5], which have yet to be discussed.

Although repeated roots are possible for planar beam grid structures having just one axis of symmetry when motions are decoupled, repeated roots are to be expected for any beam grid structure with the two perpendicular axes of symmetry. Consider the 24-beam element grid shown in Figure 6.6. The discrete masses are located at the nine beam junctures. The calculation of the natural frequencies of a beam grid of this size is best done by a commercial finite element analysis program. Using NASTRAN, with the following illustrative input data

$$E = 29 \times 10^6 \text{ psi} \quad G = 11.25 \times 10^6 \text{ psi} \quad L = 60 \text{ in.} \quad I = 22.5 \text{ in.}^4 \quad J = 58 \text{ in.}^4$$

$$m = \frac{200}{386.4} \text{ lbs.} = \text{sec}^2/\text{in.} \qquad H_x = H_y = \frac{mL^2}{20} = \frac{36{,}000}{386.4} \text{ lbs.-in.-sec}^2.$$

leads to the following first three natural frequencies in units of hertz (i.e., cycles per second):

$$f_1 = 25.551 \text{ Hz} \qquad f_2 = f_3 = 48.712 \text{ Hz}.$$

The seventh, eighth, tenth, and eleventh natural frequencies are also repeated roots.[7]

The first mode shape is often referred to as the "breathing mode." In this mode, all the beam juncture points are moving in phase. In other words, in the first mode, all the amplitudes of the beam juncture points are positive, with the maximum amplitude at the grid center, and with an obvious double symmetry for the other amplitudes. For the second and third modes, both at 48.7 Hz, the mode shapes are particularly interesting because of their lack of uniqueness, as is illustrated in the case of the previously discussed two beam grid. Again, for these two identical natural frequency values, there are an infinite number of possible pairs of mutually orthogonal mode shapes. However, there are only two possible pairs of symmetric, mutually orthogonal mode shapes at this frequency. The first pair of possibilities are ones where the nodal lines are the diagonals of the grid. That is, the line $x = y$ is a line of zero amplitudes for the first of these mode shapes, whereas the opposite diagonal is the nodal line for the second of these mode shapes. Of course, on one side of the nodal line the amplitudes are positive, whereas on the other side the amplitudes are negative. The other possible symmetric pair of mode shapes corresponding to this double root are ones where one midline, say $x = \frac{1}{2}L$, or the other, $y = \frac{1}{2}L$, is a nodal line. That is, physically, at this repeated modal frequency, the grid can vibrate in any of the following four modes: (i) its upper right half (relative to the diagonal) moving oppositely to its lower left half; (ii) its upper left half (relative to the diagonal) moving oppositely to its lower right half; (iii) its left half moving oppositely to its right half; and (iv) its top half moving oppositely to its bottom half. Numerically, the second pair of modal amplitudes can be calculated by simply combining the first pair of modal amplitudes and vice versa. It can be said that the second and third natural frequencies are the same because there is nothing inherent to the structure that distinguishes one of these two diagonals, or one of these two midlines, from the other.

Just as a matter of information, the fourth mode at 69.1 Hz has nodal lines that are both of the two midlines. The fifth mode at 77.5 Hz has nodal lines that are both of the two diagonals. The sixth mode does not have straight lines as nodal lines. In the sixth mode, at 77.9 Hz, the center joint has a positive, peak amplitude, and there are negative amplitudes of lesser magnitude at all other interior joints (a second breathing mode). Since these modes do not correspond to repeated roots, their mode shapes are unique. The mode shapes for the seventh and eighth modes, both at 90.1 Hz, have nodal lines that are midlines and other lines that divide the

[7] Calculated results by Dr. Suresh Chander.

total grid into six parts so that each of the noncentral joints has an amplitude that has a sign that differs from its noncentral neighbor.

6.5 Orthogonality and the Expansion Theorem

The determinant method was useful for introducing the process of calculating natural frequencies and their corresponding mode shapes. However, the determinant method is clearly not practical for hand calculations whenever the number of DOF exceeds three or four, which are very small numbers of DOF relative to common engineering structural models. Moreover, the determinant method is not at all efficient relative to other methods when a computer is used. Simply multiplying out a single $N \times N$ determinant takes $N*(N!)$ calculations. There are numerous other more efficient methods for calculating natural frequencies and mode shapes. Almost all of them are practical only when computer based. However, there is one method that can be used for both: (i) hand calculations when there are only several DOF and (ii) reasonably efficient computer-based calculations for structures with a moderate number of DOF (up to a few hundred). That method is called the *matrix iteration method* or the *matrix power method*. The matrix iteration method, like the determinant method, is a method that requires a separate calculation[8] for each natural frequency and its corresponding mode shape. This is in contrast to other, more modern, methods to be discussed later that calculate all the natural frequencies and mode shapes simultaneously. To understand how the matrix iteration process begins by simultaneously converging to the first modal frequency and the first mode shape, it is necessary to cover some preliminaries. First, some general nomenclature concerning matrix eigenvalue problems is introduced. Then the orthogonality of the mode shape vectors is examined. Finally, it is necessary to explain the use of the mode shapes as a "basis" for all amplitude vectors.

Reconsider the basic matrix problem to be solved for the natural frequencies and their corresponding mode shapes in the form

$$\omega^2[M]\{q\} = [K]\{q\} \quad \text{or}$$

$$[K]^{-1}[M]\{q\} = \frac{1}{\omega^2}\{q\} \quad \text{or} \quad [D]\{q\} = \lambda\{q\}. \tag{6.6}$$

In this last writing of the matrix equation, the inverse of the natural frequency squared (plus any common factors that might arise from the mass and stiffness matrices) is designated by λ. The last form of the matrix equation is the standard form for a *matrix eigenvalue problem*. The quantity λ is called the *eigenvalue*, and the unknown vector $\{q\}$ is called the *eigenvector*. The matrix $[D]$, which is the inverse of the stiffness matrix postmultiplied by the mass matrix, in this context is called the *dynamic matrix*. Since the standard form for the matrix eigenvalue problem is the starting point for the matrix iteration procedure, the price to be paid for the use of the matrix iteration method is the inversion of the stiffness matrix, or, as discussed later, a less costly preparation that is commonly used.

[8] The general meaning of the verb *to iterate* is to repeat. In engineering, it implies a trial-and-error correction process.

The first important feature of the eigenvalue problem is easily established from the standard matrix form above. Note that if $\{q\} = \{A\}$ is a nontrivial solution to the eigenvalue problem, then $\{cA\}$ is also a solution, where c is an arbitrary constant. This means that the eigenvectors are only unique up to a multiplicative constant. Again, the constant traditionally selected for most numerical work is the one that makes the largest entry in the eigenvector have the value $+1.0$, just as was done with the determinant method. Further recall that from the discussion following Eq. (6.2), where that equation can be rewritten as

$$\omega^2 = \frac{\lfloor q \rfloor [K] \{q\}}{\lfloor q \rfloor [M] \{q\}}, \tag{6.7}$$

that the natural frequency squared is always either a positive (real) number, or zero; and it can have a zero value only when the structure, like a ship or an aircraft structure, can undergo a rigid body motion.

The mode shape eigenvectors possess, or can be made to possess, the remarkable characteristic of being, in a sense to be explained, perpendicular to each other. To understand this characteristic, let ω_i be the ith natural frequency solution and ω_j be the jth natural frequency solution to the first of Eqs. (6.6). In addition, let $\{q^{(i)}\}$ and $\{q^{(j)}\}$ be the corresponding mode shape solutions. Their designation as solutions means that the following two equalities are valid

$$\omega_i^2 [M]\{q^{(i)}\} = [K]\{q^{(i)}\}$$
$$\omega_j^2 [M]\{q^{(j)}\} = [K]\{q^{(j)}\}.$$

Premultiply the first of these two equations by the row vector $\lfloor q^{(j)} \rfloor$ and the second of these two equations by $\lfloor q^{(i)} \rfloor$ to obtain

$$\omega_i^2 \lfloor q^{(j)} \rfloor [M]\{q^{(i)}\} = \lfloor q^{(j)} \rfloor [K]\{q^{(i)}\}$$
$$\omega_j^2 \lfloor q^{(i)} \rfloor [M]\{q^{(j)}\} = \lfloor q^{(i)} \rfloor [K]\{q^{(j)}\}.$$

Recall the following theorem from matrix algebra. If $[A] = [B][C]$, then $[A]^t = [C]^t[B]^t$. In other words, when taking a transpose of matrix products, it is necessary to reverse the order of the product. Further recall that the mass and stiffness matrices derived from the Lagrange equations must always be symmetric, real matrices. This symmetry means that the transpose of the mass matrix is equal to the mass matrix and the transpose of the stiffness matrix is equal to the stiffness matrix. With the above two sets of facts in mind, the transpose of the second of the above two equations leads to

$$\omega_i^2 \lfloor q^{(j)} \rfloor [M]\{q^{(i)}\} = \lfloor q^{(j)} \rfloor [K]\{q^{(i)}\}$$
$$\omega_j^2 \lfloor q^{(j)} \rfloor [M]\{q^{(i)}\} = \lfloor q^{(j)} \rfloor [K]\{q^{(i)}\}.$$

Subtracting the second of the above equations from the first yields

$$\left(\omega_i^2 - \omega_j^2\right) \lfloor q^{(j)} \rfloor [M]\{q^{(i)}\} = 0.$$

From the above equality it may be concluded that either the frequency difference must be zero or the scalar that is the triple matrix product involving the mode shapes and the mass matrix must be zero. First consider the case where the ith and jth frequency are different, which includes the case where one of the two frequencies is zero. With the two frequencies being different, the first of the above factors cannot be zero. Hence the triple matrix product involving the mass matrix must be zero for the two different[9] mode shapes. From the above equations, the triple product involving the stiffness matrix must also be zero. That is,

$$\lfloor q^{(i)} \rfloor [M] \{q^{(j)}\} = 0 = \lfloor q^{(i)} \rfloor [K] \{q^{(j)}\}. \tag{6.8}$$

The above two mathematical results are described by saying that the mode shapes for different natural frequencies are mutually *orthogonal* when "weighted" by either the mass or stiffness matrix. The use of the term "orthogonal" comes from the familiar definition for N-dimensional vectors that $\boldsymbol{A}$ and $\boldsymbol{B}$ are orthogonal if and only if $\boldsymbol{A} \cdot \boldsymbol{B} = 0$ (Recall $\boldsymbol{A} \cdot \boldsymbol{B} = A_1 B_1 + A_2 B_2 + \cdots + A_n B_n = \lfloor A \rfloor \{B\}$.) "Weighted orthogonality" only means that there is square matrix factor present in the product as set forth in Eq. (6.8). Although it is lax to do so, commonly, and thus in this textbook, the important "weighted" qualification is often omitted, and the mode shapes are simply called *orthogonal.*

However, when $\omega_i = \omega_j$, again, an uncommon case for actual structures, the corresponding eigenvectors are not necessarily orthogonal. More generally, when r natural frequencies are the same, none of the r mode shapes corresponding to that repeated frequency are necessarily orthogonal to any other mode shape corresponding to that repeated frequency. Each one is, of course, orthogonal to all mode shapes corresponding to other frequencies. Since, as previously discussed, these r mode shapes can be made to be mutually independent, they also can be made to be mutually orthogonal by the straightforward means of the Gram–Schmidt method. See Ref. [6.2] or Endnote (5). As a matter of course, it will be assumed that all mode shapes corresponding to repeated roots are rendered mutually orthogonal so that the complete set of mode shapes is an orthogonal set. For example, this was done with both sets of mode shapes deduced in Example 6.7, the illustrative example involving the twin cantilevered beams. That is, in each of those two sets, all four mode shapes are mutually orthogonal.

The final piece of preliminary information that needs to be established is that the set of N mode shape eigenvectors, each of size $N \times 1$, can be used as a *basis* for any and all $N \times 1$ deflection amplitude vectors. In this context, a basis for vectors is defined as a set of linearly independent[10] vectors of a particular size (say, $N \times 1$) that, in some linear combination, can be equated to any vector of the same size. For example, the familiar Cartesian unit vectors $\boldsymbol{i}$, $\boldsymbol{j}$, and $\boldsymbol{k}$ are one of the many possible bases for all vectors in three dimensions because it is always possible for the general three-space vector $\boldsymbol{V}$ to be written as $\boldsymbol{V} = V_x \boldsymbol{i} + V_y \boldsymbol{j} + V_z \boldsymbol{k}$. These three unit vectors are said to "span" three-space. In matrix style, these same three Cartesian unit

[9] The proof that the ith and jth mode shapes are different is from Eq. (6.7). If the ith and jth mode shapes are the same (differing only by a multiplicative constant, then the two natural frequencies would also have to be the same. Since the natural frequencies are actually different, this contradiction proves that the mode shapes connot be the same.

[10] See Endnote (1) for further discussion of linear independence.

vectors are, respectively, $\lfloor 1 \quad 0 \quad 0 \rfloor$, $\lfloor 0 \quad 1 \quad 0 \rfloor$, and $\lfloor 0 \quad 0 \quad 1 \rfloor$. In the usual vector form, quantities such as V_x (the Fourier coefficients of the above three term vector series) are determined by simply taking the dot product of $\boldsymbol{V}$ with $\boldsymbol{i}$. The equivalent matrix product of the $\lfloor V_x \quad V_y \quad V_z \rfloor$ and $\lfloor 1 \quad 0 \quad 0 \rfloor^t$ accomplishes the same result. The same exact forms hold in n-space. That is, in analogy to the above three-space sum $\boldsymbol{V} = V_x\boldsymbol{i} + V_y\boldsymbol{j} + V_z\boldsymbol{k}$, simply using the rules of matrix addition, the general N-dimensional vector $\{q\}$ can be written as

$$\begin{Bmatrix} q_1 \\ q_2 \\ q_3 \\ \cdot \\ \cdot \\ \cdot \\ q_n \end{Bmatrix} = q_1 \begin{Bmatrix} 1 \\ 0 \\ 0 \\ \cdot \\ \cdot \\ \cdot \\ 0 \end{Bmatrix} + q_2 \begin{Bmatrix} 0 \\ 1 \\ 0 \\ \cdot \\ \cdot \\ \cdot \\ 0 \end{Bmatrix} + q_3 \begin{Bmatrix} 0 \\ 0 \\ 1 \\ \cdot \\ \cdot \\ \cdot \\ 0 \end{Bmatrix} + \cdots + q_n \begin{Bmatrix} 0 \\ 0 \\ 0 \\ \cdot \\ \cdot \\ \cdot \\ 1 \end{Bmatrix}.$$

This unique representation of the general vector $\{q\}$ as a linear combination of these N different N-space Cartesian unit vectors (with obvious coefficients) is possible because these N-space unit vectors are a *complete* set of N linearly independent vectors. By definition, a complete N-space vector set, such as the Cartesian set above, allows a unique, linear sum, representation of any N-space vector. If the jth one of these unit Cartesian vectors were absent from this set, then the set would be an incomplete basis because there would be no way to duplicate the jth entry of the general N-space vector. The above Cartesian N-space vector set is more than a set of linearly independent vectors. As may be readily demonstrated, it is also a mutually (in the unweighted sense) orthogonal set, just as $\boldsymbol{i}$, $\boldsymbol{j}$, and $\boldsymbol{k}$ are a mutually orthogonal set in three-space. It is easy to prove (see Exercise 6.9) that any orthogonal set of vectors is also a set of linearly independent vectors. It is also possible to view these $N \times 1$ vectors as orthogonal in the weighted sense with the identity matrix as the square weighting matrix.

Consider the possibility of the following alternate decomposition (i.e., component formulation) of the general N-space, free vibration, amplitude vector $\{q\}$

$$\{q\} = c_1\{A^{(1)}\} + c_2\{A^{(2)}\} + c_3\{A^{(3)}\} + \cdots + c_n\{A^{(n)}\}, \tag{6.9}$$

where the c_i are real constants to be uniquely determined and the $\{A^{(i)}\}$ are mode shape amplitudes; that is, the orthogonal eigenvectors of the free vibration problem under consideration. Let the mode shapes be normalized in the usual fashion where the maximum value is $+1$. It will now be shown that because the vectors $\{A^{(i)}\}$ form a set of N (weighted) orthogonal vectors, all the c_i are unique and are not all zero unless, of course, $\{q\}$ is null. Thus it then can be concluded that the $\{A^{(i)}\}$ form a complete set of orthogonal vectors that span N-space. The first fact can be demonstrated by simply premultiplying both sides of Eq. (6.9) by the matrix product $\lfloor A^{(j)} \rfloor [M]$. The unique result is immediately $c_j = \lfloor A^{(j)} \rfloor [M]\{q\}/M_j$, where the *generalized mass* M_j is equal to the product $\lfloor A^{(j)} \rfloor [M]\{A^{(j)}\}$. Equation (6.9) is sometimes called the (eigenvector) *expansion theorem*, and it is central to understanding the eigenvalue calculation procedure that follows.

6.6 The Matrix Iteration Method

Having in place the above preliminaries concerning the (weighted) orthogonality of, and the complete basis of, the set of N mode shapes, now return to the standard N-space matrix eigenvalue problem of Eq. (6.7): $[D]\{q\} = \lambda\{q\}$. The matrix iteration method begins with an arbitrary chosen vector $\{q\} = \{q\#1\}$. As will be seen, the matrix iteration method is more efficient if the first selected numerical values of $\{q\#1\}$ are the analyst's best guess for the first eigenvector $\{A^{(1)}\}$. Continued experience with structures and mode shapes soon provides the analyst with the ability to make reasonable guesses. For example, the reader already knows that the first mode shape will not have any interior nodal points and thus have no sign changes. If by some extremely rare stroke of luck, this first guess, $\{q\#1\}$, were actually the first mode shape, then as with Eq. (6.7), substitution into $[D]\{q\} = \lambda\{q\}$ would immediately yield the eigenvalue of the first natural frequency and replicate the first mode shape. Since this choice is quite unlikely, let the first guess be other than the first mode shape. As an N-space vector, $\{q\#1\}$ can be written in terms of the yet unknown mode shapes, just as in Eq. (6.9), where, with any luck, c_1 is not zero.[11] Substitute this arbitrary vector into the left-hand side of the standard matrix eigenvalue form to obtain

$$[D]\{q\#1\} = [K^{-1}M](c_1\{A^{(1)}\} + c_2\{A^{(2)}\} + c_3\{A^{(3)}\} + \cdots + c_n\{A^{(n)}\}).$$

Since

$$[M]\{A^{(i)}\} = \frac{1}{\omega_i^2}[K]\{A^{(i)}\},$$

then

$$[D]\{q\#1\} = \frac{c_1}{\omega_1^2}\{A^{(1)}\} + \frac{c_2}{\omega_2^2}\{A^{(2)}\} + \frac{c_3}{\omega_3^2}\{A^{(3)}\} + \cdots + \frac{c_n}{\omega_n^2}\{A^{(n)}\}.$$

The right-hand side of the above equation is simply a different $N \times 1$ vector from the first guess. Call this new vector $\{q\#2\}$, and substitute it into the matrix eigenvalue problem $[D]\{q\} = \lambda\{q\}$. The result, as above, is

$$[D]\{q\#2\} = \frac{c_1}{\omega_1^4}\{A^{(1)}\} + \frac{c_2}{\omega_2^4}\{A^{(2)}\} + \frac{c_3}{\omega_3^4}\{A^{(3)}\} + \cdots + \frac{c_n}{\omega_n^4}\{A^{(n)}\} = \{q\#3\}.$$

The trend is clear. After m iterations, that is after m substitutions into the matrix eigenvalue problem, the result is

$$[D]\{q\#m\} = \frac{c_1}{\omega_1^{2m}}\{A^{(1)}\} + \frac{c_2}{\omega_2^{2m}}\{A^{(2)}\} + \frac{c_3}{\omega_3^{2m}}\{A^{(3)}\} + \cdots + \frac{c_n}{\omega_n^{2m}}\{A^{(n)}\}. \quad (6.10)$$

Since the natural frequencies are ordered from the lowest numerical value to the highest, when m is a large enough integer, the sharply decreasing size of above right-hand side coefficients c_i/ω_i^{2m} for $i > 1$ causes the right-hand side sum to be overwhelmingly composed of the first amplitude vector, regardless of the initial values of the c_i, again assuming that c_1 is not zero.

[11] If c_1 were initially zero, rough round-offs in the calculation, plus the effects of accelerating the convergence, will eventually, however slowly, produce the desired first mode shape.

In an actual numerical calculation, the conclusion of the iteration process can be viewed as one additional substitution of the last trial solution $\{q\#m\}$ into $[D]\{q\} = \lambda\{q\}$, and then the factoring out of the eigenvalue λ from the right-hand side so that there is an exact match of the right-hand side eigenvector $\{q\}$ to the left-hand side $\{q\}$ to the extent of the number of significant figures required. Of course, the factor λ contains the solution for the natural frequency. The following examples illustrate this process.

EXAMPLE 6.8 Apply the matrix iteration technique to obtain the first natural frequency and mode shape for the system of Example 6.2. Compare the result to that obtained by the determinant method.

SOLUTION The mass and stiffness matrices for that example are

$$[M] = m\begin{bmatrix} 0.50 & & \\ & 0.50 & \\ & & 1.00 \end{bmatrix} \qquad [K] = k\begin{bmatrix} 9.00 & -3.00 & -2.00 \\ -3.00 & 3.00 & 0 \\ -2.00 & 0 & 2.00 \end{bmatrix}.$$

The hand calculation of the inverse of the stiffness matrix is not difficult. The determinant of the stiffness matrix is 54 – 12 – 18 = 24. Then, from the adjoint method, Ref. [6.2], the inverse of $[K]$ is

$$[K]^{-1} = \frac{1}{24k}\begin{bmatrix} 6.00 & 6.00 & 6.00 \\ 6.00 & 14.00 & 6.00 \\ 6.00 & 6.00 & 18.00 \end{bmatrix}.$$

Postmultiplying this inverse of the stiffness matrix by the mass matrix provides the dynamic matrix, $[D]$. Making a neutral first guess of $\lfloor 1.00 \quad 1.00 \quad 1.00 \rfloor$ for the eigenvector, the eigenvalue problem $[D]\{q\} = \lambda\{q\}$ starts off being

$$\begin{bmatrix} 3.00 & 3.00 & 6.00 \\ 3.00 & 7.00 & 6.00 \\ 3.00 & 3.00 & 18.00 \end{bmatrix}\begin{Bmatrix} 1.0 \\ 1.0 \\ 1.0 \end{Bmatrix} = \frac{24.00k}{m\omega^2}\begin{Bmatrix} \\ \\ \\ \end{Bmatrix}.$$

As a comment on the first guess for the eigenvector, note that because the first mode shape is being sought, this choice of all positive ones abides by the usual rule that there should not be any sign changes that indicate interior nodes. Note that there are exceptions to this guideline of all positive values because of the qualification concerning interior nodes. A first eigenvector should have one or more sign changes if an inconsistent DOF sign convention were used, or, the structure was, for example, a straight simply supported beam with a overhang beyond one support such that as the main span of the beam vibrates upward, the overhang must travel downward so as to maintain slope continuity at that support. Another exception to the rule is when a structure is not fully constrained against rigid body motion. For example, consider a free-free beam undergoing bending vibrations in the plane of the paper. Beyond the first rigid body mode with zero nodes, and the second rigid body mode with one node, each with associated zero valued natural frequencies, the first *elastic* bending mode will have two nodes.

Furthermore, this guess of all +1 values for the eigenvector $\{q\#1\}$ says that there is no suspicion that any one of the three actual amplitudes is larger or smaller than the others. This lack of insight need not be true in this case. Examination of, say, the diagonal entries of the above flexibility matrix (the inverse of the stiffness matrix), clearly indicates that the structure is more flexible at the third DOF than at the second DOF, and more so at the second than the first. Hence a better first guess would something like $\lfloor 0.3 \quad 0.7 \quad 1.0 \rfloor$. Of course, the stiffness matrix diagonal terms tell the same story in other terms; that is, the amplitudes will be smaller where the stiffness is greater. Nevertheless, the above neutral guess of all ones will be used to clearly demonstrate the quick convergence of the matrix iteration technique even when the first guess is so poorly made.

Start the matrix iteration procedure by multiplying the dynamic matrix $[D]$ and the first guess, $\{q\#1\}$, and then factor the resulting 3×1 vector so that the largest of the three terms is 1.00 so as to match the largest term of the preceding guess. That is, obtain

$$\begin{bmatrix} 3.00 & 3.00 & 6.00 \\ 3.00 & 7.00 & 6.00 \\ 3.00 & 3.00 & 18.00 \end{bmatrix} \begin{Bmatrix} 1.0 \\ 1.0 \\ 1.0 \end{Bmatrix} = 24 \begin{Bmatrix} 0.5 \\ 0.66 \\ 1.0 \end{Bmatrix}.$$

After one iteration, the estimate for the eigenvalue $24k/m\omega^2$ is 24, and the right-hand side vector is the estimate for the first mode. However, comparison of this estimate, $\lfloor 0.50 \quad 0.66 \quad 1.0 \rfloor$, to the original guess of all positive ones shows that these would-be eigenvectors are not the same. Therefore, with different eigenvectors on each side of the equality sign, the eigenvalue problem equality $[D]\{q\} = \lambda\{q\}$ has not yet been achieved. Thus, use the right-hand side vector as an improved guess, and repeat the process. Get

$$\begin{bmatrix} 3.00 & 3.00 & 6.00 \\ 3.00 & 7.00 & 6.00 \\ 3.00 & 3.00 & 18.00 \end{bmatrix} \begin{Bmatrix} 0.50 \\ 0.66 \\ 1.0 \end{Bmatrix} = 21.5 \begin{Bmatrix} 0.44 \\ 0.56 \\ 1.0 \end{Bmatrix}.$$

Again, the right-hand side vector does not match the left-hand side vector, and another iteration is required. This time get

$$\begin{bmatrix} 3.00 & 3.00 & 6.00 \\ 3.00 & 7.00 & 6.00 \\ 3.00 & 3.00 & 18.00 \end{bmatrix} \begin{Bmatrix} 0.44 \\ 0.56 \\ 1.0 \end{Bmatrix} = 21.0 \begin{Bmatrix} 0.42 \\ 0.53 \\ 1.0 \end{Bmatrix}.$$

Notice that all rounding off is done in the direction that the values of the amplitudes are moving. (This idea can easily be carried further by "accelerating" the convergence by taking these changes in value further than the calculations show.) Also note that the process is not far from converging. Therefore it is time to start increasing the number of significant figures beyond the number of significant figures in either the mass matrix or stiffness matrix. The reasons for the need for many more significant figures are, briefly, (i) obtaining accuracy in the calculation of the higher mode shapes and (ii) obtaining accuracy when using the mode shapes to accomplish a modal transformation to be discussed in the next chapter. Another iteration produces

$$\begin{bmatrix} 3.00 & 3.00 & 6.00 \\ 3.00 & 7.00 & 6.00 \\ 3.00 & 3.00 & 18.00 \end{bmatrix} \begin{Bmatrix} 0.42 \\ 0.53 \\ 1.0 \end{Bmatrix} = 20.85 \begin{Bmatrix} 0.4244 \\ 0.5261 \\ 1.000 \end{Bmatrix}.$$

Notice that a hand calculation arithmetic error only delays the convergence. Again, repeating the process,

$$\begin{bmatrix} 3.00 & 3.00 & 6.00 \\ 3.00 & 7.00 & 6.00 \\ 3.00 & 3.00 & 18.00 \end{bmatrix} \begin{Bmatrix} 0.4244 \\ 0.5261 \\ 1.000 \end{Bmatrix} = 20.8515 \begin{Bmatrix} 0.42450 \\ 0.52543 \\ 1.0000 \end{Bmatrix}.$$

After three more iterations, convergence is secure at six significant figures. The final result is

$$\begin{bmatrix} 3.00 & 3.00 & 6.00 \\ 3.00 & 7.00 & 6.00 \\ 3.00 & 3.00 & 18.00 \end{bmatrix} \begin{Bmatrix} 0.424429 \\ 0.525190 \\ 1.00000 \end{Bmatrix} = 20.84886 \begin{Bmatrix} 0.424429 \\ 0.525190 \\ 1.00000 \end{Bmatrix}.$$

This mode shape result and the natural frequency calculation below agree quite well with those obtained via the determinant method. The first natural frequency is obtained from the first eigenvalue, 20.84886. Recall that from the original problem setup that this eigenvalue is $24k/m\omega^2$. Equating these two representations of the first eigenvalue yields $\omega_1 = 1.07291\sqrt{k/m}$. ★

In the above and subsequent matrix iteration examples, note that convergence to the eigenvalue precedes convergence to the eigenvector. This observation is reflected in *Rayleigh's principle* [see the related material of Endnote (2) in Chapter 3], which, in brief, states that a good approximation to the eigenvalue can be obtained by use of just a fair approximation to the eigenvector. In precomputer times, Rayleigh's principle, Ref. [6.6], was a quick way to obtain an upper bound for, primarily, a first natural frequency using, say, the structure's deflection pattern because of its own weight loading as an approximation to the first mode shape.

EXAMPLE 6.9 Use the matrix iteration method to calculate the first natural frequency and first mode shape for the cantilevered beam with laterally offset masses discussed in Examples 6.3 and 6.6.

SOLUTION The first task is the calculation of the dynamic matrix. The task of calculating the inverse of the stiffness matrix can be accomplished almost reasonably by hand calculator because there is no mathematical coupling between, on one hand, the third and sixth rows, and, on the other hand, the other four rows. To this end, first note that the value of the determinant of the stiffness matrix can be written as the product of the corresponding 2×2 determinant and the corresponding 4×4 determinant (Laplace's expansion rule). Then it is a matter of applying the adjugate method to the 4×4 submatrix and the 2×2 submatrix. Since the determinants to be evaluated during the process of using the adjugate method are 3×3, and because 3×3 and 2×2 matrices allow the use of simple diagonal evaluation schemes, it is, again, reasonable to do the above by hand calculator. Having said this, the result for the inverse of the stiffness matrix, from Mathematica, is

$$[K]^{-1} = \frac{1}{300k}\begin{bmatrix} 50 & 15 & 0 & 125 & 15 & 0 \\ 15 & 6 & 0 & 45 & 6 & 0 \\ 0 & 0 & 30 & 0 & 0 & 30 \\ 125 & 45 & 0 & 450 & 75 & 0 \\ 15 & 6 & 0 & 75 & 18 & 0 \\ 0 & 0 & 30 & 0 & 0 & 90 \end{bmatrix} \qquad k = \frac{EI_0}{L^3}.$$

Therefore the eigenvalue problem is

$$\begin{bmatrix} 100 & 30 & -100 & 125 & 15 & 125 \\ 30 & 12 & -30 & 45 & 6 & 45 \\ -60 & 0 & 120 & 30 & 0 & 60 \\ 250 & 90 & -250 & 450 & 75 & 450 \\ 30 & 12 & -30 & 75 & 18 & 75 \\ -60 & 0 & 120 & 90 & 0 & 180 \end{bmatrix} \left\{ \quad \right\} = \frac{300k}{m\omega^2} \left\{ \quad \right\}.$$

Since the DOF are for this structure are lateral deflections, bending slopes, and twists, the task of guessing the first mode shape from examination of the stiffness matrix is a bit more difficult. However, it should be possible to guess that the tip bending slope and tip vertical deflection are going to be, say, three or four times the corresponding midspan deflections, and the twisting DOF are going to be small values. Therefore start with the initial guess for the first eigenvector as

$$\{q\#1\}^t = \lfloor 0.3 \quad 0.1 \quad 0.3 \quad 1.0 \quad 0.3 \quad 1.0 \rfloor.$$

Postmultiplying the dynamic matrix by this initial guess and then factoring out the largest of the six entries produces the left-hand side

$$\lambda\{q\#2\}^t = 932 \lfloor 0.27 \quad 0.1 \quad 0.1 \quad 1.0 \quad 0.17 \quad 0.3 \rfloor.$$

Continuing to iterate by hand calculation leads to the following results:

$$\lambda\{q\#3\}^t = 649 \lfloor 0.29 \quad 0.102 \quad 0.067 \quad 1.0 \quad 0.16 \quad 0.21 \rfloor$$
$$\lambda\{q\#4\}^t = 621 \lfloor 0.288 \quad 0.102 \quad 0.053 \quad 1.0 \quad 0.163 \quad 0.190 \rfloor$$
$$\dots \qquad \dots.$$

After several more iterations, the six-significant-figure converged result, which agrees with the determinant method result, is

$$\lambda\{q\#12\}^t = 613.186 \lfloor 0.289089 \quad 0.102069 \quad 0.0476704 \quad 1.00000 \quad 0.163036 \quad 0.180927 \rfloor,$$

which shows that the initial guess was not a particularly good guess. From this eigenvalue result the first natural frequency is again calculated as

$$613.186 = \frac{300k}{m\omega_1^2} \quad \text{or} \quad \omega_1 = 0.70\sqrt{\frac{EI_0}{mL^3}}.$$

★

The direct calculation of the inverse of the stiffness matrix, to form the dynamic matrix, is clearly not an efficient approach to matrix iteration when the number of DOF is large. Furthermore, the software available to the analyst may only provide the eigenvalues and eigenvectors for a single symmetric matrix (in combination with the identity matrix). In that case, the above iteration approach needs modification because, as has been seen, the product of the two symmetric matrices $[K^{-1}][M]$ is not generally symmetric. If the available software provides, for example, for a *Cholesky decomposition* (or Cholesky factorization), which is a straightforward algebraic process discussed in greater detail in Endnote (2) of this chapter, then the following relatively efficient procedure will produce a symmetric dynamic matrix whose eigenvalues and eigenvectors are the same as those calculated by the determinant method.

The Cholesky decomposition technique is applicable only to symmetric,[12] positive definite matrices. Assume that the stiffness matrix is, or has been rendered, positive definite. (See Endnote (3) on how to make a stiffness matrix positive definite when the vibratory system can undergo rigid body motions). The first step is to write the stiffness matrix as the product of the matrix $[R]^t$, postmultiplied by the matrix $[R]$, which is a *r*ight (or *u*pper) triangular matrix where all the entries below the main diagonal are zero. Note that the transpose of a right triangular matrix is a *l*eft or *l*ower triangular matrix, a matrix where all entries above the main diagonal are zero; that is, $[R]^t \equiv [L]$. With the writing of $[K] = [R]^t[R]$, then the eigenvalue problem can be rewritten as

$$\left[M - \frac{1}{\omega^2} R^t R\right] \{q\} = \{0\}.$$

Substituting the linear transformation $\{q\} = [R^{-1}]\{p\}$ above, and premultiplying the above equality by the inverse of the transpose of the right triangular matrix $[R]$ (which is much easier inverse to calculate than that of the stiffness matrix), yields the same eigenvalue problem in terms of more advantageous symmetric matrices

$$\left[R^{-t} M R^{-1} - \frac{1}{\omega^2} I\right] \{p\} = \{0\}.$$

To prove that the above eigenvalue problem is the same as the original eigenvalue problem, recall again the following mathematics theorem concerning the determinants of matrices involved in a product:

$$\text{if} \quad [A][B] = [C] \quad \text{then} \quad |A||B| = |C|.$$

Applying this theorem to the above procedure with $[K] = [R^t][R]$, conclude that

$$\left|R^{-t}\right| \left|M - \frac{1}{\omega^2} K\right| \left|R^{-1}\right| = 0.$$

Recognizing that to have an inverse, the right triangular matrices must have a nonzero determinant. Then conclude that the above central determinant must be zero. This is, of course, the determinant that produced the original eigenvalues, and this concludes

[12] The LU decomposition is available for nonsymmetric matrices.

the proof. In addition, it may be concluded that the square matrix $[R^{-t}][M][R^{-1}]$ is also positive definite because the mass matrix is positive definite and the determinants of the inverse right triangular matrices are the same. The following MATLAB, Ref. [6.7], calculations illustrate the above process.

EXAMPLE 6.10 Determine the Cholesky right (or upper) triangular matrix for the stiffness matrix of Example 6.3. Then write the symmetric matrix for iteration.

SOLUTION The result of using MATLAB to obtain the right triangular matrix $[R]$ for the 6×6 matrix $[K] = [R]^t[R]$ is

$$[R] = \begin{bmatrix} 6.0000 & -5.0000 & 0 & -2.0000 & 5.0000 & 0 \\ & 16.5831 & 0 & -2.4121 & 4.5227 & 0 \\ & & 3.8730 & 0 & 0 & -1.2910 \\ & & & 1.4771 & -6.1546 & 0 \\ & & & & 4.0825 & 0 \\ & & & & & 1.8257 \end{bmatrix}.$$

The inverse of this matrix, a much simpler and thus less costly calculation than that for the inverse of the stiffness matrix, is

$$[R]^{-1} = \begin{bmatrix} 0.1667 & 0.0503 & 0 & 0.3077 & 0.2041 & 0 \\ & 0.0603 & 0 & 0.0985 & 0.0816 & 0 \\ & & 0.2582 & 0 & 0 & 0.1820 \\ & & & 0.6770 & 1.0206 & 0 \\ & & & & 2.449 & 0 \\ & & & & & 0.5477 \end{bmatrix}.$$

Then the symmetric dynamic matrix equal to $[R]^{-t}[M][R]^{-1}$ equals

$$\begin{bmatrix} 0.0556 & 0.0168 & -0.0861 & 0.1026 & 0.0680 & -0.609 \\ 0.0168 & 0.0123 & -0.0259 & 0.0428 & 0.0304 & -0.0183 \\ -0.0861 & -0.0259 & 0.2667 & -0.1589 & -0.1054 & 0.1886 \\ 0.1026 & 0.0428 & -0.1589 & 0.6671 & 0.8327 & 0.2584 \\ 0.0680 & 0.0304 & -0.1054 & 0.8327 & 1.1983 & 0.4845 \\ -0.0609 & -0.0183 & 0.1886 & 0.2584 & 0.4845 & 0.7333 \end{bmatrix}.$$

Exercise 6.6 deals with the matrix iteration of this dynamic matrix for the first natural frequency. ★

As illustrated in Ref. [6.8], p. 394, it is also possible to directly iterate using the equation form $\lambda[m]\{q_i\} = [k]\{q_{i+1}\}$; that is, without taking the inverse of the stiffness matrix. However, one price to be paid for this approach is the solution of N simultaneous equations during each iteration where only the product of the mass matrix and the previous modal vector would change from iteration to iteration. Possibly the most efficient way of carrying out those successive simultaneous solutions would involve a Cholesky decomposition of the stiffness matrix. Therefore, in comparison to the above, there is no advantage to this alternate iteration approach.

6.7 **Higher Modes by Matrix Iteration**

After completing the iteration procedure to obtain the first mode shape and first natural frequency, choosing another trial vector that now has a sign change among the amplitudes in hopes of converging to the second mode is (unless the second mode shape is guessed exactly) a path to disappointment. Such a trial vector, or any other, will generally contain, to some extent, the first mode shape as set forth in Eq. (6.9). That is, generally, c_1 will not be zero. For this reason, this trial vector will unavoidably converge to the first mode shape as in Eq. (6.10). The convergence will only take longer because of the sign change. From the discussion on convergence leading to Eq. (6.10), it should be clear that what is needed for the calculation of the second mode shape and second natural frequency is a series of trial vectors whose expansions in terms of all the mode shapes is such that the weighting coefficient for the first mode shape, c_1, is always zero to the extent of the accuracy of the calculation. Only in this way will the coefficient of the second mode dominate after many iterations, thus allowing convergence to the second mode. When the coefficient c_1 in the series expansion for the trial vector is zero, it is said that the trial vector does not contain the first mode shape as a component analogous to the vector $\boldsymbol{V} = a\boldsymbol{j} + b\boldsymbol{k}$ not having an x-direction component.

Exercise 6.10 discusses the proof that a necessary and sufficient condition for a trial vector to not have the first mode shape as a component (for c_1 of Eq (6.9) to be zero) is for the trial vector to be orthogonal to the first mode shape. (Again, this result is analogous to the fact that the vector $A_x\boldsymbol{i} + A_y\boldsymbol{j}$ has no $\boldsymbol{k}$ component and thus is orthogonal to $\mathbf{k}$.) To force (i.e., constrain) a trial vector $\{{}^cq\}$ to be orthogonal to the first mode, which is now known, simply write the equation

$$\lfloor {}^cq \rfloor [M] \{A^{(1)}\} = 0.$$

Actually carrying out the above triple product produces the result

$$q_1 \left(\Sigma M_{1j} A_j^{(1)}\right) + q_2 \left(\Sigma M_{2j} A_j^{(1)}\right) + \cdots + q_n \left(\Sigma M_{nj} A_j^{(1)}\right) = 0$$

$$\text{or} \quad q_1 = -q_2 \frac{\left(\Sigma M_{2j} A_j^{(1)}\right)}{\left(\Sigma M_{1j} A_j^{(1)}\right)} - \cdots - q_n \frac{\left(\Sigma M_{nj} A_j^{(1)}\right)}{\left(\Sigma M_{1j} A_j^{(1)}\right)}$$

$$\text{or} \quad q_1 = -s_{21}\, q_2 - s_{31}\, q_3 - \cdots - s_{n1}\, q_n,$$

where the definition of the s_{j1} ratio terms is obvious. This equation can be viewed as a single equation of constraint on the selected amplitudes of $\{{}^cq\}$, the trial vector constrained to be orthogonal to the first mode shape. Specifically, this equations says that although q_2 through q_n can be chosen arbitrarily, q_1 (for example) has to be calculated by the relationship above once those $n-1$ choices are made. To facilitate the use of this relationship in the matrix iteration scheme, that is, for each iteration, rewrite this relationship as part of the matrix equation

$$\begin{Bmatrix} q_1 \\ q_2 \\ q_3 \\ \cdot \\ \cdot \\ \cdot \\ q_n \end{Bmatrix}^C = \begin{bmatrix} 0 & -s_{21} & -s_{31} & \dots & -s_{n1} \\ & 1 & & & \\ & & 1 & & \\ & & & \dots & \\ & & & \dots & \\ & & & \dots & \\ & & & & 1 \end{bmatrix} \begin{Bmatrix} q_1 \\ q_2 \\ q_3 \\ \cdot \\ \cdot \\ \cdot \\ q_n \end{Bmatrix}^{NC}.$$

Here the right-hand vector $\{q\}$, superscripted NC(for no constraint), is without constraint; that is, it can be selected to be anything the analyst chooses it to be. However, by the above matrix equation, the left-hand vector $\{q\}$, superscripted C, is constrained to be orthogonal to the first mode shape. It is this left-hand side vector that will converge to the second mode shape. Therefore, with the above square matrix, named the first sweeping matrix $[S_1]$ for "sweeping" away the influence of the first mode shape, the matrix iteration problem for developing the second mode is simply

$$[D][S_1]\{q\} = \lambda\{q\} \quad \text{or} \quad [D_2]\{q\} = \lambda\{q\}.$$

At this point, as Eq. (6.10) shows, the matrix iteration proceeds to converge to the second mode. Of course, the convergence is still speeded along by making a good guess as to the shape of the second mode. There is an obvious factor that influences the rate of convergence. Equation (6.10) shows that when the next mode sought has a natural frequency close to that of the previous natural frequency in the ordered sequence of natural frequencies, it will take many more iterations for the lower mode coefficient in Eq. (6.10) to dominate the coefficient of the next mode. When this situation is discovered, there is one remedy in what is called a frequency shift that is discussed later in this chapter, or the simpler remedy of scaling the mass matrix. To understand this latter technique, note $[K]\{q\} = \omega^2[M]\{q\} = (10\omega)^2[\frac{1}{100}M]\{q\}$. Therefore, to multiply the natural frequencies by a factor of 10, divide the mass matrix by 100 or, alternatively, multiply the stiffness matrix by 100. Be warned that excessive scaling may lead to matrix ill-conditioning.

The process for obtaining the third and higher numbered modes follows the same idea that the trial vector for the higher numbered mode must be orthogonal to all lower numbered modes. For example, for the third mode, it is necessary to write the two equations that make the trial vector orthogonal to both the now known first and second mode shapes.

$$\lfloor^c q\rfloor[M]\{A^{(1)}\} = 0 \quad \text{and} \quad \lfloor^c q\rfloor[M]\{A^{(2)}\}.$$

These are two equations of constraint on the entries of the trial vector that will converge to the third mode. These two equations must be satisfied simultaneously. Thus it is necessary to solve these equations simultaneously for, say, q_1 and q_2 in terms of all the other $n-2$ entries of the amplitude vector. Such a solution is best done numerically. When there is a numerical solution for the first two entries of the amplitude vector, and that solution is cast in the matrix form as $\{^c q\} = [S_3]\{q\}$, then with lowercase s being a generic symbol for the above discussed numerical solutions, the result looks like

$$\begin{Bmatrix} q_1 \\ q_2 \\ q_3 \\ \cdot \\ \cdot \\ \cdot \\ q_n \end{Bmatrix}^C = \begin{bmatrix} 0 & 0 & s & \dots & s \\ 0 & 0 & s & \dots & s \\ & & 1 & & \\ & & & \dots & \\ & & & \dots & \\ & & & \dots & \\ & & & & 1 \end{bmatrix} \begin{Bmatrix} q_1 \\ q_2 \\ q_3 \\ \cdot \\ \cdot \\ \cdot \\ q_n \end{Bmatrix}^{NC}.$$

The use of this vector constrained to be orthogonal to the first and second modes, as before, leads to the matrix iteration equations

$$[D][S_2]\{q\} = \lambda\{q\} \quad \text{or} \quad [D_3]\{q\} = \lambda\{q\}.$$

EXAMPLE 6.11 Using matrix iteration, calculate the second and third mode shapes of the cantilevered beam of Examples 3.4 and 6.3.

SOLUTION To the extent of six significant figures, the first mode shape has been determined to be

$$\{A^{(1)}\}^t = \lfloor 0.289089 \quad 0.102069 \quad 0.0476704 \quad 1.000000 \quad 0.163036 \quad 0.180927 \rfloor.$$

This mode shape is now used in the single constraining equation $\lfloor {}^c q \rfloor [M]\{A^{(1)}\} = 0$. Carrying out this triple product leads to

$$0.482838q_1 + 0.204138q_2 - 0.387497q_3 + 1.18093q_4 + 0.163036q_5 + 1.36185q_6 = 0.$$

Solving the above for the first entry of the unknown vector $\lfloor {}^c q \rfloor$ in terms of the remaining five entries, and casting the result in matrix form, leads to the following relation between the amplitude vector constrained (c) to be orthogonal to the first mode and the unconstrained (nc) amplitude vector; i.e., forming $\{{}^c q\} = [S_1]\{{}^{nc} q\}$

$$\begin{Bmatrix} q_1 \\ q_2 \\ q_3 \\ q_4 \\ q_5 \\ q_6 \end{Bmatrix}^C = \begin{bmatrix} 0 & -.422788 & .802541 & -2.44581 & -.337662 & -2.82052 \\ & 1 & & & & \\ & & 1 & & & \\ & & & 1 & & \\ & & & & 1 & \\ & & & & & 1 \end{bmatrix} \begin{Bmatrix} q_1 \\ q_2 \\ q_3 \\ q_4 \\ q_5 \\ q_6 \end{Bmatrix}^{NC}.$$

The dynamic matrix for the second mode shape, $[D][S_1] = [D_2]$, is

$$\begin{bmatrix} 0 & -12.2788 & -19.7459 & -119.581 & -18.7662 & -157.052 \\ 0 & -0.683644 & -5.92376 & -28.3742 & -4.12987 & -39.6157 \\ 0 & 25.3673 & 71.8475 & 176.748 & 20.2597 & 229.231 \\ 0 & -15.6970 & -49.3646 & -161.451 & -9.41555 & -255.130 \\ 0 & -0.683644 & -5.92376 & 1.62583 & 7.87013 & -9.61566 \\ 0 & 25.6730 & 71.8475 & 236.748 & 20.2597 & 349.231 \end{bmatrix}.$$

Begin the iteration for the second mode with the not-so-clever trial vector $\lfloor -1\ \ -1\ \ -1\ \ +1\ \ +1\ \ +1 \rfloor$, which has an internal node for all three types of deflection. The iteration results are

$$\lambda\lfloor q\#2\rfloor = 509\lfloor -.5\quad -.1\quad +.6\quad -.7\quad .01\quad 1.0\rfloor$$
$$\lambda\lfloor q\#3\rfloor = 224\lfloor -.37\quad -.10\quad +.66\quad -.76\quad -.064\quad 1.0\rfloor$$
$$\lambda\lfloor q\#4\rfloor = 212\lfloor -.361\quad -.102\quad +.650\quad -.765\quad -.0713\quad 1.0\rfloor$$
$$\lambda\lfloor q\#5\rfloor = 210.8\lfloor -.3597\quad -.1015\quad +.6485\quad -.7659\quad -.07211\quad 1.0\rfloor.$$

and so forth until

$$\lambda\lfloor q\#10\rfloor = 210.402\lfloor -.359529\quad -.101481\quad +.648112\quad -.766031\quad -.0722403\quad 1.0\rfloor.$$

This mode shape and eigenvalue are in agreement with those calculated by the determinant method in Example 6.6, where, again,

$$\omega_2 = \sqrt{\frac{300}{210.402}}\sqrt{\frac{k}{m}} = 1.19\sqrt{\frac{k}{m}}.$$

The calculation for the third mode proceeds in the same manner. In addition to the first and previously used orthogonality condition $\lfloor {}^c q\rfloor[M]\{A^{(1)}\} = 0$, for the third mode there is the additional constraint $\lfloor {}^c q\rfloor[M]\{A^{(2)}\} = 0$. Carrying out the triple multiplication of the second constraint leads to

$$2.01528q_1 + 0.202961q_2 - 3.31151q_3 - 0.233970q_4 + 0.0722403q_5 - 1.23397q_6 = 0.$$

The simultaneous solution of this second constraint equation and the original constraint equation, for the first two amplitude entries in terms of the remaining four amplitudes, produces the sweeping matrix $[S_2]$

$$\begin{Bmatrix} q_1 \\ q_2 \\ q_3 \\ q_4 \\ q_5 \\ q_6 \end{Bmatrix}^{C} = \begin{bmatrix} 0 & 0 & 1.90606 & 0.917186 & 0.0585293 & 1.68573 \\ 0 & 0 & -2.61011 & -7.95432 & -0.937092 & -10.6584 \\ & & 1 & & & \\ & & & 1 & & \\ & & & & 1 & \\ & & & & & 1 \end{bmatrix} \begin{Bmatrix} q_1 \\ q_2 \\ q_3 \\ q_4 \\ q_5 \\ q_6 \end{Bmatrix}^{NC}$$

The dynamic matrix for the third mode shape, $[D][S_2] = [D_3]$, is

$$\begin{bmatrix} 0 & 0 & 12.3027 & -21.9110 & -7.25983 & -26.1790 \\ 0 & 0 & -4.13952 & -22.9363 & -3.48923 & -32.3289 \\ 0 & 0 & 5.6364 & -25.0312 & -3.51176 & -41.1438 \\ 0 & 0 & -8.3949 & -36.5923 & 5.29404 & -87.8235 \\ 0 & 0 & -4.13952 & 7.06374 & 8.51077 & -2.32890 \\ 0 & 0 & 5.63640 & 34.9688 & -3.51176 & 78.8562 \end{bmatrix}.$$

Starting the iteration with a guess that the third mode would display a twisting node between the two masses, that is, having as a first guess for the third mode $\lfloor 0.3\ \ 0.3\ \ 0.3\ \ 1.0\ \ 1.0\ \ -1.0\rfloor$, the iteration converged in fourteen steps to

$$\lambda\lfloor q\#14\rfloor = 36.7795\lfloor .0455980\quad .0615492\quad .282065\quad 1.00000\quad .278241\quad -.845635\rfloor.$$

Since the eigenvalue is $300k/m\omega^2$, the third natural frequency is again

$$\omega_3 = 2.85605\sqrt{\frac{k}{m}} = 2.86\sqrt{\frac{k}{m}}.$$

★

There is an important variation on the matrix iteration technique now to be discussed. The motivation for this variation could be either an interest in calculating just the modal frequencies and mode shapes in the vicinity of a specific frequency or the desire to improve on the numerical accuracy of one or more previously calculated modal frequencies and mode shapes in the vicinity of a specific frequency. In the first instance, an interest in a particular frequency could be prompted by a known or an anticipated excitation of significance at or near that frequency. In the second instance, for a structure with a large number of DOF, even if the analyst's interest is confined to only the first couple of dozen modes, by the time the last successive sweeping matrix and dynamic matrix are constructed, and a matrix iteration calculation is made, there is a possibility of a significant buildup of round-off error contamination. This procedure essentially restarts the round-off clock at the selected frequency. Call that analyst selected frequency of interest the *shift frequency*, ω_s. This procedure is also useful, as discussed above, when two adjacent frequencies are close in value.

The technique that accomplishes the above goals, at the price of an additional Cholesky decomposition or (hardly ever) matrix inversion, is called the *matrix iteration (or power) method with shifts*. To explain this technique, let the natural frequency closest to the frequency of interest be labeled as ω_m. Understand that this modal frequency, ω_m, and its corresponding eigenvector are not previously calculated quantities. Nevertheless, it is known that these two quantities are solutions to the eigenvalue problem $\omega_m^2\,[M]\{A^{(m)}\} = [K]\{A^{(m)}\}$. To both sides of this equality, add the unknown quantity $-\omega_s^2\,[M]\{A^{(m)}\}$. The resulting equation is $(\omega_m^2 - \omega_s^2)[M]\{A^{(m)}\} = [K - \omega_s^2 M]\{A^{(m)}\}$. If the shift frequency ω_s is not too close to the modal frequency, then the right-hand side square matrix is not close to being singular. Therefore this equation can be put into iteration form by, say for discussion purposes, inverting the known right-hand side square matrix so as to obtain

$$\left[K - \omega_s^2 M\right]^{-1}[M]\{A\}^{(m)} = \frac{1}{\omega_m^2 - \omega_s^2}\{A\}^{(m)}. \tag{6.11a}$$

Since the amplitude vector and the natural frequency for the mth mode are as yet unknown, the above equation is better written for iteration purposes as

$$\left[K - \omega_s^2 M\right]^{-1}[M]\{A\} = [D_s]\{A\} = \frac{1}{\omega^2 - \omega_s^2}\{A\}, \tag{6.11b}$$

where $[D_s]$ is the shifted dynamic matrix. Iterating upon Eq. (6.11a) will yield the mth natural frequency and mode shape because any trial vector used for that iteration will have the expansion

$$\{A\#1\} = c_1\{A\}^{(1)} + c_2\{A\}^{(2)} + c_3\{A\}^{(3)} + \cdots + c_n\{A\}^{(n)},$$

where, again, the specific values of the coefficients c_j depend on the analyst's guess and are generally not zero. Substitution of this trial vector into the above iteration equation yields

$$[D_s]\{A\#1\} = c_1[D_s]\{A\}^{(1)} + c_2[D_s]\{A\}^{(2)} + c_3[D_s]\{A\}^{(3)} + \cdots + c_n[D_s]\{A\}^{(n)}.$$

Using Eq. (6.10a,b), the above equation can be rewritten immediately as

$$[D_s]\{A\#1\} = \frac{c_1}{(\omega_1^2 - \omega_s^2)}\{A\}^{(1)} + \frac{c_2}{(\omega_2^2 - \omega_s^2)}\{A\}^{(2)} + \frac{c_3}{(\omega_3^2 - \omega_s^2)}\{A\}^{(3)} + \cdots + \frac{c_n}{(\omega_n^2 - \omega_s^2)}\{A\}^{(n)}.$$

The right-hand side vector is now designated as the second guess, $\{A\#2\}$. Following the same pattern, when this second guess is inserted into Eq. (6.11a), the result is

$$[D_s]\{A\#2\} = \frac{c_1}{(\omega_1^2 - \omega_s^2)^2}\{A\}^{(1)} + \frac{c_2}{(\omega_2^2 - \omega_s^2)^2}\{A\}^{(2)} + \frac{c_3}{(\omega_3^2 - \omega_s^2)^2}\{A\}^{(3)} + \cdots + \frac{c_n}{(\omega_n^2 - \omega_s^2)^2}\{A\}^{(n)}.$$

After k iterations, the result is

$$[D_s]\{A\#k\} = \frac{c_1}{(\omega_1^2 - \omega_s^2)^k}\{A\}^{(1)} + \frac{c_2}{(\omega_2^2 - \omega_s^2)^k}\{A\}^{(2)} + \frac{c_3}{(\omega_3^2 - \omega_s^2)^k}\{A\}^{(3)} + \cdots + \frac{c_n}{(\omega_n^2 - \omega_s^2)^k}\{A\}^{(n)}.$$

Clearly, whatever frequency is closest to the shift frequency will form, by far, the smallest denominator. Again, call that frequency the mth frequency. Hence $\{A\#(k+1)\}$ eventually will be dominated by the mth eigenvector. Thus the convergence to the mth eigenvector and natural frequency.

EXAMPLE 6.12 Using a frequency shift, recalculate the third modal frequency and mode shape of the cantilevered beam of Examples 3.4 and 6.3. Compare this result for the third mode to that of the previous example problem solution, Example 6.11.

SOLUTION Estimate the third modal frequency to be in the vicinity of $3\sqrt{k/m}$. (It was previously calculated to be $2.86\sqrt{k/m}$.) With this estimate as the selected value of the shifting frequency, the matrix $[K - \omega_s^2 M] = [K - 9.0M]$, is

$$\begin{bmatrix} 18 & -30 & 18 & -12 & 30 & 0 \\ -30 & 282 & 0 & -30 & 50 & 0 \\ 18 & 0 & -21 & 0 & 0 & -5 \\ -12 & -30 & 0 & 3 & -30 & -9 \\ 30 & 50 & 0 & -30 & 91 & 0 \\ 0 & 0 & -5 & -9 & 0 & -13 \end{bmatrix}.$$

The next step in to form the shifted dynamic matrix $[K - \omega_s^2 M]^{-1}[M]$. This matrix is

$$10[D_s] = \begin{bmatrix} .396054 & .0947883 & .265893 & -.412458 & -.505966 & .101856 \\ .292775 & .0117694 & -.490762 & -.205622 & -.279961 & .380769 \\ 2.17532 & -.197987 & -3.68869 & -.632769 & -1.04451 & 1.73682 \\ 4.15117 & -1.58403 & -10.1559 & -2.45359 & -3.12538 & 6.81994 \\ 1.07709 & -.559922 & -3.16610 & -.559922 & -.599829 & 2.00554 \\ -3.71055 & 1.17278 & 8.44973 & 1.17278 & 2.56546 & -6.92796 \end{bmatrix}.$$

The iteration of $[D_s]\{q\} = \lambda\{q\}$ proceeds as before to the largest of the eigenvalues λ and its associated eigenvector. Hopefully, in this case, that mode shape and frequency will be those of the third mode. (Therein lies the difficulty with the use of this shifting procedure *by itself* to calculate the successive modal values. It is never certain that a mode has not been skipped unless sweeping matrices are also used with the shifted dynamic matrices so that several modes are calculated at each frequency shift to form overlapping sets of modal data.) Since convergence should be relatively quick, it is not worth the effort to try and guess the third mode shape. Let the starting guess be the unimaginative set of values $\lfloor 1 \quad 1 \quad 1 \quad 1 \quad 1 \quad 1 \rfloor$. After eight iterations using Mathematica, the eigenvalue and converged mode shape are

$$-1.18626\lfloor 0.0455649 \quad 0.0615244 \quad 0.282012 \quad 0.999999 \quad 0.278266 \quad -0.845593\rfloor.$$

Comparing this mode shape with the third mode shape result of the previous example problem, which was obtained after using two sweeping matrices, shows that, on a percentage basis, the results are very close. However, the results do differ in the fourth significant digit. Since this latter result was obtained after far fewer numerical steps, it is much more likely to be the more accurate result. One check on this supposition can be made by substituting both these third mode shape results into the original eigenvalue problem and then determining which of these two vectors comes closer to replicating itself. To this end, note that the original form of the eigenvalue problem is Eq. (6.10), in Example 6.9. The sweeping matrix result for the third mode is, again,

$$\lfloor q^{(3)} \rfloor = \lfloor .0455980 \quad .0615592 \quad .282065 \quad 1.00000 \quad .278241 \quad -.845635\rfloor.$$

Substitution into the eigenvalue problem yields the result

$$[D]\{q^{(3)}\} = 36.7550\lfloor .0454092 \quad .0615005 \quad .282242 \quad 1.00000 \quad .278337 \quad -.846208\rfloor^t.$$

The shifting matrix result for the third mode is, again

$$\lfloor q^{(3)} \rfloor = \lfloor .0455649 \quad .0615244 \quad .282012 \quad 0.999999 \quad .278266 \quad -.845593\rfloor.$$

Substitution into the eigenvalue problem yields the result

$$[D]\{q^{(3)}\} = 36.7781\lfloor .0455641 \quad .0615243 \quad .282014 \quad .999999 \quad .278267 \quad -.845592\rfloor^t.$$

Comparison of the two pairs of results, where six significant figures is the common basis for all these calculations, clearly shows that this shifting matrix result is far less contaminated with other mode shapes than the sweeping matrix result, and therefore this shifting matrix result is considerably more accurate, as expected.

The frequency calculation is straightforward. The equation to be solved for the third modal frequency is simply

$$\frac{1}{\omega_3^2 - 9.0} = -1.18626 \quad \text{or} \quad \omega_3 = 2.86\sqrt{\frac{k}{m}},$$

just as before. The convergence to the eigenvalue is always superior to the convergence to the eigenvector. Exercise 6.8 provides further work on sweeping matrices and frequency shifts. ★

6.8 Other Eigenvalue Problem Procedures

As stated earlier, the primary use of the determinant method for solving eigenvalue problems is simply to introduce some of the ideas associated with such problems. Except for two or three or possibly four-DOF systems, the determinant method is a rather inefficient procedure for obtaining numerical results either by hand or by machine. The matrix iteration (with shifts) method is still widely used as a reasonably efficient procedure for obtaining the first couple of dozen modes for small or medium-sized problems. However, matrix iteration is not as widely used as a group of related methods that have been developed by mathematicians over the past several decades. There are many of these methods. They all require a knowledge of matrix theory, and quite lengthy explanations and proofs, to be fully understood. Since these more modern techniques are little suited for hand calculations, and because they usually are encountered only by engineers as a computer program menu selection, this section provides only a brief overview of the more prominent of these techniques. That is, some details, especially proofs, will be left to the references provided.

Historically, the first of these clever methods is *Jacobi's method*, see Refs. [6.9, 6.10]. Jacobi's method iterates to all the eigenvalues and eigenvectors simultaneously. This can be a substantial drawback in the use of this method because, generally, for a structure with many DOF, only a fraction, sometimes only a small fraction, of all the modes are needed for the modal transformation to be discussed in Chapter 7. Furthermore, there is also a need for a special preparation for the use of Jacobi's method. Jacobi's method requires that the dynamic matrix be a symmetric matrix. The first form of the dynamic matrix used in the previous discussion of the matrix iteration method, $[D] = [K^{-1}M]$, unless the mass matrix is proportional to the identity matrix, is never a symmetric matrix. However, there is more than one way to obtain a symmetric dynamic matrix from the matrices $[M]$ and $[K]$. Cholesky's decomposition[13] is the preferred approach to obtaining a symmetric matrix for use with the Jacobi method and the various iteration methods that spring from Jacobi's method. Since the application of Cholesky's decomposition is limited to a symmetric, positive definite matrix, in this discussion Cholesky's decomposition is applied to the

[13] See Endnote (1).

symmetric matrix $[M]$ with the assumption that $[M]$ also meets the positive definite criteria.[14]

The symmetric dynamic matrix is accomplished as follows. Return to the matrix eigenvalue problem statement in the form $[K - \lambda M]\{q\} = \{0\}$, where λ is the eigenvalue. Write the mass matrix as the product of a *l*eft or *l*ower triangular matrix and its transpose (which again is referred to as an *u*pper or *r*ight triangular matrix). That is, write $[M] = [L][L]^t$. Then introduce the transformation of coordinates $\{q\} = [L]^{-t}\{p\}$. As discussed at the end of Endnote (2), the inverse of $[L]$, another left triangular matrix, is quite cheaply obtained. Then after premultiplying by $[L]^{-1}$, the matrix eigenvalue problem becomes

$$[L]^{-1}\left([K] - \lambda[L][L]^t\right)[L]^{-t}\{p\} = \{0\}$$
$$\text{or} \quad [L^{-1}KL^{-t} - \lambda I]\{p\} = \{0\},$$

where the matrix $[L]^{-1}[K][L]^{-t} = [D]$ is clearly symmetric, as is, of course, the identity matrix $[I]$. Again, to show that this new eigenvalue problem $[D - \lambda I]\{p\} = \{0\}$ has the same eigenvalues as the original eigenvalue problem, recall again that if $[a][b] = [c]$, then the determinants of these matrices are such that $|a|\,|b| = |c|$. Then from $([D] - \lambda[I])\{p\} = \{0\}$, and the fact that $|L| \neq 0$

$$|L^{-1}KL^{-t} - \lambda I| = |L^{-1}(K - \lambda M)L^{-t}| = |L^{-1}||K - \lambda M||L^{-t}| = 0$$
$$\text{so} \quad |K - \lambda M| = 0.$$

Thus the determinant method shows that both formulations have the same eigenvalue roots. The solutions for the $\{q\}$ eigenvectors are obtained by returning to the original transformation equation, $\{q\} = [L]^{-t}\{p\}$, once the eigenvectors $\{p\}$ have been calculated.

Returning to the Jacobi method, the object is to further transform the eigenvalue problem $([D] - \lambda[I])\{p\} = \{0\}$ by postmultiplying by what is called a rotation matrix $[r]$ and premultiplying by the transpose of the same rotation matrix. In general terms, these very simple rotation matrices are chosen to render as zero just one of the pairs of off-diagonal terms of the symmetric dynamic matrix. Then this step is repeated. Unfortunately, the previously zeroed, off-diagonal terms do not stay zero. The post- and premultiplying must be repeated over and over again until all the off-diagonal terms become so small that they are less than some prechosen error limit. Then the result is, in place of the dynamic matrix, there is now a (very nearly) diagonal matrix whose diagonal terms are (very nearly) the eigenvalues, and the product of all of the rotation matrices is the matrix of eigenvectors.

In lieu of the proof of this process, which can be found in Ref. [6.9], p. 106, the following is simply a short explanation of why this process is possible. Return to the original eigenvalue solution statement for the ith eigenvalue and eigenvector and extend that statement to include all such solutions. That is

$$[D]\{q^{(i)}\} = \lambda_i\{q^{(i)}\}$$
$$\text{so} \quad [D][q^{(1)}|q^{(2)}|\ldots|q^{(n)}] = [q^{(1)}|q^{(2)}|\ldots|q^{(n)}][\backslash\lambda\backslash],$$

[14] To be positive definite, it may be necessary first to use the previously discussed static condensation technique so as to eliminate the DOF for which there are no assigned mass terms.

where $[\backslash\lambda\backslash]$ is the diagonal matrix of the N eigenvalues. Symbolize the square matrix of the eigenvectors as $[\Phi]$. Then, mimicking the result of the Jacobi process, the above equation can be written as

$$[D][\Phi] = [\Phi][\backslash\lambda\backslash]$$
$$\text{or} \quad [\Phi]^{-1}[D][\Phi] = [\backslash\lambda\backslash].$$

Since the original form of the eigenvalue problem statement is $([D] - \lambda[I])\{p\} = \{0\}$, the two weighting matrices for the mode shape orthogonality statements are the symmetric $[D]$ and $[I]$ matrices. Therefore the weighted orthogonal mode shapes which form the columns of $[\Phi]$ matrix can be normalized by dividing each mode shape by the square root of the generalized mass for that mode so that $[\Phi]^t[I][\Phi] = [\Phi]^t[\Phi] = [I]$. This last equation shows that, for this normalization, the transpose of the modal matrix is equal to its inverse. Substituting the transpose for the inverse in the above equation completes the demonstration that there exists a matrix transpose and matrix whose pre- and postmultiplication extracts the eigenvalues from the dynamic matrix, and, furthermore, the matrix that accomplishes that feat is the modal matrix. Further details can be found in Endnote (4), which illustrates this process with a short numerical calculation.

The *Givens* and *Householder* methods, Refs. [6.9,6.10,6.11], are important improvements to the Jacobi method. These two methods reduce the dynamic matrix to an exact tridiagonal matrix rather than an approximate diagonal matrix. A tridiagonal matrix is one that is zero everywhere but for the main diagonal and the entries immediately above and below the main diagonal. The advantage of going to a tridiagonal matrix is that such a matrix form can be accomplished in a finite number of steps as opposed to the open-ended, iterative procedure that is Jacobi's method. Then a follow-up procedure is required to determine the eigenvalues. Reference [6.10] offers several possible ways of calculating the eigenvalues of a tridiagonal matrix, among which are Sturm sequences (which mimic Sturm series for polynomials) and the bisection method. A procedure generally considered more effective is the QR method. This method is discussed, for example, in Refs. [6.10, 6.11].

The *QR method*, instead of using two triangular matrices, uses one triangular matrix $[R]$ and an *orthogonal* matrix $[Q]$. An orthogonal matrix[15] is a matrix whose transpose is equal to its own inverse. A very simple example of an orthogonal matrix is the matrix that relates the xy coordinates of one two-dimensional Cartesian coordinate system to those of another two-dimensional Cartesian coordinate system that is rotated through an angle θ with respect to the first. This matrix is

$$\begin{bmatrix} \cos\theta & \sin\theta \\ -\sin\theta & \cos\theta \end{bmatrix}.$$

Indeed, this simple matrix is the foundation of the rotational matrices used in the Jacobi method as discussed above.

As above, the decomposition of the dynamic matrix into the product $[Q][R]$ begins by first converting the dynamic matrix to a tridiagonal matrix, say, by use

[15] In some textbooks, such square matrices are also called *unitary*, particularly when the matrix entries are complex.

of the Householder method. The next step is to premultiply the dynamic matrix by a series of $n-1$ rotation matrices, which are orthogonal and which sequentially zero the $n-1$ entries below the main diagonal. Then the former dynamic matrix is turned into a right triangular matrix. In symbolic form, the result of the premultiplications is

$$[r_{n-1}][r_{n-2}]\ldots[r_1][D]=[R]$$
$$\text{then}\quad [D]=[Q][R]$$
$$\text{where}\quad [Q]=([r_{n-1}][r_{n-2}]\ldots[r_1])^{-1}=[r_1]^t[r_2]^t\ldots[r_{n-1}]^t.$$

An example, via Mathematica, of a QR decomposition of a symmetric matrix, after passing through the tridiagonal stage, is

$$\begin{bmatrix} 4 & 3 & 2 & 1 \\ 3 & 5 & -2 & 0 \\ 2 & -2 & 6 & 1 \\ 1 & 0 & 1 & 7 \end{bmatrix} = \begin{bmatrix} -0.730297 & 0.0147723 & 0.148337 & 0.666667 \\ -0.547723 & -0.598279 & -0.240037 & -0.533333 \\ -0.365148 & 0.782933 & -0.377586 & -0.333333 \\ -0.182574 & 0.169882 & 0.881934 & -0.400000 \end{bmatrix}$$
$$* \begin{bmatrix} -5.47723 & -4.19921 & -2.73861 & -2.37346 \\ 0.0 & -4.51294 & 6.09358 & 1.98688 \\ 0.0 & 0.0 & -0.606835 & 5.94429 \\ 0.0 & 0.0 & 0.0 & -2.46667 \end{bmatrix}.$$

Once the original decomposition is complete, call it $[D_0]=[Q_0][R_0]$. Then the QR procedure begins by writing $[D_1]=[R_0][Q_0]$. After $[D_1]$ is decomposed into $[Q_1][R_1]$, then the reversal procedure of writing $[D_2]=[R_1][Q_1]$ is repeated, and so on. If all the eigenvalues are distinct, the final $[D]$ is a diagonal matrix where the diagonal entries are the eigenvalues of the original dynamic matrix.

It can be seen that the eigenvalues of the each successive dynamic matrix are the same as the preceding dynamic matrix. First of all, from Eq. (6.9), the eigenvalues of the symmetric dynamic matrix are those of the system. Now with, say, $[D_1]=[Q_1][R_1]$ and $[D_2]=[R_1][Q_1]$, then $[D_2]=[Q_1]^{-1}[D_1][Q_1]$. Recall the theorem that says that because $[D_2]=[Q_1]^{-1}[D_1][Q_1]$, there is the determinant relationship $|D_2|=|Q_1|^{-1}\,|D_1||Q_1|$. Since the determinant value of the inverse of $[Q_1]$ is the inverse of the determinant of $[Q_1]$, the determinant and hence the eigenvalues of $[D_2]$ are the same as those of $[D_1]$.

There are several variations on this method, such as the QL method and the double QR method. A presently popular method of calculating eigenvalues and eigenvectors is the Lanczos' algorithm, Ref. [6.12, 6.13], which is another variation on the QR method. The Lanczos method is particularly suited for sparse, symmetric matrices and for the calculation of the smallest (or largest) eigenvalues and their corresponding eigenvectors. Reference [6.13] suggests a general preference for applying the Cholesky decomposition to the stiffness matrix with the inverse of the frequency term forming the eigenvalue. The effectiveness of the Lanczos method depends to some extent on the selection of frequency shifts. Frequency shifts, sometimes called spectral shifts, were discussed at the end of the section on matrix iteration.

6.9 Summary

The solution of the general structural dynamics matrix equation $[M]\{\ddot{q}\} + [C]\{\dot{q}\} + [K]\{q\} = \{Q\}$ is most often obtained by first solving the included problem, the set of homogeneous differential equations $[M]\{\ddot{q}\} + [K]\{q\} = \{0\}$. The undamped, free vibration solution is one where all the DOF solutions are proportional to a single time function that, as it turns out, must be a sinusoid. Physically, this means that for this type of motion, called harmonic motion, all the DOF are either in phase (+) or out of phase (−). Since the solution can be written as $\{q(t)\} = \{A\}\sin(\omega t + \psi)$, the acceleration vector is negatively proportional to the deflection vector, that is, $\{\ddot{q}\} = -\omega^2\{q\}$. The resulting homogeneous, algebraic, matrix equation $(-\omega^2[M] + [K])\{A\} = \{0\}$ is a matrix eigenvalue problem whose two part solution is the vibratory system's natural frequencies that are an ordered set of discrete positive values, ω_j, and a corresponding set of mode shapes, $\{A^{(j)}\}$. Each of the $N \times 1$ mode shape vectors is unique only to the extent of the relative values of its N entries; that is, $\{A^{(j)}\}$ is not a different solution from $c\{A^{(j)}\}$, where c is an arbitrary factor. The natural frequencies and mode shapes respectively are the eigenvalues and eigenvectors of the matrix equation. The quantities ω_j and $\{A^{(j)}\}$ are often called an eigenpair.

There are many ways of calculating the eigenvalues and eigenvectors of a dynamical system. The determinant method was used to introduce the fact that an N DOF system has N natural frequencies and N associated mode shapes. Once it was demonstrated that any vector of the same size as the eigenvectors could be written as a weighted sum of the eigenvectors, the matrix iteration method was established as a procedure suitable to calculate the eigenvalues and eigenvectors of small and mid-sized eigenvalue problems, particularly when frequency shifts are used with the larger problems. Eigenvalue problems of large size (i.e., with over a few hundred DOF) are generally solved using the QR method or one of its many variations. Figures 6.7(a) and (b) illustrate typical results of computer-based analyses.

The QR method requires positive definite as well as symmetric matrices for its procedures. Although the kinetic energy is a positive definite function of time, the mass matrix may only be positive semidefinite because of zero mass terms associated with some DOF such as the rotational DOF. As explained previously, these DOF can be removed from the problem statement by use of static condensation. The stiffness matrix can also be singular or positive semidefinite because the structural system is able to undergo one or more rigid body motions. Again, the effects of the rigid body motion can be removed by writing the coordinate transformations explained in Endnote (3). If these effects are not removed, then for a structural system with m possible rigid body motions ($m \leq 6$), the first m natural frequencies are zero, and the corresponding mode shapes are the deflection patterns of the rigid body motions. The presence of zero natural frequencies is illustrated in the following example problem where the simple structure is one where rigid body motion is possible.

EXAMPLE 6.13 Calculate all the natural frequencies and the first mode shape of the two-mass, one-beam element structural system shown in Figure 6.8. Note that there are unseen static lift forces counterbalancing the unseen gravitational forces

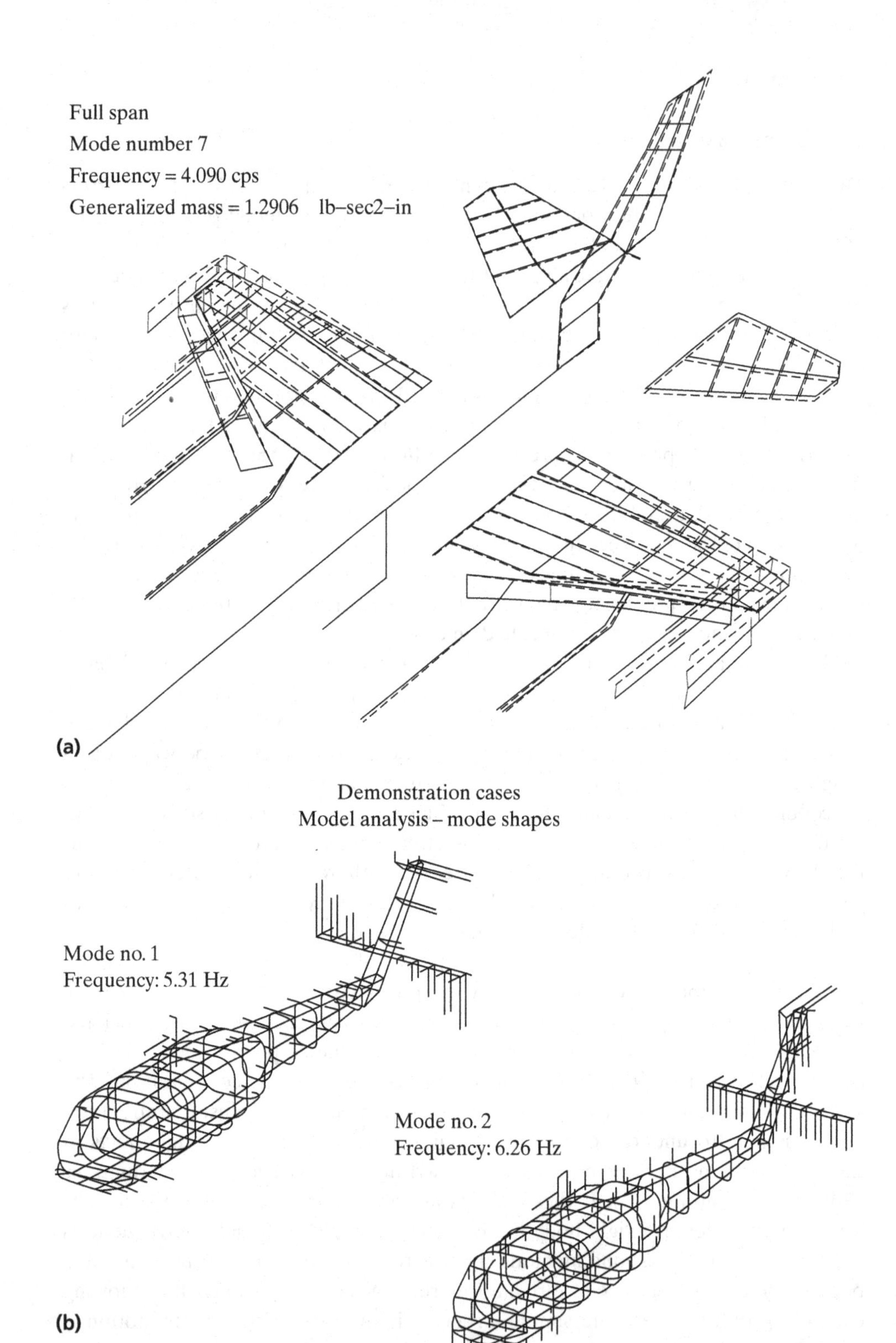

Figure 6.7. (a) Visual display of a F-16 finite element modal calculation. Courtesy of Mr. Jack A. Ellis, Lockheed-Martin Tatical Air Systems. (b) Another visual representation of mode shapes of a complicated structure. The first mode involves side bending and twisting. The second mode is mostly bending in the vertical plane. Taken from NASA CR 181975, NASTRAN result for Blackhawk helicopter.

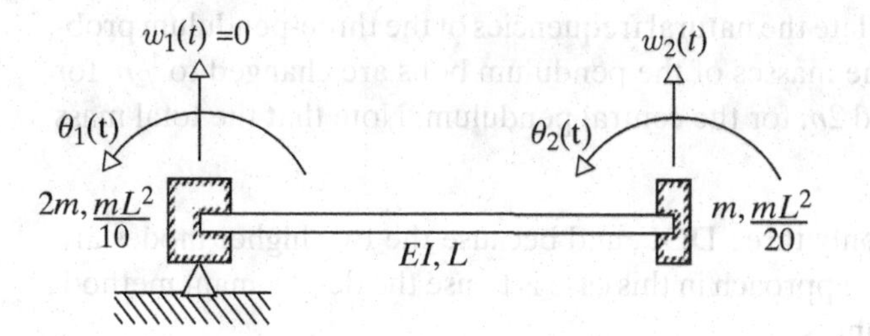

Figure 6.8. Structural system capable of one rigid body motion.

so that the SEP of the beam element is horizontal. Note the four DOF symbols shown.

SOLUTION Write the mass and FEM stiffness matrices for this three-DOF system as per usual. To achieve solely numerical values for the mass and stiffness matrix entries, divide the first and third rows by the beam element length and factor the remaining beam element lengths so that the deflection vector, mass matrix, and stiffness matrix are

$$\{q\} = \begin{Bmatrix} L\theta_1 \\ w_2 \\ L\theta_2 \end{Bmatrix}, \quad [M] = m\begin{bmatrix} 0.1 & & \\ & 1.0 & \\ & & 0.05 \end{bmatrix}, \quad [K] = \frac{EI}{L^3}\begin{bmatrix} 4 & -6 & 2 \\ -6 & 12 & -6 \\ 2 & -6 & 4 \end{bmatrix}.$$

Consider the determinant of the stiffness matrix. Note that the sum of the three rows of that determinant is a row of zeros. Thus the stiffness matrix is singular as expected and cannot be inverted to form a dynamic matrix.[16] Therefore the analysis will be carried out using the determinant method that is oblivious to the singularity of the square matrices. Form the eigenvalue λ so that $\lambda = \omega^2 mL^3/EI$ and write the determinant equation $\|[K] - \lambda[M]\| = 0$. After multiplying the first and third rows of the determinant by 10 and 20, respectively, to clear fractions without altering the value of the determinant, the determinant equation is

$$\begin{vmatrix} (40-\lambda) & -60 & 20 \\ -6 & (12-\lambda) & -6 \\ 40 & -120 & (80-\lambda) \end{vmatrix} = (\lambda - 0)(\lambda^2 - 132\lambda + 2760) = 0.$$

Therefore, the solutions for the natural frequencies are

$$\omega_1 = 0.0 \qquad \omega_2 = 5.1\sqrt{\frac{EI}{mL^3}} \qquad \omega_3 = 10.3\sqrt{\frac{EI}{mL^3}},$$

Substituting the zero value of the first eigenvalue back into the determinant equation, discarding the third row, and solving the first eigenvector is, as expected, a rigid body rotation. That is

$$\lfloor L\theta_1 \quad w_2 \quad L\theta_2 \rfloor = \lfloor 1 \quad 1 \quad 1 \rfloor \quad \text{or} \quad \lfloor \theta_1 \quad w_2 \quad \theta_2 \rfloor = \left\lfloor \frac{1}{L} \quad 1 \quad \frac{1}{L} \right\rfloor,$$

where the two rotation DOF are equal to each other and equal to the small tip deflection divided by the beam length. ★

[16] See Endnote (2) for an alternative procedure.

EXAMPLE 6.14 Recalculate the natural frequencies of the three-pendulum problem of Example 6.2 when the masses of the pendulum bobs are changed to $\frac{1}{2}m$ for the outboard pendulums and $2m$ for the central pendulum. Note that the total mass remains the same at $3m$.

SOLUTION Since there are only three DOF, and because the two higher modes are to be calculated, the simplest approach in this case is to use the determinant method. The altered 3×3 determinant is

$$\begin{vmatrix} 2\beta - \frac{1}{2}\omega^2 & -\beta & 0 \\ -\beta & 3\beta - 2\omega^2 & -\beta \\ 0 & -\beta & 2\beta - \frac{1}{2}\omega^2 \end{vmatrix} = 0.$$

Define the eigenvalue $\lambda = \omega^2/\beta$. Then the characteristic equation obtained from the above determinant is

$$27\lambda^3 - 198\lambda^2 + 432\lambda - 256 = 0.$$

After using, say, Newton–Raphson, the three ordered roots of this equation, to six significant figures, are 0.958965, 2.66667, and 3.70771. The consequent natural frequencies and mode shapes are now

$$\omega_1 = 0.979\sqrt{\beta} \quad \omega_2 = 1.633\sqrt{\beta} \quad \omega_3 = 1.926\sqrt{\beta}$$

$$\{A\}^{(1)} = \begin{Bmatrix} 0.780776 \\ 1.00000 \\ 0.780776 \end{Bmatrix} \quad \{A\}^{(2)} = \begin{Bmatrix} -1.0 \\ 0.0 \\ 1.0 \end{Bmatrix} \quad \{A\}^{(3)} = \begin{Bmatrix} 1.0 \\ -0.780776 \\ 1.0. \end{Bmatrix}.$$

Compare the above solution for the altered mass distribution to the original pendulum system solution

$$\omega_1 = 1.0\sqrt{\beta} \quad \omega_2 = 1.414\sqrt{\beta} \quad \omega_3 = 2.0\sqrt{\beta}$$

$$\{A\}^{(1)} = \begin{Bmatrix} 1.0 \\ 1.0 \\ 1.0 \end{Bmatrix} \quad \{A\}^{(2)} = \begin{Bmatrix} -1.0 \\ 0.0 \\ 1.0 \end{Bmatrix} \quad \{A\}^{(3)} = \begin{Bmatrix} 0.5 \\ -1.0 \\ 0.5 \end{Bmatrix},$$

where the last eigenvector was normalized on -1 to better facilitate comparisons. It is clear that the mode shapes have generally changed more than the natural frequencies, and yet their patterns are not all that different. Because of the symmetry of the three-pendulum system, the second mode shape has not changed at all, but its corresponding natural frequency has undergone a 15% change. This is so because the masses in motion for this mode have been reduced in magnitude by 50%. ★

It is well to repeat that natural frequencies and mode shapes are not only the first step toward quantifying a structural response, but as will be explained in the next chapter, the natural frequencies are useful for qualitatively judging the dynamic response of a structure. There are other direct uses for the natural frequencies. For example, the first natural frequency is useful for structural material fatigue estimates. When a structure is subjected to a series of intermittent pulses, such as possibly wind gusts, the structure will mostly vibrate at its first natural frequency. Thus f_1 can be use for estimating the number of significant fatigue loading cycles experienced by the structure when subject to such winds. Another immediate, and presently important, use for the natural frequencies is discussed in the next section.

6.10 **Model Tuning**

As is generally true in industrial and governmental practice, all elastic structures discussed in this textbook are modeled for finite element analyses. Although finite element analysis is a mature field, there are aspects of finite element modeling, as in any other type of structural modeling, that can be significantly in error. A good example of imprecision in mathematical modeling of physical structures is the modeling of connections and boundary conditions. For example, consider the two familiar mathematical descriptions of a physical structural joint. The hinged connection and the clamped connection represent the two extremes of relative rotational constraint, which are none and infinite, respectively. Either of these models can be a poor approximation for a riveted or bolted joint or a control system activated hinge with it finite rotational constraint. All such physical connections could be somewhat better approximated as being hinged with the addition of a linear torsional spring that would provide a finite amount of torsional restraint. However, it is seldom apparent what even the approximate magnitude of such a linear torsional spring stiffness should be. Indeed, the actual torsional resistance is likely to be a nonlinear function of relative rotation angle, and as the joint flexes over time, and the frictional resistance to rotation lessens, the torsional spring stiffness will asymptotically decrease to some nonzero value.

Of course there other types of inaccuracies that are inherent in finite element modeling. These modeling inaccuracies arise from (i) the choice of the (small) number of elements used in the model; (ii) the simplifications built into even such accurate elements as beam elements[17]; (iii) the approximation of the structural geometry using straight-line elements; and, of course, (iv) a lack of precise information on material parameters such as Young's modulus and the Poisson ratio. The difficulty in obtaining accurate values for the various Young's moduli and Poisson ratios can be a serious source of error in fiber composite materials.

Of course, the purpose of any mathematical model of a structure is to accurately predict the physical structure's response to all types of loads. The only way engineers have for determining a physical structures' actual, as opposed to predicted, responses is through testing. There are essentially two types of tests: static load tests and dynamic load tests. The latter, when the intent is to experimentally determine the structure's natural frequencies and mode shapes, are called "vibration tests." Vibration tests offer, via the experimentally determined values of the natural frequencies, and perhaps the mode shapes, too, an opportunity for the analyst to improve the accuracy of the finite element model of the structure. The complication is, of course, that experimental measurements also contain errors independent of the errors of the

[17] Beam elements are based on the strength of materials approximations that planar beam cross sections remain planar, undistorted, and perpendicular to the beam axis, after bending. Because of shearing and the Poisson effect, none of these quite reasonable approximations is precisely true. Even when shearing effects are approximated in the beam finite element, the approximation is crude.

analytical model. Since the 1970s, a group of remarkable analytical procedures has been developed that offer the structural analyst the opportunity to use the flawed experimental results to improve the flawed analytical model. There are several names given to these procedures. Modal tuning, structural validation, structural updating, and structural identification are chief among them.

There are two classes of competing procedures for modal tuning. These are the "deterministic optimization" and the "sensitivity based parameter identification" approaches. The latter approach, which currently seems to be the more fruitful, is further divided into submatrix updating and physical parameter updating. See Ref. [6.14]. The physical parameter updating procedure seems to be attracting the most attention because its results, which are the revised (i.e., corrected) system parameters are open to clear physical interpretation, whereas the submatrix updating procedure, although potentially equally useful for most structural analysis purposes, is not open to clear physical interpretation.

The physical parameter updating procedure generally begins with the construction of a quadratic objective function involving both the system test responses, such as the measured values of the natural frequencies, and the parameters of the mathematical model, such as beam element area moments of inertia. This approach is sometimes called Bayesian parameter estimation. The objective function is mostly called either the error function or the cost function in structural engineering circles, whereas mathematicians and electrical engineers may call it the estimator. The commonly used form of the error function to be minimized is written as

$$E = \lfloor \Delta R \rfloor [C_r] \{\Delta R\} + \lfloor \Delta P \rfloor [C_p] \{\Delta P\},$$

where the ΔR vector is the vector difference between the continuously corrected analytical response quantities and the fixed experimental values for the response quantities, where examples of response quantities are natural frequencies and mode shapes. Similarly, the ΔP vector is the vector difference between the continuously corrected values of the analytical model parameters and the fixed initial values of those parameters. Examples of these parameters are material constants and geometric factors, like thicknesses or area moments of inertia, which may not be accurately known. Parameters can either refer to an individual finite element, and thus be "local," or refer to a set of finite elements, and thus be "global." The values of diagonal weighting matrix $[C_r]$ are selected to reflect the confidence placed in the test results. Similarly, the values of diagonal weighting matrix $[C_p]$ are selected to reflect the confidence placed in the parameters of the analytical model. A convenient way of viewing these confidence values is to view them as the inverse of the ratio of the expected error in a given quantity to the value of the quantity itself. For example, an expected error of only 0.25% would mean a high confidence value of 400. The selection of the relative confidence values depends on the test engineer being familiar with his or her test procedures and their limitations and the analyst being familiar with the shortcomings of his or her finite element model. If the $[C_r]$ values are selected to be greater than the $[C_p]$ values, then the process of minimizing the error function places a greater emphasis on minimizing the difference between the response vectors than on minimizing the difference between the parameter vectors. This in turn means that there is more leeway in adjusting the parameters to achieve a greater coincidence for, say, the test natural frequencies and the updated analytical frequencies. Experience has shown that $[C_r]$ values that are two orders of magnitude greater than the $[C_p]$ values are necessary to appreciably change the model parameters so that, for example, the analytical values of the natural frequencies are now close to the corresponding experimental values. See, for example, Ref. [6.15].

Some details of the rather extensive calculations necessary to minimize the error function are as follows [6.16]. The first step is establishing the relationship between the difference in the responses vector, and the differences between the parameters vector, which can be written as

$$\{\Delta R\} = [S]\{\Delta P\}, \tag{6.12}$$

where $[S]$ is called the sensitivity matrix. For the purposes of discussion, it is helpful to be specific about the response quantities, but it is unnecessary to be specific about the parameters. Let the response quantities be the experimentally determined natural frequencies. It is possible to also include the measured mode shapes, but the combination of the natural frequencies and the mode shapes can lead to numerical ill-conditioning.

The sensitivity matrix is obtained from a one-term Taylor's series expansion of ΔR in terms of ΔP. This means that $[S]$ is a matrix composed of the first partial derivatives of the various response quantities, the natural frequencies, with respect to each of the parameters. These derivatives are expanded by use of the chain rule into (i) derivatives of the natural frequencies with respect to the entries of the mass and stiffness matrices and (ii) derivatives of the mass and stiffness matrix entries with respect to the selected model parameters. The latter set of derivatives are straightforward, and many times quite easy to fashion, such as, for example, the derivative of a particular stiffness matrix entry with respect to the area moment of inertial of a particular beam element or the Young's modulus of a particular plate element. The derivatives that need comment are the former set of derivatives whose development was a key step in the creation of the modal tuning process.

To consider the derivative of, say, the ith eigenvalue, λ_i with respect to a mass matrix term of the rth row and sth column, m_{rs}, start with the ith modal solution

$$\lambda_i[m]\{A\}^{(i)} = [k]\{A\}^{(i)}$$

$$\text{or} \quad \lambda_i \sum_{k=1}^{n} m_{jk}A_{ik} = \sum_{l=1}^{n} k_{jl}A_{il}. \tag{6.13}$$

Note that here the eigenvalue is directly proportional to the natural frequency squared. Make the approximation that the above typical stiffness term is independent of the above typical mass term. Then, a straightforward partial differentiation of the second of the above equations with respect to m_{rs} is as follows:

$$\frac{\partial\lambda_i}{\partial m_{rs}} \sum_{k=1}^{n} m_{jk}A_{ik} + \lambda_i \sum_{k=1}^{n} \frac{\partial m_{jk}}{\partial m_{rs}} A_{ik} + \lambda_i \sum_{k=1}^{n} m_{jk} \frac{\partial A_{ik}}{\partial m_{rs}} = \sum_{l=1}^{n} k_{jl} \frac{\partial A_{il}}{\partial m_{rs}}. \tag{6.14}$$

Now multiply the Eq. (6.14) by A_{ij} and sum all terms over j, and note that

$$\sum_{j=1}^{n}\sum_{k=1}^{n} m_{jk}A_{ij}A_{ik} = \lfloor A^{(i)} \rfloor [m]\{A^{(i)}\} = M_i,$$

where, as before, M_i is the ith modal mass. Hence the Eq. (6.14) becomes

$$\frac{\partial \lambda_i}{\partial m_{rs}} M_i + \lambda_i \sum_{j=1}^{n} \sum_{k=1}^{n} \frac{\partial m_{jk}}{\partial m_{rs}} A_{ij} A_{ik} + \lambda_i \sum_{j=1}^{n} \sum_{k=1}^{n} m_{jk} \frac{\partial A_{ik}}{\partial m_{rs}} A_{ij} = \sum_{j=1}^{n} \sum_{k=1}^{n} k_{jk} \frac{\partial A_{ik}}{\partial m_{rs}} A_{ij}. \tag{6.15}$$

With respect to the second of the above terms, the generalized coordinates are taken to be discrete deflections so that the independence of the magnitude of any one lumped mass with respect to the magnitude of any other lumped mass allows the partial derivative of the one-mass entry term with respect to the other mass entry term to be viewed as being 1.0 when $j = r$ and $k = s$ and zero otherwise. Therefore the first of the above sums can be replaced by the single term $\lambda_i A_{ir} A_{is}$.

Now consider the third of the above four terms in Eq. (6.15). Making use of the symmetry of both the mass and stiffness matrix entries, and Eq. (6.14)

$$\lambda_i \sum_{j=1}^{n} \sum_{k=1}^{n} m_{jk} \frac{\partial A_{ik}}{\partial m_{rs}} A_{ij} = \sum_{k=1}^{n} \left[\sum_{j=1}^{n} \lambda_i m_{kj} A_{ij} \right] \frac{\partial A_{ik}}{\partial m_{rs}} = \sum_{k=1}^{n} \left[\sum_{j=1}^{n} k_{kj} A_{ij} \right] \frac{\partial A_{ik}}{\partial m_{rs}}$$

$$= \sum_{j=1}^{n} \sum_{k=1}^{n} k_{jk} \frac{\partial A_{ik}}{\partial m_{rs}} A_{ij}.$$

Thus the third and fourth terms of Eq. (6.15) cancel. Therefore the conclusion that

$$\frac{\partial \lambda_i}{\partial m_{rs}} = -\frac{\lambda_i}{M_i} A_{ir} A_{is}.$$

Similarly, it can be established that

$$\frac{\partial \lambda_i}{\partial k_{rs}} = +\frac{1}{M_i} A_{ir} A_{is}.$$

Of course, for example, some of the m_{rs} terms will generally have a zero value. This fact is not upsetting because each of the above partial derivatives is to be multiplied by a partial derivative of m_{rs} with respect to some particular element parameter such as an element area moment of inertia, and this latter derivative will be zero.

Since the purpose of the system tuning process is to determine the change in the parameters given the difference between the experimental response (eigenvalue) vector and the analytical response vector, a relationship is sought in the form

$$\{\Delta P\} = [G]\{\Delta R\},$$

where $[G]$ is called the gain matrix. The above relationship can be realized by means of a matrix inverse of Eq. (6.12) only if the sensitivity matrix is a square matrix. The sensitivity matrix is rarely if ever square because, on the one hand, there is only a limited number of experimentally determined natural frequencies (typically a dozen or two), whereas, on the other hand, in a real problem, the parameters selected for updating are usually associated with quite a few finite elements of a finite element model. When the sensitivity matrix is rectangular, pseudo inverses could possibly be used. The use of such inverses is part of the first mentioned "deterministic optimization" process mentioned at the beginning of this section. However, that is not the preferred solution procedure. One of the preferred options is minimizing the error function is setting the various partial derivatives of the error function equal to zero. The acquisition of the gain matrix does not complete the determination of the change in the parameters because the process is nonlinear. The source of the nonlinearity lies in the derivatives of the eigenvalues with respect to the usual types of selected parameters. The solution for the updated parameters must be obtained by repeated iterations, and convergence must be evaluated. The iterative procedure requires considerable computation for real problems in that in each iterative cycle the new eigenvalues and their derivatives need to be computed. Commercial software is available for these calculations. Reference [6.17] is a recent textbook on this subject.

REFERENCES

6.1 Fraleigh, J. B., and R. A. Beauregard, *Linear Algebra*, 2nd ed., Addison-Wesley, Reading, MA, 1990, p. 238.

6.2 Donaldson, B., *Analysis of Aircraft Structures: An Introduction*, McGraw-Hill, New York, 1993.

6.3 Wolfram, S., *Mathematica: A System for Doing Mathematics by Computer*, 2nd ed., Addison-Wesley, Redwood City, CA, 1991.

6.4 Leissa, A. W., *Vibration of Plates*, NASA SP-160, National Aeronautics and Space Administration, 1969.

6.5 Wylie, C. R., and L. C. Barrett, *Advanced Engineering Mathematics*, McGraw-Hill, New York, 1982, Section 4.2.

6.6 Temple, G., and W. G. Bickley, *Rayleigh's Principle and Its Applications to Engineering*, Dover, New York, 1956.

6.7 PC-MATLAB for Unix Computers, Version 3.5j, The MathWorks, Inc., South Natick, MA 01760, 1991.

6.8 Chopra, A. K., *Dynamics of Structures, Theory and Applications to Earthquake Engineering*, Prentice Hall, Englewood Cliffs, NJ, 1995.

6.9 Meirovitch, L., *Analytical Methods in Vibrations*, Macmillan, New York, 1967.

6.10 Petyt, M., *Introduction to Finite Element Vibration Analysis*, Cambridge University Press, Cambridge, UK, 1990.

6.11 Kreyszig, E., *Advanced Engineering Mathematics*, 8th ed., John Wiley & Sons, New York, 1999.

6.12 Lanczos, C., "An iteration method for the solution of eigenvalue problems of linear differential and integral operators," *J. Res. Nat. Bur. Stand.*, 45, 1950, p. 255.

6.13 Grimes, R. G., J. G. Lewis, and H. Simon, "A shifted block Lanczos algorithm for solving sparse symmetric generalized eigenproblems," AMS-TR-166, Boeing Computer Services, Seattle, July 1991.

6.14 Dascotte, E., J. Strobbe, and H. Hua, "Sensitivity-based model updating using multiple types of simultaneous state variables," *Proceedings of the 13th International Modal Analysis Conference*, Feb. 1995.

6.15 Blakely, K. D., and W. B. Walton, "Selection of measurement and parameter uncertainties for finite element revision," *Proceedings of the 2nd International Modal Analysis Conference*, Feb. 1984.

6.16 Collins, J. D., G. C. Hart, T. K. Hasselman, and B. Kennedy, "Statistical identification of structures," *AIAA J.*, Vol. 12, 2, 1974.

6.17 Friswell, M. I., and J. E Mottershead, *Finite Element Model Updating in Structural Dynamics*, Kluwer, Dordrecht, The Netherlands, 1995.

6.18 Chapra, S. C., and R. P. Canale, *Numerical Methods for Engineers*, 2nd ed., McGraw-Hill, New York, 1988.

CHAPTER 6 EXERCISES

6.1 **(a)** From the determinant method, the polynomial equation for the natural frequencies of a three-DOF system is as follows:

$$\begin{vmatrix} (5-3\lambda) & -2 & 0 \\ -2 & (3-2\lambda) & -1 \\ 0 & -1 & (1-\lambda) \end{vmatrix} = 6\lambda^3 - 25\lambda^2 + 27\lambda - 6 = 0,$$

$$\text{where} \qquad \lambda = \frac{m\omega^2 L^3}{10EI}.$$

Calculate the value of the lowest natural frequency (hint: $0 < \lambda < .5$) and its corresponding mode shape. Normalize so that the largest value of the mode shape is $+1.0$. Be accurate to two or more significant figures in your final answers.

(b)(c)(d) Use the determinant method to determine all three natural frequencies of the systems whose mass and stiffness matrices are listed below. Be sure to list the frequency solutions in their proper order.

$$\text{(b)} \quad [M] = m\begin{bmatrix} 4 & & \\ & 3 & \\ & & 1 \end{bmatrix} \qquad [K] = k\begin{bmatrix} 8 & -3 & -1 \\ -3 & 5 & -2 \\ -1 & -2 & 3 \end{bmatrix}$$

$$\text{(c)} \quad [M] = m\begin{bmatrix} 3 & & \\ & 2 & \\ & & 2 \end{bmatrix} \qquad [K] = k\begin{bmatrix} 5 & -1 & -1 \\ -1 & 2 & -1 \\ -1 & -1 & 2 \end{bmatrix}$$

$$\text{(d)} \quad [M] = m\begin{bmatrix} 5 & & \\ & 3 & \\ & & 1 \end{bmatrix} \qquad [K] = k\begin{bmatrix} 12 & -4 & -2 \\ -4 & 6 & -2 \\ -2 & -2 & 4 \end{bmatrix}.$$

(e) What is the mode shape for a single-DOF system? Hint: Such a mode shape would have only a single numerical entry.

6.2 **(a)** Calculate the first mode shape for part (b) above.

(b) Calculate the first mode shape for part (c) above.

(c) Calculate the first mode shape for part (d) above.

6.3 Use matrix iteration to calculate the first natural frequency and first mode shape for the structural systems whose mass and stiffness matrices are as follows

$$\text{(a)}\quad [M] = m\begin{bmatrix} 4 & 2 & 1 \\ 2 & 2 & 1 \\ 1 & 1 & 1 \end{bmatrix} \qquad [K] = k\begin{bmatrix} 8 & & \\ & 6 & 2 \\ & 2 & 6 \end{bmatrix}$$

$$\text{(b)}\quad [M] = m\begin{bmatrix} 6 & 5 & 3 \\ 5 & 5 & 3 \\ 3 & 3 & 3 \end{bmatrix} \qquad [K] = k\begin{bmatrix} 5 & & \\ & 10 & \\ & & 10 \end{bmatrix}$$

$$\text{(c)}\quad [M] = m\begin{bmatrix} 14 & 7 & 2 \\ 7 & 7 & 2 \\ 2 & 2 & 2 \end{bmatrix} \qquad [K] = k\begin{bmatrix} 12 & & \\ & 6 & \\ & & 4 \end{bmatrix}.$$

6.4 (a) Use matrix iteration to determine the first eigenvalue and eigenvector when the three-DOF dynamic matrix is, by rows [1 1 1; 2 6 6; 2 6 14].

(b) Use matrix iteration to calculate the second eigenvalue and eigenvector when the diagonal mass matrix entries are $[\backslash \tfrac{1}{2} \;\; 1 \;\; 1 \backslash]$. Hint: It is suggested that your first guess for the second mode shape be $\lfloor 1 \;\; 1 \;\; -1 \rfloor$.

(c) From Exercise 5.7(b) the mass and stiffness matrices of the three-story building, in terms of DOF relative to the support, are

$$[M] = m\begin{bmatrix} 1 & 0 & 0 \\ 0 & 1 & 0 \\ 0 & 0 & 2 \end{bmatrix} \qquad [K] = \frac{36EI}{L^3}\begin{bmatrix} 2 & -1 & 0 \\ -1 & 2 & -1 \\ 0 & -1 & 1 \end{bmatrix}.$$

Use matrix iteration to calculate the first natural frequency and mode shape. Then use a frequency shift to search for a natural frequency near $7.0(EI/mL^3)^{1/2}$. Then use the inverse of the dynamic matrix to determine the third natural frequency.

(d) From Exercise 5.7(d) the mass and stiffness matrices of the three-story building, in terms of DOF relative to the support or the DOF located at the story immediately below, are

$$[M] = m\begin{bmatrix} 4 & 3 & 2 \\ 3 & 3 & 2 \\ 2 & 2 & 2 \end{bmatrix} \qquad [K] = \frac{36EI}{L^3}\begin{bmatrix} 1 & 0 & 0 \\ 0 & 1 & 0 \\ 0 & 0 & 1 \end{bmatrix}.$$

Follow the approach outlined in part (c) to determine the natural frequencies and mode shapes. Should the frequency solutions be the same? Should the mode shape solutions be the same?

6.5 Use matrix iteration to calculate the first natural frequency and mode shape for the structural system whose mass and stiffness matrices are those of

(a) part (b) of Exercise 6.1.

(b) part (c) of Exercise 6.1.

(c) part (d) of Exercise 6.1.

6.6 Use matrix iteration and the symmetric dynamic matrix of Example 6.10 to recalculate the first natural frequency of the cantilevered beam of Example 6.3. Note that the mode shape will not be the same as previously calculated because of the transformation $\{q\} = [R^{-1}]\{p\}$.

6.7 (a) Calculate the Cholesky factors $[L][L]^t$ for the stiffness matrix of Exercise 6.1(a).

(b) As above, for the stiffness matrix of 6.1(b).

(c) As above, for the stiffness matrix of 6.1(c).

(d) Calculate the Cholesky factors $[R]^t[R]$ for the matrix

$$[K] = k\begin{bmatrix} 3 & -1 \\ -1 & 2 \end{bmatrix}.$$

(e) Use the determinant method to prove that the eigenvalues of an upper or a right triangular matrix, U or R, are the diagonal entries of the triangular matrix.

(f) By hand calculate the inverse of the following matrix

$$\begin{bmatrix} 3 & & & \\ 2 & 4 & & \\ 1 & 5 & 9 & \\ 0 & 0 & 6 & 3 \end{bmatrix}$$

and thus show that the inverse of a lower or left triangular matrix is another lower triangular matrix whose diagonal entries are the inverses of the diagonal entries of the original matrix.

For the eager

6.8 (a) Hand calculate the second mode shape by means of matrix iteration for the structural system whose 3×3 factored diagonal mass matrix has the entries 1, 2, 4, and whose factored nonsymmetric dynamic matrix and first mode shape are

$$[D] = \begin{bmatrix} 1.00 & 2.00 & 4.00 \\ 1.00 & 4.00 & 8.00 \\ 1.00 & 4.00 & 12.0 \end{bmatrix} \quad \text{and} \quad \lfloor A^{(1)} \rfloor = \lfloor 0.3821 \quad 0.7392 \quad 1.0 \rfloor.$$

The eigenvalue is $k/(m\omega^2)$. Hint: A first step to consider is to improve the numerical accuracy of the given first mode shape by further iteration.

(b) Using computer software of your choice (since its too lengthy for a hand calculation), perform an matrix iteration method frequency shift to obtain a higher mode, hopefully the second mode, for the following four-DOF vibratory system:

$$[M] = m\begin{bmatrix} 6.00 & 0 & 0 & 0 \\ 0 & 4.00 & 0 & 0 \\ 0 & 0 & 2.00 & 0 \\ 0 & 0 & 0 & 1.00 \end{bmatrix} \quad [K] = k\begin{bmatrix} 9.00 & -3.00 & 0 & 0 \\ -3.00 & 5.00 & -2.00 & 0 \\ 0 & -2.00 & 3.00 & -1.00 \\ 0 & 0 & -1.00 & 1.00 \end{bmatrix}.$$

Let $k/m = 1.0$. The eigenvalue is $(m\omega^2/k)$, and the dynamic matrix, $[K]^{-1}[M]$, is

$$[D] = \begin{bmatrix} 1 & 2/3 & 1/3 & 1/6 \\ 1 & 2 & 1 & 1/2 \\ 1 & 2 & 2 & 1 \\ 1 & 2 & 2 & 2 \end{bmatrix}.$$

The solution for the first eigenvalue is 4.84353, which produces a first natural frequency of $0.454380\sqrt{k/m}$. (The first mode shape entries are 0.203501, 0.526474, 0.793539, 1.00000.) Guess the second modal frequency is about twice the value of the first modal frequency, rounded off so that $\omega_s = 0.9$. Expect a single sign change in the mode shape if you find the second mode shape.

6.9 If $\boldsymbol{e}_1, \boldsymbol{e}_2, \ldots, \boldsymbol{e}_n$ are a set of orthogonal unit vectors in N-space, then prove that they are linearly independent. It might be useful to consult Endnote (1).

6.10 The stiffness matrix of Example 3.4 was easily obtained using the deflection FEM. The matrix iteration technique requires the use of the inverse of the stiffness matrix, which is the flexibility matrix. The hand calculation of that 6×6 inversion would be tedious, particularly if the beneficial division of the 6×6 matrix into 4×4 and 2×2 submatrices, half of which are null, went unnoticed. An alternate approach would be to directly calculate the flexibility matrix using a force method such as Castigliano's second theorem, the unit load method, and so on.

(a) Use such a force method to calculate any *one* diagonal term and any *one* off-diagonal term of the 6×6 flexibility matrix, and make your own estimate of the relative merit of formulating and then inverting the 6×6 stiffness matrix versus directly calculating the 6×6 flexibility matrix.

(b) What checks are available on the accuracy of the calculated flexibility matrix?

6.11 Redo Example 6.7, using the determinant method to calculate the four natural frequencies, where now the parameter α has the values
(a) 0.3 (b) 0.7 (c) 0.4 (d) 0.6 (e) 1.0.

6.12 Prove that a necessary and sufficient condition for a trial vector not having the first mode shape as a component is a trial vector that is orthogonal to the first mode shape.

ENDNOTE (1): LINEARLY INDEPENDENT QUANTITIES

Again, the n quantities (functions, vectors, etc.) V_i are linearly independent if and only if the equality

$$c_1 V_1 + c_2 V_2 + \cdots + c_n V_n = 0$$

implies that all the n coefficients c_i are zero. To illustrate this concept, again consider the two functions $\sin x$ and $\cos x$. Of course, these two functions are not functionally independent because a knowledge of the value of one of them quickly leads to a knowledge of the value of the other through the mechanism of the Pythagorean theorem

$$\sin^2 x + \cos^2 x = 1.$$

Although not functionally independent, these two functions are linearly independent because there are no constants c_1 and c_2 for which, *for all x*

$$c_1 \sin x + c_2 \cos x = 0.$$

The qualification "for all x" makes the test for linear independence a test for the existence of an identity rather than the quest for a solution for what is merely an equation.

Other examples of sets of linearly independent functions are the following infinite sets of functions

$$\begin{aligned} &(1) \quad \sin x, \sin 2x, \sin 3x, \ldots, \sin nx, \ldots \\ &(2) \quad \cos x, \cos 2x, \cos 3x, \ldots, \cos nx, \ldots \\ &(3) \quad 1, x, x^2, x^3, x^4, \ldots, x^n, \ldots . \end{aligned}$$

An example of a set of three dependent functions are, from the above

$$1, \sin^2 x, \cos^2 x.$$

The same principles hold for vector quantities. The unit vectors $\boldsymbol{i}, \boldsymbol{j}, \boldsymbol{k}$ are linearly independent, as are the following four vectors written in matrix form

$$\begin{aligned} &\lfloor 0 \quad 2 \quad 3 \quad 4 \rfloor, \qquad \lfloor 1 \quad 2 \quad 3 \quad 4 \rfloor, \\ &\lfloor 2 \quad 3 \quad 4 \quad 5 \lfloor, \qquad \lfloor 0 \quad 0 \quad 0 \quad 5 \rfloor \end{aligned}$$

However, the vectors

$$\begin{aligned} &\lfloor 0 \quad 2 \quad 3 \quad 4 \rfloor, \qquad \lfloor 1 \quad 2 \quad 3 \quad 4 \rfloor, \\ &\lfloor 2 \quad 2 \quad 3 \quad 5 \rfloor, \qquad \lfloor 0 \quad 0 \quad 0 \quad 5 \rfloor \end{aligned}$$

are dependent because 10 times the difference between the second and first, plus the fourth, equals five times the difference between the third and first. The simple test for independency is to examine the determinant of the vector set. If the determinant is nonzero, then the vectors are independent. Otherwise they are dependent.

ENDNOTE (2): THE CHOLESKY DECOMPOSITION

Accomplishing the Cholesky decomposition of a positive definite, symmetric matrix is only a matter of progressively solving a succession of single algebraic equations with a single unknown. For example, consider the following matrix $[R]^t[R]$ or $[L][L]^t$ decomposition that is chosen so that all the triangular matrix entries are integers

$$\begin{bmatrix} 25 & -15 & 5 & 10 \\ -15 & 25 & -19 & 10 \\ 5 & -19 & 53 & -26 \\ 10 & 10 & -26 & 25 \end{bmatrix} = \begin{bmatrix} a & & & \\ b & c & & \\ d & e & f & \\ g & h & j & k \end{bmatrix} \begin{bmatrix} a & b & d & g \\ & c & e & h \\ & & f & j \\ & & & k \end{bmatrix}.$$

Now simply multiply the two triangular matrices, and equate the results to the corresponding entries in the original matrix. Then

$$a = 5 \quad b = -3 \quad d = 1 \quad g = 2,$$
$$\text{then} \quad c = 4 \quad e = -4 \quad h = 4 \quad f = 6 \quad j = -2 \quad k = 1.$$

Substituting these results leads to

$$\begin{bmatrix} 25 & -15 & 5 & 10 \\ -15 & 25 & -19 & 10 \\ 5 & -19 & 53 & -26 \\ 10 & 10 & -26 & 25 \end{bmatrix} = \begin{bmatrix} 5 & & & \\ -3 & 4 & & \\ 1 & -4 & 6 & \\ 2 & 4 & -2 & 1 \end{bmatrix} \begin{bmatrix} 5 & -3 & 1 & 2 \\ & 4 & -4 & 4 \\ & & 6 & -2 \\ & & & 1 \end{bmatrix}.$$

This procedure can be adapted to rapid computation on a digital computer. From Ref. [6.10], formulas for the elements of $[L]$ are

$$L_{jj} = \sqrt{M_{jj} - \sum_{k=1}^{j-1} L_{jk}^2} \quad \text{for } j = 1, 2, \ldots, n$$

$$\text{and} \quad L_{ij} = \frac{1}{L_{jj}} \left(M_{ij} - \sum_{k=1}^{j-1} L_{ik} L_{jk} \right) j = 1, 2, \ldots, (n-1); i = (j+1), \ldots, n.$$

There are a couple of advantages to triangular matrices. The first is that the inverse of an $[L]$ matrix is another left matrix, and the inverse of an $[R]$ matrix is another right matrix. For example, the hand-calculated inverse of the above $[L]$, using the usual adjoint technique (see below for a more efficient technique), is

$$\begin{bmatrix} 5 & & & \\ -3 & 4 & & \\ 1 & -4 & 6 & \\ 2 & 4 & -2 & 1 \end{bmatrix}^{-1} = \frac{1}{120} \begin{bmatrix} 24 & & & \\ 18 & 30 & & \\ 8 & 20 & 20 & \\ -104 & -80 & 40 & 120 \end{bmatrix}.$$

Another, much more important, advantage is that a coordinate transformation of the form $[L]\{q\} = \{p\}$, which has the solution that $\{p\} = [L]^{-1}\{q\}$, does not require the formal calculation of the inverse. All that is necessary to do to obtain the inverse of $[L]\{q\} = \{p\}$ is to solve these simultaneous equations successively; that is, one row at a time starting with the first row. In the first row there is only one unknown, q_1, that is immediately determined as p_1/L_{11}. After the substitution of the solution for q_1 in the second row, again there is only one unknown q_2, which is immediately determined in terms of the first two ps. This procedure is quite efficient for a digital computer. Therefore, it is clear that this is also a very effective way of calculating the inverse of either a left or right triangular matrix.

If the matrix to be decomposed is not symmetric, the matrix can still be written as the product of two triangular matrices, an $[L]$ and an $[R]$. However these two triangular matrices are not transposes of each other. See Ref. [6.18], Chapter 9.

ENDNOTE (3): CONSTANT MOMENTUM TRANSFORMATIONS

As has been seen, when a structure can undergo rigid body motions, the otherwise positive definite stiffness matrix is singular and only positive semidefinite. The degree of singularity is equal to the number of possible rigid body motions. These singularities can be removed by following reasoning. Again consider the matrix equation $[M]\{\ddot{q}\} + [C]\{\dot{q}\} + [K]\{q\} = \{Q\}$. If the externally applied loads are zero, then by Newton's second law, the system has both constant translational momentum and constant angular momentum. That is, it is possible to write an equation of the form $\lfloor M_{trans} \rfloor \{\dot{q}_{trans}\} = \text{const.}$ for both the translational momentum and the rotational rigid body motion, where $\lfloor M_{trans} \rfloor$ is a $1 \times N$ matrix that contains the mass terms of only those DOF involved in that specific rigid body translation. To illustrate this point simply, consider a mass system consisting of just two masses of magnitudes $2m$ and m. Let there initial translational momentum be zero. Then if the mass of magnitude $2m$ were to move to the left at a velocity V, then to maintain a zero net momentum vector for the mass system, the mass m would have to move to the right with a velocity $2V$. That is, the momentum equation would be $2m(-V) + m(V) = 0$. As a second, more complicated example, IF the unloaded flexible beam of Figure 4.5 were not cantilevered at its left end, but free to translate vertically at that end as well, then, following the same reasoning, the translational momentum equation would have the form

$$\text{const.} = 2m(\dot{w}_1 - e\dot{\phi}_1) + m(\dot{w}_2 + e\dot{\phi}_2) = \lfloor 2m \quad 0 \quad -2me \quad m \quad 0 \quad me \rfloor \begin{Bmatrix} \dot{w}_1 \\ \dot{\theta}_1 \\ \dot{\phi}_1 \\ \dot{w}_2 \\ \dot{\theta}_2 \\ \dot{\phi}_2 \end{Bmatrix}.$$

If the center of mass of this altered beam, or any other mass system, is not translating, then the total translational momentum of the structure, and the above constant, is zero. Therefore, it is possible to integrate $0 = (d/dt)\lfloor M_{trans} \rfloor \{q_{trans}\}$ to get $konst. = \lfloor M_{trans} \rfloor \{q_{trans}\}$. Whereas the first *const.* was related to the velocity of the center of mass, this *konst.* is a measure of the displacement of the center of mass. It too can be set equal to zero so as to keep a fixed SEP. (Of course this discussion could have begun with the lack of translation of the CG, but the constant momentum concept has enough other uses to be the preferred starting point.) Therefore, the equation $0 = \lfloor M_{trans} \rfloor \{q_{trans}\}$ represents an equation of constraint on the DOF vector allowing the elimination of one DOF. So it goes for each rigid body translation and rotation. The following example problem illustrates this DOF elimination procedure for the case of a rigid body rotation.

EXAMPLE 6.14 Prepare the free vibration system equations of the structure of Example 6.13, Figure 6.8 for matrix iteration by eliminating the singularity in that stiffness matrix.

SOLUTION To understand the angular momentum (moment of translational momentum) equation, review the relevant dynamics where again $\boldsymbol{L}$ is the angular momentum vector, $\boldsymbol{r}$ is the position vectror to the system CG, $\boldsymbol{e}_i$ is the position vector from the CG to the mass particle, m_i:

$$\boldsymbol{L} = \sum_{i=1}^{N} m_i(\boldsymbol{r} + \boldsymbol{e}_i) \times (\dot{\boldsymbol{r}} + \dot{\boldsymbol{e}}_i)$$

$$\boldsymbol{L} = \sum_{i=1}^{N} m_i(\boldsymbol{r} \times \dot{\boldsymbol{r}} + \boldsymbol{r} \times \dot{\boldsymbol{e}}_i + \boldsymbol{e}_i \times \dot{\boldsymbol{r}} + \boldsymbol{e}_i \times \dot{\boldsymbol{e}}_i)$$

$$\boldsymbol{L} = \boldsymbol{r} \times (m\dot{\boldsymbol{r}}) + \sum_{i=1}^{N} m_i \boldsymbol{e}_1 \times \dot{\boldsymbol{e}}_i .$$

The introduction of rotating unit vectors for each of the e_i vectors from the CG quickly produces the final result

$$\boldsymbol{L} = \boldsymbol{r} \times m\dot{\boldsymbol{r}} + \dot{\theta}\left(\sum m_i e_i^2\right) k.$$

Thus it is clear that, in addition to the mass moment of inertia terms, the moment of the translational momentums must also be included. Therefore, setting $\boldsymbol{L} = 0$ in this context means writing, where ℓ is the beam length

$$0 = \frac{1}{10} m\ell^2 \dot{\theta}_1 + \frac{1}{20} m\ell^2 \dot{\theta}_2 + \ell(m\dot{w}_2) = m\ell \lfloor 0.1 \quad 1.0 \quad 0.05 \rfloor \begin{Bmatrix} \ell\dot{\theta}_1 \\ \dot{w}_2 \\ \ell\dot{\theta}_2 \end{Bmatrix} .$$

Incorporating the above equation of constraint into a transformation matrix between the constrained and unconstrained coordinates leads to

$$\begin{Bmatrix} \ell\theta_1 \\ w_2 \\ \ell\theta_2 \end{Bmatrix}^C = \begin{bmatrix} 1 & 0 \\ -0.1 & -0.05 \\ 0 & 1 \end{bmatrix} \begin{Bmatrix} \ell\theta_1 \\ \ell\theta_2 \end{Bmatrix}^{NC} .$$

Making this coordinate substitution and, as per usual, premultiplying the free vibration equation of motion by the transpose of the coordinate transformation matrix yields

$$-\frac{m\omega^2}{10000} \begin{bmatrix} 1100 & 50 \\ 50 & 525 \end{bmatrix} \begin{Bmatrix} \ell\theta_1 \\ \ell\theta_2 \end{Bmatrix} + \frac{EI}{L^3} \begin{bmatrix} 5.32 & 2.96 \\ 2.96 & 4.63 \end{bmatrix} \begin{Bmatrix} \ell\theta_1 \\ \ell\theta_2 \end{Bmatrix} = \begin{Bmatrix} 0 \\ 0 \end{Bmatrix} .$$

Directly calculating the inverse of the 2 × 2 stiffness matrix, and forming the non-symmetric dynamic matrix produces the matrix eigenvalue problem prepared for matrix iteration

$$\begin{bmatrix} 49.45 & -13.225 \\ -29.945 & 26.45 \end{bmatrix} \begin{Bmatrix} \theta_1 \\ \theta_2 \end{Bmatrix} = \frac{1587EI}{\omega^2 m\ell^3} \begin{Bmatrix} \theta_1 \\ \theta_2 \end{Bmatrix} .$$

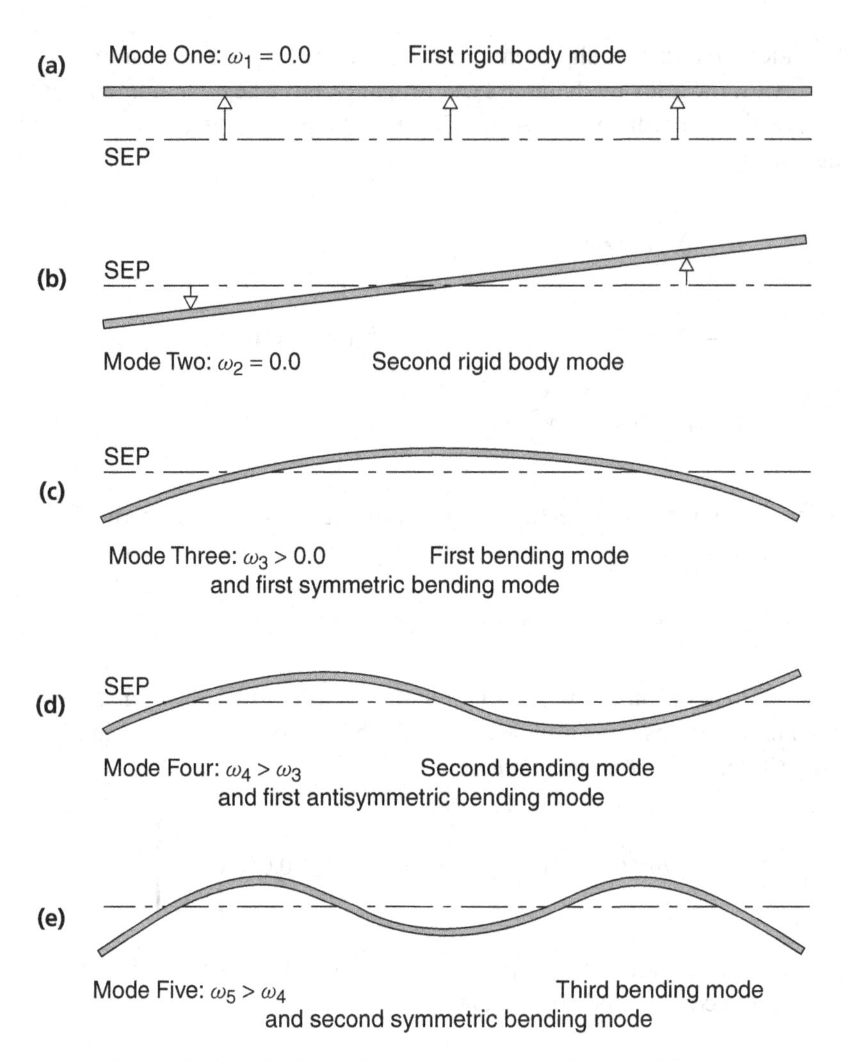

Figure 6.9. Typical mode shapes for an unsupported (free-free) beam.

Iteration produces the solution that the eigenvalue equals 60.939, and the constrained eigenvector is $\lfloor 1.0 \quad -0.8684 \rfloor$. These results lead to the conclusions that

$$\omega_2 = 5.1\sqrt{\frac{EI}{M\ell^3}} \qquad \lfloor A^{(2)} \rfloor = \lfloor 1.0 \quad -0.0566 \quad -0.8684 \rfloor$$

for the unconstrained amplitude vector. This second natural frequency result is the same as that which was determined using the determinant method. (Of course, the first natural frequency, which corresponds to the rigid body motion, is zero.) Not enough significant figures were kept to correctly calculate the third (second elastic) mode.

To further illustrate the relationship between rigid body modes and elastic modes, consider the free-free beam sketched in Figure 6.9. Let this beam, which can stand in for a simplified form of an airplane as viewed head-on, be restricted to moving only vertically. In that case, the beam has only two rigid body modes. Those modes are a vertical translation and a rotation in the plane of the paper. Of course, both of these rigid body modes have zero associated strain energy, and thus both motions have an associated zero natural frequency. The zero natural frequency can be interpreted to mean the "vibration" occurs infinitely slowly, so there are zero associated inertia forces. The rigid body mode shapes can have many forms, but the two that appear in the sketch have the advantage that these mode shapes are obviously mutually orthogonal. The first elastic mode, which is the same as the third overall mode, like all the other elastic modes in this very simplified situation, only involves bending in the plane of the paper. It is the bending mode that has the least strain energy (bending strain equals curvature) relative to the other elastic modes. Note that if the initial vertical momentum is taken to be zero (the usual viewpoint), then the CG of the beam remains stationary as the beam bends while vibrating in this first elastic mode. One way to visualize the form of this mode shape and other such mode shapes for unconstrained structures is to first recall that the vertical accelerations of the mass elements along the length of the beam are proportional to their vertical deflections. Then from the viewpoint that the inertial loads acting on the beam have to be in vertical equilibrium, the mass–deflection products all along the beam have to sum to zero. Hence, as shown for this first elastic mode, some of the beam has to travel upward while other parts have to simultaneously travel downward. Realize that even if the beam were uniform, this would not mean that half the beam would be going upward while the other half would be going downward. Each differential length of the beam is weighted by the magnitude of its deflection value (its acceleration). Hence as drawn, the unconstrained beam tips have greater deflections and greater associated inertial loads that offset more than an equal length at the center of the beam where the deflections are not as large. The higher numbered elastic beam mode shapes have greater amounts of strain energy; that is, more curvature. If the beam were symmetric about its centerline, then the first elastic mode would also be called the second symmetric mode, or first elastic symmetric mode, while the second elastic mode would be the first antisymmetric elastic mode, and so on. At this point it might be worthwhile for the reader to look again at Figures 6.7, which show free-free mode shapes for much more complicated structures.

ENDNOTE (4): ILLUSTRATION OF JACOBI'S METHOD

The following matrix, with dominant diagonal elements (for quicker results), was otherwise selected at random. All six of the 2×2 determinants along its principal diagonal, all four of the 3×3 determinants along the main diagonal, and the one 4×4 determinant have positive values. Therefore, it is verified that this matrix is positive definite and thus suitable for the Jacobi method:

$$[D_0] = \begin{bmatrix} 4 & 3 & 2 & 1 \\ 3 & 5 & -2 & 0 \\ 2 & -2 & 6 & 1 \\ 1 & 0 & 1 & 7 \end{bmatrix}.$$

The Jacobi procedure begins by selecting one pair of off-diagonal terms to be zeroed. The best choice is the largest of the off-diagonal terms. With a large matrix, computer searching for the largest such entry would be too expensive. The usual choice is to simply start with the first off-diagonal term that exceeds a selected value. However, in this small problem, finding the largest off-diagonal value can be immediately accomplished. Thus the first objective is to zero the (1,2) and (2,1) entries using the first rotation matrix

$$[r_1] = \begin{bmatrix} \cos\theta_1 & -\sin\theta_1 & & \\ \sin\theta_1 & \cos\theta_1 & & \\ & & 1 & \\ & & & 1 \end{bmatrix}.$$

Note that $[r_1]^{-1} = [r_1]^t$. The next step is determining the rotation angle for the sine and cosine terms by use of the easily derived formula

$$\tan 2\theta_1 = \frac{2d_{12}}{d_{11} - d_{22}} = -6 \quad \rightarrow \quad \theta_1 = -0.702823824 \text{ rad.}$$

Thus construct the following rotation matrix keyed to the first and second rows and columns

$$[r_1] = \begin{bmatrix} 0.763019982 & 0.646374896 & & \\ -0.646374896 & 0.763019982 & & \\ & & 1 & \\ & & & 1 \end{bmatrix}.$$

The result of the first rotation, $[D_1] = [r_1]^t[D][r_1]$, is

$$[r_1]^t[D_0][r_1] = \begin{bmatrix} 1.45862 & 0 & 2.81879 & 0.76302 \\ 0 & 7.54138 & -0.23329 & 0.646375 \\ 2.81879 & -0.23329 & 6 & 1 \\ 0.76302 & 0.646375 & 1 & 7 \end{bmatrix},$$

which demonstrates that the objective of the first rotation has been achieved. The zero in the (1,2) and (2,1) positions is actually –1.E(–9), and such subsequent zeros are E(–7) terms. There is, of course, always a round-off error at each step.

Now the largest off-diagonal term is the (1,3) or (3,1) value of 2.81879. Thus $\tan 2\theta_2 = (2)(2.81879)/(1.45862 - 6)$ or θ_2 equals –0.44633871, and the rotation matrix is

$$[r_2] = \begin{bmatrix} \cos\theta_2 & & -\sin\theta_2 & \\ & 1 & & \\ \sin\theta_2 & & \cos\theta_2 & \\ & & & 1 \end{bmatrix} = \begin{bmatrix} 0.902033594 & & 0.431665836 & \\ & 1 & & \\ -0.431665836 & & 0.902033594 & \\ & & & 1 \end{bmatrix}.$$

The result of employing this rotation matrix is $[D_2] =$

$$[r_2]^t[D_1][r_2] = \begin{bmatrix} 0.109695 & 0.100703 & 0 & 0.256604 \\ 0.100703 & 7.54138 & -0.210436 & 0.646375 \\ 0 & -0.210436 & 7.34892 & 1.2314 \\ 0.256604 & 0.646375 & 1.2314 & 7 \end{bmatrix}.$$

Now the purpose is to eliminate the (3,4) and (4,3) entries. So

$$[r_3] = \begin{bmatrix} 1 & & & \\ & 1 & & \\ & & 0.755074606 & -0.665638878 \\ & & 0.655638878 & 0.755074606 \end{bmatrix}.$$

The result of this rotation is

$$[D_3] = \begin{bmatrix} 0.109695 & 0.100703 & 0.168239 & 0.193755 \\ 0.100703 & 7.54138 & 0.264894 & 0.626031 \\ 0.168239 & 0.264894 & 8.41816 & 0 \\ 0.193755 & 0.626031 & 0 & 5.93076 \end{bmatrix}.$$

The result of the seventh rotation is

$$[D_7] = \begin{bmatrix} 0.100306 & 0.0955276 & 0.00210977 & 0 \\ 0.0955276 & 7.67195 & 0.00240573 & 0.0299617 \\ 0.00210977 & 0.00240573 & 8.50754 & -0.0776521 \\ 0 & 0.0299617 & -0.0776521 & 5.7202 \end{bmatrix}.$$

Clearly the result is converging, albeit somewhat slowly, to a diagonal matrix. From Mathematica, the eigenvalues of this matrix, in the same order as the above diagonal terms, are

$$\lambda_4 = 0.0991003 \qquad \lambda_2 = 7.67362 \qquad \lambda_1 = 8.50971 \qquad \lambda_3 = 5.71758.$$

After just seven rotations, the eigenvalues are within 1% of the correct values. The approximation to the mode shapes,

$$[\Phi] \approx [r_1][r_2][r_3][r_4][r_5][r_6][r_7]$$

$$= \begin{bmatrix} 0.688856 & 0.434032 & 0.425151 & -0.395399 \\ -0.581853 & 0.807532 & 0.0343404 & -0.0903353 \\ -0.430773 & -0.398692 & 0.573636 & -0.571333 \\ -0.0368657 & 0.0235155 & 0.699292 & 0.713497 \end{bmatrix}$$

The corresponding Mathematica calculated mode shapes, which have been normalized so that the maximum values of each mode shape are identical, are as follows. Of course, the errors here are much larger than those for the eigenvalues but not unreasonable for just seven rotations. Also note that because the original square matrix was mostly chosen at random, the nodes of these eigenvectors do not correspond to those of physically meaningful mode shapes.

$$[\Phi] = \begin{bmatrix} 0.688856 & 0.440796 & 0.449912 & -0.380207 \\ -0.596910 & 0.807532 & 0.0393502 & -0.098978 \\ -0.429416 & -0.418324 & 0.605814 & -0.534798 \\ -0.037595 & 0.0333621 & 0.699292 & 0.713497 \end{bmatrix}.$$

ENDNOTE (5): THE GRAM–SCHMIDT PROCESS FOR CREATING ORTHOGONAL VECTORS

The Gram–Schmidt process for changing a set of independent vectors into a set of orthogonal vectors is accomplished one vector at a time. As an illustration of this process, consider the randomly selected set of four vectors

$$\lfloor A_1 \rfloor = \lfloor 1 \quad 2 \quad 4 \quad 2 \rfloor \qquad \lfloor A_2 \rfloor = \lfloor 3 \quad -2 \quad 4 \quad 1 \rfloor$$
$$\lfloor A_3 \rfloor = \lfloor 3 \quad 2 \quad 6 \quad -2 \rfloor \qquad \lfloor A_4 \rfloor = \lfloor 3 \quad -2 \quad 4 \quad 6 \rfloor .$$

That these four vectors are linearly independent can be verified by calculating the 4×4 determinant of the matrix created by stacking these four vectors. If the four vectors were linearly dependent, then there would be a set of three scalar multipliers that when applied to three of these four vectors and that result summed, would yield the negative of the fourth vector. Thus the value of the determinant would be zero. Since the actual value of the determinant is $+80$, the four vectors are linearly independent. In passing, note that because the 4×4 matrix was chosen at random, while nonsingular, it is not positive definite as can be seen by examining the first 2×2 determinant on the main diagonal.

The Gram–Schmidt process begins by selecting any one of the vectors to be the first orthogonal vector. Let that first orthogonal vector $\{O_1\}$ be $\{A_1\}$. To obtain the second orthogonal vector, it is necessary only to choose another independent vector to combine with the first and then demand that this combination be orthogonal to the first orthogonal vector. That is, skipping weighting matrices, let the second orthogonal vector have the form $\{O_2\} = \{A_1\} + x\{A_2\}$ and write the orthogonality condition for the first two orthogonal vectors as

$$\lfloor A_1 \rfloor (\{A_1\} + x\{A_2\}) = 0 \quad \therefore x = -25/17$$
$$\lfloor O_2 \rfloor = \lfloor -3.41176 \quad 4.94118 \quad -1.88235 \quad 0.529412 \rfloor$$
$$\text{and} \quad \lfloor O_1 \rfloor \{O_2\} = 2.10^{-16}.$$

If it is desired to go beyond mere orthogonality to "orthonormal vectors," it is necessary only to divide, for example, $\{0_1\}$ by the square root scalar value of the result of multiplying $\{O_1\}$ by itself, which is $\sqrt{25}$.

The third of the orthogonal vectors is similarly obtained. Let $\{O_3\} = \{A_1\} + y\{A_2\} + z\{A_3\}$ and require the products of $\{O_1\}$ and $\{O_3\}$, and $\{O_2\}$ and $\{O_3\}$, be zero. This step leads to determining the values of y and z. Those calculations are as follows:

$$\lfloor A_1 \rfloor (\{A_1\} + y\{A_2\} + z\{A_3\}) = 0 \quad \text{or} \quad 25 + 17y + 27z = 0$$
$$\lfloor O_2 \rfloor (\{A_1\} + y\{A_2\} + z\{A_3\}) = 0 \quad \text{or} \quad -27.1176y - 12.7059z = 0$$
$$\therefore y = 0.615387 \quad \text{and} \quad z = -1.31339$$
$$\lfloor O_3 \rfloor = \lfloor -1.09401 \quad -1.85755 \quad -1.41879 \quad 5.24217 \rfloor$$
$$\text{and} \quad \lfloor O_1 \rfloor \{O_3\} = 0.00005 \quad \lfloor O_2 \rfloor \{O_3\} = -0.00007.$$

Clearly round-off error has accumulated rapidly, and the values for $\{O_3\}$ are good only for five significant figures rather than the six stated.

The fourth orthogonal vector is similarly obtained starting with, say, $\{O_4\} = \{A_1\} + \alpha\{A_2\} + \beta\{A_3\} + \gamma\{A_4\}$ and requiring that this vector be orthogonal to the three previously obtained orthogonal vectors. As above, those three orthogonality conditions lead to three equations to be solved simultaneously for the values of α, β, and γ.

7 The Modal Transformation

7.1 Introduction

The usual first step toward a solution of the linear, small deflection matrix equation of structural motion, $[M]\{\ddot{q}\} + [C]\{\dot{q}\} + [K]\{q\} = \{Q\}$ (called the m, c, k, Q equation), is to obtain the system natural frequencies and mode shapes. As discussed in the previous chapter, the natural frequencies and mode shapes are eigenpair solutions to the undamped, homogeneous (m, k) form of this same matrix differential equation. The focus of this chapter is on the usual final steps in the solution process. The usual step after acquiring the modal frequencies and mode shapes is to use that data to isolate the effect of each mode on the system response. The final step is to solve each of the selected modal differential equations and sum those results. For the sake of instruction and intuition building, this chapter and the next focus only on obtaining analytical solutions to those modal differential equations. To further clear the underbrush, the example problems will omit damping. Chapter 9 discusses the far more prevalent approach of numerical integration of the modal differential equations. The available commercial software packages, of course, provide numerical solutions rather than analytical solutions. Numerical solutions are almost always the only option when the original matrix differential equation is complicated by material or geometric nonlinearities.

Before introducing the specifics of the two final steps to a complete solution to the m, c, k, Q linear matrix equation of motion, it is helpful to first present the technique for incorporating the initial conditions into the free vibration, multidegree of freedom solution. This topic reinforces an understanding of the free vibration solution.

7.2 Initial Conditions

The previous chapter provided all the solution elements for $[M]\{\ddot{q}\} + [K]\{q\} = \{0\}$, the N-DOF linear matrix equation that describes an undamped system vibrating free of impressed time-varying loads. From a mathematical point of view $[M]\{\ddot{q}\} + [K]\{q\} = \{0\}$ is a set of n ordinary, second-order, homogeneous, differential equations with time as the independent variable. Recall that the previous chapter established that for such a linear N-DOF structural system: (i) every solution function is a sinusoid with time as part of its argument and (ii) there is such a sinusoid associated with each of the N discrete frequencies and orthogonal mode shapes. From the theory of differential equations, the complete solution to these N, second-order differential equations must also contain $2N$ constants of integration. Hence, with the complete solution for the undamped, free vibration displacement DOF vector is simply the following sum of all of the modal solutions

$$\{q(t)\} = a_1\{A^{(1)}\}\sin(\omega_1 t + \psi_1) + a_2\{A^{(2)}\}\sin(\omega_2 t + \psi_2) + \cdots + a_n\{A^{(n)}\}\sin(\omega_n t + \psi_n), \tag{7.1}$$

where a_i and ψ_i are the required constants of integration. The constants of integration a_i can be viewed also as the multiplicative constants (or weighting factors) associated with each mode shape once the mode shape vectors have been normalized in the customary fashion. Equation (7.1) says that, in general, the time-varying force free deflections of a multidegree of freedom system generally include all modal frequencies and mode shapes.

It is important to realize that when there are applied time-varying generalized forces, the constants of integration of the matrix differential equation complementary solution can be determined by use of the system initial conditions only after the analytical form of the particular (i.e., nonhomogeneous) solution for the motion has been obtained. In the case of an undamped, force free motion, that particular solution is simply zero. Then Eq. (7.1) alone is the required complete solution. If equivalent viscous damping is included in the structural model then, as is seen in Chapter 5 and more extensively in the latter parts of this chapter, this solution must be modified by inserting as factors decaying exponential functions. Again, if applied forces are present, the particular solutions must be added to Eq. (7.1) before applying the initial conditions.

In the case of vibrating structural systems, the initial conditions are always the system deflections and velocities at time zero. The initial deflections and velocities are, of course, expressed in terms of the system DOF, $q_i(t)$. Let these quantities be respectively symbolized as the N-dimensional vectors $\{q(0)\}$ and $\{\dot{q}(0)\}$. Substitution of these initial deflections and velocities into, for example, Eq. (7.1) leads to

$$\{q(0)\} = a_1\{A^{(1)}\}\sin\psi_1 + a_2\{A^{(2)}\}\sin\psi_2 + \cdots + a_n\{A^{(n)}\}\sin\psi_n \tag{7.2a}$$

$$\{\dot{q}(0)\} = \omega_1 a_1\{A^{(1)}\}\cos\psi_1 + \omega_1 a_2\{A^{(2)}\}\cos\psi_2 + \cdots + \omega_n a_n\{A^{(n)}\}\cos\psi_n. \tag{7.2b}$$

The task now is to fully define the motion described by Eq. (7.1) by determining the values of the $2N$ constants of integration a_i and ψ_i by use of the above two sets of N equations, Eqs. (7.2). This task is accomplished by using the orthogonality of the modal vectors. Premultiply both of Eqs. (7.2) by the row modal vector and mass weighting matrix $\lfloor A^{(j)}\rfloor[M]$. The two left-hand side results are scalars. Call these scalars D_j and V_j, respectively. Then, because of the weighted orthogonality of the terms on the right-hand side of Eqs. (7.2), the only nonzero products are the jth terms of the two series. Define again $\lfloor A^{(j)}\rfloor[M]\{A^{(j)}\} \equiv M_j$, which is called the jth generalized mass or the jth modal mass. Of course, the modal mass is not the same as m_{jj}, which is just the jth diagonal entry of the mass matrix $[M]$. Thus the result of the above premultiplication of the jth modal row vector and the mass matrix is

$$D_j = a_j M_j \sin \psi_j \qquad \text{and} \qquad V_j = \omega_j a_j M_j \cos \psi_j .$$

Again, the quantities D_j and V_j are determined by whatever the initial conditions there are for the system DOF. The above pair of equations can be solved simultaneously with the result

$$a_j = \frac{\sqrt{\omega_j^2 D_j^2 + V_j^2}}{\omega_j M_j} \qquad \text{and} \qquad \psi_j = \arctan \frac{\omega_j D_j}{V_j} .$$

Back substitution of these constants of integration into Eq. (7.1) completes the calculation of the free vibration response over time. Thus it is clear that the initial conditions completely define the force free, vibratory motion of the structural system.

As a special case of interest, let the initial displacement vector be proportional to the ith mode shape. That is, let $\{q(0)\} = c\{A^{(i)}\}$, where c is a known constant. If, in addition, the initial velocity vector is null, then all V_j terms and all the D_j terms are zero except for $D_i = cM_i$. Then all a_j and all ψ_j are zero but for $a_i = c$, and $\psi_i = \pi/2$. Therefore, the total structural system response in this case is simply

$$\{q(t)\} = c\left\{A^{(i)}\right\} \cos \omega_j t .$$

This solution says that an undamped structural system with initial deflections in the form of one of the mode shapes will vibrate indefinitely in that mode shape and that mode shape alone. Of course, this conclusion has to be altered to the extent damping is present in the system. In the case of even a lightly damped system, it can be expected that not only will the magnitude of the jth modal deflections decrease over time, but the vibration where the jth mode predominates only lasts for a short time period before lower numbered modes, particularly the fundamental mode, assert their dominance. As shown, this departure from the jth mode shape is because of two causes. One is that the higher modes generally damp out more quickly, and the second is that there will always be imperfections in setting up the initial deflections in the form of the jth mode shape. The reader can experimentally illustrate the above statements by "playing"[1] with a taught string, such as a long, coiled telephone wire. Accept for the moment that the nth mode shape of a taught string of length ℓ is $\sin(n\pi x/\ell)$. To begin your experiment, try to start the string vibrating by giving it initial deflections of, say, $\sin(3\pi x/\ell)$ and then releasing the string without an initial velocity. You would soon see that the taught string is vibrating in a shape that can be described as $\sin(\pi x/\ell)$, the fundamental mode shape where $n = 1$.

As a final comment, note that the finite series solution for the DOF vector $\{q(t)\}$ given in the form of Eq. (7.1) looks similar to a finite Fourier sine series. A Fourier series, finite or not, always produces a periodic function.[2] In a Fourier time series,

[1] The probable verb chosen by friends or spouses.

[2] A periodic function of time, $P(t)$, with period T, is such that $P(t + T) = P(t)$ for all t.

the frequencies are all integer multiples of the first frequency. This is rarely the case in Eq. (7.1), and as a result, the sum of the N deflection response sine functions is not generally a periodic function.

7.3 The Modal Transformation

The matrix differential equation of motion $[M]\{\ddot{q}\}+[C]\{\dot{q}\}+[K]\{q\}=\{Q\}$ with its three $N\times N$ symmetric, square matrices, can be viewed as merely N second-order, linear, ordinary differential equations stacked one on top of another. That is, the kth row of this matrix equation is just the ordinary differential equation

$$M_{k1}\ddot{q}_1+M_{k2}\ddot{q}_2+\cdots+M_{kn}\ddot{q}_n+C_{k1}\dot{q}_1+C_{k2}\dot{q}_2+\cdots+C_{kn}\dot{q}_n$$
$$+K_{k1}q_1+K_{k2}q_2+\cdots+K_{kn}q_k=Q_k(t).$$

Since this single equation contains n unknown functions $q_i(t)$, it is not possible to treat this equation separately from the other $N-1$ equations that are included in the matrix equation. In this circumstance, it is said that the unknown variables are "coupled." The situation would be quite different if the mass, damping, and stiffness matrices were diagonal matrices. Then the matrix product of these diagonal matrices with their corresponding DOF vector, velocity vector, or acceleration vector, as appropriate, would produce a differential equation for the kth row of the form

$$m_{kk}\ddot{q}_k+c_{kk}\dot{q}_k+k_{kk}q_k=Q_k(t).$$

Since the above equation only contains a single unknown function, $q_k(t)$, this equation is called "uncoupled," and it can be treated separately from all the other equations listed in the other matrix rows. For many mathematical models, it is possible to diagonalize the mass matrix by using DOF that originate at the SEP of the discrete masses, but then the corresponding stiffness matrix generally will not be a diagonal matrix. Correspondingly, if the structural model is so simple as to be just composed of spring elements with one fixed end, it is possible to diagonalize the stiffness matrix by using generalized coordinates that originate at the SEP of the movable spring ends. However, generally, the corresponding mass matrix will not be diagonal. It is rarely, if ever, possible to simply inspect the mathematical model of even a simple structure and thereby determine the generalized coordinates that will diagonalize both the mass and stiffness matrices. Nevertheless, there always exists a set of coordinates for which the mass matrix and the stiffness matrix are both diagonal matrices. This section focuses on using those convenient coordinates, and other devices, for the purpose of uncoupling and then solving the m, c, k, Q matrix equation.

Consider the m, c, k, Q matrix equation of motion that, of course, is in terms of the physically meaningful DOF $\{q\}$. The first step in the decoupling process is writing a coordinate transformation, called the *modal transformation*, which relates these physically meaningful DOF $\{q(t)\}$ to a new set of generalized coordinates $\{p(t)\}$ called the *modal coordinates*. This coordinate transformation is written as follows

$$\{q(t)\}=\left[A^{(1)}\middle|A^{(2)}\middle|\cdots\middle|A^{(m)}\right]\{p(t)\}=[\Phi]\{p(t)\}, \tag{7.3a}$$

where the jth column of the $N \times M$ rectangular *modal matrix* $[\Phi]$, is the eigenvector $\{A^{(j)}\}$ from the $[m]$, $[k]$ homogeneous equation, and where $M \leq N$, where N is the total number of eigenvectors. The individual terms of $\{p\}$ are called modal coordinates because the rules of matrix multiplication associate the ith coordinate p_i with the ith modal vector $\{A^{(i)}\}$, for each index i. That is, the above coordinate transformation equation can be rewritten in summation form as

$$\{q(t)\} = \{A^{(1)}\} p_1(t) + \{A^{(2)}\} p_2(t) + \cdots + \{A^{(m)}\} p_m(t), \tag{7.3b}$$

where the coordinates p_j are simply weighting factors for the mode shapes. Therefore the time function p_j determines the magnitude of the contribution of the jth mode to the total deflection vector $\{q(t)\}$. Furthermore, the above equation makes clear that any normalization factor applied to the eigenvectors is immaterial because the modal coordinate can be adjusted to provide the same contribution to the physical coordinates that would occur without the normalization factor; that is, $c_j\{A^{(j)}\}(p_j/c_j) = \{A^{(j)}\}p_j$.

Since the modal matrix is merely a collection of constants, $\{\dot{q}\} = [\Phi]\{\dot{p}\}$ and $\{\ddot{q}\} = [\Phi]\{\ddot{p}\}$. Substituting these coordinate transformations into the m, c, k, Q matrix equation of motion and then premultiplying by the transpose of the modal transformation matrix leads to

$$\begin{aligned} [\Phi]^t[M][\Phi]\{\ddot{p}\} + [\Phi]^t[C][\Phi]\{\dot{p}\} + [\Phi]^t[K][\Phi]\{p\} &= [\Phi]^t\{Q(t)\} \equiv \{P(t)\} \\ \text{or} \quad [\overline{M}]\{\ddot{p}\} + [\overline{C}]\{\dot{p}\} + [\overline{K}]\{p\} &= \{P\}, \end{aligned} \tag{7.4}$$

where the above overbars are used temporarily pending discussion in turn of each of the above triple matrix products. Consider the first of the three triple matrix products, the $M \times M$ matrix

$$[\Phi]^t[M][\Phi].$$

Using the rules of matrix multiplication, and noting that the rows of the transposed transformation matrix are also the modal vectors, the result of the triple matrix multiplication $[\Phi]^t[M][\Phi]$ for the (i, j) entry of this $M \times M$ matrix product is

$$\lfloor A^{(i)} \rfloor [M] \{A^{(j)}\} \equiv M_{ij}.$$

Due to the weighted orthogonality of the modal vectors, $M_{ij} = 0$ whenever $i \neq j$. Thus the result for $[\Phi]^t[M][\Phi]$ is a diagonal matrix. When $i = j$, which is the jth element on the diagonal, the result is the previously defined modal mass M_j. Again, the single subscript is used to help avoid any confusion over this being merely an entry in a general mass matrix. In summary, the matrix product $[\Phi]^t[M][\Phi]$ is the diagonal matrix $[\backslash M \backslash]$ whose jth diagonal element is the jth modal mass M_j.

Now consider the third of the above three triple matrix products. From premultiplying the jth eigenvalue equation solution $\omega_j^2[M]\{A^{(j)}\} = [K]\{A^{(j)}\}$ by the transpose of the modal vector $\{A^{(i)}\}$, it is again clear that the modal vectors are also mutually orthogonal when weighted by the system stiffness matrix. Hence, it can be seen by premultiplying by the transpose of the modal transformation matrix that all off-diagonal terms $\lfloor A^{(i)} \rfloor [K]\{A^{(j)}\}$ of the product $[\Phi]^t[K][\Phi]$ are zero, whereas the

jth diagonal entry, the jth generalized stiffness, $\lfloor A^{(j)} \rfloor [K] \{A^{(j)}\} \equiv K_j \equiv \omega_j^2 M_j$. Since the stiffness matrix usually has many more entries than the mass matrix, for numerical efficiency, the generalized stiffnesses are usually calculated only by multiplying the generalized masses by their corresponding frequency squared terms. Thus the triple matrix product $[\Phi]^t[K][\Phi]$ can be written as the diagonal matrix $[\backslash \omega^2 M \backslash]$, or the commutative product of the diagonal generalized mass matrix with the diagonal matrix of natural frequencies.

An actual digital computer numerical calculation of the generalized mass matrix generally will not produce off-diagonal entries that are exactly zero. The smallness of these calculated off-diagonal numbers relative to the corresponding diagonal terms is a good indication of the numerical accuracy of the mode shape calculations. Therefore, to provide the analyst with such an indication of numerical accuracy, a software package that calculates the generalized mass matrix can be expected to also print out at the very least the largest of all the off-diagonal terms, its location (so as to indicate which modes are least orthogonal), and the generalized mass terms corresponding to the row and column numbers of that off-diagonal entry in order to meaningfully scale the erroneous off-diagonal term. Many software programs go further and present something like a color-coded, two-dimensional bar graph of the off-diagonal terms relative to the diagonal terms for an easy survey of all the numerical errors up to this point in the calculation.

Now consider the modal transformation matrix triple product involving the damping matrix. A diagonal matrix result cannot be expected here. This is so because, unlike the mass and stiffness matrices, the damping matrix is not a part of the real eigenvalue problem that generates the undamped modal vectors. Recall that the damping matrix, when it actually is estimated, is often one which involves a fair amount of uncertainty. The usual course of action when a damping matrix has been constructed in terms of the physically meaningful coordinates and then transformed to the modal coordinates is to examine the result of its triple matrix product to see just how large the off-diagonal terms are relative to the diagonal terms. If, in any row, the off-diagonal terms are all somewhat "small" relative to the diagonal term, then the off-diagonal terms are simply discarded. To see why the off diagonal terms are sometimes small and thus can be discarded, recall from Chapter 4 the concept of proportional damping, sometimes called Rayleigh damping. Again, proportional or Rayleigh damping is the case where $[C] = \alpha[M] + \beta[K]$, where α and β are just scalar factors with appropriate units. Although it is rather difficult to decide on the numerical values of α and β without test data from the actual structure, the concept that the damping occurs where the mass is located (as in the case of fluid drag) and where the elastic material is located (as in the case of internal friction) is not wholly unreasonable. These associations are what the proportional damping equation states mathematically. Thus it is to be generally expected that, although the estimated damping matrix will not be modally transformed into a diagonal matrix, the triple matrix product some times will not be far from being diagonal because perfect proportional damping does indeed lead to the sum of two diagonal matrices, which, of course, can be combined into one diagonal damping matrix. If, for some reason, such as the presence of a nonstructural damping mechanism, some off-diagonal entries in the damping matrix are the same size or larger than the diagonal terms, then another solution technique is necessary for that unusual case where $[C]$ can be clearly stated. Other approaches are discussed in the next few paragraphs and in Chapter 9.

The remaining discussion in this chapter will proceed on the basis that the generalized damping matrix also has been converted into a diagonal matrix, or, as is often done, the damping matrix is entirely omitted from the original structural model and system equations. This possible alternate tactic of introducing damping later in the solution calculation will be examined later in this chapter. However, at this point, it is useful to remember that, as was done in earlier chapters, omitting damping is also a possible engineering approach where the more simply calculated undamped response is used as a close estimate and upper bound to the actual, lightly damped response.

As stated earlier, the modal transformation converts the previously coupled N differential equations in terms of the N original DOF, $q_i(t)$, into a set of M uncoupled differential equations, each of which can be solved without any reference to the other $M-1$ equations. Write the typical jth modal equation as

$$M_j \ddot{p}_j + C_j \dot{p}_j + \omega_j^2 M_j p_j = P_j(t)$$

$$\text{or} \quad \ddot{p}_j + 2\zeta_j \omega_j \dot{p}_j + \omega_j^2 p_j = \frac{P_j(t)}{M_j} \equiv \frac{1}{M_j} \sum_{i=1}^{n} A_i^{(j)} Q_i(t), \tag{7.5}$$

where the quantity $P_j(t)$ is called the *modal generalized force*. The remainder of this chapter is devoted to obtaining the solution to this ordinary, linear, differential equation with constant coefficients for special forms of the modal force, two of which are then used to formulate a solution for any generalized modal force, either in analytical or numerical form.

Before proceeding to discuss solutions of Eq. (7.5), note that this equation offers the above mentioned alternate way of introducing damping into the analysis. Damping can be introduced here, for the first time in the analysis procedure, in the form of an estimate of the jth modal damping factor, ζ_j. That is, it is possible to omit damping altogether from the analysis up to the point of Eq. (7.5), and then insert the velocity term with such an estimate of the modal damping factor.

7.4 Harmonic Loading Revisited

As Eq. (7.5) makes clear, the jth generalized modal force, $P_j(t)$, is a sum of the externally applied generalized loads each of which is weighted by an amplitude of the jth mode shape. As the first of three special cases, consider the case where the nonzero entries of $\{Q(t)\}$, the generalized forces associated with the physically meaningful DOF, are one or more harmonic loads all with a single frequency, ω_f. Let there be phase differences between these harmonic loads. The phase differences can be easily modeled by writing the amplitudes of these harmonic loads in complex algebra form. Then the jth modal load is also a harmonic load at that same frequency, and its magnitude is the sum of the complex magnitudes of all its constituent parts, where each part is weighted by the jth mode shape. For the sake of convenience in the writing of subsequent equations, write this normalized jth modal force in the following form

$$\frac{P_j(t)}{M_j} \equiv \omega_j^2 \hat{N}_j \exp(i\omega_f t),$$

where the newly defined quantity $\hat{N}_j$ has the same units, say length, as $p_j(t)$, and where ω_f is the common frequency of the physically meaningful applied loads. Again, if all the applied loads of $\{Q(t)\}$ are in or out of phase, then $\hat{N}_j$ is a real quantity. If there are phase differences between those harmonic loads, then $\hat{N}_j$ can be conveniently written as a complex constant. If there is more than one harmonic forcing frequency, then the analysis procedure that follows can be repeated for the different frequencies and then superimposed as long as the deflection response remains in the linear range.

Proceeding to a solution for the modal coordinate equation, Eq. (7.5), for this special case, let the modal deflection response, $p_j(t)$, be a complex number. As a trial solution, represent the modal deflection response as a harmonic function with frequency ω_f and a constant amplitude; that is, let

$$p_j(t) = \hat{p}_j \exp(i\omega_f t).$$

The constant amplitude of the response, $\hat{p}_j$, is also a complex number because, as is discussed below, a complex number provides the necessary phase difference between the harmonic force input and this harmonic deflection output. Substituting these complex algebra forms for the input and output into the modal equation of motion, Eq. (7.5), where the application of each time derivative to the modal deflection $p_j(t)$ is equivalent to a multiplication by $i\omega_f$, yields

$$\left(-\omega_f^2 + 2i\zeta_j\omega_j\omega_f + \omega_j^2\right)\hat{p}_j e^{i\omega_f t} = \omega_j^2 \hat{N}_j e^{i\omega_f t}$$

$$\text{or} \quad \hat{p}_j = \frac{\omega_j^2 \hat{N}_j}{\left[\left(\omega_j^2 - \omega_f^2\right) + i(2\zeta_j\omega_j\omega_f)\right]}$$

$$\text{or} \quad p_j(t) = \frac{P_j(t)/M_j}{\left[\left(\omega_j^2 - \omega_f^2\right) + i(2\zeta_j\omega_j\omega_f)\right]}$$

where again

$$P_j(t) = \sum_{i=1}^{n} A_i^{(j)} Q_i(t)$$

is a harmonic load.

The above solution for $p_j(t)$ confirms the complex nature of the response magnitude even if $\hat{N}_j$ happens to be a real constant. In this solution for the complex magnitude of the modal response, divide both the right-hand side numerator and denominator by the square of the jth natural frequency, and as before call the ratio of the forcing frequency to the jth natural frequency, ω_f/ω_j, the jth *frequency ratio*, Ω_j. Then more concise forms of the solutions for the complex magnitude of the jth modal coordinate, $\hat{p}_j$, and the modal coordinate itself, $p_j(t)$, are

$$\hat{p}_j = \frac{\hat{N}_j}{\left(1-\Omega_j^2\right)+i(2\zeta_j\Omega_j)}$$

$$\text{or} \quad p_j(t) = \frac{\hat{N}_j e^{i\omega_f t}}{\left(1-\Omega_j^2\right)+i(2\zeta_j\Omega_j)}$$

$$\text{or} \quad p_j(t) = \hat{N}_j H_j(i\Omega_j) e^{i\omega_f t}. \tag{7.6}$$

The nondimensional function $H_j(i\Omega_j)$ multiplied by the normalized harmonic force input $\hat{N}_j \exp(i\omega_f t)$ produces the harmonic deflection output. Such a complex function is called, as in Chapter 5, a frequency response function. Again, frequency response functions are forcing frequency-dependent flexibility influence coefficient for a harmonic load input. Of course, the above example of the species is called the jth modal frequency response function. The inclusion of the square root of -1 in the argument of this function is again just a reminder that this function is complex. Although the above equation only relates output to input for the case of a harmonic input, Appendix II plus the remainder of this chapter demonstrate the applicability of frequency response functions to all types of input forces.

The modal frequency response functions, of course, incorporate both the magnification of the output and the lag angle of the output. Specifically

$$|H_j(i\omega_f)| = \frac{|\hat{p}_j|}{|\hat{N}_j|} = \frac{M_j|p_j(t)|\omega_j^2}{|P_j(t)|} = \frac{1}{\sqrt{\left(1-\Omega_j^2\right)^2+(2\zeta_j\Omega_j)^2}}. \tag{7.7}$$

Thus it is apparent that there is a resonance phenomena for each and every modal frequency, and the frequency response functions describe the magnification of the response. Since the damping in the higher modes is typically much greater than that for the lower modes, then as Figure 5.8 indicates, only the resonances for the lower modes are significant.

7.5 Impulsive and Sudden Loadings

In addition to the case of harmonic loading, there are two other particular loading cases of special interest. Although these two special loadings have importance in their own right, the reason for focusing on them at this time is that they can be used to fashion an analytical solution for the general loading case, represented by Eq. (7.5), which is again

$$\ddot{p}_j(t) + 2\zeta_j\omega_j\dot{p}_j(t) + \omega_j^2 p_j(t) = \frac{P_j(t)}{M_j}, \tag{7.5}$$

where the nodal forcing function, $P_j(t)$, is an arbitrary function with units of force when the modal response, $p_j(t)$, has units of length. To begin, consider the special

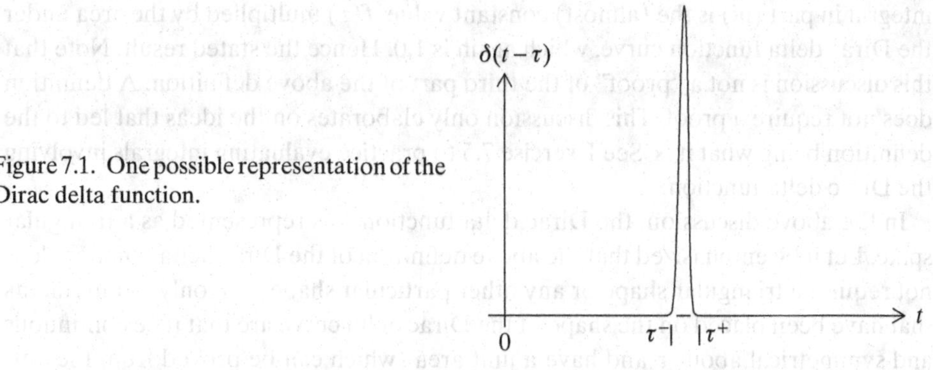

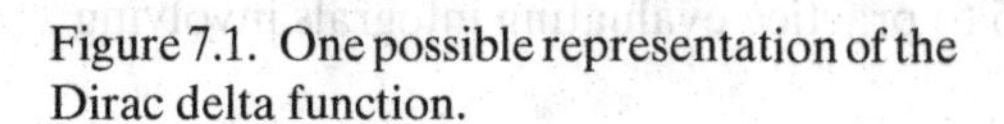
Figure 7.1. One possible representation of the Dirac delta function.

case where the time-varying modal forcing function has a very large magnitude over a very short period of time and is otherwise zero. A force of this nature is called an *impulsive force*. Examples of impulsive forces include a bat hitting a ball and a hammer striking a structure once. The time duration of the force that redirects the ball (the contact time between ball and bat or hammer and structure) is quite short relative to the duration of the subsequent flight of the ball or the first period of vibration of the structure. The mathematical tool used to describe the time variation of such a force is the *Dirac delta function*, which is symbolized by $\delta(t-\tau)$. The time t is called the variable, whereas the time τ is called the parameter.[3] The Dirac delta function is defined by the following three statements:

(i) The Dirac delta function is everywhere continuous, and an even function about τ. Thus it may be said that $\delta(t-\tau)=\delta(-(t-\tau))=\delta(\tau-t)$.

(ii) The Dirac delta function has the value zero everywhere except in the very small interval $(\tau-\frac{1}{2}\epsilon, \tau+\frac{1}{2}\epsilon)$. For the sake of brevity, this interval of very short duration ϵ is denoted as (τ^-, τ^+) and spoken of as tau minus to tau plus.

(iii) The Dirac delta function is such that for any arbitrary, bounded, continuous function $f(t)$,

$$\int_{-\infty}^{+\infty} f(\tau)\delta(t-\tau)d\tau = f(t) \quad \text{or} \quad \int_{-\infty}^{+\infty} f(t)\delta(t-\tau)dt = f(\tau).$$

The choice of the third part of the chosen definition can be explained by referring to Figure 7.1 where the Dirac delta function is represented, as one possibility, by a triangular spike with a very small base, $\tau^+-\tau^-$. Let the area within the triangular spike, the area under the Dirac delta function plot, be 1.0, without units. Since the integrand in part (iii) involves the product of the Dirac delta function and the arbitrary function $f(t)$, the integrand is zero everywhere but in the very small interval (τ^-, τ^+). Within that very small interval, which can be made as small as necessary to ensure the accuracy of the following statements, the value of the arbitrary function is little different from its apparent average value, $f(\tau)$. Recalling that the meaning of an integral is tied to measuring the area under a curve, then the value of the

[3] A parameter is merely a constant without a specific, fixed value.

integral in part (iii) is the (almost) constant value $f(\tau)$ multiplied by the area under the Dirac delta function curve, which again is 1.0. Hence the stated result. Note that this discussion is not a "proof" of the third part of the above definition. A definition does not require a proof. This discussion only elaborates on the ideas that led to the definition being what it is. See Exercise 7.5 to practice evaluating integrals involving the Dirac delta function.

In the above discussion, the Dirac delta function was represented as a triangular spike. Let it be emphasized that the above definition of the Dirac delta function does not require a triangular shape or any other particular shape. The only requirements that have been placed on the shape of the Dirac delta curve are that it be continuous and symmetrical about τ and have a unit area (which can be proved from the definition by simply setting $f = 1.0$). Hence the shape of the Dirac delta curve could just as well be parabolic or, more importantly for its use below, rectangular. Finally, the Dirac delta function, in a strict mathematical sense, is not really a function. It is not a function because its precise values are not defined in the interval (τ^-, τ^+), only its area in this interval is defined. The Dirac delta function, despite its name, is what mathematicians call a "distribution." Finally, from part (iii), note that the units associated with the Dirac delta function function are the inverse of time. More generally, the units of the Dirac delta are the inverse of whatever quantity appears as the argument of this function.

Consider a force, or an equivalent force, described as $F(t) = \mathcal{F}\delta(t)$, where the constant $\mathcal{F}$, with units of force-time, is called the magnitude of the impulse. Here the parameter τ, which indicates when in time the impulse occurs, is zero. The reason $\mathcal{F}$ is the magnitude of the impulse can be seen from the basic physics definition of an impulse as the integral of a force over a time interval. Hence, between the limits of zero minus and zero plus, $\int F(t)dt = \int \mathcal{F}\delta(t)dt = \mathcal{F}$. Let this same force be the sole force applied to a mass of fixed magnitude m. Then from Newton's second law

$$m\ddot{u} = \mathcal{F}\delta(t) \qquad \text{so} \qquad \int_{0^-}^{0^+} m d\dot{u} = \int_{0^-}^{0^+} \mathcal{F}\delta(t)dt = \mathcal{F}$$

$$\text{or} \quad m\dot{u}(0^+) = m\dot{u}(0^-) + \mathcal{F}.$$

It is clear that any impulsive force causes an increment in the velocity (or momentum) of the mass on which it acts. If the velocity of the mass is zero at time zero minus, then the velocity of the mass at time zero plus is $\mathcal{F}/m$.

Now, to parallel the above, consider a impulsive modal force $P_j(t) = \Pi_j\, \delta(t)$, where Π_j has the units of force-time when $p_j(t)$ has the units of length. In other words, Π_j is an impulsive force magnitude in modal space just as $\mathcal{F}$ above is an impulsive force magnitude in physical space. Then the jth modal equation of motion becomes

$$\ddot{p}_j + 2\zeta_j\omega_j\dot{p}_j + \omega_j^2 p_j(t) = \frac{\Pi_j}{M_j}\,\delta(t). \tag{7.8}$$

Consider the case where both the modal initial deflection and the modal initial velocity, $p_j(0^-)$ and $\dot{p}_j(0^-)$, are zero. If the initial conditions are other than null for this linear equation, then the free vibration response to those other initial conditions can simply be added to the result soon obtained. Now multiply this equation throughout by dt, so $\dot{p}dt = (dp/dt)dt = dp$, and so on. Now directly integrate this equation from time zero minus to time zero plus so that Eq. (7.8) becomes

$$\int_{0^-}^{0^+} d\dot{p}_j + 2\zeta_j\omega_j \int_{0^-}^{0^+} dp_j + \omega_j^2 \int_{0^-}^{0^+} p_j(t)dt = \frac{\Pi_j}{M_j} \int_{0^-}^{0^+} \delta(t)dt = \frac{\Pi_j}{M_j}.$$

Since the right-hand side of the above equality is a finite value, so too must be the left-hand side. The first integral, the integral of the exact differential of the modal velocity, produces just the modal velocity at time zero plus because of the presumed zero value of the modal velocity at time zero minus. The second integral, the integral of the exact differential of the modal deflection, produces the difference in the modal deflection at times zero plus and zero minus, where the latter term is also zero from the presumed initial conditions. It is important to note that the modal deflection, like any deflection, obeys the general rule that rate (velocity) times time equals distance. Here, the modal velocity is finite, and the time interval is infinitesimal. Therefore, the modal deflection at time zero plus is also infinitesimal and thus can be neglected. That is, the modal displacement at time zero plus can be said to be zero relative to any finite deflection. Thus the total result from the second integral is zero. The third integral is bounded by the product of the infinitesimal modal deflection and the infinitesimal time duration from zero minus to zero plus. Hence, it, too, is assigned a value of zero. Therefore, the final form of the above equation is simply

$$\dot{p}_j(0^+) = \frac{\Pi_j}{M_j} \quad \text{and} \quad p_j(0^+) = 0,$$

which exactly follows the previous result using Newton's second law. For the purpose of examining the modal motion *after* the application of this impulsive load, that is for the time interval starting at zero plus, the above result has the interpretation that the application of the impulsive modal load $\Pi_j\delta(t)$ produces the initial modal velocity of Π_j/M_j, and a zero initial modal deflection.

To perhaps more easily understand the argument that follows, and for the sake of versitility, it is convenient to now replace the modal equation of motion, Eq. (7.5), by the equation of motion for a single oscillator, its extact analog. Recall that an oscillator is a single-DOF system that consists of a discrete mass m, a dashpot c, and a spring k. The oscillator differential equation of motion, with $u(t)$ as the single DOF, is

$$\ddot{u}(t) + 2\zeta\omega_1\dot{u}(t) + \omega_1^2 u(t) = \frac{F(t)}{m} = \frac{\mathcal{F}}{m}\delta(t). \qquad (7.5a)$$

The recognition of the above one-to-one correspondence allows easier reference to, say, the motion of the modal mass. Since the applied (modal) force is impulsive, this external input vanishes forever after time zero plus. Thus, after time zero plus, the (modal) mass (of the oscillator) undergoes a force free vibration, and the modal deflection response, $u(t)$ or $p_j(t)$, depends exclusively on the (modal) initial conditions that are now the (modal) deflection and the (modal) velocity at time zero plus. Hence, it is now just a matter of determining the constants of integration C_0 and ψ

from Eq. (5.3b), which is the damped single-DOF, free vibration solution. That is, copying the previous single-DOF free vibration solution from Chapter 5, but writing it in terms of the modal quantities instead of the their single-DOF analogs

$$p_j(t) = C_0 e^{-\zeta_j \omega_j t} \sin\left(\omega_j t\sqrt{1-\zeta_j^2} + \psi_i\right)$$

$$\text{where} \quad p_j(t=0^+) = 0 \quad \text{and} \quad \dot{p}_j(t=0^+) = \frac{\Pi_j}{M_j}.$$

Application of the modal deflection initial condition yields $C_0 \sin\psi_j = 0$. Since C_0 cannot be zero, ψ must be either zero or π. If ψ_j equals π, (a 180° phase shift) then the effect is simply that of introducing a negative sign into the above equation which can be absorbed by C_0. Therefore choose ψ_j to be zero. Then differentiating with respect to time to obtain the expression for the modal velocity, leads to the result that $C_0 = \Pi_j/(M_j\omega_j\sqrt{1-\zeta_j^2})$. Substituting this result and the result that ψ is zero into the general, damped, free vibration solution above leads to

$$p_j(t) = \frac{\Pi_j}{M_j\omega_j\sqrt{1-\zeta_j^2}}\, e^{-\zeta_j\omega_j t} \sin\left(\omega_j t\sqrt{1-\zeta_j^2}\right)$$

$$\text{or} \quad p_j(t) = \Pi_j\, h_j(t), \tag{7.9}$$

where $h_j(t)$, which has units of length divided by force-time when the modal deflection has units of length, is called the *impulse response function* for the jth mode. Equation (7.9) says that the modal deflection response to an impulse occurring at time zero, with perhaps a rectangular shape, is simply the magnitude of the jth modal impulse, Π_j, multiplied by the impulse response function $h_j(t)$.

The solution for the modal deflection response to an arbitrarily varying modal force is quickly built on the impulse response function. Before doing so, there is a third special loading case, with its corresponding particular response function, that is worth examining even though this third response function is neither necessary nor does it provide any advantage relative to the impulse response function with respect to solving the general loading case. This case does, however, provide a clear contrast between a structural system response to a static load and a dynamic load. Consider the case where the jth modal loading[4] is the sudden application of a constant force of magnitude P_j^0. Physically, this type of force loading could be achieved by holding a fixed weight of magnitude P_j^0 just slightly above the surface of a structure and then releasing it so that it drops through a near zero distance. This third special case can be considered to be the opposite to the impulse loading case. In this case the time duration of the loading after time zero is unbounded while the magnitude of the applied force is finite. In the impulsive loading case, the smallness of the time duration of the applied force and the peak amplitude of the applied force are unbounded, whereas the magnitude of the impulse is finite. To deal with this sudden, constant loading, it is convenient to introduce another mathematical distribution called the *unit* or *Heaviside step function*, stp $(t-\tau)$, as a form for such a suddenly applied load.

[4] Or, relative to a single-DOF system, a physical loading.

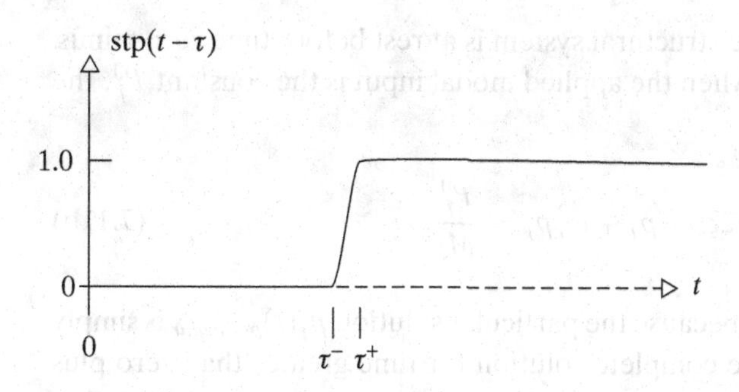

Figure 7.2. The Heaviside or unit step function.

Therefore define the Heaviside step function, where again t is the variable and τ is the parameter, as (i) everywhere differentiable; (ii) $\mathrm{stp}(t-\tau) - \frac{1}{2}$ is an odd function about τ (which makes $\mathrm{stp}(0) = \frac{1}{2}$); and, as shown in Figure 7.2, (iii)

$$\mathrm{stp}(t-\tau) = \begin{cases} 0 & t \leq \tau - \epsilon/2 = \tau^- \\ 1 & t \geq \tau + \epsilon/2 = \tau^+ \end{cases}.$$

In simple terms, the value of the nondimensional step function is zero when its argument is negative, and its value is +1 when its argument is positive. The unit step function is related to the Dirac delta function. Consider the integration of $\delta(\tilde{t} - \tau)$ from $\tilde{t}$ equal to minus infinity to $\tilde{t}$ equal to an arbitrary value of time, t. From Figure 7.1, whenever t is less than τ, the value of the integral is zero. Whenever t is greater than τ, the value of the integral is 1. Thus the values of the integral of the Dirac delta function match the values of the Heaviside step function outside the zero minus to zero plus time interval. With the smooth transition indicated in Figure 7.2, the values of the integral of the Dirac delta function also can match those of the Heaviside step function within the interval of length epsilon. Therefore write

$$\int_{-\infty}^{t} \delta(\tilde{t} - \tau)\,d\tilde{t} = \mathrm{stp}(t-\tau). \tag{7.10a}$$

Note that because $\tilde{t}$ is merely a dummy variable of integration, the left- and right-hand sides of the above equation are functions of t and τ. Differentiating the above equality with respect to t yields

$$\delta(t-\tau) = \frac{d}{dt}\mathrm{stp}(t-\tau). \tag{7.10b}$$

Clearly, the Heaviside step function has units of radians (i.e., no units). The Heaviside step function fully describes the time variation of a suddenly applied load. Thus the suddenly applied modal force of zero magnitude before time zero, and constant magnitude P_j^0 after time zero, can be written, as in Eq. (7.5a), as $P_j(t) = P_j^0\mathrm{stp}(t)$. Then the current form of the modal equation of motion, Eq. (7.5), becomes

$$\ddot{p}_j + 2\zeta_j\omega_j\dot{p}_j + \omega_j^2 p_j = \frac{P_j^0}{M_j}\mathrm{stp}(t), \tag{7.11a}$$

where again P_j^0, is a constant. The mathematical solution to this equation for time less than zero minus, that is, when the applied force is zero, is simply $p_j(t) = 0$ because, as before, it is presumed that the structural system is at rest before time zero minus. However, after time zero plus, when the applied modal input is the constant P_j^0, the equation of motion is simply

$$\ddot{p}_j + 2\zeta_j\omega_j\dot{p}_j + \omega_j^2 p_j = \frac{P_j^0}{M_j}. \tag{7.11b}$$

This is an easy equation to solve because the particular solution $p_j(t)_{particular}$ is simply the constant $P_j^0/M_j\omega_j^2$. Thus the complete solution for time greater than zero plus is this particular solution plus either standard form of the complementary solution such as

$$p_j(t) = C_j e^{-\zeta_j\omega_j t}\sin\left(\omega_j t\sqrt{1-\zeta_j^2} + \psi_j\right) + \frac{P_j^0}{M_j\omega_j^2}$$

or

$$p_j(t) = \frac{P_j^0}{M_j\omega_j^2} + e^{-\zeta_j\omega_j t}\left[A_j\sin\left(\omega_j t\sqrt{1-\zeta_j^2}\right) + B_j\cos\left(\omega_j t\sqrt{1-\zeta_j^2}\right)\right].$$

The initial conditions associated with this solution, those at time equals zero plus, can be deduced as follows. Consider the infinitesimal time interval from zero minus to zero plus. In this interval, the applied modal force behaves as does the step function in the time interval epsilon. Therefore, in this time interval, the modal force is bounded by the finite value P_j^0, which is the modal force value at time zero plus. Therefore, the acceleration of the modal mass is also finite. Hence, the modal velocity at time zero plus, which is bounded by the finite modal acceleration multiplied by the infinitesimal time interval, is infinitesimal. The only way to quantify this infinitesimal velocity is to call it zero relative to all finite values. The modal deflection at time zero plus is bounded by the product of this infinitesimal modal velocity and the length of the infinitesimal time interval. Thus, it, too, is zero. Hence, in summary, the modal velocity and modal deflection at time zero plus are both zero. Applying these initial conditions to the above complete solution yields

$$p_j(t) = \frac{P_j^0}{M_j\omega_j^2}\left[1 - e^{-\zeta_j\omega_j t}\left(\frac{\zeta_j}{\sqrt{1-\zeta_j^2}}\sin\omega_j t\sqrt{1-\zeta_j^2} + \cos\omega_j t\sqrt{1-\zeta_j^2}\right)\right]$$

or, in another form

$$p_j(t) = \frac{P_j^0}{M_j\omega_j^2}\left[1 - e^{-\zeta_j\omega_j t}\frac{\cos\left(\omega_j t\sqrt{1-\zeta_j^2} - \overline{\psi_j}\right)}{\sqrt{1-\zeta_j^2}}\right],$$

$$\text{where}\quad \cos\overline{\psi_j} = \sqrt{1-\zeta_j^2} \quad\text{or}\quad \tan\overline{\psi_j} = \frac{\zeta_j}{\sqrt{1-\zeta_j^2}}$$

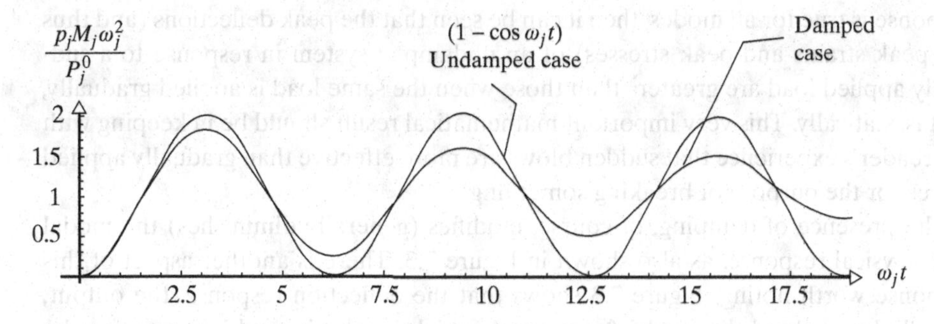

Figure 7.3. The response to an applied step force.

or, in still another form

$$p_j(t) = \frac{P_j^0}{M_j \omega_j^2}\left[1 - e^{-\zeta_j \omega_j t}\frac{\sin\left(\omega_j t\sqrt{1-\zeta_j^2} - \overline{\overline{\psi}}_j\right)}{\sqrt{1-\zeta_j^2}}\right]$$

$$\text{where} \quad \sin\overline{\overline{\psi}}_j = \sqrt{1-\zeta_j^2} \quad \text{or} \quad \text{or} \tan\overline{\overline{\psi}}_j = \frac{\sqrt{1-\zeta_j^2}}{\zeta}$$

or, for all forms

$$p_j(t) = P_j^0 g_j(t), \tag{7.12}$$

where the function $g_j(t)$ is called the *step response function*. The step response function has units of length divided by force, making it another form of a dynamic flexibility coefficient.

When the damping factor is zero, usually a worse case for the magnitude of the structural response, the nondimensionalized modal deflection becomes simply

$$\frac{p_j(t)}{\left(P_j^0 / M_j \omega_j^2\right)} = 1 - \cos\omega_j t. \tag{7.12a}$$

This equation for the modal deflection, plotted in Figure 7.3, reveals an important fact concerning dynamic loads relative to static loads. When the nondimensionalized time $\omega_j t = \pi$, or any odd, positive multiple of π, then the above nondimensional, undamped, modal deflection takes of the value 2.0. However, if the normalized modal force of magnitude P_j^0 were applied as a static load, meaning that it is applied so slowly that the acceleration and velocity terms of the modal equation of motion, Eq. (7.5) or (7.11b), are zero, then the above normalized modal deflection of Eq. (7.12a), $p_j(t)/(P_j^0/M_j\omega_j^2)$, has the constant value 1.0. (The modal mass multiplied by the modal frequency-squared term makes better sense in this static case when it is recalled that this term is equal to the modal stiffness factor K_j, which has a general force per unit displacement meaning in the static as well as dynamic load case.) Since this conclusion that the modal dynamic peak response is twice the static response is true for all modes, then it can be seen that the peak deflections (and thus the peak strains and peak stresses) of an undamped system in response to a suddenly applied load are greater[5] than those when the same load is applied gradually, that is, statically. This very important mathematical result should be in keeping with the reader's experience that sudden blows are more effective than gradually applied forces for the purpose of breaking something.

The presence of damping, of course, modifies (generally diminishes) the modal and physical response, as also shown in Figure 7.3. There is another aspect of this response worth noting. Figure 7.3 shows that the deflection response, the output, initially lags behind the modal force input, which is a horizontal line crossing the nondimensional abscissa at 1.0. Later it overshoots to reach its peak. Both the lag and the overshoot can, by analogy to a single-DOF oscillator, be ascribed to the inertia of the modal mass that causes the mass to be slow to start into motion and then to be slow to reverse direction as it vibrates.

For a dynamic load with any type of time variation, the ratio of the peak dynamic deflection to the corresponding static deflection is called the *dynamic load factor*. In this case of a single undamped mode, the ratio of the peak modal deflection response because of the dynamic load $P^0\text{stp}(t)$ to the static modal deflection response because of the equal magnitude static load P^0 produces a dynamic load factor of 2.0. If the oscillator were damped, the value would be less than 2, but still significantly greater than 1. The dynamic load factor is a convenient concept, particularly in design. If previous experience with similar structures and similar loads allows a reasonable approximation of the dynamic load factor, then the important advantage gained is that a static analysis can be used throughout the design process rather than a more expensive dynamic analysis. For example, the design code for highway bridges has 1.3 as the maximum dynamic load factor. The fact that 1.3 is considerably less than 2.0 can be explained as follows. First of all, for highway bridges, the traffic loads are not all that quickly applied relative to the first natural period of those bridges. For any structure, the dynamic load factor is reduced from 2.0 by the fact that all the dynamic modal peak amplitudes do not occur simultaneously. As is discussed in detail below, the total deflection response of the structure involves the sum of the various modal responses, where their individual peaks are spread out over time. That is, because any one modal deflection time history is only a fraction of the total deflection response, if its peak is isolated in time from the other modal peaks, then its contribution to the maximum deflection is limited. This, in addition to damping, generally lessens the overall dynamic load factor of any structure. Keep in mind, however, the closer the structural model to that of a single-DOF structure, the greater the possibility of a higher dynamic load factor.

Finally, just as the Heaviside step function and the Dirac delta function are related to each other as per Eq. (7.10), the step response function is related to the impulse response function. It is easy to show by direct differentiation that

$$\frac{d}{dt}g_j(t) = h_j(t). \tag{7.13}$$

[5] The dynamic response for even an undamped multidegree of freedom system would not be double the static response because the modal peaks do not all occur at the same time value.

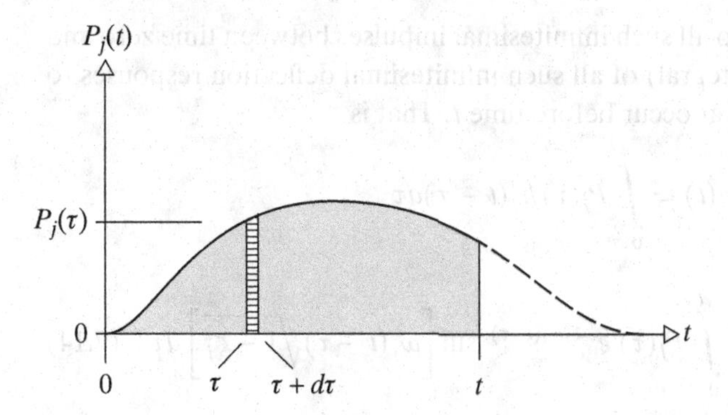

Figure 7.4. Summing infinitesmal impulses up to time t for a pulse of arbitrary shape.

Since the step response function has units of length divided by force and the impulse response function has units of length divided by force-time, the units of the above equation check.

7.6 The Modal Solution for a General Type of Loading

Now that the two special cases of impulsive and sudden loads have been investigated, it is possible to proceed to the solution for the general modal equation of motion, Eq. (7.5), which, again, is

$$\ddot{p}_j + 2\zeta_j\omega_j\dot{p}_j + \omega_j^2 p_j(t) = \frac{P_j(t)}{M_j}, \tag{7.5}$$

where $P_j(t)$ represents an arbitrary force history. There is more than one way to accomplish this task. For example, this equation with the arbitrary input can be solved using the formal mathematics of Laplace transforms. However, the same result produced by Laplace transforms can be obtained by using physical reasoning combined with either the impulse response function or the step response function. The use of one of these two special functions is preferred because the logic behind such an approach is much closer to the physical reality of applying a time-varying force to a structure, and this special function approach avoids a lengthy introduction to the sophisticated mathematics of Laplace transforms.

Consider the arbitrary applied modal force $P_j(t)$, whose typical time history is suggested in Figure 7.4, where the clock starts when the modal force first has a nonzero value. Of course, at time $t = \tau$, the modal force has the magnitude $P_j(\tau)$. Consider the infinitesimally thin rectangular area under the modal force curve between the times τ and $\tau + d\tau$ that has as its height the value $P_j(\tau)$. Recall the fact that an impulse can have any symmetric shape, including that of a rectangle. Therefore, as shown in Figure 7.4, consider this rectangular area as an impulse of infinitesimal magnitude $P_j(\tau)d\tau$. This quantity is indeed an impulse because, as has been seen, an impulse is always the area under a force-time curve during a very short interval of time. Now the modal response p_j at time t to this impulsive input at time τ can be written immediately by use of the impulse response function h as stated in Eq. (7.9), which is applicable to all impulses. That is, after an elapsed time of $t - \tau$ the deflection response is just the product of the magnitude of the impulse $P_j(\tau)d\tau$ and $h_j(t - \tau)$.

The total response at time t to all such infinitesimal impulses between time zero and time t is just the sum (i.e., integral) of all such infinitesimal deflection responses to such infinitesimal impulses that occur before time t. That is

$$p_j(t) = \int_0^t P_j(\tau)\, h_j(t-\tau)d\tau$$

or in detail

$$p_j(t) = \frac{1}{M_j\omega_j\sqrt{1-\zeta_j^2}} \int_0^t P_j(\tau)\, e^{-\zeta_j\omega_j(t-\tau)} \sin\left[\omega_j(t-\tau)\sqrt{1-\zeta_j^2}\right] d\tau, \quad (7.14)$$

where the integration is, of course, over the parameter τ. The result is a function of t, which appears in the upper limit as well as the integrand. This integral solution is named the *duhamel integral*, or the *convolution integral*, or the *superposition integral.* Endnote (1) shows by direct substitution that this integral solution satisfies the modal differential equation of motion. Exercise 7.5(g) discusses a slight variation on Eq. (7.14), which is useful when the analysis includes damping and the analytical form of the modal force is mathematically simpler than that of the impulse response function.

The advantage to a solution in the form of an integral, such as the convolution integral of Eq. (7.14), is that the input, which is the modal force, need not be written as a single smooth function for the purposes of completing the integration. For example, the integration can be performed over several subintervals if the input is known only as a series of piecewise continuous functions. Moreover, if the input is even just a tabular listing of load magnitudes at successive time points, then a numerical integration for the response can be performed.

The modal deflection response to an arbitrary modal force can also be written in terms of the step response function. In this case, think of the step function as representing an increase in the magnitude of the applied load (from zero to P^0) that lasts indefinitely over time. See Figure 7.5 where the applied modal load is described by an initial, finite load step, followed by a series of infinitesimal load steps that increase the magnitude of the applied force. In other words, where in the impulse case the load increments were vertical slices of the load history, here they are horizontal slices. Therefore, as before, consider an infinitesimal time interval. On this occasion let the interval be $[\tau^-, \tau^+]$. Using the rule that "rise equals base multiplied by slope," conclude that the infinitesimal increase in the modal force, $dP_j(\tau)$, equals $d\tau\, \dot{P}_j(\tau)$, where, of course, the dot indicates differentiation with respect to τ. Therefore, in terms of the step response function, the modal deflection response at time t to this infinitesimal step input occurring at time τ, taking into account that the elapsed time from the occurrence of the infinitesimal step is $t - \tau$, is

$$dp_j(t) = d\tau\, \dot{P}_j(\tau)\, g_j(t-\tau)$$

so summing

$$p_j(t) = P_j(0)g_j(t) + \int_0^t \dot{P}_j(\tau)\, g_j(t-\tau)d\tau$$

$$\text{where} \quad \dot{P}_j(\tau) \equiv \frac{dP_j(\tau)}{d\tau} \quad (7.15)$$

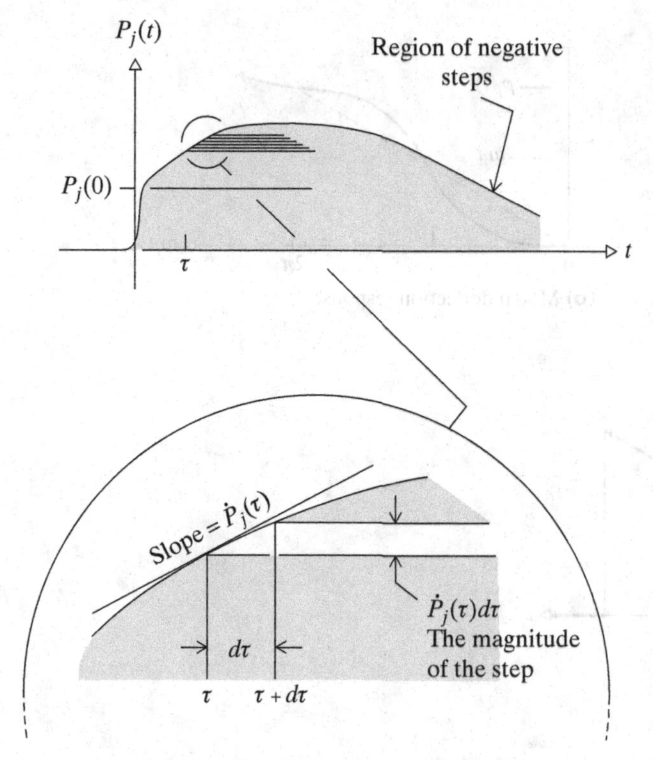

Figure 7.5. Summing infinitesimal steps, of positive and negative heights, to create a pulse of arbitrary shape.

and where $P_j(0)g_j(t)$ accounts for the initial, finite step. Exercise 7.5(h) establishes that Eq. (7.15) can also be obtained directly from Eq. (7.14) by means of integration by parts. To complete the exercise, it is necessary to take into account Eq. (7.13), the relationship between the impulse response function and the step response function. The problem with this alternate approach is that this step response form of the convolution integral seldom offers computational advantage for undamped systems, and any estimate of the slope of the applied load with respect to time is likely to be less accurate than the estimate of the modal load itself.

7.7 Example Problems

For the sake of simplicity, the problems can be considered to result from an (equivalent) dynamic load vector $\{Q(t)\}$ that only has a single nonzero entry in the ith position. This analysis situation is not uncommon. Call that single entry $Q_i(t)$. Thus jth entry of the modal load vector $\{P(t)\}$ is simply $A_i^{(j)} Q_i(t) = P_j(t)$. That is, all the modal forces have the same time variation and differ only by a constant factor. Therefore a solution for all modes is proportional to the solution for the jth mode.[6] Hence, all example problems focus on obtaining a solution for the jth mode, or, equivalently, they will be formulated for a single-DOF system. Furthermore, for the

[6] However, each term of the modal summation, Eq. (7.3), has different modal parameters, such as natural frequencies.

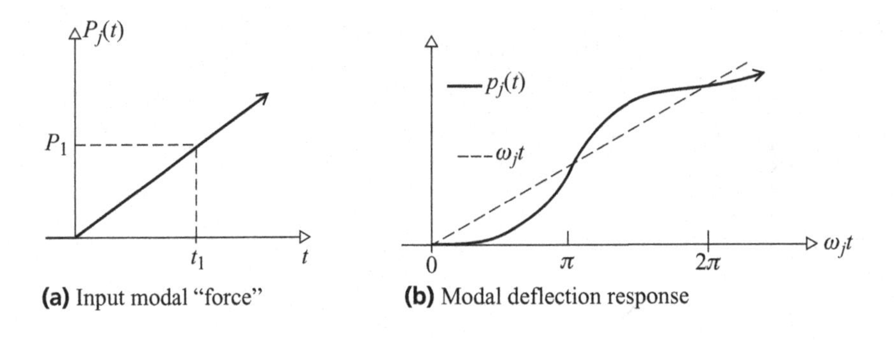

(a) Input modal "force"

(b) Modal deflection response

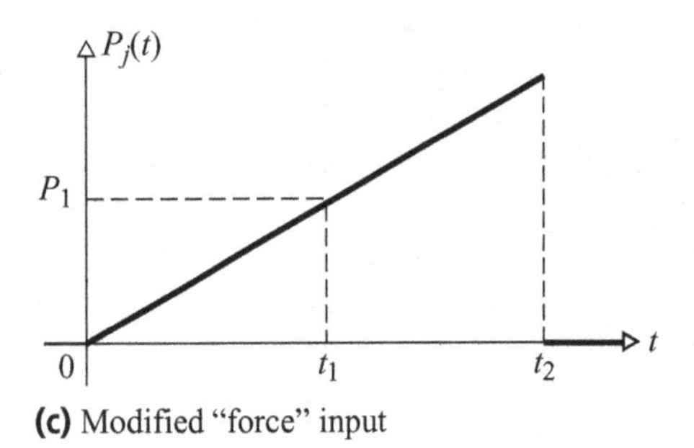

(c) Modified "force" input

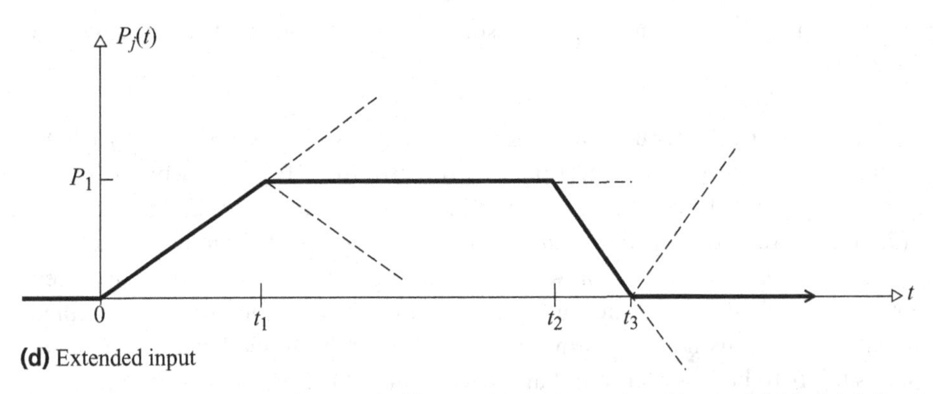

(d) Extended input

Figure 7.6. Examples 7.1, 7.2, and 7.3. (a) Model force input time history. (b) Model deflection response. (c) Modified applied force. (d) Force input extended to that of a pulse.

sake of analytical simplicity, all these example problems will be addressed on the generally worst-case basis of zero damping. The chief effect of the presence of light damping would be only to complicate the analytical integration and obscure the point of the example.

EXAMPLE 7.1 Determine the undamped jth mode response after time zero when the jth modal force $P_j(t)$ has the analytical description $(P_1 t)/t_1$. That is, $P_j(t)$ starts with a zero value and increases linearly thereafter as shown in Figure 7.6(a) with a slope P_1/t_1.

SOLUTION This is one of the very few cases where the step response form of the superposition integral is more convenient than the impulse response form. This is so because the derivative of the input happens to be a constant. Therefore, for the modal deflection output, write

$$p_j(t) = P_j(0)g_j(t) + \int_0^t \dot{P}_j(\tau)g_j(t-\tau)d\tau,$$

$$\text{where} \quad P_j(0) = 0 \quad \text{and} \quad \frac{dP_j(\tau)}{d\tau} = \frac{P_1}{t_1}.$$

Substituting for the undamped step response function, from any of the three forms before Eq. (7.12), but particularly the first, leads to the following calculation of the modal deflection response. Note that in the following calculation, the integral of the cosine function was obtained by using the trigonometric identity $\cos(\alpha-\beta) = \cos\alpha\cos\beta + \sin\alpha\sin\beta$. This is often the more convenient approach for these evaluations. However, in this case, the calculation of the integral of the cosine function could have been shortened by rewriting the differential $d\tau$ as $-d(t-\tau)$ and then integrating directly using the variable $(t-\tau)$.

$$p_j(t) = \frac{P_1}{M_j\omega_j^2 t_1}\int_0^t [1-\cos\omega_j(t-\tau)]\,d\tau$$

$$= \frac{P_1}{M_j\omega_j^2 t_1}\left[t - \cos\omega_j t\int_0^t \cos\omega_j\tau d\tau - \sin\omega_j t\int_0^t \sin\omega_j\tau d\tau\right]$$

$$= \frac{P_1}{M_j\omega_j^2 t_1}\left[t - \frac{\cos\omega_j t\sin\omega_j t}{\omega_j} + \frac{\sin\omega_j t\cos\omega_j t}{\omega_j} - \frac{\sin\omega_j t}{\omega_j}\right]$$

$$\text{or} \quad p_j(t) = \frac{P_1}{M_j\omega_j^3 t_1}[\omega_j t - \sin\omega_j t]$$

$$\text{or} \quad \frac{p_j(t)}{\left(P_1/M_j\omega_j^2\right)} = \left[\frac{t}{t_1} - \frac{\sin\omega_j t}{\omega_j t_1}\right],$$

where the latter expression contains only nondimensional quantities. A plot of this response in nondimensional terms is presented in Figure 7.6(b) where the value of $\omega_j t_1$ is selected to be $+1.0$. From the plot it can be seen that, as always, because of the inertia of the modal mass, the response of the modal mass initially lags the modal force. Later, the momentum of the modal mass carries it beyond the modal force, where after the modal spring force causes it to alternately lead and lag the modal force. ★

EXAMPLE 7.2 Redo the problem where now the linearly increasing input force drops to a constant zero magnitude at time equal to t_2 as shown in Figure 7.6(c).

SOLUTION Since the triangular force input contains a point of discontinuity at time t_2, different response expressions need to be written for the time intervals (i) $0 \leq t \leq t_2$ and (ii) $t_2 \leq t$. The solution for the first time interval is exactly that of the previous example

$$0 \leq t \leq t_2 \quad p_j(t) = \frac{P_1}{M_j\omega_j^3 t_1}\left[\omega_j t - \sin\omega_j t\right].$$

The response solution for the second time interval utilizing the step response function requires careful consideration; that is, it is easy to make a mistake. The response using the step response function can be determined by realizing that (i) the response for an arbitrary time t in this second time interval is dependent on an integration from time zero to the discontinuity time point $t = t_2$, plus an integration from time t_2 to the arbitrary time t and (ii) the slope of the zero force input is zero for all times greater than t_2. Hence it might appear that the solution for the can be adapted to this time interval by merely changing the upper limit of the integral in the solution from t to t_2. That is, it might appear that the solution can be written as

$$p_j(t) \stackrel{?}{=} \frac{P_1}{M_j\omega_j^2 t_1}\int_0^{t_2}\left[1 - \cos\omega_j(t-\tau)\right]d\tau + \int_{t_2}^{t} 0\,d\tau,$$

where the value of the second integral is, of course, zero. However, this description of the response is wrong for the following reason. This description of the response does not distinguish between a time-varying force behaving as shown in Figure 7.6(c), where the force drops to zero after time t_2 and, for example, a time-varying force that after time t_2 remains at the constant value P_1t_2/t_1, which also has a zero slope after time t_2. The necessary correction to the above expression is based on the recognition that not only is there a discontinuity in the time rate of change of the force, but that there is also a discontinuity in the force itself at t_2. It is necessary to account for the finite sized step in the value of the force at time t_2 of magnitude $-P_1t_2/t_1$ by adding to the above right-hand side the response to such a step. That is, the above right-hand side must be augmented by

$$-\frac{P_1t_2}{t_1}\left[1 - \cos\omega_j(t-t_2)\right].$$

Then the final result for all time greater than t_2 is

$$p_j(t) = \frac{P_1}{M_j\omega_j^2}\left[\frac{\sin\omega_j(t-t_2)}{\omega_j t_1} - \frac{\sin\omega_j t}{\omega_j t_1} + \frac{t_2}{t_1}\cos\omega_j(t-t_2)\right].$$

It is all too easy when using the step response form of the convolution integral solution to forget the necessity of including any finite steps at discontinuities in the applied force time history. For this reason, the step response function will no longer be used to obtain modal responses in the remainder of this textbook.

The alternative to the use of the step function formulation of the response is, of course, the use of the impulse function formulation. This approach is always more straightforward even if just occasionally more tedious. In this case, using a table of integrals,[7] the solution for $t_2 \le t$ is simply obtained as

$$p_j(t) = \frac{P_1}{M_j \omega_j t_1} \int_0^{t_2} \tau \sin \omega_j (t - \tau) d\tau + \int_{t_2}^{t} 0 \, d\tau$$

$$\text{or} \quad p_j(t) = \frac{P_1}{M_j \omega_j t_1} \left[\sin \omega_j t \int_0^{t_2} \tau \cos \omega_j \tau d\tau - \cos \omega_j t \int_0^{t_2} \tau \sin \omega_j \tau d\tau \right]$$

$$\text{or} \quad p_j(t) = \frac{P_1}{M_j \omega_j^3 t_1} \sin \omega_j t \left(\cos \omega_j t_2 + \omega_j t_2 \sin \omega_j t_2 - 1 \right)$$

$$- \frac{P_1}{M_j \omega_j^3 t_1} \cos \omega_j t \left(\sin \omega_j t_2 - \omega_j t_2 \cos \omega_j t_2 \right)$$

$$\text{or} \quad p_j(t) = \frac{P_1}{M_j \omega_j^3 t_1} \left[\sin \omega_j (t - t_2) + \omega_j t_2 \cos \omega_j (t - t_2) - \sin \omega_j t \right],$$

which is the same as above. A weak check on this solution can easily be obtained by merely checking the continuity at time t_2 between the deflection solution for the time interval $0 \le t \le t_2$, as given in Example 7.1, and that for the time interval $t_2 \le t$, as stated above. ★

EXAMPLE 7.3 Use the result of Example 7.1 to determine the modal deflection response to the *pulse*[8] shown in Figure 7.6(d).

SOLUTION The deflection response to a linearly increasing modal force $P_1 t / t_1$, as calculated in the first example problem, can be rewritten as

$$p_j(t) = \frac{P_1}{M_j \omega_j^3 t_1} (\omega_j t - \sin \omega_j t) \equiv \frac{P_1}{t_1} r_j(t),$$

where, as a matter of future convenience, $r_j(t)$, the *ramp response function* is now so defined. Again, the quantity P_1/t_1 is simply the slope of the ramp. The availability of the ramp response function greatly simplifies the writing of the modal deflection response. The superposition strategy is as follows. For the time interval $(0, t_1)$, the response is simply as above, $p_j(t) = (P_1/t_1) r_j(t)$. The modal deflection response after the first time break, t_1, is obtained by superimposing on the original applied modal force at $t = t_1$, a new applied modal force equal to the negative of the original applied modal force. That is, the sum of a positive slope force and a negative slope

[7] Again, another option for obtaining the desired analytical solution for a Duhamel integral is to use suitable software such as Mathematica. See Refs. [7.1,7.2].

[8] A *pulse* is an applied force time history where nonzero values occur over a time interval greater than the fundamental (first) period of the system. A *shock* is an applied time history where the nonzero values occur over a time interval less than the fundamental period but over a time interval too long to be called an impulse.

force of the same absolute magnitude produces the desired zero slope force, which is the same as a constant applied modal force of magnitude P_1 in the time interval (t_1, t_2) as shown in Figure 7.6(d). In other words, the modal deflection response in that time interval is the response from the continuing original load, plus the response to the new load that starts at time t_1. The mathematical description of the modal deflection response can be written as

$$\text{for} \quad 0 \le t \le t_1 \quad p_j(t) = \frac{P_1}{t_1} r_j(t)$$

$$\begin{aligned}\text{for} \quad t_1 \le t \le t_2 \quad p_j(t) &= \frac{P_1}{t_1} r_j(t) - \frac{P_1}{t_1} r_j(t - t_1) \\ &= \frac{P_1}{\omega_j^3 t_1} [\omega_j t_1 + \sin \omega_j (t - t_1) - \sin \omega_j t] \\ &= \frac{P_1}{\omega_j^3 t_1} [\omega_j t_1 - (1 - \cos \omega_j t_1) \sin \omega_j t - \sin \omega_j t_1 \cos \omega_j t].\end{aligned}$$

At time t_2, it is again a matter of superimposing another negative ramp force. That is, for the time period $t_2 \le t \le t_3$ the modal deflection response is the above response for $t_1 \le t \le t_2$ plus the following negative quantity

$$-\frac{P_1}{(t_3 - t_2)} r_j(t - t_2) = -\frac{P_1}{\omega_j^3 (t_3 - t_2)} [\omega_j (t - t_2) - \sin \omega_j (t - t_2)],$$

where the slope of the modal force input, $-P_1/(t_3 - t_2)$, is the slope used with the modal deflection output. Finally, for $t_3 \le t$, the modal deflection response is the above response for $t_2 \le t \le t_3$ plus the quantity $+P_1 r_j(t - t_3)/(t_3 - t_2)$. Again, all these deflection responses are for zero damping. Damping always mitigates the magnitude of the response to an extent depending on the amount of damping present and eventually will eliminate the response sometime after time t_3. ★

In this next example problem, for the sake of variety, there is a change in style, and style only, from the modal equation form to exactly the same equation form for a single-DOF structural model. The single-DOF equation of motion and the corresponding convolution integral solution are, of course

$$\ddot{q}(t) + 2\zeta\omega_1 \dot{q}(t) + \omega_1^2 q(t) = \frac{Q(t)}{m}$$

$$\text{so} \quad q(t) = \int_0^t Q(\tau) h(t - \tau) d\tau.$$

The generalized force $Q(t)$ can, of course, be the result of an estimated foundation motion, say $u(t)$. In that case the generalized force has the form $ku + c\dot{u}$. The reason for this change of style is that a single-DOF system can be used sometimes as a rough approximation for a simple structure for the purpose of gaining design information for dynamic responses. The rough approximation of a multidegree of freedom system by a single-DOF model can be accomplished by means of a Rayleigh analysis. See Endnote (2) for a simple demonstration of this procedure.

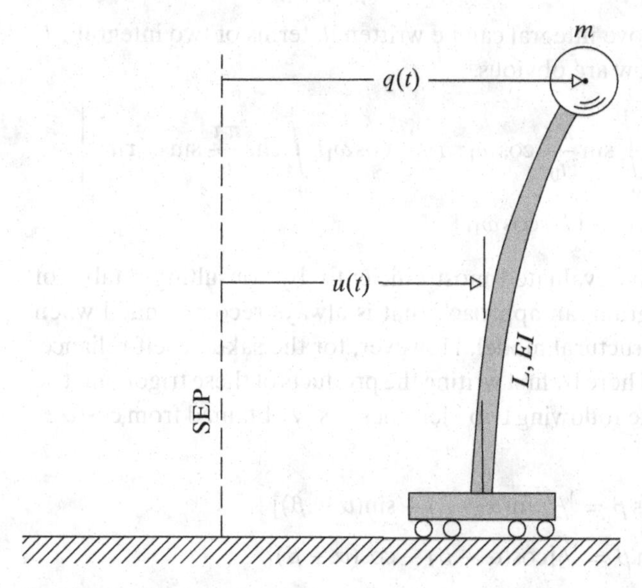

Figure 7.7. Example 7.4: Base motion activated one degree of freedom system.

EXAMPLE 7.4 The undamped, single-DOF structural system shown in Figure 7.7 is subject to a sinusoidal base motion starting at time zero, which is described as $u(t) = \Upsilon_0 \sin(\pi t/t_0)$, where Υ_0 is a constant magnitude. Use the superposition integral to determine the response $q(t)$of the mass m. The structure is at rest before time zero.

SOLUTION The generalized coordinate $q(t)$ is measured relative to the fixed axis SEP. Therefore the kinetic energy expression only involves the first time derivative of $q(t)$, whereas the strain energy expression (and the formula for the elastic force acting on the mass) involves the difference between $q(t)$ and $u(t)$. The cantilevered beam tip stiffness factor, if not already known, can be deduced from the beam element stiffness matrix as discussed in the solution to Exercise 7.6. That tip stiffness factor is $k = 3EI/L^3$. Hence the kinetic and strain energy expressions, and the resulting Lagrange equation of motion, and the integral form of its solution are

$$T = \frac{1}{2} m\dot{q}^2 \qquad U = \frac{1}{2}\frac{3EI}{L^3}(q-u)^2$$

$$\text{so} \quad \ddot{q} + \frac{3EI}{mL^3} q = \frac{3EI}{mL^3} u(t) = \frac{3EI}{mL^3} \Upsilon_0 \sin\frac{\pi t}{t_0} = \omega_1^2 \Upsilon_0 \sin\frac{\pi t}{t_0}$$

$$\text{and then} \quad q(t) = \omega_1 \Upsilon_0 \int_0^t \sin\frac{\pi\tau}{t_0} \sin(t-\tau)d\tau.$$

After checking units it is now a matter of carrying out the integration. Perhaps the most straightforward approach to evaluating this integral is to expand the second sine function. Then, after factoring out of the integral with respect to τ those functions that depend only on t, the above integral can be written in terms of two integrals, I_1 and I_2, whose definitions below are obvious.

$$q(t) = \omega_1 \Upsilon_0 \left[\sin\omega_1 t \int_0^t \sin\frac{\pi\tau}{t_0} \cos\omega_1\tau\, d\tau - \cos\omega_1 t \int_0^t \sin\frac{\pi\tau}{t_0} \sin\omega_1\tau\, d\tau \right]$$

$$\text{or} \quad q(t) = \omega_1 \Upsilon_0 [(I_1)\sin\omega_1 t - (I_2)\cos\omega_1 t].$$

The integrals I_1 and I_2 can be evaluated most efficiently by consulting a table of integrals or a computer program, an approach that is always recommended when damping is included in the structural model. However, for the sake of self-reliance, the integrations are evaluated here by first writing the products of these trigonometric functions as sums by use of the following two identities easily obtained from $\cos(\alpha \pm \beta)$ and $\sin(\alpha \pm \beta)$ formulas:

$$\sin\alpha\cos\beta = {}^1\!/\!_2[\sin(\alpha+\beta) + \sin(\alpha-\beta)]$$

$$\sin\alpha\sin\beta = {}^1\!/\!_2[\cos(\alpha-\beta) - \cos(\alpha+\beta)].$$

Then the integrals I_1 and I_2 can be written as

$$I_1 = {}^1\!/\!_2 \int_0^t \sin\left(\frac{\pi\tau}{t_0} + \omega_1\tau\right) d\tau + {}^1\!/\!_2 \int_0^t \sin\left(\frac{\pi\tau}{t_0} - \omega\tau\right) d\tau$$

$$I_2 = {}^1\!/\!_2 \int_0^t \cos\left(\omega_1 - \frac{\pi}{t_0}\right)\tau\, d\tau + {}^1\!/\!_2 \int_0^t \cos(\omega_1 + \frac{\pi}{t_0})\tau\, d\tau.$$

To carry out what is now a straightforward integration, it is necessary to first require that $\pi/t_0 \neq \omega_1$ to avoid the singularity that occurs at the undamped resonance condition $\pi/t_0 = \omega_1$. The case where $\pi/t_0 = \omega_1$ has to be treated separately. After integration and some simplification, the integrals I_1 and I_2 are

$$I_1 = \frac{1}{\left[(\pi/t_0)^2 - \omega_1^2\right]} \left[\frac{\pi}{t_0} - \frac{\pi}{t_0}\cos\frac{\pi t}{t_0}\cos\omega_1 t - \omega_1 \sin\frac{\pi t}{t_0}\sin\omega_1 t\right]$$

$$I_2 = \frac{1}{\left[(\pi/t_0)^2 - \omega_1^2\right]} \left[\omega_1 \sin\frac{\pi t}{t_0}\cos\omega_1 t - \frac{\pi}{t_0}\cos\frac{\pi t}{t_0}\sin\omega_1 t\right].$$

When these two integrals are combined, the result, after a small amount of simplification, is

$$q(t) = \frac{\omega_1 \Upsilon_0}{\left[(\pi/t_0)^2 - \omega_1^2\right]} \left(\frac{\pi}{t_0}\sin\omega_1 t - \omega_1 \sin\frac{\pi t}{t_0}\right).$$

In the same spirit that led to the defining of a ramp response function as a convenience for representing shocks and pulses, now define the sinusoid response function as

$$s_1\left(t, \frac{\pi}{t_0}\right) = \frac{1}{m\left[(\pi/t_0)^2 - \omega_1^2\right]} \left(\frac{\pi}{\omega_1 t_0}\sin\omega_1 t - \sin\frac{\pi t}{t_0}\right)$$

so that the response to a sinusoidal force of magnitude $m\,\omega_1^2 \Upsilon_0$ and period $2t_0$ is simply $q(t) = m\,\omega_1^2 \Upsilon_0\, s_1(t, \pi/t_0)$. Of course, only slight adjustments are required to adapt

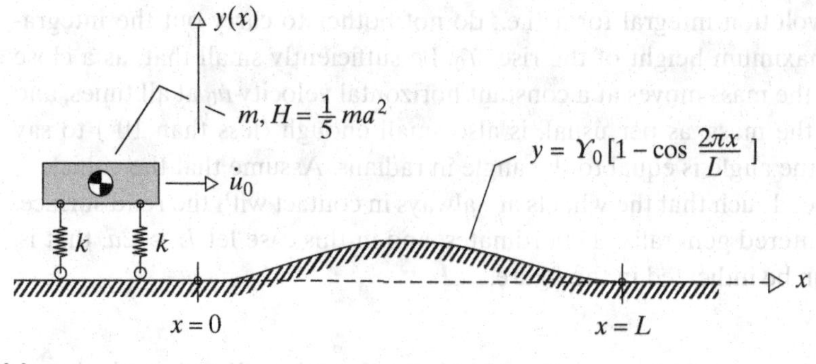

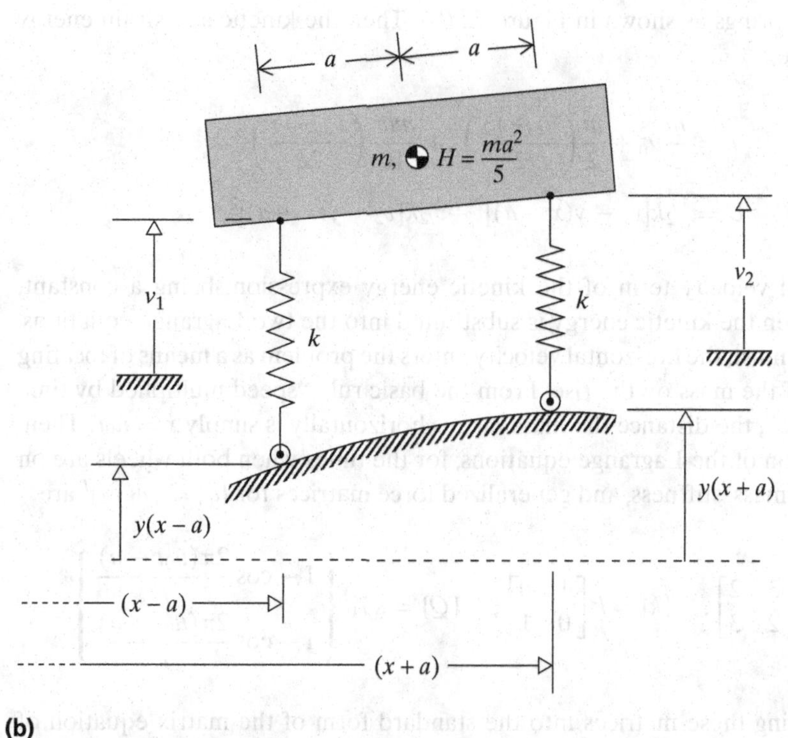

Figure 7.8. Example 7.5: Simplified vehicle traversing a speed hump at constant speed. (a) Vehicle system before encountering the hump. (b) Vehicle system in displaced configuration while on the hump.

this sinusoid response function for the single-DOF system to the sinusoid response function for the jth mode. ★

The following example problem summarizes the entire dynamic load analysis solution process. A two-DOF system is chosen to minimize the amount of calculation.

EXAMPLE 7.5 In terms of the modal coordinates, write the analytical solution for the dynamic response of the two-dimensional, undamped vehicle shown in Figure 7.8(a) during and after its encounter with a smooth, long speed hump. Leave the solution in convolution integral form (i.e., do not bother to carry out the integration). Let the maximum height of the rise, Υ_0, be sufficiently small that, as a close approximation, the mass moves at a constant horizontal velocity $\dot{u}_0$ at all times, and the rotation of the mass, as per usual, is also small enough (less than 10°) to say that the sine of the angle is equal to the angle in radians. Assume that the vehicle is traveling at a speed such that the wheels are always in contact with the road surface. Use stiffness centered generalized coordinates, and in this case let $L \gg 2a$, that is, more so than can be indicated in the figure.

SOLUTION The first task is to write the equations of motion. To this end, choose the *fixed reference* generalized coordinates, $q_1 = v_1$, and $q_2 = v_2$, which measure the stretch in the springs as shown in Figure 7.8(b). Then the kinetic and strain energy expressions are

$$T = \frac{m}{2}\dot{u}_0^2 + \frac{m}{2}\left(\frac{\dot{v}_1 + \dot{v}_2}{2}\right)^2 + \frac{ma^2}{10}\left(\frac{\dot{v}_2 - \dot{v}_1}{2a}\right)^2$$

$$U = {}^1\!/\!_2 k[v_1 - y(x-a)]^2 + {}^1\!/\!_2 k[v_2 - y(x+a)]^2.$$

The horizontal velocity term of the kinetic energy expression, being a constant, disappears when the kinetic energy is substituted into the two Lagrange equations. However, as it must, the horizontal velocity enters the problem as a means of locating the position of the mass on the rise. From the basic rule "speed multiplied by time equals distance", the distance the mass moves horizontally is simply $x = \dot{u}_0 t$. Then, after application of the Lagrange equations, for the times when both wheels are on the hump, the mass, stiffness, and generalized force matrices for $\{q\} = \lfloor v_1\ v_2 \rfloor^t$ are

$$[m] = \frac{m}{10}\begin{bmatrix} 3 & 2 \\ 2 & 3 \end{bmatrix}, \quad [k] = k\begin{bmatrix} 1 & 0 \\ 0 & 1 \end{bmatrix}, \quad \{Q\} = k\Upsilon_0 \begin{Bmatrix} 1 - \cos\dfrac{2\pi(\dot{u}_0 t - a)}{L} \\ 1 - \cos\dfrac{2\pi(\dot{u}_0 t + a)}{L} \end{Bmatrix}.$$

After inserting these matrices into the standard form of the matrix equation of motion, the next step is to solve the homogeneous equation for the two natural frequencies and mode shapes. (Contrary to the usual guidelines of necessity and accuracy as discussed in the next section, because of the small number of DOF used for this illustrative example, both of the natural modes are used to describe the system motion.) Choosing matrix iteration, first invert the stiffness matrix to get the first modal matrix equation and then, avoiding the use of a sweeping matrix, invert the mass matrix to get the last modal matrix equation. Then the dynamic matrix iteration equations for the first and second modes are, respectively,

$$\begin{bmatrix} 3 & 2 \\ 2 & 3 \end{bmatrix}\begin{Bmatrix} v_1 \\ v_2 \end{Bmatrix} = \frac{10k}{m\omega_1^2}\begin{Bmatrix} v_1 \\ v_2 \end{Bmatrix} \qquad \begin{bmatrix} 3 & -2 \\ -2 & 3 \end{bmatrix}\begin{Bmatrix} v_1 \\ v_2 \end{Bmatrix} = \frac{m\omega_2^2}{2k}\begin{Bmatrix} v_1 \\ v_2 \end{Bmatrix}.$$

This approach for the second and last mode is valid because matrix iteration always converges to the largest eigenvalue, which for this latter setup is the one containing the estimate of the second natural frequency. The results of the matrix iterations are

$$\omega_1 = \sqrt{\frac{2k}{m}} \qquad \omega_2 = \sqrt{\frac{10k}{m}}$$
$$\lfloor A^{(1)} \rfloor = \lfloor 1 \quad 1 \rfloor \qquad \lfloor A^{(2)} \rfloor = \lfloor 1 \quad -1 \rfloor$$
$$\text{thus} \quad [\Phi] = [\Phi]^t = \begin{bmatrix} 1 & 1 \\ 1 & -1 \end{bmatrix}.$$

A quick calculation confirms that these mode shapes are orthogonal when weighted by either the mass matrix or the stiffness matrix. Substituting into the matrix equations of motion the modal transformation $\{q\} = [\Phi][p]$ and premultiplying by $[\Phi]^t$ yields the following uncoupled equations of motion. (Note, as another check that for both the first and second mode, the ratio of the generalized stiffness and generalized mass terms are indeed the squares of the natural frequencies.)

$$\frac{m}{10}\begin{bmatrix} 10 & 0 \\ 0 & 2 \end{bmatrix}\begin{Bmatrix} \ddot{p}_1 \\ \ddot{p}_2 \end{Bmatrix} + k\begin{bmatrix} 2 & 0 \\ 0 & 2 \end{bmatrix}\begin{Bmatrix} p_1 \\ p_2 \end{Bmatrix}$$
$$= k\Upsilon_0 \begin{Bmatrix} 2 - \cos\dfrac{2\pi(\dot{u}_0 t - a)}{L} - \cos\dfrac{2\pi(\dot{u}_0 t + a)}{L} \\ \cos\dfrac{2\pi(\dot{u}_0 t + a)}{L} - \cos\dfrac{2\pi(\dot{u}_0 t - a)}{L} \end{Bmatrix}.$$

Ignoring the short period of time that only one of the wheels is on the rise, the solution for the first modal coordinate, for example, is

$$p_1(t) = \frac{k\Upsilon_0}{m\omega_1}\int_0^t \left[2 - \cos\frac{2\pi(\dot{u}_0\tau - a)}{L} + \cos\frac{2\pi(\dot{u}_0\tau + a)}{L}\right]\sin\omega_1(t-\tau)d\tau$$

where $k/m = \omega_1^2/2$ or $\omega_1 = \sqrt{2k/m}$, and similarly for the second modal coordinate. The response in the first mode for the time period after the vehicle leaves the rise can be obtained by simply setting the upper limit to $(L + 2a)/\dot{u}_0$, the time the vehicle CG exits the hump.

COMMENT The opposite case where $L \ll 2a$ (a bump rather than a hump) is only more complicated in the sense that the time period that the vehicle is in contact with the bump has to be broken into four parts: (i) when only the front wheel is in contact with the bump, (ii) when the wheels straddle the bump, (iii) the time when only the real wheel is in contact with the bump, and (iv) the time after the rear wheel leaves the bump. None of this involves a different treatment from the above, but clearly this case has much more detail. Another complete two-DOF problem is provided in Exercise 7.10. ★

7.8 Random Vibration Analyses

Consider formulating an engineering solution to the question of where oak leaves will land on an x, y plane after falling through outside air from a certain height above the point (0, 0) on that plane. The engineer would have to recognize that there is some variation in oak leaf sizes and shapes, as well as structure, principally leaf ribbing and thickness. If the question were restricted to a particular oak leaf falling in still air, then accurate measurements could be made of the leaf geometry, including curvatures, and its stiffness distribution. Modern computer-based numerical aerodynamic calculations could do a reasonably accurate, but expensive, prediction of the landing point of the center of gravity of that specific oak leaf and the orientation of that oak leaf relative to the Cartesian axes. It would be a large problem, but if it were of importance, it could be done for any leaf whose geometry and stiffness distribution were specified. If now the above still air condition was removed and replaced by natural wind conditions, then the already complicated problem becomes even more challenging. Again, if the wind velocities were specified as functions of time and space throughout a large volume of air surrounding the experiment, then it still might be possible, at some expense, to accurately predict the landing. However, such wind velocity specifications are simply not known for a natural landscape because of the large variation in weather, without even considering the very complicated wind patterns close to the rough Earth surface because of all the vortices generated by various objects that litter a natural surface. If now the question were reopened to refer to oak leaves in general, then the engineer would be justified in deciding to approach the question from a statistical point of view because it would be a lot less expensive to conduct a lot of experiments with a variety of oak leaves and thereby create an approximate probability distribution for the landing points of those leaves on the x, y plane in natural air. In other words, there are problems where the force inputs are so complicated in time and space that, rather than seek a deterministic model of the input forces and the structure, it is much more economical to accept some uncertainty by approaching the problem from a statistical viewpoint. When either a flexible structure or the impressed time-varying loading is described in statistical terms, then the analysis is called a *random vibration analysis*. Reference [7.3] is one of many textbooks that deal with random vibration analyses.

In aerospace engineering, random vibration analyses have been used to provide estimates on such varied questions as crew comfort and ability to function after being subject to many hours of flight through turbulent air and estimating the probability that a satellite structure would impact its aerodynamic shield of the top of a rocket as the rocket is battered by noise and aerodynamic turbulence at liftoff. A simpler aerospace engineering problem would be estimating the resulting inertial loads on a flexible aircraft wing as the aircraft taxies over different, rough, taxiway pavements on the aircraft's way to the duty runway. Mechanical engineers, of course, deal with parallel problems for land-based vehicles. Civil engineers have to concern themselves with earthquakes. For present purposes, earthquakes can serve as an example of a random motion input, even though earthquakes are more statistically complicated than some other random inputs. Earthquakes produce primarily horizontal base excitations for all sorts of very expensive and often unique structures such as buildings, bridges, dams, and so on. See Figure 7.9 for a typical time history of an earthquake base excitation. Much more so than a falling leaf, it has been necessary to view earthquakes as a random phenomenon because the time history of their ground motion cannot be predicted in advance of their occurrence. Hence, the mathematical modeling of earthquakes, in general, depends on compiling and processing the statistics of previously experienced earthquakes. Ever since the El Centro earthquake of 1940, records of earthquake excitations have been recorded in various places in the

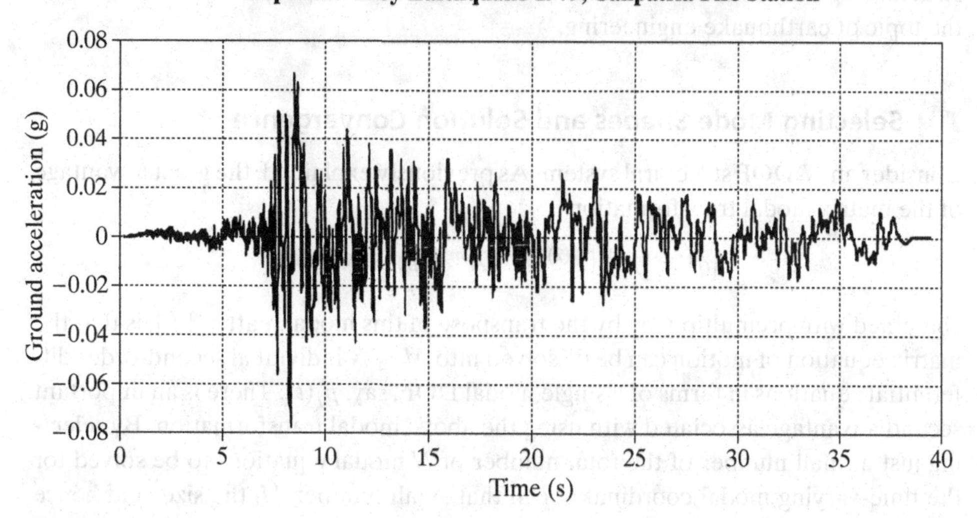

Figure 7.9. "Typical" earthquake ground motion, from Pacific Earthquake Engineering Research (PEER) center ground motion database.

United States. Hence a library of earthquake inputs from the United States and other countries is now available to the structural analyst. Again, the difficulty with using the statistics derived from that library is that these statistics are rather complicated. These statistics are characterized as *nonstationary* and *nonergodic*. Nonstationary means that the various averages associated with the earthquake data vary over the time record of the earthquake. Nonergodic means that these various averages vary from one earthquake to another. Obviously, a statistical approach, called a random vibration analysis, poses many problems for the analyst. There is an alternate approach. A structure subject to an unknown earthquake base motion can be analyzed by (i) selecting various recorded earthquake base excitations from other sites, (ii) scaling their amplitudes as appropriate for the site of interest, and (iii) using all those records as a base excitations for the structure being analyzed. Figure 7.9 shows that, say, using ramp input approximations and ramp function responses makes this approach sufficiently complicated that it needs to be done numerically. Then it is a matter of determining if any of those base excitations resulted in a structural failure as judged by whatever failure standard that was adopted for that structure. If the structural design decision was made that the criteria for success is avoidance of collapse while allowing plastic behavior, then that material nonlinearity has to be included in the computer-based analysis. However, if the design is supposed to survive a major earthquake without substantial plastic deformations, then such a stronger and stiffer structure will have larger accelerations and hence larger inertia loadings. Vibration isolation is now a common feature in large civil engineering structures. Indeed, there are enough complexities to earthquake engineering design to say that it is a subject that cannot be addressed here in detail. However, again, the structural response analysis can proceed numerically on the basis of (i) using a series of ramp or sinusoidal load functions to represent any one earthquake loading, (ii) using the corresponding response functions to craft the deflection response, and

(iii) using the responses from several earthquake records to estimate whether the structure will meet the selected failure criteria. Reference [7.4] is an introduction to the topic of earthquake engineering.

7.9 Selecting Mode Shapes and Solution Convergence

Consider an N-DOF structural system. As previously explained, the great advantage of the matrix modal transformation

$$\{q\} = \left[A^{(1)}\middle|A^{(2)}\middle|\cdots\middle|A^{(m)}\right]\{p\} \equiv [\Phi]\{p\},$$

combined with premultiplying by the transpose of this modal matrix, $[\Phi]$, is that the matrix equation of motion can be dissolved into $M \leq N$ individual second-order differential equations in terms of a single modal DOF, say, $p_j(t)$. There is an important second advantage associated with using the above modal transformation. By selecting just a small number of the total number of N modal equations to be solved for the time-varying modal coordinates, call that small number M, the size, and hence the cost, of the solving the problem can be greatly reduced.

The purpose of this section is to discuss how to choose the M modal DOF to be used in the summation for the N component deflection vector

$$\{q\} = \left[A^{(1)}\middle|A^{(2)}\middle|\cdots\middle|A^{(m)}\right]\{p\} \equiv [\Phi]\{p\},$$

where again, $M \leq N$. First of all, realize that the mode shapes of positive and negative real numbers calculated by the standard procedures of the previous chapter are not uniformly accurate relative to experimentally determined mode shapes, even after accounting for damping, ambient air, and so on. The rule of thumb is that only the lower half of the mode shapes correspond reasonable well to the experimental mode shapes of most structures.[9] The latter half of the mode shapes and natural frequencies are inaccurate for a couple of reasons. First, the actual structure is a continuum with an infinite number of DOF that is being approximated by a discrete model with a finite number of DOF. That this form of modeling produces poorer results for higher numbered modes than lower numbered modes can be surmised by considering a single beam whose mass is lumped at the beam center. Thus the beam becomes a one-DOF system, and only the first mode will be approximated, and there is no accuracy whatever for the higher numbered nodes. Second, the structural elements are described using the formulas of strength of materials theory that always include simplifying approximations. This, too, has particular importance for the higher modes with their greater number of nodes. For example, the greater number of nodes associated with the higher modes means shorter distances between nodes or nodal lines in the case of, say, plates. Therefore the effective span lengths of beams and plates are greatly reduced. This means that the use of long beam theory and thin plate theory, even with approximate shear flexibility corrections, is less accurate for the higher number nodes. There are even other, smaller effects. As the span length of the beam

[9] Another complication is that two complex structures built of thin members from the same plans, and hence having the same mathematical model, will not have exactly the same experimental higher numbered natural frequencies and mode shapes.

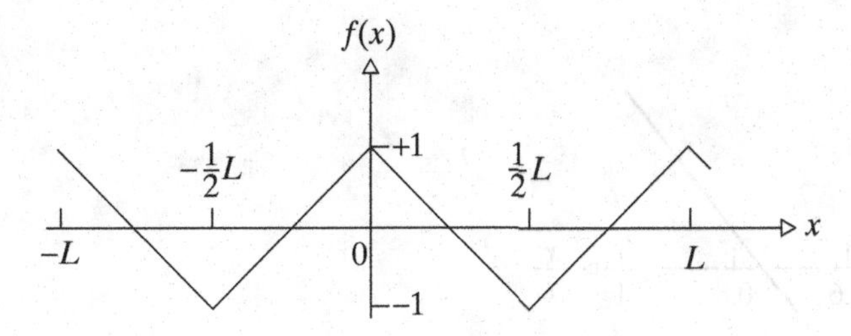

Figure 7.10. Periodic sawtooth function.

becomes smaller, the not-modeled rotary inertia of small beam segments begins to make a small difference in the experimental results.

Regardless of the accuracy of the calculated mode shapes relative to the experimentally determined mode shapes, the calculated mode shapes still decouple the mass and stiffness matrices of the (approximate) mathematical model, and the series of mode shape vectors still forms a vector basis for the general deflection vector $\{q\}$. Considering the lesser accuracy of the higher numbered modes relative to the actual structure, it is perhaps fortunate that it is never necessary to use the higher numbered mode shapes in the $[\Phi]$ coordinate transformation matrix. As may be seen more clearly in the next chapter where the eigenvectors of discrete mass models are extended to the eigenfunctions of continuous mass models, the unimportance of the higher numbered modal vectors is because series of modal vectors or modal functions are quite like a Fourier series (see below) where the lower the index number of the series term, the much larger its contribution to the sum that represents solution for the deflection function. If the number of DOF, N, is a large number (several hundred or many thousand), it is normally only necessary to use, say, the lowest numbered 15%, 10%, or 5% of all the mode shapes to accurately represent the motion of the structural system. It is not possible to suggest the use of a specific percentage of the total number of mode shapes in all circumstances. Circumstances can vary, and cases have been reported where a couple of hundred modes were necessary to achieve an accurate depiction of the physical motion. Rules of thumb on the number of mode shapes to be used in an analysis are as follows. In terms of time distributions, if f is the highest frequency component of the applied load, then use all the lower numbered modes up to the one whose natural frequency is $2f$. In terms of spatial distributions, the modes to be used have to be able, with reasonable accuracy, to duplicate the visualized dynamic deflection patterns produced by the time-varying loads. As an aid to understanding this point in a similar context, consider the sawtooth function of x shown in Figure 7.10. This piecewise linear, periodic function can be represented by either different linear expressions for each interval of length $L/2$ or by a Fourier series over all intervals of length L. That is,

$$f(x) = \frac{4x}{L} + 1 \quad \text{for} \quad -\frac{L}{2} \leq x \leq 0$$

$$\text{and} \quad f(x) = -\frac{4x}{L} + 1 \quad \text{for} \quad 0 \leq x \leq \frac{L}{2}$$

$$\text{so} \quad f(x) = \frac{8}{\pi^2} \sum_{n=1}^{\infty} \frac{\cos[2\pi(2n-1)(x/L)]}{(2n-1)^2}.$$

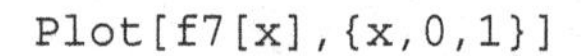

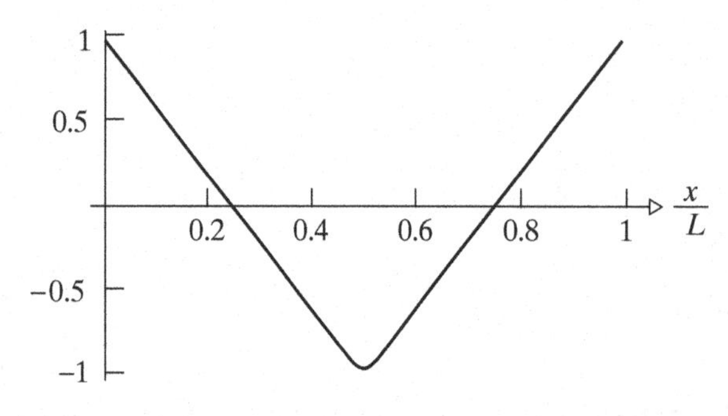

(a)

```
Plot[{f1[x],f3[x],f5[x],f7[x]},{x,0,1}]
```

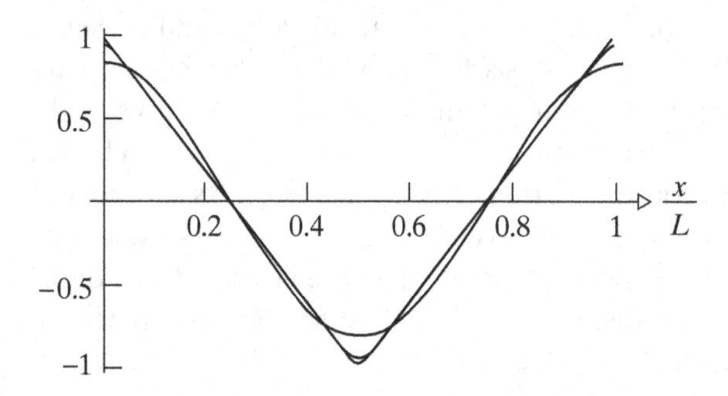

(b)

Figure 7.11. (a) Sum of the first four nonzero terms of the cosine series of Section 7.9 showing the closeness of the approximation away from the slope discontinuities at the corners. (b) Plot of the one term, the two-term sum, the three-term sum, and the four-term sum superimposed to show progressive improvement in the approx.

Consider either the interval $(-L/2, L/2)$ or the interval $(0, L)$. On either of these intervals the above-cited cosine functions are orthogonal to each other (weighting function 1.0), just as the discrete mode shapes are orthogonal to each other with the weighting factor $[m]$ or $[k]$. As is shown in Figure 7.11(a), if the above Fourier cosine series is truncated; that is, if only the first four of the infinite number of nonzero cosine terms are used as an approximation to $f(x)$, then the approximation to $f(x)$ is very good in engineering terms. Note from Figure 7.11(b) that the lower the index number of the cosine term, the much more important it is to the approximation.

Another way of looking at the process of choosing the mode shapes expected to be necessary to accurately describe the dynamic deflection is to include all mode shapes up to and including the one that has at least the same number of nodes as the imagined dynamic deflection shape in its most complicated form. Then the prudent analyst often takes an equal number of mode shapes beyond that mode shape with the same number of nodes. Of course, this estimate is not always easy to make, particularly when experience with similar structures and loads is not available. In that case the analyst can begin by making his or her best guess as to the minimum number of mode shapes necessary for an accurate deflection response and complete the analysis on that basis. Then the analyst can double the number of mode shapes and repeat the analysis. If the two answers are sufficiently close, then the original choice was satisfactory, and that number of modes can be used in further analyses of that structure and as a future guide to similar structures and loadings.[10] If the two solutions are not within some acceptable measure of accuracy, then the number of mode shapes must be increased again until two successive solutions are sufficiently close. Unfortunately, even this sort of convergence doesn't absolutely guarantee that some other mode outside the chosen few will not significantly affect the deflection solution. Returning to the above Fourier series analogy to make this point with a pathological example, let the function to be approximated, $f(x)$, by a Fourier series, be the cosine function $\cos(50\pi x/L)$, which has a lot of zeroes in the interval $(0, L)$. Then the first 49 Fourier series coefficients will be zero, the 50th will be 1.0, and all coefficients after 50 will be zero. Therefore, when comparing any two truncated series sums, where each series contains fewer than 50 terms, the series will appear to have converged because they both sum to zero, the obviously incorrect result. However, when there is some higher mode of importance, some special feature of the loading will usually clearly suggest that possibility.

There is also a simple quantitative guide to help in selecting which modes to use in an analysis. Return to the general modal equation, Eq. (7.5),

$$\ddot{p}_j + 2\zeta_j\omega_j\dot{p}_j + \omega_j^2 p_j = \frac{P_j(t)}{M_j} \equiv \frac{1}{M_j}\sum_{i=1}^{n} A_i^{(j)} Q_i(t). \tag{7.5}$$

If the magnitude of the jth modal force $P_j(t)$ divided by the modal stiffness is much smaller than other such normalized modal forces, then, if there is no resonance effect, the jth mode can be neglected. This simple idea of neglecting those modes that do not have appreciable modal force inputs is sometimes quantified in the following manner for the special case where all the applied forces have the same or similar time variations. Consider the case where each generalized force $Q_k(t)$ of the original generalized force vector $\{Q\}$ has the same nondimensional time variation, $f(t)$. That is, if the kth entry in the generalized force vector $Q_k(t) = Q_k^0 f(t)$, where Q_k^0 is the force magnitude whose spatial location is indicated by the subscript k (because it corresponds to the kth generalized coordinate), then

$$P_j(t) = f(t)\sum_{k=1}^{n} A_k^{(j)} Q_k^0 = \Gamma_j \overline{Q} f(t),$$

where $\overline{Q}$ is an arbitrarily selected force amplitude, for dimensional purposes, that is a constant for *all* selected modes. The quantity that varies from mode to mode, the nondimensional quantity Γ_j, is called the jth *participation factor*. In short, a participation factor is a nondimensional constant associated with the spatial distribution

[10] This is analogous to the Cauchy test for series convergence.

of the applied forces that is independent to the time variation of the applied forces. Here

$$\Gamma_j = \frac{1}{\overline{Q}} \sum_{k=1}^{m} A_k^{(j)} Q_k^0 = \frac{1}{\overline{Q}} \lfloor A^{(j)} \rfloor \{Q^0\}.$$

Clearly, if the spatial distribution of the applied loads as represented by $\{Q^0\}$ is near to being orthogonal to the jth mode shape, the jth participation factor will be small. Conversely, if the spatial distribution of the applied loads, $\{Q^0\}$ is near to being proportional to the jth mode shape, then the jth participation factor will relatively large. The modes selected for retention in the analysis are, of course, generally those with the larger participation factors. Another form for the participation factor is presented in the next chapter where the generalized coordinates are continuous functions of the spatial variables rather than tied to discrete spatial positions as is the case with the usual finite element analysis. Remember, although a participation factor might be small for one set of loads, the usual situation is for a structure to be subjected to more than one set of loads. Hence, usual practice is not to omit any modes up to the largest numbered mode selected.

Again, the reason that it is highly desirable to use as few mode shapes as possible in a given analysis is, of course, that the number of mode shapes is the number of modal degrees of freedom requiring solution by one method or another, and the number of mode shapes dictates the size of the matrices that require multiplication. If only, for example, 10% of the total number of modes are needed for the dynamic analysis, then the use of the modal transformation reduces the number of DOF in the core of the solution process by 90%. Thus it is worthwhile to carefully consider the question of how many modes are sufficient to obtain reasonable convergence of the solution for the original set of generalized coordinates, $\{q(t)\}$. In addition to the above-discussed procedure of comparing the solution for $\{q\}$ for M selected mode shapes and, say, $2M$ selected mode shapes, where M is small, there is another approach for testing for the convergence to the solution for $\{q\}$. This alternate procedure is called *modal acceleration*. The modal acceleration technique is as follows. First solve the original, undamped, matrix vibration equation for the vector of generalized coordinates, which again, is the objective of the vibration analysis. That result can be written as

$$\{q\} = [k]^{-1}\{Q\} - [k^{-1}m]\{\ddot{q}\} = [k]^{-1}\{Q\} - [D]\{\ddot{q}\},$$

where, again, the dynamic matrix, $[D]$, is the mass matrix premultiplied by the inverse of the stiffness matrix. Note that if the previous process of determining the system mode shapes involved calculating the inverse of the stiffness matrix, then that inverse matrix is already available at only the cost of storage in the computer. However, if the stiffness matrix is large and the inverse has not been previously calculated, then the alternative is the use of a Cholesky decomposition as on page 298. This discussion will proceed on the basis that the inverse is obtainable at an acceptable cost.

Since the modal transformation is $\{p\} = [\Phi]\{q\}$, premultiplication allows an easy inversion of this equation as $[\Phi]^t[m]\{p\} = [\Phi]^t[m][\Phi]\{q\} = [\backslash M \backslash]\{q\}$, where $[\backslash M \backslash]$ is the diagonal matrix of generalized masses. Since the inverse of any diagonal matrix is simply another diagonal matrix whose entries are the inverses of the diagonal entries of the original matrix, then

$$\{\ddot{q}\} = [\backslash M \backslash]^{-1}[\Phi]^t[m]\{\ddot{p}\}$$
$$\text{where} \quad \{\ddot{p}\} = [\backslash M \backslash]^{-1}\{P\} - 2[\backslash \zeta\omega \backslash]\{\dot{p}\} - [\backslash \omega^2 \backslash]\{p\}.$$

Substituting the second of the above equations into the first yields the solution

$$\{q\} = [k]^{-1}\{Q\} - [D][\backslash M \backslash]^{-1}[\Phi]^t[m]\left([\backslash M \backslash]^{-1}\{P\} - 2[\backslash \zeta\omega \backslash]\{\dot{p}\} - [\backslash \omega^2 \backslash]\{p\}\right),$$

where the damping is modeled only after the modal transformation. Therefore, at the expense of the above indicated matrix multiplications, the solution for the vibratory system's physical coordinates involves the full effect of the applied load vector $\{Q\}$, which is then modified using the small number of selected modal coordinates. A small difficulty with this approach is the requirement for the modal velocities, if modal damping is included in the analysis. There are several analytical or numerical ways of calculating the modal velocities once the solution for the modal deflections has been obtained. For example the convolution integral solution for p/t can be differentiated with respect to time, or a finite difference approximation can be used as discussed in Chapter 9.

7.10 Summary

Time-varying loadings can be classified as either continuing loadings or pulses. The only continuing loading worthy of special attention is the sinusoidal loading that can produce a resonance effect.[11] All other continuing loadings, such as the step load, can be treated conveniently as pulses. Pulse loadings are classified as either impulses (very short time duration relative to the first period), shocks (time durations that are a fraction of the first period), or pulses that occur over longer time durations than the first period. If the load rises to a maximum value over a time duration that is, say 10 times the length of the first period, then the loading reasonably treated as a static load, and the expense of a dynamic analysis can be avoided.

When the time variation of the loading vector of the linear matrix equation of motion is sinusoidal, so too is the vector of the steady-state deflection responses. After the decoupling the equations of motion by means of the modal coordinate transformation, and using the complex algebra procedure discussed in Chapter 5, the sinusoidal modal deflection responses can be calculated. These complex modal deflection functions are, of course, functions of the modal frequency ratio $\Omega_j = \omega_f/\omega_j$, and can be written in terms of that modal frequency response function. These modal frequency response functions incorporate the modal resonance phenomena. Specifically, where $P_j(t) = w_j^2 M_j \hat{N}_j \exp(i w_f t)$

$$\hat{p}_j e^{i\omega_f t} = \frac{\hat{N}_j e^{i\omega_f t}}{\left(1 - \Omega_j^2\right) + i(2\zeta_j \Omega_j)}$$
$$\text{or} \quad p_j(t) = \hat{N}_j H_j(i\Omega_j) e^{i\omega_f t}. \tag{7.6}$$

[11] A more general time periodic input can, via a Fourier series, be treated as a sum of many sinusoidal inputs.

Note that the peak amplitude response to a nonresonating, sinusoidal input can, and probably will, occur early during the application of the sinusoidal excitation as a result of the combining of the transient amplitudes with the steady-state amplitudes. Therefore, a sinusoidal input should also be examined for a few cycles as a pulse input. This, and all other pulse loading cases start the solution procedure the same way, that is, with a coordinate transformation to the modal coordinates. After the modal transformation is used to decouple the rows of the matrix equation of motion for light damping, the Duhamel integral (also called the convolution integral or the superposition integral) is used to obtain an analytical or numerical solution for each of the selected modal DOF. The convolution integral incorporates the impulse response function, which, along with the frequency response function, is a fundamental descriptor of the dynamic properties of a structure. Appendix II shows that the frequency response function and the impulse response function are related to each other in that they are essentially a Fourier transform pair.[12]

Carrying out the required integration of the convolution integral analytically can be facilitated by use of either tables of integrals (which can still be quite cumbersome especially when damping is included) or software such as Mathematica. Again, see Refs. [7.1,7.2]. Another option is approximating the usually imprecisely known input forces by a series of straight lines and sine curves. Since analytical expressions for the modal deflection output for straight line and sinusoidal modal acceleration inputs is known, approximate output expressions can be cobbled together, preferably using digital computer software. For easy reference, the *un*damped forms of the various response functions in modal symbols are summarized as follows:

1. For an impulsive modal loading $P_j(t) = \Pi_j \delta(t)$, the modal coordinate response is

$$p_j(t) = \Pi_j h_j(t) \quad \text{and} \quad h_j(t) = \frac{1}{M_j \omega_j} \sin \omega_j t.$$

2. For a step modal loading $P_j(t) = P_j^0 \, stp(t)$, the modal coordinate response is

$$p_j(t) = P_j^0 g_j(t) \quad \text{and} \quad g_j(t) = \frac{1}{M_j \omega_j^2}[1 - \cos \omega_j t].$$

3. For a ramp loading $P_j(t) = (t/t_1)P_j^0$, the modal coordinate response is

$$p_j(t) = \frac{P_j^0}{t_1} r_j(t) \quad \text{and} \quad r_j(t) = \frac{1}{M_j \omega_j^3}[\omega_j t - \sin \omega_j t].$$

4. For a sine loading $P_j(t) = P_j^0 \sin(\pi t/t_1)$, where $\omega_f = \pi/t_1$, the modal coordinate response is

$$p_j(t) = P_j^0 s_j\left(t, \frac{\pi}{t_1}\right) \quad \text{and} \quad s_j\left(t, \frac{\pi}{t_1}\right) = \frac{1}{M_j \omega_j} \left\{ \frac{(\pi/t_1) \sin \omega_j t - \omega_j \sin \pi (t/t_1)}{(\pi/t_1)^2 - \omega_j^2} \right\},$$

[12] Fourier transforms play an important role when, for example, the applied loads can only be described in probabilistic terms.

where of course, to avoid an indeterminate form, the forceing frequency (π/t_1) cannot equal the modal frequency ω_j. If, instead of a force excitation, the excitation is a base motion of amplitude Υ_j^0, then the force amplitude P_j^0 in the above expressions needs to be replaced by $M_j\omega_j^2\Upsilon_j^0$, which has the same units.

The final option for evaluating a superposition integral is just to use numerical integration and form a list of modal deflection outputs at a series of closely selected time points. This latter process is best done using commercial software. The transformation back to the physically meaningful DOF, $\{q\} = [\Phi]\{p\}$ completes the solution procedure. Of course, most often for commercial applications, the solution for the deflections requires further processing such as plotting and color animation for ease of understanding, and using the deflections to calculate, and similarly plot, for example, stresses.

EXAMPLE 7.6 In the jth mode, the time history of the modal force $P_j(t)$ has an isosceles triangular shape that starts at time zero, rises to a peak value of P_0 at time t_0, and falls back to a zero value at time $2t_0$ and remains at a zero force level thereafter. Determine the jth modal deflection as a function of time.

SOLUTION The ramp response function, $r_j(t)$, is defined in Example 7.2. This function can be used to immediately write the modal deflection response for each time interval after time zero from superposition of the input ramp force so as to achieve the stated force time history. Therefore,

$$\text{for} \quad 0 \le t \le t_0 \quad p_j(t) = r_j(t) = \frac{P_0}{\omega_j^3 t_0}(\omega_j t - \sin\omega_j t)$$

$$\begin{aligned}\text{for} \quad t_0 \le t \le 2t_0 \quad p_j(t) &= r_j(t) - 2r_j(t - t_0)\\ &= \frac{P_0}{\omega_j^3 t_0}[\omega_j(2t_0 - t) + 2\sin\omega_j(t - t_0) - \sin\omega_j t]\end{aligned}$$

$$\begin{aligned}\text{for} \quad 2t_0 \le t \quad p_j(t) &= r_j(t) - 2r_j(t - t_0) + r_j(t - 2t_0)\\ &= \frac{P_0}{\omega_j^3 t_0}[2\sin\omega_j(t - t_0) - \sin\omega_j t - \sin\omega_j(t - 2t_0)].\end{aligned}$$

COMMENT One weak check on this solution is that the deflection response in the third time interval should not be increasing as time proceeds, as are the solutions in the first two time intervals. It is not, because this part of the solution contains only sine functions rather than powers of the time variable. This expectation is so because, in this time interval, there is no applied "force" to provide the energy necessary to drive the deflection response away from its undamped mean value.

7.11 **Aeroelasticity**

As mentioned before, the dynamic interaction of structures and fluids in motion, either fluids contained by the structure or fluids surrounding the structure, can pose a significant challenge to an analyst. This section discusses relatively simple examples of the latter type of interaction. Before proceeding to do that, understand that the former type of interaction, that of an enclosed fluid, is not so simple that the enclosed fluid can be treated as if it were a solid. The reader can prove that statement to himself or herself by placing both a raw egg and a hard-boiled egg on their sides on a horizontal surface and spinning them with one's fingers. The hard-boiled egg will spin much more rapidly because it takes time for the viscosity of the fluid in the raw egg to transmit the initial spinning moment throughout that fluid and thereby overcome the inertia of the fluid.

Returning to the case of the moving fluid external to the structure, the selected examples illustrate a situation where the external dynamic forces and moments applied to the structure by the surrounding fluid depend on the motion of the structure as well as the far field velocity of the external fluid. In other words, there is a feedback mechanism between the fluid and the structure. The motion of the structure, in general, is described by the generalized coordinates and their first and second time derivatives. In such circumstances, the applied fluid forces can be grouped together with the inertial, damping, and elastic forces with the result that the equations of motion are mathematically homogeneous. Homogeneous differential equations suggest an eigenvalue problem, which in this case, similar to buckling problems, leads to a stability analysis.

When the fluid of the fluid–structure interaction is air, the above interaction falls under the topic heading *aeroelasticity*. This main purpose of this section is to discuss one aspect of dynamic aeroelasticity called low-speed airfoil flutter.[13] Broadly speaking, flutter is a dynamic instability resulting from the (fluid induced) forces and moments acting on the structure doing positive work on the structure. This work done by the airflow is converted into increased kinetic energy, which in turn leads to increased vibratory velocities and deflections. To understand this point, recall that from Newton's second law $F = m(dV/dt)$, where V symbolizes velocity of the body. Multiply both sides of this equation by dx. After shifting the dt of the acceleration to the dx of the path integration

$$\int_{x_1}^{x_2} F dx = \int_{t_1}^{t_2} m dV \frac{dx}{dt} \quad \text{or} \quad \Delta W = \int_{t_1}^{t_2} mV dV = \Delta\left(\frac{1}{2} mV^2\right) = \Delta T.$$

The increased kinetic energy of the structural motion, T, results in increased deflections, increased strains, increased stresses, and, if unchecked, structural failure. The time to structural failure for, say an aircraft wing, can be so short that no pilot can react sufficiently quickly to avert the failure of the wing. Furthermore, the possibility of airfoil flutter is, and has been, so common among aircraft, that even today most high-performance aircraft are speed constrained so as to avoid that possibility. Thus the flutter problem is of historic as well as present concern.

Since only this section addresses the topic of aeroelasticity, it is appropriate to provide a very brief overview of the topic as a whole. Although the flutter phenomenon is its most dramatic aspect, the topic of aeroelasticity covers several types

[13] A description of the fluid forces acting on an airfoil at high speeds requires consideration of compressibility effects as characterized by a Mach number and, if necessary, the location and motion of shock waves if they occur.

of instabilities, both static and dynamic. With respect to flight vehicles, a static instability is one that only involves elastic and aerodynamic forces, and the deflections of the structure typically continue to increase in a single direction. A dynamic instability is one that involves elastic, aerodynamic, and inertial forces. The unstable deflections of the structure typically involve a back-and-forth motion with increasing amplitudes. The most critical airfoil static instability is called *airfoil divergence.*[14] The reader has probably had the experience of riding in an automobile at highway speeds and sticking his or her arm out the window, into the airstream, with his or her hand flattened so as to be parallel to the road surface. If so, the reader may have also slightly rotated that flattened hand one way or the other and experienced the effect of the airstream that was to further sweep the entire arm backward in the same direction of the rotated hand. Airfoil divergence is very much the same thing.

To gain a mathematical insight into airfoil divergence, consider the greatly simplified mathematical model of a three-dimensional wing shown in Figure 7.12(a). Since (i) the width (into the paper) of the uniform airfoil is unspecified, (ii) the airfoil thickness does not directly enter into the analysis, and (iii) the only specific overall dimension of this airfoil is its chord length c, this is termed a one-dimensional model. Of course, the effectiveness of a one-dimensional model for representing a three-dimensional reality is somewhat limited. Fortunately, the mathematics of the one-dimensional model is sufficient to illuminate the physics of airfoil divergence phenomena, which is the present purpose.[15]

Such a one-dimensional model was used extensively before digital computers were available because this type of model was the only model suitable for most hand calculations. Also such one-dimensional models were possible because most wings at that time were high aspect ratio wings, meaning that their span-to-chord length ratio was 10 or more to 1. Thus the wings could be roughly viewed as nonuniform beams. The translational spring and the rotational spring stiffnesses of this one-dimensional model were selected, as a rule of thumb, as the beam bending and torsional stiffnesses at 70–75% of the wing semispan, outboard from the wing root at the fuselage centerline. Again viewing the wing as a nonuniform beam, the fore and aft location of the springs along the airfoil chord is at the shear center of the structural portion of the wing cross section. This is so because the shear center is the point about which the cross section rotates if the beam is subjected to a pure torque and the point where a vertical force will not produce a twisting of the beam. In other words, the shear center is the unique point on the cross section where, for small deflections, the twisting and bending motions of the beam are decoupled. For such high aspect ratio wings, the major portion of the wing-beam structure is a box beam with a thin, nearly rectangular, cross section. It is not difficult to calculate the location of the shear center for such a thin beam cross section.

The two traditional symbols for the deflections of the airfoil segment of Figure 7.12(a) are h, here the positive upward vertical translation of the airfoil, and

[14] Recall Section 2.10 for an earlier discussion of system stability.

[15] Unfortunately, the one-dimensional model is too simple to reveal the mathematical procedure for a wing with a finite aspect ratio (the ratio of the length of the wing span to the average airfoil chord length), taper, sweep back, and so on. For that purpose, see Ref. [7.5], p. 816 ff.

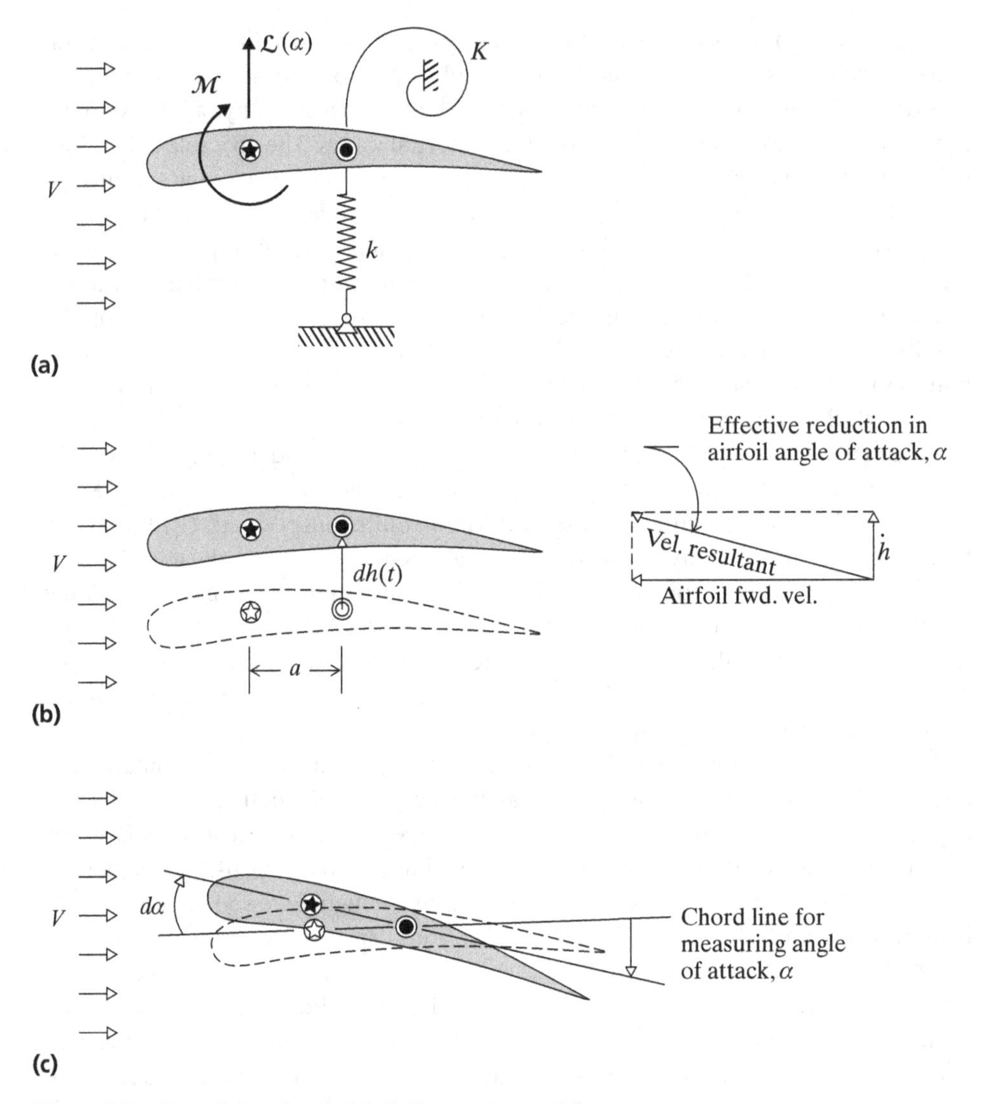

Figure 7.12. One-dimensional airfoil divergence model.

α, the rotation about an axis perpendicular to the paper, positive leading edge up and trailing edge down. Let these two deflections be measured at the juncture point of the springs and the airfoil segment. The aerodynamic loads impressed on the airfoil segment by the airstream are the static[16] (also called steady state) lift force $\mathcal{L}$ and the static aerodynamic moment $\mathcal{M}$, which act very close to the quarterchord (one-quarter of the distance from the leading edge to the trailing edge). The steady-state aerodynamic moment is very little affected by changes in either deflection and thus it is essentially a constant in this analysis. However, over a broad range of rotations, the steady-state aerodynamic lift is directly proportional to the airfoil rotation. With

[16] The force and moment are described as "static" because the deflections of the airfoil are viewed as occurring slowly, so slowly that kinetic energy has no part in this analysis.

C_l being the nondimensional lift coefficient,[17] ρ being the mass density of the air, V being the airstream velocity a long way from the airfoil, and S being the planform area of the airfoil segment (chord length multiplied by wing segment width), then from basic aerodynamics, the lift force acting on the one-dimensional airfoil model is $\mathcal{L} = \frac{1}{2}C_l\rho V^2 S$. For a wide range of values of α, the lift coefficient C_l varies linearly with α. Therefore, if α is measured from the rotation orientation where the lift force is zero, analogous to "rise" equals "slope" multiplied by "run," the lift coefficient (rise) can be written as the lift curve slope multiplied by the rotation (run); that is, $C_l = C_{l\alpha}\alpha$, where the constant $C_{l\alpha} \equiv dC_l/d\alpha$. Substituting into the lift expression

$$\mathcal{L} = \tfrac{1}{2}C_{l\alpha}\rho V^2 S\alpha. \tag{7.16}$$

Return now to the airfoil of Figure 7.12(a). This airfoil is statically stable with regard to the h motion regardless of the airstream velocity. To understand this point, consider Figure 7.12(b), where the airfoil is translated upward from its static equilibrium position a distance dh, without any rotation α. The upward airfoil velocity vector dh/dt plus the forward airfoil velocity V add vectorially to produce an airfoil velocity vector that is slightly rotated clockwise from the original airfoil velocity vector that was just V. The velocity of the airstream as seen from the airfoil is a velocity vector that is equal and opposite to airfoil velocity vector. Thus, because of the rotation of the velocity vector only, the effective angle of attack of the airfoil has decreased with a resulting decrease in the lift force. A decrease in the lift force is the same as adding a downward directed incremental lift force to the original lift force that is equilibrated by the original spring force. Therefore an upward translation of the airfoil results in both an aerodynamic force tending to return the airfoil to its original position as well as a spring force doing the same. If the airfoil translates downward, both the incremental aerodynamic force and the spring force are directed upward. Since any translational motion of the airfoil is opposed by both the aerodynamic and elastic forces, the airfoil is unconditionally stable in vertical translation. That is, for all fluid velocities, the airfoil tends to return to its (stable) equilibrium position.

Now consider a positive increment in the airfoil rotational angle as shown in Figure 7.12(c). The increase in the rotation angle $d\alpha$, causes an increase in the lift force, which can be written as $d\mathcal{L} = \frac{1}{2}C_{l\alpha}\rho V^2\, S d\alpha$. This in turn causes a clockwise moment about the elastic center of magnitude $a d\mathcal{L}$. The increment in the rotational angle also causes an increase in the counterclockwise moment produced by the torsional spring of magnitude $K d\alpha$. When the increase in the moment produced by the spring is greater than the increase in the aerodynamic moment about the elastic center, then the airfoil will rotate back toward its original position and the system will be stable. If the reverse is true, that is, if the increase in the aerodynamic moment $a\, d\mathcal{L}$ is greater than the increase in the opposing spring moment, then the rotation angle will increase further. The further increase in α will lead to a further increase in that aerodynamic moment that will again be greater than the increase in the spring moment, leading to a still greater increase in α, and so on, as the system diverges from its

[17] For a given rotation, the lift coefficient depends on the thickness variation of the airfoil and varies with Reynolds number, but only gradually for the airspeeds considered in this discussion. Hence it can be considered to be nearly a constant.

original, unstable, equilibrium position. Between these two possibilities there is a third possibility. The third possibility is where the increases in the aerodynamic and spring moments are equal and thus balance each other. Here is the neutral stability point between stability and instability. This is equivalent to the glass marble being displaced on a horizontal plane as opposed to a convex or concave surface. The marble can be moved from one point on the plane to another, but such a motion itself does not induce further motion. To mathematically determine this balance point for the airfoil, keep in mind that the equilibrium position from which the differential increments in the angle of attack occur is one where the moment about the elastic axis of the aerodynamic lift force, $a\mathcal{L}$, and the spring moment, $K\alpha$, are in balance; that is, static equilibrium. Hence it is only necessary to equate the increases in the moments, $ad\mathcal{L}$ to $Kd\alpha$, and determine the velocity for which this second equality exists. Doing so, and canceling the differential of alpha leads to $\frac{1}{2}aC_{l\alpha}\rho V^2 S = K$ or the divergence velocity solution

$$V_{div} = \sqrt{\frac{2K}{aC_{l\alpha}\rho S}}.$$

As a final comment on airfoil divergence, a two-dimensional analysis for an unswept wing is not too different from the above one-dimensional analysis. See Ref. [7.5], p. 816. When a wing is swept forward, the moment arms of outboard lifting forces relative to inboard airfoil sections are greater, sometimes much greater, than they would be if the wing were unswept. Hence, the swept-forward wing will suffer divergence at a lower airspeed than the same wing would suffer if it were unswept. The same wing swept back would have a higher, perhaps a much higher, divergence airspeed. Thus, despite the small aeronautical and structural advantages to having a swept forward wing, the divergence phenomenon is why swept-forward wings are very rare, whereas swept-back wings are common for high-speed flight. The advent of tailored, carbon fiber composite materials permitted the selective stiffening of wings against divergence. The appeal of this concept led to the building of the XF-29. See Ref. [7.6]. Nevertheless, the experience developed from the XF-29 was that the advantages of forward sweep didn't offset the risks.

Turn now to the more challenging problem of low-speed airfoil flutter. To understand how such a dynamic instability could possibly occur, consider Figure 7.13(a) and 7.13(b). These figures show an airfoil undergoing a sinusoidal vibration as it moves through the air. The vibration could be, for example, in response to a vertical gust or control motion. In both sketches, the vibrating airfoil is both translating up and down and rotating. In the first sketch, as the airfoil moves upward, the leading edge rotates downward, and as the airfoil moves downward, the leading edge rotates upward. In these circumstances, as the airfoil moves upward, the downward rotation of the leading edge means that the angle of attack is decreasing and consequently the lift force is decreasing. Thus the incremental lift force is opposing the upward motion of the airfoil. Similarly, when the airfoil moves downward, the lift force increases and, again, the incremental lift force opposes the motion. Since the elastic forces of the wing structure (not shown) also oppose the motion, the airfoil

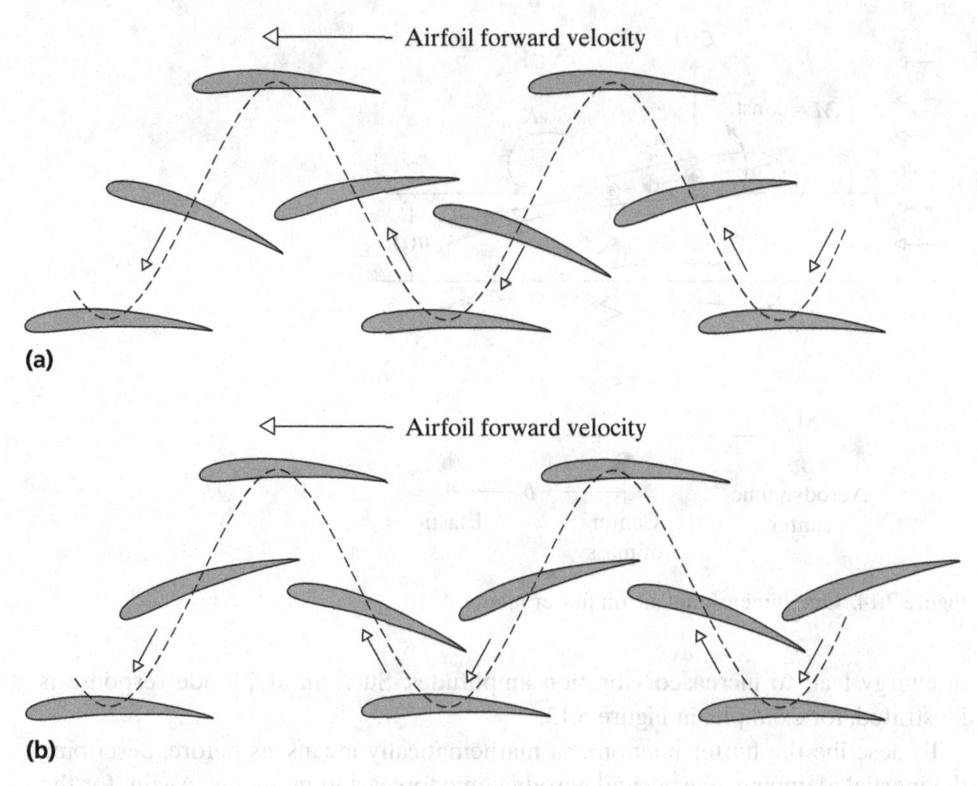

Figure 7.13. (a) Vibratory phasing that causes the lift force to diminish the vibratory amplitude. (b) Vibratory phasing that causes the lift force to increase the vibratory amplitude and thus destabilize the system.

will always tend to return to its equilibrium position and thus is dynamically stable in this situation. The opposite is true in the second sketch. Here the relationship between the two independent DOF are such that as the airfoil moves, for example, upward, the lift force increases. This tends to make the airfoil move up even further. The elastic forces still oppose the motion, so it is a question of whether the increments in the aerodynamic forces are greater than or less than the increases in the elastic forces. Since the aerodynamic forces depend approximately on the airspeed squared, and the elastic forces are independent of the airspeed, it is a question of whether the airspeed is greater or equal to that flutter airspeed where there is neutral stability. To be more specific, neutral stability occurs when the time-varying aerodynamic lift resulting from the vibratory motion is exactly balanced by the time-varying elastic and inertial generalized forces that also result from the vibratory motion, as the amplitudes of the vibratory motion neither increase or decrease. Another way of looking at these possible examples of phase differences between the upward translation $h(t)$ and the leading edge upward rotation $\alpha(t)$ is, for example, to say that in the first case the incremental lift force does negative work on the elastically supported airfoil, whereas in the second case, the work is positive. This positive work is stored in the airfoil as increased strain energy and increased kinetic energy. These increases

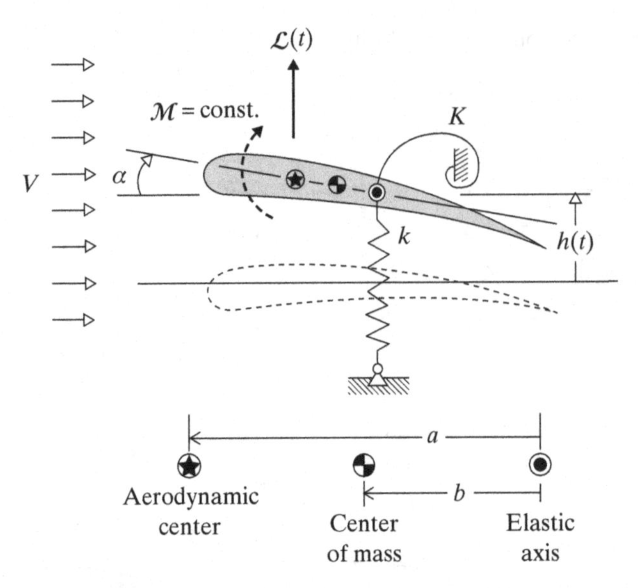

Figure 7.14. One-dimensional airfoil flutter model.

in energy lead to increased vibration amplitudes. Such an amplitude response is illustrated, for example, in Figure 5.13.

To describe the flutter phenomena mathematically means, as before, describing the inertial, damping, elastic, and aerodynamic forces and moments. Again, for the sake of computational simplicity, only the case of the one-dimensional airfoil of Figure 7.14 is considered here. Again there are only two DOF, and it is a simple process to write the equations of motion. As is commonplace in flutter analyses, damping (to be discussed later) is ignored. Then the quantities necessary to write the Lagrange equations of motion are

$$T = \tfrac{1}{2} H_{cg}\dot{\alpha}^2 + \tfrac{1}{2} m(\dot{h} + b\dot{\alpha})^2$$

$$\text{and} \quad U = \tfrac{1}{2} kh^2 + \tfrac{1}{2} K\alpha^2 \qquad \delta W = \mathcal{L}\delta h + (\mathcal{M} + a\mathcal{L})\delta\alpha, \tag{7.17}$$

where, of course, H_{cg}, and later, H_{ea}, are the airfoil mass moments of inertia about the airfoil center of mass and the elastic axis, respectively. The difficulty with any flutter analysis is entirely associated with writing the mathematical expressions for the time-varying aerodynamic lift $\mathcal{L}$ and the time-varying aerodynamic moment $\mathcal{M}$. These expressions can be quite complicated even in the present case of a one-dimensional airfoil in a low-speed air stream. The usual approach in engineering practice is to use a numerical scheme to describe the airflow. Here, however, for the sake of explanation, only analytical approaches are discussed.

The simplest approach for describing the air loads analytically is to adapt the planar flow lift force and aerodynamic moment expressions used above in the divergence analysis. The adaptation is replacing the slowly varying (i.e., static) rotational angle α by a fully time-varying $\alpha(t)$. Such air loads are called *quasisteady*, because the previously used time-invariant (static) air loads are called *steady-state* air loads. Considerable testing has demonstrated that solutions based on the use of these planar flow, quasisteady air loads are in poor agreement with experimental results. The shortcomings of quasisteady air loads, compared to the planar flow, *unsteady* air loads that are discussed later, is that the latter account for the effect of the motion of the air foil on the air flow, which in turn affects the airfoil. Nevertheless, quasisteady air loads are a convenient first step toward getting a grip on the concept of aerodynamic flutter.

As per usual, the analysis begins with writing the Lagrange equations of motion for the degrees of freedom $\alpha(t)$ and $h(t)$ from the energy and virtual work expressions set forth in Eq. (7.17). The only nonroutine aspect of this task is remembering that the quasisteady aerodynamic moment $\mathcal{M}$ is independent of both these DOF. Thus the aerodynamic moment is a constant and, as such, has no place in this dynamic analysis. Thus, the matrix equation of motion for this airfoil segment in planar flow is

$$\begin{bmatrix} m & mb \\ mb & (H_{cg} + mb^2) \end{bmatrix} \begin{Bmatrix} \ddot{h} \\ \ddot{\alpha} \end{Bmatrix} + \begin{bmatrix} k & 0 \\ 0 & K \end{bmatrix} \begin{Bmatrix} h \\ \alpha \end{Bmatrix} = \begin{Bmatrix} \mathcal{L} \\ a\mathcal{L} \end{Bmatrix}$$

As discussed above, when the airfoil is moving up and down as well as rotating, the effective angle of attack is not just α, but, for the adopted sign convention, is $\alpha - (\dot{h}/V)$. Then the lift force becomes $\mathcal{L} = \frac{1}{2} C_{l\alpha} \rho V^2 S[\alpha(t) - \dot{h}(t)/V] = A_0 V^2[\alpha - \dot{h}/V]$, where the definition of the aerodynamic coefficient A_0 is obvious. Substituting this latter expression for the quasisteady lift force into the above matrix equation of motion yields

$$\begin{bmatrix} m & mb \\ mb & H_{ea} \end{bmatrix} \begin{Bmatrix} \ddot{h} \\ \ddot{\alpha} \end{Bmatrix} + \begin{bmatrix} k & 0 \\ 0 & K \end{bmatrix} \begin{Bmatrix} h \\ \alpha \end{Bmatrix} = -A_0 V \begin{bmatrix} 1 & 0 \\ a & 0 \end{bmatrix} \begin{Bmatrix} \dot{h} \\ \dot{\alpha} \end{Bmatrix} + A_0 V^2 \begin{bmatrix} 0 & 1 \\ 0 & a \end{bmatrix} \begin{Bmatrix} h \\ \alpha \end{Bmatrix}. \tag{7.18a}$$

Clearly, the aerodynamic coefficient matrices are not symmetric. The first step in solving Eq. (7.18a) is to recognize that every differential equation term involves one or the other of the two DOF or their derivatives. That is, because the generalized forces depend on the deflections, velocities, and accelerations, the above matrix equation can be written in the homogeneous equation form as

$$\begin{bmatrix} m & mb \\ mb & H_{ea} \end{bmatrix} \begin{Bmatrix} \ddot{h} \\ \ddot{\alpha} \end{Bmatrix} + \begin{bmatrix} A_0 V & 0 \\ a A_0 V & 0 \end{bmatrix} \begin{Bmatrix} \dot{h} \\ \dot{\alpha} \end{Bmatrix} + \begin{bmatrix} k & -A_0 V^2 \\ 0 & K - a A_0 V^2 \end{bmatrix} \begin{Bmatrix} h \\ \alpha \end{Bmatrix} = \begin{Bmatrix} 0 \\ 0 \end{Bmatrix}. \tag{7.18b}$$

The homogeneous form itself and the fact that there are but two equations (rows) and three unknown quantities (the two DOF and V) suggest that the above equation is a differential equation eigenvalue, problem with V as an eigenvalue, while the two DOF compose the eigenvector. This proves to be the case. The next step in the solution process is to deal with the time variation of the two DOF.

The solution to Eq. (7.18) that is now sought is the one associated with the neutral stability point, the point on the airspeed scale between a region of airspeeds where the vibratory deflections increase without bound (i.e., where the airfoil motion is unstable) and the region of lower airspeeds where the vibratory motion of the airfoil decreases over time (i.e., where the airfoil motion is stable). At the neutral stability point the amplitudes of the deflections are constant, and no net work is done on the airfoil by the airstream. Again, the airspeed associated with this neutral point is called the flutter speed of the airfoil. Since the vibratory airfoil motion will persist indefinitely at constant amplitudes, the adopted trial solution for each DOF will be harmonic in form. The last aspect of the trial solution recognizes that, as was explained above, phase differences between the vertical motion DOF and the rotational DOF are crucial to the explanation of the flutter phenomenon. Therefore, to reflect mathematically the constant-amplitude, harmonic motion with phase differences, write the trial solution for the two DOF as either

$$\begin{aligned} &\alpha(t) = A_1 \sin\omega t + A_2 \cos\omega t \quad && h(t) = B_1 \sin\omega t \\ \text{or} \quad &\alpha(t) = (A_1 + iA_2)e^{i\omega t} && h(t) = B_1 e^{i\omega t} \end{aligned} \tag{7.19}$$

where the real amplitude components A_1, A_2, and B_1 are unknown, and therefore the phase angle between the up-and-down translational motion and the rotational motion is also unknown. Using the complex form choice, as per usual, differentiating and substituting into Eq. (7.18), after some organizing, yields

$$\begin{bmatrix} m\omega^2 b + A_0 V^2 & m\omega^2 - k \\ K - H_{ea}\omega^2 - aA_0V^2 & -m\omega^2 b \end{bmatrix} \begin{Bmatrix} A_1 \\ B_1 \end{Bmatrix} + i \begin{bmatrix} -\omega A_0 V & m\omega^2 b + A_0 V^2 \\ a\omega A_0 V & K - H_{ea}\omega^2 - aA_0V^2 \end{bmatrix} \begin{Bmatrix} B_1 \\ A_2 \end{Bmatrix} = \begin{Bmatrix} 0 \\ 0 \end{Bmatrix}.$$

When a complex number is zero, both the real part and the imaginary part are zero. Thus[18]

$$\begin{bmatrix} m\omega^2 b + A_0 V^2 & m\omega^2 - k \\ K - H_{ea}\omega^2 - aA_0V^2 & -m\omega^2 b \end{bmatrix} \begin{Bmatrix} A_1 \\ B_1 \end{Bmatrix} = \begin{Bmatrix} 0 \\ 0 \end{Bmatrix}$$

$$\begin{bmatrix} -\omega A_0 V & m\omega^2 b + A_0 V^2 \\ a\omega A_0 V & K - H_{ea}\omega^2 - aA_0V^2 \end{bmatrix} \begin{Bmatrix} B_1 \\ A_2 \end{Bmatrix} = \begin{Bmatrix} 0 \\ 0 \end{Bmatrix}.$$

These are two coupled matrix eigenvalue problems where, in addition to the unknown amplitudes A_1, A_2, and B_1 of the two eigenvectors, there are the two unknown quantities V and ω at the neutral stability point, which are now recognized as the two eigenvalues. Again, nontrivial solutions exist only if the coefficient matrices are singular. Setting the second determinant equal to zero yields the solution for the flutter frequency

$$\omega_f^2 = \frac{K}{H_{ea} - abm}.$$

This solution for the flutter frequency, the frequency of the vibration at the flutter airspeed, is independent of the airspeed, and closely related to the uncoupled natural frequency for a torsional vibration. The first observation casts doubt on the

[18] If sines and cosines were used rather than complex notation, the linear independence of those two functions would allow the same separation into two matrix equations.

completeness of this solution. Solving the equation produced by the determinant of the first of the above matrix equations yields

$$V_f^2 = \frac{(k - m\omega_f^2)(K - H_{ea}\omega_f^2) - (mb\omega_f^2)^2}{A_0\left[a(k - m\omega_f^2) + bm\omega_f^2\right]}.$$

where the value of the flutter frequency is taken from the previous equation. Clearly this simple solution for the flutter airspeed is not so simple that the effects of any parameter other than those present in A_0 are easily discerned. The dubious nature of this approach suggest that this solution does not deserve further attention.

A more accurate answer can be obtained using the unsteady airloads developed in the mid-1930s. Still these air loads do not account for the effect of viscosity (no Reynolds number), the effect of compressibility (no Mach number), the effects of finite wing aspect ratio,[19] or airfoil thickness. Within these limitations, however, the unsteady air loads differ from the quasisteady air loads by accounting for the effects on the airfoil of the entire flow field, particularly the oscillating wake that affects the airflow at the airfoil. As a result, these air loads depend not only on the instantaneous angle of attack, $\alpha(t)$, but also on the first and second time derivatives of both $\alpha(t)$ and $h(t)$. The expressions for the unsteady aerodynamic lift per unit of span length and the unsteady moment per unit of span length, both acting at the elastic axis, are, from Ref. [7.7], p. 199,

$$\mathcal{L} = -\pi\rho\frac{c^3}{8}\left[\frac{2\ddot{h}}{c} + 8\frac{V\dot{h}}{c^2}\mathcal{C}(\kappa) + \frac{2e}{c}\ddot{\alpha} + \frac{2V}{c}\left[\left(\frac{4e}{c} - 1\right)\mathcal{C}(\kappa) - 1\right]\dot{\alpha} + 8\frac{V^2}{c^2}\mathcal{C}(\kappa)\alpha\right]$$

$$\mathcal{M} = -\frac{\pi\rho c^4}{16}\left[\frac{4e\ddot{h}}{c^2} + \frac{4V}{c^2}\left(1 + \frac{4e}{c}\right)\mathcal{C}(\kappa)\,\dot{h} + \left(\frac{1}{8} + \frac{4e^2}{c^2}\right)\ddot{\alpha}\right.$$
$$\left. - \frac{2V}{c}\left[\frac{2e}{c} - \frac{1}{2} + \left(\frac{1}{2} - \frac{8e^2}{c^2}\right)\mathcal{C}(\kappa)\right]\dot{\alpha} - \frac{4V^2}{c^2}\left(1 + \frac{4e}{c}\right)\mathcal{C}(\kappa)\alpha\right], \quad (7.20)$$

where c is the chord length (the distance in the direction of the airstream between the airfoil leading edge and trailing edge); κ, the reduced frequency,[20] is equal to the nondimensional quantity $c\omega/(2V)$; $e = (a/c) - (1/4)$ is a nondimensional parameter that locates the elastic axis aft of the midchord position; V is the airspeed; and ρ is the zero-velocity mass density of the air. The quantity $\mathcal{C}(\kappa)$, which is a factor for each of the DOF and their first time derivatives, is called the Theodorsen function. The Theodorsen function is equal to the complex quantity $\mathcal{F}(\kappa) + i\mathcal{G}(\kappa)$, where

$$\mathcal{F}(\kappa) = \frac{J_1(\kappa)[J_1(\kappa) + Y_0(\kappa)] + Y_1(\kappa)[Y_1(\kappa) - J_0(\kappa)]}{[J_1(\kappa) + Y_0(\kappa)]^2 + [Y_1(\kappa) - J_0(\kappa)]^2}$$

$$\mathcal{G}(\kappa) = -\frac{Y_0(\kappa)Y_1(\kappa) + J_0(\kappa)J_1(\kappa)}{[J_1(\kappa) + Y_0(\kappa)]^2 + [Y_1(\kappa) - J_0(\kappa)]^2},$$

where J_0 and J_1 are Bessel functions of the first kind of order zero and 1, respectively, and Y_0 and Y_1 are Bessel functions of the second kind of order zero and 1, respectively.

[19] See Ref. [7.8] for the quite complicated low-speed airloads that account for the effect of finite aspect ratio. See Refs. [7.9,7.10] for some of the history of the engineering understanding of aircraft flutter.

[20] The common symbol for the reduced frequency is k. Since k is being used here to represent the translational spring stiffness, κ is chosen instead to represent the reduced frequency.

Here the moment is not static, and unlike the quasisteady case, this moment must be included in the equations of motion. Since these unsteady airloads act at the airfoil elastic axis that is the location of the two DOF, the generalized forces are simply the above lift and moment multiplied by the airfoil width. Thus for the simple case of the airfoil of Figure 7.14 with width ℓ, the equations of motion are

$$\begin{bmatrix} m & mb \\ mb & H_{ea} \end{bmatrix} \begin{Bmatrix} \ddot{h} \\ \ddot{\alpha} \end{Bmatrix} + \begin{bmatrix} k & 0 \\ 0 & K \end{bmatrix} \begin{Bmatrix} h \\ \alpha \end{Bmatrix} = \begin{Bmatrix} \ell\mathcal{L}(\ddot{h}, \dot{h}, \ddot{\alpha}, \dot{\alpha}, \alpha) \\ \ell\mathcal{M}(\ddot{h}, \dot{h}, \ddot{\alpha}, \dot{\alpha}, \alpha) \end{Bmatrix}. \tag{7.21}$$

Again, inspection of the above unsteady lift and moment expressions listed above shows that the resulting differential equations are homogeneous, and they constitute a complex algebra eigenvalue problem that, again, is solved at the neutral stability point where the airfoil motion is harmonic in both DOF with a phase angle difference between the two DOF as described in Eq. (7.19).

The remainder of the solution process is similar to, but not quite the same as, that employed above for the case of the quasisteady airloads. It is the same procedure in that, in addition to the same three unknown deflection amplitudes A_1, A_2, and B_1, the two unknown quantities of the resulting real and imaginary determinant equations are the same eigenvalues V_f and ω_f. However, in the unsteady airload case there is an important complication that was not present in the quasisteady airload case. That difficulty is that the ratio of the two unknowns, V_f and ω_f, in the form of the reduced frequency, forms the argument of the complicated Theodorsen function. Most analysts would prefer not to deal with trying to solve equations involving the Theodorsen function with an unknown argument. For example, if the powerful Newton–Raphson method were used, it would be necessary to differentiate the Theodorsen function and its ratio of the squares of Bessel functions. Hence, other solution techniques have been sought and found. A rather clever and common approach is to begin the remainder of the solution process by specifying a value of the reduced frequency, κ which, again, is equal to $c\omega/(2V)$. This step, of course, makes the Theodorsen function into merely a constant for each calculation associated with that selected value of the reduced frequency. However, this specification of a reduced frequency increases the number of equations by 1, from 2 to 3, whereas the number of unknowns remains at 2. To restore the correct number of unknowns, an unknown, artificial, material damping factor, g, is introduced into the problem in the form of the factor $(1 + ig)$ for the stiffness matrix, $[k]$, as shown below:

$$\begin{bmatrix} m & mb \\ mb & H_{ea} \end{bmatrix} \begin{Bmatrix} \ddot{h} \\ \ddot{\alpha} \end{Bmatrix} + (1+ig) \begin{bmatrix} k & 0 \\ 0 & K \end{bmatrix} \begin{Bmatrix} h \\ \alpha \end{Bmatrix} = \begin{Bmatrix} \ell\mathcal{L}(\ddot{h}, \dot{h}, \ddot{\alpha}, \dot{\alpha}, \alpha) \\ \ell\mathcal{M}(\ddot{h}, \dot{h}, \ddot{\alpha}, \dot{\alpha}, \alpha) \end{Bmatrix}.$$

This introduction of an artificial material damping factor is strictly a mathematical convenience. Thus, for each selected reduced frequency, there can now be a straightforward solution from the real and imaginary parts of the homogeneous matrix equation of motion for the two unknowns, V and g. To be on the safe side, the actual material damping and all the other types of actual damping, that are actually present in the wing system, are generally ignored. Hence when the value of the selected reduced frequency is found that results in the artificial material damping being zero, as it should be, then that value is the flutter value of the reduced frequency. The flutter value of the reduced frequency, along with the velocity solution, immediately yields solutions for V_f and ω_f. A simple graph of the artificial damping factor g versus (usually the inverse of) the selected reduced frequency, identifies which value of the reduced frequency corresponds to a zero damping factor. See Figure 7.15(a) for a typical example of a plot of the artificial material damping factor versus the reciprocal of the reduced frequency. Note that each point on the curve is a neutral stability point. Therefore, at the lower values of $(1/\kappa)$, which correspond to lower values of airspeed, negative values of damping are required for constant amplitude vibrations, and thus this is a stable region of reduced frequency. At the higher values of $(1/\kappa)$, which correspond to higher values of airspeed, positive values of damping are required to maintain constant amplitude vibrations, and thus this is an unstable region of reduced frequency.

It is evident that even in this simple case of low speed flow, the unsteady aerodynamic lift and moment require computational effort that is better suited to the use of a computer. Computer processing is particular important when the above unsteady air loads are applied to, say, a high aspect ratio wing rather than a single segment of a wing. Consider a high aspect ratio, tapered wing without sweepback of the elastic axis as shown in Figure 7.16. For simplicity of discussion, the wing is cantilevered at the fuselage centerline as would essentially be the case if the mass of the loaded fuselage and the tail is much larger than that of the wing. (As an alternative, a full unsupported wing can be dealt with as outlined in Endnote (3) of the previous chapter.) To facilitate the use of the unsteady airloads discussed above, the wing planform is divided into a series of adjacent wing strips, where the edges of each strip parallel the direction of the planar airstream. Note each wing strip has its own chord length. Since the aspect ratio is large for such a wing, the wing can be viewed structurally as a beam whose elastic axis is the wing's loci of shear centers of the beam's cross-sections. Beam finite elements, each with its own stiffness coefficients GJ and EI, join the finite element method nodes located at the centers of each wing strip. The beam DOF required to account for the motion of jth strip are the vertical deflection, h_j, the beam bending slope θ_j, and the angle of twist α_j. The beam stiffness matrix in effect joins together the various strips, and there are mass matrix terms associated with each DOF. The unsteady airloads are associated only with each $h_j(t)$ and $\alpha_j(t)$, and zeros would be entered in the generalized force matrix corresponding to each $\theta_j(t)$. Such an application of the above unsteady airloads to each wing strip is called *basic strip theory.* For a wing with sweepback, see Ref. [7.5], p. 824 and p. 817. Basically, a sweepback angle Λ for the wing elastic axis merely means that the relationship between the angle of attack of the aerodynamic cross section, α, and the twist, ϕ, and bending slope, θ, of the beam cross section becomes $\alpha = \phi \cos \Lambda - \theta \sin \Lambda$. Basic strip theory was useful for many years because it generally produced safe estimates of the experimental low-speed flutter speed. See Endnote (3) for a comparison of a typical low-speed experimental result and the corresponding basic strip theory result.

If the calculated flutter speed of a wing is so low that, subject to flight testing, the operation of an aircraft would need to be unacceptably restricted, there are remedies available at the cost of generally greater weight, or greater wind drag, and hence less overall aircraft performance. First of all, for example, a high aspect ratio straight wing, modeled as a tapered beam, the wing torsional stiffness coefficient GJ, is the most important parameter influencing flutter speed. For such a wing,

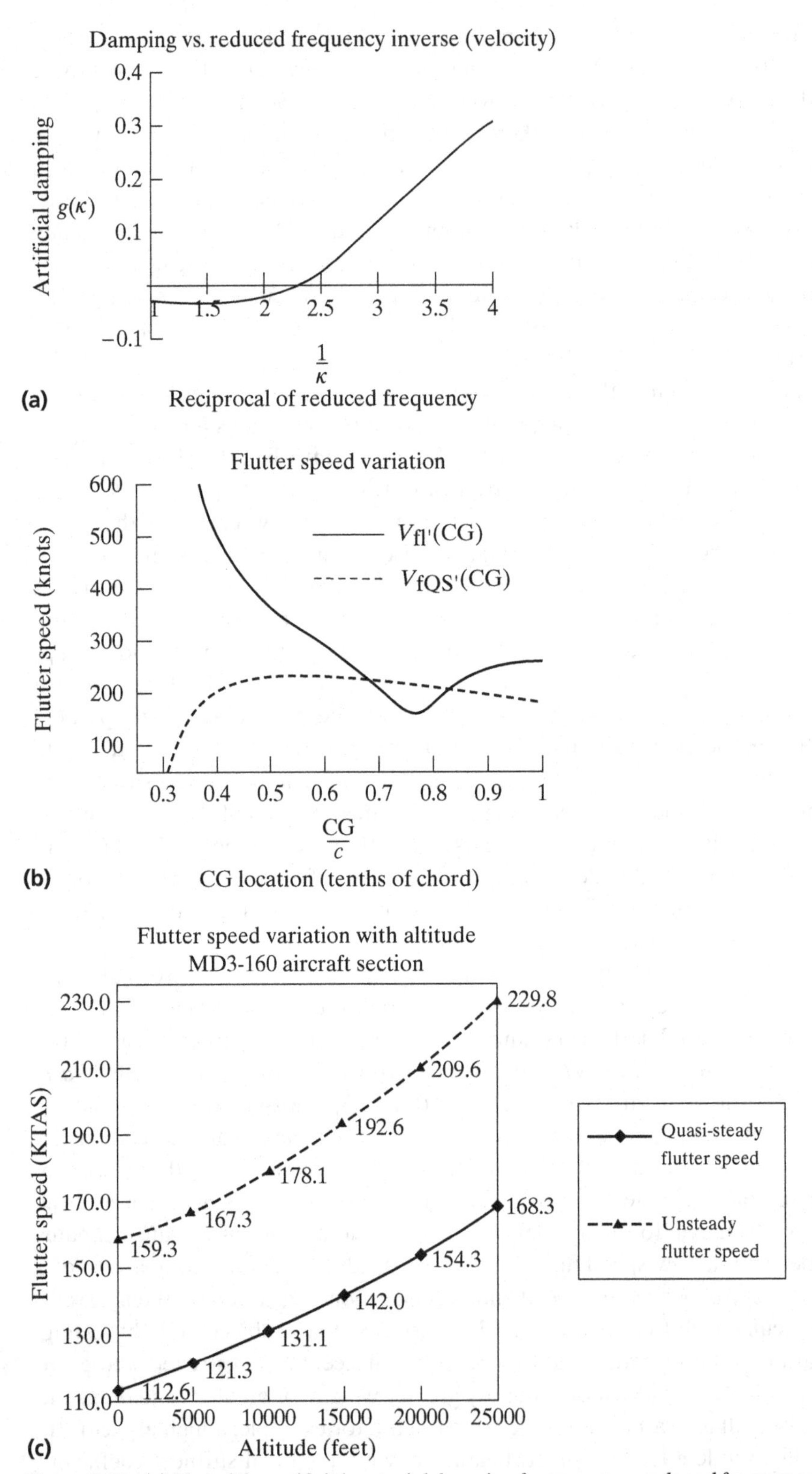

Figure 7.15. (a) Plot of the artificial material damping factor versus reduced frequency reciprocal. (b) Plot of flutter airspeed versus CG location along airfoil chord calculated using unsteady and quasisteady airload theories. (c) Calculated flutter airspeed versus altitude, Ref. [7.11].

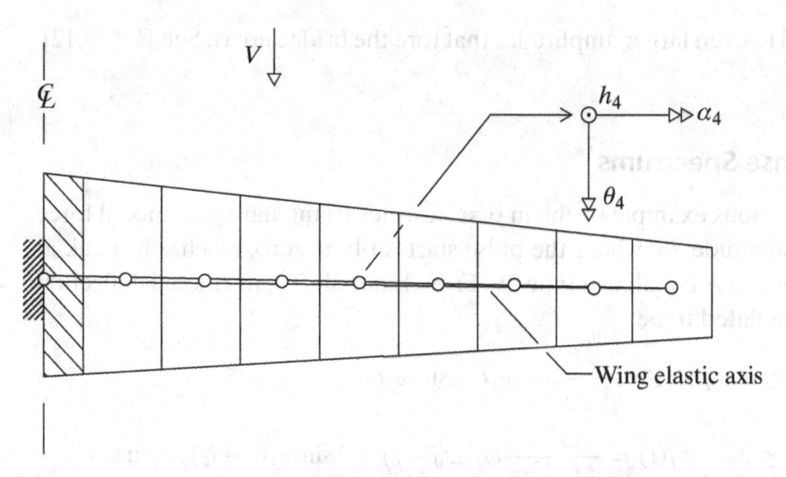

Figure 7.16. High aspect ratio wing model suitable for application of basic strip theory.

the usual construction consists of a forward spar near the wing's leading edge and a main spar near the middle of the wing, where the spars run through the fuselage from wing tip to wing tip. Viewing such a wing as a tapered beam, the St. Venant constant for uniform torsion, J, is somewhat proportional to the cross-sectional area enclosed by the centerlines of the forward spar, the top wing skin, the main spar, and the bottom wing skin. See Ref. [7.5], p. 419ff. Therefore, increasing the torsional stiffness coefficient of the wing can be achieved by further increasing the beam cross-sectional area between the wing front spar and the main spar, called the wing's torsion box. This can be done by moving the spars a bit and making the wing a bit thicker.

The second remedy to consider for increasing the flutter airspeed is that of shifting forward on the wing chord the location of the wing center of gravity. This is sometimes even accomplished by simply adding mass ahead of the leading edge on a forward protruding boon. The effectiveness of this approach is illustrated by Figure 7.15(b), which is taken from Ref. [7.11]. This plot is based on calculations using the above unsteady air loads, Eqs. (7.20), as applied to the simple airfoil strip model of Figure 7.14. Also from Ref. [7.11], Figure 7.15(c) shows how flutter speed varies with altitude.

The interaction of blunt bodies and air streams is more complicated than that of smooth airfoils, but that interaction can also be quite dramatic. The flow around a blunt body, at all but the very slowest of airspeeds, results in flow separation from the blunt body, and therefore all manner of vortices and turbulence. The classical example of blunt body flutter is the collapse of the highly flexible Takoma Narrows bridge in 1940. The cross section of the bridge was an extended H where the horizontal bar of the H was the road bed. The air flow that collapsed the bridge was a steady, low-speed wind blowing down the river valley spanned by the bridge. That air flow interacted with bridge oscillations in a way that resulted in the alternating shedding of clockwise and then counterclockwise vortices (called von Karman vortices). These vortices resulted in up and then down aerodynamic loads on the bridge road bed that were in phase with the bridge oscillations; constantly increasing them until nonlinear effects limited the oscillatory amplitudes to large values. The large amplitudes weakened the bridge, which led to even larger amplitudes that tore the bridge apart. See Ref. [7.12], p. 17.

7.12 **Response Spectrums**

Return to the previous example problem of a isosceles triangular spike modal force input of peak magnitude P_0, where the pulse starts at time zero, reaches its peak at time t_0, and drops to a zero value at time $2t_0$. From Example 7.6, the modal deflection response was calculated to be

$$\text{for} \quad 0 \le t \le t_0 \quad p_j(t) = \frac{P_0}{M_j\,\omega_j^3 t_0}(\omega_j t - \sin\omega_j t)$$

$$\text{for} \quad t_0 \le t \le 2t_0 \quad p_j(t) = \frac{P_0}{M_j\,\omega_j^3 t_0}[\omega_j(2t_0 - t) + 2\sin\omega_j(t - t_0) - \sin\omega_j t]$$

$$\text{for} \quad 2t_0 \le t \quad p_j(t) = \frac{P_0}{M_j\,\omega_j^3 t_0}[2\sin\omega_j(t - t_0) - \sin\omega_j t - \sin\omega_j(t - 2t_0)]$$

it is clear that for any given input magnitude P_0, the magnitude of the deflection response, and hence the stress response, depends entirely on the two parameters ω_j and t_0, which respectively characterize the period of the jth mode of the structural system and the duration of the pulse loading. This is typical of all pulses. The engineering design question is as follows: for a given ω_j (or, for a given value of the period $T_j = 2\pi/\omega_j$), what nondimensional values of $\omega_j t_0$ or t_0/T_j will maximize the deflection response? In this manner, the worst possible case can be anticipated. This question can be answered in a straightforward manner by simply plotting the nondimensional response $p_j M_j \omega_j^3 t_0/P_0$ versus t_0/T_j or $\omega_j t_0$ for each of the three time intervals discussed. Such a plot is called a *response spectrum* or a *shock response spectrum*. To illustrate the process of calculating a response spectrum, first consider the simpler load input case where the jth modal force is a rectangular pulse. In this case there are only two time intervals of concern rather than the three time intervals of the triangular spike loading. The mathematical description of the modal force input is

$$P_j(t) = P_j^0[\text{stp}(t) - \text{stp}(t - t_0)].$$

Using step response functions, the undamped modal deflection response is

$$\text{for} \quad 0 \le t \le t_0 \quad p_j(t) = \frac{P_j^0}{M_j\omega_j^2}[1 - \cos\omega_j t]$$

$$\text{for} \quad t \ge t_0 \quad p_j(t) = \frac{P_j^0}{M_j\omega_j^2}[\cos\omega_j(t - t_0) - \cos\omega_j t].$$

In the first time interval, the maximum values of the deflection response, up to a time value where $\omega_j t = \pi$, are controlled by the value of t_0, the length of the time interval of the applied force. That is, the ever increasing peak values obtained in this initial time interval $(0, t_0)$ are only limited by the available time in the pulse interval. The ever-increasing response is described by the $(1 - \cos\omega_j t)$expression until that expression reaches the value of +2.0. After $\omega_j t = \pi$, the maximum value of 2.0 is never exceeded. Hence 2.0 remains the maximum value thereafter for this case of

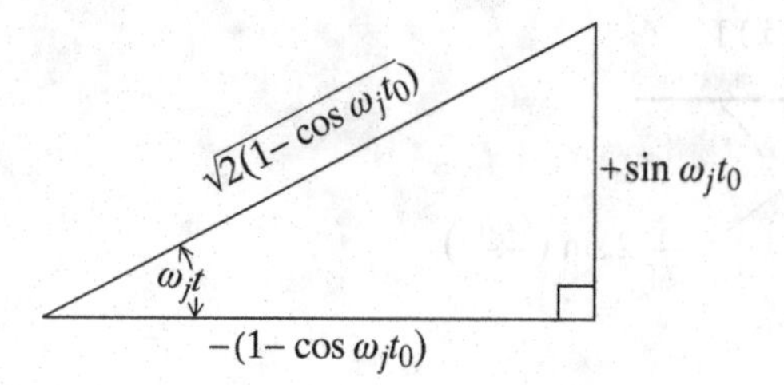

Figure 7.17. Schematic for calculating sines and cosines from tangents.

$t < t_0$. These maximums must be compared with the maximum modal deflections obtained for the second time interval.

To determine the maximum modal response in the time interval where $t > t_0$, first use the usual calculus routine for determining the time value for the maximums of the second time interval response. (The minimum deflection response occurs when $t_0 = 0$.) That is, from the second of the above modal deflection solutions, write

$$\frac{d}{dt}[\cos\omega_j(t-t_0) - \cos\omega_j t] = -\omega_j[\sin\omega_j(t-t_0) - \sin\omega_j t] = 0.$$

To solve for the times for the maximums expand the above expression for $\sin\omega_j(t-t_0)$ to obtain the following solution for $t_{\max/\min}$

$$\tan\omega_j t_{max} = \frac{-\sin\omega_j t_0}{1+\cos\omega_j t_0}.$$

To facilitate the substitution of the above times for the maximum deflections into the above second solution expression for $p_j(t)$, where the function $\cos\omega_j(t-t_0)$ is expanded to be $\cos\omega_j t\cos\omega_j t_0 + \sin\omega_j t\sin\omega_j t_0$, draw the right angle triangle diagram of Figure 7.17, which embodies the above solution for the tangent. Now the cosine and sine are easily determined with the result, after some algebra,

$$p_{j:max} = \sqrt{2}\frac{P_j^0}{M_j\omega_j^2}\sqrt{1-\cos\omega_j t}.$$

To more easily interpret this solution, write

$$\cos\omega_j t_0 = \cos\omega_j\left(\frac{t_0}{2}+\frac{t_0}{2}\right) = \cos^2\omega_j\frac{t_0}{2} - \sin^2\omega_j\frac{t_0}{2}$$

$$\text{and}\quad 1 = \cos^2\omega_j\frac{t_0}{2} + \sin^2\omega_j\frac{t_0}{2}.$$

Then the solution for the maximum values of the deflection response has the simpler form

$$p_{j:max/min} = \frac{2P_j^0}{M_j\omega_j^2}\sin\frac{\omega_j t_0}{2}.$$

Comparing this solution to the solution for the first time interval shows that the absolute value of this solution for the maximum response is always larger than the absolute value of the first solution for the maximum responses. That is, for the time period $0 < \omega_j t_0 < 2\pi$, $2\sin(\omega_j t_0/2) > 1 - \cos\omega_j t_0$ except at the point $\omega_j t_0 = \pi$ where the two expressions are equal. Thus this second solution determines the response spectrum, which can be plotted as shown in Figure 7.18.

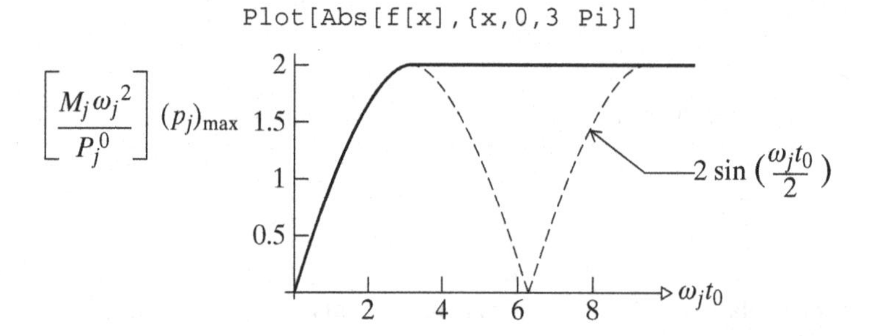

Figure 7.18. Response spectrum for a rectangularly shaped pulse of magnitude P_j^0 and duration t_0 after time zero.

The response spectrum for the originally considered isosceles triangular shaped pulse can be found in Ref. [7.13], p. 107. That reference also shows the response spectrum for a half-wave sinusoidal pulse on that same page, and on p. 125, it shows the response spectrum for a ramp input that levels off at time t_1. Clearly these response spectra are more valuable when the number of modes needing consideration is small, and they are most valuable when the structure, or, for example, the item to be packaged, can be, at least crudely, modeled as a single-DOF system.

REFERENCES

7.1 Skeel, R. D., and J. B. Keiper, *Elementary Numerical Computing with Mathematica*, McGraw-Hill, New York, 1993.

7.2 Wolfram, S., *Mathematica*, 2nd ed., Addison-Wesley, Reading, MA, 1991.

7.3 Lin, Y. K., *Probabilistic Theory of Structural Dynamics*, McGraw-Hill, New York, 1967; Krieger, Melbourne, FL, 1976.

7.4 Chopra, A. K., *Dynamics of Structures, Theory and Applications to Earthquake Engineering*, Prentice Hall, Englewood Cliffs, NJ, 1995.

7.5 Donaldson, B. K., *Analysis of Aircraft Structures: An Introduction*, McGraw-Hill New York, 1993.

7.6 Krone, N. J., "Divergence elimination with advanced composites," AIAA 1975 Aircraft Systems and Technology Meeting, Los Angeles, CA, paper no. 75-1009.

7.7 Scanlan, R. H., and R. Rosenbaum, *Introduction to the Study of Aircraft Vibration and Flutter*, Macmillan, New York, 1951.

7.8 Reissner, E., *Effect of finite span on the airload distributions for oscillating wings, Parts I and II* (with Stevens, J. E.), NACA TN 1194 and 1195, 1947.

7.9 Garrick, I. E., and W. H. Reed, "Historical development of aircraft flutter," *AIAA* J. Aircraft, vol. 18, 11, 1981, p. 897.

7.10 Pines, S., in "An elementary explanation of the flutter mechanism," Flomenhoft, H. I. (ed.), *The Revolution in Structural Dynamics*, Dynaflo Press, Palm Beach Gardens, FL, 1997.

7.11 Wheeler, P. C., *An Explication of Airfoil Section Bending-Torsion Flutter*, MS thesis in Civil Engineering, University of Maryland, College Park, 2004.

7.12 Gimsing, N. J., *Cable Supported Bridges*, John Wiley & Sons, Chichester, UK, 1983.

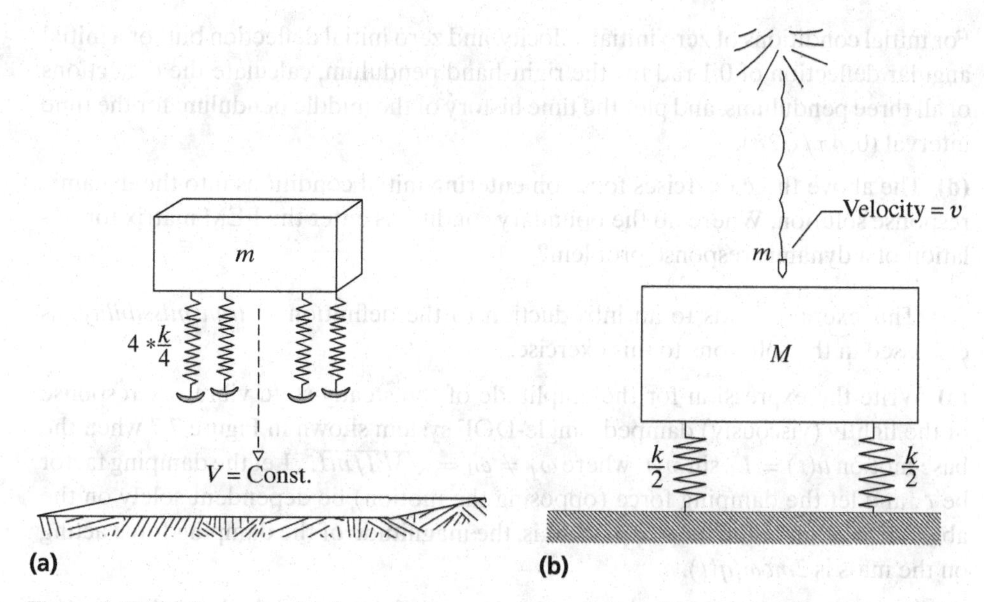

Figure 7.19. (a) Lunar lander descending at constant velocity. (b) Measuring muzzle velocity.

7.13 Thomson, W. T., *Theory of Vibration with Applications*, 4th ed., Prentice Hall, Englewood Cliffs, NJ, 1993.

7.14 7.14 Meirovitch, L., *Analytical Methods in Vibrations*, Macmillan, New York, Co., 1967.

7.15 Donaldson, B. K., "Evaluation of Reissner's correction for finite span aerodynamic effects," *J. Aircraft*, vol. 9, 7, 1972.

CHAPTER 7 EXERCISES

7.1 (a) The undamped m, k moon-landing vehicle shown in Figure 7.19(a) descends to the lunar surface at a constant descent velocity V. Write the expression for the harmonic motion of the mass m after initial contact with the lunar surface, time zero. Let g be the lunar acceleration of gravity.

(b) Consider the single-DOF, undamped M, K system shown in Figure 7.19(b). At time zero, the system is impacted by a bullet of mass m traveling a short distance through air at a constant velocity v. With the constant quantities m, M, K, and U (U being the deflection amplitude of the vibrating target plus imbedded bullet) being measured directly, calculate the (muzzle) velocity of the bullet, v. Could a pendulum impacted at its center of mass serve as another means of determining muzzle velocity?

(c) Consider the three-DOF pendulum system shown in Figure 2.15, for Exercise 2.4. From Examples 5.1 and 5.4, the natural frequencies and mode shapes of this system are

$$\begin{aligned} \omega_1 &= \sqrt{\beta} & A^{(1)} &= \lfloor 1.0 \quad 1.0 \quad 1.0 \rfloor \\ \omega_2 &= \sqrt{2\beta} & A^{(2)} &= \lfloor -1.0 \quad 0.0 \quad 1.0 \rfloor \\ \omega_3 &= 2\sqrt{\beta} & A^{(3)} &= \lfloor -0.5 \quad +1.0 \quad -0.5 \rfloor. \end{aligned}$$

For initial conditions of zero initial velocity, and zero initial deflection but for a initial angular deflection of 0.1 rad for the right-hand pendulum, calculate the deflections of all three pendulums, and plot the time history of the middle pendulum for the time interval $(0, 4\pi/\sqrt{2\beta})$.

(d) The above three exercises focus on entering initial conditions into the dynamic response solution. Where do the boundary conditions enter the FEM matrix formulation of a dynamic response problem?

7.2 This exercise leads to an introduction to the definition of *transmissibility*, as discussed in the solutions to this exercise.

(a) Write the expression for the amplitude of the steady-state vibratory response of the lightly (viscously) damped, single-DOF system shown in Figure 7.7 when the base motion $u(t) = U_0 \sin\omega_f t$, where $\omega_f \neq \omega_1 = \sqrt{3EI/mL^3}$. Let the damping factor be ζ, and let the damping force (opposing the motion) be dependent solely on the absolute motion of the mass, $q(t)$; that is, the magnitude of the damping force acting on the mass is $2m\zeta\omega_1\dot{q}(t)$.

(b) Repeat part (a), but this time let the damping force be solely dependent on the motion of the mass relative to the base, $q(t) - u(t)$; that is, the magnitude of the damping force acting on the mass is $2m\zeta\omega_1(\dot{q} - \dot{u})$.

(c) It is not unusual for a structure to elastically support a large, relatively rigid mass that contains a component that rotates at high angular velocity. Two examples are a piece of heavy machinery used for manufacturing supported by a factory building structure, and an aircraft engine supported by an airframe structure. The centrifugal force associated with a slight imbalance of the rotating component impresses (in any one direction) a harmonic force on the relatively "rigid" mass that is the piece of machinery or aircraft engine. The design of the machine or engine mounting, that is, the connection or interface between the rigid mass the supporting structure, often seeks to minimize the transmission of the impressed harmonic force from the rigid mass to the supporting structure. The design model is that of a one-DOF system where the rigid body has mass m, the elastic elements forming the support interface between the rigid body and the structure are grouped together as a single spring k, and the inherent damping is represented by a dashpot c, where again the spring and dashpot connect the mass to the supporting structure. The imbalance induced harmonic force acting on the structure has the form $F_0 \sin\omega_f t$. Determine the ratio of the magnitudes of the transmitted force (i.e., the sum of the forces in the spring and dashpot) to the impressed force, F_0. Hint: Use as your generalized coordinate, $q(t)$, the absolute motion of the rigid mass, and approximate the small motion of the supporting structure as nonexistent.

7.3 As shown in Figure 7.20, a building of mass m is supported vertically by a series of rollers and against horizontal motion by springs of total stiffness k and damping idealized as dashpots with a total coefficient c. The building is subjected to a horizontal ground motion $\Upsilon_0 \sin\omega_f t \Rightarrow \Upsilon_0 \exp(i\omega_f t)$.

(a) Write the building's equation of motion and then determine the building's steady-state response $u_0 \exp(i\omega_f t)$. Form the ratio of the amplitude of the building response to the amplitude of the ground motion.

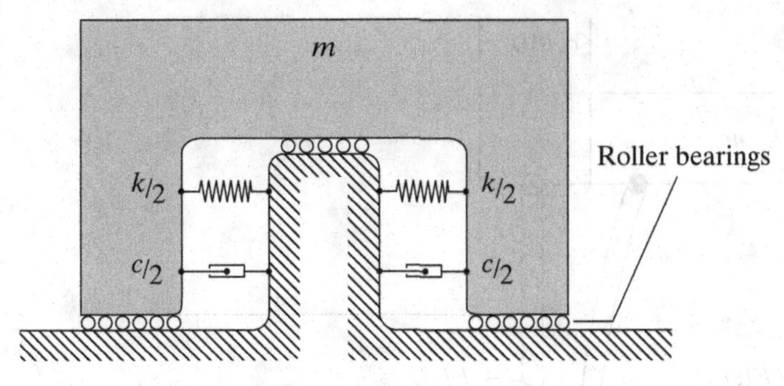

Figure 7.20. Exercise 7.3.

(b) If the frequency of the ground motion is estimated (known), then what would you recommend with regard to the design of the horizontal stiffeners, k, and what would you recommend with regard to system friction?

7.4 (a) Calculate the value of the frequency ratio for which Eq. (7.7) has its maximum value and that maximum value.

(b) Show that IF the amplitudes of the harmonic loading vector $\{Q(t)\} = \{Q_0\}e^{i\omega t}$ are proportional to the jth modal vector weighted by the mass matrix, that is, IF the amplitude vector $\{Q_0\} = \tilde{c}[M]\{\Phi^{(j)}\}$, then only the jth mode is excited (i.e., the deflection response is limited to $p_j(t)$ deflections).

7.5 (a) Evaluate the following integrals involving the Dirac delta function

$$(a)\quad \int_{-1}^{+1} x^2\delta(x-2)dx \qquad (b)\quad \int_{-1}^{+5} x^2\delta(x-2)dx$$

$$(c)\quad \int_{2(precisely)}^{3} x^2\delta(x-2)dx \qquad (d)\quad \int_{2^-}^{3} x^2\delta(x-2)dx$$

$$(e)\quad \int_{-\pi/2}^{\pi/2} \tan x\,\delta(x)\,dx \qquad (f)\quad \int_{-\pi}^{\pi} \delta(x-\tfrac{1}{2}\pi)\sin x dx.$$

(g) Show, by means of the coordinate transformation, that the impulse response function form of the Duhamel integral can also be written as

$$p(t) = \int_0^t P(\tau)h(t-\tau)d\tau = \int_0^t P(t-\bar{\tau})h(\bar{\tau})d\bar{\tau}.$$

(h) Show, by means of integration by parts, that the impulse response function form of Duhamel's integral and the step response form of Duhamel's integral are the same. That is, show

$$p_j(t) = \int_0^t P_j(t-\tau)h_j(\tau)d\tau = P_j(0)g_j(t) + \int_0^t \dot{P}_j(t-\tau)g_j(\tau)d\tau.$$

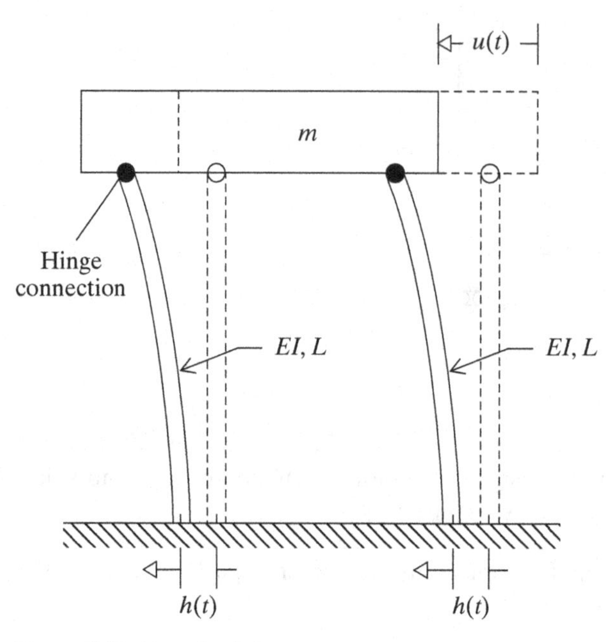

Figure 7.21. Exercise 7.6.

7.6 The undamped, planar structure sketched in Figure 7.21 moves only in the plane of the paper in response to a time-varying foundation motion (very simplified earthquake motion) that is

$$0 \leq t \leq t_0 \qquad h(t) = \frac{t}{t_0} h_0 \quad \text{and} \quad t_0 \leq t \qquad h(t) = h_0.$$

(a) In terms of the sideways displacement of the mass, $u(t)$, write the equation of motion for this structural system.

(b) Write, in integral form, the solution for the motion $u(t)$ in the time interval $0 \leq t \leq t_0$. Do not carry out any integrations.

(c) Write the solution for the motion $u(t)$ in the time interval $t \geq t_0$. Do not integrate.

7.7 **(a)** Use the undamped impulse response form of the Duhamel integral, Eq. (7.14), to obtain the response to a normalized step modal force written as $P_j^0 \text{stp}(t)$.

(b) Use the undamped step response form the Duhamel integral, Eq. (7.15), to obtain the response to a normalized impulsive acceleration written as $\Pi_j \delta(t)$.

(c) Repeat part (a), but this time use Eq. (7.14), the damped impulse response form of the convolution integral.

7.8 **(a)** Write, for an undamped system model, in terms of the ramp response function and the step response function, the response for the jth mode when the normalized modal force time history is, at time zero, a vertical jump to a magnitude P_j^0, followed by a linear decline to zero at time t_1, followed by a constant zero value thereafter.

(b) Write, for an undamped system, in terms of the modal form of the sine response function, the response for the jth mode when the normalized modal acceleration is a positive half sine wave in the time interval $(0, t_0)$ with a maximum value of P_j^0/M_j.

7.9 Consider a lightly damped single-DOF system whose equation of motion is

$$\ddot{q} + 2\zeta\omega_1\dot{q} + \omega_1^2 q = 2\zeta\omega\dot{v} + \omega_1^2 v,$$

where $v(t)$, the base motion, is zero everywhere but in the finite interval $(0, t_0)$ and in that interval the base motion is described by the parabolic shape $v(t) = +(4\Upsilon/t_0^2)(tt_0 - t^2)$.

(a) Write the convolution integral solution for the deflection response of the mass in the time interval $(0, t_0)$. Do NOT carry out the integration.

(b) As above, set up the integral for the deflection solution after time t_0. Do NOT integrate.

For the eager

7.10 A cantilevered beam with a uniform stiffness coefficient EI and total length L supports nonstructural mass such that the mass properties of the beam are modeled as a discrete mass of magnitude $2m$ at the beam center, and m at the beam tip. The mass moments of inertia of the lumped masses are negligible. At time zero, the tip mass is subjected only to a pulse loading that is a positive force described analytically by the function $F(t)$.

(a) Write the equations of motion for this system using only the vertical deflections at the masses as DOF.

(b) Use the modal transformation to write the superposition integral solutions for modal deflections for this simple model.

7.11 Substitute the step function form of the Duhamel integral into the modal equation of motion so as to prove that this integral is indeed a solution to the modal equation of motion. Hint: See Endnote (1).

7.12 When pushing a child on a playground swing, the person pushing is careful to time his or her pushes to coincide with the natural period of the child and swing. To examine the effects of selecting the time duration and timing of continuous pulses, consider an undamped one-DOF system subjected to a base motion that has the form of a pulse of constant magnitude Υ for a nondimensional time duration of $\omega_1 t_0$ and is zero thereafter. If $q(t)$ is the deflection response, then that response can be written as

$$\begin{aligned} &\text{for} \quad 0 \le t \le t_0 \quad q(t) = \Upsilon(1 - \cos\omega_1 t) \\ &\text{for} \quad t \ge t_0 \qquad\;\; q(t) = \Upsilon[\cos\omega_1(t - t_0) - \cos\omega_1 t]. \end{aligned}$$

(a) If $\omega_1 t_0 = 2\pi$, sketch the response for all time.

(b) If $\omega_1 t_0 = \pi$, sketch the response for all time.

(c) What would happen in case (b) if another rectangular pulse having the same $\omega_1 t_0 = \pi$ duration were applied at time $\omega_1 t_0 = 2\pi$? Another way of viewing this is to consider the work done on the system by the applied, or equivalent applied, force.

$$W = \int F(t)dq(t) = \int F(t)\frac{dq(t)}{dt}dt.$$

Recall that the total mechanical energy, that is, the sum of the kinetic and potential energies (and thus the amplitude of the motion) of the system is increased whenever positive work is being done on the system. The above formula shows that work done on the system is positive whenever the applied force and the velocity of the system are in phase (i.e., have the same sign). Recall further that the velocity of the mass is just the slope of the deflection time history. Thus it is a simple matter to decide when and how long to apply additional pulses to either increase or decrease the amplitude of a motion.

7.13 Using the modal transformation to diagonalize a multidegree of freedom structural system modeled as having proportional or Rayleigh damping was discussed in Section 7.4. There is a slightly more general form of the damping matrix that allows diagonalization by the modal transformation that was developed by T. K. Caughey [7.5]. The Caughey damping matrix can be written as follows:

$$[c] = [m]\sum_{j=0}^{N-1} a_j\left([m]^{-1}[k]\right)^j.$$

To diagonalize the Caughey damping matrix, it is first necessary to scale (i.e., normalize) the modal amplitude vectors differently than previously where a positive unit value was assigned to the largest, in absolute value, entry in the vector. This previous normalization led to the results

$$[\Phi]^t[m][\Phi] = [\backslash M\backslash] \quad \text{and} \quad [\Phi]^t[k][\Phi] = [\backslash\omega^2 M\backslash].$$

Now alter each jth modal vector in the modal matrix by dividing all its entries by the square root of its corresponding modal mass term, M_j. Then the above equations simplify to

$$[\Phi]^t[m][\Phi] = [\backslash I\backslash] \quad \text{and} \quad [\Phi]^t[k][\Phi] = [\backslash\omega^2\backslash].$$

Your task is to show that the Caughey damping matrix is indeed diagonalized by the modal transformation by considering, say, the first four terms in the Caughey series.

ENDNOTE (1): VERIFICATION OF THE DUHAMEL INTEGRAL SOLUTION

The validity of a purported solution to a differential equation can be tested by substitution of that proposed solution into the original differential equation to determine if the equation equality is truly satisfied. To show that Eq. (7.12), the Duhamel integral solution does indeed satisfy the modal equation of motion, Eq. (7.5a), recall the mechanics of differentiating an integral. It is not difficult to prove that when an integral has a parametric limit or a parameter appears in the integrand, the easily proved rules for differentiation are

$$\frac{d}{dy}\int_a^y f(x)dx = f(y) \quad \text{and} \quad \frac{d}{dy}\int_a^b f(x,y)dx = \int_a^b \frac{\partial f(x,y)}{\partial y}dx.$$

When an integral, such as the Duhamel integral, has both a parametric limit and the same parameter in the integrand, then the derivative is obtained from use of the chain rule for derivatives. To be more explicit, consider a general integral where the dummy variable of integration is τ, and the parameter is t. Let the parameter t appear in both of the limits of the integral as well as the integrand. Then the derivative with respect to the parameter t is obtained as follows:

$$\frac{d}{dt}G(u(t),v(t),w(t)) = \frac{d}{dt}\int_{u(t)}^{v(t)} f(w(t),\tau)d\tau = \frac{\partial G}{\partial u}\frac{du}{dt} + \frac{\partial G}{\partial v}\frac{dv}{dt} + \frac{\partial G}{\partial w}\frac{dw}{dt}$$

$$\therefore \frac{dG}{dt} = -f(w(t),u(t))\times\dot{u}(t) + f(w(t),v(t))\times\dot{v}(t) + \dot{w}(t)\times\int_{u(t)}^{v(t)}\frac{\partial f(w,\tau)}{\partial w}d\tau.$$

Now apply the above formula to the Duhamel integral solution of the modal equation of motion

$$\ddot{p}_j + 2\zeta_j\omega_j\dot{p}_j + \omega_j^2 p_j(t) = \frac{P_j(t)}{M_j},$$

which is

$$p_j(t) = \int_0^t P_j(\tau)h_j(t-\tau)d\tau,$$

where from Eq. (7.14)

$$h_j(t) = \frac{1}{M_j\omega_j\sqrt{1-\zeta_j^2}}e^{-\zeta_j\omega_j t}\sin\left[\omega_j t\sqrt{1-\zeta_j^2}\right].$$

Note that for the Duhamel integral, $u = 0$ and $v = w = t$. Thus

$$\dot{p}_j(t) = P_j(t)h_j(0) + \int_0^t P_j(\tau)\dot{h}_j(t-\tau)d\tau,$$

$$\text{where} \quad h_j(0) = 0$$

$$\text{and} \quad \ddot{p}_j(t) = P_j(t)\dot{h}_j(0) + \int_0^t P_j(\tau)\ddot{h}_j(t-\tau)d\tau,$$

$$\text{where} \quad \dot{h}_j(0) = 1/M_j.$$

Substitution of the above results into the modal equation of motion produces

$$P_j/M_j(t) + \int_0^t P_j(\tau)[\ddot{h}_j + 2\zeta_j\omega_j\dot{h}_j + \omega_j^2 h_j(t-\tau)]d\tau \stackrel{?}{=} P_j(\tau)/M_j.$$

The square-bracketed quantity in the integrand above is zero because the impulse response function, $h_j(t-\tau)$, is a solution to the force free (homogeneous) differential equation of motion. Hence, the equality is satisfied, and the Duhamel integral is verified as a solution to the modal equation of motion. This same process can be applied to the step response function form of Duhamel's integral.

ENDNOTE (2): A RAYLEIGH ANALYSIS EXAMPLE

A Rayleigh analysis approximates a multidegree of freedom structure as a single-DOF structure by guessing at the structure's first mode shape and using only that assumed first mode shape to describe the motion of the structure. Thus a Rayleigh analysis is a special case of a Ritz analysis, which uses more than one assumed mode shape and is sometimes called a Rayleigh–Ritz analysis. A person who has no experience with Rayleigh analyses would understandably question the usefulness of any analysis that begins with a guess. Since the Rayleigh–Ritz procedure was one of the very few general analysis tools available to engineers in the pre-computer era, and certainly the most elegant one, it was used and studied extensively. It has been proven that a good estimation of the first natural frequency of the structure can be had for any reasonable estimate of the first mode shape. See Ref. [7.14], p. 118.

A simple example suffices for a brief explanation of the Rayleigh procedure for the present purpose of obtaining a good estimate of the first natural frequency of a structure, and an estimate of the effective mass associated with the first natural frequency. Consider a structure that reasonably can be modeled as a single, nonuniform cantilevered beam. Note that a Rayleigh analysis can be applied to any type of structure, but choosing a beam simplifies the discussion. Let the beam structure undergo a lateral vibration. Then, temporarily viewing this nonuniform beam as a continuum, the kinetic energy of this beam is

$$T = \frac{1}{2}\int_0^L m(x)[\dot{w}(x,t)]^2 dx,$$

where $m(x)$ is the mass per unit length along the length of the beam and $w(x,t)$ is the lateral deflection of the beam axis. From engineering beam theory, the strain in a beam that is bending only in the x, y plane is $-z(d^2w/dx^2)$. In this circumstance, the strain energy (the energy stored because of deformation) of the nonuniform beam can be written as

$$U = \frac{1}{2}\int_L \int\int_A E(x)[\epsilon(x,t)]^2\, dA\, dx = \frac{1}{2}\int_L EI(x)[w''(x,t)]^2 dx.$$

To initiate the Rayleigh method, guess that the first mode shape of this cantilevered beam has the shape of a cubic polynomial. Then for a free vibration of the cantilevered beam, with the coordinate x starting at the clamped end, write $w(x,t) = W_{max}\,(x/L)^3 \sin\omega_1 t$, where W_{max} is the tip deflection of the cantilevered beam. Substituting this guess for the lateral deflections into the expressions for the kinetic energy and the strain energy leads to

$$T = \frac{1}{2}[W_{max}\omega_1]^2 \cos^2\omega_1 t \int_0^L m(x)\frac{x^6}{L^6}dx$$

$$U = \frac{36}{2}[W_{max}]^2 \sin^2\omega_1 t \int_0^L \frac{EI(x)}{L^4}\frac{x^2}{L^2}dx.$$

As the beam vibrates, every point along the beam axis passes through the beam's undeflected position simultaneously. At that point in time, the beam velocities, and hence the kinetic energy of the beam as a whole, are a maximum. At this same point in time, the strain energy is zero because the beam is undeflected. At the peak of its deflections, just as the motion of the beam is about to change direction, the velocities, and hence the kinetic energy, are zero. However, at the peak deflections, the strain energy is a maximum. Since, in this model, the structure vibrates without damping, its total energy, which is the sum of the kinetic and strain energies, is a constant. From the above discussion, this constant is equal to the the maximum kinetic energy or the maximum strain energy. Hence, the maximum kinetic energy equals the maximum strain energy. The points in time where the above energy expressions are a maximum is clearly when the two trigonometric functions have the value of 1. Hence, equating these maximum energies leads to the following approximation for the first natural frequency for this structure

$$\omega_1^2 = \frac{36\displaystyle\int_L \frac{EI(x)}{L^4}\left(\frac{x}{L}\right)^2 dx}{\displaystyle\int_0^L m(x)\frac{x^6}{L^6}dx} = \frac{K_{eff}}{M_{eff}}.$$

The integrations are easily carried out numerically by selecting 8, 10, or more nodes along the length of the beam and using beam stiffness coefficients, EI, at those nodes, and lumping the surrounding mass at those nodes. Hence an estimate for the first natural frequency is obtained. Note that the mass near the beam tip, where the value of x is greater, has a much greater influence on the result than the mass near the base of the nonuniform beam. The natural frequency and the effective mass, which is the denominator of the above fraction, are the only parameters required to now treat the multidegree of freedom structure as a single-DOF system in the forced vibration case.

ENDNOTE (3): AN EXAMPLE OF THE ACCURACY OF BASIC STRIP THEORY

Since flutter calculations based on basic strip theory do not account for such effects as those produced by airfoil thickness, airstream viscosity and compressibility, and, particularly, a finite wing aspect ratio with its resulting wing tip vortices (a three- rather than two-dimensional flow field), fully accurate calculated flutter speeds and flutter frequencies should not be expected. The aircraft industry response to these limitations was extensive experimentation with aircraft models and carefully monitored flight testing. Both flutter model testing and flight testing are expensive operations requiring careful preparation. The testing of a cantilevered wing flutter model is the subject of this endnote. First note that a "flutter model" requires the difficult task of scaling not just the aerodynamic shape of the model, but also the elastic and inertial properties of the system being modeled. It is the latter challenge that makes the models so expensive to build.

Reference 7.15 discusses a wind tunnel flutter model wing of 37.65-in. semispan length. The model wing had a low aspect ratio of 3.5, a taper ratio of 0.2, a root chord length of 35.857 in., and a leading edge sweep back angle of 45°. The scaled elastic portion of the structure consisted entirely of a finely milled, tapered aluminum beam that was cantilevered at 5% of the semispan. The beam was perpendicular to the support (wind tunnel wall) until 20% of the semispan and then was swept back at an angle of 32.54°. To be as consistent as possible with the simplifications associated with the unsteady aerodynamic loads of basic strip theory, the balsa wood wing segments were connected to the aluminum beam only at the centers of those segments, which were connections at every one-tenth of the semispan length starting at 5% of the semispan length. Small lead weights were used to scale the mass properties of the wing design. The wind tunnel experimental flutter speed was 95.9 mph with a flutter frequency of 12.5 Hz. Calculations using the low-speed, unsteady airloads discussed above estimated the flutter airspeed at 68.9 mph and the flutter frequency at 14.5 Hz. When the more sophisticated low-speed aerodynamics developed by Eric Reissner, Ref. [7.8], which accounted for a finite aspect ratio, were used, the calculated flutter airspeed and frequency were 99.6 mph and 12.0 Hz, a much more accurate result, but, alas, not on the safe side. Again see Ref. [7.15]. Unfortunately, the use of the Reissner aerodynamics applied to other such model wings didn't always result in such close estimates of the experimental results, just better results than the basic strip theory aerodynamics, which, of course, are without the finite aspect ratio correction.

ENDNOTE (4): NONLINEAR VIBRATIONS

Modern computing power makes possible the routine investigation of nonlinear behavior by numerical methods. The presence of nonlinear behavior in a structural system is to be expected if the amplitudes of the vibration exceed whatever are "small" deflections for the system of interest. For example, small beam deflections are deflections that do not exceed approximately one-quarter of the depth of the beam. Otherwise such deflections are called "finite" deflections. "Large" deflections exceed the depth of the beam. What often happens when the deflections are finite is that there is significant coupling between the axial forces and those associated with beam bending, as well as coupling between the bending moments in the two orthogonal planes of bending, and the twisting moment. The result of this coupling is that the structure becomes stiffer as the amplitudes of vibration increase. Greater stiffness means higher natural frequencies. Thus the situation where the natural frequencies depend on the amplitude of the vibration. This low-damping-value situation can be depicted graphically, say, by altering the straight-up resonant peak of Figure 5.8 by bending it to the right so that the resonant peak that starts out at the frequency ratio $\Omega = 1.0$ is only reached by increasing the frequency ratio. This phenomena can lead to surprising results. In this case, an applied harmonic load of increasing frequency, sometimes called a frequency sweep, will lead to increasing amplitudes as the first natural frequency is approached from below just as in the wholly linear case. However, the increase in amplitudes will not be rapid as in the linear case. Then, as the frequency continues to slowly increase, the amplitude will suddenly drop to the low intermode value. The upper path of the bent over resonance diagram became unstable, and the amplitude drops past the lower portion of the resonance curve to the intermode value. If the sweep is from higher frequencies to lower frequencies, then as the lower natural frequency is approached, the amplitudes will jump up from the lowest, intermode curve past the lower path of the bent over resonance peak to the upper path. That is the amplitudes will suddenly jump in magnitude.

Strange things happen not only with the amplitudes but also can happen with a response frequency. A simple example of such a phenomenon is the case of a long, uniform, thin beam-column undergoing fully elastic Euler buckling. Although the static Euler beam-buckling ordinary differential equation has a linear form, it is actually a nonlinear problem because the derivation of that ordinary differential equation requires consideration of a deformed beam element. From another viewpoint, the static beam buckling equation (the dynamic form of beam bending equations is discussed in the next chapter) for a compressive axial force P is

$$EIw''''(x) + Pw''(x) = 0.$$

The axial compressive force is related to the axial deflections $u(x)$ according to the formula

$$P = -EAu'(x).$$

If the second of these two formulas is substituted into the first, then the second term of the equation involves the product of the two deflections, which is a nonlinear term. Now let this long, thin beam-column be subjected to a harmonic axial force whose minimum magnitude is the Euler buckling load for that beam-column, and the maximum amplitude is some greater value that still results in fully elastic behavior in the beam-column. Start the applied harmonic force at its minimum value. As the force increases in magnitude to its peak value and then recedes again to its minimum value, the beam-column will bend out from its straight position, say to the left, and then return to its straight position. Then the beam-column's inertia loads will carry the beam through its straight position to the right, and the still-increasing values of the applied axial force will abet the motion to the right. Thus a peak deflection will be achieved on the right side and will subside to the original equilibrium position, and so on. Note the periods of the applied axial force and the beam-column response. The applied harmonic force goes through two periods of its motion, whereas the beam-column goes through only one. Therefore the forced harmonic response frequency of the nonlinear beam-column is one-half that of the applied harmonic load. Recall that in all the previously considered linear cases, response harmonic frequency is the same as the applied load's harmonic frequency.

Chapter 9 provides examples of numerical calculations of vibratory responses for relatively simple, nonlinear structural systems.

8 Continuous Dynamic Models

8.1 Introduction

The previous four chapters emphasized the advantages of using discrete mass mathematical models wherein both the structural mass and the nonstructural mass is "lumped" at selected (usually a relatively few) finite element nodes or at short distances from those finite element nodes. The alternative in mass modeling is the seemingly more realistic mathematical model where the mass is distributed throughout each structural element. Such distributed or continuous mass models are not nearly as useful as discrete mass models. However, continuous mass models do have enough instructional value and occasional engineering value that they cannot be wholly ignored. Their instructional value resides in (i) seeing the results of dealing with what is essentially an infinite DOF system; (ii) the reinforcement, and perhaps deeper understanding, obtained through repetition of the same analysis procedures used with discrete mass systems in a different context; and (iii) discovering the very few types of structures which can be usefully described by this much more concise type of modeling. Therefore the purpose of this chapter is to discuss some of those situations where the use of continuous mass models is of some, albeit small, value in the study of structural dynamics.

Again, continuous mass models are practical only in quite restricted circumstances. All cases examined here are limited to structures that are modeled as a single structural element (e.g., one beam or one plate). Furthermore, each structural element must have either a uniform geometry and mass distribution, or a geometry that varies in such a simple and smooth manner that it can be described by use of a low-order polynomial function.

8.2 Derivation of the Beam Bending Equation

The matrix form for the equations of motion for the general discrete mass structural model was derived using the Lagrange equations of motion. Since the Lagrange equations of motion utilize generalized coordinates that can represent deflections at specific points in the structure, the Lagrange equations are very compatible with the discrete mass finite element model of a structure. However, to derive, for example, the beam bending equation of motion in terms of a lateral deflection at any point along the axis of a beam with continuous mass modeling, it is much more convenient to use Hamilton's principle. From the first chapter, after separating the total virtual work into the sum of the virtual work of the internal loads and the virtual work of the external loads, Hamilton's principle can be written as

$$\int_{t_1}^{t_2} [\delta T + \delta W_{in} + \delta W_{ex}]\, dt = 0 = \int_{t_1}^{t_2} [\delta T - \delta U - \delta V]\, dt,$$

where, again, the limits of integration are wholly arbitrary. Now it is a matter of detailing each of the three components of either integrand. For the sake of initial simplicity, let the beam vibrations be limited to lateral vibrations in the x, z plane. The placement of the x, y, and z Cartesian coordinates for the beam is as follows. The y and z axes lie in the plane of the beam cross section and originate at the centroid of the beam cross section. The x axis runs the length of the beam and is the loci of those centroids. For this limitation of vibrating only in the x, z plane to be realized, it is necessary that (i) the z-direction lateral loading per unit of beam length, $f_z(x, t)$, act only along the x axis and thus have lines of action that pass through the centroid of the beam cross section and (ii) the shear center[1] of each cross section also lie on each z axis. This alignment of the cross-section centroid and shear center will occur if the z axis is an axis of symmetry of the beam cross section. Then the only beam lateral deflection is $w(x, t)$, which is positive in the z direction, and the product of inertia, I_{yz}, is zero. Therefore the kinetic energy (one-half the mass multiplied by velocity squared) of a differential length of the beam located at the point x along the length of the beam is

$$dT = \frac{1}{2}[\rho A(x) dx] \left[\frac{\partial w(x, t)}{\partial t}\right]^2,$$

where ρ is the mass density, and A is the cross-sectional area. In this case, $w(x, t)$ can be measured from either the shear center or the centroid of the cross section. Hence, the kinetic energy for the entire beam when it undergoes lateral bending vibrations is

$$T = \frac{1}{2} \int_0^L \rho A(x) \left[\frac{\partial w(x, t)}{\partial t}\right]^2 dx.$$

Therefore, the variation on the kinetic energy is

$$\delta T = \int_0^L \rho A(x) \frac{\partial w(x, t)}{\partial t} \frac{\partial [\delta w(x, t)]}{\partial t} dx.$$

In this limited circumstance, the virtual work of the external loads involves only the lateral loading per unit length f_z. The positive upward work producing force acting at

[1] For small displacements, lateral loads applied along the loci of shear centers do not cause the beam to twist. The shear center is the center of twist for a beam cross section.

any point x along the length of the beam is $f_z(x,t)dx$ and the corresponding virtual displacement is $\delta w(x,t)$. Hence, the external virtual work for the entire beam is

$$\delta W_{ex} = \int_0^L f_z(x,t)\delta w(x,t)\,dx.$$

Again, the virtual work of the internal forces is equal to the negative of the variation of its corresponding potential function, which is the elastic strain energy. That is, $\delta W_{in} = -\delta U$. The use of the elastic strain energy alone, as is done here, limits the validity of all subsequent beam analyses based on this derivation to the elastic range of material behavior. Since engineering beam bending theory hypothesizes that the only significant bending strain is the normal strain ϵ_{xx}, the small deflection beam bending strain energy is, from Ref. [8.1]

$$U = \frac{1}{2}\iiint \sigma_{xx}\epsilon_{xx}\,dx\,dy\,dz = \frac{1}{2}\iiint E\epsilon_{xx}^2\,dx\,dy\,dz,$$

where Hooke's law (linear elasticity) for no temperature change is used to obtain the second integral. Any elementary strength of materials textbook explains that, for beam bending, the above-normal strain at any point on the beam cross section is approximated, with good accuracy, as the negative of the distance of that point on the cross section above the centroid, z, multiplied by the local beam curvature, $\partial^2 w(x,t)/\partial x^2$. Thus

$$U = \frac{1}{2}\int_0^L E\left[\frac{\partial^2 w(x,t)}{\partial x^2}\right]^2\left(\int_A\int z^2\,dy\,dz\right)dx.$$

The above area integral is called the area moment of inertia about the y axis, symbolized as I_{yy}. Since this is the only area moment of inertia to enter this development, the symbol is shortened to just I. Applying the variational operator to both sides of the above expression

$$-\delta W_{in} = \delta U = \int_0^L EI\frac{\partial^2 w(x,t)}{\partial x^2}\frac{\partial^2}{\partial x^2}[\delta w(x,t)]\,dx.$$

Now that all three terms of the integrand of Hamilton's principle have been detailed, they can be substituted into that integrand to obtain

$$\int_{t_1}^{t_2}\int_0^L\left\{\rho A(x)\frac{\partial w(x,t)}{\partial t}\frac{\partial[\delta w(x,t)]}{\partial t} + f_z(x,t)\delta w(x,t)\right.$$
$$\left. - EI(x)\frac{\partial^2 w(x,t)}{\partial x^2}\frac{\partial^2}{\partial x^2}[\delta w(x,t)]\right\}dx\,dt = 0.$$

To have the same varied deflection, $\delta w(x,t)$, as a common factor for all three terms of the integrand, it is necessary to integrate by parts over both time and space. For the first term of the integrand, the time and distance integrals are first reordered, and one integration by parts over time is performed. The second term is fine the way it is. The third term requires two integrations by parts over distance. When the integration by parts is completed, the result is

$$\int_{t_1}^{t_2}\int_0^L \left\{ -\rho A(x)\frac{\partial^2 w(x,t)}{\partial t^2}\delta w(x,t) + f_z(x,t)\delta w(x,t) - \frac{\partial^2}{\partial x^2}\left[EI(x)\frac{\partial^2 w(x,t)}{\partial x^2}\right]\delta w(x,t)\right\} dx\,dt$$

$$+ \int_0^L \rho A \frac{\partial w}{\partial t}\delta w\Big|_{t_1}^{t_2} dx - \int_{t_1}^{t_2} EI\frac{\partial^2 w(x,t)}{\partial x^2}\delta\left[\frac{\partial w(x,t)}{\partial x}\right]\Big|_0^L dt$$

$$+ \int_{t_1}^{t_2} \frac{\partial}{\partial x}\left[EI(x)\frac{\partial^2 w(x,t)}{\partial x^2}\right]\delta w(x,t)\Big|_0^L dt = 0.$$

The next step is to recall that the values of the limits of integration, t_1 and t_2, are wholly arbitrary. For the first *single* integral to be individually zero, it is sufficient that the virtual displacement be chosen to have, as per usual, a zero value at those time limits everywhere along the length of the beam. The second of the single integrals can be disposed of by saying that for all values of time, either the bending moment (EIw'') is zero or the bending slope (w') is a constant at both ends of the beam. The third integral is zero for all values of time if either the shear force $(EIw'')'$ is zero or the displacement (w) is a constant at both ends of the beam. These latter two integrals provide the general beam boundary conditions. With these requirements, the above result reduces to the one double integral

$$\int_{t_1}^{t_2}\int_0^L -\left\{\rho A(x)\frac{\partial^2 w(x,t)}{\partial t^2} + f_z(x,t) - \frac{\partial^2}{\partial x^2}\left[EI(x)\frac{\partial^2 w(x,t)}{\partial x^2}\right]\right\}\delta w(x,t)\,dx\,dt = 0.$$

Note that this double integral has a zero value regardless of the arbitrary choice for the values of the continuous virtual displacement function between the time limits. For example, for any fixed value of time t, whether the value of δw is positive over the left-hand part of the beam and zero over the right-hand portion of the beam, or vice versa, the integral is always zero. The only way that this can happen is for the quantity within braces to be zero for all values of x and all values of t within the time limits. This result is the partial differential equation for beam bending

$$\frac{\partial^2}{\partial x^2}\left[EI(x)\frac{\partial^2 w(x,t)}{\partial x^2}\right] + \rho A(x)\frac{\partial^2 w(x,t)}{\partial t^2} = f_z(x,t). \tag{8.1}$$

This simplified equation is sufficient to illustrate the process of determining natural frequencies and mode shapes for beam bending and then the determination of deflection responses to applied forces. Note if damping were part of the beam model, it would have appeared in the external virtual work expression. Alternately, it could be entered here as a velocity-dependent term as part of $f_z(x,t)$. More extensive beam bending equations are set forth in Endnote (1), along with a beam twisting equation and a beam axial deflection equation. The plate bending equation is also included in Endnote (1). Endnote (2) repeats much of the above for the case where there are also point masses and springs included in the mathematical model of the structure.

8.3 Modal Frequencies and Mode Shapes for Continuous Models

When a discrete/lumped mass model is used for a vibration analysis, the initial result is a set of many coupled, ordinary differential equations, with time as the independent variable. These equations are best solved in matrix form. Whenever a single structural element with a continuously distributed mass is used as the model for a vibration analysis, the linear governing differential equation is always a partial differential equation with time and at least one spatial variable acting as the independent variables. This typifies the usual circumstance of a large set of ordinary differential equations being an approximation to one or more partial differential equations. In general, partial differential equations are a significantly greater challenge than ordinary differential equations. However, it can be established, as below, and as was the case for the discrete mass models, that when the structure is undergoing a force free vibration, the time variation of the motion must be sinusoidal. As will be seen, this information allows the elimination of the independent time variable from the free vibration differential equation. Consider, for example, a beam deflection partial differential equation that has time, t, and distance along the beam axis, x, as its two independent variables. After the introduction of harmonic motion, a known variation of the time variable, this partial differential equation becomes a single ordinary differential equation in terms of x alone. Hence, it is possible that a particular distributed-mass beam vibration equation is open to a manageable analytic solution, as well as to a numerical solution.

As will be seen in general, the linear partial differential equations that describe the force free, small deflection, *un*damped motion of distributed mass structural elements (beams, plates, etc.), and thus structures, vibrating about their static equilibrium configuration all have the form

$$\mathcal{P}[w(\boldsymbol{x}, t)] + \mathcal{H}[w(\boldsymbol{x}, t)] = 0,$$

where w represents a deflection of any type, $\boldsymbol{x}$ is one or more spatial variables, and $\mathcal{P}$ is an even-order, partial derivative operator in the relevant spatial coordinates only. This operator also involves quantities that describe the geometry and the elastic properties of the structure. The partial derivative operator $\mathcal{H}$ describes the inertial loading. As such, it involves two time differentiations, the mass density of the structure, and quantities that further describe the geometry of the structure. There are various techniques for determining general solutions for the above partial differential equation, such as those set forth, for example, in Ref. [8.2]. These general solutions involve arbitrary functions just as ordinary differential equation solutions involve arbitrary constants. With a few exceptions, such as that for the wave equation, these solutions in the form of arbitrary functions are seldom useful in structural engineering applications. The most prominent mathematical technique for finding useful solutions for those partial differential equations associated with structural engineering is called *separation of variables.* As the name implies, this technique applied to the above partial differential equation would result in the unknown function $w(\mathbf{x}, t)$ written as either the sum of functions each with fewer variables than w or the product of such functions. In this case a useful solution is obtained using a product. Specifically, let $w(x, t) = W(x)T(t)$. Substituting this trial variables separable solution recalling that $\mathcal{P}$ involves only spatial derivatives while $\mathcal{H}$ involves two temporal derivatives, the above governing differential equation becomes

$$T(t)\,\mathcal{P}[W(x)] + \ddot{T}(t)\,\overline{\mathcal{H}}[W(x)] = 0$$

$$\text{or} \quad \frac{\mathcal{P}[W(x)]}{\overline{\mathcal{H}}[W(x)]} = -\frac{\ddot{T}(t)}{T(t)} = +\omega^2,$$

a constant, where the two time derivatives of the $\mathcal{H}$ operator were removed and applied to the time function, leaving only the spatial operator or factor $\overline{\mathcal{H}}$, which involves material density and geometric factors. The reason the above two ratios are equal to the same constant is as follows. Note that the first ratio involves only the spatial coordinates, whereas the second ratio only involves time. Fix all the spatial coordinates and vary time. Since the spatial coordinates are fixed, the ratio involving the derivatives of $W(x)$ does not vary. Therefore, the ratio involving $T(t)$ also does not vary despite the fact that t is varying. Thus the ratio involving $T(t)$ must be equal to a constant and so, too, the ratio involving $W(x)$. A positive constant is selected because a negative constant would require the deflections to increase exponentially with increasing time. Such a deflection time history is contrary to the nature of a force free vibratory motion and thus is rejected. With the choice of a positive constant, the function $T(t)$ must satisfy the following now very familiar differential equation

$$\ddot{T} + \omega^2 T = 0 \qquad \text{so} \qquad T(t) = A\,\sin(\omega t + \psi).$$

Therefore the engineering solution to the original partial differential equation becomes $w(x, t) = W(x)\sin(\omega t + \psi)$ with the amplitude A absorbed into $W(x)$. This separation of the time and spatial variables, which is valid for all free vibrations, allows the elimination of the time variable from the original differential equation. That is, after canceling the function $T(t)$, the original partial differential equation with time and spatial variables becomes either a partial differential equation only in terms of spatial variables or, if there is just one spatial variable, the ordinary differential equation

$$\mathcal{P}[W(x)] - \omega^2\overline{\mathcal{H}}[W(x)] = 0.$$

With the above background in place, consider the following beam example problems that illustrate this process.

EXAMPLE 8.1 Write the mathematical description of the force free motion of the single, undamped, uniform, simply supported beam shown in Figure 8.1(a) as it undergoes bending vibrations in the x, z plane. Again let $w(x, t)$ and $W(x)$ be respectively the z-direction bending deflection and bending deflection amplitude beyond the beam's static equilibrium position (SEP). To confine the motion to that

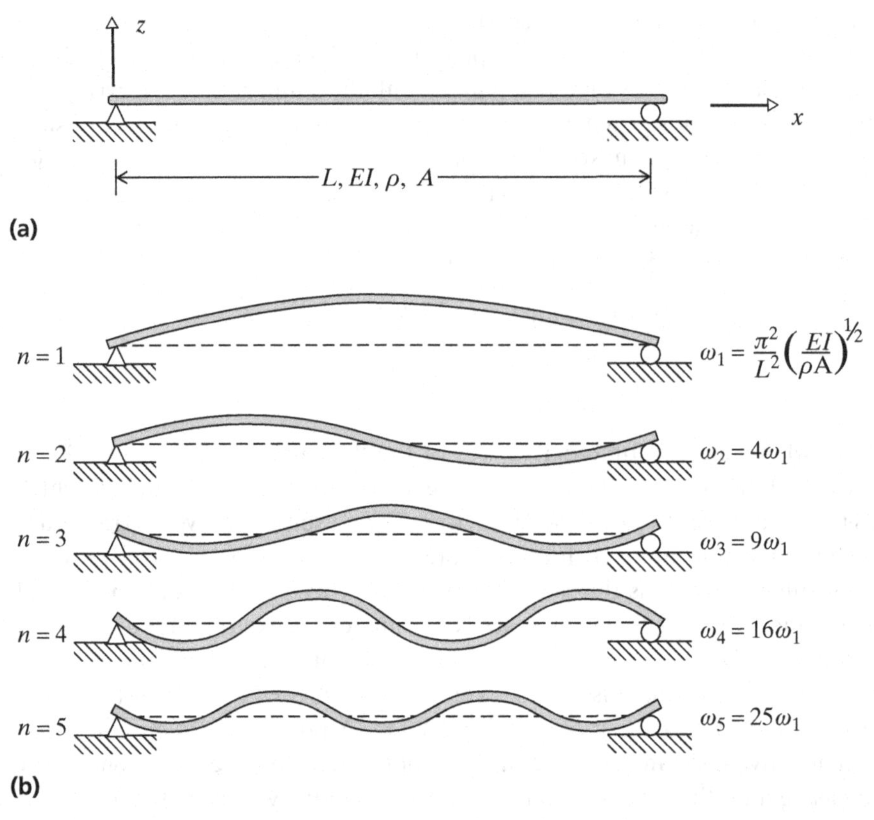

Figure 8.1. Example 8.1. Natural frequencies and mode shapes for a simply supported, uniform beam.

of beam bending in the plane of the paper, let, as discussed previously, the y and z axes be the principal axes of the beam and let both the shear center (the elastic center) and centroid (the mass center) of each cross section lie in the x, z plane so that the beam does not twist as it deflects [8.1].

SOLUTION Equation (8.1) provides the explicit form of the governing differential equation that can easily be adapted to describe the free vibration of this beam. However, just for the purpose of demonstrating a different point of view, note that because there is no axial force, N, along the length of the beam, from Endnote (1), the fourth-order[2] beam bending differential equation that describes the z-direction, lateral deflection of this uniform beam can be written in the form

$$EI\frac{\partial^4 w(x,t)}{\partial x^4} = f_z(x,t),$$

where $f_z(x, t)$ is the z-direction load per unit of beam length. In this free vibration case, this distributed loading is entirely the result of the vibratory accelerations of

[2] Briefly, this form of the fourth-order beam bending equation can be obtained from the second-order equation $EI(\partial^2 w/\partial x^2) = M(x,t)$ by differentiating twice with respect to x while recalling that the derivative of the internal moment is the internal shear force, and the derivative of the internal shear force is the externally applied force per unit length of beam axis.

the distributed mass. That is, the lateral load is the inertial force per unit length. Since any inertia force is the negative of mass multiplied by acceleration, the lateral inertial force per unit of beam length is the negative of the mass per unit length, ρA, multiplied by the second time derivative of the lateral deflection. Therefore, from this argument, or the detailed energy approach of Endnote (2), the beam fourth-order, governing differential equation for a free vibration in the x, z plane is

$$EI\frac{\partial^4 w(x,t)}{\partial x^4} = -\rho A\frac{\partial^2 w(x,t)}{\partial t^2}.$$

Since this beam is simply supported, the boundary conditions (BCs) are those of zero deflection and zero bending moment at both beam ends. The mathematical description of these BCs reduces to $w(0,t) = w''(0,t) = w(L,t) = w''(L,t) = 0$, for all times t, where primes indicate partial derivatives with respect to x. After employing the free vibration (variables separable) substitution of $w(x,t) = W(x)\sin(\omega t + \psi)$, and after cancellation of the $\sin(\omega t + \psi)$ term, the above partial derivative equation and its associated BCs reduce to the following ordinary differential equation and its BCs

$$W''''(x) - \lambda^4 W(x) = 0,$$
$$\text{where} \quad \lambda^4 \equiv \frac{\rho A\omega^2}{EI}$$
$$W(0) = W''(0) = W(L) = W''(L) = 0. \tag{8.2}$$

The differential equation and BCs of Eq. (8.2) are an example of the ordinary differential equation form of an eigenvalue problem. Note that, just like the matrix eigenvalue problem discussed previously, (i) there is a trivial solution for both the differential equation and the boundary conditions, which is, again $W(x) = 0$; (ii) if $W(x)$ is a solution, then so too is $cW(x)$, where c is an arbitrary factor; and (iii) the single governing differential equation contains two unknowns, $W(x)$ and λ. Hence, when this fourth-order equation is integrated, there will be four unknown constants of integration plus the unknown parameter λ; that is, five quantities to be determined by the four BC equations. Thus, just as was true for the $n \times 1$ amplitude vector of the matrix eigenvalue problem, it will be possible to determine only the infinite dimensioned vector (i.e., the function) $W(x)$ up to a multiplicative constant. In other words, for this formulation for beam bending, whereas the shape of $W(x)$ will be determined, its magnitude will be indefinite. This fact is evident because, again, if $W(x)$ is a solution to this differential equation, then $cW(x)$ is also a solution to the same equation.

Rather than just state the four linearly independent functions of x that satisfy Eq. (8.1), the algebraic dexterity required makes it worthwhile to derive the solution. The solution to this equation, as it is for all linear, ordinary differential equations with constant coefficients, is obtained from the always successful use of the trial function solution $W(x) = B\exp(rx)$, where here B is a constant of integration and the constant r is to be determined. Substitution of the trial solution into the differential equation leads to the algebraic equation

$$B\exp(rx)[r^4 - \lambda^4] = 0.$$

At least one of these three factors must be zero. The exponential function is never zero regardless of the value of r, real or complex. The amplitude B must be excluded from being zero because then the trial function $B\exp(rx)$ always would be zero, which is the above-rejected trivial SEP case. Therefore, the only useful possibility is when $r^4 = +\lambda^4$. There are four distinct roots to this algebraic equation, which is called the *characteristic equation* of the structural system. These roots are

$$r_1 = +\lambda \qquad r_2 = -\lambda \qquad r_3 = +i\lambda \qquad r_4 = -i\lambda.$$

Substitution of each of these four possible distinct roots into the trial function $B\exp(rx)$ leads to the following form of the general solution

$$W(x) = B_1 e^{\lambda x} + B_2 e^{-\lambda x} + B_3 e^{i\lambda x} + B_4 e^{-i\lambda x}.$$

Recalling that[3]

$$\begin{aligned}\exp(\pm i\lambda x) &= \cos(\lambda x) \pm i\sin(\lambda x)\\ \exp(\pm\lambda x) &= \cosh(\lambda x) \pm \sinh(\lambda x)\end{aligned}$$

the solution for $W(x)$ can be rewritten in more convenient form as

$$\begin{aligned}W(x) &= (B_1 + B_2)\cosh\lambda x + (B_1 - B_2)\sinh\lambda x\\ &\quad + (B_3 + B_4)\cos\lambda x + i(B_3 - B_4)\sin\lambda x\\ \text{or}\quad W(x) &= C_1\sinh\lambda x + C_2\cosh\lambda x + C_3\sin\lambda x + C_4\cos\lambda x. \qquad (8.3)\end{aligned}$$

Note that the constants C_3 and C_4 are real quantities. All these constants of integration can be proven to be real simply by noting (i) that λ is, by definition, real and (ii) $W(x)$ is real for all real values of x. Then select four real values of λx that produce the following four linearly independent equations[4] where $j = 1, 2, 3, 4$

$$W(x_j) = C_1\sinh(\lambda x_j) + C_2\cosh(\lambda x_j) + C_3\sin(\lambda x_j) + C_4\cos(\lambda x_j)$$

or in matrix form

$$\{W_j\} = [\textit{real constants}]\{C_j\}.$$

Since each of the above equations is linearly independent of the others, the coefficient matrix of "real constants," derived from the sinh through cos functions, can be inverted[5] to obtain another matrix of strictly real quantities. Since the product of this real square matrix and the vector $\{W_j\}$ must also be real, so, too, are the constants of integration. Thus the original constants of integration B_3 and B_4 are complex conjugates.

[3] The definitions of the hyperbolic sine and hyperbolic cosine functions are $\sinh z = (e^z - e^{-z})/2$ and $\cosh z = (e^z + e^{-z})/2$.

[4] One possible choice for the four such values of λx is $(0, \pi/2, -\pi/2, \pi)$. The resulting second and third equation prove that C_2 is real. Then the first equation proves C_4 is real. Then the fourth equation proves C_1 is real, and hence C_3 is real.

[5] See Endnote (3) for confirmation of this assertion.

Now it is a matter of substituting the above solution for the vibration amplitude function, $W(x)$, into the previously stated boundary conditions. When dealing with differential equation eigenvalue problems, usually it is best to apply the BCs in the systematic way of writing a matrix equation. The four equations $W(0) = W(L) = W''(0) = W''(L) = 0$ respectively yield the four rows of the matrix equation

$$\begin{bmatrix} 0 & 1 & 0 & 1 \\ \sinh\lambda L & \cosh\lambda L & \sin\lambda L & \cos\lambda L \\ 0 & 1 & 0 & -1 \\ \lambda^2\sinh\lambda L & \lambda^2\cosh\lambda L & -\lambda^2\sin\lambda L & -\lambda^2\cos\lambda L \end{bmatrix} \begin{Bmatrix} C_1 \\ C_2 \\ C_3 \\ C_4 \end{Bmatrix} = \begin{Bmatrix} 0 \\ 0 \\ 0 \\ 0 \end{Bmatrix}.$$

Using the same argument used in the discrete mass case that this square matrix must be singular to avoid the otherwise inescapable trivial solution where all the coefficients of integration are zero, conclude that the determinant of the square matrix must be zero. That is, conclude

$$\begin{vmatrix} 0 & 1 & 0 & 1 \\ \sinh\lambda L & \cosh\lambda L & \sin\lambda L & \cos\lambda L \\ 0 & 1 & 0 & -1 \\ \sinh\lambda L & \cosh\lambda L & -\sin\lambda L & -\cos\lambda L \end{vmatrix} = 0,$$

where the nonzero factor of λ^2 was factored out of the last row. This determinant is easily simplified by adding the third row to the first row. This determinant equation can be further simplified by adding or subtracting the fourth row from the second row. Expansion by minors quickly leads to the eigenvalue equation

$$\sinh\lambda L \sin\lambda L = 0.$$

The factor $\sinh\lambda L$ is never zero unless λ is zero. If λ were zero, that would imply that the natural frequency is zero. A zero natural frequency in a stable system constrained against rigid body motion, such as this simply supported beam, identifies the nonvibratory (static) trivial case. That possibility is rejected. Thus the zeros must be found in the second factor, $\sin\lambda L$. Since the zero value for λ has already been rejected, conclude that the necessarily positive quantity $\lambda L = n\pi$. Using the above definition of λ, the solution for the beam natural frequencies is

$$\omega_n = \frac{n^2\pi^2}{L^2}\sqrt{\frac{EI}{\rho A}} \quad \text{for} \quad n = 1, 2, 3, \ldots. \tag{8.4}$$

Just as was the case for the discrete mass model, of course, natural frequencies vary as the square root of the stiffness term divided by the mass term. It is quite unusual to have all the theoretical values of the natural frequencies be integer products of the fundamental natural frequency.

Substitution of the solution for $\lambda = n\pi/L$ into the matrix equation for the constants of integration yields, after some algebra, $C_1 = C_2 = C_4 = 0$, whereas C_3 remains indeterminate for each value of n. Therefore the mode shapes are

$$W_n(x) = C_{3n}\sin\frac{n\pi x}{L} \equiv A_n\sin\frac{n\pi x}{L}.$$

Figure 8.1(b) shows sketches of the first five mode shapes. These sketches show that the number of humps, or number of nodes, in the modal deflection pattern increases with the modal number. At some point the effective beam length, which is the distance between nodes, will decrease to the point that long beam theory,[6] the basis of this analysis, will no longer be valid. Another way of saying the same thing is to say that the distance between nodes will eventually become so short that shearing deflections will become a significant portion of the total deflection, with the result that the above estimates of the natural frequencies and mode shapes will be increasingly in error. The calculated, undamped, natural frequencies would be larger than the "actual" undamped natural frequencies (for vibration in a vacuum) because the above strength of materials differential equation incorporates the previously mentioned constraint against shearing deformations, and any such mathematical constraint artificially stiffens a structure. Simply on the basis that natural frequencies squared are proportional to stiffness divided by mass, overestimating the stiffness overestimates the values of the natural frequencies. Furthermore, this analysis also ignores the mass moment of inertia of each differential length of beam about the y axis. This is a very small effect, but underestimating the effective mass also contributes to overestimating the natural frequencies. Any ambient fluid, such as air, adds effective mass to the beam as it vibrates and thus ignoring that effect also results in these calculated estimates for the natural frequencies to be on the high side.

The complete solution for the beam's free vibration deflections is the linear combination of all the linearly independent modal deflection shapes. Thus the solution for the original partial differential equation is

$$w(x,t) = \sum_{n=1}^{\infty} A_n \sin\frac{n\pi x}{L} \sin(\omega_n t + \phi_n),$$

$$\text{where} \quad \omega_n = \frac{n^2\pi^2}{L^2}\sqrt{\frac{EI}{\rho A}}. \tag{8.5}$$

★

EXAMPLE 8.2 Adapt the complete deflection solution of the previous example, as given by Eq. (8.5), to the initial conditions of a symmetric, parabolic initial deflection with peak amplitude w_0 and zero initial velocity.

COMMENT An initial parabolic deflection pattern for the beam length could result from equal and oppositely directed externally applied moments at each end of the simply supported beam that would result in a constant internal moment at every point along the length of the beam.

SOLUTION Mathematically, the given initial conditions (ICs) can be described as

$$w(x,0) = 4w_0\left(\frac{x}{L}\right)\left(1-\frac{x}{L}\right), \quad \dot{w}(x,0) = 0.$$

[6] The rule of thumb for long beam theory is that the beam length is at least 10 times the beam depth.

For the sake of the second of these ICs, it is necessary to differentiate Eq. (8.5) with respect to time. Then in both the deflection and velocity expressions, setting time equal to zero, and making the above two correspondences, leads to

$$\sum_{n=1}^{\infty} A_n \sin\frac{n\pi x}{L}\sin\phi_n = 4w_0\frac{x}{L}\left(1-\frac{x}{L}\right)$$

$$\sum_{n=1}^{\infty} \omega_n A_n \sin\frac{n\pi x}{L}\cos\phi_n = 0. \tag{8.6}$$

Now it is just a matter of solving these two simultaneous equations for the two sets of unknowns, $\{A_n\}$ and $\{\phi_n\}$. Once their values have been determined, the undamped motion of the beam will be completely specified.

The key to solving for these two sets of unknowns is the weighted orthogonality of the mode shapes. The proof of the orthogonality of the mode shape functions very closely follows that for mode shape vectors. To that end, return to the beam deflection amplitude governing differential equation where the amplitudes A_n and the cross-sectional area A are not to be confused

$$EI\,W''''(x) = \omega^2\rho A\,W(x).$$

Let the eigenpairs ω_n, $\Phi_n(x)$ and ω_m, $\Phi_m(x)$ be any two distinct solutions for the beam natural frequency and the deflection amplitude function $W(x)$, where the ϕ of the phase angle and the Φ of the mode shape are also not to be confused. That is, let

$$EI\,\Phi_n''''(x) = \omega_n^2\rho A\,\Phi_n(x) \quad \text{and} \quad EI\,\Phi_m''''(x) = \omega_m^2\rho A\,\Phi_m(x).$$

For example, in the case of the above beam, $\Phi_n(x) = A_n\sin(n\pi x/L)$. Now multiply the first of these equations by $\Phi_m(x)$ and multiply the second of the above pair of equations by $\Phi_n(x)$. Then integrate both equations over the length of the beam. (This integration of the product of two functions is what is called an "inner product" just as the dot-product of two vectors or a row matrix postmultiplied by a column matrix of the same size is called an inner product.) The result is

$$\int_0^L EI\,\Phi_m(x)\Phi_n''''(x)\,dx = \omega_n^2\int_0^L \rho A\Phi_m(x)\Phi_n(x)\,dx$$

$$\int_0^L EI\,\Phi_n(x)\Phi_m''''(x)\,dx = \omega_m^2\int_0^L \rho A\Phi_n(x)\Phi_m(x)\,dx.$$

If the left-hand sides are equal, the structural system is called *self-adjoint*. The left-hand side integrals can be integrated by parts twice using the boundary conditions that, in this case, $\Phi_n(0) = \Phi_n(L) = 0$ with the first integration and $\Phi_n''(0) = \Phi_n''(L) = 0$ with the second integration. (The BCs will always eliminate the nonintegral "uv"

term of the integration by parts procedure.) The result is the following equal left-hand sides

$$\int_0^L EI\,\Phi_m''(x)\Phi_n''(x)\,dx = \omega_n^2 \int_0^L \rho A \Phi_m(x)\Phi_n(x)\,dx$$

$$\int_0^L EI\,\Phi_n''(x)\Phi_m''(x)\,dx = \omega_m^2 \int_0^L \rho A \Phi_n(x)\Phi_m(x)\,dx.$$

Now subtract the second equation from the first to obtain

$$\left(\omega_n^2 - \omega_m^2\right) \int_0^L \rho A \Phi_m(x)\Phi_n(x)\,dx = 0.$$

Since the two natural frequencies are stipulated to be distinct, then

$$\int_0^L \rho A \Phi_m(x)\Phi_n(x)\,dx = \int_0^L EI\,\Phi_n''(x)\Phi_m''(x)\,dx = 0.$$

These two inner products establish the weighted orthogonality of the modal functions themselves and the weighted orthogonality of their second derivatives. Of course, this result closely parallels the discrete model orthogonality statements, and, as is seen in Endnote (2), the first integral is associated with the system kinetic energy and the system mass matrix, whereas the second integral is associated with the system strain energy[7] and the system stiffness matrix.

Return to the determination of the two sets of unknown constants, $\{A_n\}$ and $\{\phi_n\}$. Multiply both of Eq. (8.6) by the orthogonal mode shape $\sin(k\pi x/L)$, where k is an arbitrary integer. Since the mass density and cross-sectional area are constants in this example, they can be omitted. Now integrate over the length of the beam to obtain the following two equations to be solved for the constants A_k and ϕ_k

$$\sum_{n=1}^{\infty} A_n \sin\phi_n \int_0^L \sin\frac{k\pi x}{L} \sin\frac{n\pi x}{L}\,dx = 4w_0 \int_0^L \frac{x}{L}\left(1 - \frac{x}{L}\right) \sin\frac{k\pi x}{L}\,dx$$

$$\sum_{n=1}^{\infty} \omega_n A_n \cos\phi_n \int_0^L \sin\frac{k\pi x}{L} \sin\frac{n\pi x}{L}\,dx = 0.$$

The integrals involving the product of the sine functions (the mode shapes) are, of course, zero whenever $n \neq k$. When $n = k$, the value of that integral is simply $L/2$. The value of the integral involving the polynomial and the sine function has to be evaluated by either using suitable software, or a table of integrals, or, as a last resort, integration by parts. The result is

$$\text{for } k \text{ even:} \quad A_k \sin\phi_k = 0 \quad \text{and} \quad \omega_k A_k \cos\phi_k = 0$$

$$\text{for } k \text{ odd:} \quad A_k \sin\phi_k = \frac{32 w_0}{k^3\pi^3} \quad \text{and} \quad \omega_k A_k \cos\phi_k = 0.$$

[7] Recall that the bending strain in a beam is $-zw''$, and Φ is a special case of w.

Hence A_k is zero when k is even, and when k is odd, the simultaneous solution is

$$\phi_k = \frac{\pi}{2} \qquad A_k = \frac{32w_0}{k^3\pi^3}.$$

Substituting these results into Eq. (8.1) yields the complete description of the force free, undamped motion.

$$w(x,t) = \frac{32w_0}{\pi^3} \sum_{n=1,3,5,\ldots}^{\infty} \frac{1}{n^3} \sin\frac{n\pi x}{L} \cos\omega_n t,$$

$$\text{where} \quad \omega_n = \frac{n^2\pi^2}{L^2}\sqrt{\frac{EI}{\rho A}}.$$

Since the first mode shape, that is, the sine function with $n = 1$, closely resembles the parabolic initial deflection, it is not surprising that the first term of the sum dominates all other terms in the sum by being 27 times larger than the next largest term. Further, note that each term of the sum has a different time function. This means that the undamped, free vibration motion is not simply harmonic. However, if the motion is damped, as it always is, then the necessary additional exponential factors with negative arguments (i.e., $\exp(-\zeta\omega_n t)$) modify the above result. In each series term, as the index integer n increases, the value of the natural frequency in the argument of the exponential function increases rapidly. One result is that the negative exponential function in each term with a higher index number n goes to zero much quicker than the lower numbered terms. The end result is that soon the damped beam is vibrating harmonically only in the first mode at the first modal frequency. ★

EXAMPLE 8.3 Extend the problem of Example 8.1 to the forced vibration problem where there is an applied load that is an upwardly directed concentrated force of constant magnitude F_0 that is moving from left to right along the length of the beam at a constant velocity v_0. Let the force start at $x = 0$ at time $t = 0$. (This is called the moving load problem.) Let the initial conditions for the beam be zero initial deflection and zero initial velocity.

SOLUTION The first step is to devise an analytical description for the applied load. Since the applied load is a concentrated force acting at the distance $x = v_0 t$, this force can be described using the Dirac delta function as $F_0\delta(x - v_0 t)$. Since the force is acting upward, it is assigned a positive value. Therefore the governing differential equation and the boundary conditions are

$$EIw''''(x,t) + \rho A\ddot{w}(x,t) = F_0\,\delta(x - v_0 t)$$

$$\text{and} \quad w(0,t) = w(L,t) = w''(0,t) = w''(L,t) = 0,$$

where again primes indicate partial derivatives with respect to the spatial variable x and dots indicate partial derivatives with respect to the temporal variable t. The key step toward solving this nonhomogeneous, ordinary differential equation is to now seek a modal series solution to the above equations where each term in the series is the product of an unknown amplitude (a generalized coordinate) and a vibratory mode shape. That is, take as a trial solution the modal expansion

$$w(x,t) = \sum_{n=1}^{N} p_n(t)\Phi_n(x) = \sum_{n=1}^{N} p_n(t)\sin\frac{n\pi x}{L}, \tag{8.7}$$

where N is the number of modes necessary for an accurate solution, and the generalized modal coordinates $p_n(t)$ are again the weighting factors for the mode shapes. This multiplication of each mode shape by a modal coordinate, of course, exactly parallels the discrete mass modal coordinate transformation $\{q\} = [\Phi]\{p\}$. The modal generalized coordinates $p_n(t)$ are also called *distributed coordinates* because they are not associated with geometric points on the structure but rather are factors for functions that distribute deflections over the entire structure. Substituting the above trial solution into the governing differential equation leads to

$$EI\sum_{n=1}^{N}\left(\frac{n\pi}{L}\right)^4 p_n(t)\sin\frac{n\pi x}{L} + \rho A\sum_{n=1}^{N}\ddot{p}_n(t)\sin\frac{n\pi x}{L} = F_0\,\delta(x - v_0 t).$$

To use mode shape orthogonality to uncouple the modal coordinates $p_n(t)$ in the above equation, multiply both sides of the above equality by the mode shape function with arbitrary integer index m and then integrate over the beam length; that is, apply the following operator to both sides of the above equation

$$\frac{2}{L}\int_0^L [\cdots]\sin\frac{m\pi x}{L}dx = \frac{\int_0^L \rho A(x)[\cdots]\Phi_m(x)\,dx}{\int_0^L \rho A(x)\Phi_k(x)\Phi_m(x)\,dx},$$

where the integral in the denominator produces the normalizing factor $L/2$. This step parallels the premultiplication of the discrete mass matrix equation by the transpose of the modal matrix. The result, after dividing by the mass density and cross-sectional area, is the following single, ordinary differential equation in the variable t, where the coefficient of the modal deflection, $p_m(t)$, is, from Example 8.1, the squared mth natural frequency of this simply supported beam

$$\ddot{p}_m(t) + \frac{EI}{\rho A}\left(\frac{m\pi}{L}\right)^4 p_m(t) = \frac{2F_0}{\rho AL}\int_0^L \delta(x - v_0 t)\sin\frac{m\pi x}{L}dx$$

$$\ddot{p}_m(t) + \omega_m^2 p_m(t) = \frac{2F_0}{\rho AL}\sin\frac{m\pi v_0 t}{L}.$$

Since this ordinary differential equation is so simple, the convolution integral solution is set aside to follow earlier procedures. The complementary solution to this equation is familiar as

$$p_m(t) = A_m\sin\omega_m t + B_m\cos\omega_m t,$$

$$\text{where again}\quad \omega_m = (m\pi)^2\sqrt{\frac{EI}{\rho AL^4}}.$$

As long as the speed of the moving force is not equal to $\omega_m L/(m\pi)$, the particular solution for this undamped case is easily obtained by using the method of undetermined coefficients and the single trial function $\sin(m\pi v_0 t/L)$. Then the particular solution result is

$$p_m(t) = \frac{2F_0 L}{m^2\pi^2\left[m^2\pi^2(EI/L^2) - \rho A v_0^2\right]} \sin\frac{m\pi v_0 t}{L}.$$

Thus the complete solution is

$$p_m(t) = A_m \sin\omega_m t + B_m \cos\omega_m t + \frac{2F_0 L}{m^2\pi^2\left[m^2\pi^2\left(\frac{EI}{L^2}\right) - \rho A v_0^2\right]} \sin\frac{m\pi v_0 t}{L}.$$

Note that regardless of the initial conditions and therefore regardless of the values of the constants of integration, A_m and B_m, there is a resonant effect coming from the denominator of the coefficient of the particular solution. The solution for the modal deflection becomes unbounded when the speed of the applied force is

$$v_0 = m\pi\sqrt{\frac{EI}{\rho A L^2}}, \quad \text{where} \quad m = 1, 2, 3, \ldots.$$

This is not normally an immediate problem. For example, for a steel beam with a value for the radius of gyration squared, (I/A), of 60 in.^{2}, a weight density of 0.283 lbs./in.^{3}, and with a length of 50 feet, the resonant velocity for the index value of $m = 1$ is approximately 466 mph or 750 km/hr. Clearly, such speeds are not common for vehicular traffic. As far as the resonant effect is concerned, a vehicle traveling at that speed, 680 ft/sec, traverses the beam in less than one-tenth of a second. This is insufficient time for the resonance effect to be realized.

Application of the given initial conditions, and on this occasion simply utilizing the linear independence of the $\sin n\pi x/L$ functions rather than modal orthogonality, and solving for the constants of integration, A_m and B_m, yields the final form of the complete solution

$$w(x,t) = \sum_{n=1}^{N} \frac{2F_0 L}{n\pi\left[n^2\pi^2(EI/L^2) - \rho A v_0^2\right]} \left[\frac{1}{n\pi}\sin\frac{m\pi v_0 t}{L} - \left(\frac{v_0}{\omega_n L}\right)\sin\omega_n t\right] \sin\frac{n\pi x}{L}.$$

Again, this undamped, and hence approximate, solution is valid only when the single force is still traveling the length of the simply supported beam. After the time the force leaves the beam, L/v_0, the beam undergoes a free vibration with initial conditions that are the deflections and velocities of the beam when the force exited the beam. If another such force enters onto the beam as the first force is exiting, at time L/v_0, then the above analysis can be repeated with a time t being replaced by time $t - L/v_0$.

COMMENT Exercise 8.7 discusses a variation on the above example, where the moving force is because of the weight of a moving mass. ★

EXAMPLE 8.4 Redo the first example problem involving the uniform, simply supported beam, but here let there be an externally applied lateral loading per unit of span length that is constant over the entire length of the beam. With respect to time, let the applied loading have the magnitude zero before time zero, and after time zero, let the applied loading suddenly achieve a constant magnitude f_0 acting downward. Let the initial conditions again be zero initial lateral deflection and zero initial velocity.

COMMENT Exercise 8.3 considers the similar problem where the magnitude of the lateral load per unit length acting on the simply supported beam varies sinusoidally along the span and sinusoidally with respect to time.

SOLUTION This time the beam equation of motion is

$$EIw''''(x,t) + \rho A\ddot{w}(x,t) = -f_0 stp(t).$$

Once again, the solution process begins with writing Eq. (8.7), the modal solution with unknown modal generalized coordinates. This time, because the step function is simply the value 1 after time zero, the result is

$$EI\sum_{n=1}^{N}\left(\frac{n\pi}{L}\right)^4 p_n(t)\sin\frac{n\pi x}{L} + \rho A\sum_{n=1}^{N}\ddot{p}_n(t)\sin\frac{n\pi x}{L} = -f_0.$$

Dividing by the mass density and the cross-sectional area and applying the modal orthogonality yields

$$\ddot{p}_m(t) + \frac{EI}{\rho A}\left(\frac{m\pi}{L}\right)^4 p_m(t) = \frac{2f_0}{\rho AL}\int_0^L \sin\frac{m\pi x}{L}dx = \frac{2f_0}{\rho AL}\left(\frac{L}{m\pi}\right)[1-(-1)^m]$$

$$= \frac{4f_0}{m\pi\rho A} \quad \text{for} \quad m = odd; \qquad \text{and} = 0 \quad \text{for} \quad m = even.$$

Hence p_m is zero for these initial conditions when m is even. The lack of even valued indices just says that the resulting vibration is symmetric about the beam center as it should be. The complete solution for m odd is

$$p_m(t) = A_m \sin\omega_m t + B_m\cos\omega_m t + \frac{4f_0}{m\pi EI}\left(\frac{L}{m\pi}\right)^4.$$

Applying the zero initial conditions and using the linear independence of the $\sin m\pi x/L$ functions leads to the complete solution

$$w(x,t) = \sum_{n=1,odd}^{N}\frac{4f_0}{n\pi EI}\left(\frac{L}{n\pi}\right)^4[1-\cos\omega t]\sin\frac{n\pi x}{L}.$$

This very quickly converging series, like the previous forced vibration solution, shows that the response is composed of many mode shapes. However, because of the factor $1/n^5$, the first mode dominates the deflection. Moreover, if damping were part of the model, then the higher modes would damp out sooner, leaving the vibration wholly dominated by the first mode. ★

EXAMPLE 8.5 The uniform, cantilevered beam shown in Figure 8.2(a) is homogeneous, linearly elastic, and possesses a doubly symmetric cross section. At its tip, at $x = L$, the beam supports a nonstructural mass of magnitude M. That mass M has a mass moment of inertia at its center of mass (at the beam tip) about the y axis

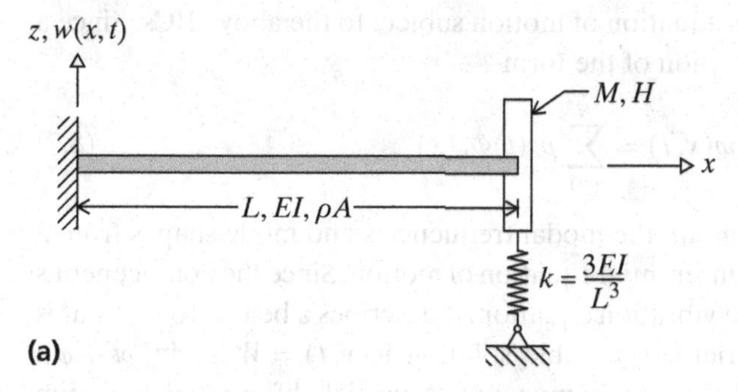

Figure 8.2. (a). Example 8.5. Cantilevered uniform beam with a tip support and a tip rigid mass.

of magnitude H. The discrete mass is also supported by a spring that has a stiffness factor of $3EI/L^3$. The beam is loaded by a spatially constant but time-varying downward acting load per unit length of magnitude $f_z = f_0 \sin \omega_f t$ that starts at time zero. Proceed toward determining the vibratory deflection response of the beam in the plane of the paper.

COMMENT The purpose of this example problem is to present a seemingly simple problem where the necessary calculations are far more tedious than those of the simply supported beam and thus remedy the possibly false impression from the previous examples that continuous mass models provide easily obtained solutions. In fact, the distributed mass approach to this problem becomes sufficiently tedious that it is not worth completing. Hence, the above choice of the words *proceed toward determining* rather than just *determine*.

SOLUTION As in the previous examples, the first step is to adapt the fourth-order beam bending differential equation, with its force per unit length loading term, to this beam vibration. Again, if the reader is not sufficiently familiar with the fourth-order beam bending equation to confidently use that equation in this circumstance, Endnote (1) provides a summary of this equation, and several others, whereas Endnote (2) provides guidance by means of a derivation of both the beam bending equation of motion and its associated BCs. From any source, the partial differential equation of motion is

$$EIw''''(x,t) + \rho A\ddot{w}(x,t) = -f_0 \sin \omega_f t.$$

The four required boundary conditions are the two deflection boundary conditions $w(0,t) = w'(0,t) = 0$ and, say, from a free body diagram of the beam tip and end mass using the inertia loads associated with its positive (upward) deflection and its positive (counterclockwise) rotation

$$EIw'''(L,t) - M\ddot{w}(L,t) - \frac{3EI}{L^3} w(L,t) = 0$$

$$\text{and} \quad EIw''(L,t) + H\ddot{w}'(L,t) = 0.$$

Note that, as is appropriate, the tip mass M and the elastic spring enter the beam equation only through the BCs at $x = L$ in the form of tip shear forces and a tip bending moment.

Now the task is to solve this equation of motion subject to the above BCs. Since a (variables separate) modal solution of the form

$$w(x,t) = \sum_{n=1}^{N} p_n(t)\Phi_n(x) \tag{8.7}$$

is sought, the next step is to obtain the modal frequencies and mode shapes from a solution to the homogeneous differential equation of motion. Since the homogeneous equation has the form of a free vibration equation, it describes a beam motion that is harmonic. Thus write the variable separable solution $w(x,t) = W(x)\sin(\omega t + \phi)$. Then, using that solution, the beam homogeneous partial differential equation becomes the ordinary differential equation

$$W''''(x) - \frac{\rho A \omega^2}{EI} W(x) = W''''(x) - \lambda^4 W(x) = 0$$

because this is the same GDE encountered in Example 8.1, it has the same solution

$$W(x) = C_1 \sinh \lambda x + C_2 \cosh \lambda x + C_3 \sin \lambda x + C_4 \cos \lambda x.$$

The four amplitude beam boundary conditions are $W(0) = W'(0) = 0$,

$$EI\, W'''(L) + M\omega^2 W(L) - \frac{3EI}{L^3} W(L) = 0,$$

$$\text{and} \quad EI\, W''(L) - H\omega^2 W'(L) = 0$$

or, after division by EI and defining the usual eigenvalue and the following non-dimensional mass ratios,

$$\lambda^4 = \frac{\rho A \omega^2}{EI} \qquad \mu_1 = \frac{M}{\rho AL} \qquad \mu_2 = \frac{H}{\rho AL^3}$$

the beam tip BCs are

$$L^3 W'''(L) + [\mu_1(\lambda L)^4 - 3]W(L) = 0 \qquad LW''(L) - \mu_2(\lambda L)^4 W'(L) = 0.$$

Since a solution of the modal frequency equation for arbitrary values of the mass ratios is unnecessary, as well as inconvenient, let the tip mass have, say, the mass values of a rigid beam identical in its mass properties to the cantilevered beam when this second beam's center of mass is fastened to the cantilevered beam at the point $x = L$ on the cantilevered beam so as to form a T-shaped structure. This choice was made so that the mass moment of inertial would be substantial. Then μ_1 has the value 1, whereas μ_2 has the value 1/12. Then the full set of BCs reduce to

$$W(0) = 0 \quad L^3 W''(L) + [(\lambda L)^4 - 3]W(L) = 0.$$

$$W''(0) = 0 \quad LW''(L) - \frac{(\lambda L)^4}{12} W'(L) = 0.$$

Now the above complementary solution $W(x) = C_1 \sinh \lambda x + \cdots$ can be substituted into these BCs to obtain the transcendental determinant equation for the eigenvalue and hence the natural frequencies. The wall BCs produce $C_3 = -C_1$ and $C_4 = -C_2$.

Then, where $z = \lambda L$ (which is nondimensional), after much algebraic hand work, the remaining 2×2 by two determinant equation (see below) reduces to

$$\frac{z^4+9}{z^4-3} - \frac{z^4-15}{z^4-3}\cosh(z)\cos(z) - \frac{z^6-12z^4+36}{z^7-3z^3}\sinh(z)\cos(z)$$
$$-\frac{z^6+z^4-36}{z^7-3z^3}\cosh(z)\sin(z) = 0.$$

This transcendental equation, solvable by, say, Mathematica (or by hand using Newton–Raphson after a lot of work), establishes the point of this example, which again is that continuous models do not generally lead to simple solutions. Indeed solving the above equation for the natural frequencies is but the first step in the solution process for the originally stated problem. Those frequency solutions would have to be substituted into the boundary condition equations to determine the mode shapes, which can be expected to be algebraically cumbersome. Only then can the process of determining the forced response begin.

The situation is a bit simpler if the nonstructural tip mass and the spring at the beam tip are removed from the structure, leaving just a uniform cantilevered beam. In this case the BCs are

$$W(0) = W'(0) = W''(L) = W'''(L) = 0.$$

After again using the result of the first two BCs that $C_3 = -C_1$ and $C_4 = -C_2$, the determinant from the matrix solution for C_1 and C_2 is

$$\begin{vmatrix} (\sinh\lambda L + \sin\lambda L) & (\cosh\lambda L + \cos\lambda L) \\ (\cosh\lambda L + \cos\lambda L) & (\sinh\lambda L - \sin\lambda L) \end{vmatrix} = 0, \tag{8.8}$$

which quickly reduces to the characteristic equation

$$1 + \cosh\lambda L \cos\lambda L = 0.$$

This equation is easily solved because all its infinity of roots alternate between being just past the odd-numbered zeros of $\cos\lambda L$ and then being just before the even-numbered zeros of $\cos\lambda L$, with the differences between the zeros of the cosine function and these roots rapidly diminishing as the root number increases. Thus using Newton–Raphson, the first three hand-determined solutions, which are approximately $(\pi/2)++$, $(3\pi/2)-$, $(5\pi/2)+$, are

$$(\lambda L)_1 = 1.8751, \quad (\lambda L)_2 = 4.6941, \quad (\lambda L)_3 = 7.85476.$$

As before, these three values can be substituted into the matrix precursor of Eq. (8.8), meaning the matrix equation for C_1 and C_2 corresponding to the above determinant equation, which, with $C_3 = -C_1$ and $C_4 = -C_2$, determines the first three mode shapes. For example, using the first and then the second eigenvalues, the solution for the first and second mode shapes are

$$\Phi_1(x) = -0.7341\sinh\left(1.8751\frac{x}{L}\right) + 1.0000\cosh\left(1.8751\frac{x}{L}\right)$$
$$+0.7341\sin\left(1.8751\frac{x}{L}\right) - 1.0000\cos\left(1.8751\frac{x}{L}\right)$$

$$\Phi_2(x) = 1.0000 \sinh\left(4.6941\frac{x}{L}\right) - 0.98187 \cosh\left(4.6941\frac{x}{L}\right)$$
$$- 1.0000 \sin\left(4.6941\frac{x}{L}\right) + 0.98187 \cos\left(4.6941\frac{x}{L}\right).$$

A quick check shows that these functions and their slopes are zero at $x = 0$. For the sake of some consistency in normalizing the mode shapes, the final step in writing the mode shapes is determining their maximum value and dividing by that maximum value so that the resulting maximum value becomes 1.000. In the above cases of the first and second mode shapes, it would mean dividing by 2.000. See Figures 8.2(b) and (c) for graphs of these normalized mode shapes.

The next step after determining sufficient mode shapes is to use Eq. (8.6) to obtain the particular solution. This is to be combined with the complementary (free vibration) solution and the initial conditions to obtain the complete solution. Since this type of lengthy procedure is completed in a following example, this final step is left uncompleted in this example. ★

The previous example problem clearly demonstrates that using continuous models to calculate deflection responses is not an efficient approach when there is more than one structural element. Although this is generally true for single structural elements, hand calculating natural frequencies, particularly if the algebraic characteristic equation is solved on a digital computer, is sometimes feasible and may even be advantageous. The next two example problems (along with the simplicity of the first example) suggest that, in certain limited circumstances such as those of a single element structure with smoothly varying elastic properties, solving the continuous model may be even superior to creating and solving a FEM model. Of course, some experience in dealing with differential equations is necessary to fully take advantage of these opportunities. Hopefully this chapter does provide some help toward reconstructing those differential equation solution skills, if they need bolstering. Toward that end, Endnotes (3) and (4) provide a mathematical overview of the next group of problems that involve second-order differential equations. Endnote (3) also offers another general guarantee of orthogonal mode shapes.

EXAMPLE 8.6 Determine the natural frequencies for an axial vibratory deflections $u(x, t)$ of the long, cantilevered beam with the tapered, square cross section shown in Figure 8.3(a). With the origin of the spanwise coordinate located at the beam tip, the analytical description of the cross-sectional area is $A(x) = A_0(x/L)^2$, where A_0 is the value of the cross-sectional area at the clamped end.

SOLUTION Let $u(x, t) = U(x) \sin \omega_n t$, where $U(x)$ is the beam theory axial deflection amplitude. Then from Endnote (1), where the axial force per unit length is the axial inertia force $-\rho A(x)\, \ddot{u}(x, t)$, the general beam axial free vibration equation is $[EA(x)U'(x)]' + \rho\omega^2 A(x)U(x) = 0$. Define the eigenvalue squared, λ^2, as $\rho\omega^2/E$. After substitution of the cross-sectional area function, the equation for the axial vibratory amplitude distribution $U(x)$ reduces to

$$x^2U''(x) + 2xU'(x) + \lambda^2x^2U(x) = 0. \tag{8.9}$$

```
In[26]:= Plot[0.5(-0.7341 Sinh[1.8751 z]+
         Cosh[1.8751 z] + 0.7341 Sin[1.8751 z]-
         Cos[1.8751 z]),{z,0,1}]
```

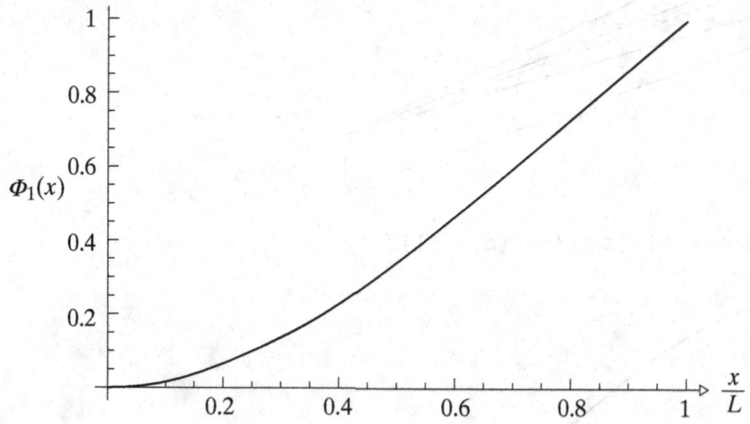

```
Out[26]= -Graphics-
```

(b)

```
In[1]:= a = 4.6941
Out[1]= 4.6941
In[2]:= c = -(Sinh[a] + Sin[a])/(Cosh[a] + Cos[a])
Out[2]= -0.981868
In[4]:= Plot[0.5(Sinh[a z] + c Cosh[a z] - Sin[a z]
       -c Cos[a z]),{z,0,1,}]
```

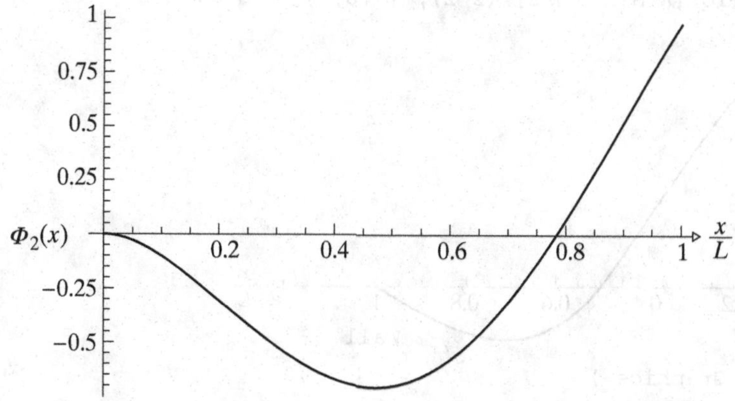

```
Out[4]= -Graphics-
```

(c)

Figure 8.2. (b) First mode shape of a cantilevered, uniform beam. (c) Second mode shape of a uniform, cantilevered beam.

The step following the writing the amplitude equation is dependent on the reader's familiarity with ordinary differential equations. Hopefully the reader has already acquired sufficient familiarity with the subject of differential equations to note that this amplitude differential equation appears as something like a Bessel function

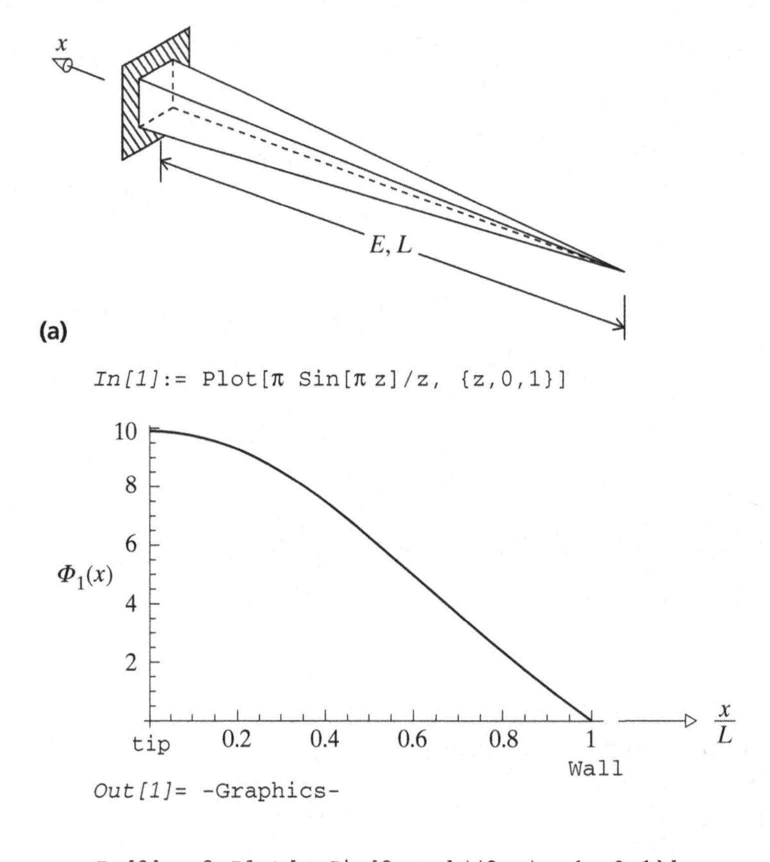

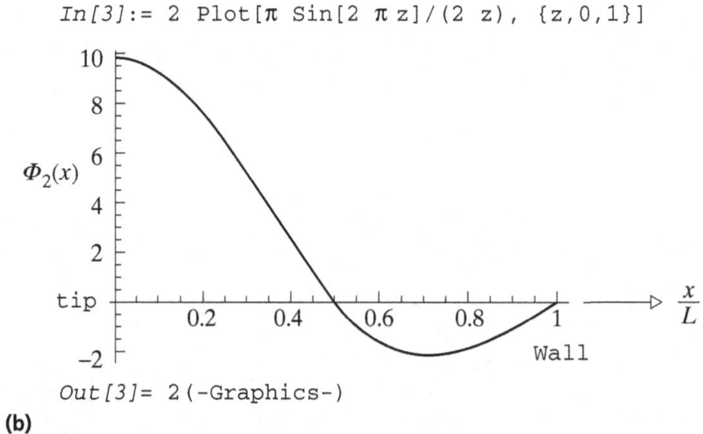

Figure 8.3. (a) Example 8.6: Solid, tapered beam undergoing axial free vibration. (b) First two axial mode shapes of a tapered, cantilevered beam.

equation, but not quite in the standard Bessel equation form.[8] After consultation with a differential equations textbook such as Refs. [8.3,8.4], or Endnote (4), it can be learned that the transformation necessary to put this equation in the standard Bessel

[8] See the corollary at the end of Endnote (4) for a helpful template for solving linear, ordinary differential equations with polynomial coefficients.

function equation form is $U(x) = V(x)/\sqrt{x}$, where, of course, $V(x)$ is a function to be determined. Substituting

$$\frac{dU}{dx} = x^{-\frac{1}{2}}\frac{dV}{dx} - \frac{1}{2}x^{-\frac{3}{2}}V$$

$$\text{and} \quad \frac{d^2U}{dx^2} = x^{-\frac{1}{2}}\frac{d^2V}{dx^2} - x^{-\frac{3}{2}}\frac{dV}{dx} + \frac{3}{4}x^{-\frac{5}{2}}V$$

into the given equation yields

$$x^2V''(x) + xV'(x) + \left(\lambda^2x^2 - \frac{1}{4}\right)V(x) = 0.$$

This is the desired standard form for the Bessel equation of order one-half for which the solution can immediately be written as

$$V(x) = C_1J_{1/2}(\lambda x) + C_2J_{-1/2}(\lambda x)$$

$$\text{or} \quad U(x) = \frac{C_1}{\sqrt{x}}J_{1/2}(\lambda x) + \frac{C_2}{\sqrt{x}}J_{-1/2}(\lambda x).$$

From Ref. [8.5], the half order Bessel functions, the only such Bessel functions, can be expressed in terms of "simple" functions. That is, the solution to Eq. (8.8) can be written in the more convenient form

$$U(x) = \frac{A_1}{x}\sin\lambda x + \frac{A_2}{x}\cos\lambda x. \tag{8.10}$$

The reader is urged to verify this latter form of the solution by directly testing it in Eq. (8.8). The natural frequencies are determined by application of the BCs, which are zero axial force at $x = 0$ and zero deflection at $x = L$. In terms of the deflection function, the second BC is $U(L) = 0$, but the first boundary condition at $x = 0$, $EA(0)U'(0) = 0$, cannot be usefully written because the cross-sectional area at the beam tip, $A(0)$, is zero. That is, this BC just becomes the identity $0 = 0$. All that can be said for the beam tip is that the deflection amplitude, U, must be finite there at $x = 0$. This statement, however, is sufficient to conclude that A_2 must be zero so as to avoid the singularity. The boundary condition at $x = L$ leads to the conclusion that

$$\sin\lambda L = 0 \quad \text{or} \quad \omega_n = \sqrt{\frac{n\pi E}{\rho L}} \quad \text{for} \quad n = 1, 2, 3, \ldots.$$

Notice how the frequencies crowd each other as n increases. The mode shapes are simply

$$\Phi_n(x) = \frac{A_n}{x}\sin\frac{n\pi x}{L},$$

where the undetermined constant of integration A_2 is now a set of modal factors, A_n. Plots of the first two normalized mode shapes are shown in Figure 8.3(b). Notice that the nodal locations for these and all other modes are evenly spaced because of the sine function.

COMMENT Consider the contrast between the above analysis and a finite element model of the same beam. To reasonably approximate the actual variation in the bar

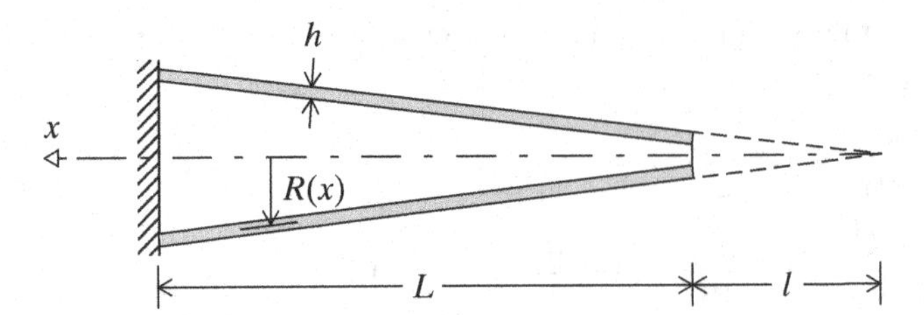

Figure 8.4. Example 8.7: Hollow, tapered beam undergoing torsional free vibration.

cross-sectional properties with the (usually) constant cross-sectional properties of the bar elements, the analyst would have to use at least 8 to 10 bar elements with as many axial deflection DOF. Hence, in addition to not being open to a hand solution, the numerically determined mode shapes and frequencies would not readily reveal the pattern for all mode shapes and frequencies. ★

The previous example problems involved beam bending and beam extension. Therefore, to cover all three of the important types of beam motion, the last of these example problems involves beam twisting. This problem is presented to further contrast: (i) the corresponding finite element model with its many discrete masses and its (usually) uniform elastic elements that can provide only a step-type approximation to a varying stiffness and (ii) the use of the differential equation approach. From the point of view of simplicity, this contrast is not particularly to the advantage of the differential equation approach even in these limited circumstances. However, the differential equation solution can be viewed as a bit more reliable because it better describes the system geometry, and it certainly is more concise for reporting purposes.

As is the case for a finite element analysis, solving this particular differential equation, as a practical matter, also requires the use of sophisticated software. Mathematica has been used here. Even so, this task is challenging even to those who are not new to Bessel functions. The reader can judge if all that can be said in favor of this differential equation approach is that it is an alternate approach, one that can be used to validate a finite element model.

EXAMPLE 8.7 Consider the long, uniformly tapered, truncated conical structure shown in Figure 8.4 as it undergoes free torsional vibrations. The cylinder material is homogeneous, linearly elastic, and isotropic. Treating the structure as a beam[9] rather than as a conical shell, determine the first few natural frequencies for the twisting of this structure.

SOLUTION The general, free vibration, beam twisting equation, without warping restraint, is

$$\frac{\partial}{\partial x}\left[GJ(x)\frac{\partial\phi(x,t)}{\partial x}\right] - \rho I_p(x)\frac{\partial^2\phi(x,t)}{\partial t^2} = 0,$$

[9] Since the cross section is annular, there is no warping as the structure twists. Thus, if the structure is, say, 10 times as long as it is wide, and not too thin, then beam theory will provide a reasonable approximation for calculating torsional modes.

where ϕ is the angular twist of the cross section at a spanwise position x, G is the shear modulus of the beam material, $J(x)$ is the St. Venant torsion constant (a constant for any one cross section) for uniform torsion, and I_p is the area polar moment of inertia about the cross-sectional center. Annular and circular cross sections have the special property that the St. Venant constant for uniform torsion is the same as the polar moment of inertia. In this thin-walled case, $J = I_p = 2\pi R^3 h$. For the sake of simplifying the use of the BCs later in this example, let the length coordinate x originate where the beam radius has the minimum value R_0. Then $R(x) = R_0(x + \ell)/\ell$. Substituting these expressions for J, I_p, and R and canceling common terms leads to

$$\frac{\partial}{\partial x}\left[(x+\ell)^3 \frac{\partial \phi(x,t)}{\partial x}\right] = \frac{\rho}{G}(x+\ell)^3 \frac{\partial^2 \phi(x,t)}{\partial t^2}.$$

Again a solution of the form $\phi(x,t) = \Phi(x)\sin\omega_n t$ is sought, where $\phi(x)$ is the twist amplitude, not to be confused with the Φ_n, symbol for mode shapes, which always has at least one subscript. Substituting for ϕ produces the ordinary spatial differential equation

$$\frac{d}{dx}\left[(x+\ell)^3 \frac{d\Phi(x)}{dx}\right] + \lambda^2 (x+\ell)^3 \Phi(x) = 0,$$

$$\text{where} \quad \lambda^2 = \frac{\rho\omega^2}{G}.$$

Now the task is to solve this GDE subject to the BCs that the torque at $x = 0$ is zero [i.e., $GJ(0)\Phi'(0) = 0$] and the twist at $x = L$ is zero [i.e., $\Phi(L) = 0$]. Simplifying any equation is usually a worthwhile step, so let $x + \ell = \xi$ and note that

$$\frac{d\Phi(x)}{dx} = \frac{d\Phi(\xi)}{d\xi}\frac{d\xi}{dx} = \frac{d\Phi(\xi)}{d\xi}, \quad \text{and so on.}$$

Substituting and carrying out the differentiation on the left-hand side produces

$$\xi^3\Phi''(\xi) + 3\xi^2\Phi'(\xi) + \lambda^2\xi^3\Phi(\xi) = 0$$

$$\text{or} \quad \xi^2\Phi''(\xi) + 3\xi\Phi'(\xi) + \lambda^2\xi^2\Phi(\xi) = 0.$$

If the coefficient 3 were not there, the latter equation would be the Bessel equation of order zero. However, the 3 is there, and the latter equation is something that is just close in appearance to being a Bessel equation. As the material pertaining to Eq. (8.10) in Endnote (4) explains, there is a transformation, this time on the dependent variable Φ, that converts this type of equation into a Bessel equation. That transformation in this case is $\Phi(\xi) = \Psi(\xi)/\xi$. Noting that

$$\Phi'(\xi) = \frac{\Psi'(\xi)}{\xi} - \frac{\Psi(\xi)}{\xi^2} \quad \text{and} \quad \Phi''(\xi) = \frac{\Psi''(\xi)}{\xi} - 2\frac{\Psi'(\xi)}{\xi^2} + 2\frac{\Psi(\xi)}{\xi^3}$$

this transformation results in the Bessel equation of order 1

$$\xi^2\Psi'' + \xi\Psi' + (\lambda^2\xi^2 - 1)\Psi(\xi) = 0.$$

Since the order of the Bessel equation is an integer, in this case 1, the solution must be written in terms of a Bessel function of the second kind of order 1 as well as a Bessel function of the first kind of order 1. That is, the solution is

$$\Psi(\xi) = C_1 J_1(\lambda\xi) + C_2 Y_1(\lambda\xi)$$
$$\text{or} \quad \Phi(x) = \frac{C_1}{x+\ell} J_1(\lambda(x+\ell)) + \frac{C_2}{x+\ell} Y_1(\lambda(x+\ell)).$$

To apply the BC at $x = 0$, it is first necessary to differentiate the solution for $\Phi(x)$. To carry out that differentiation using the formulas offered in Endnote (4), it is best to reintroduce $\xi = x + \ell$ so as to write the above solution as

$$\Phi(\xi) = C_1 \frac{1}{\xi} J_1(\lambda\xi) + C_2 \frac{1}{\xi} Y_1(\lambda\xi).$$

Then using the following derivative formula from Endnote (4) that is applicable to both kinds of Bessel functions

$$\frac{d[x^{-\nu} J_\nu(x)]}{dx} = -x^{-\nu} J_{\nu+1}(x)$$

the result is

$$\frac{d\Phi(\xi)}{d\xi} = C_1 \frac{d}{d\xi}\left[\frac{1}{\xi} J_1(\lambda\xi)\right] + C_2 \frac{d}{d\xi}\left[\frac{1}{\xi} Y_1(\lambda\xi)\right] = -\lambda \frac{C_1}{\xi} J_2(\lambda\xi) - \lambda \frac{C_2}{\xi} Y_2(\lambda\xi).$$

Using the recurrence relations stated in Endnote (4) that

$$J_{\nu+1}(x) = \frac{2\nu}{x} J_\nu(x) - J_{\nu-1}(x) \quad \text{or} \quad J_2(x) = \frac{2}{x} J_1(x) - J_0(x)$$

the derivative of the twist amplitude can be written as

$$\frac{d\Phi(\xi)}{d\xi} = C_1 \left[\frac{\lambda}{\xi} J_0(\lambda\xi) - \frac{2}{\xi^2} J_1(\lambda\xi)\right] + C_2 \left[\frac{\lambda}{\xi} Y_0(\lambda\xi) - \frac{2}{\xi^2} Y_1(\lambda\xi)\right]$$

or, returning to the original spatial coordinate, x,

$$\frac{d\Phi(x)}{dx} = C_1 \left[\frac{\lambda}{x+\ell} J_0(\lambda(x+\ell)) - \frac{2}{(x+\ell)^2} J_1(\lambda(x+\ell))\right]$$
$$+ C_2 \left[\frac{\lambda}{x+\ell} Y_0(\lambda(x+\ell)) - \frac{2}{(x+\ell)^2} Y_1(\lambda(x+\ell))\right].$$

Now all is prepared to write the BCs $GJ(0)\Phi'(0) = 0$ and $\Phi(L) = 0$. Since $GJ(0)$ is not zero, it may be canceled. Therefore, the first of these two BCs is

$$0 = C_1 \left[\frac{\lambda}{\ell} J_0(\lambda\ell) - \frac{2}{\ell^2} J_1(\lambda\ell)\right] + C_2 \left[\frac{\lambda}{\ell} Y_0(\lambda\ell) - \frac{2}{\ell^2} Y_1(\lambda\ell)\right].$$

The second BC is

$$0 = \frac{C_1}{L+\ell} J_1(\lambda(L+\ell)) + \frac{C_2}{L+\ell} Y_1(\lambda(L+\ell)).$$

Casting these equations in matrix form and requiring that the determinant of the square coefficient matrix be zero yields the characteristic equation. To reduce the number of parameters, let $\ell = L$, and for the sake of simplifying the writing of the characteristic equation, let $\eta = \lambda L$. Then the equation to be solved for the eigenvalue, and hence the natural frequency, is

$$\eta J_0(\eta) Y_1(2\eta) - 4 J_1(\eta) Y_1(2\eta) - \eta J_1(2\eta) Y_0(\eta) + 4 J_1(2\eta) Y_1(\eta) = 0.$$

This is an equation that, given a month or more, can be solved by hand using, say, Newton–Raphson, along with tables of Bessel functions values. However, Mathematica does this calculation in the blink of an eye.[10] The first seven roots of the above equation are

2.59803	5.33476	8.27884	11.3131
14.3893	17.4873	20.5979	

which clearly approach a regular spacing as the root number increases. From the local definitions that $\eta = \lambda L$ and $\lambda = \omega\sqrt{\rho/G}$, the first seven natural frequencies are the above values multiplied by the square root of $G/\rho L^2$.

To obtain the mode shapes, return to the zero rotation boundary condition at the clamped end of the beam. Solving that equation for

$$C_2 = -C_1 \frac{J_1(2\lambda L)}{Y_1(2\lambda L)}$$

and substituting into the general solution, the jth mode shape has the form

$$\Phi_j(x) = \frac{C_j}{x+L} J_1(\lambda_j(x+L)) - \frac{C_j}{x+L}\left[\frac{J_1(2\lambda_j L)}{Y_1(2\lambda_j L)}\right] Y_1(\lambda_j(x+L)),$$

where $C_j L$ has replaced the unknowable C_1. To conveniently write the expressions for the various individual mode shapes, let $\zeta = x/L$. Then the first mode shape in terms of the nondimensional ζ coordinate, where $\xi = L(1+\zeta)$, is

$$\Phi_1(\zeta) = \frac{C_1^*}{1+\zeta} J_1\,(2.59803(1+\zeta)) - \frac{C_1^*}{1+\zeta}\left[\frac{J_1(5.19606)}{Y_1(5.19606)}\right] Y_1\,(2.59803(1+\zeta))$$

$$\text{or} \quad \Phi_1(\zeta) = C_1^*\left[\frac{J_1\,(2.59803(1+\zeta))}{1+\zeta} + 4.25851\frac{Y_1\,(2.59803(1+\zeta))}{1+\zeta}\right],$$

where the factor C_1^* is simply the normalizing factor for the first mode shape. In this example, it is not worthwhile calculating the values of these mode shape factors that, for example, cause the maximum mode shape value to be 1.0. However, a good approximation to such values for the first four modes can be determined from the mode shape graphs presented in Figures 8.5(a), (b), (c), and (d). For example, for the first mode shape, the normalizing factor is approximately a little more than 1.5. Recall that having the maximum value equal to 1.0 is merely a convention. The next few mode shapes, without the normalizing factors, are

$$\Phi_2(\zeta) = \frac{J_1\,(5.33476(1+\zeta))}{1+\zeta} + 0.538999\frac{Y_1\,(5.33476(1+\zeta))}{1+\zeta}$$

$$\Phi_3(\zeta) = \frac{J_1\,(8.27884(1+\zeta))}{1+\zeta} + 0.0871505\frac{Y_1\,(8.27884(1+\zeta))}{1+\zeta}$$

$$\Phi_4(\zeta) = \frac{J_1\,(11.3131(1+\zeta))}{1+\zeta} - 0.134591\frac{Y_1\,(11.3131(1+\zeta))}{1+\zeta}$$

$$\Phi_5(\zeta) = \frac{J_1\,(14.3893(1+\zeta))}{1+\zeta} - 0.274723\frac{Y_1\,(14.3893(1+\zeta))}{1+\zeta}.$$

[10] One must be careful that a variety of initial guesses are tried so that no roots are overlooked.

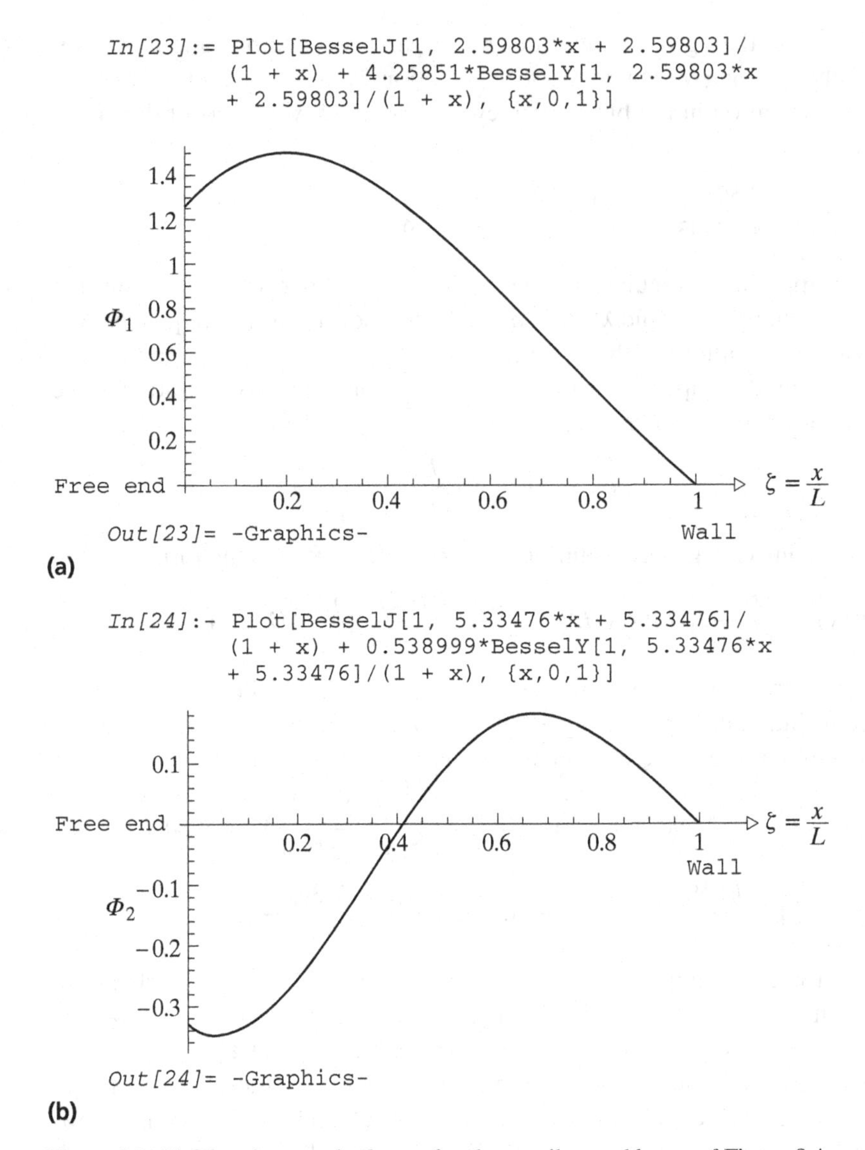

Figure 8.5. The first four mode shapes for the cantilevered beam of Figure 8.4.

Remember that the beam length coordinate, ζ, is zero at the beam tip and 1 at the wall where the beam is clamped. Therefore the mode shapes are oriented backwards from the original beam diagram, and each mode shape twist has a zero value at $\zeta = 1$. Note again that the number of nodes increases by 1 as the modal number increases by 1. Further, note that for any one of the mode shape numbers greater than 2, the distance between nodes is very close to being the same. It also is interesting to note, and perhaps counterintuitive, that, at least for these first four modes, the beam tip modal deflection is never a maximum, although the trend is clearly that it is becoming closer to being a maximum as the modal number increases. ★

```
In[25]:= Plot[BesselJ[1, 8.27884*x + 8.27884]/
          (1 + x) + 0.0871505*BesselY[1, 8.27884*x
          + 8.27884]/(1 + x), {x,0,1}]
```

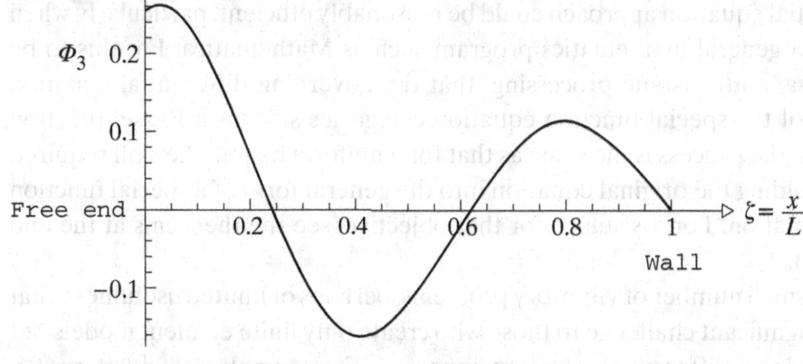


```
Out[25]= -Graphics-
```

(c)

```
In[26]:= Plot[BesselJ[1, 11.3131*x + 11.3131]/
          (1 + x) + 0.134591*BesselY[1, 11.3131*x
          + 11.3131]/(1 + x), {x,0,1}]
```

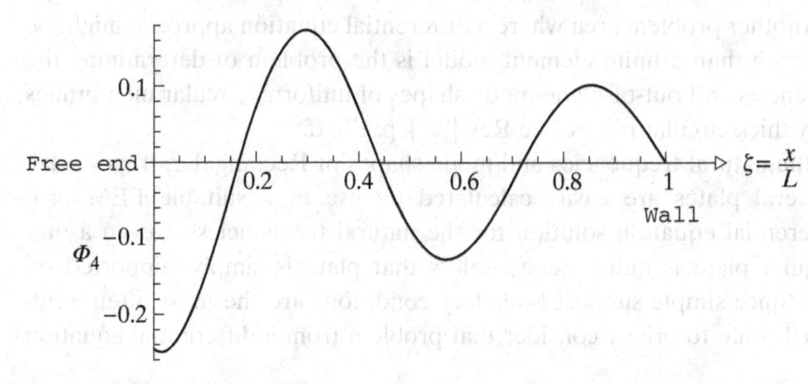


```
Out[26]= -Graphics-
```

(d)

Figure 8.5 (*continued*)

8.4 Conclusion

The example problems offered above dealt with single beams that were separately bending, twisting, and extending as they vibrated. Regardless of the rigid support boundary conditions, if the beam is uniform and without the complications of nonstructural mass or elastic support, then calculating the first few natural frequencies sometimes can be done somewhat efficiently (relative to a computer-based finite element program) using a differential equation approach. However, the mode shapes may involve the sum of two or four terms, making their hand manipulation rather cumbersome. However, this difficulty can be overcome using a general math program or a modern spread sheet program.

When the beam stiffness varies in a fashion that is representable by a low-order polynomial (the usual engineering approximation regardless of whether the structural model mass is modeled as continuous or discrete), it is still possible, just possible, that a differential equation approach could be reasonably efficient, particularly when supported by a general mathematics program such as Mathematica. For this to be so, it is necessary, after some processing, that the governing differential equation falls into one of the special function equation categories such as a Bessel function equation. Then the process is the same as that for a uniform beam. The skill required is mostly in molding the original equation into the general form of a special function differential equation. For assistance for that objective, see the theorems at the end of Endnote (4).

There are a small number of vibratory problems, perhaps of limited usefulness, that might pose a significant challenge to those who create only finite element models but are solvable using a differential equation approach. For example, Ref. [8.4], p. 606, presents the differential equation solution for the natural frequencies and lateral deflection mode shapes of a cable (zero EI bending stiffness) hanging vertically from one end in a gravity field and free at the other end. The tension in the cable because of its weight provides the resistance to bending in this infinitely flexible pendulum. The solution to this problem involves Bessel functions of the first and second kind of order zero with arguments that involve the square root of the length coordinate. Another problem area where a differential equation approach might be a better approach than a finite element model is the problem of determining the natural frequencies and out-of-plane mode shapes of uniform, circular membranes and uniformly thick circular plates. See Ref. [8.5], p. 172 ff.

However, the natural frequencies and mode shapes of Rectangular, trapezoidal, and quadrilateral plates are easily calculated by use of a suitable FEM program. A differential equation solution for the natural frequencies of even a uniform rectangular plate is quite messy, unless that plate is simply supported on all four sides. Since simple support boundary conditions are the most often modeled, it is worthwhile to briefly consider that problem from a differential equation viewpoint.

EXAMPLE 8.8 Calculate the natural frequencies and mode shapes for the uniform, isotropic, rectangular, thin plate shown in Figure 8.6.

SOLUTION Let $w(x, y, t)$ be the plate midplane lateral deflection, positive up as shown in Figure 8.6. Let $f_z(x, y, t)$ be the lateral load per unit area on the midplane of the thin plate, also positive up. Then, from Ref. [8.1], p. 769, the plate bending equation is

$$D\nabla^4 w(x, y, t) = f_z(x, y, t) \quad \text{or} \quad D\left[\frac{\partial^4 w}{\partial x^4} + 2\frac{\partial^4 w}{\partial x^2 \partial y^2} + \frac{\partial^4 w}{\partial y^4}\right] = -\rho h \frac{\partial^2 w}{\partial t^2},$$

where the plate stiffness coefficient $D = Eh^3/[12(1 - \nu^2)]$, h is the constant plate thickness, ν is the Poisson ratio, and ρ is the plate mass density. Along each of the simple supported plate edges, the lateral deflection, and the moment per unit length

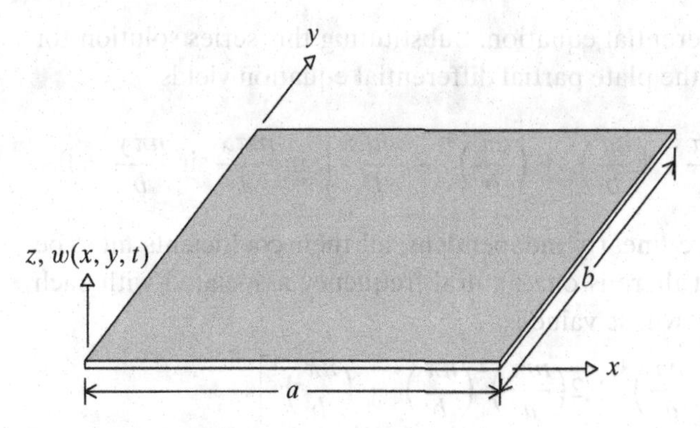

Figure 8.6. Thin, rectangular plate.

in the direction normal to the edge, are zero. In mathematical terms these BCs come down to the following eight equations [8.1]

$$w(0, y, t) = w(a, y, t) = w(x, 0, t) = w(x, b, t) = 0$$
$$w_{,xx}(0, y, t) = w_{,xx}(a, y, t) = w_{,yy}(x, 0, t) = w_{,yy}(x, b, t) = 0,$$

where each coordinate subscript following a comma indicates one partial differentiation with respect to that variable.

The complete solution for the natural frequencies and mode shapes associated with the above partial differential equation and BCs can be obtained, as per usual, by first writing

$$w(x, y, t) = W(x, y) \sin(\omega t + \phi),$$

where $W(x, y)$ is the plate lateral deflection amplitude. Then, with the time variable so removed, the partial differential equation reduces to

$$\left[\frac{\partial^4 W(x, y)}{\partial x^4} + 2\frac{\partial^4 W(x, y)}{\partial x^2 \partial y^2} + \frac{\partial^4 W(x, y)}{\partial y^4}\right] - \rho h \omega^2 W(x, y) = 0$$

and the BCs reduce to

$$W(0, y) = W(a, y) = W(x, 0) = W(x, b) = 0$$
$$W_{,xx}(0, y) = W_{,xx}(a, y) = W_{,yy}(x, 0) = W_{,yy}(x, b) = 0.$$

The next step in the solution process is write a Navier series (a double Fourier series) for the unknown deflection amplitude function. Write

$$W(x, y) = \sum_{m=1}^{\infty}\sum_{n=1}^{\infty} q_{mn} \sin\frac{m\pi x}{a} \sin\frac{n\pi y}{b},$$

where q_{mn} are unknown multiplicative constants that when multiplied by $\sin(\omega t + \phi)$ are the generalized coordinates of the plate's lateral deflection. This expression for the deflection amplitude has the singular virtue that it already satisfies all eight of the deflection amplitude BCs. Thus the only equation that remains to be satisfied is the (governing) partial differential equation. Substituting this series solution for the deflection amplitude into the plate partial differential equation yields

$$\sum_{m=1}^{\infty}\sum_{n=1}^{\infty} q_{mn}\left[\left(\frac{m\pi}{a}\right)^4 + 2\left(\frac{m\pi}{a}\right)^2\left(\frac{n\pi}{b}\right)^2 + \left(\frac{n\pi}{b}\right)^4 - \frac{\rho h\omega^2}{D}\right]\sin\frac{m\pi x}{a}\sin\frac{n\pi y}{b} = 0.$$

Since all the sine functions are linearly independent, all their coefficients must be zero. Therefore conclude that there is one natural frequency associated with each pair of integer indices m and n whose value is

$$\omega_{mn}^2 = \frac{D}{\rho h}\left[\left(\frac{m\pi}{a}\right)^4 + 2\left(\frac{m\pi}{a}\right)^2\left(\frac{n\pi}{b}\right)^2 + \left(\frac{n\pi}{b}\right)^4\right].$$

Of course, the fundamental natural frequency corresponds to $m = n = 1$. The case of a square plate ($a = b$) produces a particularly concise solution for the natural frequencies that is

$$\omega_{mn}^2 = \frac{\pi^4 D}{\rho h a^4}(m^2 + n^2)^2,$$

where there would be many repeated roots.

The rectangular plate mode shapes associated with each of the m, n natural frequencies are simply

$$\Phi_{mn}(x, y) = q_{mn}\sin\frac{m\pi x}{a}\sin\frac{n\pi y}{b},$$

where each q_{mn} turns out to be a modal weighting factor that can be assigned the value 1.0. ★

As illustrated above, the simply supported, thin rectangular plate free vibration solution is easily obtained. However, when the BCs are other than those of simple support at all four edges, the differential equation solution is not nearly as neat and tidy. To illustrate this point, a slight variation on the above problem is briefly considered in the next example problem.

EXAMPLE 8.9 Redo the above example problem, but this time let the boundary conditions at $x = 0$ and $x = a$ be those of a clamped support.

SOLUTION After introducing harmonic motion and thereby eliminating the time variable, the partial differential equation is again

$$\left[\frac{\partial^4 W(x, y)}{\partial x^4} + 2\frac{\partial^4 W(x, y)}{\partial x^2 \partial y^2} + \frac{\partial^4 W(x, y)}{\partial y^4}\right] - \rho h\omega^2 W(x, y) = 0.$$

The BCs are altered by the change from simple support to fixed support at two of the four edges. The BCs are now

$$W(0, y) = W(a, y) = W(x, 0) = W(x, b) = 0$$
$$W_{,x}(0, y) = W_{,x}(a, y) = W_{,yy}(x, 0) = W_{,yy}(x, b) = 0.$$

The form of the solution for the previous example problem, the Navier series solution, does not satisfy these zero slope boundary conditions at $x = 0, a$. However the sine function still works well for the BCs at $y = 0, b$. Therefore the trial solution for this situation is manufactured from sine functions for the y-direction variation of the deflection amplitude and unknown general functions for the x-direction variation of the deflection amplitude. That is, write the Lévy series trial solution

$$W(x, y) = \sum_{n=1}^{\infty} X_n(x) \sin \frac{n\pi y}{b}. \tag{8.11}$$

Again, this trial solution satisfies the BCs at $y = 0, b$, but the unknown functions $X_n(x)$ need to be chosen so as to satisfy the plate bending equation and then be adjusted so that the BCs at $x = 0, a$ are also satisfied. Substitution of the trial solution into the plate amplitude differential equation leads to

$$\sum_{n=1}^{\infty} \left\{ X_n''''(x) - 2\left(\frac{n\pi}{b}\right)^2 X_n''(x) + \left[\left(\frac{n\pi}{b}\right)^4 - \frac{\rho h\omega^2}{D}\right] X_n(x) \right\} \sin \frac{n\pi y}{b} = 0.$$

Again, because of the linear independence of the sine functions, each of the sine function coefficients in the above sum must be zero. Therefore, the next step is to solve the following ordinary differential equation for each value of n

$$X_n''''(x) - 2\left(\frac{n\pi}{b}\right)^2 X_n''(x) + \left[\left(\frac{n\pi}{b}\right)^4 - \frac{\rho h\omega^2}{D}\right] X_n(x) = 0.$$

Similarly, substitution of the above Lévy series trial solution into the BC equations yields the BCs on the function $X_n(x)$, which are

$$X_n(0) = X_n(a) = X_n'(0) = X_n'(a) = 0.$$

At this point it might seem that the remainder of the solution process will be a simple matter. This is not so because the sign of the coefficient of the zeroth order derivative depends on the relative values of the integer index n and the magnitude of the unknown value of the natural frequency. For this point to be clear, it is necessary to examine the details of the solution development.

As ever the solution to the above linear, ordinary differential equation with constant coefficients begins with its trial function solution

$$X_n(x) = A_n e^{r_n x}.$$

Substitution of this trial function into the ordinary differential equation, and canceling the nonzero common factors $A_n \exp(r_n x)$, yields the characteristic equation

$$r_n^4 - 2\left(\frac{n\pi}{b}\right)^2 r_n^2 + \left[\left(\frac{n\pi}{b}\right)^4 - \frac{\rho h\omega^2}{D}\right] = 0.$$

The two roots for this quadratic equation are

$$r_n^2 = \left(\frac{n\pi}{b}\right)^2 \pm \omega\sqrt{\frac{\rho h}{D}}.$$

One more step is necessary because the trial solution requires values of r_n. The two square roots of the above expression will be real or imaginary depending, again, on the values of n and ω. To be clear on when the roots are real, define a noninteger critical value n^* such that

$$\left(\frac{n^*\pi}{b}\right)^2 = \omega\sqrt{\frac{\rho h}{D}} \equiv \omega\gamma.$$

Then for $n < n^*$, the four roots of the characteristic equation are

$$r_n = +\sqrt{\omega\gamma + (n\pi/b)^2} \qquad r_n = -\sqrt{\omega\gamma + (n\pi/b)^2}$$

$$r_n = +i\sqrt{\omega\gamma - (n\pi/b)^2} \qquad r_n = -i\sqrt{\omega\gamma - (n\pi/b)^2}.$$

Then, as before, converting the exponential functions with real arguments to hyperbolic functions, and those with imaginary arguments to circular functions, the total solution for $n < n^*$ is

$$X_n(x) = C_{1n}\sinh r_n x + C_{2n}\cosh r_n x + C_{3n}\sin s_n x + C_{4n}\cos s_n x$$

with the definitions

$$r_n \equiv \sqrt{\omega\gamma + (n\pi/b)^2} \qquad s_n \equiv \sqrt{\omega\gamma - (n\pi/b)^2}$$

For the case $n > n^*$, there are no imaginary roots. The four linearly independent terms that comprise the complete solution can be written as

$$X_n(x) = C_{1n}\sinh r_n x + C_{2n}\cosh r_n x + C_{3n}\sinh s_n x + C_{4n}\cosh s_n x, \tag{8.12}$$

where

$$r_n \equiv \sqrt{(n\pi/b)^2 + \omega\gamma} \qquad s_n \equiv \sqrt{(n\pi/b)^2 - \omega\gamma}$$

Now, for both ranges of the integer index n, it is necessary to use the boundary conditions to determine the relationship between the constants of integration and thereby develop the characteristic equation that determines the natural frequencies. To simplify and condense the algebra associated with these steps, it is convenient to combine the above two solutions into one by defining

$$\text{for } n < n^* \quad \sin(h)s_n x \equiv \sin s_n x \qquad \cos(h)s_n x \equiv \cos s_n x$$
$$\text{for } n > n^* \quad \sin(h)s_n x \equiv \sinh s_n x \qquad \cos(h)s_n x \equiv \cosh s_n x.$$

Then the combined solution is

$$X_n(x) = C_{1n}\sinh r_n x + C_{2n}\cosh r_n x + C_{3n}\sin(h)s_n x + C_{4n}\cos(h)s_n x.$$

Substituting the combined solution into the four boundary conditions leads, after some algebra, to the following characteristic equation

$$1 = \cosh r_n a \cos(h)s_n a - \frac{r_n^2 + s_n^2 \operatorname{sgn}(n^* - n)}{2r_n s_n}\sinh r_n a \sin(h)s_n a,$$

$$\text{where} \quad \operatorname{sgn}(\theta) = -1 \quad \text{if} \quad \theta < 0 \quad \text{and} \quad \operatorname{sgn}(\theta) = +1 \quad \text{if} \quad \theta > 0.$$

For solution purposes, the above characteristic equation is conveniently separated into its two parts. For $n < n^*$

$$1 = \cosh r_n a \cos s_n a - \frac{\left(\dfrac{n\pi}{b}\right)^2}{\sqrt{\dfrac{\omega^2 \rho h}{D} - \left(\dfrac{n\pi}{b}\right)^4}}\sinh r_n a \sin s_n a.$$

For $n > n^*$

$$1 = \cosh r_n a \cosh s_n a - \frac{\left(\frac{n\pi}{b}\right)^2}{\sqrt{\left(\frac{n\pi}{b}\right)^4 - \frac{\omega^2 \rho h}{D}}} \sinh r_n a \sinh s_n a.$$

After specifying the plate geometric and material properties, the above two equations can be solved for the natural frequency that corresponds to each chosen value of n. These are not easy equations to solve accurately. For the sake of a numerical calculation, a 1-in.-thick steel plate ($E = 29{,}000{,}000$ psi, $\nu = 0.3$, $\rho = 490$ lbs./ft^3), 50×50 in. was chosen. In the case of $n < n^*$, selecting $n = 1$, which determines the shape of the mode shape in the y direction as $\sin(\pi y/b)$, the first two solutions for ω, the natural frequency in radians per second, are 35.44 and 84.86. These solutions for ω can now be substituted into Eq. (8.12), and then the constants of integration $C_{1n}, C_{2n}, C_{3n}, C_{4n}$ can be determined up to a multiplicative constant. When $n = 2$ and the mode shape between the simple supports is $\sin(2\pi y/b)$, the Newton–Raphson solution for ω barely converged to a first root of 74.65. Again, this solution procedure for this uniform square plate does not appear to offer the slightest advantage relative to a finite element solution. ★

Solutions for other rectangular plate problems with two opposite edges simply supported proceed in the same fashion as above. If the rectangular plate boundary conditions are not such that there are two opposite edges that are simply supported, or the straight-edge planform geometry is other than a rectangle, then superposition is necessary. See Ref. [8.6]

REFERENCES

8.1 Donaldson, B. K., *Analysis of Aircraft Structures: An Introduction*, McGraw-Hill, New York, 1993, p. 287.

8.2 Miller, F. H., *Partial Differential Equations*, John Wiley & Sons, New York, 1941.

8.3 Kreyszig, E., *Advanced Engineering Mathematics*, 7th ed., John Wiley & Sons, New York, 1993, Chapter 5.

8.4 Wylie, C. R., and L. C. Barrett, *Advanced Engineering Mathematics*, 5th ed., McGraw-Hill, New York, 1982, Chapter 10.

8.5 Meirovitch, L., *Analytical Methods in Vibrations*, Macmillan, New York, 1967.

8.6 Chander, S., B. K. Donaldson, and H. M. Negm, "Improved extended field method numerical results," *J. Sound Vibrat.* vol. **66**, 1, 1979, pp. 39–51.

8.7 Hodgman, C. D. (ed.), *C.R.C. Standard Mathematical Tables*, 12th ed., Chemical Rubber, Cleveland, 1959, p. 319.

8.8 Brush, D. O., and B. O. Almroth, *Buckling of Bars, Plates, and Shells*, McGraw-Hill, New York, 1975.

8.9 Wilf, H. S., *Mathematics for the Physical Sciences*, John Wiley & Sons, New York, 1962.

CHAPTER 8 EXERCISES

8.1 By means of drawing free body diagrams of the beam tip and the tip nonstructural mass that include inertia forces, derive the tip boundary conditions for the cantilevered beam of Example 8.5, which are

$$EIw'''(L,t) - M\ddot{w}(L,t) - \frac{3EI}{L^3}w(L,t) = 0$$

$$\text{and} \quad EIw''(L,t) + H\ddot{w}'(L,t) = 0.$$

Recall that for this doubly symmetric cross section (the area product of inertia is zero) the internal elastic shear force is EIw''' and the internal elastic bending moment is EIw''.

8.2 **(a)** Redo Example 8.4 when the applied downward loading per unit length is spatially uniform but increases linearly with time. That is, let the simply supported, undamped beam be subject to the following force per unit of beam length

$$f_z(x,t) = -\frac{f_0 t}{t_1},$$

where f_0 has units of force per length.

(b) Use the result of part (a) to get the response to a triangular-shaped pulse that peaks at time $t = t_1$ and that is symmetrical about $t = t_1$.

8.3 Determine the complete response of a uniform, simply supported beam of length L and stiffness coefficient EI, originally at rest, when it is subjected to an x, z plane lateral loading per unit of beam length that varies sinusoidally in both space and time. In other words, solve the beam bending equation when the upwardly acting, externally applied loading per unit length for the time period $0 \le t \le t_1$ is

$$f_z(x,t) = f_0 \sin\frac{\pi x}{L} \sin\frac{\pi t}{t_1}$$

and is zero otherwise. Hint: Be sure to include the initial conditions.

8.4 After consulting Endnote (5), find the response to the simply supported beam of Example 8.1 when there is no applied lateral loading over the length of the beam ($f_z(x,t) = 0$), and the initial conditions are zero, but there are following upward foundation movements:

(a) At $x = 0$, the support motion is $w_0 \sin(\pi t/t_0)$ for $0 < t < t_0$ and is zero otherwise.

(b) In addition to the above-discussed support motion at $x = 0$, there is also at $x = L$, the support motion $2w_0 \sin[\pi(t - t_0)/t_0]$ for $t_0 < t < 2t_0$ and is zero otherwise.

8.5 Write the equation of motion for the cantilevered beam of Example 8.5, if the lateral loading along the length of the beam is removed and the beam-spring-discrete mass system is driven by a foundation motion that is:

(a) A vertical wall motion $W_0 \sin \omega_f t$ starting at time zero.

(b) A motion at the base of the spring of magnitude $W_0 \sin \omega_f t$ starting at time zero.

8.6 Consider a uniform, taut wire of length L, mass density ρ, and cross-sectional area A that is clamped at both of its ends. Let the tensile force, N, in the wire be sufficiently great that its value is not significantly affected by the wire's lateral deflections.

(a) Consulting Endnote (1) if necessary, consider the wire's bending stiffness coefficients (EI) to be negligible and thus adapt the beam bending equation to show that the wire's governing differential equation for free vibration along with its two boundary conditions are

$$c^2 \frac{\partial^2 w(x,t)}{\partial x^2} = \frac{\partial^2 w(x,t)}{\partial t^2},$$

$$\text{where} \quad c^2 = \frac{N}{\rho A}$$

and $w(0,t) = w(L,t) = 0$. Note that the quantity c has units of velocity. This equation is called the wave equation.

(b) Just as general solutions for ordinary differential equations involve arbitrary constants of integration, general solutions for partial differential equations involve arbitrary functions [8.7]. There are not very many occasions where general solutions to partial differential equations have proven useful for engineering purposes. This is an exception. Let $F(x - ct)$ and $G(x + ct)$ be any functions of their single, compound variables. Show that the following is the general solution for the freely vibrating wire

$$w(x,t) = F(x - ct) + G(x + ct).$$

In this solution, called the *wave solution*, c is the wave velocity. When the axial coordinate x originates at the left end of the wire and is positive to the right, the first function represents a fixed deflection shape moving to the right. This is so because when x increases in fixed proportion to ct, which means moving to the right along the wire, the argument of F is constant and thus so is the form of the contribution of F to w. Similarly, $G(x + ct)$ represents a wave of fixed form moving to the left. This undamped solution is particularly useful when the initial conditions are just those of a deflection; see Ref. [8.5], Chapter 8.

(c) For the symmetric, triangular, initial deflection $w(x,0) = 2W_0x/L$ for $0 < x < L/2$, and $w(x,0) = 2W_0[1 - (x/L)]$ for $L/2 < x < L$ and for zero initial velocity, write the variables separable trial solution $w(x,t) = W(x)\sin(\omega t + \phi)$ and thereby obtain the complete solution for an undamped wire undergoing a force free vibration.

8.7 Set up the differential equation of motion, the boundary conditions, and the initial conditions for the beam of Example 8.3 where now the beam is subjected to a moving mass of magnitude m_0. Hint: As is done in Endnote (2), rather than trying to adapt the beam bending equation, write the expressions for the kinetic and strain energies.

ENDNOTE (1): THE LONG BEAM AND THIN PLATE DIFFERENTIAL EQUATIONS

Let the x axis of the beam always be directed along the beam length, and let the y and z beam cross-sectional axes always originate at the cross section's centroid. Then, in the absence of temperature changes, for a homogeneous, isotropic,[11] linearly elastic material, the bending portion of the finite deflection, combined beam bending and extension equation for the x, z plane,[12] in terms of the lateral deflection in the z direction at the beam cross section's coincident shear center and centroid, $w(x,t)$, is, from Ref. [8.8], p. 41, and Ref. [8.1], p. 287,

$$\frac{\partial^2}{\partial x^2}\left[EI_{yy}(x)\frac{\partial^2 w(x,t)}{\partial x^2}\right] + \frac{\partial^2}{\partial x^2}\left[EI_{yz}(x)\frac{\partial^2 v(x,t)}{\partial x^2}\right] - N\frac{\partial^2 w(x,t)}{\partial x^2}$$
$$= f_z(x,t) + \frac{\partial m_y(x,t)}{\partial x},$$

where E is Young's modulus, I_{yy} is the area moment of inertia about the y axis, I_{yz} is the area product of inertia, N is the internal axial force (positive when tensile), $f_z(x,t)$ is the external force per unit of beam length (positive in the z direction), and $m_y(x,t)$ is the externally applied bending moment about the positive y axis per unit of beam length. There is, of course, a similar equation for bending in the x, y plane

$$\frac{\partial^2}{\partial x^2}\left[EI_{zz}(x)\frac{\partial^2 v(x,t)}{\partial x^2}\right] + \frac{\partial^2}{\partial x^2}\left[EI_{yz}(x)\frac{\partial^2 w(x,t)}{\partial x^2}\right] - N\frac{\partial^2 v(x,t)}{\partial x^2}$$
$$= f_y(x,t) - \frac{\partial m_z(x,t)}{\partial x}.$$

In the case of a free vibration, the moment per unit length terms $m_y(x,t)$ and $m_z(x,t)$ would result from the beam's mass moment of inertia, per unit of beam length, about the y and z axes, respectively. This term is always quite small and almost always neglected. Another recommended step when there is an axial force N, when the BCs permit, is to rotate the beam cross sectional axes to those of the cross section's principal axes. When this is done, the area product of inertia becomes zero, and the second of the above terms disappears. The internal bending stress resultants at the centroid, which may be needed for the purpose of writing boundary conditions, are as follows when the area product of inertia and temperature change are zero

$$\text{bending moment:}\quad EI_{yy}(x)\frac{\partial^2 w(x,t)}{\partial x^2} = M_y(x,t)$$

$$\text{shear force:}\quad \frac{\partial}{\partial x}\left[EI_{yy}(x)\frac{\partial^2 w(x,t)}{\partial x^2}\right] = V_z(x,t) + m_y(x,t).$$

When there is no boundary constraint against the beam stretching as it bends, the differential equation that governs beam axial extension or axial contraction under

[11] The beam equation has the same form if the beam has a nonhomogeneous cross section or the material is orthotropic [8.1]. If the beam were modeled as nonlinearly elastic, then the form of the beam equations would change.

[12] The cross-sectional coordinates y and z originate at the centroid of the cross section. The internal beam axial and shearing forces act at the centroid. However, the loci of shear centers is the axis about which the beam twists. Requiring the shear center and centroid to be coincident, as is done here, eliminates coupling between bending and twisting and thus simplifies these equations.

the same conditions stated for beam bending above, but for small axial deflections only, is

$$\frac{\partial}{\partial x}\left[EA(x)\frac{\partial u(x,t)}{\partial x}\right] = -f_x(x,t),$$

where $A(x)$ is the cross-sectional area, $u(x,t)$ is the elastic axis deflection in the positive x direction, and $f_x(x,t)$ is the externally applied force, per unit of beam length, in the positive x direction. The axial force is related to the axial deflection by the relationship

$$N(x,t) = EA(x)\frac{\partial u(x,t)}{\partial x}.$$

If the beam is long, as has been required throughout these discussions, the beam twisting equation term involving the beam warping constant can be neglected. Then the small deflection beam twisting equation has the same simple form as the beam extension equation; that is,

$$\frac{\partial}{\partial x}\left[GJ(x)\frac{\partial \phi(x,t)}{\partial x}\right] = -m_t(x,t),$$

where G is the shear modulus, $J(x)$ is the St. Venant constant for uniform torsion (only equal to the polar moment of inertia in the case of a circular or annular cross section), ϕ is the beam angle of twist, and $m_t(x,t)$ is the external twisting moment per unit of beam length. The internal twisting moment is related to the angle of twist by the relationship

$$M_t(x,t) = GJ(x)\frac{\partial \phi(x,t)}{\partial x}.$$

Shearing deflections in long beams are negligible.

There is no value here in writing the finite deflection plate differential equation corresponding to the above combined beam bending and extension equation because during a vibration, the three unknown, in-plane plate forces (corresponding to the beam's single axial force N) are never constant. The presence of such unknown, nonconstant terms greatly complicates the differential equation governing the thin plate's lateral, finite sized deflections. Therefore, for the sake of having a relatively simple differential equation, let the plate's lateral deflections be limited to being small; that is, less than one-quarter or one-third of the plate depth. In this case, with the x and y axes imbedded in the plate's midplane, and in the absence of a temperature change,[13] the bending and twisting of the linearly elastic, isotropic, homogeneous, uniform thin plate is described by

$$D\left[\frac{\partial^4 w(x,y,t)}{\partial x^4} + 2\frac{\partial^4 w(x,y,t)}{\partial x^2 \partial y^2} + \frac{\partial^4 w(x,y,t)}{\partial y^4}\right] = f_z(x,y,t),$$

$$\text{where} \quad D = \frac{Eh^3}{12(1-\nu^2)}$$

[13] Temperature changes usually occur so slowly relative to the fundamental period of the structure that there is essentially no interaction between the temperature change as an input and the deflection or any other dynamic output. However, this is not always the case for some highly flexible spacecraft.

and where $w(x, y, t)$ is the z-direction deflection of the plate midplane, $f_z(x, y, t)$ is external applied load per unit of midplane area acting in the z direction, h is the constant plate thickness, and ν is Poisson's ratio. The rigid support deflection-type boundary conditions for plates, just like those for beams, involve only fixed values of the edge deflections and slopes normal to the edges. However, the force-type boundary conditions for plates are more complicated and not worth discussing here. Those boundary conditions are discussed in such references as [8.1,8.5].

ENDNOTE (2): DERIVATION OF THE BEAM EQUATION OF MOTION USING HAMILTON'S PRINCIPLE

With specific reference to the cantilevered beam of Example 8.5, consider that beam vibrating without twisting in the x, z plane. Consider a segment of that beam of length dx at a typical point x along the length of the beam. The mass of this infinitesimal segment is $\rho A dx$, where ρ is the mass density of the beam material. Since $w(x, t)$ is the beam lateral deflection, the quantity $\partial w(x, t)/\partial t$ is the lateral velocity of this beam segment. Hence the kinetic energy of this infinitesimal beam segment is $\frac{1}{2}\rho A\, dx\, (\partial w/\partial t)^2$. Integrating along the entire length of the beam and adding the kinetic energy of the beam tip rigid mass leads to the total system kinetic energy and its first variation

$$T = \frac{1}{2}\int_0^L \rho A[\dot{w}(x, t)]^2 dx + \frac{1}{2}M\dot{w}(L, t)^2 + \frac{1}{2}H\dot{w}'(L, t)^2$$

$$\delta T = \rho A \int_0^L \dot{w}(x, t)\delta\dot{w}(x, t)dx + M\dot{w}(L, t)\delta\dot{w}(L, t) + H\dot{w}'(L, t)\delta\dot{w}'(L, t).$$

Since the beam's lateral deflections are measured from the SEP, the potential energy V is zero. Since the beam tip rigid body has no strain energy, the elastic strain energy of the system is entirely that of the beam and the spring supporting the beam tip. From Ref. [8.1], p. 520 and p. 640, the beam and spring system strain energy[14] and its first variation are

$$U = \frac{1}{2}\int_0^L EI[w''(x, t)]^2 dx + \frac{1}{2}\left(\frac{3EI}{L^3}\right)w(L, t)^2$$

$$\delta U = EI\int_0^L w''(x, t)\delta w''(x, t)dx + \left(\frac{3EI}{L^3}\right)w(L, t)\delta w(L, t).$$

[14] The elastic strain energy (the recoverable, internally stored work) per unit of beam volume is the triangular area beneath the linear stress–strain curve. Hence, the total strain energy is one-half the integral over both the beam length and the beam cross-sectional area, of the beam bending stress multiplied by the beam strain. In the absence of a temperature change, and for $N = I_{yz} = 0$, this beam stress is the moment (M_y) multiplied by the vertical distance from the centroid z divided by the area moment of inertia (I_{yy}). The strain is the stress divided by Young's modulus (E). Recalling that $\iint z^2 dA = I_{yy}$ and replacing the bending moment by the familiar formula $M_y = EI_{yy}w''(x, t)$ leads to the given result.

Since there is no damping, the expression for the virtual work of the external forces only involves the downward acting applied force per unit length. (Recall that the lateral deflection $w(x, t)$ is positive up.) The virtual work expression is determined from consideration of the total external force $f_0 dx$ acting on the same infinitesimal beam segment, multiplied by the oppositely directed positive virtual deflection of that infinitesimal segment, $\delta w(x, t)$. After integrating over the length of the beam

$$\delta W_{ex} = -\int_0^L f_0(t)\delta w(x, t)\, dx.$$

Hence, in Hamilton's principle, where $\delta W = \delta W_{ex} + \delta W_{in} = \delta W_{ex} - \delta U$,

$$\int_{t_1}^{t_2} \{\delta T + \delta W\} dt = 0 \quad \text{or}$$

$$\int_{t_1}^{t_2} \left\{ \int_0^L \rho A\dot{w}(x, t)\delta\dot{w}(x, t)\, dx + M\dot{w}(L, t)\delta\dot{w}(L, t) + H\dot{w}'(L, t)\delta\dot{w}'(L, t) \right.$$

$$\left. - \frac{3EI}{L^3} w(L, t)\delta w(L, t) - \int_0^L f_0(t)\delta w(x, t)\, dx - \int_0^L EIw''(x, t)\delta w''(x, t) dx \right\} dt = 0.$$

Now the task is to extract the governing differential equation and boundary conditions from this variational statement by converting the above integral expressions into other integral expressions that only have $\delta w(x, t)$ as a factor in the integrands. This step is accomplished by integration by parts, either with respect to the time variable or the spatial variable, as appropriate. Only the terms involving the spring constant and the applied load integral do not require integration by parts because they already have the independent quantity δw as a factor. For the first integral, the kinetic energy integral, interchange the order of integration (which is always possible because both integrations are over finite intervals) and integrate by parts with respect to t. The uv term of this integration involves the factors $\delta w(x, t_2)$ and $\delta w(x, t_1)$, which are set to zero because all variations at the arbitrary time limits are set to zero in the derivation of Hamilton's principle. The same thing happens with the integration by parts over time of the terms involving M and H. Note that the term involving H ends up having $\delta w'(L, t)$ as a factor. Since this term is not included in a spatial integral, it cannot be further integrated, nor is it necessary to do so because at a specific point, a deflection and a bending slope are independent quantities, that is, $\delta w'(L, t)$ is independent of $\delta w(L, t)$. The strain energy integral needs to be integrated by parts with respect to x twice. After applying the requirements from the beam BCs at the clamped end that $\delta w'(0, t) = \delta w(0, t) = 0$, the remaining uv terms, for this case where

$EI = \text{constant}$, are $-EIw''(L,t)\delta w'(L,t)$ and $+EIw'''(L,t)\delta w(L,t)$. Therefore, the total result of the various integrations is

$$\int_{t_1}^{t_2} \Bigg\{ -\int_0^L \rho A\ddot{w}(x,t)\delta w(x,t)\,dx - M\ddot{w}(L,t)\delta w(L,t) - H\ddot{w}'(L,t)\delta w'(L,t)$$

$$-\frac{3EI}{L^3}w(L,t)\delta w(L,t) - \int_0^L f_0(t)\delta w(x,t)\,dx - EIw''(L,t)\delta w'(L,t)$$

$$+ EIw'''(L,t)\delta w(L,t) - \int_0^L EIw''''(x,t)\delta w(x,t)\,dx \Bigg\}\, dt = 0.$$

Regardless of the choice made for the arbitrary time interval limits, t_1 and t_2, the time integral is always zero. This can be true only if the integrand is zero. Therefore, conclude

$$\int_0^L [-\rho A\ddot{w}(x,t) - EIw''''(x,t) - f_0(t)]\,\delta w(x,t)\,dx$$

$$+ \left[-M\ddot{w}(L,t) - \frac{3EI}{L^3}w(L,t) + EIw'''(L,t)\right]\delta w(L,t)$$

$$+ [-H\ddot{w}'(L,t) - EIw''(L,t)]\,\delta w'(L,t) = 0.$$

Whatever the choice for the variation for the lateral deflection over the length of the beam, the zero sum of all terms is unaffected. This can be true only if the factor multiplying the varied deflection function inside the spatial integral is zero. This leaves the terms involving the variation of beam tip deflection and the variation of the bending slope at the beam tip. Since these two variations are independent of each other, and because changes in one while the other is fixed also leave the zero sum unaffected, their coefficients must also be zero. Therefore conclude in the first instance that

$$+EIw''''(x,t) + \rho A\ddot{w}(x,t) = -f_0(t)$$

and in the second instance

$$EIw'''(L,t) - M\ddot{w}(L,t) - \frac{3EI}{L^3}w(L,t) = 0$$

$$\text{and} \quad EIw''(L,t) + H\ddot{w}'(L,t) = 0.$$

The first equation is, of course, the beam differential equation of motion. The second and third equations are the force boundary conditions at $x = L$. (The two other BCs for this fourth-order differential equation are the previously used kinematic BCs of zero deflection and zero bending slope at $x = 0$.) This has been a somewhat long process. The equation of motion and the BCs could have been obtained more quickly and easily in this case from the beam bending equation and free body diagrams of the beam tip that involved the inertia force and inertia moment associated with the tip mass.

ENDNOTE (3): STURM–LIOUVILLE PROBLEMS

From Ref. [8.3], the Sturm–Liouville problem is defined as a differential equation eigenvalue problem on the interval $a \leq x \leq b$ that consists of the following variable coefficient differential equation and associated boundary conditions

$$[r(x)y'(x)]' + [q(x) + \lambda p(x)]y(x) = 0$$
$$\alpha_1 y(a) + \alpha_2 y'(a) = 0 \quad \text{and} \quad \beta_1 y(b) + \beta_2 y'(b) = 0,$$

where the αs and βs are constants. If the real valued functions $r'(x), r(x), q(x)$, and $p(x)$ are continuous, and $p(x) > 0$ on the given interval, then the eigenfunctions $y_m(x)$ and $y_n(x)$, corresponding to the eigenvalues λ_m and λ_n, respectively, are orthogonal on $a \leq x \leq b$ with weighting function $p(x)$. That is,

$$\int_a^b y_m(x)y_n(x)p(x)dx = 0 \quad \text{if} \quad m \neq n.$$

All the second-order differential equations considered in this chapter fit this mold. For example, the relatively simple case of torsional vibrations of a uniform, cantilevered beam has the following differential equation and boundary conditions

$$\phi''(x) + \frac{\omega^2 \rho I_p}{GJ}\phi(x) = 0,$$
$$\text{where} \quad \phi(0) = 0 \quad \text{and} \quad \phi'(L) = 0.$$

If $\omega^2 \rho I_p / GJ = \lambda^2$, then the three coefficient functions of the Sturm–Liouville statement are simply $r(x) = 1, q(x) = 0$, and $p(x) = 1$. Since these functions fulfill the continuity and sign requirements stated above, then it may be concluded that the eigenfunction solutions $\sin(n\pi x/2L)$ for $n = 1, 3, 5, \ldots$ possess the orthogonality relationship

$$\int_0^L \sin\frac{m\pi x}{2L} \sin\frac{n\pi x}{2L}(1)dx = 0.$$

The Bessel equation and its boundary conditions, discussed in the next endnote, is another example of a Sturm–Liouville problem.

ENDNOTE (4): THE BESSEL EQUATION AND ITS SOLUTIONS

The Bessel equation is one of the more important ordinary differential equations with variable coefficients. From, for example, Ref. [8.4], the standard form of the Bessel equation of order ν with a parameter (eigenvalue) λ is

$$x^2\frac{d^2y}{dx^2} + x\frac{dy}{dx} + (\lambda^2 x^2 - \nu^2)y(x) = 0.$$

The transformation $\xi = \lambda x$ leads to the similar Bessel equation form

$$\xi^2\frac{d^2y}{d\xi^2} + \xi\frac{dy}{d\xi} + (\xi^2 - \nu^2)y(\xi) = 0.$$

The infinite series solution to the above equation, defined for all positive values of x, is obtained using the Frobenius method [8.3]. When ν is not an integer, the solution to the above equation can be written as

$$y(\xi) = C_1 J_\nu(\xi) + C_2 J_{-\nu}(\xi),$$

where the first of these two Bessel functions of the first kind, J_ν, is everywhere bounded, but the other Bessel function of the first kind, $J_{-\nu}$, is singular at zero. When ν is an integer, the above solution must be written in the form

$$y(\xi) = C_1 J_\nu(\xi) + C_2 Y_\nu(\xi),$$

where Y_ν is called the Bessel function of the second kind of order ν. It too is singular at the origin. Therefore, from an engineering viewpoint, once a differential equation has been written in one of the above standard forms for a Bessel equation, the solution procedure is just a matter of writing one or the other of the above solutions, depending on whether ν is an integer.

The catch may be writing the engineering equation in the standard Bessel equation form, or the standard form of any of the other occasionally encountered named equations, such as those covered in the last paragraph of this endnote. If, for example, the engineering equation ends up having the form

$$x\frac{d^2y}{dx^2} + \alpha\frac{dy}{dx} + xy(x) = 0,$$

then, unless α has the value 1, this is not the necessary standard form that allows the immediate statement of the solution. (If α is 1, then the solution is written in terms of the zero-order Bessel functions $J_0(x)$ and $Y_0(x)$.) However, the dependent variable transformation

$$y(x) = x^\beta u(x), \qquad \text{where} \quad \beta = \frac{1}{2} - \frac{\alpha}{2} \tag{8.13}$$

will change the above equation into one with the standard Bessel equation form in terms of $u(x)$. Sometimes it is necessary to transform the independent variable as well as the dependent variable. See Ref. [8.4], p. 231. Thus, whenever the engineering equation is something like a Bessel equation, there may be a relatively simple transformation that will convert it to a standard form for either a Bessel function solution, or one of the standard polynomial solutions discussed in the last paragraph.

There are a few additional useful facts concerning Bessel functions worth mentioning. Higher order Bessel functions can be written in terms of lower order Bessel functions of the first kind by means of the recurrence relation, from Ref. [8.3]

$$J_{\nu-1}(x) + J_{\nu+1}(x) = \frac{2\nu}{x} J_\nu(x).$$

In general experience, the above equation means that other than half order Bessel functions, Bessel functions of order zero and one are most often used to express solutions. Graphs of the functions $J_0(x)$, $J_1(x)$, $Y_0(x)$, and $Y_1(x)$ are shown in Figure 8.7. Derivatives of Bessel functions of the first kind can be expressed by use of

```
In[4]:= Plot[{BesselJ[0, x],BesselJ[1, x]},{x,0,22}]
```

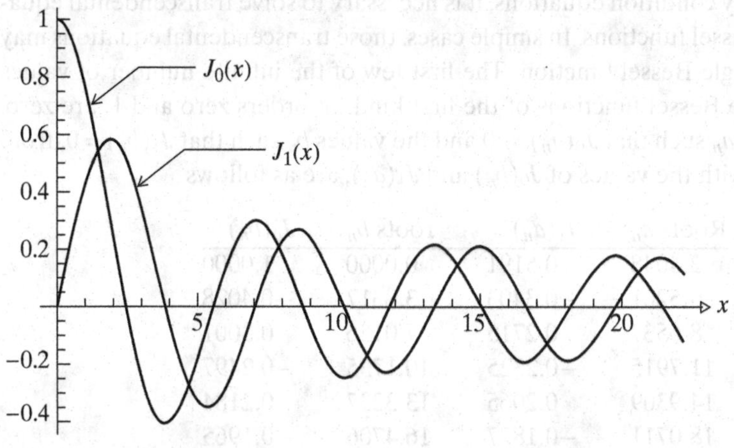


```
Out[4]= -Graphics-
```

(a)

```
In[5]:= Plot[BesselY[0, x],BesselY[1, x]},
        {x,0.5,22.0}]
```

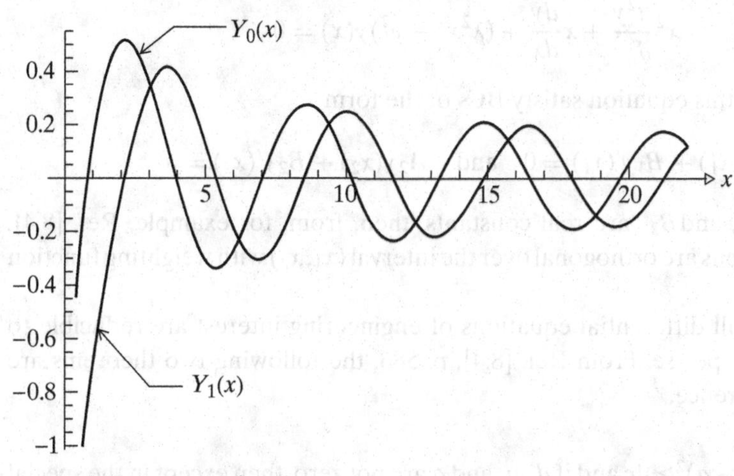


```
Out[5]= -Graphics-
```

(b)

Figure 8.7. Plots of Bessel functions of the first and second kind of orders zero and one.

the following formulas, also from Ref. [8.3]

$$J_{\nu-1}(x) - J_{\nu+1}(x) = 2J'_{\nu}(x)$$

$$\frac{d}{dx}[x^{\nu} J_{\nu}(x)] = x^{\nu} J_{\nu-1}(x) \quad \text{and} \quad \frac{d}{dx}[x^{-\nu} J_{\nu}(x)] = -x^{-\nu} J_{\nu+1}(x).$$

From Ref. [8.4], derivatives of Bessel functions of the second kind follow the same forms as those of Bessel functions of the first kind. These recurrence and differentiation formulas are useful for integration of Bessel functions.

When pursuing the process of determining the values of the system eigenvalues from the boundary condition equations, it is necessary to solve transcendental equations involving Bessel functions. In simple cases, those transcendental equations may involve only a single Bessel function. The first few of the infinite number of values of x for which the Bessel functions of the first kind, of orders zero and 1, are zero, that is, the values a_n such that $J_0(a_n) = 0$ and the values b_n such that $J_1(b_n) = 0$, from Ref. [8.7], along with the values of $J_0(b_n)$ and $J_1(a_n)$, are as follows

Roots a_n	$J_1(a_n)$	roots b_n	$J_0(b_n)$
2.4048	0.5191	0.0000	1.0000
5.5201	−0.3403	3.8317	−0.4028
8.6537	0.2715	7.0156	0.3001
11.7915	−0.2325	10.1735	−0.2497
14.9309	0.2065	13.3237	0.2184
18.0711	−0.1877	16.4706	−0.1965
21.2116	0.1733	19.6159	0.1801

Finally, Bessel functions of the first kind and the same order ν are orthogonal to each other over certain intervals. Return to the Bessel equation with a parameter λ

$$x^2\frac{d^2y}{dx^2} + x\frac{dy}{dx} + (\lambda^2x^2 - \nu^2)y(x) = 0.$$

If the solutions to this equation satisfy BCs of the form

$$A_1y(x_1) + B_1y'(x_1) = 0 \quad \text{and} \quad A_2y(x_2) + B_2y'(x_2) = 0,$$

where A_1, A_2, B_1, andB_2 are real constants, then, from, for example, Ref. [8.4], p. 599, those solutions are orthogonal over the interval (x_1, x_2) with weighting function x.

Of course, not all differential equations of engineering interest are reducible to a Bessel equation, per se. From Ref. [8.4], p. 586, the following two theorems are presented for reference.

THEOREM 1. If $(1 - a)^2 \geq 4c$ and if d, p, and q are not zero, then except in the special cases when it reduces to Euler's equation,[15] the differential equation

$$x^2y''(x) + x(a + 2bx^p)y'(x) + [c + dx^{2q} + b(a + p - 1)x^p + b^2x^{2p}]y(x) = 0.$$

has as a complete solution

$$y(x) = x^\alpha \exp(-\beta x^p)[C_1J_\nu(\lambda x^q) + C_2Y_\nu(\lambda x^q)],$$

$$\text{where} \quad \alpha = \frac{1-a}{2} \quad \beta = \frac{b}{p} \quad \lambda = \frac{\sqrt{|d|}}{q} \quad \nu = \frac{\sqrt{(1-a)^2 - 4c}}{2q}.$$

[15] The Euler equation, also called the Euler-Cauchy equation, has the form $x^n y^{(n)}(x) + a_1x^{n-1}y^{(n-1)}(x) + \cdots + a_{n-1}xy'(x) + a_ny(x) = 0$. It is converted to one with constant coefficients by the transformation of the independent variable: $z = \ln|x|$.

If $d < 0$, then J_ν and Y_ν are to be replaced by I_ν and K_ν, respectively. If ν is not an integer, Y_ν and K_ν can be replaced by $J_{-\nu}$ and $I_{-\nu}$. A corollary of this theorem is as follows.

COROLLARY 1. If $(1-r)^2 \geq 4b$, then except in the special case $a = 0$ $r = 2$, and $s = b = 0$, when it reduces to Euler's equation, the differential equation

$$(x^r y')' + (ax^s + bx^{r-2})y = 0.$$

has the complete solution

$$y(x) = x^\alpha [C_1 J_\nu(\lambda x^\gamma) + C_2 Y_\nu(\lambda x^\gamma)],$$

$$\text{where} \quad \alpha = \frac{1-r}{2} \quad \gamma = \frac{2-r+s}{2} \quad \lambda = \frac{2\sqrt{|a|}}{2-r+s} \quad \nu = \frac{\sqrt{(1-r)^2-4b}}{2-r+s}.$$

If $a < 0$, then J_ν and Y_ν are to be replaced by I_ν and K_ν, respectively. If ν is not an integer, Y_ν and K_ν can be replaced by $J_{-\nu}$ and $I_{-\nu}$ respectively.

For the sake of more completeness, it is necessary to mention some of the many other standard equations. From Ref. [8.3], p. 224, Gauss' hypergeometric equation has the standard form

$$x(1-x)y''(x) + [c - (a+b+1)x]y(x) - aby(x) = 0.$$

The hypergeometric function solution is given at the above reference. Again from Ref. [8.3], p. 209, the Legendre differential equation, whose solutions are Legendre functions, is

$$(1-x^2)y''(x) - 2xy'(x) + n(n+1)y(x) = 0.$$

When n is a nonnegative integer, one of the two independent Legendre functions will be a Legendre polynomial $P_n(x)$. A convenient summary of the differential equations, recurrence formulas, and so on, for Legendre, Tschebycheff, Laguerre, and Hermite polynomials is found in Ref. [8.9], p. 67.

ENDNOTE (5): NONHOMOGENEOUS BOUNDARY CONDITIONS

In Exercise 8.4 the governing differential equation for the bending vibration of a homogeneous, orthotropic, linearly elastic, uniform beam always has the same form, which is

$$EIw''''(x,t) + \rho A\ddot{w}(x,t) = f_z(x,t),$$

where in this exercise the load per unit length is zero. The initial conditions are again

$$w(x,0) = \dot{w}(x,0) = 0.$$

However, instead of four BCs that have the form

$$\mathfrak{F}_1[w(0,t)] = \mathfrak{F}_2[w(0,t)] = \mathfrak{F}_3[w(L,t)] = \mathfrak{F}_4[w(L,t)] = 0,$$

where each of the various $\mathfrak{F}_j$ terms are possibly differential operators, now consider the case where the BCs have the form

$$\mathfrak{F}_1[w(0,t)] = g_1(t) \quad \mathfrak{F}_2[w(0,t)] = g_2(t)$$
$$\mathfrak{F}_3[w(L,t)] = g_3(t) \quad \mathfrak{F}_4[w(L,t)] = g_4(t).$$

This is called the nonhomogeneous BC case. See Ref. [8.5], p. 300. This BC case, of course, corresponds to beam support motions.

The previously discussed procedure of writing the deflection solution as a series expansion in terms of the mode shapes for $w(x,t)$ will not suffice when the BCs are not homogeneous. The insufficiency of the previous procedure is quite evident in this simply supported beam case where the mode shapes are simply sine functions. That is, if a solution were attempted here using the previous modal expansion

$$w(x,t) = \sum_{n=1}^{N} p_n(t) \sin \frac{n\pi x}{L}$$

then setting $x = 0$ always leads to $w(0,t) = 0$, which is a contradiction of the stated BC. Clearly the previously successful procedure needs to be modified in the face of nonhomogeneous BCs. A successful procedure is to write a transformation on the dependent variable that turns this nonhomogeneous boundary value problem into a homogeneous boundary value problem. That transformation has the form

$$w(x,t) = v(x,t) + h_1(x)g_1(t) + h_2(x)g_2(t) + h_3(x)g_3(t) + h_4(x)g_4(t),$$

where the functions $g_j(t)$ are the nonhomogeneous functions of the BCs, and the generally nonunique $h_j(x)$ are chosen so as to render the new BCs for $v(x,t)$ homogeneous. For example, if the beam BCs were, say,

$$w(0,t) = w_0 \frac{t}{t_0} \quad \text{and} \quad w'(0,t) = s_0 \frac{t^2}{t_0^2},$$

then write

$$w(x,t) = v(x,t) + \frac{t}{t_0} h_1(x) + \frac{t^2}{t_0^2} h_2(x). \tag{8.14}$$

Then

$$w(0,t) = v(0,t) + \frac{t}{t_0} h_1(0) + \frac{t^2}{t_0^2} h_2(0)$$

$$w'(0,t) = v'(0,t) + \frac{t}{t_0} h_1'(0) + \frac{t^2}{t_0^2} h_2'(0).$$

To have $v(0,t) = v'(0,t) = 0$, and thus achieve homogeneous BCs for the new dependent variable $v(x,t)$, it is necessary in this case to chose the functions $h_1(x)$ and $h_2(x)$ such that $h_1(0) = w_0$, $h_2(0) = 0$, $h_1{}'(0) = 0$, and $h_2{}'(0) = s_0$. One possible choice is simply $h_1(x) = w_0$ and $h_2(x)' = s_0 x$. Once homogeneous BCs have been achieved, then the previously used modal expansion will work as before. The only other thing to note is that the transformation of Eq. (8.14) will alter the beam governing differential equation by adding additional equivalent loading terms to the right-hand side.

9 Numerical Integration of the Equations of Motion

9.1 Introduction

As discussed in the last part of Chapter 5, digital computer software capabilities have currently reached a point where numerical solutions to very large, linear, structural dynamics problems can be successfully achieved. As an indication of the growth in size of structural models being used in dynamic analyses, note that it is now not uncommon for structural dynamic analyses to employ the same detailed FEM models prepared for the purposes of static stress analyses. As a result of this marked increase in the number of DOF used in analyses, and just as importantly, as part of the clear trend toward automating everything, the integration of the equations of motion is rarely done by any means other than by digital computer-based numerical methods. Although these reasons are sufficient for looking at numerical integration techniques, there are still other important reasons. The foremost of these other reasons is that numerical integration is the only practical approach when material nonlinearities (e.g., plasticity) or geometric nonlinearities are part of the system's mathematical model.

Today, numerical integration is a well-developed field with many textbooks available to provide a comprehensive overview on both simplistic and sophisticated levels. See, for example, Refs. [9.1,9.2]. Therefore it is appropriate for this textbook to provide only a brief introduction to the popular numerical integration techniques that are particularly suitable for the numerical integration of the ordinary differential equations that result from the modal transformation applied to a finite element model or are suitable for the direct integration of the matrix equation of motion in terms of the original generalized coordinates. Of course, either of these equation sets have time as the only independent variable. Throughout the subsequent discussion, the type of dynamic loading associated with these equations of motion is limited to that of a "pulse"; that is, a loading having a duration, T_{load}, that does not exceed a few natural periods.[1] One reason for this focus on a shock loading is that any long-acting,

[1] If the characteristic time duration of the time-dependent smoothly varying load exceeds about six times the first natural period of the structure, then the time-varying load can be effectively treated as a static load, meaning that the kinetic energy of the structure can be ignored.

time-varying load is generally periodic; that is, the result of the superposition of several harmonic loadings. Harmonic loadings are simply treated by the methods of Sections 5.6 and 5.7.

9.2 The Finite Difference Method

One of the oldest of general numerical methods, and still a reasonably accurate method for the integration of second or first order, ordinary differential equations, is the finite difference method (FDM). The FDM gets its name from the fact that this method of numerical integration replaces all the total derivatives of a differential equation, where the derivatives can be viewed as ratios of two differential increments, by the approximating ratio of two finite sized increments. In other words, this FDM approximation may be understood by viewing, for example, a first-order derivative dw/dt as the ratio of two differentials $(dw)/(dt)$ and then approximating this ratio of infinitesimals by the ratio of the two finite increments $(\Delta w)/(\Delta t)$. Furthermore, instead of approximating derivatives at every point in some interval of interest, the approximate derivatives are evaluated only at a representative finite number of discrete points on the time interval of interest. These solution points are separated by various values of Δt, and this finite set of N points is chosen to represent all the points on the time interval of interest. The quantity Δt is called the *time step*.

The replacement of the actual derivatives by approximations can be accomplished in the following way. Since the physical nature of the deflections, velocities, and accelerations of vibrating structures is such that these quantities are always bounded, continuous functions, they can be deemed to be analytical functions. As such, these functions are representable by means of a Taylor's series. Recall that a Taylor's series can be written in either of the following forms

$$f(b) = f(a) + (b-a)f'(a) + {}^1\!/\!_2(b-a)^2 f''(a) + \frac{1}{3!}(b-a)^3 f'''(a) + \frac{1}{4!}(b-a)^4 f''''(a) + \cdots$$

$$\text{or}\quad f(t_j + \Delta t_j) = f(t_j) + \Delta t_j f'(t_j) + \frac{1}{2}\Delta t_j^2 f''(t_j) + \frac{1}{3!}\Delta t_j^3 f'''(t_j) + \frac{1}{4!}\Delta t_j^4 f''''(t_j) + \cdots,$$

where here, only because several derivatives are needed, primes rather than dots are used temporarily to indicate derivatives with respect to time.

IF: (i) $p(t)$ represents one the various modal degrees of freedom whose magnitude is to be determined over several periods after the loading begins; (ii) $\Delta t_j = \Delta t$ represents a small increment in the value of the dependent variable t; (iii) t_j designates the jth selected discrete value of the independent variable, time, after time zero; and (iv) $p(t_j) \equiv p_j$ and $p(t_j{+}\Delta t_j) \equiv p(t_{j+1}) \equiv p_{j+1}$, etc.;

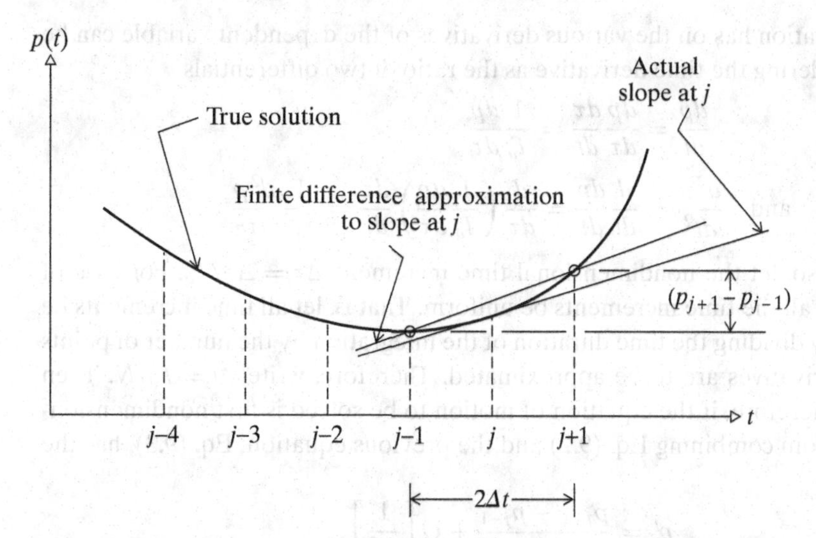

Figure 9.1. Geometric interpretation of the finite difference approximation to the first derivative at time step j.

THEN the Taylor's series for this modal coordinate function can be written for the two time points adjacent to time t_j as follows

$$p_{j+1} = p_j + (\Delta t)p'_j + \frac{1}{2}(\Delta t)^2 p''_j + \frac{1}{(3!)}(\Delta t)^3 p'''_j + \frac{1}{(4!)}(\Delta t)^4 p''''+\cdots$$

$$p_{j-1} = p_j - \Delta t p'_j + \frac{1}{2}(\Delta t)^2 p''_j - \frac{1}{(3!)}(\Delta t)^3 p'''_j + \frac{1}{(4!)}(\Delta t)^4 p''''-\cdots, \quad (9.1)$$

where, in the second of these equations, the increment in time going from t_j to t_{j-1} is negative. If the second equation is subtracted from the first, then the following approximating expression for the first derivative is obtained

$$p'_j = \frac{p_{j+1} - p_{j-1}}{2\Delta t} - \frac{1}{3}(\Delta t)^2 p'''_j - \frac{1}{60}(\Delta t)^4 p'''''_j - \cdots$$

$$= \frac{p_{j+1} - p_{j-1}}{2\Delta t} + O[\Delta t]^2, \quad (9.2)$$

where the last term is read as "terms of order delta tee squared," meaning the largest of these terms is proportional to the small quantity $(\Delta t)^2$. Thus, if the approximation

$$p'_j = \frac{p_{j+1} - p_{j-1}}{2\Delta t} + \text{<one-step approximation error>} \quad (9.3)$$

is used, then the error of the approximation at the jth time point is a collection of terms that are, at worst, proportional to the square of the small time increment. The geometric interpretation of Eq. (9.3) is illustrated in Figure 9.1 for an arbitrary solution function.

To be clear why the largest term of this approximation error, the term involving $(\Delta t)^2$, is a small quantity, it is instructive to temporarily nondimensionalize the modal equation of motion over the total period of time of the integration, T_a. Introduce the nondimensional time variable τ that is such that $0 \leq \tau = t/T_a \leq 1$. The effect that this transformation has on the various derivatives of the dependent variable can be seen by considering the time derivative as the ratio of two differentials

$$\frac{dp}{dt} = \frac{dp}{d\tau}\frac{d\tau}{dt} = \frac{1}{T_a}\frac{dp}{d\tau}$$

and $$\frac{d^2p}{dt^2} = \frac{d}{dt}\frac{dp}{dt} = \frac{d}{d\tau}\left(\frac{1}{T_a}\frac{dp}{d\tau}\right)\frac{d\tau}{dt} = \frac{1}{T_a{}^2}\frac{d^2p}{d\tau^2},$$

and so on. Also, let the nondimensional time increment $\Delta\tau_j = \Delta t_j/T_a$. For ease of discussion, let all the time increments be uniform. That is, let all time increments be determined by dividing the time duration of the integration by the number of points where the derivatives are to be approximated. Therefore, write $\Delta t = T_a/N$. Then $\Delta\tau = 1/N$. Therefore, if the equation of motion to be solved is first nondimensionalized, then from combining Eq. (9.2) and the previous equation, Eq. (9.3), has the form

$$p'_j = \frac{p_{j+1} - p_{j-1}}{2\Delta\tau} + O\left[\frac{1}{N^2}\right],$$

where the above derivative is now with respect to τ. It is immediately evident that since $\Delta\tau = 1/N$ (i) the smaller the quantity $\Delta\tau$ is (the larger the number N is), the smaller is the error for this approximation of the first derivative at that typical time point, and (ii) terms in the original Taylor's series involving Δt raised to exponents greater than 2 are of lesser importance than those just raised to the second power. Thus, the largest part of the error included in the quantity indicated by the $O[\Delta t^2]$ symbol, read as "of order delta tee squared," is the term associated with Δt^2 itself. As an aside note that although the error associated with the Eq. (9.2) approximation of the first derivative at any time point is proportional to the inverse of N^2, the total error is proportional to $1/N$. This is so because there would be N calculations at the N time points spaced over the time interval T_a, each involving an error of $1/N^2$, which adds up to a total error of $1/N$. Thus N must be a large number, and the total error would decrease slowly as N is increased.

As is soon illustrated, individual errors of order Δt^2 for derivatives are sufficient for routine numerical integrations. However, it is possible to obtain a still more accurate approximation for the first and higher derivatives at the expense of a greater number of computer calculations. Such a more accurate approximation for the first derivative can be obtained by using two more adjacent time points in the Taylor series

$$p_{j+2} = p_j + 2\Delta t\, p'_j + \frac{4}{2}(\Delta t)^2 p''_j + \frac{8}{(3!)}(\Delta t)^3 p'''_j + \frac{16}{(4!)}(\Delta t)^4 p''''_j + \cdots$$

$$p_{j-2} = p_j - 2\Delta t\, p'_j + \frac{4}{2}(\Delta t)^2 p''_j - \frac{8}{(3!)}(\Delta t)^3 p'''_j + \frac{16}{(4!)}(\Delta t)^4 p''''_j - \cdots. \tag{9.4}$$

To isolate the first-order derivative by first eliminating the second-order derivative, subtract the second of Eqs. (9.4) from the first. Listing that result with the similar subtraction in Eqs. (9.1) yields

$$p_{j+2} - p_{j-2} = 4\Delta t\, p'_j + \frac{8}{3}(\Delta t)^3 p'''_j + O[\Delta t^5]$$

$$p_{j+1} - p_{j-1} = 2\Delta t\, p'_j + \frac{1}{3}(\Delta t)^3 p'''_j + O[\Delta t^5].$$

Now multiply the second of these equations by 8 and subtract the first to further isolate the first derivative. After solving for the first derivative, the result is

$$p'_j = \frac{p_{j-2} - 8p_{j-1} + 8p_{j+1} - p_{j+2}}{12\Delta t} + O[\Delta t^4].$$

This fourth-order approximation is clearly more accurate than the second-order approximation because the error involves a much smaller factor. However, it involves twice as many sums. Again, the second-order approximation is satisfactory unless the dependent variable is changing rapidly.

A second-order approximation for the second derivative can also be obtained from Eqs. (9.1) simply by adding those two equations. After solving for the second derivative the result is

$$p''_j = \frac{p_{j-1} - 2p_j + p_{j+1}}{\Delta t^2} + O[\Delta t^2]. \tag{9.5}$$

Although other higher order finite difference total and partial derivatives can be approximated in the same fashion as above, they are not of present concern. As a final comment, note that the above approximations for derivatives use equal numbers of time points on each side of the time point under consideration. Such approximations are called *central differences*, as opposed to *forward* and *backward differences*, which respectively employ only time points ahead or behind the time point of interest. Since central differences are generally more accurate, the forward and backward difference approximations are generally used only with spatially independent variables to facilitate the expression of boundary conditions.

EXAMPLE 9.1 In Example 9.4, an undamped, one-DOF, linear vibratory system was subjected to a sinusoidal base motion input $Y_0 \sin(\pi t/t_0)$ starting at time zero. With ω_1 being the system natural frequency, the system differential equation of motion is

$$\ddot{q}(t) + \omega_1^2 q(t) = \omega_1^2 Y_0 \sin \frac{\pi t}{t_0}. \tag{9.6}$$

For the case of zero initial conditions, the deflection response solution was determined to be

$$q(t) = \frac{\omega_1 Y_0}{(\pi/t_0)^2 - \omega_1^2} \left[(\pi/t_0) \sin \omega_1 t - \omega_1 \sin(\pi t/t_0)\right].$$

A numerical integration of the equation of motion does not produce a solution such as the above, which is an analytical expression in terms of arbitrary values of the system parameters. Any numerical calculation requires specific choices for most, if not all, system parameters. Thus, for the purposes of the following numerical calculations, choose t_0 to be $\pi/2$ seconds and ω_1 to be 1 rad/sec. Therefore, the nondimensional form of the above solution reduces to

$$\frac{q(t)}{Y_0} \equiv \overline{q}(t) = {}^1\!/\!_3 \left[2 \sin t - \sin 2t\right].$$

This will be referred to as the "exact" solution (in the numerical sense) for this problem.

Now, for the purpose of comparison to the above analytical solution for this vibratory system, examine the finite difference approach to numerically integrating the system differential equation, Eq. (9.6). That equation, in terms of the selected values for the system parameters, is, where again $\overline{q} = q(t)/Y_0$

$$\overline{q}''(t) + \overline{q}(t) = \sin 2t.$$

Let the numerical integration be carried out using the moderate-sized time step $\Delta t = 0.2$ sec. A selected value for Y_0 is not necessary for this calculation. However, because such a selection will be necessary for the next calculation, also choose Y_0 to have the value 1.0, which makes $\overline{q}$ the same as q.

SOLUTION Using the finite difference approximation for the second derivative of $q(t)$, the original differential equation of motion, Eq. (9.1), is, after some algebra, converted into the following finite difference equation for a typical time point t_j

$$q_{j-1} - [2.0 - (0.2)^2]q_j + q_{j+1} = (0.2)^2 \sin(0.4j)$$

$$\text{or} \quad q_{j-1} - 1.96q_j + q_{j+1} = 0.04 \sin(0.4j)$$

$$\text{or} \quad q_{j+1} = 0.04 \sin(0.4j) + 1.96q_j - 1.0q_{j-1}, \tag{9.6a}$$

where $t_j = j\Delta t = 0.2j$. The first initial condition of zero deflection at time equals zero (i.e., at $j = 0$) leads immediately to $q_0 = 0$. The second initial condition of zero velocity at time zero, after using the finite difference expression for the first derivative, Eq. (9.3), leads to $q_{-1} = q_1$. Application of $q_{-1} = q_1$ to the above recurrence relationship, Eq. (9.6a), at the time point $j = 0$ leads immediately to $q_1 = 0$. Substitution of these solutions q_0 and q_1 to Eq. (9.6a) at the time point $j = 1$ leads to $q_2 = 0.04(0.3894183) = 0.0155767$. Continuing step by step allows the calculation of the value of q_{j+1} at the jth time step according to the above recursion formula. The first few results using a hand calculator are as follows

j	q_j	j	q_j
$j = 0$	$q_j = 0.0000000$	$j = 6$	$q_j = 0.3901883$
1	0.0000000	7	0.5409705
2	0.0155767	8	0.6835134
3	0.0592246	9	0.7963808
4	0.1377851	10	0.8596922
5	0.2508171	11	0.8583438

A plot of the Mathematica numerical results superimposed on the exact solution is shown in Figure 9.2. From Figure 9.2, it is evident that, with the chosen step size, the numerical solution is a very close fit to the exact solution, especially at the peak responses, which are of greatest interest. The slow increase in numerical error is evident close to the baseline where the deviation from the true solution continually increases as n increases from the low 30s to the low 60s to the low 90s. Also keep in mind that the input load is expected to act only for several periods, after which it goes to zero. Therefore, this slow buildup of error is not particularly consequential in this case. ★

One of the advantages of the FDM, like other most other numerical methods, is that it is a straightforward matter to apply the FDM to a nonlinear equation of motion. Recall that numerical integration of one form or another is usually the only practical procedure for obtaining solutions to nonlinear differential equations. As an

```
In[14]:= Show[PlotExactSolution,PlotFDMSolution]
```

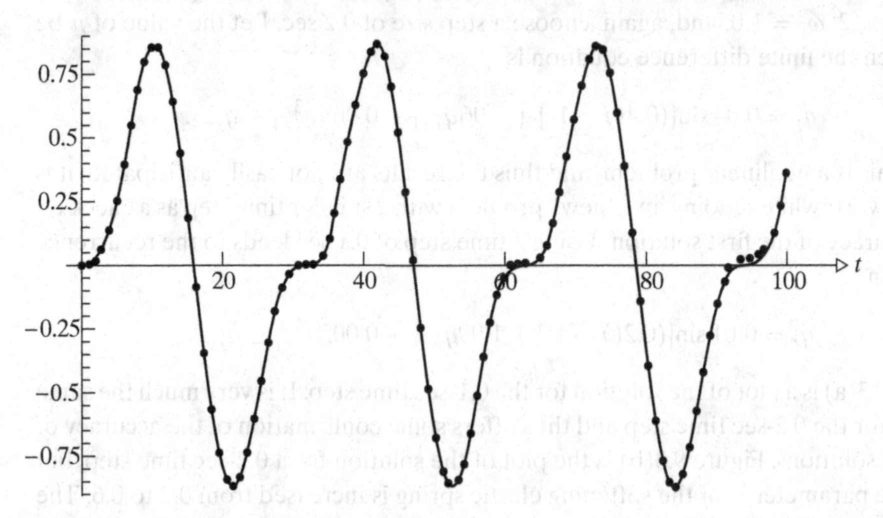

Figure 9.2. Finite difference numerical solution for an oscillator subjected to base excitation superimposed on the analytical solution.

example, consider the following undamped single-DOF equation of motion that is a form of *Duffing's equation* [9.3]

$$\ddot{q}(t) + \omega_1^2 q(t) + \mu q^3(t) = \frac{Q(t)}{M},$$

where μ is a small, positive number. Comparing this equation with its cubic deflection term to the standard m, k, Q single-DOF linear differential equation shows that the above second and third term together represent an elastic spring that is "hardening"; that is, the restoring spring force has a positive curvature on the spring force versus spring deflection curve that in turn means a larger increment in the applied force is required for an additional increment of spring deflection than was required for the previous increment in spring deflection. If there were a minus sign before the cubic term, the force–deflection curve would turn down from the original linear tangent, and the elastic spring would be labeled *softening*. The application of the FDM to this *quasilinear equation*[2] proceeds in exactly the same way as with a linear equation. In this case, the resulting recursive, algebraic equation is

$$q_{j+1} = \Delta t^2 \frac{Q_j}{M} + q_j \left[2 - \omega_1^2 \Delta t^2\right] - \mu \Delta t^2 q_j^3 - q_{j-1}, \tag{9.6b}$$

where again the initial deflection and the initial velocity provide the means to start this recursive relationship.

EXAMPLE 9.2 Consider an (undamped) Duffing's equation with a softening elastic spring; that is, where the sign before the transposed term containing the parameter μ in Eq. (9.6b) is changed from a minus sign to a plus sign. Use the finite difference method to numerically integrate this equation to obtain the time history of the motion, $q(t)$. As in Example 9.1, let the externally applied generalized force result

[2] A quasilinear equation is one where the nonlinearities do not involve the highest derivative.

from a harmonic base motion described as $Y_0 \sin(\pi t/t_0)$. As in Example 9.1, choose t_0 to be $\pi/2$, $\omega_1 = 1.0$, and, again, choose a step size of 0.2 sec. Let the value of μ be 0.2. Then the finite difference equation is

$$q_j = 0.04 \sin[(0.4(j-1)] + 1.96 q_{j-1} + 0.008 q_{j-1}^3 - q_{j-2}.$$

Since this is a nonlinear problem, and thus the results are not easily anticipated, it is always worthwhile redoing any "new" problem with a smaller time step as a check on the accuracy of the first solution. Using a time step of 0.1 sec leads to the recurrence equation

$$q_j = 0.01 \sin[(0.2(j-1)] + 1.99 q_{j-1} + 0.002 q_{j-1}^3 - q_{j-2}.$$

Figure 9.3(a) is a plot of the solution for the 0.1-sec time step. It is very much the same as that for the 0.2-sec time step and thus offers some confirmation of the accuracy of the two solutions. Figure 9.3(b) is the plot of the solution for a 0.2-sec time step, but with the parameter μ of the softening elastic spring is increased from 0.2 to 0.6. The shape of the response with the still softer spring as shown in Figure 9.3(b) is, after the first peak, quite different from that of Figure 9.3(a). The only difference between to the two plots that is easily anticipated is that the peak magnitude responses with the softer spring, μ equal to 0.6, are slightly larger than those with μ equal to 0.2. Even though the cubic term coefficient was tripled, it seems that the amplitudes are only slightly larger because the quantity being cubed, q_j, is mostly less than 1.0. ★

In addition to applying the finite difference method to individual modal equations, the FDM also can be applied directly to the original $[m]$, $[c]$, $[k]$, $\{Q\}$ matrix equation for the system under study. This choice, for example, might be prompted by stiffness nonlinearities or a modal damping matrix that had significant off-diagonal terms. In the case of stiffness nonlinearities, the entries of the stiffness matrix would not be constants but would depend on the values of the generalized coordinates, just as the Duffing equation 1×1 stiffness matrix entry, divided by the mass term, is $(\omega_1^2 \pm \mu q^2)$. In the discussion that follows, the stiffness matrix entries are constants.

If, as before, $\{q(t)\}$ is the N by one vector of unknown system generalized coordinates, then each time derivative of the individual DOF that form the entries of this vector can be approximated by use of the above order Δt^2 FDM equations. That is, for the jth time step

$$\{q_j'\} = -\frac{1}{2\Delta t}\{q_{j-1}\} + \frac{1}{2\Delta t}\{q_{j+1}\}$$

$$\text{and} \quad \{q_j''\} = \frac{1}{\Delta t^2}\{q_{j-1}\} - \frac{2}{\Delta t^2}\{q_j\} + \frac{1}{\Delta t^2}\{q_{j+1}\}.$$

Substitution into the original $[m]$, $[c]$, $[k]$ differential equation and solving for the system DOF at the following $(j+1)$th time step leads to

$$\left(\frac{1}{\Delta t^2}[m] + \frac{1}{2\Delta t}[c]\right)\{q_{j+1}\}$$
$$= \{Q_j\} + \left(\frac{2}{\Delta t^2}[m] - [k]\right)\{q_j\} - \left(\frac{1}{\Delta t^2}[m] - \frac{1}{2\Delta t}[c]\right)\{q_{j-1}\}.$$

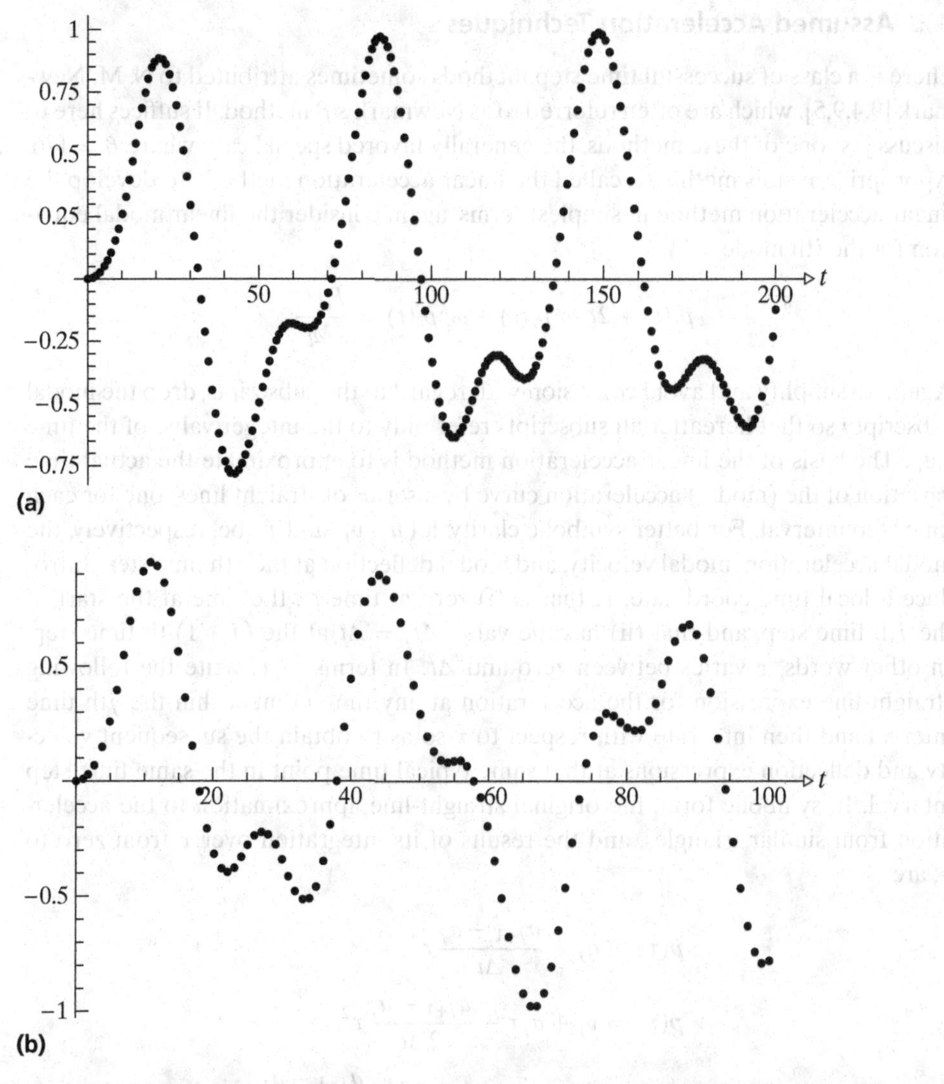

Figure 9.3. (a) FDM numerical solution for an undamped single-DOF system with a softening spring (Duffing's equation with $\mu = 0.2$) for a 0.1-sec time step. (b) FDM solution to Duffing's equation with $\mu = 0.6$ (a still softer spring) using a time step of 0.2 sec.

The coefficient matrix on the left-hand side of the above equation, which is within parentheses, temporarily call it $[a]$, does not change when j is changed. The right-hand side, which forms a single vector, temporarily call it $\{b_j\}$, needs to be updated with every change in j. Therefore, this is a set of simultaneous equations of the form $[a]\{q_{j+1}\} = \{b_j\}$ even if the stiffness matrix were dependent on q_j values. Therefore these simultaneous equations can be solved using any appropriately efficient numerical technique to obtain $\{q_{j+1}\}$. Again, for any particular chosen time step, the left-hand side coefficient matrix needs be calculated only once. Hence this is a case where it may be worthwhile calculating that inverse of that coefficient matrix for $\{q_{j+1}\}$. Just as is true for the modal equation case, the two vectors needed to start the procedure, $\{q_0\}$ and $\{q_1\}$, are obtained from the initial conditions cast in matrix form.

9.3 Assumed Acceleration Techniques

There is a class of successful time step methods sometimes attributed to N. M. Newmark [9.4,9.5], which are often referred to as Newmark's β method. It suffices here to discuss just one of these methods, the generally favored special case where $\beta = 1/6$. Appropriately, this method is called the linear acceleration method. To develop the linear acceleration method in simplest terms, again consider the linear modal equation for the ith mode

$$\ddot{p}_i(t) + 2\zeta_i \omega_i \dot{p}_i(t) + \omega_i^2 p_i(t) = \frac{P_i(t)}{M_i}.$$

Again, to simplify and avoid confusion with regard to the subscripts, drop the modal subscript i so that hereafter all subscripts refer only to the integer value of the time step. The basis of the linear acceleration method is to approximate the actual time variation of the (modal) acceleration curve by a series of straight lines, one for each time step interval. For better symbolic clarity, let a_j, v_j, and p_j be, respectively, the modal acceleration, modal velocity, and modal deflection at the jth time step. Introduce a local time coordinate, τ, that is (i) zero at time t_j, the time at the start of the jth time step, and that (ii) has the value $\Delta t_j = \Delta t$ at the $(j+1)$ th time step. In other words, τ varies between zero and Δt. In terms of τ, write the following straight-line expression for the acceleration at any time point within the jth time interval and then integrate with respect to τ so as to obtain the subsequent velocity and deflection expressions at that same typical time point in the same time step interval. In symbolic form, the original straight-line approximation to the acceleration from similar triangles, and the results of its integration over τ from zero to τ, are

$$\ddot{p}(\tau) = a_j + \frac{a_{j+1} - a_j}{\Delta t}\tau$$

$$\dot{p}(\tau) = v_j + a_j\tau + \frac{a_{j+1} - a_j}{2\Delta t}\tau^2$$

and $$p(\tau) = p_j + v_j\tau + {}^1\!/\!_2 a_j\tau^2 + \frac{a_{j+1} - a_j}{6\Delta t}\tau^3.$$

Note the one-sixth factor in the last term. Specializing these general velocity and deflection expressions within the time interval to the $(j+1)$th time step by setting $\tau = \Delta t$, produces

$$v_{j+1} = v_j + \frac{\Delta t}{2}(a_{j+1} + a_j)$$

$$p_{j+1} = p_j + \Delta t v_j + \frac{\Delta t^2}{3}a_j + \frac{\Delta t^2}{6}a_{j+1}. \tag{9.7}$$

Equations (9.7) provide solutions for the velocity and deflection at the end of the time step, but these solutions are in terms of the accelerations at the beginning and at the end of the time step. Fortunately there are two more relevant equations to be combined with Eqs. (9.7). They are the dynamic equilibrium equations at these two adjacent time steps.[3] Those equations are

$$a_j + 2\zeta\omega v_j + \omega^2 p_j = \frac{P_j}{M}$$

$$a_{j+1} + 2\zeta\omega v_{j+1} + \omega^2 p_{j+1} = \frac{P_{j+1}}{M}.$$

These four equations are now to be manipulated to obtain algebraic expressions for the unknown deflection and velocity values at the end of the time step entirely in terms of the known deflection and velocity values at the beginning of the time step. Note that the force input is considered known at all time steps. To this end, the first and the fourth equations are to be solved simultaneously for a_{j+1} and v_{j+1}, where the velocity solution is used later. These two solutions are

$$(1 + \zeta\omega\Delta t)a_{j+1} = \frac{P_{j+1}}{M} - \omega^2 p_{j+1} - 2\zeta\omega v_j - \zeta\omega\Delta t a_j$$

$$(1 + \zeta\omega\Delta t)v_{j+1} = \frac{P_{j+1}\Delta t}{2M} - \frac{\omega^2\Delta t}{2}p_{j+1} + v_j + \frac{\Delta t}{2}a_j, \tag{9.8}$$

where, again, the load input value P_{j+1} is a known quantity and the forward value of the modal deflection, p_{j+1}, will not be a difficulty because it will be combined with a like term in the second equation of the original four equations. Now it is just a matter of substituting the first of the above two equations and the third of the original four equations into the original second equation so as to eliminate the two acceleration terms. After some algebra

$$p_{j+1}\left[1 + \frac{\omega^2\Delta t^2}{6(1+\zeta\omega\Delta t)}\right] = p_j\left[1 - \frac{\omega^2\Delta t^2}{3} + \frac{\omega^2\Delta t^2(\zeta\omega\Delta t)}{6(1+\zeta\omega\Delta t)}\right]$$
$$+ \Delta t\, v_j\left[1 - \zeta\omega\Delta t\frac{3+\zeta\omega\Delta t}{3(1+\zeta\omega\Delta t)}\right]$$
$$+ \frac{P_j\Delta t^2}{6M}\left[2 - \frac{\zeta\omega\Delta t}{(1+\zeta\omega\Delta t)}\right] + \frac{P_{j+1}\Delta t^2}{6M}\left[\frac{1}{1+\zeta\omega\Delta t}\right]. \tag{9.9a}$$

Similarly, substituting the jth time step equilibrium equation into the Eq. (9.8) solution for the velocity term so as to again eliminate a_j, yields

$$(1+\zeta\omega\Delta t)v_{j+1} = v_j(1 - \zeta\omega\Delta t) - \omega^2\Delta t\frac{p_{j+1}+p_j}{2} + \frac{\Delta t}{2}\frac{P_{j+1}+P_j}{M}. \tag{9.9b}$$

Since all the right-hand side deflection and velocity terms are known at the jth time step, and the applied force terms are known at all time steps, the above two equations, (9.8a) and (9.8b), can be solved first for p_{j+1} and then for v_{j+1} which requires the value of p_{j+1} for its solution. Note also that there is no difficulty starting this calculation because all that these equations require for a successful start are the known initial conditions p_0 and v_0.

[3] The use of these two equations means that at every discrete time step, dynamic equilibrium is being enforced, as was done for the FDM.

EXAMPLE 9.3 Redo Example 9.1, but this time use the linear acceleration technique, as expressed in Eqs. (9.9), to accomplish a numerical integration of the linear, undamped, equation of motion

$$\ddot{q}(t) + \omega_1^2 q(t) = \omega_1^2 Y_0 \sin \frac{\pi t}{t_0}$$

$$\text{or} \quad \frac{\ddot{q}(t)}{Y_0} + \frac{q(t)}{Y_0} = \sin 2t,$$

where $P_j(t)/M = \omega_1^2 Y_0 \sin(\pi t/t_0)$. Again, the selected value of the time step is 0.2 sec, the natural frequency is 1.0 rad/sec, and t_0 is chosen to be a nonresonant $\pi/2$ sec. As before, now let $q(t)$ represent the previous nondimensional deflection $q(t)/Y_0$ or, equivalently, let Y_0 have a unit value.

COMMENT This numerical integration technique is stable only when the chosen time step is less than 55.1% of the natural period associated with the equation being integrated [9.4]. This is a very loose requirement in that the accuracy requirement demands a much smaller time step. Here, because $\omega = 1.0$ rad/sec, the natural period is 2π sec, and the time step of 0.2 sec is only 3% of the natural period.

SOLUTION Since the single-DOF system of Example 9.1 is undamped, the damping factor ζ equals zero. This choice, along with the other parameter choices reduces the constant acceleration numerical integration equations to

$$q_{j+1} = 0.98013245 q_j + 0.198675496 v_j + 0.0066225159 \sin[0.4(j+1)] + 0.0132450328 \sin[0.4j]$$

$$\text{and} \quad v_{j+1} = v_j - 0.1(q_{j+1} + q_j) + 0.1 \sin[0.4(j+1)] + 0.1 \sin[0.4j].$$

These calculations were made using a spreadsheet program. Two time steps were used. The first time step was 0.2 sec, whereas the second time step was one-third of the first time step. The choice of the second time interval was made so that every third time step of the second calculation falls exactly on a time step of the first calculation. Thus at time 2.00 secs, steps 10 and 30, respectively, the corresponding deflections are 0.845 and 0.857, 1.4% difference at the first positive peak. At the first negative peak, at 4.2 sec and steps 21 and 63, the corresponding values are −.849 and −.864, which is a 1.8% difference. Therefore, it seems that the 0.2 sec time step is satisfactory for this simple problem.

Figure 9.4 is a plot of those results superimposed on the exact solution. The plot shows that this more complicated numerical integration scheme, in this simple case, has only slightly better accuracy than the finite difference method as judged by looking at time steps in the vicinity of time step 90. However, at the more important peak responses, there is no apparent difference in accuracy in this simple case. ★

EXAMPLE 9.4 Derive the linear acceleration numerical equations necessary to determine the response of a single-DOF system whose motion is determined by Duffing's stiffening equation for a value of μ equal to 0.2 as in Example 9.2. Comment on the feasibility of this approach.

```
In[15]:= Show[draw1,draw2]
```

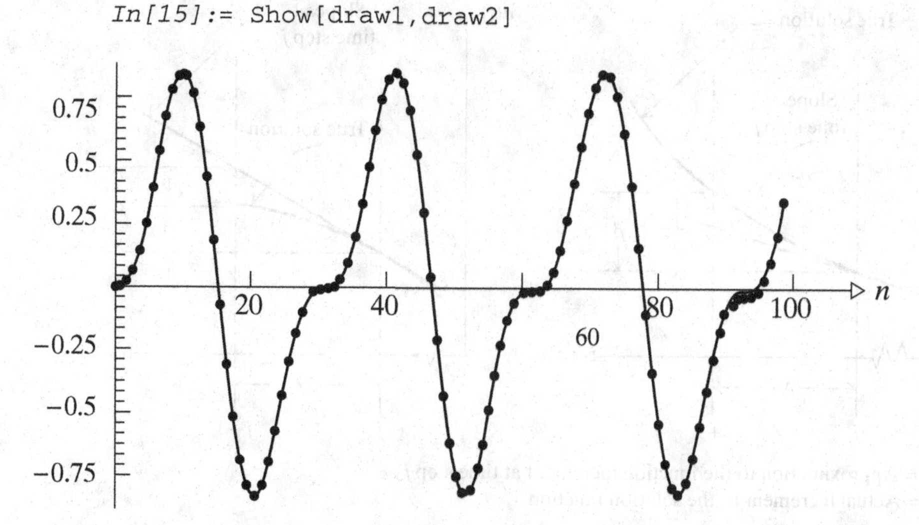

Figure 9.4. Example 9.3: Linear acceleration solution superimposed on exact solution.

SOLUTION In terms of the nondimensional deflection, which, again, is the actual deflection divided by the amplitude of the base motion, the equation of motion at the mth and $(m+1)$th time step, with $\omega = 1.0$ rad/sec and $\mu = 0.2$, are

$$\ddot{q}_m + q_m + 0.2q_m^3 = \sin 2t_m \qquad \ddot{q}_{m+1} + q_{m+1} + 0.2q_{m+1}^3 = \sin 2t_{m+1}$$

$$\text{or } a_m = \sin 2t_m - q_m - 0.2q_m^3 \qquad a_{m+1} = \sin 2t_{m+1} - q_{m+1} - 0.2q_{m+1}^3.$$

Substituting these latter two equations into the linear acceleration equations, Eqs. (9.7), yields

$$\begin{aligned} 0.0013333333q_{m+1}^3 + 1.00666666q_{m+1} &= 0.98666666q_m - 0.026666666q_m^3 + 0.2v_m \\ &\quad + 0.0133333333 \sin 0.4m \\ &\quad + 0.0066666666 \sin 0.4(m+1) \end{aligned}$$

$$\text{and} \quad \begin{aligned} v_{m+1} &= v_m + 0.1 \sin 0.4m + 0.1 \sin 0.4(m+1) - 0.1q_m - 0.1q_{m+1} \\ &\quad - 0.02q_m^3 - 0.02q_{m+1}^3. \end{aligned}$$

There is no difficulty starting these two equations with $q_0 = v_0 = 0.0$. The difficulty is that the first equation, which must be dealt with first, is a cubic equation that must be solved for the real root at each m step. Since the cubic term of this equation is much smaller than the linear term, it is possible to solve the cubic equation iteratively. Thus this approach is feasible, but of limited appeal. ★

9.4 Predictor-Corrector Methods

To provide an insight to predictor-corrector methods, first consider the simplest of all numerical integration schemes, which is known as Euler's method or the Euler-Cauchy method. Euler's method uses a time-stepping approach to address the integration of first order, *quasilinear*, differential equations of the form

$$\frac{dq(t)}{dt} = f(t, q(t)) \quad \text{or} \quad dq(t) = f(t, q(t))dt, \tag{9.10}$$

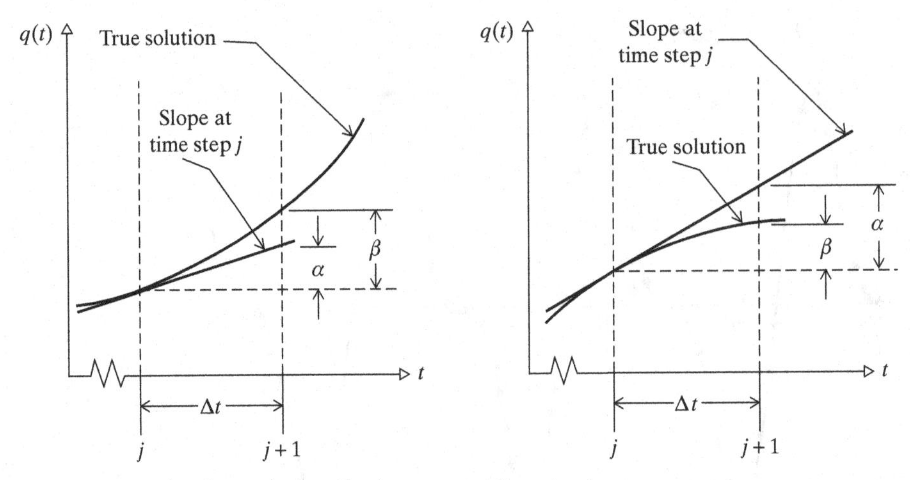

α = Approximation to the function increment at time step j
β = Actual increment to the solution function

Figure 9.5. Graphical view of Euler's method.

where f is *any* piecewise smooth function of its arguments. The Euler numerical solution is based on the function f being the slope of the function q. That is, the Euler solution for the small, finite, increment Δq_j in the function $q(t)$ at time step j, used to obtain the approximation for q_{j+1}, is simply

$$\Delta q_j = f(t_j, q_j)\Delta t.$$

This is, of course, just the finite increment corresponding to the infinitesimal increment form stated in Eq. (9.10). For the sake of simplicity in the following discussion, let the value of the function $q(t)$ be known precisely at the jth time step, as would be the case for precise initial conditions at the zeroth time step. Then the fundamental reason that this approach is unacceptably inaccurate, as illustrated in Figure 9.5, is that the approximate slope used to obtain the increment in $q(t)$ is just the slope at the beginning of the time step, Δt. Since the slope at the beginning of the interval, $f(t_j, q_j)$, can be quite unrepresentative of the average slope on the finite interval, inaccuracy is the result. To clarify this point, note that the average slope of an arbitrary function $g(x)$ on the interval (a, b) is simply

$$\text{average slope} \equiv \frac{1}{b-a}\int_a^b g'(x)dx = \frac{g(b)-g(a)}{b-a}.$$

Rearranging the first and third parts of this equation shows that the value of the function at the beginning of the time increment, $g(a)$, plus the average slope multiplied by the length of the interval, $b-a$, yields the exact value of the function at the end of the interval, $g(b)$. Thus, what is clearly needed for a more accurate numerical determination of the value of the function at the end of a time increment is a better approximation to the average slope over the time increment. The quest for a good approximation to the average slope has spawned all manner of methods, including the predictor-corrector methods discussed in this section and the Runge-Kutta methods discussed in the next section. Of course, in any numerical integration procedure, the exact value of the average slope is not available any more than is the exact value of the function at the end of the interval, $g(b)$.

Before going on to the predictor-corrector and Runge-Kutta methods as superior approaches, note that the focus of this discussion, a second-order ordinary differential equation in time such as the damped version of Duffing's equation, can easily be converted to two first-order equations to be solved simultaneously as follows. Let

$$\dot{q}(t) \equiv v(t),$$

then the single second-order differential equation

$$\ddot{q}(t) + 2\zeta\omega\dot{q}(t) + \omega^2 q(t) + \mu q^3 = Q(t)$$

becomes the following two first-order differential equations

$$\dot{v}(t) = Q(t) - 2\zeta\omega\, v(t) - \omega^2 q(t) - \mu q^3 \quad \text{and} \quad \dot{q}(t) = v(t). \tag{9.11}$$

Return to the basic problem statement $\dot{q}(t) = f(t, q)$. One reasonable way to approximate the true average slope of $q(t)$ over the time increment Δt is to average (i) the slope at the beginning of the time increment and (ii) the slope at the end of the time increment. The slope at the beginning of the time increment at time step j can be calculated directly, as before, as simply $f(t_j, q_j)$. The slope at the end of the time interval, $f(t_{j+1}, q_{j+1})$, cannot be calculated directly because the true value of q_{j+1} is not available. However, an Euler approximation can be made to obtain a first estimate of q_{j+1}. That is, write $q_{j+1}(\#1) = q_j + f(t_j, q_j)\Delta t$ and then use the average of the initial slope and the approximate end slope, based on this first estimate, to obtain a better approximation of the functional value at the end of the time step. That is, write

$$q_{j+1}(\#2) = q_j + \frac{\Delta t}{2}[f(t_j, q_j) + f(t_j + \Delta t, q_{j+1}(\#1))]. \tag{9.12}$$

Clearly it is possible to continue this process of improving the prediction for the slope at the end of the time interval, $f(t_{j+1}, q_{j+1})$, by improving the estimate for q_{j+1} and then correcting that prediction in an iterative fashion for a still better resulting estimate of q_{j+1}. The formulas given above are often referred to as Heun's method [9.1].

The error associated with the Euler method-based predictor portion of Heun's method can be reduced at the expense of a more complicated computation. To follow this path, first substitute $dq(t)/dt = f(t, q)$ into the chain rule formulation for the second derivative that is obtained from this same expression to get $d^2q(t)/dt^2 = \partial f(t, q)/\partial t + (f)(\partial f/\partial q)$ where $\dot{q}$ has been replaced by f. Then write the Taylor series expansion for the desired function $q(t)$, which can be extended one more term as follows

$$q_{j+1} = q_j + \Delta t\, q_j' + {}^1\!/\!_2(\Delta t)^2 q_j'' + O(\Delta t^3)$$

$$\text{or} \quad q_{j+1} = q_j + \Delta t\, f(t_j, q_j) + {}^1\!/\!_2(\Delta t)^2 \left[\frac{\partial f(t_j, q_j)}{\partial t} + f(t_j, q_j)\frac{\partial f(t_j, q_j)}{\partial q}\right] + O(\Delta t^3),$$

where now the local error term is of order Δt^3 rather than just Δt^2. The price to be paid for this increase in accuracy in the predictor portion of the calculation is the calculation of the partial derivatives of the given function f. If the partial derivatives of f are complicated, this procedure is not recommended.

EXAMPLE 9.5 Using Heun's method, numerically integrate the second-order ordinary differential equation of Example 9.1, which is, again,

$$\ddot{q}(t) + q(t) = \sin 2t.$$

Use, for the sake of comparison, the same time step of 0.2 sec, and the same zero initial conditions. For the sake of simplicity, use the Euler predictor and limit the number of iterations to one rather than use a percentage difference criteria.

SOLUTION The first step is to reduce this second-order equation to two first-order equations that can be written as

$$\dot{q}(t) = v(t) \quad \text{and} \quad \dot{v}(t) = -q(t) + \sin 2t.$$

Then, with subscripts referring as usual to the time step, and the number in parentheses referring to the iteration number, the Euler predictor portion of the Heun's method equations start out as follows, where use is made of the initial conditions $q_0 = v_0 = 0$

$$q_1(\#0) = q_0 + \Delta t\, \dot{q}_0 = q_0 + \Delta t\, v_0 = 0$$
$$v_1(\#0) = v_0 + \Delta t\, \dot{v}_0 = v_0 + \Delta t[-q_0 + \sin 2t_0] = 0.$$

The corrector portion of the equations for this first time step are

$$q_1(\#1) = q_0 + \frac{\Delta t}{2}[\dot{q}_0 + \dot{q}_1(\#0)] = q_0 + \frac{\Delta t}{2}[v_0(\#0) + v_1(\#0)] = 0$$

$$\begin{aligned} v_1(\#1) &= v_0 + \frac{\Delta t}{2}[\dot{v}_0 + \dot{v}_1(\#0)] \\ &= q_0 + \frac{\Delta t}{2}[-q_0 - q_1(\#0) + \sin 2t_0 + \sin 2t_1] \\ &= 0 + 0.1\sin(0.4) = 0.038941834. \end{aligned}$$

Iterating once, where, again, $t_0 = 0$ and $t_j = j\Delta t = 0.2j$

$$q_1(\#2) = q_0 + \frac{\Delta t}{2}[\dot{q}_0 + \dot{q}_1(\#1)] = q_0 + \frac{\Delta t}{2}[v_0 + v_1(\#1)] = 0 + 0.0038941834$$

$$\begin{aligned} v_1(\#2) &= v_0 + \frac{\Delta t}{2}[\dot{v}_0 + \dot{v}_1(\#1)] = v_0 + \frac{\Delta t}{2}[-q_0 - q_1(\#1) + \sin 2t_0 + \sin 2t_1] \\ &= 0 + 0.1\sin(0.4) = 0.038941834. \end{aligned}$$

It is immediately evident that because these two calculations are done sequentially, the calculation for v_j can be improved by making use of the previous calculation for q_j. Therefore, revise the second of the above two equations as follows:

$$\begin{aligned} v_1(\#2) &= v_0 + \frac{\Delta t}{2}[-q_0 - q_1(\#2) + \sin 2t_0 + \sin 2t_1] \\ &= 0 + 0.1[-0.00389418 + \sin(0.4)] = 0.03855242. \end{aligned}$$

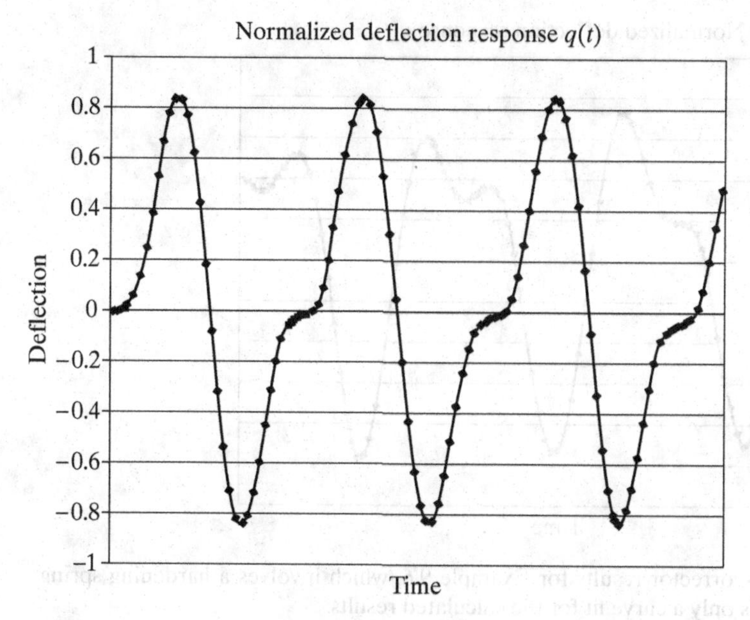

Figure 9.6. Predictor-corrector results for Example 9.5 (the solid line is merely a curve fit).

The calculation pattern is now established. The predictor equations are

$$q_{j+1} = q_j + \Delta t\, v_j \quad \text{and} \quad v_{j+1} = v_j + \Delta t[-q_j + \sin(0.4j)]$$

and the corrector equations for the final values of q_{j+1} and v_{j+1} are where n indicates the number of the iteration

$$q_{j+1}(n+1) = q_j + \frac{\Delta t}{2}[v_j + v_{j+1}(n)] \quad \text{and}$$

$$v_{j+1}(n+1) = v_j + \frac{\Delta t}{2}[-q_j - q_{j+1}(n+1) + \sin(0.4j) + \sin(0.4j + 0.4)].$$

A spreadsheet program is well suited to these calculations. Figure 9.6 shows a plot of the spreadsheet results, where, again, each q_j and v_j value were recalculated twice and where the solid line here is not the analytical (i.e., not the exact solution which is unavailable) but merely a line that connects the data points. As the graph shows, from examining the results near the zero deflection line, this result is no more or less accurate than the preceding finite difference and linear acceleration methods. ★

EXAMPLE 9.6 As in Example 9.2, using the above predictor-corrector algorithm, solve the undamped Duffing's equation, but this time with a hardening spring, as represented by the equation

$$\ddot{q}(t) + \omega_1^2 q(t) + \mu q^3(t) = \omega_1^2 Y_0 \sin\frac{\pi t}{t_0},$$

where $\omega_1 = 1.0$, $\mu = 0.2$, and $t_0 = \pi/2$.

SOLUTION Once again, after nondimensionalizing the deflection response $q(t)$ by dividing by the input amplitude Y_0 (while retaining the same symbol), it is necessary

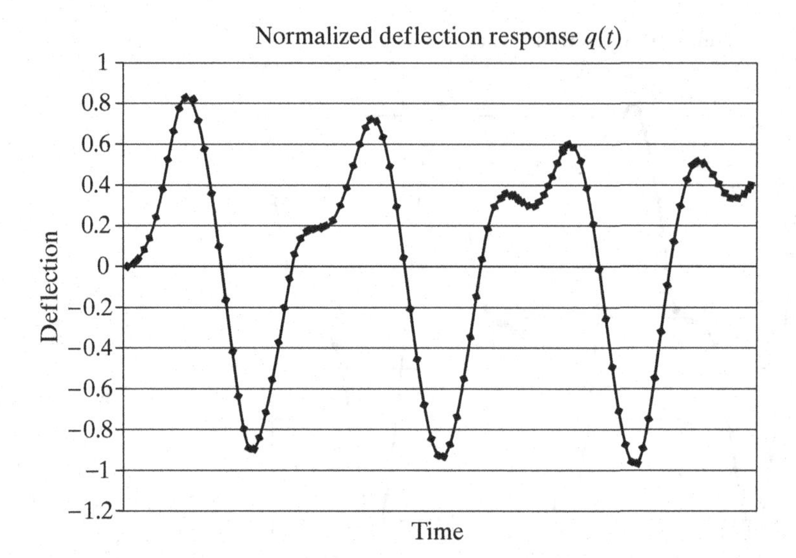

Figure 9.7. Predictor-corrector results for Example 9.6, which involves a hardening spring. Again, the solid line is only a curve fit for the calculated results.

to break the one second-order differential equation into two, first-order differential equations. That result is

$$\dot{q}(t) = v(t) \quad \text{and} \quad \dot{v}(t) = -q(t) - 0.2q^3(t) + \sin 2t.$$

The close similarity of these equations with those of the previous example, the single nonlinear term being the only difference, permits the above equations to be modified to the present circumstances as follows. The predictor equations are

$$q_{j+1} = q_j + \Delta t\, v_j \quad \text{and} \quad v_{j+1} = v_j + \Delta t[-q_j - 0.2q_j^3 + \sin(0.4j)],$$

whereas the two corrector equations, where again n is the number of the iteration, are

$$q_{j+1}(n+1) = q_j + \frac{\Delta t}{2}\left[v_j + v_{j+1}(n)\right],$$

which is exactly the same as in the linear case, and

$$v_{j+1}(n+1) = v_j + \frac{\Delta t}{2}\left[-q_j - q_{j+1}(n+1) - 0.2q_j^3 - 0.2q_{j+1}^3(n+1) + \sin(0.4j) + \sin(0.4j + 0.4)\right].$$

The spreadsheet implementation of these Duffing equations is little different from that for the linear equation of the previous example. The numerical result is presented in Figure 9.7. ★

9.5 The Runge-Kutta Method

One of the two most popular, general purpose, numerical integration techniques is the fourth-order Runge-Kutta method (RK method). It, like predictor-corrector methods, is a sophisticated version of Euler's method and differs from other methods only in its approach to calculating the average slope between time steps. There are many different classifications of Runge-Kutta methods. The primary classification refers to the "order" of the method. The order number is related to the number of calculations in the time increment between the integer numbered steps, Δt. For example, a second-order RK method will employ two calculations (one each at two different points in this case) in the interval of the time increment, and a fourth-order RK method will employ four calculations (in this case at three different points). From Ref. [9.1], a common choice for a fourth-order RK method solution for the unknown step value q_{i+1} of the deflection function $q(t)$ described by a first-order quasilinear equation of the form

$$\dot{q}(t) = F(t, q)$$

is the following sum of a present (ith) value plus the time increment multiplied by a four-term approximation for the average velocity

$$q_{i+1} = q_i + \frac{\Delta t}{6}(k_1 + 2k_2 + 2k_3 + k_4),$$

$$\text{where} \quad k_1 = F(t_i, q_i) \qquad k_2 = F\left(t_i + \frac{\Delta t}{2}, q_i + \frac{\Delta t}{2}k_1\right)$$

$$k_3 = F\left(t_i + \frac{\Delta t}{2}, q_i + \frac{\Delta t}{2}k_2\right) \qquad k_4 = F(t_i + \Delta t, q_i + \Delta t k_3). \quad (9.13)$$

Note that the slope terms k_1, k_2, k_3, and k_4 terms are calculated sequentially, and that the k_1 and the k_4 terms are respectively, the approximations to the slope at the beginning and end of time step interval. The double-weighted k_2 and k_3 terms are two different approximations to the slope at the midpoint of the time interval.

EXAMPLE 9.7 Using the above version of the fourth-order RK method, redo Example 9.1, the single-DOF, forced, linear vibration problem whose differential equation is

$$\ddot{q}(t) + \omega_1^2 q(t) = \omega_1^2 Y_0 \sin\frac{\pi t}{t_0}.$$

Again let the natural frequency be 1 rad/sec and t_0 be $\pi/2$ sec. Therefore, as before, dividing the deflection response by the input amplitude, Y_0 (or setting the input amplitude to a unit value), the normalized differential equation is simply

$$\ddot{q}(t) + q(t) = \sin 2t.$$

To avoid a round-off error in the numerical integration, slightly change the previously used time step from 0.2 to 0.198 sec, which, when divided by 6 as required by the above equations, is exactly 0.033 sec.

SOLUTION As before, first rewrite the above second-order differential equation as the following two first-order differential equations

$$\dot{v}(t) = \omega_1^2 Y_0 \sin\frac{\pi t}{t_0} - \omega_1^2 q(t),$$

$$\text{and} \quad \dot{q}(t) \equiv v(t)$$

where the parameters ω_1, Y_0, and t_0 are temporarily retained for the purpose of checking units in each of the RK method expressions. Now write each of these two equations in increment form as an initial value plus a time increment multiplied by an RK averaged slope

$$v_{n+1} = v_n + \frac{\Delta t}{6}[k_1 + 2k_2 + 2k_3 + k_4]$$

$$q_{n+1} = q_n + \frac{\Delta t}{6}[h_1 + 2h_2 + 2h_3 + h_4],$$

where the symbol h is used to distinguish the slope terms in the deflection expression from those in the velocity expression. Of course the ks are acceleration terms, whereas the hs are velocity terms. Adapting the formulas of Eqs. (9.13) to this example problem, the explicit values of the RK method slope terms are, where $t_n = n\Delta t$

$$k_1 = \omega_1^2 Y_0 \sin\frac{n\pi\,\Delta t}{t_0} - \omega_1^2 q_n$$

$$k_2 = \omega_1^2 Y_0 \sin\frac{\left(n+\frac{1}{2}\right)\pi\,\Delta t}{t_0} - \omega_1^2\left(q_n + \frac{\Delta t}{2}h_1\right)$$

$$k_3 = \omega_1^2 Y_0 \sin\frac{\left(n+\frac{1}{2}\right)\pi\,\Delta t}{t_0} - \omega_1^2\left(q_n + \frac{\Delta t}{2}h_2\right)$$

$$k_4 = \omega_1^2 Y_0 \sin\frac{(n+1)\pi\,\Delta t}{t_0} - \omega_1^2(q_n + \Delta t\, h_3)$$

$$h_1 = v_n \quad h_2 = v_n + \frac{\Delta t}{2}k_1 \quad h_3 = v_n + \frac{\Delta t}{2}k_2 \quad h_4 = v_n + \Delta t\, k_3.$$

Now, as before, substituting the selected values of the natural frequency and the time scale value for the forced motion, t_0, normalizing the deflection response by the input amplitude, and noting that $2(n + {}^1\!/\!_2) = 2n + 1$, the above equations reduce to

$$k_1 = \sin(2n\Delta t) - q_n$$

$$k_2 = \sin[(2n+1)\Delta t] - \left[q_n + \frac{\Delta t}{2}h_1\right]$$

$$k_3 = \sin[(2n+1)\Delta t] - \left[q_n + \frac{\Delta t}{2}h_2\right]$$

$$k_4 = \sin[(2n+2)\Delta t] - [q_n + \Delta t\, h_3]$$

$$h_1 = v_n \quad h_2 = v_n + \frac{\Delta t}{2}k_1 \quad h_3 = v_n + \frac{\Delta t}{2}k_2 \quad h_4 = v_n + \Delta t\, k_3.$$

It may appear that there is a difficulty that the two sets of slope terms are interdependent; that is, the slope terms for the deflection depend on the slope terms for the velocity and vice versa. This difficulty can be resolved with proper sequencing. A successful procedure to both start and continue the time step calculation is to calculate the k_1 and h_1 values first using the initial or previous step values of the deflection and velocity. Then use the h_1 and k_1 quantities to calculate the h_2 and k_2 quantities. Then use h_2 and k_2 to calculate h_3 and k_3, and these latter two quantities to calculate h_4 and k_4. Thus the next time step values for velocity and deflection are obtained without difficulty. In this manner the calculation can proceed from time step to time step.

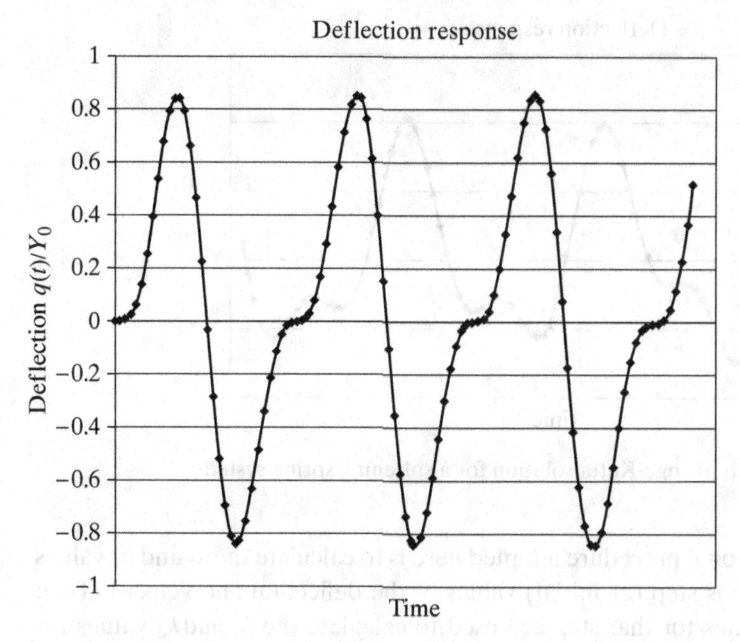

Figure 9.8. Example 9.7: Runge-Kutta results for the linear vibratory system.

The plot of the spreadsheet calculation results for this example problem are shown in Figure 9.8. Judging by the result near the zero deflection line, this is the most accurate of the four-linear calculations that have been carried out. Of course, the computational effort was also slightly greater. ★

EXAMPLE 9.8 Apply the fourth-order RK method to the Duffing equation with a softening spring as first treated in Example 9.2 ($\mu = 0.2$) and the Duffing equation with a hardening spring as first treated in Example 9.6. Compare the results of the previous calculations with the RK result.

SOLUTION Using the same parameter selections of Examples 9.2 and 9.4, the resulting softening and hardening spring equations to be numerically integrated for the nondimensional deflection q/Y_0 (represented as q) are, respectively,

$$\ddot{q}(t) + q(t) - 0.2q^3(t) = \sin 2t \quad \text{and} \quad \ddot{q}(t) + q(t) + 0.2\, q^3(t) = \sin 2t.$$

Beginning with the case of the softening spring, the first-order RK equations are

$$h_1 = v_n \quad h_2 = v_n + \frac{\Delta t}{2}k_1 \quad h_3 = v_n + \frac{\Delta t}{2}k_2 \quad h_4 = v_n + \Delta t\, k_3$$

and

$$k_1 = -q_n + 0.2q_n^3 + \sin 2n\Delta t$$

$$k_2 = -\left(q_n + \frac{\Delta t}{2}h_1\right) + 0.2\left(q_n + \frac{\Delta t}{2}h_1\right)^3 + \sin(2n+1)\Delta t$$

$$k_3 = -\left(q_n + \frac{\Delta t}{2}h_2\right) + 0.2\left(q_n + \frac{\Delta t}{2}h_2\right)^3 + \sin(2n+1)\Delta t$$

$$k_4 = -(q_n + \Delta t\, h_3) + 0.2(q_n + \Delta t\, h_3)^3 + \sin 2(n+1)\Delta t.$$

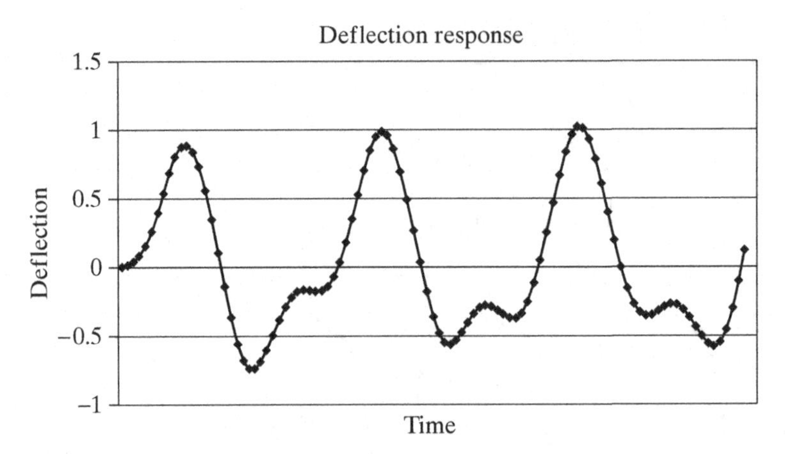

Figure 9.9. Example 9.8: Runge-Kutta solution for a softening spring system.

Again the computational procedure adopted here is to calculate the k_1 and h_1 values first using the previous step (or initial) values of the deflection and velocity. Then the h_1 and k_1 quantities for that step are used to calculate the h_2 and k_2 values for that step, etc. The spreadsheet plot of the numerical results are shown in Figure 9.9. Again, the solid line on the plot is a curve fit only for the numerical data. Comparison to Figure 9.3(a), which is for a time increment of only 0.1 sec, shows good agreement between the two different types of solution through the end of the third cycle. This good agreement serves to generally validate both solutions.

The equations for the case of the hardening spring are the same as those above with the sole exception the sign before the cubic terms is changed from a plus sign to a minus sign. The calculated results from a spreadsheet are shown in Figure 9.10, where the time step was again set at 0.198 sec. Again, the solid line in the plot merely connects the calculated deflection response. The Figure 9.10 response compares quite closely with the predictor-corrector result displayed in Figure 9.7. As a final confirmation of the accuracy of these calculations, a Mathematica derived result for the same Duffing equation with a hardening spring is shown in Figure 9.11. If it were not

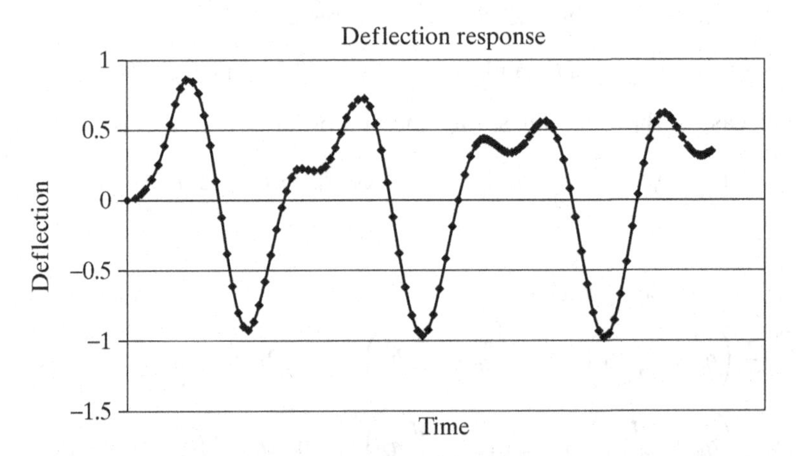

Figure 9.10. Example 9.8: Runge-Kutta solution of the Duffing equation for a hardening spring.

```
In[4]:= z = NDSolve[{y''[x] + y[x] + 0.2*y[x]^3 ==
        Sin[2*x],y[0] == 0.0,y'[0] == 0.0},y,
        {x,0,6*Pi}]

Out[4]= {{y→InterpolatingFunction[{{0.,18.8496}},
         <>]}}

In[5]:= F[x_] = y[x]/. First[z]

Out[5]= InterpolatingFunction[{{0.,18.8496}},
         <>][x]

In[7]:= Plot[F[x],{x,0,18.8}]
```

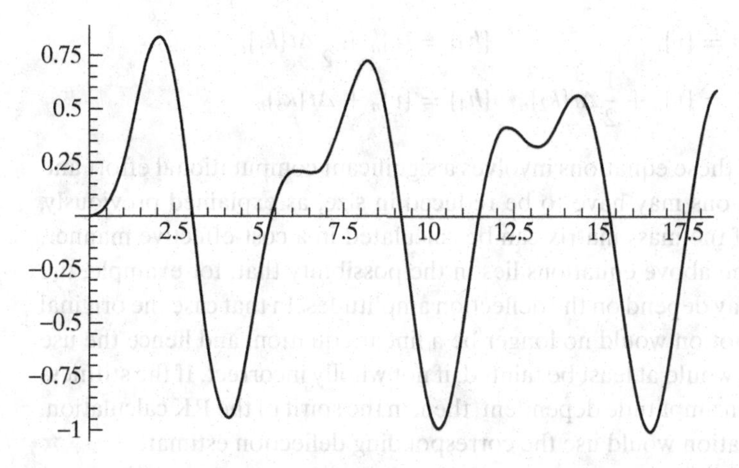


```
Out[7]= -Graphics-
```

Figure 9.11. Mathematica solution to the same Duffing equation whose solution is presented in Figures 9.7 and 9.10. Note the irregularity of the "zero crossings".

clear that the previous RK and Heun calculations provided a sufficient number of calculated deflection responses per bump in the deflection response curve for the purpose of accurately estimating the deflection response curve, this Mathematica result confirms that the choice of the time step, 0.2 or 0.198 sec, was a satisfactory choice.

The application of the Runge-Kutta method need not be constrained to equations with a single DOF. The RK method can be directly applied to, say, the nth-order linear matrix equation as follows. Let $\{q(t_n)\} = \{q\}_n$, with the same notation for the velocity vector. Then, undertaking, say, the expense of an efficient inverse of the mass matrix, the second-order matrix equation of motion can be rewritten as

$$\{\dot{q}\}_n = \{v\}_n \quad \text{and} \quad \{\dot{v}\}_n = [m]^{-1}\left(-[c]\{v\}_n - [k]\{q\}_n + \{Q(t_n)\}\right).$$

Then the RK matrix equations become

$$\{v\}_{n+1} = \{v\}_n + \frac{\Delta t}{6}\left(\{k_1\}_n + 2\{k_2\}_n + 2\{k_3\}_n + \{k_4\}_n\right)$$

$$\{q\}_{n+1} = \{q\}_n + \frac{\Delta t}{6}\left(\{h_1\}_n + 2\{h_2\}_n + 2\{h_3\}_n + \{h_4\}_n\right),$$

where the various slope vectors, from Eqs. (9.13), are calculated as follows:

$$\{k_1\}_n = [m]^{-1}\left(\{Q(n\Delta t)\} - [c]\{v\}_n - [k]\{q\}_n\right)$$

$$\{k_2\}_n = [m]^{-1}\left(\left\{Q\left(n\Delta t + \frac{\Delta t}{2}\right)\right\} - [c]\left(\{v\}_n + \frac{\Delta t}{2}\{k_1\}_n\right) - [k]\left(\{q\}_n + \frac{\Delta t}{2}\{h_1\}_n\right)\right)$$

$$\{k_3\}_n = [m]^{-1}\left(\left\{Q\left(n\Delta t + \frac{\Delta t}{2}\right)\right\} - [c]\left(\{v\}_n + \frac{\Delta t}{2}\{k_2\}_n\right) - [k]\left(\{q\}_n + \frac{\Delta t}{2}\{h_2\}_n\right)\right)$$

$$\{k_4\}_n = [m]^{-1}\left(\{Q(n\Delta t + \Delta t)\} - [c]\left(\{v\}_n + \Delta t\{k_3\}_n\right) - [k]\left(\{q\}_n + \Delta t\{h_3\}_n\right)\right)$$

and

$$\{h_1\} = \{v\}_n \qquad \{h_2\} = \{v\}_n + \frac{1}{2}\Delta t\{k_1\}_n$$

$$\{h_3\} = \{v\}_n + \frac{1}{2}\Delta t\{k_2\}_n \qquad \{h_4\} = \{v\}_n + \Delta t\{k_3\}_n.$$

Clearly, (i) the use of these equations involves a significant computational effort and (ii) the matrix equations may have to be reduced in size, as explained previously, so that the inverse of the mass matrix can be calculated in a cost-effective manner. The importance of the above equations lies in the possibility that, for example, the stiffness matrix $[k]$ may depend on the deflection amplitudes. In that case the original matrix equation of motion would no longer be a linear equation, and hence the use of the modal method would at least be tainted, if not wholly incorrect. If the stiffness matrix were deflection amplitude dependent, then, in the spirit of the RK calculation, the each slope calculation would use the corresponding deflection estimate. ★

9.6 Summary

According to Ref. [9.1], tests performed on numerical methods tend to favor the Runge-Kutta methods, particularly the fourth-order method relative to the Heun predictor-corrector method. Butcher's method, which is the fifth-order RK method, provides better accuracy, but the increased numerical effort is such that the usual judgement is that the fourth-order RK method is optimum.

There are additional aspects of numerical integration that, although not mentioned previously, are of importance. Foremost among these additional techniques is *adaptive step size control.* Adaptive step size control is simply increasing the size of the time step when the solution values are not changing much when the previous step size is used and decreasing the time step when the calculated slopes exceed a preset value. As has been seen by the results shown for the example problems, the solutions to dynamic problems usually involve significant changes most of the time. Thus adaptive step size control is not all that valuable for dynamic problems. Another adaptation that will only be mentioned is a class of solutions called *multistep methods*. For example, in the case of the Heun method and the RK method, the solution value at time step $n + 1$ was calculated entirely on the basis of the previously obtained solution at time step n. Multistep methods also use the solution result at the $n - 1$ time step and perhaps other previous time steps. The finite difference method fits this category. A possible drawback of these methods is that they often require the use of a different method to start the calculation. Among these methods, *Adams-Bashforth*, which uses a combination of a Taylor's series expansion and a backward finite difference equation, is popular. Again, see Ref. [9.1], p. 618.

9.7 **Matrix Function Solutions**

Although not a numerical method in the sense of the above direct time integration methods discussed in the previous sections of this chapter, there is still another approach for computing the deflection solution to the $N \times N$ forced vibration, linear, matrix equation of motion

$$[m]\{\ddot{q}(t)\} + [c]\{\dot{q}(t)\} + [k]\{q(t)\} = \{Q(t)\}.$$

It must be emphasized that this additional approach bears mentioning just for the sake of a more complete discussion of solution techniques. This approach, that of using functions of matrices, is merely an additional topic because it has not been found to offer computational advantage relative to the previously discussed solution methods. Hence, this approach, at present, is more of a curiosity than a fruitful alternate computational approach.

The first thing to do is to explain what is meant by a function of a matrix. To this end, consider the mathematics of an arbitrary, real $N \times N$ symmetric matrix $[B]$, and the associated $[B]$, $[I]$ matrix eigenvalue equation

$$[B]\{u\} = \lambda\{u\} = \lambda[I]\{u\}.$$

Since the identity matrix $[I]$ is one of the two weighting matrices of this eigenvalue problem, the $N \times 1$ modal vectors, $\{A^{(n)}\}$, are mutually orthogonal is the usual vector sense; that is, without a weighting matrix

$$\lfloor A^{(n)} \rfloor [I] \{A^{(n)}\} = \lfloor A^{(n)} \rfloor \{A^{(n)}\} = 0 \quad \text{if } m \neq n.$$

For present purposes, normalize these eigenvectors so that

$$\lfloor A^{(n)} \rfloor [I] \{A^{(n)}\} = \lfloor A^{(n)} \rfloor \{A^{(n)}\} = 1,$$

which is easily done by dividing each entry of an eigenvector by the square root of the original result of the product of the eigenvector with itself. Let the N eigenvalue and eigenvector solutions to this equation be arranged, as before, as

$$[B][\Phi] = [\Phi][\backslash \Lambda \backslash] \quad \text{or} \quad [B] = [\Phi][\backslash \Lambda \backslash][\Phi]^t,$$

where the $[\backslash \Lambda \backslash]$ matrix is the diagonal matrix of the eigenvalues and where the second of the above equations is obtained by postmultiplying by the transpose of the matrix of eigenvectors. That is, after normalizing the eigenvectors as above, the modal matrix inverse equals the transpose of the modal matrix,

$$[\Phi]^t[\Phi] = [\Phi][\Phi]^t = [I].$$

Now consider, for example, the square and the cube of the symmetric matrix $[B]$, where those terms mean what they should mean.

$$\begin{aligned}[B]^2 &= [B][B] = [\Phi][\backslash\Lambda\backslash][\Phi]^t[\Phi][\backslash\Lambda\backslash][\Phi]^t = [\Phi][\backslash\Lambda\backslash][\backslash\Lambda\backslash][\Phi]^t \\ &= [\Phi][\backslash\Lambda\backslash]^2[\Phi]^t = [\Phi][\backslash\Lambda^2\backslash][\Phi]^t \\ [B]^3 &= [\Phi][\backslash\Lambda\backslash][\Phi]^t[\Phi][\backslash\Lambda\backslash][\Phi]^t[\Phi][\backslash\Lambda\backslash][\Phi]^t = [\Phi][\backslash\Lambda^3\backslash][\Phi]^t.\end{aligned}$$

Note that whereas the product of a symmetric matrix with another, different, symmetric matrix is generally not symmetric, the product of a symmetric matrix with itself is always symmetric.

Recall that a scalar analytical function can be expanded as power series. For example, these common scalar functions have the following power series expansions

$$\begin{aligned}\sin x &= x - \frac{x^3}{3!} + \frac{x^5}{5!} - \frac{x^7}{7!} + \cdots \\ \cos x &= 1 - \frac{x^2}{2!} + \frac{x^4}{4!} - \frac{x^6}{6!} + \cdots \\ \exp x &= 1 + x + \frac{x^2}{2!} + \frac{x^3}{3!} + \frac{x^4}{4!} + \cdots.\end{aligned}$$

Using, for example, the first of these series expansions as a model, define the sine function of the product of the constant, symmetric matrix $[B]$ and the scalar variable t as follows:

$$\begin{aligned}\sin([B]t) &= [B]t - \frac{1}{3!}[B]^3t^3 + \frac{1}{5!}[B]^5t^5 - \frac{1}{7!}[B]^7t^7 + \cdots \\ &= [\Phi][\backslash(\Lambda t - \frac{1}{3!}\Lambda^3t^3 + \frac{1}{5!}\Lambda^5t^5 - \cdots)\backslash][\Phi]^t = [\Phi][\backslash \sin\lambda t\backslash][\Phi]^t,\end{aligned}$$

where the diagonal entries of the matrix $[\backslash\sin\lambda t\backslash]$ are the N ordered, individual sines of the product of the eigenvalues and the variable t. This definition of the sine of a symmetric matrix is useful because, just as is the case with the scalar function, the derivative of the sine of a matrix is essentially the cosine of that matrix. To demonstrate this fact, differentiate both sides of the above equation with respect to time. Using the last part of the above equality

$$\begin{aligned}\frac{d}{dt}\sin([B]t) &= \frac{d}{dt}[\Phi][\backslash\sin\lambda t\backslash][\Phi]^t = [\Phi][\backslash\lambda\cos\lambda t\backslash][\Phi]^t \\ &= [\Phi][\backslash\Lambda\backslash][\backslash\cos\lambda t\backslash][\Phi]^t = [\Phi][\backslash\Lambda\backslash][\Phi]^t[\Phi][\backslash\cos\lambda t\backslash][\Phi]^t \\ &= [B]\cos([B]t) = \cos([B]t)[B].\end{aligned}$$

The added bonus is the above-indicated commutativity that is easily proved by reversing the order of the scalar factors Λ and $\cos\lambda t$ on the above second equation line. It is also easy to show that the derivative with respect to time of the $\cos[B]t$ is similarly the negative of the sine of that matrix, and the derivative of the exponential function of a symmetric matrix is similarly the exponential function of that matrix, and so on. The only other result that will be needed below is

$$\begin{aligned}[B] &= [B^{1/2}][B^{1/2}] \\ \text{and}\quad [B^{1/2}] &= [\Phi][\backslash\Lambda^{1/2}\backslash][\Phi]^t.\end{aligned}$$

There are basically two ways of using functions of matrices to write solutions to the matrix equation of vibratory motion. The first approach is to create a matrix form for the Duhamel integral. To simplify that discussion by keeping the algebra as simple as possible, only the undamped form of the matrix equation of motion is considered in the first approach. The damping matrix, estimated as best it can be, is retained in the second approach, which is called the phase state approach. For the first approach, let the applied forces be a collection of impulses written as

$$[m]\{\ddot{u}(t)\} + [k]\{u(t)\} = \{\mathcal{F}_0\}\delta(t),$$

where the right-hand side vector is a vector of fixed magnitudes. Note that all the impulses are centered at time zero. The first step in this procedure, as per Section 5.8, is to use a Cholesky decomposition to obtain a symmetric dynamic matrix. Let

$$[m] = [L_m][L_m]^t \quad \text{and} \quad \{u(t)\} = [L_m]^{-t}\{r(t)\},$$

where $\{r\}$ has the strange units of deflection divided by the square root of mass (or mass moment of inertia). Making the above substitutions and premultiplying by the inverse of $[L_m]$ leads to

$$[I]\{\ddot{r}(t)\} + [L_m]^{-1}[k][L_m]^{-t}\{r(t)\} = [L_m]^{-1}\{\mathcal{F}_0\}\delta(t)$$

or, after consolidating the above terms, where $[D]$ is a symmetric matrix

$$\{\ddot{r}(t)\} + [D]\{r(t)\} = \{R_0\}\delta(t).$$

The first step in solving this equation is to directly integrate this equation over time between the limits of 0^- and 0^+, where, without loss of generality, it is specified that all deflections and velocities are zero at time 0^-. Therefore, the integral of the acceleration vector leads to the velocity vector at time zero plus. As for the second integral, because $[D]$ is a matrix of constants, it comes out of the time integral, leaving only the vector of deflections. Reusing the argument first made in Section 7.5 that the time duration between zero minus and zero plus is so short that the finite velocities generated by the impulsive loading can produce only a negligible deflections, the result of this time integration are

$$\{\dot{r}(0^+)\} = \{R_0\} \quad \text{and} \quad \{r(0^+)\} = \{0\}.$$

These two results become initial conditions for the vibration after time zero plus. Again, after time zero plus, there is no further force input. Hence, after time zero plus, the system undergoes a force free vibration. As can be proved by direct substitution into the above differential equation, the free vibration solution can be expressed in the following form

$$\{r(t)\} = \sin([D]^{1/2}t)\{C_1\} + \cos([D]^{1/2}t)\{C_2\},$$

where, concerning the square root of the dynamic matrix, the solution to Exercise 9.8 shows that all the (selected) natural frequencies are represented in this solution. The initial conditions, those at time zero plus, can be used to determine the entries in the constant of integration vectors. To this end, note that, where [0] is the null matrix, an $N \times N$ by n matrix of zeros

$$\sin[0] = [\Phi_D][0][\Phi_D] = [0] \quad \text{and} \quad \cos[0] = [\Phi_D][I][\Phi_D] = [I].$$

Then

$$\{C_2\} = \{0\} \quad \text{and} \quad \{C_1\} = [D]^{-1/2}\{R_0\}.$$

Therefore, the free vibration solution is

$$\{r(t)\} = \sin([D]^{1/2}t)[D]^{-1/2}\{R_0\}$$
$$\text{or} \quad \{u(t)\} = [L_m]^{-t}\sin([D]^{1/2}t)[D]^{-1/2}[L_m]^{-1}\{\mathfrak{F}_0\}.$$

Just as was done in Section 7.6, the above impulse response solution can be used to build the response solution to an arbitrary load vector $\{F(t)\}$ at time t by breaking each of those force–time histories into a continuous series of infinitesimal impulses, which at the representative time τ, have the magnitudes $\{F(\tau)\}d\tau$. Then summing all the resulting responses to the infinitesimal impulses over the elapsed time $t - \tau$ leads to

$$\{u(t)\} = \int_0^t [L_m]^{-t}\sin([D]^{1/2}(t-\tau))[D]^{-1/2}[L_m]^{-1}\{F(\tau)\}d\tau,$$

where the square matrix of impulse response functions is everything in the integrand but the applied force vector, $\{F(\tau)\}$. Since the above integrand has the form of a $N \times 1$ vector, the integration should not pose any unusual problems. The challenge here is simply obtaining the triangular, square root, and sine matrices.

The second approach to a nonmodal solution of the matrix equation of vibratory motion using functions of matrices begins by writing the equation of motion in the first-order differential equation form where the first column vector is the first derivative of the second column vector

$$\begin{bmatrix} m & 0 \\ 0 & -k \end{bmatrix}\begin{Bmatrix} \ddot{u}(t) \\ \dot{u}(t) \end{Bmatrix} + \begin{bmatrix} c & k \\ k & 0 \end{bmatrix}\begin{Bmatrix} \dot{u}(t) \\ u(t) \end{Bmatrix} = \begin{Bmatrix} F(t) \\ 0 \end{Bmatrix}.$$

The other possible symmetric arrangement for a first-order differential equation

$$\begin{bmatrix} 0 & m \\ m & c \end{bmatrix}\begin{Bmatrix} \ddot{u}(t) \\ \dot{u}(t) \end{Bmatrix} + \begin{bmatrix} -m & 0 \\ 0 & k \end{bmatrix}\begin{Bmatrix} \dot{u}(t) \\ u(t) \end{Bmatrix} = \begin{Bmatrix} 0 \\ F(t) \end{Bmatrix}$$

is not suitable here because of the singularity of the leading square matrix due to the zero submatrix on the main diagonal. This double-sized matrix approach has the advantage relative to the preceding matrix approach that the presence of the estimated damping matrix does not complicate this approach. Indeed, the damping matrix need only be symmetric, which of course it should be if $[c]$ only involves damping terms.[4] Again, adopting the first possibility for symmetric matricies assures that the leading $2N \times 2N$ matrix is nonsingular.

[4] From Ref. [9.6], gyroscopic terms produce a skew-symmetric coefficient matrix for the velocity terms.

The next step is to accomplish a Cholesky decomposition of the leading square matrix

$$\begin{bmatrix} m & 0 \\ 0 & -k \end{bmatrix} = [L_l][L_l]^t.$$

Then write the coordinate transformation $\lfloor \dot{u} \quad u \rfloor^t = [L_l]^{-t} \quad \lfloor \dot{q} \quad q \rfloor^t$ and premultiply by the matrix $[L_l]^{-1}$ for the result

$$\begin{Bmatrix} \ddot{q}(t) \\ \dot{q}(t) \end{Bmatrix} + [L]^{-1} \begin{bmatrix} c & k \\ k & 0 \end{bmatrix} [L]^{-t} \begin{Bmatrix} \dot{q}(t) \\ q(t) \end{Bmatrix} = [L]^{-1} \begin{Bmatrix} F(t) \\ 0 \end{Bmatrix}$$

$$\text{or} \quad \{\dot{r}(t)\} + [B]\{r(t)\} = \{R(t)\},$$

where $[B]$ is a symmetric matrix. In passing, note that all attempts to date to apply modal procedures to this first-order equation have been unsuccessful because, unlike the case for the modes for the second-order form of the equation of motion, there is no diminishing in the importance of the mode shapes to the solution as the modal number increases. Indeed, in one test for heat transfer equations, the highest numbered mode proved to be the most important.

As might be expected for such a first-order equation as $\{\dot{r}(t)\} + [B]\{r(t)\} = \{R(t)\}$, the matrix form of the solution involves the exponential function of the matrix $[B]$ and follows the same pattern as the solution to the first-order scalar equation. To quickly review the meaning of $\exp[B]t$, write the eigenpair solutions for the matrices $[B]$ and $[I]$ as

$$[B][\Phi_B] = [\Phi_B][\backslash \Lambda_B \backslash] \quad \text{or} \quad [B] = [\Phi_B][\backslash \Lambda_B \backslash][\Phi_B]^t.$$

Then, mimicking the Taylor's series expansion for

$$\exp(x) = 1 + x + x^2/2! + x^3/3! + x^4/4! + \cdots, \text{ define}$$

$$\exp([B]t) = [I] + [B]t + (1/2!)[B]^2t^2 + (1/3!)[B]^3t^3 + (1/4!)[B]^4t^4 + \cdots.$$

This definition provides the result that the derivative of $\exp([B]t)$ with respect to t is $[B]\exp([B]t)$ or $\exp([B]t)[B]$. Now the complete solution to $\{r(t)\} + [B]\{r(t)\} = \{R(t)\}$ can be obtained by multiplying both sides by the integrating factor $\exp([B]t)$. The result is

$$\exp(+[B]t)\{\dot{r}(t)\} + \exp(+[B]t)[B]\{r(t)\} = \exp(+[B]t)\{R(t)\}$$

$$\text{or} \quad \frac{d}{dt}(\exp(+[B]t)\{r(t)\}) = \exp(+[B]t)\{R(t)\}.$$

Since the integral of a matrix is the matrix of the integrals of the matrix entries, integrating the above equation from time equals 0 to time equals t yields

$$\exp(+[B]t)\{r(t)\} - \{r(0)\} = \int_0^t \exp(+[B]t)\{R(t)\}\, dt.$$

Hence the complete solution for the phase state vector containing both the velocities and the deflections, and their initial conditions, is

$$\{r(t)\} = \exp(-[B]t)\{r(0)\} + \exp(-[B]t)\int_0^t \exp(+[B]t)\{R(t)\}\,dt.$$

This is the result that provides a means of dealing with a nonproportional damping matrix that cannot be approximated by a diagonal matrix after a modal transformation.

REFERENCES

9.1 Chapra, S. C., and R. P. Canale, *Numerical Methods for Engineers*, 2nd ed., McGraw-Hill, New York, 1988.

9.2 Hoffman, J. D., *Numerical Methods for Engineers and Scientists*, McGraw-Hill, New York, 1992.

9.3 Thomson, W. T., *Vibration Theory and Applications*, Prentice Hall, Englewood Cliffs, NJ, 1965, p. 140.

9.4 Newmark, N. M., "A method for computation for structural dynamics," *J. Eng. Mech. ASCE*, **85**, 1959, pp. 67–94.

9.5 Chopra, A. K., *Dynamics of Structures, Theory and Applications to Earthquake Engineering*, Prentice Hall, Englewood Cliffs, NJ, 1995, p. 164.

9.6 Ziegler, Hans, *Principles of Structural Stability*, Blaisdell, Waltham, MA, 1968, p. 29ff.

CHAPTER 9 EXERCISES

9.1 **(a)** Using Taylor series expansions, derive a central finite difference approximation for a second total derivative at the jth time step where the error is of order Δt^4.

(b) Using Taylor series expansions, derive a central finite difference approximation for a fourth total derivative at the jth spatial step where the error is of order Δx^2.

9.2 **(a)** Using spreadsheet software, repeat the step time integration of Example 9.1, but this time use the larger time step of 0.04 sec for 50 time steps. Compare this result with that obtained in Example 9.1.

(b) As in part (a), use a time step of 0.01 sec for 200 time steps.

9.3 Using spreadsheet software and (a) Heun's method; (b) the fourth order Runge-Kutta method; numerically integrate the linear, undamped, one-DOF equation of Example 9.1, but here let the base motion input be a step function. That is, numerically solve

$$\ddot{q}(t) + \omega_1^2 q(t) = \omega_1^2 Y_0 stp(t - 0).$$

Again, let the natural frequency have a unit value, and extend the integration from time zero to $t = 10$ sec. Choose your own time step. Compare your numerical result with the correct analytical solution.

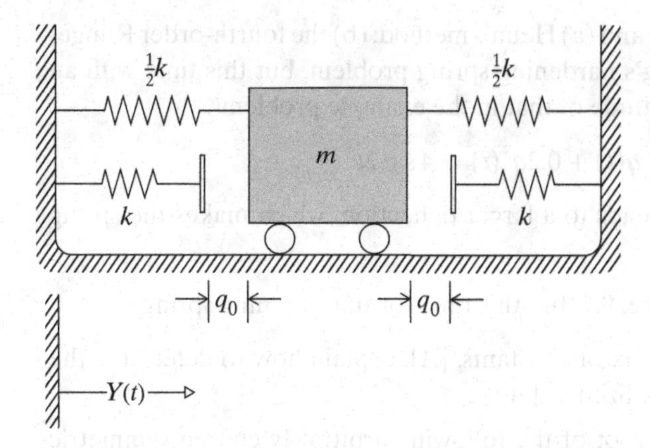

Figure 9.12. Exercise 9.4: A piecewise linear, one-DOF, vibratory system.

9.4 (a) Using spreadsheet software and (a) Heun's method; (b) the fourth order Runge-Kutta method; find a numerical solution for the horizontal, one-DOF motion of the mass shown in Figure 9.12 for the time period (0. 6.28 sec). Note that the second set of springs do not engage the mass unless the mass' deflection exceeds q_0, and then only the spring on one side of the mass resists the motion to that side. Assume the mass maintains constant contact with the second spring whenever its deflection exceeds q_0. (This is one way to approximate a nonlinear spring by use of two linear segments for the force–displacement curve.) The differential equation that describes this motion is called piecewise linear. Let the small deflection natural frequency of the system have a unit value, and let the base motion be described by the equation

$$Y(t) = Y_0 \sin 2t.$$

Hint: For example, in Excel, the "logical if" function can be used to replicate the effect of the initial gap between the mass and the second set of springs.

9.5 (a) Using spreadsheet software and either (i) Heun's method or (ii) the fourth-order Runge-Kutta method, numerically integrate the following single-DOF equation with velocity-squared damping

$$\ddot{q}(t) + 0.1[\dot{q}(t)]^2 + q(t) = \sin 2t.$$

Integrate from time zero to $t = 6.28$ sec. Choose your own time step, and then halve that time step for a check on your first calculation for this nonlinear equation.

(b) As above, letting $g/L = \pi^2$, corresponding to a 2-sec period, numerically integrate the following nonlinear pendulum differential equation subject to the stated BCs

$$\ddot{\theta}(t) + \frac{1}{2}\frac{g}{L}\sin\theta = 0 \qquad \theta(0) = \frac{\pi}{3}, \qquad \dot{\theta}(0) = 0$$

through three full swings of the pendulum. Is the period of the vibration larger or smaller than that predicted by the linear equation?

9.6 Using spreadsheet software and (a) Heun's method; (b) the fourth-order Runge-Kutta method; redo the Duffing's hardening spring problem, but this time with an input force four times the magnitude of that in the example problems.

$$\ddot{q}(t) + q(t) + 0.2q^3(t) = 4\sin 2t.$$

A force of a larger magnitude leads to a larger deflection, which makes the spring nonlinearity more prominent.

9.7 Repeat the previous exercise, 9.6, but this time for the softening spring.

9.8 **(a)** Given a symmetric matrix of constants, $[A]$, explain how to determine the square root of this matrix, symbolized as $[A]^{1/2}$.

(b) Find and check the square root of the following arbitrarily chosen symmetric matrix. (From examination of the main diagonal determinants of various sizes, it can be concluded that this matrix is positive definite. Hence the eigenvalues will all be positive, but since it was arbitrarily chosen do not expect the eigenvectors to follow the normal pattern for nodes.)

$$[A] = \begin{bmatrix} 6 & 2 & 0 \\ 2 & 8 & 1 \\ 0 & 1 & 8 \end{bmatrix}.$$

APPENDIX I

Answers to Exercises

CHAPTER 1 SOLUTIONS

1.1 **(a)** Draw an arbitrary velocity vector for the mass, $\boldsymbol{v}$, and another arbitrary velocity vector, $\boldsymbol{v}_2$, to represent the motion of the second coordinate system relative to the first coordinate system. Placing these two vectors tail to tail shows that the velocity of the mass relative to the second coordinate system is $\boldsymbol{v} - \boldsymbol{v}_2$, which is the vector connecting the heads of the first two vectors. Since $\boldsymbol{v}_2$ is a constant, the time derivative of $\boldsymbol{v} - \boldsymbol{v}_2$, the acceleration of the mass relative to (i.e., "in") the second coordinate system is the same as that for mass relative to the first coordinate system. Then Newton's second law is exactly the same in both coordinate systems.

(b) The angular momentum of any one of the particles is simply the cross product of the position vector and the linear momentum vector. The latter vector is, of course, the product of the particle mass and the time rate of change of the position vector. Then the time rate of change of the angular momentum involves two terms that arise from differentiating with respect to time. The first term is zero because it is the cross product of the time rate of change of the position vector with itself. The second term is also zero because the second time derivative of the position vector is zero in this case of constant velocity:

$$\sum_i^n \frac{d}{dt}(\boldsymbol{r}_i \times m_i\dot{\boldsymbol{r}}_i) = \sum_i^n [(\dot{\boldsymbol{r}}_i \times m_i\dot{\boldsymbol{r}}_i) + (\boldsymbol{r}_i \times m_i\ddot{\boldsymbol{r}}_i)] = 0 + 0.$$

(c) Draw a diagram of two collinear and equal but opposite forces $\boldsymbol{f}$. From the arbitrarily selected point that represents the moment center, draw the position vectors to the base points of the above two forces. Call these two position vectors $\boldsymbol{r}_1$ and $\boldsymbol{r}_2$. The cross product of these position vectors with their respective forces is equal to the product of the magnitudes of the position vectors, the magnitudes of the forces, and the sine of the angle between the positively directed position vectors and the forces. In the case where there is an obtuse angle, write the product as one with the negative value of the force and the sine of the acute angle that is the complement to the obtuse angle. From the sketch it is possible to see that the value of the product of each position vector magnitude and its respective sine term is equal to the perpendicular distance between the line of action of the forces and the arbitrarily selected moment center. Thus with the above introduction of the minus sign, the sum of the two moments is zero.

1.2 The center of mass is located at the geometric center of the rectangular parallelepiped. Locate the origin of the Cartesian coordinate system at the center of mass. The differential volume can be written as $c\,dx\,dy$, which is equivalent to immediately carrying out the integration over the z variable. Thus

$$H_{CG} = c\rho \int_{-b/2}^{b/2} \int_{-a/2}^{a/2} (x^2 + y^2) dx\, dy$$

$$= \frac{c\rho}{12}(a^3 b + ab^3) = \frac{m}{12}(a^2 + b^2),$$

where the total mass $m = \rho abc$.

1.3 **(a)** Two. One way to view this task is to break the motion into a rolling motion and a slipping motion. When the cylinder rolls without slipping, the distance the cylinder rolls is the same as the distance around the circumference that defines the total angle of the cylinder rotation. Thus either the distance the cylinder rolls or the angle of rotation could be chosen as the one generalized coordinate that defines the location of all mass particles in the cylinder for the rolling portion of the motion. When the cylinder slips, another coordinate measures how far the cylinder moves without rotating. A simpler and more basic approach is simply to have one DOF for the distance from the datum to the point of contact between the cylinder and the plane and another to define the rotation of the cylinder with respect to its original orientation.

(b) Five. Two DOF (e.g., polar-type coordinates) are needed to locate on the plane the point of contact between the sphere and the plane. Two more DOF (e.g., a colatitude angle and an east longitude angle) are required to locate that point of contact on the surface of the sphere. One further angle is required to define the rotation of the sphere around the instantaneous axis passing through the point of contact and the center of the sphere. In other words, this sphere can be viewed as a rigid body in space with those six spatial DOF (say three rectilinear coordinates to locate the CG, plus pitch, roll, and yaw angles) subjected to the one constraint equation that is that the sphere's center must be the distance of its radius above the plane. Six coordinates less one constraint leaves five unconstrained, independent coordinates that are the five generalized coordinates.

1.4 **(a)** The variation of the deflection function is obtained by remembering that the generalized coordinates are the dependent quantities. As such, only they have a nonzero variation. The application of the variational operator to the independent quantities that are part of the deflection function (e.g., f, X) and constants (e.g., ℓ, EI) only produces zero. Thus

$$\delta w(x) = (2X^3 - 3X^2 + 1)\delta w_1 + \ell(X^3 - 2X^2 + X)\delta\theta_1$$
$$+ (-2X^3 + 3X^2)\delta w_2 + \ell(X^3 - X^2)\delta\theta_2.$$

(b) Give each generalized coordinate a virtual change in turn and note the movement of the externally applied forces. When q_1 is augmented by the positive quantity δq_1, the sole external force F doesn't move because its point of application is fixed by the constant value of q_2. When q_2 is augmented, the virtual word done is $+F\delta q_2$. Hence, $Q_1 = 0$, whereas $Q_2 = +F$.

(c) The virtual work done by the external loads is

$$\delta W = -f_0 \int_0^L [\delta q_1 \sin(\pi x/L) + \delta q_2 \sin(2\pi x/L)]dx$$

$$= -f_0 \delta q_1 \int_0^L \sin(\pi x/L)\, dx - f_0 \delta q_2 \int_0^L \sin(2\pi x/L) dx$$

$$= -\frac{2}{\pi} f_0 L \delta q_1$$

Carrying out the integrations yields

$$Q_1 = -\frac{2}{\pi} f_0 L, \quad Q_2 = 0.$$

1.5 Let F_r and F_θ be the conservative or nonconservative forces at the CG in the r and θ directions, respectively. Then,

$$T = {}^1\!/\!_2 m(\dot{r}^2 + r^2\dot{\theta}^2) \quad \text{and} \quad \delta W = F_r \delta r + F_\theta r \delta\theta.$$

Substitution into the Lagrange equations with $q_1 = r$, $q_2 = \theta$ yields

$$F_r = m\ddot{r} - mr\dot{\theta}^2 \quad \text{and} \quad F_\theta = mr\ddot{\theta} + 2m\dot{r}\dot{\theta}.$$

1.6 **(a)** Draw a diagram of the Earth's equatorial plane with the projection of the north pole at the center of the equatorial disk. Let x, y be the (valid) fixed Cartesian coordinate system originating at the center of the equatorial disk, and XY be the (invalid) coordinate system fixed at a general point on the rotating equator that has a counterclockwise rotation from the x axis of magnitude ωt. Let X be the above-the-equator altitude coordinate, whereas Y is the equatorial tangential coordinate. This latter coordinate system is perfectly fine for everything but writing Newton's law or equations derived from Newton's laws such as the Lagrange equations. Let u, v be the generalized coordinates of the particle. Let these two DOF originate at the origin of the XY coordinate system and parallel those two axes respectively. Now, it is correct to write the virtual work equation in terms of those generalized coordinates and, in effect, the invalid coordinate system, because virtual work has nothing to do with Newton's laws. Therefore, write

$$\delta W = F_X \delta u + F_Y \delta v.$$

However, the kinetic energy, which is part of the Lagrange equation, must be written in terms of the fixed coordinate system. However, once written in terms of the fixed system, valid substitutions can be made freely. That is, write

$$T = {}^1\!/\!_2 m(\dot{x}^2 + \dot{y}^2)$$
$$\text{where } x = (R+u)\cos\omega t - v\sin\omega t$$
$$y = (R+u)\sin\omega t + v\cos\omega t.$$

Now it is simply a matter of differentiating x and y with respect to time and substituting the result into T, and the T and the δW result into the Lagrange equations for u and v.

(b) The derivative of the position vector $r(t)$ is

$$\dot{\boldsymbol{r}} = \dot{r}\boldsymbol{p} + r\dot{\boldsymbol{p}}.$$

Since there is no rotational kinetic energy for the particle, the total kinetic energy is

$$T = {}^1\!/\!_2 m(\dot{r}^2 + r^2\dot{\phi}^2 + r^2\dot{\theta}^2\sin^2\phi),$$

which shows that the radial arm for the motion in the θ direction (just as was true for the spherical pendulum problem) is $r \sin\phi$. Thus, with the gravitational force being just one part of $\boldsymbol{F}_r$, the virtual work is

$$\delta W = F_r \delta r + F_\phi r \delta\phi + F_\theta r \sin\phi \delta\theta .$$

Substituting into the Lagrange equations of motion, where again the gravitation force is simply one part of $\boldsymbol{F}_r$, and then simplifying, leads to

$$\begin{aligned} m\ddot{r} &= F_r + mr\dot{\phi}^2 + mr\dot{\theta}^2 \sin^2\phi \\ mr\ddot{\phi} &= F_\phi - 2m\dot{r}\dot{\phi} + {}^1\!/\!_2 mr\dot{\theta}^2 \sin 2\phi \\ mr\ddot{\theta} &= F_\theta \csc\phi - 2m\dot{r}\dot{\theta} - 2mr\dot{\theta}\dot{\phi}\cot\phi . \end{aligned}$$

(d) Again the first task is to obtain orthogonal rectilinear velocity components for the center of mass. Since the horizontal displacement of the center of mass is $u - (L/2)(1 - \cos\psi)$, and the vertical displacement is ${}^1\!/\!_2 L \sin\psi$, differentiating with respect to time provides

$$T = \frac{1}{2} m \left[\left(\dot{u} - \frac{L}{2}\dot{\psi} \sin\psi \right)^2 + {}^1\!/\!_4 L^2 \dot{\psi}^2 \cos^2\psi \right] + \frac{1}{2}\left(\frac{mL^2}{12}\right)\dot{\psi}^2$$

$$V = \frac{1}{2} mg L \sin\psi \qquad \text{thus} \qquad m(\dot{u} - L\dot{\psi}\sin\psi) = C_1$$

$$\text{and} \qquad \frac{7mL^2}{12}\ddot{\psi} - \frac{mL^2}{2}\ddot{u}\sin\psi + \frac{1}{2} mg L \cos\psi = 0,$$

where the wall and floor reaction forces do no virtual work and the angle ψ is not small. Clearly the first of these two equations of motion can be use to eliminate $u(t)$ in the second equation of motion with the result of one nonlinear equation in the one unknown time function, ψ.

1.8 Calculate the first Lagrange equation term with the correct order of differentiation, and then do the same calculation reversing the order of differentiation to show that there is an additional term when the order is reversed

$$\begin{aligned} \frac{d}{dt}\left(\frac{\partial T}{\partial \dot{q}}\right) &= \frac{\partial^2 T}{\partial \dot{q}^2}\ddot{q} + \frac{\partial^2 T}{\partial q \partial \dot{q}}\dot{q} + \frac{\partial^2 T}{\partial t \partial \dot{q}} \\ \frac{\partial}{\partial \dot{q}}\left(\frac{dT}{dt}\right) &= \frac{\partial}{\partial \dot{q}}\left(\frac{\partial T}{\partial \dot{q}}\ddot{q} + \frac{\partial T}{\partial q}\dot{q} + \frac{\partial T}{\partial t}\right) \\ &= \frac{\partial^2 T}{\partial \dot{q}^2}\ddot{q} + \frac{\partial^2 T}{\partial \dot{q} \partial q}\dot{q} + \frac{\partial T}{\partial q} + \frac{\partial^2 T}{\partial \dot{q} \partial t} . \end{aligned}$$

CHAPTER 2 SOLUTIONS

2.1 **(a)** This is a single-DOF pendulum problem where the motion is about a fixed axis. After drawing the system in its displaced configuration, write the equation of motion. If the fixed axis form of Newton's law is used, where weight $= mg$, obtain

$$H_{FA}\ddot{\theta} = -mgh \sin\theta,$$

$$\text{where} \qquad H_{FA} = H_{CG} + mh^2$$

$$\text{so} \quad \ddot{\theta} + \left(\frac{mgh}{H_{CG} + mh^2}\right)\theta = 0$$

$$\text{therefore} \quad \frac{2\pi}{T_1} = \omega_1 = \sqrt{\frac{mgh}{H_{CG} + mh^2}}$$

$$\text{then} \quad H_{CG} = wh\left(\frac{T_1^2}{4\pi^2} - \frac{h}{g}\right) > 0.$$

Note that the units for the mass moment of inertia check.

(b) One relatively cheap design concept is as follows. The mass of the long object can be determined by any or all of various types of measurements. The location of the center of mass can be determined by balancing the object. The design of the object can be adapted to facilitate the balancing of the object and simplifying the labeling of the CG for measuring distances from the suspension points. As indicated in the hint, measuring the periods of oscillation for the first and second suspension points (i.e., for two different pendulums with the same mass characteristics) allows forming the ratio of the natural frequency squared for the first suspension point over that for the second suspension point. That measured ratio is $\mathfrak{R} = (h_1/h_2)[(H_{cg} + mh_2^2)/(H_{cg} + mh_1^2)]$, which can be solved for the mass moment of inertia about the center of mass. Unfortunately, the mathematical result involves a ratio of accuracy losing differences between numbers with close numerical values. However, the good news is that the design can make the denominator the larger of the two differences. Then it is a simple matter of calculating the acceleration of gravity from either (use both) period measurements. There are other ways of measuring the acceleration of gravity. How is your way better, and how is it worse?

2.2 **(a)** $[mr^2(R-r)^2 + H(R-r)^2]\ddot{\theta} + mgr^2(R-r)\theta = 0$. Thus the square of the natural frequency is $\omega^2 = mgr^2(R-r)/[mr^2(R-r)^2 + H(R-r)^2]$. The natural period is $2\pi/\omega$

(b) $(\pi - 2)R\ddot{\theta} + g\theta = 0$. The natural frequency is the square root of $g/[R(\pi - 2)]$.

2.3 **(a)** It is quite possible to solve this one-DOF problem by working with each of the two rigidly connected arms individually. However, it is much easier to locate the center of mass of the system and thereby deal with the entire system in terms of this single CG. Of course, the system center of mass lies halfway between the centers of mass for each arm, which puts it at a distance $\sqrt{3}$ L/4 below the pivot point. Mimicking the analysis that lead to Eq. (2.1), leads to the following equation of motion when the mass moment of inertia of the arms is ignored

$$2M\left(\frac{\sqrt{3}L}{4}\right)^2 \ddot{\theta} + 2Mg\left(\frac{\sqrt{3}L}{4}\right)\theta = 0$$

$$\text{thus} \quad \omega^2 = \frac{4g}{\sqrt{3}L}$$

$$\text{and} \quad T = \pi\sqrt{\frac{\sqrt{3}L}{g}},$$

where, as always, the square of the circular frequency is equal to the coefficient of the θ term over the coefficient of the $\ddot{\theta}$.

(b) Since the given mass moment of inertia value for each of the two pendulum arms is already stated for an axis parallel to the desired axis passing through the system center of mass, only a transfer through a distance $L/4$ for each arm is necessary to obtain the system value of $7ML^2/48$. Again referring to Eq. (2.1), this adjustment leads to

$$\omega^2 = \frac{3\sqrt{3}g}{4L} \qquad \text{and} \qquad T = 4\pi\sqrt{\frac{L}{3\sqrt{3}g}}.$$

Thus it can be seen that the effect of the mass moment of inertia, in this case where the mass distribution is not compact about the center of mass, is 25%.

2.4 This is, of course, a three-DOF system. Arbitrarily assume that θ_3 is greater than θ_2, which in turn is greater than θ_1. With this assumption in mind, the energy quantities required for the three Lagrange equations are

$$\begin{aligned}
&T = {}^1\!/\!_2 mL^2\dot{\theta}_1^2 + {}^1\!/\!_2 mL^2\dot{\theta}_2^2 + {}^1\!/\!_2 mL^2\dot{\theta}_3^2 \\
&U = {}^1\!/\!_2 k[\alpha L(\sin\theta_2 - \sin\theta_1)]^2 + {}^1\!/\!_2 k[\alpha L(\sin\theta_3 - \sin\theta_2)]^2 \\
&\text{or} \quad U \approx {}^1\!/\!_2 \alpha^2 kL^2[(\theta_2 - \theta_1)^2 + (\theta_3 - \theta_2)^2] \\
&V = mgL(1 - \cos\theta_1) + mgL(1 - \cos\theta_2) + mgL(1 - \cos\theta_3) \\
&\text{or} \qquad V \approx {}^1\!/\!_2 mgL\,(\theta_1^2 + \theta_2^2 + \theta_3^2) \\
&\text{and} \qquad \delta W = 0.
\end{aligned}$$

Before substituting the above energy quantities into the three Lagrange equations (one for each θ), note that in this case where there is a fixed axis of rotation for each pendulum, the direct application of Newton's fixed axis rotation equation is particularly simple. After setting the sines of angles equal to the angles, the result is

$$\begin{aligned}
mL^2\ddot{\theta}_1 &= -mgL\theta_1 + \alpha Lk[\alpha L(\theta_2 - \theta_1)] \\
mL^2\ddot{\theta}_2 &= -mgL\theta_2 - k\alpha^2L^2(\theta_2 - \theta_1) + k\alpha^2L^2(\theta_3 - \theta_2) \\
mL^2\ddot{\theta}_3 &= -mgL\theta_3 - k\alpha^2L^2(\theta_3 - \theta_2).
\end{aligned}$$

The application of the Lagrange equations, of course, produces the same result. Using the stated relation $g/L = \alpha^2 k/m$ and casting the above equations in matrix form (i.e., writing the first equation in the first row, the second equation in the second row, etc.) leads to the following matrix equation, where the first square matrix is the inertia matrix divided by the mass value and the second square matrix is the stiffness matrix, also divided by the mass value

$$\begin{bmatrix} 1 & 0 & 0 \\ 0 & 1 & 0 \\ 0 & 0 & 1 \end{bmatrix} \begin{Bmatrix} \ddot{\theta}_1 \\ \ddot{\theta}_2 \\ \ddot{\theta}_3 \end{Bmatrix} + \frac{\alpha^2 k}{m} \begin{bmatrix} 2 & -1 & 0 \\ -1 & 3 & -1 \\ 0 & -1 & 2 \end{bmatrix} \begin{Bmatrix} \theta_1 \\ \theta_2 \\ \theta_3 \end{Bmatrix} = \begin{Bmatrix} 0 \\ 0 \\ 0 \end{Bmatrix},$$

Once again, it is not possible in a linear multidegree of freedom system, as it is possible in a linear one-DOF system, to determine the system natural frequencies by merely inspecting the constant coefficients of the equations of motion. How those natural frequencies are determined is explained later.

2.5 **(a)** Note that the total velocity of the center of mass is *not* simply the product of the offset distance a multiplied by the angular velocity $\dot{\theta}$. This is so because the center of mass is not moving about a fixed axis. Rather, the rectilinear portion of the kinetic energy is obtained by first obtaining the horizontal and vertical components of the total movement of the center of mass as the cylinder moves through and angle θ. These components are, respectively, $R\theta - a\sin\theta$ and $a(1-\cos\theta)$. Thus,

$$T = \tfrac{1}{2}m[(R\dot{\theta} - a\dot{\theta}\cos\theta)^2 + (a\dot{\theta}\sin\theta)^2] + \tfrac{1}{2}H\dot{\theta}^2$$
$$U = \tfrac{1}{2}k(R\theta)^2 \qquad V = mga(1-\cos\theta).$$

Substitution into the Lagrange equations produces

$$[H + m(R^2 - 2Ra\cos\theta + a^2)]\ddot{\theta} + mRa\dot{\theta}^2\sin\theta + mga\sin\theta + kR^2\theta = 0.$$

This equation linearizes to

$$[H + m(R-a)^2]\ddot{\theta} + (kR^2 + mga)\theta = 0$$

$$\text{so} \qquad T = 2\pi\sqrt{\frac{H + m(R-a)^2}{kR^2 + mga}}$$

(b) This is clearly a pendulum problem because gravity causes a restoring torque to act on the swinging gate. As always, the (constant) natural frequency is determined by first writing the (linearized) equation of motion for this single-DOF system. Using the Lagrange equation, the kinetic energy of the gate swinging through an angle θ is simply

$$T = \tfrac{1}{2}m(\tfrac{1}{2}b\dot{\theta})^2 + \tfrac{1}{2}\left(\frac{mb^2}{12}\right)\dot{\theta}^2 = \left(\frac{mb^2}{6}\right)\dot{\theta}^2 = \tfrac{1}{2}H_{FA}\dot{\theta}^2.$$

The potential energy expression is the challenge in this problem. As always, the potential energy is the weight, mg, multiplied by the height the center of mass rises vertically above its datum point, which, as per usual, is the lowest point attained by the center of mass. For the sake of greater clarity, redraw Fig. 2.16, the front view of the gate, with an angle α approximately equal to somewhere between 30° and 45°. Picture the gate rotating out of the plane of the paper, say, toward you. From this front view, the center of mass appears to move along the perpendicular line between the hinge line and the datum point for the center of mass. *In the plane of the swinging center of mass*, the center of mass moves up in that plane a distance $(\tfrac{1}{2}b)(1-\cos\Theta)$. Since the plane of the swinging center of mass is not vertical, the vertical rise of the center of mass is not this quantity. With this quantity as the hypotenuse (edge view of that swinging CG plane) of a triangle drawn on the front view, the rise of the CG is seen to be $(\tfrac{1}{2}b)(1-\cos\Theta)$ multiplied by $\sin\alpha$. Thus $V = \tfrac{1}{2}mb(1-\cos\Theta)\sin\alpha$. Note that if α is zero, then, as it should be, there is no potential energy. However, if α is $\pi/2$, then the gate is an ordinary pendulum, and the potential energy has the correct value for an ordinary pendulum. From the completed equation of motion

$$\omega_1 = \sqrt{\frac{3g\sin\alpha}{2b}}.$$

(c) The required quantities for the Lagrange equations are

$$T = \tfrac{1}{2}m[\dot{v}^2 + (L+v)^2\dot{\theta}^2] + \tfrac{1}{2}H\dot{\theta}^2 \qquad U = \tfrac{1}{2}kv^2$$
$$V = mg(L - L\cos\theta - v\cos\theta)$$
$$\text{and} \quad \delta W = F\cos\theta(L+v)\delta\theta + F\sin\theta\,\delta v$$
$$\text{therefore} \quad m\ddot{v} - m(L+v)\dot{\theta}^2 + kv = mg\cos\theta + F\sin\theta$$
$$[H + m(L+v)^2]\ddot{\theta} + 2m(L+v)\dot{v}\dot{\theta} + mg(L+v)\sin\theta = F\sin\theta.$$

2.6 **(a)** This exercise can be more easily completed by writing Newton's second law, particularly after obtaining the energy solution. However, it is important to practice using the Lagrange equation. The first step of the analysis, the sketch of the manometer tube with the fluid columns displaced; that is, with one higher than the other, is provided in the problem statement. The up or down displacement of either column is $u(t)$. Thus one column top is a distance $2u$ above the other column top. Since the entire body of the fluid moves together, the kinetic energy is $T = \frac{1}{2}(\rho AL)\dot{u}^2$. The potential energy of the displaced configuration can best be understood by visualizing what masses have to be moved within the gravitational field to reach the displaced configuration from the static equilibrium configuration, which is the datum configuration. Picture altering the static equilibrium configuration by removing the mercury at the top of one of the two equal-height columns to a depth u, without (somehow) allowing the remainder of the fluid to move. Place this removed mercury on top of the other column. Now the static equilibrium configuration has become the displaced configuration, and this change has been achieved by moving a weight ρgAu upward a distance u. Hence $V = \rho gAu^2$. Therefore

$$\rho AL\ddot{u} + 2\rho gAu = 0 \quad \text{so} \quad \omega_1 = \sqrt{\frac{2g}{L}}.$$

(b) No.

2.8 **(a)** First of all, the mass moment of inertia of the thin rod about its own center of mass about an axis perpendicular to the plane of the paper is $mL^2/12$. The key to the geometry of the problem is realizing that the distance along the rod from the center of mass to the point of contact between the rod and the cylinder is $R\theta$. Then it is relatively simple to conclude that

$$\ell = R\theta\cos\theta \quad \text{and} \quad h = R\theta\sin\theta$$

$$u = R\sin\theta - \ell \quad \text{and} \quad v = h + R\cos\theta - R$$

$$\text{thus} \quad \dot{u} = R\dot{\theta}(\cos\theta - \cos\theta + \theta\sin\theta)$$

$$\text{and} \quad \dot{v} = R\dot{\theta}(\sin\theta + \theta\cos\theta - \sin\theta)$$

$$\text{hence} \quad T = \frac{mL^2\dot{\theta}^2}{24} + \frac{mR^2\theta^2\dot{\theta}^2}{2}$$

$$\text{and} \quad V = mgR(\theta\sin\theta + \cos\theta - 1)$$

$$\text{therefore} \quad m\left(\frac{L^2}{12} + R^2\theta^2\right)\ddot{\theta} + mR^2\theta\dot{\theta}^2 + mgR\theta\cos\theta = 0$$

$$\text{linearizing: } \ddot{\theta} + 12\left(\frac{R}{L}\right)\left(\frac{g}{L}\right)\theta = 0 \qquad \omega_1 = \sqrt{12\left(\frac{R}{L}\right)\left(\frac{g}{L}\right)}.$$

(c) This is a two-DOF system. As generalized coordinates, choose the usual θ (positive counterclockwise) for the mass M, and u, positive to the right, which measures the position of the mass m *relative* to the axis of the pendulum arm. The motion of the mass m is a bit complicated. Therefore, it is best to go back to basics. Draw a large diagram of the displaced system and then calculate the horizontal and vertical deflections of the mass m:

$$\text{ver. displ.} = L(1 - \cos\theta) + u\sin\theta \quad \text{and} \quad \text{hor.displ.} = L\sin\theta + u\cos\theta$$

$$\therefore \quad \text{vertical vel.} = L\dot{\theta}\sin\theta + u\dot{\theta}\cos\theta + \dot{u}\sin\theta$$

$$\therefore \quad \text{horizon. vel.} = L\dot{\theta}\cos\theta - u\dot{\theta}\sin\theta + \dot{u}\cos\theta$$

$$\text{vel.}^2 = (L^2 + u^2)\dot{\theta}^2 + \dot{u}^2 + 2L\dot{\theta}\dot{u} = (L\dot{\theta} + \dot{u})^2 + u^2\dot{\theta}^2.$$

Then the energy expressions are

$$T = \tfrac{1}{2}M\ell^2\dot{\theta}^2 + \tfrac{1}{2}H\dot{\theta}^2 + \tfrac{1}{2}m(\text{vel.}^2) + \tfrac{1}{2}h\dot{\theta}^2 \quad \text{and} \quad U = ku^2$$
$$\text{and} \quad V = Mg\ell(1 - \cos\theta) + mgL(1 - \cos\theta) + mgu\sin\theta.$$

Now it is just a matter of substitution into the Lagrange equations of motion. That result, after some algebra, is

$$[H + h + M\ell^2 + m(L^2 + u^2)]\ddot{\theta} + 2mu\dot{u}\dot{\theta} + mL\ddot{u} + (M\ell + mL)g\sin\theta + mgu\cos\theta = 0$$
$$mL\ddot{\theta} - mu\dot{\theta}^2 + m\ddot{u} + mg\sin\theta + 2ku = 0.$$

Linearizing here begins with replacing the sine of θ by θ and the cosine of θ by 1. Linearization is completed by ruthlessly eliminating all products of the small displacements and their derivatives. Then the matrix form of the linearized equations of motion is

$$\begin{bmatrix} (H + h + M\ell^2 + mL^2) & mL \\ mL & m \end{bmatrix} \begin{Bmatrix} \ddot{\theta} \\ \ddot{u} \end{Bmatrix} + \begin{bmatrix} (M\ell + mL)g & mg \\ mg & 2k \end{bmatrix} \begin{Bmatrix} \theta \\ u \end{Bmatrix} = \begin{Bmatrix} 0 \\ 0 \end{Bmatrix}.$$

(d) With infinitely stiff springs, the two masses become one rigidly connected mass, and thus the system becomes one simple pendulum with just one DOF. An analysis of the new system would require the location of the combined center of mass of the previously distinct two masses and, using the parallel axis theorem, their combined mass moment of inertia about the combined center of mass.

(e) In the zero stiffness case, the springs disappear. Then, in the absence of friction, there would be no forces (just small moments) acting on the smaller mass. Therefore, the smaller mass would travel back and forth within the pendulum bob, but it would not translate relative to an outside observer. The straight track for the smaller mass would cause the smaller mass to rock back and forth through the angle θ as the pendulum swings through that same angle. If the track were a circular curve with radius L, then the smaller mass would not move at all.

2.9 It is not immediately apparent which of the possibilities is the best choice for the single generalized coordinate necessary to describe the instantaneous position of the trapeze bar as it twists about its axis of symmetry. In a situation like this, the best course of action is usually to first write the expressions for the kinetic and potential energies in terms of whatever (perhaps) temporary variables most simplify those expressions and then to relate those various variables. Last, select whichever generalized coordinate(s) offer the greater simplicity in the final differential equation of motion.

The kinetic energy, as per usual, has two parts: the kinetic energy of translation and that of rotation:

$$T = \tfrac{1}{2}m\dot{v}^2 + \tfrac{1}{2}\left(\frac{mL^2}{12}\right)\dot{\theta}^2 \quad \text{and} \quad V = mgv.$$

To work with the deflected trapeze geometry, introduce the angle ϕ, which is the angle between the support wire in its static equilibrium vertical position and the same wire in its deflected position. The slightly rotated plane in which this angle resides can be seen in (top) edge view in Fig. 2.19(c) by augmenting that drawing as follows. Redraw the right half of that top view where the undeflected trapeze bar is projected on the plane of the deflected trapeze bar. By means of a straight line, connect the two wire connection points of the deflected and projected bars to form an isosceles triangle in the plane of the deflected bar. The angle θ as the smaller, unequal angle. Then bisecting the angle θ leads to the relationship

$$\sin\left(\frac{\theta}{2}\right) = \frac{\tfrac{1}{2}h\sin\phi}{\tfrac{1}{2}\ell} = \left(\frac{h}{\ell}\right)\sin\phi.$$

From Fig. 2.19(b), the angle ϕ can also be related to v as

$$v = h - h\cos\phi.$$

Thus θ is related to v via ϕ, but in a complicated fashion. However, because this analysis is to be limited to small deflections, it is quite convenient to choose ϕ as the single generalized coordinate with the following geometric approximations, where the latter is based on the series expansion for the cosine

$$\theta \approx \left(2\frac{h}{\ell}\right)\phi$$

$$\text{and} \quad v \approx h\frac{\phi^2}{2}$$

$$\therefore \quad T = \tfrac{1}{2}mh^2\phi^2\dot{\phi}^2 + \left(\frac{mL^2}{24}\right)\left(2\frac{h}{\ell}\right)^2\dot{\phi}^2$$

$$\text{and} \quad V = \tfrac{1}{2}mgh\phi^2.$$

Substituting into the Lagrange equation and linearizing leads to the result

$$\ddot{\phi} + 3\left(\frac{g}{h}\right)\left(\frac{\ell^2}{L^2}\right)\phi = 0$$

$$\text{so} \quad \omega_1 = \frac{\ell}{L}\sqrt{\frac{3g}{h}}.$$

2.10 **(a)** To understand the geometry of the motion of this horizontal pendulum, first consider point B. Point B must always lie on a sphere of radius a centered at point C. Point B must also lie on a sphere of radius h centered at point D. Therefore, point B must move on the circular arc that is the intersection of those two spheres. The edge view of that circular arc is obtained by first drawing a straight line from point C to point D and beyond and then drawing the perpendicular to this line from point B. That perpendicular is the edge view of the circular arc followed by point B. Similarly, drawing the perpendicular from point A to this extended line CD (call that intersection point, point F) provides the edge view of the circular arc on which the mass m moves. In other words, the line AC traces out a conical surface as the pendulum arm moves in and out of the plane of the paper, where the apex is, of course, point C and the axis of the cone is the line CF. From the geometry of the triangles ACF and BCD, the radius of the circular arc along which the mass moves is

$$\text{dist}\{AF\} = \frac{bh}{\sqrt{a^2 + h^2}}.$$

Since this is a case of rotation about a fixed axis, CF, the simplest approach is to use the fixed axis version of Newton's law for rotational motion

$$H_{FA}\ddot{\theta} = M_{FA},$$

$$\text{where} \quad H_{FA} = \frac{mb^2h^2}{a^2 + h^2},$$

where θ is the angular DOF of the mass in the skewed plane of motion of the mass (the plane with edge view AF). The gravitational moment in that same plane is constructed by combining the component acting in that plane of the vertical weight vector with the usual pendulum moment arm. That is,

$$\left(\frac{mb^2h^2}{a^2+h^2}\right)\ddot{\theta} = -\left(\frac{mga}{\sqrt{a^2+h^2}}\right)\left(\frac{bh}{\sqrt{a^2+h^2}}\right)\sin\theta$$

$$\text{or}\quad \ddot{\theta} + \left(\frac{g}{h}\frac{a}{b}\right)\theta = 0$$

$$\text{so}\quad \omega_1 = \sqrt{\frac{g}{h}\frac{a}{b}}.$$

(b) Looking first at the effect of the circular arc on the pendulum arm and then the result of the rotated straight line, the vertical rise and horizontal displacement of the pendulum bob are, respectively,

$$\Delta y = R(\theta - \sin\theta) + (L - R\theta)(1 - \cos\theta)$$
$$\Delta x = (R - R\cos\theta) + (L - R\theta)\sin\theta.$$

Multiplying the first of the above expressions by the bob weight provides the potential energy expression. Differentiating both expressions with respect to time provides the velocities required for the kinetic energy expression. The result is

$$T = \tfrac{1}{2}m[(L - R\theta)^2\dot{\theta}^2\sin^2\theta + (L - R\theta)^2\dot{\theta}^2\cos^2\theta] = \tfrac{1}{2}m(L - R\theta)^2\dot{\theta}^2.$$

The interesting aspect of this kinetic energy expression is that it illustrates the fact that the total velocity is merely the tangential velocity with a moment arm measured from the instantaneous beginning of the straight-line portion of the pendulum arm. Since these energy expressions are nonlinear, and the system model does not include friction, it would be better to write the constant energy equation rather than write out the Lagrange equation.

2.12 **(a)** The task is to relate $\dot{\phi}$ to $\dot{\theta}$ in this one-DOF system. The strategy to be followed for this purpose is to focus on the distance the roller moves over the supporting surface, $s(t)$, which is the distance from where θ has the value zero to its instantaneous point of contact with the support surface. This distance in terms of the angle ϕ is simply $r\phi$. This same distance in terms of the angle θ is a bit more complicated. The quantity $s(\theta)$ can be determined from integrating ds.

First note that because R is not constant, it is *not* true that along the arc of the supporting surface, $ds = R(\theta)d\theta$. One way of obtaining the correct value of ds is to introduce the quantity $v(\theta)$, which is the vertical rise of the support surface above the point where θ is zero. Also introduce $h(\theta)$, which is the horizontal projection of the point on the support surface at the angle θ, again measured from the point where θ is zero. Therefore,

$$v(\theta) = R_0 - R(\theta)\cos\theta \quad\text{and}\quad h(\theta) = R(\theta)\sin\theta$$

$$\text{and}\quad ds = \sqrt{(dv)^2 + (dh)^2} = d\theta\sqrt{\left(\frac{dv}{d\theta}\right)^2 + \left(\frac{dh}{d\theta}\right)^2}$$

$$\text{since}\quad \frac{dv}{d\theta} = R\sin\theta - R'\cos\theta$$

$$\text{and}\quad \frac{dh}{d\theta} = R\cos\theta + R'\sin\theta$$

$$\text{then}\quad ds = [R^2 + (R')^2]^{1/2}d\theta.$$

Integrating over θ and differentiating with respect to time yields the result stated. There is no differentiation of the integrand with respect to time because the θ of the integrand is not a function of time. That is, these intermediate values of θ are just geometric values up to the limiting generalized coordinate value of θ that appears as the upper limit. Of course, the generalized coordinate is a function of time and, as such, has a time derivative. This result is confirmed simply by dividing both sides of the last of the above equations by the differential dt.

(b) The kinetic energy and potential energy expressions for the one-DOF circular cylinder rolling on a very smooth, concave circular surface whose radius varies with the angle θ are, where $R_0 = R(0)$ and again $R' = dR/d\theta$

$$T = \tfrac{1}{2} H\,[d/dt \text{ of angle of rotation w.r.t. fixed axis}]^2$$

$$+\tfrac{1}{2} m\left(\dot{s} + \tfrac{1}{2} r\dot{\phi}\right)^2$$

$$T = \tfrac{1}{2} H\left[\dot{\phi} - \frac{d}{dt}\left(\frac{dv}{dh}\right)\right]^2 + \tfrac{1}{2} m\dot{\theta}^2[(R')^2 + R^2]$$

$$+\tfrac{1}{2} mr\dot{\theta}\dot{\phi}[(R')^2 + R^2]^{\frac{1}{2}} + \tfrac{1}{2} m\left(\tfrac{1}{4} r^2\dot{\phi}^2\right)$$

$$V = mg\{v - r[1 - \cos(dv/dh)]\}$$

$$V = mg[R_0 - R\cos\theta - r + r\cos(dv/dh)].$$

From part (a),

$$\frac{dv}{dh} = \frac{dv/d\theta}{dh/d\theta} = \frac{R\sin\theta - R'\cos\theta}{R\cos\theta + R'\sin\theta}.$$

Differentiating this expression with respect to θ, in route to obtaining its time derivative, is best left undone. The general complexity of these expressions strongly suggests that this problem is not best formulated in terms of the varying radius R. A formulation directly in terms of $v(h)$ also has its difficulties.

CHAPTER 3 SOLUTIONS

3.1 **(a)** Let the left-hand beam element be numbered 10 and the right-hand beam element be numbered 20. With these two beam elements merely bending in the x, z plane this is a six-DOF system. Name the six global DOF $w_1 = w(0)$, $\theta_1 = w'(0)$, $w_2 = w(L/2)$, $\theta_1 = w'(L/2)$, $w_3 = w(L)$, and $\theta_3 = w'(L)$. Since each beam element is of length $L/2$, the two element stiffness matrices with the corresponding global DOF are

$$[k_{10}]\{q\}^{(10)} = \frac{EI}{L^3}\begin{bmatrix} 96 & 24L & -96 & 24L \\ 24L & 8L^2 & -24L & 4L^2 \\ -96 & -24L & 96 & -24L \\ 24L & 4L^2 & -24L & 8L^2 \end{bmatrix}\begin{Bmatrix} w_1 \\ \theta_1 \\ w_2 \\ \theta_2 \end{Bmatrix}$$

and

$$[k_{20}]\{q\}^{(20)} = \frac{EI}{L^3}\begin{bmatrix} 96 & 24L & -96 & 24L \\ 24L & 8L^2 & -24L & 4L^2 \\ -96 & -24L & 96 & -24L \\ 24L & 4L^2 & -24L & 8L^2 \end{bmatrix}\begin{Bmatrix} w_2 \\ \theta_2 \\ w_3 \\ \theta_3 \end{Bmatrix}.$$

The result of combining the two element stiffness matrices to obtain the system stiffness matrix equation is

$$\begin{Bmatrix} R \\ 0 \\ F \\ 0 \\ R \\ 0 \end{Bmatrix} = \frac{EI}{L^3}\begin{bmatrix} 96 & 24L & -96 & 24L & 0 & 0 \\ 24L & 8L^2 & -24L & 4L^2 & 0 & 0 \\ -96 & -24L & 192 & 0 & -96 & 24L \\ 24L & 4L^2 & 0 & 16L^2 & -24L & 4L^2 \\ 0 & 0 & -96 & -24L & 96 & -24L \\ 0 & 0 & 24L & 4L^2 & -24L & 8L^2 \end{bmatrix}\begin{Bmatrix} w_1 \\ \theta_1 \\ w_2 \\ \theta_2 \\ w_3 \\ \theta_3 \end{Bmatrix}.$$

Insert the boundary conditions that $w_1 = w_3 = 0$. Now note that everywhere there is a known deflection DOF, there is an unknown external generalized force and vice versa. Now the first and fifth rows can be set aside and the first and fifth columns deleted. The remaining four simultaneous equations can then for solved for the remaining DOF. However, based on the symmetry of the structure, there are important simplifications possible. The first is to note that θ_2 is zero, which deletes the fourth of the original six columns. After deleting the fourth row, the matrix equation is

$$\begin{Bmatrix} 0 \\ F \\ 0 \end{Bmatrix} = \frac{EI}{L^3}\begin{bmatrix} 8L^2 & -24L & 0 \\ -24L & 192 & 24L \\ 0 & 24L & 8L^2 \end{bmatrix}\begin{Bmatrix} \theta_1 \\ w_2 \\ \theta_3 \end{Bmatrix}.$$

From the first and third rows

$$\theta_1 = -\theta_3 = 3\frac{W_2}{L}$$

Substituting this result into the middle equation yields the perhaps familiar result

$$w_2 \equiv w\left(\frac{L}{2}\right) = \frac{FL^3}{48\,EI}.$$

(b) In the 6×6 system matrix equation stated above, the new boundary and symmetry conditions are such that all system DOF are zero except w_2. Therefore, deleting the first, second, fourth, fifth, and sixth columns and setting aside the corresponding rows leads immediately to

$$w_2 \equiv w\left(\frac{L}{2}\right) = \frac{FL^3}{192\,EI}.$$

These two solutions provide the stiffness factors, or flexibility factors, at the midspans of simply supported and clamped–clamped beams.

3.2 **(a)** Since this is a small problem, it is not necessary to show the intermediate steps. (Part (b) shows intermediate steps.) Thus the matrix equation involving all seven DOF is

$$\begin{Bmatrix} R \\ R \\ 0 \\ M_0 \\ R \\ R \\ R \end{Bmatrix} = \frac{EI}{\ell^3}\begin{bmatrix} 12 & 6\ell & -12 & 6\ell & 0 & 0 & 0 \\ 6\ell & 4\ell^2 & -6\ell & 2\ell^2 & 0 & 0 & 0 \\ -12 & 6\ell & 41 & 6\ell & -5 & -24 & 12\ell \\ 6\ell & 2\ell^2 & 6\ell & 12\ell^2 & 0 & -12\ell & 4\ell^2 \\ 0 & 0 & -5 & 0 & 5 & 0 & 0 \\ 0 & 0 & -24 & -12\ell & 0 & 24 & -12\ell \\ 0 & 0 & 12\ell & 4\ell^2 & 0 & -12\ell & 8\ell^2 \end{bmatrix}\begin{Bmatrix} w_1 \\ \theta_1 \\ w_2 \\ \theta_2 \\ w_3 \\ w_4 \\ \theta_4 \end{Bmatrix}.$$

After the application of the boundary conditions where all the DOF but those associated with node 2 are zero, the final FEM equation is just

$$\begin{Bmatrix} 0 \\ M_0 \end{Bmatrix} = \frac{EI}{\ell^3}\begin{bmatrix} 41 & 6\ell \\ 6\ell & 12\ell^2 \end{bmatrix}\begin{Bmatrix} w_2 \\ \theta_2 \end{Bmatrix}.$$

(b) Since this problem is a fairly large hand calculation, it is useful to show the intermediate steps. Start with beam 12 bending about the x axis, then the y axis, and then twisting. After applying the clamped support boundary conditions, those element matrices are, respectively,

$$\frac{EI}{\ell^3}\begin{bmatrix} 3 & 3\ell \\ 3\ell & 4\ell^2 \end{bmatrix}\begin{Bmatrix} v_2 \\ \psi_2 \end{Bmatrix} \quad \frac{EI}{\ell^3}\begin{bmatrix} 41 & 6\ell \\ 6\ell & 12\ell^2 \end{bmatrix}\begin{Bmatrix} w_2 \\ \theta_2 \end{Bmatrix} \quad \frac{EI}{\ell^3}\begin{bmatrix} 0.25\ell^2 \end{bmatrix}\begin{Bmatrix} \phi_2 \end{Bmatrix}.$$

Turning now to beam 23, where the small displacement DOF w_2 is zero, the element matrices for bending about the x axis and twisting about the y axis are, respectively,

$$\frac{EI}{\ell^3}\begin{bmatrix} 4\ell^2 & -6\ell & 2\ell^2 \\ -6\ell & 12 & -6\ell \\ 2\ell^2 & -6\ell & 4\ell^2 \end{bmatrix}\begin{Bmatrix} \psi_2 \\ w_3 \\ \psi_3 \end{Bmatrix} \quad \frac{EI}{\ell^3}\begin{bmatrix} 0.5\ell^2 & -0.5\ell^2 \\ -0.5\ell^2 & 0.5\ell^2 \end{bmatrix}\begin{Bmatrix} \theta_2 \\ \theta_3 \end{Bmatrix}.$$

The element stiffness matrix for bending about the z axis for beam 23 is

$$\frac{EI}{\ell^3}\begin{bmatrix} 12 & -6\ell & -12 & -6\ell \\ -6\ell & 4\ell^2 & 6\ell & 2\ell^2 \\ -12 & 6\ell & 12 & 6\ell \\ -6\ell & 2\ell^2 & 6\ell & 4\ell^2 \end{bmatrix}\begin{Bmatrix} u_2 \\ \phi_2 \\ u_3 \\ \phi_3 \end{Bmatrix}.$$

3.3 The key to this problem is understanding that, because the displacements are small, the deflections in one direction do not disturb the symmetry of the deflections in the other direction, and there is no significant relative twisting of any of the beam-columns. In other words, the small deflections in the x direction are independent of those in the y direction. Thus, because of the symmetry of the structure and the loading, $u_1 = u_2$ and $v_1 = v_2$. Furthermore, the symmetry extends to the nodal rotations. Therefore, $\theta_1 = \theta_3$, $\phi_1 = \phi_2$, and, as before, all the ψs are zero. Furthermore, as can be seen from a side-view sketch of the deflection pattern, because the upper beams are inextensible, all the θ DOF are the same and all the ϕ DOF are the same. Thus the number of independent DOF is just 4: u, v, θ, and ϕ. Consider deflections in the x direction. The two beams between nodes 1 and 3 and between 2 and 4 provide no resistance to motion in the x direction. For each of the four beam-columns and for the two upper beams whose ends are rotated when there are deflections in the x direction (the beams between nodes 1 and 2, and between nodes 3 and 4), the element matrix equations are respectively

$$\frac{EI}{\ell^3}\begin{bmatrix} 12 & -6\ell \\ -6\ell & 4\ell^2 \end{bmatrix}\begin{Bmatrix} u \\ \theta \end{Bmatrix} \quad \frac{EI}{\ell^3}\begin{bmatrix} 4\ell^2 & 2\ell^2 \\ 2\ell & 4\ell^2 \end{bmatrix}\begin{Bmatrix} \theta \\ \theta \end{Bmatrix}.$$

The latter $[k]\{q\}$ product reduces to just having the equal nodal bending moments equal to simply $(EI/\ell^3)(6\ell^2)\theta$. Of course there is a similar result for y-direction deflections. The assembly of the global stiffness matrix leads to

$$\begin{Bmatrix} 2F_1 \\ 0 \\ 2F_2 \\ 0 \end{Bmatrix} = \frac{EI}{\ell^3}\begin{bmatrix} 48 & -24\ell & 0 & 0 \\ -24\ell & 28\ell^2 & 0 & 0 \\ 0 & 0 & 48 & 24\ell \\ 0 & 0 & 24\ell & 28\ell^2 \end{bmatrix}\begin{Bmatrix} u \\ \theta \\ v \\ \phi \end{Bmatrix},$$

which is easily solved for the four unknown DOF.

3.4 Assembling the 3×3 element stiffness matrices for beams 10 and 30, and the 6×6 element matrix for beam 20, and writing the virtual work expression leads to

$$\begin{Bmatrix} F_1 \\ 0 \\ 0 \\ 0 \\ -M_2 \\ -M_1 \end{Bmatrix} = \frac{EI}{\ell^3}\begin{bmatrix} 18.5 & -6\ell & 1.5\ell & -1.5\ell & 0 & 1.5\ell \\ -6\ell & 4.25\ell^2 & 0 & 0 & -0.25\ell^2 & 0 \\ 1.5\ell & 0 & 2.5\ell & -1.5\ell & 0 & \ell^2 \\ -1.5 & 0 & -1.5\ell & 16.5 & -6\ell & -1.5\ell \\ 0 & -0.25\ell^2 & 0 & -6\ell & 4.25\ell^2 & 0 \\ 1.5\ell & 0 & \ell^2 & -1.5\ell & 0 & 2.5\ell^2 \end{bmatrix}\begin{Bmatrix} v_2 \\ \theta_2 \\ \phi_2 \\ v_3 \\ \theta_2 \\ \phi_3 \end{Bmatrix}.$$

3.5 There are three generalized coordinates requiring solution: u_1, θ_1, and θ_2. However, the known displacement, u_0, must also be included in the stiffness matrix equations. The element stiffness equations are

$$k_{10}u_1 = \frac{EI}{L^3}[3]u_1 \qquad [k_{20}]\{q\} = \frac{EI}{L^3}\begin{bmatrix} 12 & -12 & 6L \\ -12 & 12 & -6L \\ 6L & -6L & 4L^2 \end{bmatrix}\begin{Bmatrix} -u_0 \\ u_1 \\ \theta_1 \end{Bmatrix}$$

$$[k_{30}]\{q\} = \frac{EI}{L^3}\begin{bmatrix} 2L^2 & L^2 \\ L^2 & 2L^2 \end{bmatrix}\begin{Bmatrix} \theta_1 \\ \theta_2 \end{Bmatrix}$$

$$[k_{40}]\{q\} = \frac{EI}{L^3}\begin{bmatrix} 24 & -12L \\ -12L & 8L^2 \end{bmatrix}\begin{Bmatrix} u_1 \\ \theta_2 \end{Bmatrix}.$$

The fact that the global DOF for the spring is oppositely directed relative to the spring element DOF does not result in a sign change for any diagonal element or, for that matter, for any element of a 2×2 stiffness matrix. Note that the stiffness matrix for beam 30 reflects its length of $2L$, whereas that for beam 40 reflects its larger stiffness coefficient. Assembling the global stiffness matrix, partitioning and removing the column associated with the input, and setting aside the corresponding row lead to the following equation to be solved for the unknown DOF:

$$\frac{EI}{L^3}\begin{bmatrix} 39 & -6L & -12L \\ -6L & 6L^2 & L^2 \\ -12L & L^2 & 10L^2 \end{bmatrix}\begin{Bmatrix} u_1 \\ \theta_1 \\ \theta_2 \end{Bmatrix} = \frac{EI}{L^3}u_0\begin{Bmatrix} -12 \\ 6L \\ 0 \end{Bmatrix},$$

where the stiffness coefficients can be canceled. In the case of only three algebraic simultaneous equations to be solved by hand, Cramer's rule would be suitable. From that perspective, each of the unknown DOF is proportional to the input u_0.

CHAPTER 4 SOLUTIONS

4.1 **(a)** The augmented stiffness matrix is

$$[k]\{w\} = \frac{EI_0}{L^3}\begin{bmatrix} 36 & -6L & & -12 & 6L & \\ -6L & 12L^2 & & -6L & 2L^2 & \\ & & 3\beta L^2 & & & -\beta L^2 \\ -12 & -6L & & 36 & +6L & \\ 6L & 2L^2 & & +6L & 12L^2 & \\ & & -\beta L^2 & & & 3\beta L^2 \end{bmatrix}\begin{Bmatrix} w_1 \\ \theta_1 \\ \phi_1 \\ w_2 \\ \theta_2 \\ \phi_2 \end{Bmatrix}.$$

(b) No and no. The mass matrix and applied load vectors are the same because the DOF are the same, and the loads and inertial properties are also the same. Only the stiffness matrix changes.

(c) The first change would be that there would now be three more required DOF, the three bending slope rotations at the bases of the columns. Hence the mass and stiffness matrices would now be 7×7. It likely would be appropriate to estimate mass moments of inertia corresponding to these DOF.

4.2 **(a)** The free vibration matrix equation of motion is

$$\begin{bmatrix} m & & & \\ & H & & \\ & & 2m & \\ & & & 2H \end{bmatrix}\begin{Bmatrix} \ddot{w}_1 \\ \ddot{\theta}_1 \\ \ddot{w}_2 \\ \ddot{\theta}_2 \end{Bmatrix} + \frac{EI_0}{L^3}\begin{bmatrix} 19 & 6L & -12 & 6L \\ 6L & 4L^2 & -6L & 2L^2 \\ -12 & -6L & 25 & -6L \\ 6L & 2L^2 & -6L & 7L^2 \end{bmatrix}\begin{Bmatrix} w_1 \\ \theta_1 \\ w_2 \\ \theta_2 \end{Bmatrix} = \begin{Bmatrix} 0 \\ 0 \\ 0 \\ 0 \end{Bmatrix}.$$

(b) The only part of the above equation that needs to be altered is the previously null applied load vector. It now must be

$$\lfloor Q \rfloor = \left\lfloor \frac{7EI_0 w_0(t)}{L^3} \quad + M_0(t) \quad - F_0(t) \quad 0 \right\rfloor.$$

(c) If Newton's second law is used to write the equation of motion, the beam supplied contact force is the only force to be equated to the mass multiplied by its acceleration. For small beam tip rotations, that contact force is the beam tip shear force. This shear force, V, can be determined from the beam element stiffness matrix in terms of the beam tip lateral deflection $u(t)$in the form $V = ku$, where k is beam stiffness at its tip. If the Lagrange equation is used to write the equation of motion, the simplest form for the strain energy also involves this stiffness factor. Turning to the beam element stiffness matrix where R entries again indicate beam support reactions without present interest and the (relative) element deflections have been entered in terms of the quantities of the system under discussion:

$$\begin{Bmatrix} R \\ R \\ V \\ 0 \end{Bmatrix} = \frac{EI}{L^3}\begin{bmatrix} 12 & 6L & -12 & 6L \\ 6L & 4L^2 & -6L & 2L^2 \\ -12 & -6L & 12 & -6L \\ 6L & 2L^2 & -6L & 4L^2 \end{bmatrix}\begin{Bmatrix} 0 \\ 0 \\ u - v \\ \theta_2 \end{Bmatrix}.$$

The equation of the last row yields $\theta_2 = 3(u - v)/2L$. Substituting into the equation of the third row provides the solution that $V = 3EI(u - v)/L^3$. Therefore, of course, the stiffness factor is $k = 3EI/L^3$. Of course, there are many other ways to determine the stiffness factor at the beam tip. Hence, the equation of motion in terms of $u(t)$, and the natural frequency, are

$$m\ddot{u}(t) + \frac{3EI}{L^3}u(t) = \frac{3EI}{L^3}v(t) \quad \text{and} \quad \omega = \sqrt{\frac{3EI}{mL^3}}.$$

(d) The use of the relative deflection DOF $w(t)$ leads to the same left-hand side as above and hence the same natural frequency. In this case, the driving force on the right-hand side is $-m\ddot{v}$.

4.3 **(a)** Only four global DOF are required: w_2, w_3, θ_3, w_4, positive up and counterclockwise, at the nodes indicated by the subscripts. The two beam element stiffness matrices and the two spring element stiffness matrices are

$$[k_{10}]\{q_{10}\} = \frac{EI_0}{L^3}\begin{bmatrix} 12 & -6L \\ -6L & 4L^2 \end{bmatrix}\begin{Bmatrix} w_3 \\ \theta_3 \end{Bmatrix} \qquad [k_{20}]\{q_{20}\} = \frac{EI_0}{L^3}\begin{bmatrix} 12 & 6L \\ 6L & 4L^2 \end{bmatrix}\begin{Bmatrix} w_3 \\ \theta_3 \end{Bmatrix}$$

$$[k_{up}]\{q_{up}\} = \frac{6EI_0}{L^3}\begin{bmatrix} +1 & -1 \\ -1 & +1 \end{bmatrix}\begin{Bmatrix} w_2 \\ w_3 \end{Bmatrix} \qquad [k_{lw}]\{q_{lw}\} = \frac{3EI_0}{L^3}\begin{bmatrix} +1 & -1 \\ -1 & +1 \end{bmatrix}\begin{Bmatrix} w_3 \\ w_4 \end{Bmatrix}.$$

Therefore the matrix equation of motion is

$$\begin{bmatrix} 2 & & & \\ & 1 & & \\ & & 0 & \\ & & & 1 \end{bmatrix}\begin{Bmatrix} \ddot{w}_2 \\ \ddot{w}_3 \\ \ddot{\theta}_3 \\ \ddot{w}_4 \end{Bmatrix} + \frac{EI_0}{mL^3}\begin{bmatrix} 6 & -6 & & \\ -6 & 33 & & -3 \\ & & 8L^2 & \\ & -3 & & 3 \end{bmatrix}\begin{Bmatrix} w_2 \\ w_3 \\ \theta_3 \\ w_4 \end{Bmatrix} = \frac{F(t)}{m}\begin{Bmatrix} 0 \\ 0 \\ 0 \\ 1 \end{Bmatrix}.$$

It is important to note that inspection of the third of these four simultaneous equations immediately yields the result that $\theta_3(t)$ equals zero. This means that with this perfectly symmetrical structure and perfectly symmetrical force excitation, the system can undergo only symmetric vibrations. Furthermore, this rotational DOF and its corresponding third equation can now be removed from the above matrix equation, thus reducing its size to only that of 3×3 matrices.

(b) In this case, the DOF θ_1 and θ_5 have to be added to the previous four DOF for a total of six DOF. Now the beam element stiffness matrices are

$$\frac{EI_0}{L^3}\begin{bmatrix} 4L^2 & -6L & 2L^2 \\ -6L & 12 & -6L \\ 2L^2 & -6L & 4L^2 \end{bmatrix}\begin{Bmatrix} \theta_1 \\ w_3 \\ \theta_3 \end{Bmatrix} \qquad \frac{EI_0}{L^3}\begin{bmatrix} 12 & 6L & 6L \\ 6L & 4L^2 & 2L^2 \\ 6L & 2L^2 & 4L^2 \end{bmatrix}\begin{Bmatrix} w_3 \\ \theta_3 \\ \theta_5 \end{Bmatrix}.$$

Therefore the matrix equations of motion are

$$\begin{bmatrix} 0 & & & & & \\ & 2 & & & & \\ & & 1 & & & \\ & & & 0 & & \\ & & & & 1 & \\ & & & & & 0 \end{bmatrix}\begin{Bmatrix} \ddot{\theta}_1 \\ \ddot{w}_2 \\ \ddot{w}_3 \\ \ddot{\theta}_3 \\ \ddot{w}_4 \\ \ddot{\theta}_5 \end{Bmatrix}$$

$$+\frac{EI_0}{mL^3}\begin{bmatrix} 4L^2 & & -6L & 2L^2 & & \\ & 6 & -6 & & & \\ -6L & -6 & 33 & & -3 & 6L \\ 2L^2 & & & 8L^2 & & 2L^2 \\ & & -3 & & 3 & \\ & & 6L & 2L^2 & & 4L^2 \end{bmatrix}\begin{Bmatrix} \theta_1 \\ w_2 \\ w_3 \\ \theta_3 \\ w_4 \\ \theta_5 \end{Bmatrix} = \frac{F(t)}{m}\begin{Bmatrix} 0 \\ 0 \\ 0 \\ 0 \\ 1 \\ 0 \end{Bmatrix}.$$

Although it is still possible to guess partial solutions, such as the antisymmetrical result $\theta_3 = -2\theta_1 = -2\theta_5$, the by-hand reduction in the size of the matrices in this case is not easily accomplished.

(c) The hinge at the center of the structure introduces the possibility of two different bending slopes at node 3, while the deflections on both sides of the node remain the same. Call the bending slope at the left of the node 3 θ_3 and the bending slope at the right of the node 3 ψ_3. Thus the DOF of part (a) are now augmented by the additional DOF of ψ_3 and the matrix equation of motion is

$$\begin{bmatrix} 2 & & & & \\ & 1 & & & \\ & & 0 & & \\ & & & 0 & \\ & & & & 1 \end{bmatrix} \begin{Bmatrix} \ddot{w}_2 \\ \ddot{w}_3 \\ \ddot{\theta}_3 \\ \ddot{\psi}_3 \\ \ddot{w}_4 \end{Bmatrix} + \frac{EI_0}{mL^3} \begin{bmatrix} 6 & -6 & & & \\ -6 & 33 & -6L & 6L & -3 \\ & -6L & 4L^2 & & \\ & 6L & & 4L^2 & \\ & -3 & & & 3 \end{bmatrix} \begin{Bmatrix} w_2 \\ w_3 \\ \theta_3 \\ \psi_3 \\ w_4 \end{Bmatrix} = \frac{F(t)}{m} \begin{Bmatrix} 0 \\ 0 \\ 0 \\ 0 \\ 1 \end{Bmatrix}.$$

Note that the third and the fourth of the above simultaneous equations can be used to immediately determine the values of θ_3 and ψ_3 in terms of w_3. That is, $\theta_3 = -\psi_3 = 3w_3/2L$. In this case it is not difficult to reduce the size of the coefficient matrices to be 3×3 and the vectors to 3×1. It is only a matter of using the above partial solution and then removing the third and fourth columns and rows.

(d) The left-hand beam element stiffness matrix is now

$$\frac{EI_0}{L^3} \begin{bmatrix} 12 & -12 & 6L \\ -12 & 12 & -6L \\ 6L & -6L & 4L^2 \end{bmatrix} \begin{Bmatrix} w_1 \\ w_3 \\ \theta_3 \end{Bmatrix}.$$

Setting aside the first row, and separating the known applied motion function w_1 from the unknown DOF, leads to the new matrix equation of motion

$$\begin{bmatrix} 2 & & & \\ & 1 & & \\ & & 0 & \\ & & & 1 \end{bmatrix} \begin{Bmatrix} \ddot{w}_2 \\ \ddot{w}_3 \\ \ddot{\theta}_3 \\ \ddot{w}_4 \end{Bmatrix} + \frac{EI_0}{mL^3} \begin{bmatrix} 6 & -6 & & \\ -6 & 33 & & -3 \\ & & 8L^2 & \\ & -3 & & 3 \end{bmatrix} \begin{Bmatrix} w_2 \\ w_3 \\ \theta_3 \\ w_4 \end{Bmatrix} = \frac{EI_0 w_1(t)}{mL^3} \begin{Bmatrix} 0 \\ 12 \\ -6L \\ 0 \end{Bmatrix}.$$

Note that, as would be expected, θ_3 is not zero. Moreover, it is easily determined to be

$$\theta_3(t) = -\frac{3w_1(t)}{4L}.$$

4.4 **(a)** The only difference in the element stiffness matrices is that the stiffness matrix for beam element 40 has to be altered so that now the bending slope, θ_4 is no longer zero, but the center line deflection is zero. Thus for the antisymmetrical vibration, the beam element 40 stiffness matrix becomes

$$[k_{40}]\{q_{40}\} = \frac{EI}{L^3} \begin{bmatrix} 12 & 6L & 6L \\ 6L & 4L^2 & 2L^2 \\ 6L & 2L^2 & 4L^2 \end{bmatrix} \begin{Bmatrix} w_3 \\ \theta_3 \\ \theta_4 \end{Bmatrix}.$$

Using this stiffness matrix for element 40, the assembled global stiffness matrix and deflection vector are

$$[K]\{q\} = \frac{EI}{L^3}\begin{bmatrix} 4L^2 & -6L & 2L^2 & 0 & 0 & 0 & 0 & 0 \\ -6L & 24 & 0 & -12 & 6L & 0 & 0 & 0 \\ 2L^2 & 0 & 8L^2 & -6L & 2L^2 & 0 & 0 & 0 \\ 0 & -12 & -6L & 24 & 0 & -12 & 6L & 0 \\ 0 & 6L & 2L^2 & 0 & 8L^2 & -6L & 2L^2 & 0 \\ 0 & 0 & 0 & -12 & -6L & 24 & 0 & 6L \\ 0 & 0 & 0 & 6L & 2L^2 & 0 & 8L^2 & 2L^2 \\ 0 & 0 & 0 & 0 & 0 & 6L & 2L^2 & 4L^2 \end{bmatrix}\begin{Bmatrix} \theta_0 \\ w_1 \\ \theta_1 \\ w_2 \\ \theta_2 \\ w_3 \\ \theta_3 \\ \theta_4 \end{Bmatrix}.$$

(b) Yes. The mass matrix needs to be adjusted for the different acceleration vector where now the second time derivative of w_4 is zero and the second time derivative of θ_4 is not zero.

4.5 Since the beam elements have infinite axial stiffness, the only possible motions within the plane of the frame at the nodes are rotations in the plane (i.e., rotations about the z axis). Call these rotations, the two DOF of the system, ψ_1 and ψ_2. The matrix equation of motion is

$$H_z\begin{bmatrix} 1 & 0 \\ 0 & 1 \end{bmatrix}\begin{Bmatrix} \ddot{\psi}_1 \\ \ddot{\psi}_2 \end{Bmatrix} + \frac{4EI_0}{L}\begin{bmatrix} 10 & 1 \\ 1 & 10 \end{bmatrix}\begin{Bmatrix} \psi_1 \\ \psi_2 \end{Bmatrix} = M_1(t)\begin{Bmatrix} 0 \\ 1 \end{Bmatrix}.$$

4.6 **(a)** The only necessary alteration in the previous solution is the determination of a new equivalent load vector. The equivalent load vector resulting from an enforced motion $w_0(t)$ is

$$\lfloor Q \rfloor = \frac{EI_0 w_0(t)}{L^3}\lfloor 24 \quad -12L \quad 0 \quad 0 \quad 0 \quad 0 \rfloor.$$

(b) The equivalent load vector resulting from an enforced $\theta_0(t)$ is

$$\lfloor Q \rfloor = \frac{EI_0 \theta_0(t)}{L^3}\lfloor 12L \quad -4L^2 \quad 0 \quad 0 \quad 0 \quad 0 \rfloor.$$

(c) From the element stiffness matrix for the right-hand column, the new equivalent load vector is

$$\lfloor Q \rfloor = \frac{EI_0 \theta_0(t)}{L^3}\lfloor +3L \quad 0 \quad 0 \quad -2L^2 \rfloor.$$

4.7 **(a, b)** In general terms, natural frequencies squared are proportional to the ratio of stiffness over mass. Thus an increase in mass generally lowers the natural frequencies, whereas an increase in stiffness generally raises the natural frequencies. More specifically, the natural frequencies squared of any structure are inversely proportional to what are called the modal masses. The modal masses, to be more fully discussed later, are the lumped masses of the structure weighted by (multiplied by) the squares of the values at the lumped masses of the deflection amplitudes for that vibration mode (i.e., the amplitudes of the deflection pattern associated with that natural frequency). Thus the addition of mass at a point of zero deflection for that natural frequency does not decrease that natural frequency. However, because the points of zero deflection (also called nodes) are generally different from natural frequency to natural frequency, it is quite likely that the addition of mass anywhere lowers most, if not all, natural frequencies. Similarly, the stiffer a structure is made, the higher most of the natural frequencies. Note, however, as the question is posed in part (b), increasing I_{yy} without changing the cross-sectional area (i.e., without changing the mass per unit of beam length) not always, but generally means that I_{zz} must decrease if the same thicknesses are maintained. Hence, the out-of-paper natural frequencies must decrease in this circumstance, whereas those for deflections in the plane of the paper increase.

4.8 **(a)** The fundamental natural frequency decreases. In this case where the total mass remains the same, but is redistributed toward the regions of greater vibratory amplitude, the first natural frequency squared decreases. Later, when the modal solution method is discussed, it is mathematically shown that the square of any natural frequency is equal to the modal stiffness term for that frequency divided by the modal mass term for that frequency. These modal masses and stiffnesses are terms that not only depend on the mass and stiffness distributions, but they weight (multiply) those mass and stiffness distributions by the squares of the vibratory amplitudes. With a bit of experience, an analyst generally finds it easy to roughly imagine the distribution of vibratory amplitudes for at least the first force free vibratory frequency. In this question, the equivalent stiffness is (unrealistically) roughly unaltered, but the effective mass, the mass multiplied by (relative) deflection squared, is markly increased because the mass is being moved where the associated deflections are greatest. Thus the denominator is markly increasing, whereas the numerator is roughly stationary.

(b) This is a difficult question to respond to theoretically. However, the answer is that the first natural frequency does decrease slightly with only half the total mass lumped at the beam midspan. See the result listed in the first column of the table of Endnote (1).

4.9 **(a)** When in doubt, always go back to basics. In this case, going back to basics means writing the expressions for T, U, and δW. The kinetic energy is

$$T = \tfrac{1}{2}H_1\dot{\phi}_1^2 + \tfrac{1}{2}H_2\left(\frac{R_1}{R_2}\right)^2\dot{\phi}_1^2 + \tfrac{1}{2}H_3\dot{\phi}_2^2.$$

The strain energy of the total structure is the sum of the strain energies of its two bar element components. Their strain energies are

$$U_{10} = \tfrac{1}{2}\lfloor 0 \quad \phi_1 \rfloor \frac{2GJ_0}{L}\begin{bmatrix} +1 & -1 \\ -1 & +1 \end{bmatrix}\begin{Bmatrix} 0 \\ \phi_1 \end{Bmatrix}$$

$$U_{20} = \tfrac{1}{2}\left\lfloor \frac{R_1}{R_2}\phi_1 \quad \phi_2 \right\rfloor \frac{GJ_0}{L}\begin{bmatrix} +1 & -1 \\ -1 & +1 \end{bmatrix}\begin{Bmatrix} \frac{R_1}{R_2}\phi_1 \\ \phi_2 \end{Bmatrix}.$$

Thus the matrix equations of motion are

$$\begin{bmatrix} H_1 + \frac{R_1^2}{R_2^2}H_2 & \\ & H_3 \end{bmatrix}\begin{Bmatrix} \ddot{\phi}_1 \\ \ddot{\phi}_2 \end{Bmatrix} + \frac{GJ_0}{L}\begin{bmatrix} 2+\frac{R_1^2}{R_2^2} & -\frac{R_1}{R_2} \\ -\frac{R_1}{R_2} & 1 \end{bmatrix}\begin{Bmatrix} \phi_1 \\ \phi_2 \end{Bmatrix} = M_t\begin{Bmatrix} 0 \\ -1 \end{Bmatrix}.$$

(b) Writing the Lagrange matrix equation of motion in this case requires writing the expression for the kinetic energy and assembling the system stiffness matrix. Since there are neither applied forces nor explicit damping present, a virtual work statement is not necessary:

$$T = \frac{1}{2}\left[m\dot{w}_1^2 + H_y\dot{\theta}_1^2 + H_x\dot{\phi}_1^2 + 2m\dot{w}_2^2 + 2H_y\dot{\theta}_2^2 + 2H_x\dot{\phi}_2^2\right].$$

This leads to the following diagonal mass matrix and the acceleration vector:

$$[M]\{\ddot{q}\} = \begin{bmatrix} m & & & & & \\ & H_y & & & & \\ & & H_x & & & \\ & & & 2m & & \\ & & & & 2H_y & \\ & & & & & 2H_x \end{bmatrix} \begin{Bmatrix} \ddot{w}_1 \\ \ddot{\theta}_1 \\ \ddot{\phi}_1 \\ \ddot{w}_2 \\ \ddot{\theta}_2 \\ \ddot{\phi}_2 \end{Bmatrix}.$$

The challenge of this problem lies in writing and assembling the element stiffness matrices. From the standard beam element stiffness matrix template, for beam element number 10:

$$[k_{10}]\{q\} = \frac{EI}{L^3}\begin{Bmatrix} -12 \\ 6L \\ 0 \end{Bmatrix} w_0(t) + \frac{EI}{L^3}\begin{bmatrix} 12 & -6L & 0 \\ -6L & 4L^2 & 0 \\ 0 & 0 & L^2 \end{bmatrix}\begin{Bmatrix} w_1 \\ \theta_1 \\ \phi_1 \end{Bmatrix}$$

$$[k_{20}]\{q\} = +\frac{EI}{L^3}\begin{bmatrix} 1.5 & 0 & 1.5L & -1.5 & 0 & 1.5L \\ 0 & 0.5L^2 & 0 & 0 & -0.5L^2 & 0 \\ 1.5L & 0 & 2L^2 & -1.5L & 0 & L^2 \\ -1.5 & 0 & -1.5L & 1.5 & 0 & -1.5L \\ 0 & -0.5L^2 & 0 & 0 & 0.5L^2 & 0 \\ 1.5L & 0 & L^2 & -1.5L & 0 & 2L^2 \end{bmatrix}\begin{Bmatrix} w_1 \\ \theta_1 \\ \phi_1 \\ w_2 \\ \theta_2 \\ \phi_2 \end{Bmatrix}$$

$$[k_{30}]\{q\} = +\frac{EI}{L^3}\begin{bmatrix} 24 & -12L & 0 \\ -12L & 8L^2 & 0 \\ 0 & 0 & 2L^2 \end{bmatrix}\begin{Bmatrix} w_2 \\ \theta_2 \\ \phi_2 \end{Bmatrix}$$

with the final result that

$$[K]\{q\} = \frac{EI}{L^3}\begin{bmatrix} 13.5 & -6L & 1.5L & -1.5 & 0 & 1.5L \\ -6L & 4.5L^2 & 0 & 0 & -0.5L^2 & 0 \\ 1.5L & 0 & 3L^2 & -1.5L & 0 & L^2 \\ -1.5 & 0 & -1.5L & 25.5 & -12L & -1.5L \\ 0 & -0.5L^2 & 0 & -12L & 8.5L^2 & 0 \\ 1.5L & 0 & L^2 & -1.5L & 0 & 4L^2 \end{bmatrix}\begin{Bmatrix} w_1 \\ \theta_1 \\ \phi_1 \\ w_2 \\ \theta_2 \\ \phi_2 \end{Bmatrix}$$

and

$$\lfloor Q \rfloor = \frac{EI}{L^3} w_0(t) \lfloor 12 \quad -6L \quad 0 \quad 0 \quad 0 \quad 0 \rfloor.$$

(c) All DOF are needed but u_2. Hence the stiffness matrix will be 5×5 as follows:

$$[K]\{q\} = \frac{EI_0}{L^3}\begin{bmatrix} 24 & -12L & 0 & 0 & 0 \\ -12L & 8L^2 & 0 & 0 & 0 \\ 0 & 0 & 36 & -18L & 0 \\ 0 & 0 & -18L & 12L^2 & 0 \\ 0 & 0 & 0 & 0 & L^2 \end{bmatrix}\begin{Bmatrix} v_2 \\ \psi_2 \\ w_2 \\ \theta_2 \\ \phi_2 \end{Bmatrix}.$$

$\{Q\}^t$ is obtained from the stiffness matrix as follows

$$\{Q\} = \frac{EI_0 h(t)}{L^3} \lfloor 0 \quad 0 \quad 36 \quad -18L \quad 0 \rfloor^t.$$

Note that there is no coupling in the stiffness matrix between bending in the y plane and bending in the z plane and no coupling between torsion and bending in either plane. Thus, in the case of a static loading, the beam will respond only with deflections in the direction of the tip forces.

(d) The writing of the mass matrix begins with writing the kinetic energy expression and ends with factoring that expression into matrix form. Using the right-hand rule

$$T = \tfrac{1}{2}m_1 \left[\dot{v}_2^2 + \dot{w}_2^2\right] + \tfrac{1}{2}m_2 \left[(e_z\dot{\theta}_2)^2 + (\dot{v}_2 - e_z\dot{\phi}_2)^2 + \dot{w}_2^2\right] \\ + \tfrac{1}{2}m_3 \left[(e_y\dot{\psi}_2)^2 + (\dot{v}_2 + e_x\dot{\psi}_2)^2 + (\dot{w}_2 + e_x\dot{\theta}_2 - e_y\dot{\phi})^2\right].$$

Now assemble the mass matrix in the same fashion that the stiffness matrix is assembled to get

$$[M]\{\ddot{q}\} \\ = \begin{bmatrix} m_1+m_2+m_3 & m_3e_x & 0 & 0 & -m_2e_z \\ m_3e_x & m_3\left(e_x^2+e_y^2\right) & 0 & 0 & 0 \\ 0 & 0 & m_1+m_2+m_3 & m_3e_x & -m_3e_y \\ 0 & 0 & m_3e_x & m_2e_z^2+m_3e_x^2 & -m_3e_xe_y \\ -m_2e_z & 0 & -m_3e_y & -m_3e_xe_y & m_2e_z^2+m_3e_y^2 \end{bmatrix} \begin{Bmatrix} \ddot{v}_2 \\ \ddot{\psi}_2 \\ \ddot{w}_2 \\ \ddot{\theta}_2 \\ \ddot{\phi}_2 \end{Bmatrix}.$$

Note that all DOF are coupled because of nonzero, off-diagonal terms in the mass matrix. This means that, although the beam is excited only in the z direction, it will nevertheless also undergo bending vibrations in the y direction and twist about the x axis.

4.10 **(a)** In this motion, and the other two motions, the roof mass remains horizontal. Thus the kinetic energy is simply $T = \tfrac{1}{2}m\dot{u}^2$. The ends of each bent beam-column are constrained against a bending slope rotation, and the lower deflection is zero and the upper deflection is u. Thus, taking the value 12 from the general beam element stiffness matrix, for each of the eight beams, $U_e = \tfrac{1}{2}(EI_0/L^3)(12)u^2$. Multiplying the latter quantity by 8 and substituting into the Lagrange equation of motion leads to the result

$$\omega_u = \omega_1 = 4\sqrt{\frac{6EI_0}{mL^3}}.$$

(b) The procedure is as above, noting the greater beam bending stiffness coefficient for motion in the y direction. The result is

$$\omega_v = \omega_2 = 8\sqrt{\frac{3EI_0}{mL^3}}.$$

(c) Since this motion is a bit more complicated, it is worthwhile to draw a top view figure showing the location of the tops of the eight columns relative to the axis of rotation. As above, kinetic energy expression is a single term. The strain energy expression also can be created as above, by (i) calculating the separate x- and y-direction lateral deflections at the tops of the beam-columns because of the rotation about the z axis and (ii) using the general strain energy expression for beam-bending $\tfrac{1}{2}q^tkq$. Another, and wholly equivalent, way to do this exercise is to anticipate the approach stressed in the next chapter and do the following: determine the torsional stiffness factor of this structure, K, by (imagining) twisting it through the small but arbitrary angle ϕ, and then calculating the elastic, resisting torque, $M_t = K\phi$. Then, as is true for any torsional spring, rotary mass system, regardless of what it looks like, the torsional natural frequency is the square root of the ratio of the torsional stiffness factor divided by the value of the rotary mass moment of inertia. The details of the calculation are as follows. First, for convenience of discussion, divide the eight beam-columns into two groups: the outboard (end) four and the inboard (middle) four. In response to a rotation about the z axis through an angle ϕ, the tops of the inboard columns move a distance $\frac{1}{2}L\phi$ in both the x and the y directions. At each column top, these motions produce a resisting x-direction force of magnitude $(EI_0/L^3)(12)(\frac{1}{2}L\phi)$ and twice that for the y-direction resisting force. Similarly, the outboard column tops move the distances $3L\phi/2$ in the y direction, and $\frac{1}{2}L\phi$ in the x direction. These deflections create a resisting x-direction force of magnitude $(EI_0/L^3)(12)(\frac{1}{2}L\phi)$ and a resisting beam top y-direction force of magnitude $(EI_0/L^3)(12)(3L\phi/2)$. Now it is just a matter of summing the moments about the z axis produced by these 16 force components.

$$M_t = 4\left[\frac{L}{2}\left(\frac{6EI_0}{L^2}\right)\phi + \frac{L}{2}\left(\frac{12\,EI_0}{L^2}\right)\phi + \frac{L}{2}\left(\frac{6EI_0}{L^2}\right)\phi + \frac{3L}{2}\left(\frac{36EI_0}{L^2}\right)\phi\right] = 264\frac{EI_0}{L}\phi$$

$$\text{so} \quad \omega_\phi = \omega_3 = 2\sqrt{\frac{66\,EI_0}{H_2 L}}.$$

Note the three natural frequencies are always numbered and subscripted so that the lowest frequency value is the first (or fundamental) natural frequency, and the second lowest value is the second natural frequency, and so on. As will be seen, only three natural frequencies are possible for a structural system such as this which has only three DOF.

(d) No. There is only one mass with three possible independent motions.

4.11 **(a)** Take the mass of each half beam and lump it at its nearest joint. Then the discrete masses at the corners have the magnitudes $1.5mL$, whereas those at the first level joints have the magnitudes $2.5mL$.

(b) Taking into account the positive directions assigned to the selected DOF in Fig. 3.13(b), the expression for the kinetic energy is

$$T = \tfrac{1}{2}(1.5mL + 1.5mL)\dot{u}_1^2 + \tfrac{1}{2}(5mL)\dot{u}_2^2 + \tfrac{1}{2}(2.5mL)(\dot{u}_1 - e\dot{\theta}_1)^2 + \tfrac{1}{2}(2.5mL)(\dot{u}_1 - e\dot{\theta}_3)^2.$$

Upon arranging the kinetic energy in the matrix form $T = \frac{1}{2}\lfloor q \rfloor [M]\{q\}$, the mass matrix and acceleration vector are seen to be

$$mL\begin{bmatrix} 8 & -2.5e & & & -2.5e & \\ -2.5e & 2.5e^2 & & & & \\ & & 5 & & & \\ & & & 0 & & \\ -2.5e & & & & 2.5e^2 & \\ & & & & & 0 \end{bmatrix}\begin{Bmatrix} \ddot{u}_1 \\ \ddot{\theta}_1 \\ \ddot{u}_2 \\ \ddot{\theta}_2 \\ \ddot{\theta}_3 \\ \ddot{\theta}_4 \end{Bmatrix}.$$

(c) The assembled stiffness matrix, for a deflection vector having the same DOF ordering as the acceleration vector, is

$$[K] = \frac{EI}{L^3}\begin{bmatrix} 24 & 6L & -24 & 6L & 6L & 6L \\ 6L & 6L^2 & -6L & 2L^2 & L^2 & 0 \\ -24 & -6L & 72 & 6L & -6L & 6L \\ 6L & 2L^2 & 6L & 14L^2 & 0 & L^2 \\ 6L & L^2 & -6L & 0 & 6L^2 & 2L^2 \\ 6L & 0 & 6L & L^2 & 2L^2 & 14L^2 \end{bmatrix}.$$

(d) $\lfloor Q \rfloor = \lfloor 2(F_1 + F_2) \quad 0 \quad (F_1 + F_2) \quad 0 \quad 0 \quad 0 \rfloor.$

4.12 **(a)** The motion of the single mass is described entirely by the four DOF at node 2. Since there are no products of inertia included in the mathematical model, the kinetic energy expression is

$$T = \tfrac{1}{2}m(\dot{v}_2 + e_x\dot{\theta}_2)^2 + \tfrac{1}{2}m(e_y\dot{\theta}_2)^2 + \tfrac{1}{2}m(e_y\dot{\phi}_2 - e_x\dot{\psi}_2)^2 + \tfrac{1}{2}H_x\dot{\phi}_2^2 + \tfrac{1}{2}H_y\dot{\psi}_2^2 + \tfrac{1}{2}H_z\dot{\theta}_2^2.$$

Therefore the mass and acceleration vector are

$$[M]\{\ddot{q}\} = \begin{bmatrix} m & me_x & 0 & 0 \\ me_x & [H_z + m(e_x^2 + e_y^2)] & 0 & 0 \\ 0 & 0 & H_x + me_y^2 & -me_xe_y \\ 0 & 0 & -me_ye_x & H_y + me_x^2 \end{bmatrix}\begin{Bmatrix} \ddot{v}_2 \\ \ddot{\theta}_2 \\ \ddot{\phi}_2 \\ \ddot{\psi}_2 \end{Bmatrix}.$$

(b) It is suggested that the individual beam stiffness matrices be written out so as to make later assembly a simple matter. Starting with beam element 10, accounting for the different bending stiffness coefficients and the element length $2L$, the stiffness matrix is

$$[k_{10}]\{q\} = \frac{EI_0}{L^3}\begin{bmatrix} 4.5 & -4.5L & 0 & 0 \\ -4.5L & 6L^2 & 0 & 0 \\ 0 & 0 & 0.5L^2 & 0 \\ 0 & 0 & 0 & 10L^2 \end{bmatrix}\begin{Bmatrix} v_2 \\ \theta_2 \\ \phi_2 \\ \psi_2 \end{Bmatrix}.$$

The stiffness matrix for beam 20 needs to include the two base motion components so that the equivalent generalized force vector can be identified. The result is

$$[k_{20}]\begin{Bmatrix} q \\ u_3 \\ v_3 \end{Bmatrix} = \frac{EI_0}{L^3}\begin{bmatrix} 12 & 0 & 6L & 0 & 0 & -12 \\ 0 & L^2 & 0 & 0 & 0 & 0 \\ 6L & 0 & 4L^2 & 0 & 0 & -6L \\ 0 & 0 & 0 & 8L^2 & -12L & 0 \\ 0 & 0 & 0 & -12L & 24 & 0 \\ -12 & 0 & -6L & 0 & 0 & 12 \end{bmatrix}\begin{Bmatrix} v_2 \\ \theta_2 \\ \phi_2 \\ \psi_2 \\ u_3 \\ v_3 \end{Bmatrix}.$$

Therefore the assembled stiffness matrix and generalized force vectors are

$$[K]\{q\} = \frac{EI_0}{L^3}\begin{bmatrix} 16.5 & -4.5L & 6L & 0 \\ -4.5L & 7L^2 & 0 & 0 \\ 6L & 0 & 4.5L^2 & 0 \\ 0 & 0 & 0 & 18L^2 \end{bmatrix}\begin{Bmatrix} v_2 \\ \theta_2 \\ \phi_2 \\ \psi_2 \end{Bmatrix}.$$

and

$$\{Q\} = \frac{EI_0}{L^3}\begin{Bmatrix} 0 \\ 0 \\ 0 \\ 12L \end{Bmatrix} u_3(t) + \frac{EI_0}{L^3}\begin{Bmatrix} 12 \\ 0 \\ 6L \\ 0 \end{Bmatrix} v_3(t).$$

Of course, the complete equations of motion are

$$[M]\{\ddot{q}\} + [K]\{q\} = \{Q\}.$$

Inspection of this matrix equation of motion makes clear that all the DOF are coupled by either the mass matrix or the stiffness matrix. That is, none of there DOF can be separated from the other DOF.

4.13 Since there are no mass moments of inertia specified, the kinetic energy is simply one-half the mass multiplied by the total velocity squared. The total velocity squared is simply the sum of the squares of each of the three orthogonal velocity components. Each of the rotations, in combination with the offsets, produces a velocity component in two orthogonal directions. Thus the kinetic energy is

$$T = \tfrac{1}{2}m\left[(-e_z\dot{\theta} - e_y\dot{\psi})^2 + (\dot{v} - e_z\dot{\phi} + e_x\dot{\psi})^2 + (\dot{w} + e_x\dot{\theta} + e_y\dot{\phi})^2\right].$$

After substituting the above kinetic energy expression into the Lagrange equation and factoring into matrix form, the mass matrix and acceleration vector becomes

$$[m]\{\ddot{q}\} = m\begin{bmatrix} 1 & 0 & 0 & -e_z & e_x \\ 0 & 1 & e_x & e_y & 0 \\ 0 & e_x & (e_x^2 + e_z^2) & e_x e_y & e_y e_z \\ -e_z & e_y & e_x e_y & (e_y^2 + e_z^2) & -e_x e_z \\ e_x & 0 & e_y e_z & -e_x e_z & (e_x^2 + e_y^2) \end{bmatrix}\begin{Bmatrix} \ddot{v} \\ \ddot{w} \\ \ddot{\theta} \\ \ddot{\phi} \\ \ddot{\psi} \end{Bmatrix}.$$

4.14. There are only two nonzero DOF, u_4 and v_4. The element stiffness matrices, and then the global stiffness matrix equation, are as follows:

$$[k_{14}] = \frac{\sqrt{3}EA}{2h}\begin{bmatrix} 1/4 & \sqrt{3}/4 \\ \sqrt{3}/4 & 3/4 \end{bmatrix} = \frac{EA}{h}\begin{bmatrix} 0.2165 & 0.3750 \\ 0.3750 & 0.6495 \end{bmatrix}$$

$$[k_{24}] = \frac{\sqrt{3}EA}{2h}\begin{bmatrix} 1/4 & -\sqrt{3}/4 \\ -\sqrt{3}/4 & 3/4 \end{bmatrix} = \frac{EA}{h}\begin{bmatrix} 0.2165 & -0.3750 \\ -0.3750 & 0.6495 \end{bmatrix}$$

$$[k_{34}] = \frac{2EA}{\sqrt{2}h}\begin{bmatrix} 1/2 & -1/2 \\ -1/2 & 1/2 \end{bmatrix} = \frac{EA}{h}\begin{bmatrix} 0.7071 & -0.7071 \\ -0.7071 & 0.7071 \end{bmatrix}$$

$$\therefore \begin{Bmatrix} F \\ 0 \end{Bmatrix} = \frac{EA}{h}\begin{bmatrix} 1.140 & -0.7071 \\ -0.7071 & 2.006 \end{bmatrix}\begin{Bmatrix} u_4 \\ v_4 \end{Bmatrix}.$$

4.15 Using the three DOF shown in the sketch, the kinetic energy of the system is

$$T = \tfrac{1}{2}m(\dot{w} - e\dot{\phi})^2 + \tfrac{1}{2}H_y\dot{\theta}^2 + \tfrac{1}{2}H_x\dot{\phi}^2$$

This leads to the following mass and acceleration vector for the Lagrange matrix equation:

$$[M]\{\ddot{q}\} = \begin{bmatrix} m & 0 & -me \\ 0 & H_y & 0 \\ -me & 0 & H_x + me^2 \end{bmatrix}\begin{Bmatrix} \ddot{w} \\ \ddot{\theta} \\ \ddot{\phi} \end{Bmatrix}.$$

Again, writing the expression for the strain energy amounts to assembling the system stiffness matrix from the two element stiffness matrices. The element stiffness matrices and corresponding deflection vectors for the near and far beam elements, respectively, are

$$[k_{near}]\{q\} = \frac{EI}{L^3}\begin{bmatrix} 12 & 6L & 0 \\ 6L & 4L^2 & 0 \\ 0 & 0 & L^2 \end{bmatrix}\begin{Bmatrix} w \\ \theta \\ \phi \end{Bmatrix}$$

$$[k_{far}]\{q\} = \frac{EI}{L^3}\begin{bmatrix} 12 & 6L & 0 \\ 6L & 4L^2 & 0 \\ 0 & 0 & L^2 \end{bmatrix}\begin{Bmatrix} w - 2e\phi \\ \theta \\ \phi \end{Bmatrix}.$$

Recall that the strain energy has the form $\frac{1}{2}\lfloor q \rfloor [k]\{q\}$. Multiplying out the latter matrix product as a triple product and regrouping terms leads to the following stiffness matrix for the far beam

$$[k_{far}]\{q\} = \frac{EI}{L^3}\begin{bmatrix} 12 & 6L & -24e \\ 6L & 4L^2 & -12Le \\ -24e & -12Le & (L^2 + 48e^2) \end{bmatrix}\begin{Bmatrix} w \\ \theta \\ \phi \end{Bmatrix}.$$

Thus the second part of the free vibration equation $[M]\{q\} + [K]\{q\} = \{0\}$ is

$$[K]\{q\} = \frac{EI}{L^3}\begin{bmatrix} 24 & 12L & -24e \\ 12L & 8L^2 & -12Le \\ -24e & -12Le & (2L^2 + 48e^2) \end{bmatrix}\begin{Bmatrix} w \\ \theta \\ \phi \end{Bmatrix}.$$

4.16 Since the original partial derivative equation is linear, the separation of the total equation into static and dynamic parts proceeds exactly as it did for the linear matrix equations of motion. That is, again start with $\mathcal{F} = \mathcal{F}_{stat} + \mathcal{F}_{dyn}$, $w = w_{stat} + w_{dyn}$. Complete the process by subtracting the static equation load-deflection equation, $\mathcal{P}[w_{stat}(x)] = \mathcal{L}_{stat}(x)$, from the original partial derivative equation to obtain the desired result because $\mathcal{H}[w_{stat}] = 0$.

4.17 **(a)** Equation (1.17), which applies to a single lumped mass (call it the jth mass), can be rewritten as

$$2T_j = \lfloor \dot{u} \quad \dot{v} \quad \dot{w} \quad \dot{\theta}_x \quad \dot{\theta}_y \quad \dot{\theta}_z \rfloor^{(j)} \begin{bmatrix} m & & & & & \\ & m & & & & \\ & & m & & & \\ & & & H_{xx} & H_{xy} & H_{xz} \\ & & & H_{xy} & H_{yy} & H_{yz} \\ & & & H_{xz} & H_{yz} & H_{zz} \end{bmatrix}^{(j)} \begin{Bmatrix} \dot{u} \\ \dot{v} \\ \dot{w} \\ \dot{\theta}_x \\ \dot{\theta}_y \\ \dot{\theta}_z \end{Bmatrix}^{(j)},$$

where u through θ_z are the generalized coordinates of this particular lumped mass. In the same style that the FEM elastic element DOF for the jth element are related to the FEM global DOF, that is, through a constant transformation matrix $[\Xi_j]$, these mass element DOF can be related to the global DOF. Thus the kinetic energy of this particular mass element can be written in terms of the global DOF, $\{q\}$, as

$$T_j = \tfrac{1}{2}\lfloor q \rfloor [\Xi_j]^t [m_j][\Xi_j]\{q\}.$$

The triple product $[\Xi_j]^t[m_j][\Xi_j]$ is the jth mass contribution to the global mass matrix $[M]$. This typical triple product is a symmetric square matrix because $[m_j]$ is a symmetric square matrix. Since the sum of symmetric matrices is a symmetric matrix, the global mass matrix $[M]$ is symmetric.

(b) The global mass matrix is positive definite because (i) the kinetic energy T is always a positive value whenever there are any velocities, positive or negative, associated with the mass of the system (i.e., there is no such thing as negative kinetic energy) and (ii) the very definition of positive definiteness that says a matrix $[m]$ is positive definite whenever

$$\lfloor v \rfloor [\overline{m}] \{v\} > 0$$

for all nonnull vectors $\{v\}$, combined with the standard global kinetic energy expression

$$2T = \lfloor \dot{q} \rfloor [M] \{\dot{q}\}.$$

4.18 The "proof" breaks down at the point where it is claimed that because the velocity vector is arbitrary, the central square matrix of the triple product is zero. If the two velocity vectors were different arbitrary vectors, then that conclusion would be valid. However, they are the same arbitrary vector, and thus all that can be concluded is that the central square matrix is skew-symmetric (i.e., $m_{ij} = -m_{ji}$). Test this idea with a 2×2 square matrix with arbitrary entries.

CHAPTER 5 SOLUTIONS

5.1 Write

$$C_1 = \pm C_0 \sin\chi \quad \text{and} \quad C_2 = C_0 \cos\chi$$
$$\text{then} \quad u(t) = C_0 e^{-\zeta\omega t} \cos(\omega_d t \pm \chi).$$

5.2 **(a,b)** With either approach, the result is

$$\begin{bmatrix} m & 0 \\ 0 & H \end{bmatrix} \begin{Bmatrix} \ddot{u} \\ \ddot{\theta} \end{Bmatrix} + \begin{bmatrix} (c_1 + c_2 + c_3) & (-ac_1 + bc_2) \\ (-ac_1 + bc_2) & (a^2 c_1 + b^2 c_2) \end{bmatrix} \begin{Bmatrix} \dot{u} \\ \dot{\theta} \end{Bmatrix}$$
$$+ \begin{bmatrix} (k_1 + k_2) & (-ak_1 + bk_2) \\ (-ak_1 + bk_2) & (a^2 k_1 + b^2 k_2) \end{bmatrix} \begin{Bmatrix} u \\ \theta \end{Bmatrix} = \begin{Bmatrix} F \\ M \end{Bmatrix}.$$

(c) The use of the Lagrange equations produces the result

$$\begin{bmatrix} m & ma \\ ma & H + ma^2 \end{bmatrix} \begin{Bmatrix} \ddot{v} \\ \ddot{\theta} \end{Bmatrix} + \begin{bmatrix} (c_1 + c_2 + c_3) & (Lc_2 + ac_3) \\ (Lc_2 + ac_3) & (L^2 c_2 + a^2 c_3) \end{bmatrix} \begin{Bmatrix} \dot{v} \\ \dot{\theta} \end{Bmatrix}$$
$$+ \begin{bmatrix} k_1 + k_2 & Lk_2 \\ Lk_2 & L^2 k_2 \end{bmatrix} \begin{Bmatrix} v \\ \theta \end{Bmatrix} = \begin{Bmatrix} F \\ M + aF \end{Bmatrix}.$$

(d) The use of Newton's second law with this particular set of generalized coordinates produces the (highly undesirable) nonsymmetrical matrix equations of motion

$$\begin{bmatrix} m & ma \\ 0 & H \end{bmatrix} \begin{Bmatrix} \ddot{v} \\ \ddot{\theta} \end{Bmatrix} + \begin{bmatrix} (c_1 + c_2 + c_3) & (Lc_2 + ac_3) \\ (bc_2 - ac_1) & Lbc_2 \end{bmatrix} \begin{Bmatrix} \dot{v} \\ \dot{\theta} \end{Bmatrix}$$
$$+ \begin{bmatrix} k_1 + k_2 & Lk_2 \\ bk_2 - ak_1 & Lbk_2 \end{bmatrix} \begin{Bmatrix} v \\ \theta \end{Bmatrix} = \begin{Bmatrix} F \\ M \end{Bmatrix}.$$

5.3 **(a)** The tedious approach to this problem is to write the Coulomb differential equation for a one-DOF system subjected to a harmonic applied force with any frequency other than the system natural frequency and obtain its solution for a quarter or half cycle; that is, write and obtain

$$m\ddot{u} + ku = F_0 \sin \omega_f t - \mu mg$$

$$u(t) = C_1 \sin \omega t + C_2 \cos \omega t + \frac{(F_0/k)}{1 - (\omega_f/\omega)^2} \sin \omega_f t - \mu(mg/k)$$

and then apply BCs. (Note that the response can grow very large as the forcing frequency approaches the natural frequency; see Exercises 4.8 and 4.9 for consideration of the situation where the forcing frequency equals the natural frequency.) A far simpler approach is to realize that because the Coulomb damping force is a constant, the work done by this force over a quarter cycle is simply $-\mu mgA$, where A is the amplitude of the vibration. Thus the work done over a full cycle is four times that amount. Therefore,

$$c_{eq} = \frac{4\mu mg}{\pi \omega_1 A}.$$

There are some problems with this solution. From the solution to the differential equation of motion, there is no steady-state response if the frequency of the applied force is equal to the natural frequency of the oscillator, and the solution may be beyond the bounds of linearity if the frequency of the applied force is near to the value of the natural frequency. (In the first case, more energy is being put into the system per cycle by the applied force than can be removed by the friction force.) The solution for the equivalent damping coefficient itself is unsettling at first glance because the amplitude term is in the denominator. However, this is not unreasonable in itself. If the damping coefficient is small, then the amplitude should be large and vice versa. Nevertheless, the concept of equivalent viscous damping simply doesn't work very well in this case of Coulomb damping.

5.4 **(a)** Let θ be the counterclockwise rotation of the bar and $u(t)$ be the upward deflection of the right-hand mass. Then

$$T = \tfrac{1}{2}H\dot{\theta}^2 + \tfrac{1}{2}m\dot{u}^2, \quad U = \tfrac{1}{2}k(L\theta)^2 + \tfrac{1}{2}k(L\theta - u)^2,$$

$$\text{and} \quad \delta W = -(cL\dot{\theta})(L\delta\theta) - c\dot{u}\delta u.$$

Therefore the matrix equations of motion are

$$\begin{bmatrix} m & 0 \\ 0 & H \end{bmatrix} \begin{Bmatrix} \ddot{u} \\ \ddot{\theta} \end{Bmatrix} + \begin{bmatrix} c & 0 \\ 0 & cL^2 \end{bmatrix} \begin{Bmatrix} \dot{u} \\ \dot{\theta} \end{Bmatrix} + \begin{bmatrix} k & -kL \\ -kL & 2kL^2 \end{bmatrix} \begin{Bmatrix} u \\ \theta \end{Bmatrix} = \begin{Bmatrix} 0 \\ 0 \end{Bmatrix}.$$

(b) Since the rotations are small, the displacement of the smaller mass is just $u - L\theta$. The strain energy for the right-hand spring is $\tfrac{1}{2}(2k)(u + L\theta - v)^2$, and so on. The matrix equation is

$$m\begin{bmatrix} 3 & -1 \\ -1 & 1.5 \end{bmatrix} \begin{Bmatrix} \ddot{u} \\ L\ddot{\theta} \end{Bmatrix} + c\begin{bmatrix} 3 & -2 \\ -2 & 4 \end{bmatrix} \begin{Bmatrix} \dot{u} \\ L\dot{\theta} \end{Bmatrix} + k\begin{bmatrix} 3 & 1 \\ 1 & 3 \end{bmatrix} \begin{Bmatrix} u \\ L\theta \end{Bmatrix} = 2kv(t)\begin{Bmatrix} 1 \\ 1 \end{Bmatrix},$$

where all the elements of the coefficient matrices have been reduced to plain numbers by factoring and dividing the second equation (second row) by L. Note that this is not a case of proportional damping regardless of the positive values of k and c.

(c) This system is so straightforward that Newton's second law and the Lagrange equations are equally convenient. Assuming that $u_2 > u_1$, or $u_1 > u_2$, either resulting matrix equation is

$$\begin{bmatrix} m_1 & 0 \\ 0 & m_2 \end{bmatrix} \begin{Bmatrix} \ddot{u}_1 \\ \ddot{u}_2 \end{Bmatrix} + \begin{bmatrix} (c_1 + c_2) & -c_2 \\ -c_2 & (c_2 + c_3) \end{bmatrix} \begin{Bmatrix} \dot{u}_1 \\ \dot{u}_2 \end{Bmatrix} + \begin{bmatrix} (k_1 + k_2) & -k_2 \\ -k_2 & k_2 \end{bmatrix} \begin{Bmatrix} u_1 \\ u_2 \end{Bmatrix} = \begin{Bmatrix} 0 \\ F(t) \end{Bmatrix}.$$

(e) The tangential velocity of the left bar center of mass is merely $\tfrac{1}{2}L\dot{\theta}$. The right bar center of mass has an upward deflection equal to $\tfrac{1}{2}L\sin\theta$ and a leftward deflection equal to $2L(1-\cos\theta) - \tfrac{1}{2}L(1-\cos\theta)$. This latter expression can be understood by moving the right bar center of mass along the following path. Momentarily detach the two bars at their common hinge before they both rotate through the angle θ. The right bar center of mass moves upward the $\tfrac{1}{2}L\sin\theta$ mentioned above and rightward $\tfrac{1}{2}L(1-\cos\theta)$. Now translate the right bar leftward through the gap $2L(1-\cos\theta)$ that opened when the two bars were momentarily detached, until the two bars are again in a position where they can be reattached. Thus the horizontal component of the right bar center of mass velocity is $(3L\dot{\theta}/2)\sin\theta$. Furthermore, the rotational spring is deflected through the relative rotation angle 2θ. The rightward force in the dashpot is $c(d/dt)2L[1-\cos\theta]$. The leftward movement of this equivalent viscous damping force during the virtual displacement $\delta\theta$ is $2L[1-\cos(\theta+\delta\theta)] - 2L(1-\cos\theta) = 2L\delta\theta\sin\theta$. Thus the kinetic energy, strain energy, and virtual work are

$$T = 2(\tfrac{1}{2})\left(\frac{mL^2}{12}\right)\dot{\theta}^2 + \tfrac{1}{2}m\left(\frac{L}{2}\dot{\theta}\right)^2 + \tfrac{1}{2}m\left(\tfrac{1}{4}L^2\dot{\theta}^2 + 2L^2\dot{\theta}^2\sin^2\theta\right)$$

$$\text{so} \quad T = \tfrac{1}{3}mL^2\dot{\theta}^2 + mL^2\dot{\theta}^2\sin^2\theta \qquad U = \tfrac{1}{2}K(2\theta)^2$$

$$\text{and} \quad \delta W = +M(t)\delta\theta - 2Lc\dot{\theta}\sin\theta(2L\delta\theta\sin\theta).$$

Substituting into the Lagrange equations of motion the nonlinear and linear equations of motion and natural frequency are respectively

$$2mL^2(\tfrac{1}{3} + \sin^2\theta)\ddot{\theta} + mL^2\dot{\theta}^2\sin 2\theta + 4cL^2\dot{\theta}\sin^2\theta + 4K\theta = M(t)$$

$$\text{for} \quad \theta^2 \ll 1, \quad \ddot{\theta} + \left(\frac{6K}{mL^2}\right)\theta = M(t)$$

$$\text{so} \quad \omega = \sqrt{\frac{6K}{mL^2}}.$$

The effect of the dashpot disappears in the linearization.

(h) Let the two DOF be the indicated θ and $u(t)$. Using the parallel axis theorem to get the mass moment of inertia of the rigid rod about its base, the expressions for the kinetic energy, strain energy, virtural work, and the resulting Lagrange equation are

$$T = \frac{1}{2}m\dot{u}^2 + \frac{1}{2}\left(\frac{2}{3}mL^2 + 2mL^2\right)\dot{\theta}^2$$

$$U = \frac{1}{2}k(u - 2L\theta)^2$$

$$\delta W = -c\dot{u}\delta u - cL\dot{\theta}(L\delta\theta) + M(t)\delta\theta$$

$$m\begin{bmatrix}1 & 0\\ 0 & \frac{8}{3}\end{bmatrix}\begin{Bmatrix}\ddot{u}\\ L\ddot{\theta}\end{Bmatrix}+c\begin{bmatrix}1 & 0\\ 0 & 1\end{bmatrix}\begin{Bmatrix}\dot{u}\\ L\dot{\theta}\end{Bmatrix}+k\begin{bmatrix}1 & -2\\ -2 & 4\end{bmatrix}\begin{Bmatrix}u\\ L\theta\end{Bmatrix}=\begin{Bmatrix}0\\ M(t)/L\end{Bmatrix}.$$

This is another case where the damping is not proportional regardless of the values of m, c, and k.

5.5 Using Eq. (4.5), with $A_1/A_{11}=3$, the result rounds off to ζ equal to 0.02.

5.6 **(a)** Since the beam-columns are fixed at both ends, it is convenient to use the FEM beam element stiffness matrix because all but one of the element DOF are zero. Specifically, with the beam element DOF being $w_1=u$, $\theta_1=w_2=\theta_2=0$, the shear force acting on the mass from just one beam-column is, from the $(1,1)$ element of the stiffness matrix, $(12EI/L^3)u$. For three beam-columns, the stiffness factor is three times the above quantity in parantheses. With the damping factor being 0.2, the factor that converts the undamped natural frequency to the damped natural frequency, the $\sqrt{1-\zeta^2}$, has a value of the $\sqrt{0.96}$. Hence the damped natural frequency is

$$\omega_d=\sqrt{\frac{(36)(0.96)EI}{mL^3}}=5.88\sqrt{\frac{EI}{mL^3}}.$$

(b) The stiffness at the mass for a single beam-column of this structure can be determined using the beam element stiffness matrix in the following way. First note that what is being sought is again the relation between the shear force at the top of the beam and the lateral deflection at that same beam end. Call the top of the beam-column the 1 end of the beam element. Then with $w_1=u$, $w_2=\theta_2=M_1=0$, from writing the first two rows of the beam element stiffness matrix and solving the second of those two equations for θ in terms of u and substituting in the first of those two equations leads to

$$\frac{VL^3}{EI}=12u+6L\theta \quad \text{and} \quad 0=6Lu+4L^2\theta$$

$$\text{or} \quad V=\left(\frac{3EI}{L^3}\right)u.$$

Alternately, the unit load method or the differential equaton method could be used to directly obtain the same result. Either of these latter two approaches would start by placing an arbitrary force V at the free end of the cantilevered beam-column. Then the free end deflection, u, is calculated in terms of V. After the arrangement of this solution in the form $V=(k)u$, the stiffness coefficient for the cantilevered beam-column at its free end is evident. Taking into account that there are three beam-columns, the (equivalent viscous) damped natural frequency of the system is

$$\omega_d=\sqrt{\frac{(3)(3)(0.96)EI}{mL^3}}=2.94\sqrt{\frac{EI}{mL^3}}.$$

(c) This is exactly the same stiffness and mass situation as the previous case, so the undamped and damped natural frequencies are also the same.

(d) This is exactly the same situation as those of the above parts (b), and (c), except here there are four beams rather than three. Thus the damped natural frequency is

$$\omega_d=\sqrt{\frac{(4)(3)(0.96)EI}{mL^3}}=3.4\sqrt{\frac{EI}{mL^3}}.$$

(e) This is exactly like part (a) except that here there are four beams. That is, because all other DOF but the first (or third) lateral deflection are zero, from the 1, 1 (i.e., row 1, column 1) entry of the beam element stiffness matrix

$$\omega_d = \sqrt{\frac{(4)(12)(0.96)EI}{mL^3}} = 6.8\sqrt{\frac{EI}{mL^3}}.$$

(f) In this case $T = \frac{1}{2} H_z \dot{\psi}^2$. Since the beams also have fixed end conditions at the mass, as a first approximation, the elastic resisting torque acting on the mass is a result of the bending of the beam ends at the mass in such a manner that they do not deflect laterally but do undergo a bending slope. This is an approximation because when the mass rotates through the angle ψ, there is a tip deflection equal to this angle multiplied by the radius from the center of the mass to the point of connection between the mass and the beam. Hence, using the approximation of zero radius and zero deflection, from the 2,2 entry of the beam element stiffness matrix, $U = \frac{1}{2}(4)(EI/L^3)(4L^2)\psi^2$. Thus

$$\omega_d = \sqrt{\frac{(4)(4)(0.96)EI}{H_z L}} = 3.9\sqrt{\frac{EI}{H_z L}}.$$

A more accurate linear deflection answer incorporating the nonzero radius of the mass can be made using the beam element stiffness matrix. With $w_1 = \theta_1 = 0$, $w_2 = -r\psi$, $\theta_2 = +\psi$, and with the total resisting torque on the mass being $M_{tot} = M_2 + Vr$, from the last two rows of the stiffness matrix

$$\frac{V_2 L^3}{EI} = +12r\psi + 6L\psi$$

$$\text{and} \quad \frac{M_2 L^3}{EI} = +6Lr\psi + 4L^2\psi$$

$$\text{so} \quad M_{tot} = \frac{4EI}{L^3}(L^2 + 3rL + 3r^2)\psi$$

$$\text{and} \quad \omega_d = 4.0\sqrt{\frac{(L^2 + 3rL + 3r^2)EI(0.96)}{HL^3}}.$$

It is clear that only if, $L > 50r$, is the first approximation sufficiently accurate for engineering purposes. As a check, note that when $r = 0$, this answer reduces to the previous answer.

(g) Here the elastic resisting moment comes from the twisting of two of the beams to the extent of $2GJ\phi/L = 2\alpha EI\phi/L$, and the bending of the other two beams to the extent of $2(EI/L^3)(4L^2\phi)$ for $r = 0$. Thus the resisting moment is $2(4 + \alpha)EI\phi/L$ and

$$\omega_d = 1.39\sqrt{\frac{(4 + \alpha)EI}{H_x L}}.$$

(h) To answer this question, consider dividing the angular rotation vector into two perpendicular components where each component is aligned with a beam axis. Then the resisting moment for each component, 0.707 of the total angular deflection, is 0.707 of the answer to part (g), and the total vector form of the moment is the same as that of part (g). Thus the damped natural frequency is the same.

5.7 **(a)** For use in Newton's second law, the free body diagram for, say, the middle mass consists of the middle mass deflected to the left a distance $u_2(t)$ with a positive (rightward) horizontal shearing force acting on its top, and a negative shearing force acting on its bottom. For example, the shearing force resulting from the deformation of each of the three upper beam columns is, from the beam element stiffness matrix, $+12EI/L^3(u_3 - u_2)$. Since none of the masses rotate, the matrix equation of motion is

$$\begin{bmatrix} 1 & & \\ & 1 & \\ & & \alpha \end{bmatrix} \begin{Bmatrix} \ddot{u}_1 \\ \ddot{u}_2 \\ \ddot{u}_3 \end{Bmatrix} + 12\zeta\sqrt{\frac{EI}{mL^3}} \begin{bmatrix} 2 & -1 & 0 \\ -1 & 2 & -1 \\ 0 & -1 & 1 \end{bmatrix} \begin{Bmatrix} \dot{u}_1 \\ \dot{u}_2 \\ \dot{u}_3 \end{Bmatrix}$$
$$+ \frac{36EI}{mL^3} \begin{bmatrix} 2 & -1 & 0 \\ -1 & 2 & -1 \\ 0 & -1 & 1 \end{bmatrix} \begin{Bmatrix} u_1 \\ u_2 \\ u_3 \end{Bmatrix} = \frac{F(t)}{m} \begin{Bmatrix} 1 \\ 2 \\ 2 \end{Bmatrix}.$$

Here is a case of proportional damping.

(b) With the mass parameter α having the value 2, the matrix equation of motion in terms of DOF referenced to the fixed vertical axis, is

$$\begin{bmatrix} 1 & & \\ & 1 & \\ & & 2 \end{bmatrix} \begin{Bmatrix} \ddot{u}_1 \\ \ddot{u}_2 \\ \ddot{u}_3 \end{Bmatrix} + \frac{c}{m} \begin{bmatrix} 2 & -1 & 0 \\ -1 & 2 & -1 \\ 0 & -1 & 1 \end{bmatrix} \begin{Bmatrix} \dot{u}_1 \\ \dot{u}_2 \\ \dot{u}_3 \end{Bmatrix}$$
$$+ \frac{36EI}{mL^3} \begin{bmatrix} 2 & -1 & 0 \\ -1 & 2 & -1 \\ 0 & -1 & 1 \end{bmatrix} \begin{Bmatrix} u_1 \\ u_2 \\ u_3 \end{Bmatrix} = \frac{36EI}{mL^3} \begin{Bmatrix} 1 \\ 0 \\ 0 \end{Bmatrix} h(t) + \frac{c}{m} \begin{Bmatrix} 1 \\ 0 \\ 0 \end{Bmatrix} \dot{h}(t).$$

(c) In terms of the relative DOF, the matrix equation of motion is

$$\begin{bmatrix} 4 & 3 & 2 \\ 3 & 3 & 2 \\ 2 & 2 & 2 \end{bmatrix} \begin{Bmatrix} \ddot{u}_1 \\ \ddot{u}_2 \\ \ddot{u}_3 \end{Bmatrix} + \frac{c}{m} \begin{bmatrix} 1 & & \\ & 1 & \\ & & 1 \end{bmatrix} \begin{Bmatrix} \dot{u}_1 \\ \dot{u}_2 \\ \dot{u}_3 \end{Bmatrix} + \frac{36EI}{mL^3} \begin{bmatrix} 1 & & \\ & 1 & \\ & & 1 \end{bmatrix} \begin{Bmatrix} u_1 \\ u_2 \\ u_3 \end{Bmatrix} = \begin{Bmatrix} 4 \\ 3 \\ 2 \end{Bmatrix} \ddot{h}(t).$$

This too is a case of proportional damping.

5.8 **(d)** Since the one possible solution for the time when the velocity is zero, $w_1 t = 2\mu mg/F_0 > \pi$. the factor $\sin\omega t$ must supply the first occuring stopping point. Thus the time duration of the first half period is π/ω, and the distance traveled over the first half cycle is

$$u\left(\frac{\pi}{\omega}\right) = \frac{\pi u_{stat}}{2} - 2\mu u_g = \frac{\pi F_0}{2k} - 2\mu \frac{mg}{k}.$$

(e) Yes. The amplitude reduction per half cycle is $2\mu u_g$.

5.9 **(a)** The solution for the motion for the first half cycle is

$$u(t) = \tfrac{1}{2}\left(\frac{F_0}{k}\right)\omega_1 t \sin\omega_1 t + \frac{k\ell - \mu mg}{k}(1 - \cos\omega_1 t).$$

(b) The transcendental equation to be solved for the time at which the motion stops is

$$\omega_1 t = -\left(1 - \frac{2\mu mg}{F_0}\right)\tan\omega_1 t = -0.5\tan\omega_1 t.$$

A quick sketch of the functions $y = \omega t$, $y = -0.5\tan\omega t$ shows that the two curves cross somewhere between $\omega = 0$ and $\pi/2$.

5.10 **(a)** Let $u(t)$ be the motion of the block to the left after time zero. The initial conditions are those of zero initial deflection and zero initial velocity. After time zero, the stretch in the spring is such that Newton's second law is

$$m\ddot{u} = +k(V_0t + d_0 - u) - mg\mu_d$$
$$\text{or} \quad \ddot{u} + \omega_1^2 u = (\omega_1^2 V_0)\,t + (\omega_1^2 d_0 - g\mu_d)$$

The complete solution to this differential equation is

$$u(t) = A\sin\omega_1 t + B\cos\omega_1 t + V_0 t + d_0 - \frac{\mu_d g}{\omega_1^2}.$$

Applying the initial conditions leads to

$$B = \frac{\mu_d g}{\omega_1^2} - d_0 \qquad A = -\frac{V_0}{\omega_1}$$

so that the solution for the first part of the motion can be written as

$$u(t) = \frac{V_0}{\omega_1}[\omega_1 t - \sin\omega_1 t] - \left(\frac{\mu_d g}{\omega_1^2} - d_0\right)[1 - \cos\omega_1 t].$$

(b) Therefore the velocity of the block mass is

$$\dot{u}(t) = V_0[1 - \cos\omega_1 t] - \left(\frac{\mu_d g}{\omega_1} - \omega_1 d_0\right)\sin\omega_1 t.$$

The time, if any, at which the block comes to a stop is determined by setting the velocity equal to zero and solving explicitly for the corresponding time

$$\frac{\omega_1 V_0}{\mu_d g - \omega_1^2 d_0} = \frac{\sin\omega_1 t}{1 - \cos\omega_1 t} = \frac{2\sin\frac{\omega_1 t}{2}\cos\frac{\omega_1 t}{2}}{2\sin^2\frac{\omega_1 t}{2}} = \cot\frac{\omega_1 t}{2}$$
$$\text{or} \quad \tan\frac{\omega_1 t}{2} = \frac{\mu_d g - \omega_1^2 d_0}{\omega_1 V_0}.$$

The constant numerator of the right-hand fraction is not zero unless the dynamic friction coefficient is equal to the static friction coefficient, which is not the case. Since the dynamic friction coefficient is less than the static friction coefficient, the numerator is less than zero, making the left-hand side a negative quantity. Specifically, from the given relationship that here the dynamic Coulomb friction coefficient is $-4kd_0/(5mg)$, so that

$$\omega_1 t_{stop} = 2\arctan\left(\frac{-\omega_1 d_0}{5V_0}\right).$$

Thus a sketch of the tangent function shows that the nondimensional time (natural frequency multiplied by time) at which the block first comes to a stop must lie between π and 2π, even if the vehicle velocity is very small (yet is large enough to reach the distance d_0 in a finite time) and it increases with the speed of the vehicle asymtopically to the value of 2π.

(c) For the given choice of parameters, the nondimensional time that the block stops is $3\pi/2$. Now the question concerns the force in the spring. Is the spring compressed sufficiently at this time to overcome the static friction and send the mass moving to the right? For this selection of parameters, after evaluation of the block displacement and some other algebra, the spring force is

$$F_{spring} = k\left(d_0 - \frac{2V_0}{\omega_1}\right) = \mu_d mg.$$

Since this is a tensile force, the mass will not be traveling to the right. On the contrary, it is three-quarters of the way toward restarting its motion to the left.

CHAPTER 6 SOLUTIONS

6.1 **(a)** Recall the Newton–Raphson iterative (tangent) method which provides a second approximation from a first guess according to $r_2 = r_1 - f(r_1)/f(r_1)$. Start with λ equal to 0.25 and get a second approximation of 0.30. Use 0.30 to get a third approximation 0.299. The solution to four digits is 0.2991. Using this value for λ in the first and third equations yields the result that the first mode shape is $\lfloor 0.3417 \quad 0.7009 \quad 1.0 \rfloor^t$, whereas the first frequency is $\omega_1 = 1.73\sqrt{(EI/mL^3)}$.

6.4 **(a)** The first eigenvalue is 17.433, and the first eigenvector is $\lfloor 0.09377 \quad 0.5411 \quad 1.000 \rfloor$.

(b) The equation of constraint is $0.04689q_1 + 0.05411q_2 + q_3 = 0$. Dividing by the largest coefficient for better accuracy, then the product of the dynamic matrix $[D]$ and the sweeping matrix $[S_1]$ is

$$\begin{bmatrix} 1 & 1 & 1 \\ 2 & 6 & 6 \\ 2 & 6 & 14 \end{bmatrix} \begin{bmatrix} 1 & 0 & 0 \\ 0 & 1 & 0 \\ -0.04689 & -0.5411 & 0 \end{bmatrix} = \begin{bmatrix} 0.95311 & 0.4589 & 0 \\ 1.71866 & 2.7534 & 0 \\ 1.34354 & -1.5754 & 0 \end{bmatrix}.$$

Iterating this second dynamic matrix leads to an eigenvalue of 3.117 and an eigenvector $\lfloor 0.212 \quad 1.00 \quad -0.414 \rfloor$.

(c) The initial dynamic matrix, $[K]^{-1}[M]$, is $[(1,1,2),(1,2,4),(1,2,6)]$. The eigenvalues are 7.89167, 0.785825, and 0.322504. Hence, for example, the second natural frequency is $6.77(EI/mL^3)^{1/2}$. The first, second, and third mode shapes arranged in rows for amplitudes A_1, A_2, and A_3 (bottom to top) are

0.398534	0.746568	1.000
1.000	0.727455	−0.470814
−0.908484	1.000	−0.192253

(d) Here the dynamic matrix coincides with the mass matrix. The eigenvalues, and hence the natural frequencies, are the same because they are characteristics of the physical system and thus unaffected by the analyst's choice for DOF. However, the mode shapes are dependent on the choice of the DOF. In this case of relative coordinates, the mode shapes as above are

1.000	0.873282	0.635910
−0.834540	0.227452	1.000
−0.476022	1.000	−0.624711

6.6 The eigenvalue is 2.04396, which relates to the inverse of the first natural frequency. Thus, again, the first natural frequency is $0.70\sqrt{k/m}$.

6.7 **(d)** With $[K] = [R^t][R]$, which is the same as $[L][L]^t$, working with the elementary rules of matrix multiplication only leads to the right triangular matrix result

$$[R] = \frac{1}{\sqrt{3}}\begin{bmatrix} 3 & -1 \\ 0 & \sqrt{5} \end{bmatrix}.$$

As a check, the product $[R^t][R]$ returns $[K]$.

(e) From, for example, the characteristic determinant $|U - \lambda I| = 0$, expand the determinant using the rule of minors by using the first column, then second, and so on, to obtain the equation

$$(u_{11} - \lambda)(u_{22} - \lambda)\ldots(u_{nn} - \lambda) = 0$$

from whence the proposed conclusion.

(f) The inverse of the given matrix is

$$[L]^{-1} = \begin{bmatrix} \frac{1}{3} & 0 & 0 & 0 \\ -\frac{1}{6} & \frac{1}{4} & 0 & 0 \\ \frac{1}{18} & -\frac{5}{36} & \frac{1}{9} & 0 \\ -\frac{1}{9} & \frac{5}{18} & -\frac{2}{9} & \frac{1}{3} \end{bmatrix}.$$

6.8 **(a)** Applying the equation of constraint that $\lfloor A^{(1)} \rfloor [M]\{q\}^c = 0$ and, as before, solving for the first DOF (solving for the third DOF would have been a better choice for numerical accuracy) leads to the following transformation matrix equation $\{q\}^c = [S_1]\{q\}^{nc}$ and the dynamic matrix

$$[S_1] = \begin{bmatrix} 0 & -3.869 & -10.468 \\ 0 & 1.00 & 0 \\ 0 & 0 & 1.00 \end{bmatrix}$$

$$\text{and} \quad [D_2] = \begin{bmatrix} 0 & -1.869 & -6.468 \\ 0 & 0.131 & -2.468 \\ 0 & 0.131 & 1.532 \end{bmatrix}.$$

Notice the additional loss of accuracy in $[D_2]$ because of entries such as 0.131 being the small difference of two larger numbers. The second eigenvalue and mode shape from this calculation are, respectively, 1.24 and $\lfloor 0.838 \quad 1.00 \quad -0.450 \rfloor$.

(b) With $(k/m) = 1.0$, the modified dynamic matrix is formed from the premultiplication of $[M]$ by the inverse of $[1.0K - 0.81M]$. The new, shifted dynamic matrix is

$$[D_{shift}] = \begin{bmatrix} 6.56242 & 4.70409 & -1.21141 & -3.18792 \\ 7.05614 & 6.49165 & -1.67174 & -4.39933 \\ -3.63423 & -3.34349 & 0.345978 & 0.910469 \\ -19.1275 & -17.5873 & 1.82094 & 10.0551 \end{bmatrix}.$$

The first and largest eigenvalue is $1/(\omega^2 - \omega_s^2) = 23.4226$. The corresponding mode shape is as follows:

$$\lfloor -0.312564 \quad -0.404647 \quad 0.147306 \quad 1.00000 \rfloor.$$

Since this mode shape evidences a single node, it is expected that it is the second mode shape. This is confirmed when the software results for all the modes are inspected.

6.9 To test for linear independence, form the sum and equality

$$c_1\boldsymbol{e}_1 + c_2\boldsymbol{e}_2 + c_3\boldsymbol{e}_3 + \cdots + c_n\boldsymbol{e}_n = 0$$

and test whether any of the coefficients c_i can be other than zero. This is accomplished by multiplying both sides of the equality by any one of the orthogonal vectors, say $\boldsymbol{e}_j$. As a result of their orthogonality, the sum becomes the single term $c_j(\boldsymbol{e}_j \cdot \boldsymbol{e}_j) = c_j = 0$. Thus this coefficient and all like it must be zero. Hence the orthogonal vectors are linearly independent.

6.10 **(a)** In the author's opinion, when the matrix size is 4 × 4 or less, it is clearly better to invert the stiffness matrix by hand than calculate the flexibility matrix by hand in that it is both less work and less likelihood of arithmetic error. At larger sizes, either approach by hand is too lengthy, but the direct calculation of the symmetric flexibility matrix is a somewhat more rewarding experience. The complete flexibility matrix is

$$\begin{Bmatrix} w_2 \\ \theta_2 L \\ \phi_2 L \\ w_3 \\ \theta_3 L \\ \phi_3 L \end{Bmatrix} = \frac{L^3}{12EI} \begin{bmatrix} 2 & 3 & 0 & 2 & 3 & 0 \\ 3 & 6 & 0 & 3 & 6 & 0 \\ 0 & 0 & 12 & -12 & 0 & 12 \\ 2 & 3 & -12 & 18 & 3 & -18 \\ 3 & 6 & 0 & 3 & 30 & 0 \\ 0 & 0 & 12 & -18 & 0 & 24 \end{bmatrix} \begin{Bmatrix} F_2 \\ M_2/L \\ T_2/L \\ F_3 \\ M_3/L \\ T_3/L \end{Bmatrix}$$

The above calculation was done using the virtual load method, which again is essentially the same as the unit load method or the dummy load method or Castiglianp's second theorem. The setup for this calculation is

$$\delta W^*_{ex} = \int_0^L \frac{M(x)\delta M(x)}{2EI}dx + \int_0^L \frac{M(y)\delta M(y)}{EI}dy + \int_0^L \frac{T(x)\delta T(x)}{EI}dx + \int_0^L \frac{2T(y)\delta T(y)}{EI}dy$$

Note that because the virtual load method is a force or flexibility method, as opposed to the deflection or stiffness finite element method used everywhere else in this textbook, the varied quantities here are forces and moments, while the actual quantities are displacements, bending slopes, and twists. Furthermore, the above equation is an application of the principle of complementary virtual work which states that the complementary virtual work of the external loads is equal to the complementary strain energy. To use the principle to calculate, say, the entry in the second row and first column, using this approach, it is necessary to first apply the real force F_2 at node two in the positive direction of w_2, and then apply a virtual moment δM_2 in the positive direction of θ_2. Then the external complementary virtual work done on the beam grid is the virtual moment multiplied by the real rotation, $\delta M_2\,\theta_2$. Then it is simply a matter of calculating the actual bending and twisting moments due to F_2, and the virtual bending and twisting moments due to δM_2. Note that the arbitrary valued δM_2 will also appear in all terms on the right-hand side of the above equation, and thus be cancelled. The result of the calculation is $\theta_2 = F_2 L^2/(4EI)$.

(b) The flexibility matrix must be a symmetric matrix. In this case where the beam is cantilevered, the flexibility coefficients for w_2, θ_2, and ϕ_2 must by larger than the same quantities with subscript 1. Finally, the product of the flexibility matrix and stiffness matrix, in either order, must be the 6 × 6 identity matrix.

6.11 The first and second pairs of repeated natural frequencies in units of radians per second are

(a) 1.36 and 6.08

(b) 0.97 and 5.60

(c) 1.22 and 5.87

(d) 1.03 and 5.66

(e) 0.82 and 5.49.

Notice that with increasing tip mass and constant stiffness, both the first and second pairs of natural frequencies drop in magnitude.

6.12 First consider sufficiency; that is, given the vector $\{v\}$ is orthogonal to the first mode shape, show that c_1 is zero. By way of contradiction, presume c_1 is not zero in the expansion of Eq. (5.8) for $\{v\}$. Then $\lfloor v \rfloor [M]\{A^{(1)}\} = 0$ requires that c_1 is zero. Hence, the contradiction and the conclusion that $\{v\}$ does not contain the first mode. Now for necessity; that is, given that $c_1 = 0$, show that $\{v\}$ is orthogonal to the first mode shape. This is a simple matter of considering the matrix product $\lfloor v \rfloor [M]\{A^{(1)}\}$ and seeing that this must be zero.

CHAPTER 7 SOLUTIONS

7.1 **(a)** As the lunar lander approaches the surface, its massless springs are unstretched. Therefore, at the instant of contact, the initial deflection of this single-DOF system is $+mg/k$ (the static deflection of the mass weight acting on the springs), and the initial velocity is $-V$. From the second chapter, the formula for free vibration is

$$u(t) = u(0)\cos\omega_1 t + \frac{\dot{u}(0)}{\omega_1}\sin\omega_1 t, \text{ where } \quad \omega_1 = \sqrt{\frac{k}{m}}.$$

Applying the above initial conditions leads to

$$u(t) = \frac{mg}{k}\cos\omega_1 t - \frac{V}{\omega_1}\sin\omega_1 t.$$

(b) Let the bullet and the single-DOF system together constitute the dynamic system under study relative to Newton's laws. Since at no time are there forces external to this two part system acting on this system, the total momentum of this system is conserved (i.e., constant), despite the fact that the energy of the system is far from conserved. Thus, with V being the velocity of the bullet lodged in the target and the target itself, writing the expressions for the momentum before and after impact, and equating them, leads to

$$mv = (M+m)V \quad \text{or} \quad V = \frac{m}{(M+m)}v.$$

At time zero plus, after the impact of the bullet, the target and the bullet begin an undamped harmonic motion with initial conditions that are a zero initial deflection and an initial velocity V. Therefore, the motion of the target and bullet is simply $u(t) = U\sin\omega_1 t$, where $\omega_1^2 = K/(M+m)$, $U = (V/\omega_1)$. Thus the muzzle velocity of the bullet is $v = (M+m)\omega_1 U/m$.

(c) The solution for the free vibration motion of this pendulum system is stated as Eq. (6.1), where $n = 3$. Thus it is necessary only to calculate the three coefficients a_j

and the three phase angles ψ_j. To this end, note that the mass matrix is simply m multiplying the 3×3 identity matrix. Since the initial velocities are all zero

$$V_j = \lfloor A^{(j)} \rfloor [m]\{\dot{q}(0)\} = 0.$$

Therefore all the phase angles $\psi_j =$ arctan $\omega_j D_j / V_j$ are 90°. Hence, in Eq. (6.1), the sines with 90° phase angles can be replaced by cosines with 0° phase angles. The solutions for the coefficients a_j now simplify to D_j / M_j. The numerators and denominators of these ratios are

$$D_1 = m\lfloor 1 \quad 1 \quad 1 \rfloor \begin{Bmatrix} 0 \\ 0 \\ 0.1 \end{Bmatrix} = 0.1m \qquad M_1 = m\lfloor 1 \quad 1 \quad 1 \rfloor \begin{Bmatrix} 1 \\ 1 \\ 1 \end{Bmatrix} = 3m$$

$$D_2 = m\lfloor -1 \quad 0 \quad 1 \rfloor \begin{Bmatrix} 0 \\ 0 \\ 0.1 \end{Bmatrix} = 0.1m \qquad M_2 = m\lfloor -1 \quad 0 \quad 1 \rfloor \begin{Bmatrix} -1 \\ 0 \\ 1 \end{Bmatrix} = 2m$$

$$D_3 = m\lfloor -0.5 \quad 1.0 \quad -0.5 \rfloor \begin{Bmatrix} 0 \\ 0 \\ 0.1 \end{Bmatrix} = -0.05m$$

$$M_3 = m\lfloor -0.5 \quad 1.0 \quad -0.5 \rfloor \begin{Bmatrix} -0.5 \\ 1.0 \\ -0.5 \end{Bmatrix} = 1.5m.$$

Therefore

$$\begin{aligned} q_1(t) = \theta_1(t) &= \frac{1}{30} \cos \sqrt{\beta} t - \frac{1}{20} \cos \sqrt{2\beta} t + \frac{1}{60} \cos 2\sqrt{\beta} t \\ q_2(t) = \theta_2(t) &= \frac{1}{30} \cos \sqrt{\beta} t - \frac{1}{30} \cos 2\sqrt{\beta} t \\ q_3(t) = \theta_3(t) &= \frac{1}{30} \cos \sqrt{\beta} t + \frac{1}{20} \cos \sqrt{2\beta} t + \frac{1}{60} \cos 2\sqrt{\beta} t. \end{aligned}$$

These answers check at time zero.

(d) As always for matrix equations of motion, the BCs enter the problem in the stiffness matrix.

7.2 **(a)** $T = \frac{1}{2} m\dot{q}^2$, $U = \frac{1}{2}(3EI/L^3)(q-u)^2$, and $\delta W = -(\text{damping force})\delta q$. Substituting into the Lagrange equation of motion and dividing by the mass term m leads to the equation of motion

$$\ddot{q} + 2\zeta\omega_1\dot{q} + \omega_1^2 q = \omega_1^2 U_0 \sin \omega_f t \rightarrow \omega_1^2 U_0 e^{i\omega_f t},$$

where the real harmonic function has again been replaced by its complex equivalent so that the solution may be written as $q(t) = q_0 e^{i\omega t}$. Substituting this solution and simplifying

$$\frac{q_0}{U_0} = \frac{1}{(1-\Omega_1^2) + 2i\zeta\Omega_1} \quad \text{or} \quad \frac{|q_0|}{U_0} = \frac{1}{\sqrt{(1-\Omega_1^2)^2 + (2\zeta\Omega_1)^2}}.$$

This, of course, is the same form of solution as that obtained for the modal equation of motion.

(b) With the damping force depending on the relative motion of the mass and base, the simplified equation of motion is, after writing the base velocity in terms of its deflection,

$$\ddot{q} + 2\zeta\omega_1\dot{q} + \omega_1^2 q = \omega_1^2 U_0 e^{i\omega_f t} + 2\zeta\omega_1\omega_f U_0 e^{i\omega_f t}.$$

Again, writing and substituting the solution for the mass motion in complex form as $q(t) = q_0 e^{i\omega t}$, the result is

$$\frac{q_0}{U_0} = \frac{1 + 2i\zeta\Omega_1}{(1 - \Omega_1^2) + 2i\zeta\Omega_1} \quad \text{or} \quad \frac{|q_0|}{U_0} = \frac{\sqrt{1 + (2\zeta\Omega_1)^2}}{\sqrt{(1 - \Omega_1^2)^2 + (2\zeta\Omega_1)^2}}.$$

The ratio of the absolute value (amplitude) of the mass deflection to the amplitude of the base motion is called the *transmissibility*. It is, of course, a measure of the amount of steady-state base motion transmitted to the mass. Examination of this function shows two facts of importance. The first is that there is a resonance phenomenon at $\Omega_1 = 1.0$, and the second is that the value of this amplitude ratio is increasingly less than 1 as Ω_1 increases beyond the value $\sqrt{2}$. Furthermore, the less the damping, the less the transmissibility after Ω_1 exceeds $\sqrt{2}$.

(c) In the case of zero support motion, the equation of motion of the rigid mass and its solution for the mass motion are

$$m\ddot{q} + c\dot{q} + kq = F_0 e^{i\omega_f t}$$

$$\text{or} \quad \ddot{q} + 2\zeta\omega_1\dot{q} + \omega_1^2 q(t) = (F_0/m)(k/k)e^{i\omega_f t} = \frac{\omega_1^2 F_0}{k} e^{i\omega f t}$$

$$\text{so} \quad q(t) = q_0 e^{i\omega_f t}$$

$$\text{and} \quad q_0 = \frac{F_0/k}{(1 - \Omega_1^2) + 2i\zeta\Omega_1}.$$

The force resultant transmitted to the structure is the force in the spring plus the force in the dashpot, which is

$$F_{tr} = kq + c\dot{q} \quad \text{or} \quad \frac{F_{tr}}{k} = q + \frac{2\zeta}{\omega_1}\dot{q} = [1 + 2i\zeta\Omega_1]q_0 e^{i\omega_f t}.$$

Using the above two results to form the ratio F_{tr}/F_0 yields

$$\frac{F_{tr}}{F_0} = \frac{1 + 2i\zeta\Omega_1}{(1 - \Omega_1^2) + 2i\zeta\Omega_1} \quad \text{or} \quad \frac{|F_{tr}|}{F_0} = \frac{\sqrt{1 + 2\zeta\Omega_1}}{\sqrt{(1 - \Omega_1^2)^2 + (2\zeta\Omega_1)^2}},$$

which is the same transmissibility factor defined in the previous part of this exercise.

7.3 If $v(t)$ is the horizontal ground motion, then the equation of motion for the building is

$$m\ddot{u} = -k(u - v) - c(\dot{u} - \dot{v})$$

$$\text{or} \quad \ddot{u} + 2i\zeta\omega_1\dot{u} + \omega_1^2 u = (\omega_1^2 + 2i\zeta\omega_1\omega_f)\,\Upsilon_0 e^{i\omega_f t}.$$

Introducing the solution form $u(t) = u_0 \exp(i\omega_f t)$, obtain the complex form of the transmissibility

$$\frac{u_0}{\Upsilon_0} = \frac{1 + 2i\zeta\Omega_1}{1 - \Omega_1^2 + 2i\zeta\Omega_1}.$$

Taking absolute values provides the real form of the transmissibility

$$\frac{|u_0|}{\Upsilon_0} = \frac{\sqrt{1+(2\zeta\Omega_1)^2}}{\sqrt{(1-\Omega_1^2)^2+(2\zeta\Omega_1)^2}}.$$

7.5 (**a**) zero
(**b**) 4
(**c**) ½ 4 = 2
(**d**) 4
(**e**) zero
(**f**) 1
(**g**) Write the coordinate transformation $t-\tau=\eta$. The desired form of the integrand is immediately achieved. As for the differential, because τ, η are the variables of integration, whereas t is only a parameter, $d(t-\tau)=d\eta \Rightarrow d\tau=-d\eta$. Now use the transformation to determine the limits of integration with respect to η. When $\tau=0$, $\eta=t$. When $\tau=t, \eta=0$. Reversing the limits cancels the negative sign, and the desired result is achieved.

7.6 This is a single-DOF system with the sole generalized coordinate $u(t)$ measuring the lateral displacement of the mass m with respect to a fixed coordinate axis. Since the system is so uncomplicated, Newton's second law is easy to apply, and the Lagrange equation is no more complicated. Using the Lagrange equation

$$\frac{d}{dt}\frac{\partial T}{\partial \dot{u}} + \frac{\partial U}{\partial u} = 0$$

$$\text{where} \quad T=\frac{1}{2}m\dot{u}^2 \quad U=\frac{1}{2}(2k)(u-h)^2$$

$$\therefore \quad m\ddot{u}(t)+2ku(t)=2kh(t)$$

$$\text{or} \quad \ddot{u}(t)+\omega_1^2u(t)=\omega_1^2h(t).$$

The cantilevered beam tip stiffness factor k, if not remembered, can be deduced as follows. Consider the beam element stiffness matrix equation for just one of the two beam columns when the left-hand DOF (here the DOF at the bottom of a column) are zero (fixed end), and there is no bending moment, M, at the top of the column where a hinge is located. Then V, which is the shear force that acts at the top of the beam-column, is the restoring spring force for the mass. Then the bottom two rows of this beam element FEM equation reduce to

$$\begin{Bmatrix} V \\ M=0 \end{Bmatrix} = \frac{EI}{L^3}\begin{bmatrix} 12 & -6L \\ -6L & 4L^2 \end{bmatrix}\begin{Bmatrix} u \\ \theta \end{Bmatrix} \rightarrow \theta = \frac{3u}{2L} \rightarrow V = \frac{3EI}{L^3}u = ku.$$

Hence, the total stiffness factor for the mass is $(6EI/L^3)$, and the system natural frequency squared is equal to $2k/m = 6EI/mL^3$. Therefore the solution for the above equation of motion for the mass is

$$0 \le t \le t_0: \quad u(t) = \frac{\omega_1 h_0}{t_0}\int_0^t \tau \sin\omega_1(t-\tau)d\tau$$

$$\text{and for} \quad t \ge t_0: \quad u(t) = \frac{\omega_1 h_0}{t_0}\int_0^{t_0} \tau \sin\omega_1(t-\tau)d\tau + \omega_1 h_0 \int_{t_0}^{t} \sin\omega_1(t-\tau)d\tau.$$

7.8 **(a)** Since the applied modal acceleration time history can be described in terms of a step function and a (negative) ramp function, then the modal deflection response of this linear system can be described in terms of the individual responses to these components of the loading. Therefore the modal deflection response can be written immediately as

$$\text{for} \quad t \leq 0 \qquad p_j(t) = 0 \quad (\textit{as per usual})$$

$$0 \leq t \leq t_1 \qquad p_j(t) = P_j^0 g_j(t) - \frac{P_j^0}{t_1} r_j(t)$$

$$t_1 \leq t \qquad p_j(t) = P_j^0 g_j(t) - \frac{P_j^0}{t_1}[r_j(t) - r_j(t - t_1)].$$

(b) The sine response function, as defined in Example 6.4, needs to be adapted to the modal case. This is simply done by replacing the first natural frequency by the jth natural frequency and the base acceleration amplitude coefficient of the original sine response function by the modal acceleration amplitude. Then the complete solution in terms of the sine response function

$$s_j(t) = \frac{1}{(\pi/t_0)^2 - \omega_j^2}\left(\frac{\pi}{\omega_j t_0}\sin\omega_j t - \sin\frac{\pi t}{t_0}\right)$$

is

$$t \leq 0: \quad p_j(t) = 0; \qquad 0 \leq t \leq t_0: \quad p_j(t) = P_j^0 s_j(t);$$

and for

$$t_0 \leq t: \quad p_j(t) = P_j^0[s_j(t) + s_j(t - t_0)].$$

7.9 **(a)** First calculate the velocity of the base motion, and thereby form the expression for the equivalent input acceleration term of the convolution integral

$$\dot{u} = \frac{4\Upsilon_0}{t_0^2}(t_0 - 2t)$$

$$\text{thus} \quad P(t) = \frac{4\Upsilon}{t_0^2}\left[\omega_1^2(tt_0 - t^2) + 2\zeta\omega_1(t_0 - 2t)\right].$$

Then for the time interval $(0, t_0)$, the deflection solution is, in integral form,

$$q(t) = \frac{1}{\omega_d}\int_0^t \left\{\frac{4\Upsilon}{t_0^2}\left[\omega_1^2(\tau t_0 - \tau^2) + 2\zeta\omega_1(t_0 - 2\tau)\right]\exp[-\zeta\omega_1(t - \tau)]\sin\omega_d(t - \tau)\right\}d\tau.$$

(b) The deflection solution here is as in part (a) but for the single change that is replacing the upper limit t of the integral by t_0.

7.10 The object is to describe this system using only two DOF, the vertical deflections at the lumped masses, which are here labeled w_1 and w_2. To accomplish this time-saving approach, either the 2×2 flexibility matrix must be determined and inverted or the 4×4 stiffness matrix, which also includes the two bending slope DOF, must be partitioned so as to eliminate the bending slope DOF. Since the cantilevered beam is statically determinate, the former approach is reasonable. However, to continue the focus on the finite element method, the latter approach is used here. Rearranging rows and columns, and partitioning the mass, stiffness, and applied force vector, the original 4×4 matrix equation of motion becomes the following two sets of 2×2 equations:

$$m\begin{bmatrix} 2 & 0 \\ 0 & 1 \end{bmatrix}\begin{Bmatrix} \ddot{w}_1 \\ \ddot{w}_2 \end{Bmatrix} + \frac{8EI}{L^3}\begin{bmatrix} 24 & -12 \\ -12 & 12 \end{bmatrix}\begin{Bmatrix} w_1 \\ w_2 \end{Bmatrix} + \frac{8EI}{L^3}\begin{bmatrix} 0 & 3L \\ -3L & -3L \end{bmatrix}\begin{Bmatrix} \theta_1 \\ \theta_2 \end{Bmatrix} = \begin{Bmatrix} 0 \\ F \end{Bmatrix}$$

$$\frac{8EI}{L^3}\begin{bmatrix} 0 & -3L \\ 3L & -3L \end{bmatrix}\begin{Bmatrix} w_1 \\ w_2 \end{Bmatrix} + \frac{8EI}{L^3}\begin{bmatrix} 2L^2 & \frac{1}{2}L^2 \\ \frac{1}{2}L^2 & L^2 \end{bmatrix}\begin{Bmatrix} \theta_1 \\ \theta_2 \end{Bmatrix} = \begin{Bmatrix} 0 \\ 0 \end{Bmatrix}.$$

In the latter of the above two equations, multiplying through by the inverse of the 2×2 matrix that multiplies the bending slope vector (easily done), allows for a quick solution for the bending slope vector in terms of the lateral deflection vector. When that result for the bending slope vector is substituted into the previous equation, and the matrices multiplying the lateral deflection vector are combined, the result is

$$m\begin{bmatrix} 2 & 0 \\ 0 & 1 \end{bmatrix}\begin{Bmatrix} \ddot{w}_1 \\ \ddot{w}_2 \end{Bmatrix} + \frac{48EI}{7L^3}\begin{bmatrix} 16 & -5 \\ -5 & 2 \end{bmatrix}\begin{Bmatrix} w_1 \\ w_2 \end{Bmatrix} = \begin{Bmatrix} 0 \\ F \end{Bmatrix}.$$

For this 2×2 matrix equation, the solution for the modal frequencies and mode shapes can easily be done using the determinant method or the iteration method. The results are

$$\omega_1^2 = 2.4905\left(\frac{EI}{mL^2}\right) \qquad \lfloor A^{(1)} \rfloor = \lfloor 0.32736 \quad 1.0 \rfloor$$

$$\omega_2^2 = 66.08\left(\frac{EI}{mL^3}\right) \qquad \lfloor A^{(2)} \rfloor = \lfloor 1.0 \quad -0.6547 \rfloor.$$

Now write and substitute the modal transformation $\{w\} = [\Phi]\{p\}$ to obtain the uncoupled matrix equations, where each row is the corresponding modal equation

$$m\begin{bmatrix} 1.214 & 0 \\ 0 & 2.429 \end{bmatrix}\begin{Bmatrix} \ddot{p}_1 \\ \ddot{p}_2 \end{Bmatrix} + \frac{48EI}{7L^3}\begin{bmatrix} 0.4411 & 0 \\ 0 & 23.41 \end{bmatrix}\begin{Bmatrix} p_1 \\ p_2 \end{Bmatrix} = \begin{Bmatrix} F \\ -0.6547F \end{Bmatrix}.$$

The ratios of the generalized stiffness entries to the generalized mass entries are the squares of the natural frequencies as they should be. Therefore, the first and second modal equations reduce to

$$\ddot{p}_1 + \omega_1^2 p_1 = \frac{F(t)}{m} = P_1(t)$$

$$\text{and} \qquad \ddot{p}_2 + \omega_2^2 p_2 = -0.6547\frac{F(t)}{m} = P_2(t).$$

Now it is just a matter of substituting into the superposition integral.

$$p_1(t) = \frac{1}{\omega_1 m}\int_0^t F(\tau)\sin\omega_1(t-\tau)d\tau$$

$$\text{and} \quad p_2(t) = \frac{-0.6547}{\omega_2 m}\int_0^t F(\tau)\sin\omega_2(t-\tau)d\tau.$$

7.12 **(a)** After the nondimensional time value of 2π, the response is evermore zero.

(b) After the nondimensional time value of π, the response is a continuous harmonic vibration of amplitude 2Υ.

7.13 The first four terms of the Caughey damping matrix series are

$$[c] = a_0[m] + a_1[k] + a_2[km^{-1}k] + a_3[km^{-1}km^{-1}k].$$

Note that forming the inverse of the mass matrix is often not costly. Now that the modified modal vector normalization process for the mass matrix produces the identity matrix, the following conclusions can be drawn. That is, because

$$[\Phi]^t[m][\Phi] = [I] \to [m] = [\Phi]^{-t}[\Phi]^{-1} \to [m]^{-1} = [\Phi][\Phi]^t.$$

Substituting for the inverse of the mass matrix in the damping matrix series above yields

$$[c] = a_0[m] + a_1[k] + a_2[k\Phi\Phi^t k] + a_3[k\Phi\Phi^t k\Phi\Phi^t k].$$

Premultiplying and postmultiplying, respectively, by the transpose of the modal matrix and the modal matrix leads to

$$[\Phi^t c\Phi] = a_0[\backslash I\backslash] + a_1[\backslash\omega^2\backslash] + a_2[\backslash\omega^4\backslash] + a_3[\backslash\omega^6\backslash],$$

where, of course, the sum of diagonal matrices is a diagonal matrix. This clever idea has had no impact on engineering practice. Reference [6.5] says that the calculation of the a_j coefficients often leads to ill-conditioning, and the inclusion of more than two terms in the series is not useful.

CHAPTER 8 SOLUTIONS

8.3 Although not strictly necessary for this uniform geometry and set of simple support boundary conditions that have already been examined in three example problems, it is nevertheless suggested that the first step toward a solution is that of writing the equation of motion. The governing partial differential equation for the time period $(0, t_1)$ is

$$EIw''''(x,t) + \rho A\ddot{w}(x,t) = f_0 \sin\frac{\pi x}{L}\sin\frac{\pi t}{t_1}.$$

Seeking the solution in terms of the mode shapes, write

$$w(x,t) = \sum_{n=1}^{N} p_n(t)\sin\frac{n\pi x}{L}$$

$$\text{so that}\quad w''''(x,t) = \sum_{n=1}^{N}\left(\frac{n\pi}{L}\right)^4 p_n(t)\sin\frac{n\pi x}{L}$$

$$\text{and}\quad \ddot{w}(x,t) = \sum_{n=1}^{N}\ddot{p}_n(t)\sin\frac{n\pi x}{L}.$$

Substituting the above into the equation of motion yields

$$\sum_{n=1}^{N}\left[\ddot{p}_n(t) + \left(\frac{n\pi}{L}\right)^4\left(\frac{EI}{\rho A}\right)p_n(t)\right]\sin\frac{n\pi x}{L} = \frac{f_0}{\rho A}\sin\frac{\pi t}{t_1}\sin\frac{\pi x}{L}.$$

Note that the coefficient of $p_n(t)$ is the nth natural frequency squared. The N unknown functions $p_n(t)$ in the above single equation can now be uncoupled by applying to both sides the following modal multiplication and integration

$$\frac{2}{L}\int_0^L [\ldots]\sin\frac{m\pi x}{L}\,dx.$$

From the orthogonality of the mode shapes, or from the orthogonality of the sine functions in this simple case

$$\text{for} \quad m \neq 1: \qquad \ddot{p}_m(t) + \omega_m^2 p_m(t) = 0$$
$$\text{for} \quad m = 1: \qquad \ddot{p}_1(t) + \omega_1^2 p_1(t) = \frac{f_0}{\rho A} \sin \frac{\pi t}{t_1}.$$

Since the initial conditions are zero deflection and zero velocity, the solution for the first of the above two equations is simply $p_m(t) = 0$ for all $m \neq 1$. For the case where $m = 1$, the total solution could be obtained using the convolution integral. However, it is simpler to just use the method of undetermined coefficients to obtain a particular solution and then (i) combine the particular solution with the familiar complementary solution and (ii) apply the initial conditions. Write the trial solution as

$$p_1(t) = B_1 \sin \frac{\pi t}{t_1}$$
$$\text{so that} \qquad \ddot{p}_1(t) = -\frac{\pi^2}{t_1^2} B_1 \sin \frac{\pi t}{t_1}.$$

Substituting the above into the $m = 1$ equation and solving for the constant B_1 yields

$$B_1 = \frac{f_0 L^4 t_1^2}{\pi^4 E I t_1^2 - \pi^2 \rho a L^4} = \frac{f_0 t_1^2}{\rho A [(\omega_1 t_1)^2 - \pi^2]}.$$

Hence, the solution for $p_1(t)$ is

$$p_1(t) = C_1 \sin \omega_1 t + C_2 \cos \omega_1 t + \frac{f_0 t_1^2}{\rho A [(\omega_1 t_1)^2 - \pi^2]} \sin \frac{\pi t}{t_1}.$$

Since the factor $\sin(\pi x/L)$ is nowhere zero on the interior of the beam span, the initial conditions $w(x, 0) = \dot{w}(x, 0) = 0$, lead to the conclusion that $p_1(0) = \dot{p}_1(0) = 0$. Therefore, the above constants of integration have the values

$$C_2 = 0 \qquad \text{and} \quad C_1 = -\frac{\pi \omega_1 t_1 f_0}{A \omega_1^2 [(\omega_1 t_1)^2 - \pi^2]}.$$

Therefore, the complete deflection solution for $0 \leq t \leq t_1$ is

$$w(x, t) = \frac{f_0 \omega_1 t_1}{\rho A \omega_1^2 [(\omega_1 t_1)^2 - \pi^2]} \left(\omega_1 t_1 \sin \frac{\pi t}{t_1} - \pi \sin \omega_1 t \right) \sin \frac{\pi x}{L}.$$

Thus when the beam is excited by a loading that is only proportional to the first mode shape, it vibrates only in the first mode shape. Furthermore, there is a resonance effect when $t_1 = \pi/\omega_1$. Also note the similarity of this result to that of the half wave sine response function developed in the previous chapter. The above solution can be used to define the simply supported beam sine response function.

The response for $t_1 \leq t$ can be constructed from that for $0 \leq t \leq t_1$ by viewing the time history of the applied force in this second time interval as the superposition of the original applied force onto itself, starting at time t_1. In this way the now descending portion of the first sine function is canceled by the ascending portion of the second sine function, resulting in the specified zero applied force after time t_1. In other words, adding $w(x, t - t_1)$ to the above response yields for $t_1 \leq t$

$$w(x, t) = \frac{-\pi f_0 \omega_1 t_1}{\rho A \omega_1^2 [(\omega_1 t_1)^2 - \pi^2]} [\sin \omega_1 t + \sin \omega_1 (t - t_1)] \sin \frac{\pi x}{L}.$$

8.4 **(a)** The governing differential equation and BCs for $0 < t < t_0$ are

$$w''''(x,t) + \frac{\rho A}{EI}\ddot{w}(x,t) = 0$$

$$w(0,t) = w_0 \sin\left(\frac{\pi t}{t_0}\right)$$

$$w''(0,t) = w(L,t) = w''(L,t) = 0.$$

Introduce the transformation suggested in Endnote (4). Let

$$w(x,t) = v(x,t) + h(x)\sin\frac{\pi t}{t_0}$$

$$\text{so that } w(0,t) = v(0,t) + h(0)\sin\frac{\pi t}{t_0} = w_0 \sin\frac{\pi t}{t_0}.$$

To make $v(0,t) = 0$, let $h(0) = w_0$. When the above transformation involving $h(x)$ is substituted into the above governing differential equation, it will be differentiated four times. Therefore, it is best to select a smooth, as well as simple, function for $h(x)$. Let $h(x) = w_0(1 - x/L)$. Then substitution of the above transformation leads to the new governing differential equation in terms of $v(x,t)$ and an equivalent loading per unit length. This result is

$$v''''(x,t) + \frac{\rho A}{EI}\ddot{v}(x,t) = \frac{\pi^2 \rho A w_0}{EI t_0^2}\left(1 - \frac{x}{L}\right)\sin\frac{\pi t}{t_0}.$$

Since the homogeneous form of the above equation is not different from the corresponding, and previously solved, equation in $w(x,t)$, then it can be concluded that the natural frequencies and mode shapes are the same as before, which is not a surprise. Therefore, write the modal expansion for the particular solution to the above equation as

$$v(x,t) = \sum_{n=1}^{N} p_n(t)\sin\frac{n\pi x}{L}.$$

Substituting into the governing differential equation yields

$$\left[\ddot{p}_n(t) + \frac{EI}{\rho A}\left(\frac{n\pi}{L}\right)^4 p_n(t)\right]\sin\frac{n\pi x}{L} = \frac{\pi^2 w_0}{t_0^2}\left(1 - \frac{x}{L}\right)\sin\frac{\pi t}{t_0}.$$

Employing the orthogonality of the mode shapes over the beam length; that is, multiplying both sides by $(2/L)\sin(m\pi x/L)dx$ and integrating from 0 to L, yields

$$\ddot{p}_n(t) + \frac{EI}{\rho A}\left(\frac{n\pi}{L}\right)^4 p_n(t) = \frac{2\pi w_0}{m t_0^2}\sin\frac{\pi t}{t_0}.$$

This equation can be easily solved using the method of undetermined coefficients. The result is

$$p_m(t) = \frac{2\pi w_0}{m\left[\left(\frac{m\pi}{L}\right)^4 \frac{EI t_0^2}{\rho A} - m\pi^2\right]}\sin\left(\frac{\pi t}{t_0}\right).$$

Thus the particular solution for this simply supported beam is

$$w(x,t) = w_0\left(1 - \frac{x}{L}\right)\sin\frac{\pi t}{t_0} + \sum_{n=1}^{N}\frac{2\pi w_0}{n\left[\left(\frac{n\pi}{L}\right)^4 \frac{EI t_0^2}{\rho A} - n\pi^2\right]}\sin\frac{\pi t}{t_0}\sin\frac{n\pi x}{L},$$

whenever the denominator in the above sum is not zero. Actually, because damping, which severely limits a resonance response, was ignored in this analysis, the above solution will be inaccurate when the denominator is just close to zero. The complete solution consists of the homogeneous solution

$$w(x,t) = \sum_{n=1}^{N} C_n \sin\frac{n\pi x}{L} \sin\omega_n t$$

and the above particular solution. Note, that the complete solution satisfies all BCs and the zero initial deflection condition. The zero initial velocity condition determines the values of the constants of integration C_n, which, as per usual, are calculated using the orthogonality of the mode shapes.

(b) Since this is a small deflection problem, one described by linear equations, superimpose the result from the second support motion upon the above result for the first support motion.

8.6 **(b)** To verify the proposed solution, substitute into given partial differential equation. Carry out the required differentiations using the chain rule. For example, for the function $F(x - ct)$ these differentiations are as follows

$$\frac{\partial F}{\partial x} = \frac{\partial F}{\partial(x-ct)}\frac{\partial(x-ct)}{\partial x} = \frac{\partial F}{\partial(x-ct)} = F'$$
$$\frac{\partial^2 F}{\partial x^2} = \frac{\partial F'}{\partial x} = \frac{\partial F'}{\partial(x-ct)}\frac{\partial(x-ct)}{\partial x} = \frac{\partial F'}{\partial(x-ct)} = F''$$
$$\frac{\partial F}{\partial t} = \frac{\partial F}{\partial(x-ct)}\frac{\partial(x-ct)}{\partial t} = -c\frac{\partial F}{\partial(x-ct)} = -cF'$$
$$\frac{\partial^2 F}{\partial t^2} = -c\frac{\partial F'}{\partial t} = -c\frac{\partial F'}{\partial(x-ct)}\frac{\partial(x-ct)}{\partial t} = +c^2\frac{\partial F'}{\partial(x-ct)} = c^2 F''.$$

Substitution of the above results show that the wave equation is identically satisfied and thus this part of the proposed solution is indeed one half of the general solution.

(c) This solution path is the one emphasized in this chapter, which is that of obtaining a complete rather than a general solution. Substitution of the proposed solution form immediately yields

$$W''(x) + \lambda^2 W(x) = 0,$$
$$\text{where} \quad \lambda^2 = \frac{\rho A\omega^2}{N}$$

Application of the BCs $W(0) = W(L) = 0$ to the sine and cosine solutions to the above differential equation quickly show that the natural frequencies and mode shapes for the wire are

$$\omega_n = \frac{n\pi}{L}\sqrt{\frac{N}{\rho A}}$$
$$\text{and} \quad \Phi_n(x) = \sin\frac{n\pi x}{L}.$$

Therefore, write the complete solution in modal terms as

$$w(x,t) = \sum_n p_n \sin\frac{n\pi x}{L}\sin(\omega_n t + \phi_n),$$

where the p_n is an unknown constant associated with the nth mode that is to be determined by use of the initial conditions. Respectively, the initial deflection and velocity condition equations reduce to

for $0 \le x \le L/2$:

$$\sum_n p_n \sin \frac{n\pi x}{L} \sin \phi_n = 2w_0 x/L$$

$$\sum_n \omega_n p_n \sin \frac{n\pi x}{L} \cos \phi_n = 0$$

and a similar pair of equations for the right half of the wire. Multiplying both sides of each equation by $(2/L)\sin(m\pi x/L)$ and taking into account the even and odd characters of both the initial deflection function and these sine functions yields $p_m = 0$ for all even values of m. For odd values of m,

$$p_m = \frac{8w_0}{m^2\pi^2}(-1)^{\frac{m-1}{2}} \quad \text{and} \quad \phi_m = \frac{\pi}{2}.$$

Insertion of these values into the modal solution form completes the undamped free vibration solution as

$$w(x,t) = \frac{8w_0}{\pi^2} \sum_{m=1,odd} \frac{(-1)^{\frac{m-1}{2}}}{m^2} \sin \frac{m\pi x}{L} \cos(\omega_m t).$$

8.7 The kinetic and strain energy expressions are as follows:

$$T = \frac{1}{2} m_0 [\dot{w}(v_0 t, t)]^2 + \frac{1}{2}\int_0^L \rho A[\dot{w}(x,t)]^2 dx$$

$$\text{and} \quad U = \frac{1}{2}\int_0^L EI[w''(x,t)]^2 dx.$$

Again write and substitute the modal transformation

$$w(x,t) = \sum p_n(t) \sin \frac{n\pi x}{L}$$

$$\text{and} \quad w(v_0 t, t) = \sum p_n(t) \sin \frac{n\pi v_0 t}{L}$$

Then the use of the orthogonality of the sine functions produces the requested differential equation in terms of the modal coordinate.

CHAPTER 9 SOLUTIONS

9.1 **(a)** First add Eqs. (8.1) and then add Eqs. (8.4) to obtain

$$p_{j-1} - 2p_j + p_{j+1} = (\Delta t)^2 p_j{}'' + \frac{1}{12}(\delta t)^4 p_j{}'''' + O[\Delta t^6]$$

$$p_{j-2} - 2p_j + p_{j+2} = 4(\Delta t)^2 p_j{}'' + \frac{16}{12}(\Delta t)^4 p_j{}'''' + O[\Delta t^6].$$

Multiply the first of these equations by 16 and then subtract the second to obtain

$$p_j{}'' = \frac{-p_{j-2} + 16p_{j-1} - 30p_j + 16p_{j+1} - p_{j+2}}{12\Delta t^2} + O[\Delta t^4].$$

9.4 The single piecewise linear equation of motion of this single-DOF system can be written by combining the equations for the two linear portions of motion, which are

$$\begin{aligned} &\text{for} \quad q < q_0 \qquad \ddot{q}(t) + \omega_1^2 q(t) = \omega_1^2 Y_0 \sin 2t \\ &\text{for} \quad q > q_0 \qquad \ddot{q}(t) + 2\omega_1^2 q(t) = 2\omega_1^2 Y_0 \sin 2t. \end{aligned}$$

Combined as a single equation for convenience, the result is

$$\ddot{q}(t) + q(t) + q(t)\text{stp}(q - q_0) = Y_0 \sin 2t + Y_0 \,\text{stp}(q - q_0) \sin 2t.$$

Now it is a matter of setting up a spreadsheet solution.

9.8 **(a)** To obtain the square root of a matrix, just as finding any function of that matrix, first solve the eigenvalue problem associated with that matrix. Then

$$\begin{aligned} [A] &= [\Phi][\backslash \Lambda \backslash][\Phi]^t = [\Phi][\backslash \sqrt{\lambda} \backslash][\backslash \sqrt{\lambda} \backslash][\Phi]^t \\ &= [\Phi][\backslash \sqrt{\lambda} \backslash][\Phi]^t[\Phi][\backslash \sqrt{\lambda} \backslash][\Phi]^t = [A]^{1/2}[A]^{1/2}. \end{aligned}$$

This procedure, because of the need for an eigenvalue solution, is usually not as efficient as a Cholesky decompositon, unless, of course, the eigenvalue solution is to be obtained for other purposes.

(b) The eigenvalues and eigenvectors of this matrix are such that the original matrix can be expanded as

$$\begin{bmatrix} 6 & 2 & 0 \\ 2 & 8 & 1 \\ 0 & 1 & 8 \end{bmatrix} = \begin{bmatrix} -0.423085 & 0.379706 & 0.822692 \\ -0.778758 & 0.311752 & -0.544377 \\ -0.463179 & -0.870996 & -0.163801 \end{bmatrix}$$

$$* \begin{bmatrix} 9.68133 & 0 & 0 \\ 0 & 7.64207 & 0 \\ 0 & 0 & 4.67660 \end{bmatrix} \begin{bmatrix} -0.423085 & -0.778758 & -0.463179 \\ 0.379706 & 0.311752 & -0.870996 \\ 0.822692 & -0.544377 & -0.163801 \end{bmatrix}.$$

Then, replacing the eigenvalues by their square roots, the square root matrix is

$$\begin{bmatrix} 2.41918 & 0.383906 & -0.01310 \\ 0.383906 & 2.79654 & 0.178853 \\ -0.01310 & 0.178853 & 2.82274 \end{bmatrix} = \begin{bmatrix} -0.423085 & 0.379706 & 0.822692 \\ -0.778758 & 0.311752 & -0.544377 \\ -0.463179 & -0.870996 & -0.163801 \end{bmatrix}$$

$$* \begin{bmatrix} 3.11148 & 0 & 0 \\ 0 & 2.76443 & 0 \\ 0 & 0 & 2.16254 \end{bmatrix} \begin{bmatrix} -0.423085 & -0.778758 & -0.463179 \\ 0.379706 & 0.311752 & -0.870996 \\ 0.822692 & -0.544377 & -0.163801 \end{bmatrix}.$$

APPENDIX II

Fourier Transform Pairs

II.1 Introduction to Fourier[1] Transforms

Consider a function $f(t)$ that is periodic with period[2] T. Let the function $f(t)$ satisfy the Dirichlet[3] conditions. From Ref. [II.1], these conditions are that, in addition to being periodic, $f(t)$ is a bounded function that in any one period has at most a finite number of local maxima and local minima and a finite number of points of discontinuity. Then the following trigonometric series for $f(t)$ converges to that function at all points where the function $f(t)$ is continuous and converges to the average of the right- and left-hand limits of $f(t)$ at each point where $f(t)$ is discontinuous:

$$f(t) = a_0 + \sum_{n=1}^{\infty} a_n \cos\left(\frac{2n\pi t}{T}\right) + b_n \sin\left(\frac{2n\pi t}{T}\right). \tag{AII.1}$$

On the basis of the orthogonality of these sine and cosine functions over any t-interval of length T, the series coefficients of the sine and cosine functions, a_0, a_n, b_n, for all n, can be determined from the relations

$$a_0 = \frac{1}{T}\int_{-T/2}^{+T/2} f(t)\,dt \qquad a_n = \frac{2}{T}\int_{-T/2}^{+T/2} f(t)\cos\left(\frac{2n\pi t}{T}\right)dt$$

$$b_n = \frac{2}{T}\int_{-T/2}^{+T/2} f(t)\sin\left(\frac{2n\pi t}{T}\right)dt. \tag{AII.2}$$

In the above integrations, the intervals of integration can be any time interval of total duration T, such as $(0, T)$.

The above Fourier series can also be written in terms of spatial variables, that is, with x and L, respectively, replacing t and T. Although their importance per se has faded in the digital age, such Fourier series can have many uses. For example, a Fourier series can be used with a differential equation to describe a periodic loading and/or the unknown response over a

[1] From Ref. [II.1], Jean Fourier (1768–1830), a confidant of Napoleon, first undertook the systematic study of the expansions that bear his name in "Theorie analytique de la chaleur" in 1822. The use of such series dates back to the time of Daniel Bernoulli (1700–1782), Swiss physicist and mathematician.

[2] A function $f(t)$ has a period T if and only if $f(t + T) = f(t)$ for all t.

[3] Peter Dirichlet (1805–1859), German mathematician.

fixed interval. For present purposes, rewrite Eq. (AII.1) using complex algebra. Recall the result

$$e^{i\theta} = \cos\theta + i\sin\theta$$
$$e^{-i\theta} = \cos\theta - i\sin\theta.$$

Solving simultaneously for the values of the sine and cosine functions leads to

$$\cos\theta = \frac{(e^{i\theta} + e^{-i\theta})}{2}$$

$$\text{and} \quad \sin\theta = \frac{(e^{i\theta} - e^{-i\theta})}{2i} = -i\frac{(e^{i\theta} - e^{-i\theta})}{2}.$$

Substituting these complex forms for the cosine and sine functions into the original form for the Fourier series, Eq. (AII.1), leads to

$$f(t) = a_0 + \sum_{n=1}^{\infty} \frac{(a_n - ib_n)}{2} e^{+i2n\pi t/T} + \frac{(a_n + ib_n)}{2} e^{-i2n\pi t/T}. \qquad \text{(AII.3)}$$

At this point, for $n = 0, 1, 2, 3, \ldots$, define $b_0 = 0$, $b_{-n} = -b_n$, and $a_{-n} = +a_n$. Then define the complex coefficients

$$c_n \equiv \frac{(a_n - ib_n)}{2} \quad \text{so that} \quad c_{-n} = \frac{(a_n + ib_n)}{2}.$$

Further define $c_0 \equiv a_0$. After these definitions, the series can be rewritten as

$$f(t) = \sum_{n=0}^{\infty} c_n e^{(in2\pi t/T)} + \sum_{n=+1}^{+\infty} c_{-n} e^{(-in2\pi t/T)} = \sum_{n=0}^{\infty} c_n e^{(in2\pi t/T)} + \sum_{n=-\infty}^{-1} c_{+n} e^{(+in2\pi t/T)}.$$

Note that if in the first group of terms in the exponential form of the above sum, the index n is replaced by $-n$, then this first part of the sum becomes the second part and vice versa, except for the term c_0, which can be transferred back and forth between sums because its exponential factor is simply the value $\exp(0) = 1.0$. The above two parts of the sum to be combined into a single sum as

$$f(t) = \sum_{n=-\infty}^{\infty} c_n e^{i(2n\pi t/T)}. \qquad \text{(AII.4)}$$

This is the complex form for the Fourier series for $f(t)$. From Eq. (AII.2), for any positive or negative or zero value of n, the complex coefficients can be evaluated as

$$c_n = \frac{1}{T}\int_{-T/2}^{+T/2} f(t)\left[\cos\left(\frac{2n\pi t}{T}\right) - i\sin\left(\frac{2n\pi t}{T}\right)\right] dt$$

$$\text{or} \quad c_n = \frac{1}{T}\int_{-T/2}^{+T/2} f(t)\, e^{-i(2n\pi t/T)} dt. \qquad \text{(AII.5)}$$

To progress from the complex algebra forms for a Fourier series, Eqs. (AII.4) and (AII.5), to the Fourier transform equations, it is necessary to resort to the idea of frequency as introduced in the second chapter. Again, the circular frequency ω is defined as $2\pi/T$. Also define $n\omega \equiv \omega_n = 2n\pi/T$ and $\omega_{n+1} - \omega_n = \Delta\omega_n$ for all values of n, positive and negative. Further, define the following function of those now-introduced discrete values of frequency ω_n that cover the range of frequencies from $-\infty$ to $+\infty$. Let $F(\omega_n) \equiv (T/2\pi)c_n$ so that $c_n = F(\omega_n)\Delta\omega_n$. These definitions allow Eqs. (AII.4) and (AII.5) to be rewritten as

$$f(t) = \frac{2\pi}{T}\sum_{-\infty}^{+\infty} F(\omega_n)\, e^{+i\omega_n t} = \sum_{-\infty}^{+\infty} F(\omega_n)\, e^{+i\omega_n t}\, \Delta\omega_n$$

$$\text{and} \quad F(\omega_n) = \frac{1}{2\pi}\int_{-T/2}^{+T/2} f(t)\, e^{-i\omega_n t}\, dt.$$

Each of the infinite number of ω_ns can be regarded as a discrete value of a continuous frequency variable ω, that is, a continuous frequency variable that covers the interval from $\omega = -\infty$ to $\omega = +\infty$, an interval called the frequency spectrum. The second of the above sums has the exact appearance of that sum (local ordinate multiplied by an increment in the abscissa) used to evaluate a Riemann[4] definite integral in terms of the continuous variable ω over the interval between $-\infty$ and $+\infty$. Thus, if at this juncture it is required that $T \to \infty$, which implies $\Delta\omega_n = 2\pi/T \to d\omega_n = d\omega$, then the differences between adjacent values of the discrete variable ω_n become infinitesimal, allowing the replacement of the discrete variable by the continuous variable ω. The same process allows, as above, the replacement of $\Delta\omega_n$ by $d\omega$, and the replacement of the sum over n by an integration over the continuous variable ω. Thus, under these limiting conditions, the joint expressions for $f(t)$ and $F(\omega)$ become

$$f(t) = \int_{-\infty}^{+\infty} F(\omega)\, e^{i\omega t}\, d\omega$$

$$F(\omega) = \frac{1}{2\pi}\int_{-\infty}^{+\infty} f(t)\, e^{-i\omega t}\, dt. \qquad \text{(AII.6)}$$

Equations (AII.6) are known as the Fourier transform pair for the time domain function $f(t)$ and the frequency domain function $F(\omega)$. For further explanation, consider the second of these two integrals. This is an integral over the dummy variable t. It is also an integral that contains the parameter ω. Thus the second of these two integrals represents a transformation of a time function into a frequency function. This is sometimes described as a transformation from the time domain to the frequency domain. The first of the Fourier transform pair integrals does exactly the opposite. The first integral expression is often viewed as a summing of all the frequency components of $f(t)$. From Ref. [II.2], only two conditions are sufficient for the existence of the Fourier transform of a function $f(x)$. These conditions are that (i) $f(x)$ must be piecewise continuous on every finite interval and (ii) $f(x)$ must be absolutely integrable on $(-\infty,+\infty)$.

The near symmetry of the above pair of transformations is unmistakable. That symmetry can be further strengthened by defining a new $F(\omega)$ that is equal to the old $F(\omega)$ multiplied by the factor $\sqrt{2\pi}$. Then the result is

$$f(t) = \frac{1}{\sqrt{2\pi}}\int_{-\infty}^{+\infty} F(\omega)\, e^{i\omega t}\, d\omega$$

$$F(\omega) = \frac{1}{\sqrt{2\pi}}\int_{-\infty}^{+\infty} f(t)\, e^{-i\omega t}\, dt. \qquad \text{(AII.7)}$$

[4] Georg Riemann (1826–1866), German mathematician [1].

Further note that letting $T \to \infty$ has the effect of making the period of the originally periodic function $f(t)$ infinite. Since any nonperiodic function can be viewed as periodic with an infinite period, imposing the limit $T \to \infty$ has the effect of eliminating the requirement that $f(t)$ be periodic.

To show that the two primary response functions of structural dynamics, the frequency response function and the impulse response function, are essentially a Fourier transform pair, consider a damped single-DOF system subjected to an applied force $F(t)$. Let the displacement response of the mass be $u(t)$. Then, from Chapter 6, using the superposition/Duhamel/convolution integral, the output response can be expressed in terms of the force input as

$$u(t) = \int_0^t F(\tau)h(t-\tau)\,d\tau. \tag{AII.8}$$

Since this solution is predicated on zero initial conditions, let the input force be zero before time zero. Since the input force is zero from time equals minus infinity to time equals zero, the lower limit of the integral can be changed to minus infinity without altering the deflection response. Now let the original input force $F(t)$ be replaced by a new input force with the same symbol that is everywhere equal to the original input force from time equals minus infinity to time t, but differs from the original input after the arbitrary time t stated in $F(t)$ in that the new force is zero thereafter. This change also does not affect the above equality. Moreover, it allows the upper limit to be increased to plus infinity without altering the equality. Hence Eq. (AII.8) can be written as

$$u(t) = \int_{-\infty}^{+\infty} F(\tau)h(t-\tau)\,d\tau.$$

Now, to move in the direction of the frequency response function, let the input force be harmonic. That is, let

$$F(t) = F_0 \sin \omega_f t \to F_0 e^{i\omega_f t}.$$

Substituting this special case into the previous general response formulation for the output deflection yields

$$u_h(t) = \int_{-\infty}^{+\infty} F_0 e^{i\omega_f \tau} h(t-\tau)\,d\tau, \tag{AII.9}$$

where $u_h(t)$ is specifically the response to the harmonic input. Recall that the harmonic response for this same single-DOF (m, c, k) system can be written in terms of the frequency response function. In review, the differential equation of motion of this one-DOF system subjected to this same harmonic force, where, again, $\omega_1^2 = k/m$ and $2\zeta\omega_1 = c/m$, is

$$\ddot{u}_h(t) + 2\zeta\omega_1\dot{u}_h(t) + \omega_1^2 u_h(t) = \left(\frac{F_0}{m}\right) e^{i\omega_f t} = \omega_1^2\left(\frac{F_0}{k}\right) e^{i\omega_f t}.$$

As demonstrated in Chapter 6, the solution for the harmonic deflection output can be written as $u_h(t) = u_0 \exp(i\omega_f t)$, where u_0 is a complex number depending on the frequency and damping factor. Substituting this form of the solution yields

$$u_0 = \frac{F_0/k}{(1-\omega_f^2/\omega_1^2) + 2i\zeta\omega_f/\omega_1} \quad \text{or} \quad u_h(t) = F_0 H(\omega_f) e^{i\omega_f t},$$

where, for the sake of convenience, the stiffness factor has been combined with the other elements of the frequency response function. Combining the above result with Eq. (AII.9) yields

$$F_0\, H(\omega_f)\, e^{i\omega_f t} = \int_{-\infty}^{+\infty} F_0\, e^{i\omega_f \tau}\, h(t-\tau)\, d\tau.$$

After canceling the force amplitude constant, introduce the following transformation from τ to θ. Let $\theta = t - \tau$. Then the above equation can be written as

$$H(\omega_f)\, e^{i\omega_f t} = -\int_{+\infty}^{-\infty} e^{i\omega_f (t-\theta)}\, h(\theta)\, d\theta = \int_{-\infty}^{+\infty} e^{i\omega_f (t-\theta)}\, h(\theta)\, d\theta.$$

Since the integration is over θ, the exponential term $\exp(i\omega_f t)$ can be factored out of the integral and canceled. Since t no longer appears in the result, make the cosmetic change from θ to t with the result

$$H(\omega_f) = \int_{+\infty}^{-\infty} e^{-i\omega_f t}\, h(t)\, dt.$$

Except for the factor of $1/(2\pi)$, this is identical with the second of Eqs. (AII.6). This difference of a constant factor explains the previous reference to the impulse response function and the frequency response function as being "essentially" a Fourier transform pair. Of course, this slight discrepancy could easily be removed by simply redefining the frequency response function, which will not be done here. Simply for reference, the reverse transformation is

$$h(t) = 2\pi \int_{-\infty}^{+\infty} H(\omega_f)\, e^{i\omega_f t} d\omega_f.$$

REFERENCES

II.1 Wylie, C. R., and L. C. Barrett, *Advanced Engineering Mathematics*, 5th ed., McGraw-Hill, New York, 1982.

II.2 Kreyszig, D., *Advanced Engineering Mathematics*, 7th ed., John Wiley & Sons, New York, 1993.

Index

Printed in the United States
By Bookmasters